MW01235741

Methods in Molecular Biology

For further volumes:
http://www.springer.com/series/7651

Nanoscale Imaging

Methods and Protocols

Edited by

Yuri L. Lyubchenko

Department of Pharmaceutical Sciences, University of Nebraska Medical Center, Omaha, NE, USA

Editor
Yuri L. Lyubchenko
Department of Pharmaceutical Sciences
University of Nebraska Medical Center
Omaha, NE, USA

ISSN 1064-3745 ISSN 1940-6029 (electronic)
Methods in Molecular Biology
ISBN 978-1-4939-8590-6 ISBN 978-1-4939-8591-3 (eBook)
https://doi.org/10.1007/978-1-4939-8591-3

Library of Congress Control Number: 2018944128

Printed on acid-free paper

This Humana Press imprint is published by the registered company Springer Science+Business Media, LLC part of Springer Nature.
The registered company address is: 233 Spring Street, New York, NY 10013, U.S.A.

Preface

Nanoimaging: Emerging Field in Biomedicine

Nanoimaging combines a set of experimental approaches which are the foundation for single-molecule biophysics (SMB). The significance of SMB is defined by the two major features. First, the biological systems, including cells, consist of a limited number of biological molecules or complexes, so their study required techniques capable of probing single molecules. Second, upon functioning, biological molecules and systems assemble in transient states and characterization of these transient states is critical for understanding molecular mechanism of the biological process under study. Currently, SMB nanoimaging is applied to numerous biological problems associated with understanding molecular mechanisms of DNA replication, transcription, and translation as well as functioning of molecular machines. The main nanoimaging techniques are associated with Nobel Prizes. Historically, the first of them is atomic force microscopy (AFM) that belongs to the family of scanning probe microscopes, for which the Nobel Prize in Physics in 1986 was awarded to G. Binnig and H. Rohrer at IBM, Zurich. Importantly, their first publication was made only 4 years prior to the award. Single-molecule fluorescence methods originated from the 1989 publication of W. Moerner and L. Kador, and a Nobel Prize for this discovery was given to W. Moerner along with Eric Betzig and Stefan W. Hell in 2014. Optical traps or tweezers utilize the phenomenon of manipulating of atoms and small particles by the focused light beam. Applications of this method in physics to cooling and trapping atoms with laser light, enabling the authors to construct a Bose-Einstein condensate, led to Nobel Prizes in physics in 1997 and 2001.

Protocols utilizing a set of nanoimaging techniques are presented in this book and are illustrated with applications to several important biological systems and questions while also selectively highlighting the most useful and exciting strengths of these novel and evolving methods. The primary target audience for this book includes biophysicists, biochemists, and molecular biologists who want to learn more about the field and potentially utilize these techniques in understanding molecular mechanisms of major cellular processes.

Following the preface are five major parts (Parts I–V) organized primarily on the complexity of the system. These parts comprise several chapters with each describing the protocol for a specific method. The first of these major parts is Part I: Imaging and Probing of Biomolecules, which is comprised of nine chapters. The first chapter (Chapter 1) provides protocols for high-resolution AFM imaging of DNA with a focus on imaging in aqueous solutions. The chapter is complemented with a discussion of the factors that influence the high-resolution imaging. Chapter 2 describes AFM force spectroscopy and its application to receptor-ligand systems; specifically, the interaction of the serotonin transporter in the membrane of the cell culture in the presence of S-citalopram, a drug used for the treatment of depression. Two binding sites for the drug on the transporter were revealed and characterized, and the protocol provides details for how such measurements can be performed. The application of AFM force spectroscopy to elucidating the mechanical properties of proteins is described in Chapter 3. Folding of proteins in a complex structure is defined by interactions between the protein segments and these interactions can be characterized by AFM in experiments in which the protein is stretched under defined force applied to the AFM tip. The authors of this chapter provide a detailed protocol on how the experiments are performed and how the data are analyzed. Chapter 4 explains how AFM force spectroscopy data with sub-pN resolution can be obtained. The key in the described protocol is the modification of the AFM cantilever. The authors describe two methods enabling one to prepare the required cantilevers. Successful probing of intermolecular interactions depends on the immobilization of the macromolecules. Chapter 5 describes the immobilization approach, termed flexible nano array (FNA) in which interacting macromolecules are covalently tethered inside a polymer at a predefined position. Importantly, the same pair of macromolecules can be probed multiple times allowing one to repeatedly follow unfolding-folding events. The approach is illustrated by measuring the interaction within amyloid dimers. Chapter 6 outlines an approach for the application of magnetic tweezers (MT) to measure twist and torque in DNA and RNA. The authors focus on two recently developed methodologies in MT, freely orbiting magnetic tweezers (FOMT) and magnetic torque tweezers (MTT), and provide detailed protocols for precise measurements with the use of both the FOMT and MTT approaches. Additionally, the methodology for multiplexed experiments in which multiple tethered DNA molecules are probed in one experiment is described. AFM is a tool that enables various types of measurements and the authors of Chapter 7 describe the methodology for measurements with AFM local electrical properties of the sample. The chapter is accompanied by an extensive introduction in which the basics for measurements of electrostatic properties of materials using AFM operating in contact and oscillation modes are given. Detailed protocols for electrostatic measurements including critical instrument-related issues are provided. The chapter is accompanied by a few examples illustrating the capabilities of this modality of AFM. Many biological processes occur on a membrane and AFM as a topographic technique is capable of characterization of this process at the nanoscale. Supported lipid bilayers are widely used experimental systems for cellular membrane, but such issues as stability of assembled bilayers and their smoothness are critical for successful AFM applications. Protocols for the preparation of such bilayers are given in Chapter 8. Assembled bilayers are stable for hours which allowed the authors to perform time-lapse AFM observation over several hours. This methodology is illustrated by time-lapse observation of the self-assembly of amyloid proteins on lipid bilayers surfaces. DNA nanotechnology is a largely growing area of nanotechnology that allows one to build with DNA duplexes various three-dimensional structures. Application of such nanostructures for numerous applications, including a drug delivery vehicle, requires a thorough characterization of a DNA structure with a selected small molecule. The authors of Chapter 9 describe their approach for probing the interaction of small molecules with DNA nanostructures. The developed methodology combines computational modeling (molecular docking simulation) with a set of experimental methods that provide different properties of the ligand-DNA complexes.

Part II consists of six chapters describing a set of applications of high-speed AFM (HS-AFM) instrumentation, the key feature of which being the capability to visualize and probe interactions at the millisecond timescale that exceed the temporal capabilities of a traditional AFM by three orders of magnitude. Substrate preparation enabling reliable and reproducible observation of dynamics of molecules observed with HS-AFM instrumentation is the main topic of Chapter 10. The authors cover a wide range of substrates including bare mica and a set of functionalized mica substrates, supported lipid bilayers, graphite, gold substrate, and polydimethylsiloxane (PDMS). Additional recommendations for operation of the HS-AFM instrument are provided. The authors of Chapter 11 outline detailed procedures for HS-AFM imaging enabling one to obtain high-resolution topographic images.

Special attention is given to data analysis and processing and displaying HS-AFM videos. The protocols are illustrated by high-resolution imaging of Outer Membrane Porin F (OmpF) protein. Methods and protocols for HS-AFM direct observation of self-assembly of amyloidogenic protein Aβ42 are presented in Chapter 12. This method allowed the authors to follow the time-dependent growth of individual Aβ42 fibrils. In Chapter 13, the HS-AFM based methodology that enabled the authors to directly observe hybridization dynamics of DNA is described. This method utilizes the assembled DNA origami approach by which the two DNA molecules were end-immobilized inside the DNA origami rectangular. The application of HS-AFM to study the dynamics of DNA nucleosomes is described in Chapter 14. The method is illustrated by the use of centromere nucleosome and the developed approach allowed the authors to directly visualize sliding of individual nucleosomes over one hundred base pairs and transferring nucleosomes from one DNA template to another. The ability to operate HS-AFM in the force spectroscopy mode is described in Chapter 15. The authors provide a step-by-step protocol of HS-AFM force spectroscopy experiments including sample preparation, measurements, and analysis of the acquired data. The approach is illustrated by unfolding of a concatemer made of 8 repeats of the titin I91 domain.

Methods in which nanoimaging and probing is applied to large biomolecular complexes are assembled as Part III of the book which is comprised of five chapters. A single-molecule fluorescence approach developed for probing of protein-protein interactions is described in Chapter 16. The authors provide a step-by-step protocol that includes RNA labeling, purification of protein complexes, and imaging per se. The approach is illustrated by the demonstration of how human Dicer and its cofactor TRBP orchestrate the biogenesis of microRNA in real time. Cryo-electron microscopy is currently applied to high-resolution imaging of various biological samples. Chapter 17 provides protocols for negative-stain grid and cryo-grid preparation for studying protein-nucleic acid complexes. Probing of chromatin dynamics on the reconstructed chromatin fibrils with the use of magnetic tweezers (MT) is described in Chapter 18. The authors provide detailed protocols for the preparation of chromatin fibrils for MT studies, performing probing experiments and the data analysis of the force extension data. Chapter 19 presents a methodology for probing of protein-RNA complexes assembled at initial stages of RNP granules assembly. The approach combines a single-molecule FRET (smFRET) assay and an electrophoretic mobility shift assay (EMSA). Probing of complex biomolecular systems with the use of an instrument that integrates AFM in the fluorescence microscope is descried in Chapter 20. The authors provide step-by-step protocols for sample preparation, data acquisition, and data processing that yield nanoscale topographic resolution, high image registration accuracy, and single fluorophore sensitivity. The description of factors influencing the data quality is discussed. The methodology is illustrated by simultaneous characterization of mesoscale geometry and composition in a nucleoprotein complex.

Part IV comprises 13 chapters in which nanoimaging methods are applied to prokaryotic and eukaryotic cells, tissues, and embryos. Chapter 21 provides a protocol for probing of bacterial surfaces. The developed sample preparation method allowed the authors to acquire topographic images of the bacterial surface with nanometer resolution and to perform force spectroscopy experiments on living bacterial cells. Application of AFM to characterization of large protein bacterial microcompartments complexes (BMP) is described in Chapter 22. The authors describe the procedure to prepare samples for AFM and outline the high-resolution AFM imaging and nanoindentation experiments on BMC shell proteins and intact BMCs, as well as data analysis. The method can be modified to study other self-assembling biological systems as well. Chapter 23 presents time-lapse imaging with an integrated AFM-optical microscopy instrument. With such instrumentation, relatively slow processes as division of bacterial cells were imaged. The chapter describes the protocols for the assembly and operation of the integrated instrument along with the methodology for data acquisition and analysis. Chapter 24 is dedicated to single-molecule AFM measurements of bacteria adhesion using functionalized AFM probes and presents protocols to graft molecules and bacteria to AFM cantilevers. The bioprobes enable one to measure forces down to the single-cell and single-molecule levels. Chapter 25 presents protocols for conducting quantitative FCS measurements on DNA inside living eukaryotic and prokaryotic cells. The protocols include sample preparation, dye selection and characterization, FCS data acquisition, and data analysis. Tracking of individual particles inside living cells is widely applied to the characterization of cellular machinery such as molecular motors and membrane proteins, and the methodology of intracellular 3D tracking of single molecules is presented in Chapter 26. The authors use quantum dots conjugated with transmembrane proteins and provide protocols for single-molecule tracking of protein-conjugated QDs on the membrane and in the cytosol of living cells. Chapter 27 presents protocols for AFM probing of cells in which mechanical properties are characterized by the analysis of the indentation by the AFM tip. The authors describe a step-by-step method to analyze the AFM force-indentation curves to derive cell mechanics and the parameters of the pericellular coat (density and thickness of the coat layer). Chapter 28 describes the application of ringing mode AFM, a mode recently developed by the authors, to the imaging of surfaces of such soft biological samples as cells and tissues. The authors describe a step-by-step approach to collect images in ringing mode applied to both biological samples and soft materials in general. Technical details, potential difficulties, and points of special attention are described. Chapter 29 describes methods for probing single virus binding sites on living mammalian cells using AFM. The authors provide a step-by-step description of experiments to investigate interactions between viral particles and cell surface receptors including the preparation of the required biological materials. The methodology enables one to characterize virus binding to living mammalian cells and to quantify the kinetic and thermodynamic parameters of binding events. The authors in the introduction provide a detailed description of AFM force spectroscopy in application to such complex biological systems. Probing of interactions on living cells including cellular receptors and a ligand and cell-cell interaction is described in Chapter 30. Probe functionalization is a critical component of the protocol. For experiments designed to examine cell-cell adhesions, a single cell is attached to a tip-less cantilever which is then brought into contact with other cultured cells to measure adhesions. The method is illustrated by measuring adhesions between fibronectin and $\alpha 5\beta 1$ integrin, and breast cancer cells and bone marrow endothelial cells. Chapter 31 is dedicated to topographic imaging of cells. Two main precautions must be considered to obtain high-resolution morphological and nanomechanical characterization of biological specimens with an atomic force microscope: the tip-sample interaction and the sample-substrate adhesion. The necessary steps for a correct preparation of three types of biological samples, erythrocytes, bacteria, and osteoblasts, are provided. Chapter 32 focuses on imaging in live embryos using lattice light-sheet microscopy. Living *Drosophila Melanogaster* embryos were primary systems in illustrating the method. This protocol allowed the authors to directly observe both transcription factor diffusion and binding dynamics. The application of the same methodology to live mouse embryos extends the developed method to thick samples. Chapter 33 presents the method for multidimensional scanning electron microscopy imaging of tissues. The described approach improves overall sample conductivity, resulting in the minimization of charging under high-vacuum conditions and an improvement in lateral resolution and image contrast. The sample preparation approach is illustrated by the data obtained for the liver tissue.

Part V contains Chapter 34 in which the computer modeling of complex biological systems is described. The chapter presents techniques for classical simulations of protein and protein-nucleic complexes, revealing their dynamics and protein-substrate binding energies. The approach is based on classical atomistic molecular dynamics (MD) simulations of the experimentally determined structures of the complexes. The application is illustrated by examples of snapshots of proteins and their complexes with DNA and RNA.

In summary, this volume presents modern methods and protocols utilizing nanoimaging and nanoprobing techniques in application to a broad range of biological systems. The set of techniques assembled in this book includes the most widely used in nanoimaging experiments including single-molecule biophysics. These methods have already led to a large number of discoveries and some of them are described in the book. The major trend in the nanoimaging instrumentatation is currently made to the combination of the techniques in one integrated instrument and few such integrations are described in the book. However even these methods are in flux and undergo constant improvements and modifications. Novel methods appear as well and these advances set the base for future editions of such a book.

I would like to thank all of the authors for their enthusiasm and effort in putting together this set of methods, their help in reviewing other chapters, as well as the members of my laboratory for additionally reviewing and correcting of the chapters. I would also like to thank my wife Luda for her help with the book assembly, keeping me organized, and her patience.

Omaha, NE, USA ***Yuri L. Lyubchenko***

Contents

Contributors

JOHN ALEXANDER • *SPM Labs LLC, Tempe, AZ, USA*
DAVID ALSTEENS • *Louvain Institute of Biomolecular Science and Technology, Université catholique de Louvain, Louvain-la-Neuve, Belgium*
PABLO ARES • *Department of Condensed Matter Physics, Universidad Autónoma de Madrid, Madrid, Spain*
SIDDHARTHA BANERJEE • *Department of Pharmaceutical Sciences, University of Nebraska Medical Center, Omaha, NE, USA*
AUDREY BEAUSSART • *CNRS, LIEC (Laboratoire Interdisciplinaire des Environnements Continentaux), UMR7360, Nancy, France; Université de Lorraine, CNRS, LIEC, F-54000, Nancy, France*
SERGEY BELIKOV • *SPM Labs LLC, Tempe, AZ, USA*
FILIP BRAET • *School of Medical Sciences (Discipline of Anatomy and Histology)—The Bosch Institute, The University of Sydney, Camperdown, NSW, Australia; Australian Centre for Microscopy and Microanalysis (ACMM), The University of Sydney, Camperdown, NSW, Australia; Cellular Imaging Facility, Charles Perkins Centre, The University of Sydney, Camperdown, NSW, Australia*
THOMAS B. BROUWER • *Huygens-Kamerlingh Onnes Laboratory, Leiden Institute of Physics, Leiden University, Leiden, The Netherlands*
MARTINO CALAMAI • *LENS—European Laboratory for Non-linear Spectroscopy, Sesto Fiorentino, Italy; INO—National Institute of Optics, Sesto Fiorentino, Italy*
MARCO CAPITANIO • *LENS—European Laboraotry for Non-Linear Spectroscopy, Sesto Fiorentino, Italy; Department of Physics and Astronomy, University of Florence, Sesto Fiorentino, Italy*
IGNACIO CASUSO • *LAI, Aix-Marseille Université, INSERM UMR_S 1067, CNRS UMR 7333, Marseille, France*
DELFINE CHENG • *School of Medical Sciences (Discipline of Anatomy and Histology)—The Bosch Institute, The University of Sydney, Camperdown, NSW, Australia*
ALESSANDRO COSTA • *Macromolecular Machines Laboratory, The Francis Crick Institute, London, UK*
XAVIER DARZACQ • *Department of Molecular and Cell Biology, University of California, Berkeley, CA, USA*
NYNKE H. DEKKER • *Department of Bionanoscience, Kavli Institute of Nanoscience, Delft University of Technology, Delft, The Netherlands*
MARTIN DELGUSTE • *Louvain Institute of Biomolecular Science and Technology, Université catholique de Louvain, Louvain-la-Neuve, Belgium*
SIMONE DINARELLI • *Istituto di Struttura della Materia ISM—CNR, Via del Fosso del Cavaliere 100, Rome, Italy*
MAXIM E. DOKUKIN • *Department of Mechanical Engineering, Tufts University, Medford, MA, USA*
MICHAEL B. EISEN • *Department of Molecular and Cell Biology, University of California, Berkeley, CA, USA; Biophysics Graduate Group, University of California, Berkeley, CA,*

USA; Howard Hughes Medical Institute, University of California, Berkeley, CA, USA; Department of Integrative Biology, University of California, Berkeley, CA, USA

SOFIANE EL-KIRAT-CHATEL • *CNRS, Laboratoire de Chimie Physique et Microbiologie pour les Matériaux et l'Environnement, LCPME, UMR 7564, Nancy, France; Université de Lorraine, CNRS, LCPME, F-54000, Nancy, France*

MASAYUKI ENDO • *Department of Chemistry, Graduate School of Science, Kyoto University, Sakyo-ku, Kyoto, Japan; Institute for Integrated Cell-Material Sciences, Kyoto University, Sakyo-ku, Kyoto, Japan*

HAIG ALEXANDER ESKANDARIAN • *School of Engineering, École Polytechnique Fédérale de Lausanne, Lausanne, Switzerland; School of Life Sciences, École Polytechnique Fédérale de Lausanne, Lausanne, Switzerland*

GEORG ERNEST FANTNER • *School of Engineering, École Polytechnique Fédérale de Lausanne, Lausanne, Switzerland*

MOHAMED FAREH • *Department of BioNanoScience, Kavli Institute of NanoScience, Delft University of Technology, Delft, The Netherlands; Cancer Immunology Program, Peter MacCallum Cancer Center, East Melbourne, Victoria, Australia. Sir Peter MacCallum Department of Oncology, University of Melbourne, Parkville, Victoria, Australia*

MATTHEW FAULKNER • *Institute of Integrative Biology, University of Liverpool, Liverpool, UK*

STEVEN DE FEYTER • *Division of Molecular Imaging and Photonics, Department of Chemistry, KU Leuven-University of Leuven, Leuven, Belgium*

WOUT FREDERICKX • *Division of Molecular Imaging and Photonics, Department of Chemistry, KU Leuven-University of Leuven, Leuven, Belgium*

YASUHIKO FUJITA • *Division of Molecular Imaging and Photonics, Department of Chemistry, KU Leuven-University of Leuven, Leuven, Belgium*

ALEXANDER GALL • *Cepheid, Inc., Bothell, WA, USA*

YINGNING GAO • *Chemical and Environmental Engineering, University of California Riverside, Riverside, CA, USA*

HERNAN GARCIA • *Department of Molecular and Cell Biology, University of California, Berkeley, CA, USA; Biophysics Graduate Group, University of California, Berkeley, CA, USA; Department of Physics, University of California, Berkeley, CA, USA*

LUCIA GARDINI • *LENS—European Laboratory for Non-linear Spectroscopy, Sesto Fiorentino, Italy; INO—National Institute of Optics, Sesto Fiorentino, Italy*

MARCO GIRASOLE • *Istituto di Struttura della Materia ISM—CNR, Rome, Italy*

VLADISLAV V. GLINSKY • *Dalton Cardiovascular Research Center, University of Missouri, Columbia, MO, USA; Department of Pathology and Anatomical Sciences, University of Missouri, Columbia, MO, USA; Research Service, Harry S. Truman Memorial Veterans Hospital, Columbia, MO, USA*

JULIO GOMEZ-HERRERO • *Department of Condensed Matter Physics, Universidad Autónoma de Madrid, Madrid, Spain; Condensed Matter Physics Center (IFIMAC), Universidad Autónoma de Madrid, Madrid, Spain*

SURESH GORLE • *Department of Chemistry and Biochemistry, University of Texas at El Paso, El Paso, TX, USA*

PANCHALI GOSWAMI • *Macromolecular Machines Laboratory, The Francis Crick Institute, London, UK*

HERMANN J. GRUBER • *Institute of Biophysics, Johannes Kepler University Linz, Linz, Austria*

ANDERS S. HANSEN • *Department of Molecular and Cell Biology, University of California, Berkeley, CA, USA*

HIROYASU HATAKEYAMA • *Frontier Research Institute for Interdisciplinary Sciences, Tohoku University, Sendai, Japan*
PETER HINTERDORFER • *Institute of Biophysics, Johannes Kepler University, Linz, Austria*
DIRK HOCKEMEYER • *Department of Molecular and Cell Biology, University of California, Berkeley, CA, USA*
CAMERON HODGES • *Department of Biophysics, University of Michigan, Ann Arbor, MI, USA*
HUANG HUANG • *Dalton Cardiovascular Research Center, University of Missouri, Columbia, MO, USA; Department of Medical Pharmacology and Physiology, University of Missouri, Columbia, MO, USA*
CHIRLMIN JOO • *Department of BioNanoScience, Kavli Institute of NanoScience, Delft University of Technology, Delft, The Netherlands*
ARTUR KACZMARCZYK • *Huygens-Kamerlingh Onnes Laboratory, Leiden Institute of Physics, Leiden University, Leiden, The Netherlands; Department of Bionanoscience, Kavli Institute of Nanoscience, Delft University of Technology, Delft, The Netherlands*
MAKOTO KANZAKI • *Graduate School of Biomedical Engineering, Tohoku University, Sendai, Japan*
ANGELIKA KARDINAL • *Department of Physics, Nanosystems Initiative Munich, and Center for Nanoscience, LMU Munich, Munich, Germany*
HERLINDE DE KEERSMAECKER • *Division of Molecular Imaging and Photonics, Department of Chemistry, KU Leuven-University of Leuven, Leuven, Belgium*
GAVIN M. KING • *Department of Physics and Astronomy, University of Missouri, Columbia, MO, USA; Department of Biochemistry, University of Missouri, Columbia, MO, USA*
NORIYUKI KODERA • *Nano Life Science Institute (WPI-NanoLSI), Kanazawa University, Kakuma, Kanazawa, Ishikawa, Japan*
MELANIE KOEHLER • *Louvain Institute of Biomolecular Science and Technology, Université catholique de Louvain, Louvain-la-Neuve, Belgium*
FRANZISKA KRIEGEL • *Department of Physics, Nanosystems Initiative Munich, and Center for Nanoscience, LMU Munich, Munich, Germany*
XUYE LANG • *Chemical and Environmental Engineering, University of California Riverside, Riverside, CA, USA*
LEIKE XIE • *Dalton Cardiovascular Research Center, University of Missouri, Columbia, MO, USA; Department of Pathology and Anatomical Sciences, University of Missouri, Columbia, MO, USA*
JAN LIPFERT • *Department of Physics, Nanosystems Initiative Munich, and Center for Nanoscience, LMU Munich, Munich, Germany*
LU-NING LIU • *Institute of Integrative Biology, University of Liverpool, Liverpool, UK*
JULIA LOCKE • *Macromolecular Machines Laboratory, The Francis Crick Institute, London, UK*
GIOVANNI LONGO • *Istituto di Struttura della Materia ISM—CNR, Rome, Italy*
ZHENGJIAN LV • *Department of Pharmaceutical Sciences, University of Nebraska Medical Center, Omaha, NE, USA; Bruker Nano Surfaces Division, Santa Barbara, CA, USA*
YURI L. LYUBCHENKO • *Department of Pharmaceutical Sciences, University of Nebraska Medical Center, Omaha, NE, USA*
SERGEI MAGONOV • *SPM Labs LLC, Tempe, AZ, USA*
SIBAPRASAD MAITY • *Department of Pharmaceutical Sciences, College of Pharmacy, University of Nebraska Medical Center, Omaha, NE, USA*

ARIN MARCHESI • *U1006 INSERM, Université Aix-Marseille, Parc Scientifique et Technologique de Luminy, Marseille, France; LAI, Aix-Marseille Université, INSERM UMR_S 1067, CNRS UMR 7333, Marseille, France*

PIOTR E. MARSZALEK • *Department of MEMS, Duke University, Durham, NC, USA*

TINA R. MATIN • *Department of Anesthesiology, Weill Cornell Medicine, New York, NY, USA; Department of Physics and Astronomy, University of Missouri, Columbia, MO, USA*

JENS-CHRISTIAN MEINERS • *Departments of Biophysics and Physics, University of Michigan, Ann Arbor, MI, USA*

GERALD A. MEININGER • *Dalton Cardiovascular Research Center, University of Missouri, Columbia, MO, USA; Department of Pathology and Anatomical Sciences, University of Missouri, Columbia, MO, USA; Department of Medical Pharmacology and Physiology, University of Missouri, Columbia, MO, USA*

MUSTAFA MIR • *Department of Molecular and Cell Biology, University of California, Berkeley, CA, USA*

FERNANDO MORENO-HERRERO • *Department of Macromolecular Structures, Centro Nacional de Biotecnología, Consejo Superior de Investigaciones Científicas, Madrid, Spain*

SUA MYONG • *Department of Biophysics, Johns Hopkins University, Baltimore, MD, USA*

THOMAS NICOLAUS • *Department of Physics, Nanosystems Initiative Munich, and Center for Nanoscience, LMU Munich, Munich, Germany*

ADRIAN PASCAL NIEVERGELT • *School of Engineering, École Polytechnique Fédérale de Lausanne, Lausanne, Switzerland*

JOHN VAN NOORT • *Huygens-Kamerlingh Onnes Laboratory, Leiden Institute of Physics, Leiden University, Leiden, The Netherlands*

YOO JIN OH • *Institute of Biophysics, Johannes Kepler University, Linz, Austria; Keysight Technologies Austria GmbH, Linz, Austria*

KENJIRO ONO • *Department of Neurology, School of Medicine, Showa University, Tokyo, Japan*

FRANCESCO SAVERIO PAVONE • *LENS-European Laboraotry for Non-Linear Spectroscopy, Sesto Fiorentino, Italy; INO- National Institute of Optics, Sesto Fiorentino, Italy; Department of Physics and Astronomy, University of Florence, Sesto Fiorentino, Italy*

CHI PHAM • *Huygens-Kamerlingh Onnes Laboratory, Leiden Institute of Physics, Leiden University, Leiden, The Netherlands*

ANNA E. PITTMAN • *Department of Physics and Astronomy, University of Missouri, Columbia, MO, USA*

LORENA REDONDO-MORATA • *Inserm U1019, Institut Pasteur de Lille, Center for Infection and Immunity of Lille, Lille, France*

ARMANDO REIMER • *Biophysics Graduate Group, University of California, Berkeley, CA, USA*

FELIX RICO • *LAI, Aix-Marseille Université, INSERM UMR_S 1067, CNRS UMR 7333, Marseille, France*

SUSANA ROCHA • *Division of Molecular Imaging and Photonics, Department of Chemistry, KU Leuven-University of Leuven, Leuven, Belgium*

JORGE RODRIGUEZ-RAMOS • *Institute of Integrative Biology, University of Liverpool, Liverpool, UK*

JAYA SARKAR • *Department of Biophysics, Johns Hopkins University, Baltimore, MD, USA*

SIMON SCHEURING • *Department of Anesthesiology, Department of Physiology and Biophysics, Weill Cornell Medical College, New York, NY, USA*

ZACKARY N. SCHOLL • *Department of Physics, University of Alberta, Edmonton, AB, Canada*

GERALD J. SHAMI • *School of Medical Sciences (Discipline of Anatomy and Histology)—The Bosch Institute, The University of Sydney, Camperdown, NSW, Australia*
KRISHNA P. SIGDEL • *Department of Physics and Astronomy, University of Missouri, Columbia, MO, USA*
IGOR SOKOLOV • *Department of Mechanical Engineering, Tufts University, Medford, MA, USA; Department of Biomedical Engineering, Tufts University, Medford, MA, USA; Department of Physics, Tufts University, Medford, MA, USA*
MICHAEL STADLER • *Department of Molecular and Cell Biology, University of California, Berkeley, CA, USA*
MICAH P. STUMME-DIERS • *Department of Pharmaceutical Sciences, University of Nebraska Medical Center, Omaha, NE, USA*
HIROSHI SUGIYAMA • *Department of Chemistry, Graduate School of Science, Kyoto University, Sakyo-ku, Kyoto, Japan; Institute for Integrated Cell-Material Sciences, Kyoto University, Sakyo-ku, Kyoto, Japan*
FIDAN SUMBUL • *LAI, Aix-Marseille Université, INSERM UMR_S 1067, CNRS UMR 7333, Marseille, France*
ZHIQIANG SUN • *Department of Pharmaceutical Sciences, University of Nebraska Medical Center, Omaha, NE, USA*
HIROHIDE TAKAHASHI • *Department of Anesthesiology, Department of Physiology and Biophysics, Weill Cornell Medical College, New York, NY, USA*
ASTOU TANGARA • *Department of Molecular and Cell Biology, University of California, Berkeley, CA, USA*
ANDREEA TRACHE • *Department of Medical Physiology, Texas A&M Health Science Center, College Station, TX, USA; Department of Biomedical Engineering, Texas A&M University, College Station, TX, USA*
TAKAYUKI UCHIHASHI • *Nagoya, Nagoya University, Nagoya, Aichi, Japan*
HIROSHI UJI-I • *Division of Molecular Imaging and Photonics, Department of Chemistry, KU Leuven-University of Leuven, Leuven, Belgium; Research Institute for Electronic Science, Nanomaterials and Nanoscopy, Hokkaido University, Sapporo, Japan*
WILLEM VANDERLINDEN • *Department of Physics, Nanosystems Initiative Munich, and Center for NanoScience, LMU Munich, Munich, Germany; Division of Molecular Imaging and Photonics, Department of Chemistry, KU Leuven-University of Leuven, Leuven, Belgium*
EKATERINA VIAZOVKINA • *Cepheid, Inc., Bothell, WA, USA*
LELA VUKOVIĆ • *Department of Chemistry and Biochemistry, University of Texas at El Paso, El Paso, TX, USA*
HIROKI WATANABE • *Research Institute of Biomolecule Metrology Co. Ltd, Tsukuba, Ibaraki, Japan*
TAKAHIRO WATANABE-NAKAYAMA • *WPI Nano Life Science Institute (WPI-NanoLSI), Kanazawa University Kakuma-machi, Kanazawa, Japan*
IAN WHEELDON • *Chemical and Environmental Engineering, University of California Riverside, Riverside, CA, USA*
KAREN ZAGORSKI • *Department of Pharmaceutical Sciences, University of Nebraska Medical Center, Omaha, NE, USA*
RONG ZHU • *Institute of Biophysics, Johannes Kepler University Linz, Linz, Austria*
FRANCESCA ZUTTION • *LAI, Aix-Marseille Université, INSERM UMR_S 1067, CNRS UMR 7333, Marseille, France*

Part I

Imaging and Probing of Biomolecules

Chapter 1

High-Resolution Atomic Force Microscopy Imaging of Nucleic Acids

Pablo Ares, Julio Gomez-Herrero, and Fernando Moreno-Herrero

Abstract

Exploring the limits of spatial resolution has been a constant in the history of atomic force microscopy imaging. Since its invention in 1986, the AFM has beaten the barrier of resolution continuously, thanks to technical developments, miniaturization of tips, and implementation of new imaging modes. The double helix structure of DNA has been always at the horizon of resolution. Today, this milestone has been reached, not only imaging DNA but also its close relative double-stranded RNA. Here, we provide a comprehensive description of the methods employed and the steps required to image the helical periodicity of these two nucleic acids with the sample immersed in a buffer solution.

Key words Atomic force microscopy, Double-stranded DNA, Double-stranded RNA, AFM imaging methods

1 Introduction

In atomic force microscopy, a sharp tip supported by a micrometer-size cantilever is employed to scan and probe the topography of a surface. Tip-sample interaction is transduced into a change of a parameter of the supporting cantilever (typically deflection, oscillation amplitude, or frequency), which is used as control in a feedback loop to obtain a topographic image. The high sensitivity of the technique in the vertical dimension relies on the strong dependence of the tip-sample interaction with the tip-sample distance. However, the lateral resolution of AFM has been always punished by the finite size of the tip giving rise to the so-called tip-dilation problem when imaging objects smaller than the size of the tip. The development of new imaging modes and, most importantly, the use of cantilevers with resonance frequency of tens of kHz in liquid and low stiffness have been critical to the improvement of the level of control on the tip-sample distance and, consequently, on the force applied to soft materials such as nucleic acids [1–3]. Since high-imaging speed is not a requirement here, cantilevers do not need to

Yuri L. Lyubchenko (ed.), *Nanoscale Imaging: Methods and Protocols*, Methods in Molecular Biology, vol. 1814,
https://doi.org/10.1007/978-1-4939-8591-3_1,

be ultrasmall, and regular commercial cantilevers available for most AFM setups are valid for high resolution. Nevertheless, in principle, high-speed imaging may be beneficial to improve spatial resolution as low-frequency noise effects are reduced. Minimization of the force exerted by the tip has proven to be fundamental to achieve high resolution of soft matter, but still in order to probe narrow and deep trenches, a sharp tip is required. We have found that the combination of small forces and sharp point tips is the key factor to image with enough resolution to resolve the helical structure of nucleic acids. On the contrary, the imaging mode chosen to scan the sample was not critical.

Despite the seminal work by Mou et al. [4] in contact mode, it has been only recently when the double helix of DNA has been clearly resolved using dynamic modes [1, 2, 5–7]. High-resolution AFM imaging of double-stranded RNA has also been recently reported [3]. The more commonly used AFM dynamic modes are amplitude modulation AFM (AM-AFM, also known as tapping™ or intermittent contact) [8, 9] and frequency modulation AFM (FM-AFM) [10–13]. Other advanced dynamic modes include drive amplitude modulation (DAM-AFM) [14], Kelvin probe force microscopy [15], and multifrequency-based imaging modes [16]. In dynamic modes, the cantilever is oscillated at or near its resonance frequency, and either the cantilever oscillation amplitude (AM-AFM) or the frequency shift (FM-AFM) is used as control parameter for topographic imaging. Force-distance-based imaging modes include peakforce tapping® [17] or jumping mode plus (JM +) [18–20]. These imaging modes use the force given by the deflection of the cantilever as control parameter. In references [2, 3], it is also proved that force-distance-based imaging modes can be used for high-resolution imaging of nucleic acids.

A list of requirements for high-resolution imaging is given below.

A soft tip-sample interaction. In dynamic modes, this is achieved using cantilevers with a high resonance frequency (e.g., ~110 kHz in air conditions, typically 25 kHz in liquid) and a low constant force (e.g., ~0.09 N m^{-1}), and this necessarily requires relatively small cantilevers. In force-distance-based modes, the cantilever is not driven at its resonance frequency. However, in order to minimize drag forces, which result in a hysteresis deflection loop, small cantilevers are also desirable. Fine-tuning of parameters for each imaging mode is essential (*see* below).

Sharp tips. Most commercially available AFM tips have a nominal radius (<10 nm) larger than the size of a nucleic acid groove. However, this showed not to be essential to resolve the helical periodicity of dsDNA [2] and dsRNA [3]. A plausible explanation is that the small features of the biological specimens are probed by a tiny irregularity at the end of the tip, which effectively is able to get inside the grooves. Note, however, that probing at the grooves is possible only if low forces are maintained while imaging. This also introduces a random factor in the experiments: not all cantilevers perform the same in terms of high resolution.

Minimization of cantilever spurious resonance peaks in liquid. In dynamic modes, the cantilever is oscillated at or near its resonance frequency. Due to the low-quality factor of the cantilever resonance in liquids, its response in these media is broad, and acoustic excitation of the tip holder can produce multiple spurious peaks. Not all of these peaks, which could hide the real one, work satisfactorily for dynamic modes operation. For high-resolution imaging, it is important to have a clean resonance peak. Different AFM designs deal with this problem using different strategies, which include photothermal [21] and magnetic excitation [22] of the tip, or optimized acoustic excitation [23].

High sensitivity of the detection system. The relation between the distance the cantilever moves (in nanometers) and the output it generates at the detection system (in volts) is known as the deflection sensitivity (in nm V^{-1}). A low value of this magnitude (<20 nm V^{-1}) has proven to be crucial to control the low amplitudes and forces required for high-resolution imaging.

High mechanical and thermal stability. The microscope has to be stable over image acquisition. These systems are often inside an acoustic isolation enclosure placed over an active or passive vibration isolation system. An RMS noise in vertical direction lower than ~0.5 Å must be achieved. Regarding thermal stability, it is desirable to keep the microscope in a constant temperature environment to minimize thermal tip-sample drift. Low temperatures (15–18 °C) will also reduce liquid evaporation and changes in buffer composition. Rotation of the image scan area to make DNA/RNA molecules to coincide with the fast scanning direction also minimizes drift effects (*see* below).

Precise scanner calibration and low electronic noise. AFM piezo scanners are factory calibrated to adapt the piezo-driving voltages to the required scanning sizes. The specifications of the scanner should be appropriate for high-resolution imaging (noises must be lower than 1 Å in the X–Y directions and 0.5 Å in the Z direction). Additionally, for high-resolution measurements, it is advisable to recalibrate the piezo scanner using AFM test grids with pitch distances as close as possible to the distances to be measured. Otherwise, the helical features of the nucleic acids might be imaged at wrong sizes. For instance, atomic steps (0.34 nm) in a graphite surface can be used to accurately calibrate along the vertical direction and its atomic periodicity (0.25 nm) to calibrate along the X–Y directions.

2 Materials

1. Cantilevers: Biolever mini BL-AC40TS-C2 available from Olympus.
2. V-4 grade mica sheets (SPI supplies). Mica substrates are fixed with double-sided tape to magnetic AFM specimen discs of 15 mm diameter (Ted Pella, INC).
3. Plasmid pGEM3Z (Promega) at 1 μg/μL.
4. Restriction enzyme BamHI (New England Biolabs).
5. RNase-free DNase I (Roche).
6. QIAquick PCR purification Kit (Qiagen).
7. RNeasy MinElute Cleanup Kit (Qiagen).
8. HiScribe T7 in vitro transcription kit (New England Biolabs).
9. Chemicals to prepare buffers: $NiCl_2$, Tris base, and HCl to adjust the pH to 8.0.
10. Milli-Q quality water.
11. Buffer A: 10 mM $NiCl_2$, 10 mM Tris–HCl pH 8.0.
12. Buffer B: 10 mM Tris–HCl pH 8.0.
13. Double-stranded DNA and RNA molecules produced following the methods described below and stored in TE buffer at 4 °C at a concentration of ~0.6 ng/μL.
14. Atomic force microscope instrument from Nanotec Electronica S.L.

3 Methods

3.1 Fabrication of dsDNA Molecules for High-Resolution AFM Imaging

dsDNA molecules in their linear form are typically produced by PCR or restriction digestion of plasmids that contain sites of interest or particular structures. Multiple DNA plasmids are commercially available but can be also produced by cloning in bacteria. DNA molecules produced from plasmid restriction are often more homogeneous in size than those fabricated by PCR.

1. Cleave the pGem3Z plasmid with BamHI, mix 1 μg of pGem3Z (1 μL from stock at 1 μg/μL) with 1 μL BamHI (stock at 20 units/μL) in 1× NEBuffer 3.1 for a total volume of 20 μL, and incubate the mixture for 1 h at 37 °C.
2. Purify the linearized DNA using the QIAquick PCR purification kit following the instructions of the manufacturer. DNA is eluted in 20 μL of TE buffer (10 mM Tris–HCl pH 8.0, 1 mM EDTA).
3. Measure the concentration of DNA and prepare a dilution of 0.6 ng/μL for AFM measurements.

3.2 Fabrication of dsRNA Molecules for High-Resolution AFM Imaging

dsRNA molecules are fabricated by annealing two complementary single-stranded RNA molecules produced by transcription using T7 RNA polymerase on an appropriate dsDNA template containing the T7 RNA polymerase promoter. After transcription and annealing, the dsRNA is purified by RNeasy MinElute Cleanup Kit (Qiagen) and stored in TE buffer [24, 25].

1. Fabricate complementary ssRNA strands by transcription. Transcription is performed at 42 °C for 3 h, in RNase-free PCR tubes using the HiScribe T7 in vitro transcription kit, with 1500 U of T7 RNA polymerase in a total reaction volume of 60 μL.
2. Anneal the complementary ssRNA strands. After 3 h of transcription reaction, pause the reactions and add EDTA to reach a final concentration of 30 mM EDTA (10.6 μL of 0.2 M EDTA in 500 mM Tris, pH 8.0, in RNase-free water). Mix both transcription reactions to anneal the complementary ssRNA strands. Annealing is achieved by heating for 1 h at 65 °C and slowly cooling down to room temperature at 1.2 °C/5 min rate. Resulting dsRNA molecules are purified using the QiaGen RNeasy Purification Kit and eluted in RNase-free water.
3. Eliminate the template DNA by digestion with 5 U of RNase-free DNase I (Roche) for 1 h at 37 °C.
4. Purify the dsRNA molecules by using the QiaGen RNeasy MinElute Cleanup kit, eluting in RNase-free water or TE buffer.
5. Measure the concentration of dsRNA and prepare a dilution of 0.6 ng/μL for AFM measurements.

3.3 dsDNA and dsRNA Sample Preparation for High-Resolution AFM Imaging

For imaging of nucleic acids, mica is the substrate of preference because it exposes a clean and atomically flat surface after mechanical cleavage that can be recycled multiple times. Importantly, immediately after its cleavage, the mica surface is always negatively charged, and this allows the use of different divalent cations to attach negatively charged molecules such as DNA and RNA. These include Mg^{2+}, Ca^{2+}, Mn^{2+}, and Ni^{2+}, which electrostatically bridge the negatively charged surface of mica with DNA or RNA. We found that a millimolar concentration of $NiCl_2$ was enough to firmly attach both dsDNA and dsRNA to the mica surface [25].

1. Functionalize the mica surface. First, cleave the mica and then add 10 μL of buffer A. Leave for 1 min.
2. Add 1 μL (0.6 ng) of dsDNA or dsRNA to the droplet and wait for up to 15 min. To avoid evaporation, the sample can be

covered by small box and a beaker with water placed inside to increase the local humidity.

3. Add 40 μL of buffer A and then add 50 μL of buffer B to reach a final volume of 100 μL.
4. Place the sample on the stage of the AFM (*see* Subheading 3.4, **step 9**).

3.4 AFM Setup

Most of the AFMs use the optical beam deflection method to detect the deflection of the cantilever [26] due to its simplicity, reliability, low noise, and ability to be applied to a variety of cantilevers. Hence, we will only consider AFMs of this kind. The following **steps 1–8** should be performed before or during sample preparation. **Steps 9–14** should be applied after sample preparation:

1. Turn on the AFM control unit and run the acquisition software. This will allow the electronics of the AFM to warm up.
2. Insert a cantilever into the cantilever holder and place it into the AFM head.
3. Align the laser beam to be focused on the cantilever.
4. Adjust the photodiode position maximizing the total intensity of light.
5. Calibrate the cantilevers' spring constant. Thermal [27] and Sader's [28] methods are the commonest. If your AFM control software includes a calibration method, follow the manufacturer's instructions. Otherwise, you can use either thermal or Sader's method (*see* **Note 1**).
6. Switch on the AM-AFM mode and tune the cantilever resonance frequency.
7. Place a dummy mica substrate (with the same thickness of the substrate employed in the nucleic-acid specimen) into the sample holder and mount it on the AFM scanner.
8. Coarse approach the cantilever to the substrate until a distance of less than 1 mm between the cantilever chip and the mica is observed. Remove the AFM head from the acquisition position. This will prevent from crashing the tip later when placing the AFM head on the DNA/RNA sample with liquid.
9. Load the sample (mica substrate + nucleic acids) onto the AFM scanner. Sample should be prepared just before loading into the AFM. Special care should be taken to prevent that sample gets dry.
10. Pre-wet the cantilever chip by placing a drop of ~30–40 μL of buffer B directly on the cantilever holder (*see* **Note 2**). This volume may change depending on the cantilever holder design.
11. Return the AFM head to its acquisition position.

12. Due to the presence of liquid, the position of the laser on the cantilever has moved in the longitudinal direction of the cantilever. Realign it.
13. Readjust the photodiode position.
14. Retune the cantilever resonance frequency. Due to the presence of liquid, the new value should be ~1/3–1/4 of the value in air and the quality factor much lower.

3.5 Tip-Sample Approach

Independently of the acquisition mode, tip-sample approach is done in AM-AFM mode, since it is the easiest mode among the suitable ones for high resolution. Modern AFMs include easy-to-use routines for the tip-sample approach where no parameter selection is necessary. If this is not the case, follow the next steps.

1. Select a relatively high free oscillation amplitude, to a value of ~7–10 nm (*see* **Note 3**). By using a value in this range, false engagements during approach in liquids will be minimized.
2. Select a set point value of about 75% of the free amplitude.
3. Select a value of about 15 μm/s for the approach speed.
4. Tip-sample approach routines typically use an approach control parameter (i.e., amplitude reduction) to stop the motion. Select an initial conservative value.
5. Start the tip-sample approach motion. As the tip approaches the sample, the amplitude will decrease until reaching the approach control parameter condition.
6. If the approach motion stops quickly after motor movement, false engagement is occurring. If this happens, increase the value of the approach control parameter and start the approach procedure again. Repeat this step increasing the control parameter value until the tip engages the sample.
7. At this point, the amplitude equals the set point at 75% of the free amplitude, but since we are using low-spring constant cantilevers, the tip might not be in imaging range. Change the set point, so the piezo scanner will carefully reduce the tip-sample gap until the tip gets in imaging range (*see* **Note 4**).

3.6 Amplitude Modulation, AM-AFM Mode

AM-AFM operates by using the amplitude of the cantilever oscillation as the control parameter for topography acquisition. A feedback diagram of this mode can be found in Fig. 1a.

1. Once the tip is in imaging range, reduce the drive amplitude to about 1 nm and the set point value accordingly to get back to imaging range (*see* **Note 5**). Keep a set point value just below the amplitude value at which the cantilever lifts off the sample, typically 0.5–0.8 nm. Using low oscillation amplitude is a key factor to obtain high resolution in liquid.

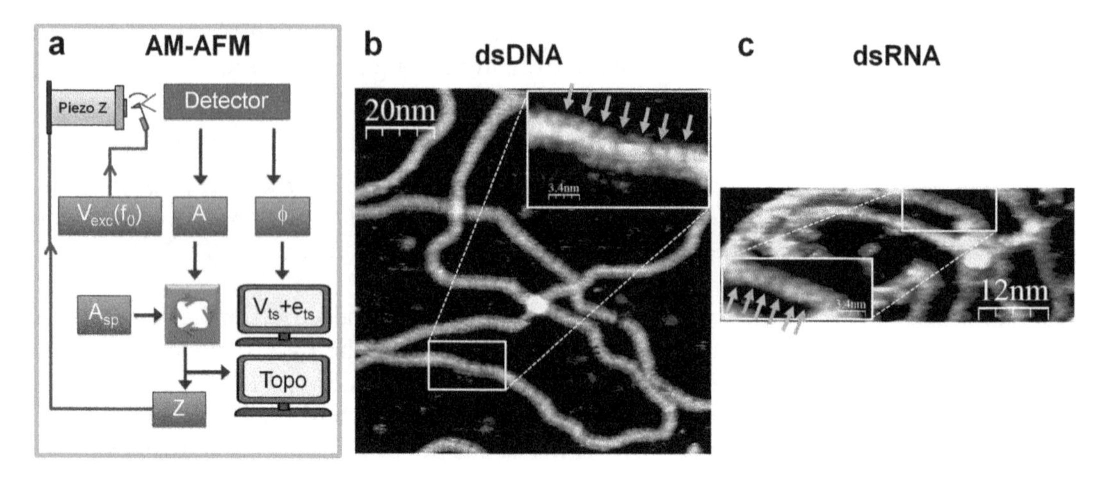

Fig. 1 Amplitude modulation AFM mode and high-resolution imaging of dsDNA and dsRNA. (**a**) Diagram of AM-AFM mode experimental setup. (**b**) AM-AFM high-resolution image of dsDNA. (**c**) AM-AFM high-resolution image of dsRNA. Insets show fragments of dsDNA and dsRNA molecules, where their major and minor grooves are resolved. Panels (**a**) and (**c**) [3]—Adapted with permission from The Royal Society of Chemistry

2. Select an area of $1 \times 1\ \mu m^2$ with a resolution of 128×128 pixels and a scanning frequency of ~2 lines per s and start the scanning operation. A wide field is preferred at this moment to pinpoint molecules with straight segments over several tens of nm.
3. Reduce the scan sizes to 50×50–$150 \times 150\ nm^2$ and zoom in over straight DNA/RNA segments. Acquire at 512×512 pixels resolution and scanning frequencies of 3–5 lines per s. To visualize the helical features of the nucleic acids, set the fast scan direction preferably parallel to the straight segments (*see* **Note 6**).
4. Initially scan each line in the image from left to right (trace direction) and from right to left (retrace direction). Then, if the topographic features are conserved in both trace and retrace images, you may acquire in trace direction only to double the scan rate.
5. In AM-AFM, there is only one feedback loop where the amplitude is used as the controlled input for the topography feedback. Adjust the feedback gains to optimize topography acquisition (*see* **Note 7**).
6. If the effective tip radius is small enough, lower than ~2.5 nm according to [3], double-stranded nucleic acids major and minor grooves should be visible. Figure 1b, c shows examples of high-resolution images of dsDNA and dsRNA, respectively.

DAM-AFM operates by using three different feedback loops. A phase-locked loop (PLL) is used to track the resonance frequency.

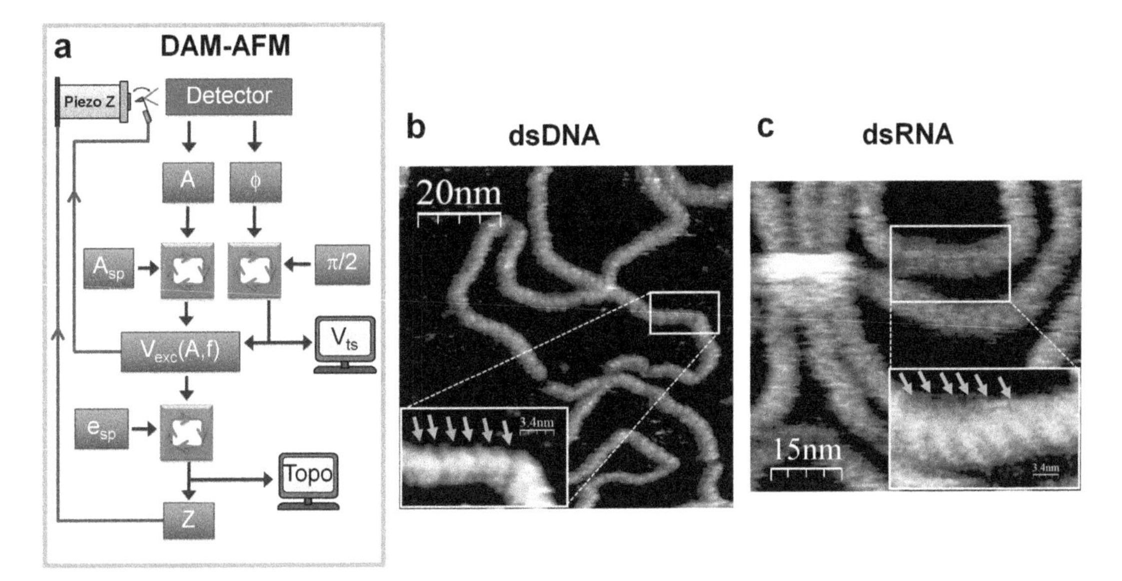

Fig. 2 Drive amplitude modulation AFM mode and high-resolution imaging of dsDNA and dsRNA. (**a**) Diagram of DAM-AFM mode experimental setup. (**b**) DAM-AFM high-resolution image of dsDNA. Inset shows a fragment of dsDNA, where its major and minor grooves are resolved. (**c**) DAM-AFM high-resolution image of dsRNA, where the helical periodicity was resolved (inset). Panels (**a**) and (**c**) [3]—Adapted with permission from The Royal Society of Chemistry

3.7 Drive Amplitude Modulation, DAM-AFM Mode

A second feedback loop adjusts the drive amplitude in order to maintain the cantilever oscillation amplitude at a fixed value. The drive amplitude channel is related to the power dissipated by the tip-sample interaction. Finally, a third feedback loop controls the topography acquisition adjusting the position of the scanner in the Z direction by keeping the drive amplitude constant at a set point value. A feedback diagram of this mode can be found in Fig. 2a.

1. Approach in AM-AFM mode (*see* Subheading 3.6 above). Once the tip is in imaging range, withdraw the tip from the sample for about 1 μm. By doing this, we avoid any possible damage that may occur while performing the procedure to engage all the different feedback loops. One micrometer distance is also convenient to easily get back to imaging range.
2. Enable the PLL and adjust its feedback gains (*see* **Note 7**).
3. Enable an extra feedback loop. The input of this feedback is the cantilever amplitude and the output is the drive amplitude. Select a set point for this feedback of 0.5–0.8 nm. Adjust its feedback gains (*see* **Note 7**).
4. Change the topography feedback channel, selecting the drive amplitude as input channel. Check the numerical value of the drive amplitude and select a topography set point about 10% higher than this drive amplitude value.

5. Start the tip-sample approach motion as in Subheading 3.5, **step 5**. As the tip approaches the sample, the drive amplitude will increase until the drive amplitude set point is reached. If it does not reach the region of tip-sample interaction (imaging range), increase the drive amplitude set point. As already mentioned, drive amplitude is related to the tip-sample power dissipation. Typical values of drive amplitude set point for imaging nucleic acids are in the range of 0.2–0.5 fW (*see* **Note 8**).
6. Start image acquisition following the same steps as for AM-AFM mode (*see* Subheading 3.6, **steps 2–4**). Adjust the topography feedback gains, and readjust the other feedback gains if necessary to optimize topography acquisition (*see* **Note 7**).
7. Double-stranded nucleic acids helical structure should be resolved. Figure 2b, c show examples of dsDNA and dsRNA, respectively, where the helical pitch can be clearly resolved.

3.8 Force-Distance-Based Modes

Force-distance-based modes operate by performing a quick FZ curve at each pixel of the scanned area. Depending on the AFM manufacturer, the name of the mode varies (jumping mode, pulsed mode, peakforce tapping®, QI™ mode, force-distance mapping mode, HybriD™ mode, etc.), and there are slight differences on the way the movement of the tip is performed. In this section, we will consider jumping mode plus (JM+) as it was described in reference [3]. JM+ moves the tip from pixel to pixel at the farthest tip-sample distance, thus minimizing lateral forces. JM+ uses the peak force referenced to the force baseline as the control parameter for the topography acquisition, allowing a precise direct control on the applied forces. A feedback diagram of this mode can be found in Fig. 3a.

1. Once the tip is in imaging range (*see* approach in AM-AFM, Subheading 3.6), remove the cantilever oscillation and select a low set point in contact mode. This will bring the tip to a gentle contact with the substrate (*see* **Note 9**).
2. Select the proper values for the amplitude and frequency of the Z excursion movement. Amplitudes of 15–35 nm and frequencies of 0.5–1 kHz should work fine for JM+ (*see* **Note 10**).
3. Select the option to compensate for the dragging force (if it is available at your AFM control software).
4. Select the set point with respect to the force baseline. Typical values for the force set point are of the order 30–50 pN (*see* **Note 11**).
5. Start the image acquisition following the same steps as for AM-AFM mode (*see* Subheading 3.6, **steps 2–4**). In the case

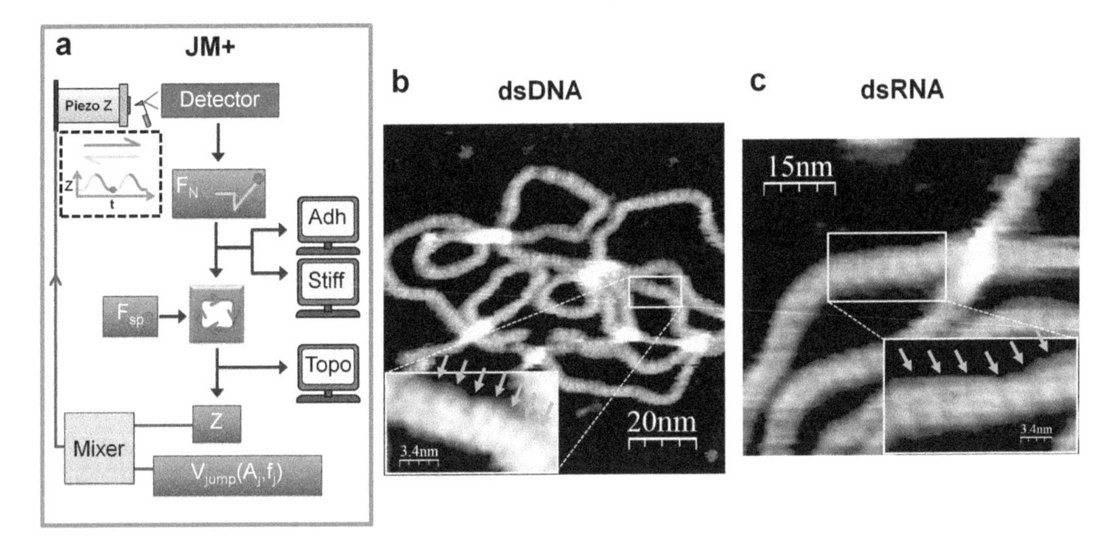

Fig. 3 Jumping mode plus and high-resolution imaging of dsDNA and dsRNA. (**a**) Diagram of JM+ mode experimental setup. (**b**) JM+ high-resolution image of dsDNA. Inset shows a fragment of dsDNA, where its major and minor grooves are resolved. (**c**) JM+ high-resolution image of dsRNA, where the helical periodicity was resolved (inset). Panels (**a**) and (**c**) [3]—Adapted with permission from The Royal Society of Chemistry

of JM+, the scan rate will be related to the frequency of the Z excursion. The abovementioned parameters will give typical scan rates of 3–4 lines per s. Adjust the feedback gains to optimize topography acquisition (*see* **Note 7**).

6. Double-stranded nucleic acids helical structure should be resolved. Figure 3b, c show examples on dsDNA and dsRNA, respectively, where the helical pitch can be clearly resolved.

4 Notes

1. The "thermal method" is based on the cantilever's thermal distribution spectrum (square of the fluctuations in amplitude as a function of frequency). According to the equipartition theorem, the mean square amplitude of the cantilever's thermal fluctuation in the vertical direction $\langle z^2 \rangle$ can be expressed as $\langle z^2 \rangle = (k_B\ T)/k_N$, where k_B is Boltzmann's constant, T is the temperature of the cantilever, and k_N is the cantilever spring constant.

   Sader's method can be applied to rectangular cantilevers. It incorporates the viscosity and density of the medium in which the cantilever is immersed, along with experimentally determined values of the resonance frequency and quality factor, together with the length and width of the cantilever. Thus it can be considered a geometric approach, although it does not need the cantilever thickness. The spring constant can be easily calculated online at the University of Melbourne webpage (http://www.ampc.ms.unimelb.edu.au/afm/calibration.html) or directly on some AFM control software.

2. Pre-wetting the cantilever is useful to preserve its integrity when forming a liquid meniscus between the cantilever holder and the sample. It is recommended to pre-wet the whole cantilever chip, as this will ensure a more stable liquid meniscus.

3. Some AFM systems display the cantilever amplitude in volts. To convert the cantilever amplitude to nm, multiply the value in volts by the deflection sensitivity of the detection system, taking also into account possible gains that may be applied to the amplitude channel. For example, the typical sensitivity of our system is ~10 nm V^{-1}, and therefore the approach amplitude of 10 nm corresponds to 1 V.

4. It might happen that the piezo scanner fully extends before reaching the imaging position. If this happens, perform a coarse approach using motor steps until the vertical scanner position is within its elongation range. Change again the set point to extend the piezo until the tip gets in imaging range or repeat the coarse approach procedure. One way to check if the AFM is in imaging range is by performing a force vs. distance (FZ) curve, monitoring the amplitude as the tip moves away from the sample. If the tip is in imaging range, the amplitude should increase with tip-sample distance as far as the tip-sample interaction is relevant. Beyond this point, the amplitude remains constant.

5. To reduce the oscillation of the cantilever, first reduce the drive amplitude and then change the set point to drive the system back to imaging range. The cantilever oscillation is typically proportional to the drive amplitude. For example, a decrease of the drive amplitude by a factor of 10 implies an equivalent change of set point.

6. All scanning microscopies suffer from the so-called $1/f$ noise (being f the noise frequency). Thus, low-frequency signals contribute with high noise amplitudes. The $1/f$ noise mainly includes contributions from thermal drift and mechanical noise. AFM is a scanning technique where the tip is moved along to orthogonal directions. In general, the fast scan direction is about two orders of magnitude faster than the other orthogonal direction. A common way to minimize the $1/f$ noise is to increase the acquisition frequency, and this is easily done by measuring along the fast scan axis. By setting the fast scan direction to be parallel to the straight segments of the molecules, low-frequency noise is minimized. Note that if the molecule is not aligned with the fast axis, points in close

proximity will be probed within a long lag time contributing to $1/f$ noise that may result in loss of spatial resolution. Therefore, low-frequency noise is much likely to have an effect on data points acquired along the slow scan direction, and this is the reason why we rotate the image to have the molecules preferably aligned with fast scan axis (the horizontal axis of the image). Nevertheless, under conditions of very low thermal drift and mechanical noise, it is possible to observe the periodicity along molecules not aligned to the horizontal direction of the image [3].

7. The feedback gain values that optimize the acquisition without damaging the sample need to be determined for each AFM setup. In general, the higher the gain, the faster will be the response. However, a too high gain value will produce instabilities and oscillations in the feedback.

8. The drive amplitude (in volts) can be related to the power (watts) dissipated between tip and sample. To convert volts into watts, the Cleveland formula can be used [29]:

$$\overline{P_{\text{tip}}} = 2\pi f_0 \frac{\frac{1}{2}kA^2}{Q}\left(\frac{V_{\text{drive}}}{V_{\text{drive},0}} - \frac{f}{f_0}\right)$$

   where P_{tip} is the dissipation power, f_0 is the resonance frequency of the free cantilever, f is the resonance frequency in imaging range, k is its spring constant, A is its amplitude, Q is its quality factor, $V_{\text{drive},0}$ is the driving voltage applied to the free cantilever, and V_{drive} is the driving voltage of the cantilever in imaging range.

9. When changing from AM-AFM to contact mode, check the value of the normal force before removing the cantilever oscillation and use a slightly higher value for the topography set point to establish a very gentle contact.

10. In JM+, the amplitude of the Z excursion is called jump off, which typically ranges from 15 to 35 nm for this kind of experiments. The frequency of the Z excursion is controlled through two parameters, jump sample and control cycles. Values of 15–30 and 5–10, respectively, will provide a frequency of the Z excursion in the 0.5–1 kHz range. In JM+, the frequency of the Z excursion coincides with the frequency of pixel acquisition. Images of 256 × 256 pixels obtained by scanning only in the trace direction are typically taken in 120 s. Other force-based modes may operate using different frequencies.

11. These forces correspond to set points of 0.02–0.07 V, for Olympus BioLever mini cantilevers, and considering a deflection sensitivity of <20 nm V^{-1}.

In summary, high-resolution imaging of nucleic acids can be achieved by different imaging modes, provided the list of requirements given in the introduction is fulfilled. Therefore, the choice of a particular one should not be motivated by its resolution to image DNA or RNA but by the additional information of the sample that each particular mode may provide. For instance, AM-AFM can give information on material composition via the phase/frequency shift channel, DAM-AFM allows the quantification of the power dissipated in the process of imaging; and force-distance-based imaging modes give information on the mechanical properties of the sample.

Acknowledgments

We thank E. Herrero-Galan and C. Aicart for providing details of the protocol for dsRNA and dsDNA fabrication and A. Gil for critical reading of the manuscript. We thank the financial support from the Spanish MINECO/FEDER (projects MAT2016-77608-C3-3-P and The "María de Maeztu" Programme for Units of Excellence in R&D (MDM-2014-0377) to J.G.-H. and FIS2014-58328-P to F.M.-H.). F.M.-H. also acknowledges support from European Research Council (ERC) under the European Union's Horizon 2020 research and innovation (grant agreement No 681299).

References

1. Ido S, Kimura K, Oyabu N, Kobayashi K, Tsukada M, Matsushige K, Yamada H (2013) Beyond the helix pitch: direct visualization of native DNA in aqueous solution. ACS Nano 7 (2):1817–1822. https://doi.org/10.1021/nn400071n
2. Pyne A, Thompson R, Leung C, Roy D, Hoogenboom BW (2014) Single-molecule reconstruction of oligonucleotide secondary structure by atomic force microscopy. Small 10(16):3257–3261. https://doi.org/10.1002/smll.201400265
3. Ares P, Fuentes-Perez ME, Herrero-Galan E, Valpuesta JM, Gil A, Gomez-Herrero J, Moreno-Herrero F (2016) High resolution atomic force microscopy of double-stranded RNA. Nanoscale 8(23):11818–11826. https://doi.org/10.1039/c5nr07445b
4. Mou J, Czajkowsky DM, Zhang Y, Shao Z (1995) High-resolution atomic-force microscopy of DNA: the pitch of the double helix. FEBS Lett 371(3):279–282
5. Maaloum M, Beker AF, Muller P (2011) Secondary structure of double-stranded DNA under stretching: elucidation of the stretched form. Phys Rev E Stat Nonlinear Soft Matter Phys 83(3 Pt 1):031903
6. Kitazawa M, Ito S, Yagi A, Sakay N, Uekusa Y, Ohta R, Inaba K, Hayashi A, Hayashi Y, Tanemura M (2011) High-resolution imaging of plasmid DNA in liquids in dynamic mode atomic force microscopy using a carbon nanofiber tip. Jpn J Appl Phys 50(8):S3
7. Leung C, Bestembayeva A, Thorogate R, Stinson J, Pyne A, Marcovich C, Yang J, Drechsler U, Despont M, Jankowski T, Tschope M, Hoogenboom BW (2012) Atomic force microscopy with nanoscale cantilevers resolves different structural conformations of the DNA double helix. Nano Lett 12

(7):3846–3850. https://doi.org/10.1021/nl301857p
8. Lyubchenko YL, Shlyakhtenko LS (2009) AFM for analysis of structure and dynamics of DNA and protein-DNA complexes. Methods 47(3):206–213. https://doi.org/10.1016/j.ymeth.2008.09.002
9. Cassina V, Manghi M, Salerno D, Tempestini A, Iadarola V, Nardo L, Brioschi S, Mantegazza F (2015) Effects of cytosine methylation on DNA morphology: an atomic force microscopy study. Biochim Biophys Acta 1860(1 Pt A):1–7. https://doi.org/10.1016/j.bbagen.2015.10.006
10. Marti O, Drake B, Hansma PK (1987) Atomic force microscopy of liquid-covered surfaces: atomic resolution images. Appl Phys Lett 51 (7):484–486
11. Martinez-Martin D, Carrasco C, Hernando-Perez M, de Pablo PJ, Gomez-Herrero J, Perez R, Mateu MG, Carrascosa JL, Kiracofe D, Melcher J, Raman A (2012) Resolving structure and mechanical properties at the nanoscale of viruses with frequency modulation atomic force microscopy. PLoS One 7 (1):e30204. https://doi.org/10.1371/journal.pone.0030204
12. Fukuma T, Jarvis SP (2006) Development of liquid-environment frequency modulation atomic force microscope with low noise deflection sensor for cantilevers of various dimensions. Rev Sci Instrum 77:043701
13. Yamada H, Kobayashi K, Fukuma T, Hirata Y, Kajita T, Matsushige K (2009) Molecular resolution imaging of protein molecules in liquid using frequency modulation atomic force microscopy. Appl Phys Express 2(9):095007
14. Jaafar M, Martinez-Martin D, Cuenca M, Melcher J, Raman A, Gomez-Herrero J (2012) Drive-amplitude-modulation atomic force microscopy: from vacuum to liquids. Beilstein J Nanotechnol 3:336–344. https://doi.org/10.3762/bjnano.3.38
15. Nonnenmacher M, O'Boyle MP, Wickramasinghe HK (1991) Kelvin probe force microscopy. Appl Phys Lett 58(25):2921–2923
16. Garcia R, Herruzo ET (2012) The emergence of multifrequency force microscopy. Nat Nanotechnol 7(4):217–226. https://doi.org/10.1038/nnano.2012.38
17. Pittenger BB, Erina N (2012) Application note #128. Quantitative mechanical property mapping at the nanoscale with PeakForce QNM. Bruker, Santa Barbara, CA
18. de Pablo PJ, Colchero J, Gómez-Herrero J, Baro AM (1998) Jumping mode scanning force microscopy. Appl Phys Lett 73 (22):3300–3302
19. Ortega-Esteban A, Horcas I, Hernando-Perez-M, Ares P, Perez-Berna AJ, San Martin C, Carrascosa JL, de Pablo PJ, Gomez-Herrero J (2012) Minimizing tip-sample forces in jumping mode atomic force microscopy in liquid. Ultramicroscopy 114:56–61. https://doi.org/10.1016/j.ultramic.2012.01.007
20. Rosa-Zeise A, Weilandt E, Hild S, Marti O (1997) The simultaneous measurement of elastic, electrostatic and adhesive properties by scanning force microscopy: pulsed-force mode operation. Meas Sci Technol 8:1333–1338
21. Kiracofe D, Kobayashi K, Labuda A, Raman A, Yamada H (2011) High efficiency laser photothermal excitation of microcantilever vibrations in air and liquids. Rev Sci Instrum 82 (1):013702. https://doi.org/10.1063/1.3518965
22. Han W, Lindsay SM, Jing T (1996) A magnetically driven oscillating probe microscope for operation in liquids. Appl Phys Lett 69 (26):4111–4113
23. Carrasco C, Ares P, de Pablo PJ, Gomez-Herrero J (2008) Cutting down the forest of peaks in acoustic dynamic atomic force microscopy in liquid. Rev Sci Instrum 79 (12):126106. https://doi.org/10.1063/1.3053369
24. Dekker NH, Abels JA, Veenhuizen PT, Bruinink MM, Dekker C (2004) Joining of long double-stranded RNA molecules through controlled overhangs. Nucleic Acids Res 32(18): e140. https://doi.org/10.1093/nar/gnh138
25. Herrero-Galan E, Fuentes-Perez ME, Carrasco C, Valpuesta JM, Carrascosa JL, Moreno-Herrero F, Arias-Gonzalez JR (2013) Mechanical identities of RNA and DNA double helices unveiled at the single-molecule level. J Am Chem Soc 135 (1):122–131. https://doi.org/10.1021/ja3054755
26. Meyer G, Amer NM (1988) Novel optical approach to atomic force microscopy. Appl Phys Lett 53:1045
27. Butt H-J, Jaschke M (1995) Calculation of thermal noise in atomic force microscopy. Nanotechnology 6:1–7
28. Sader JE (1998) Frequency response of cantilever beams immersed in viscous fluids with applications to the atomic force microscope. J Appl Phys 84:64
29. Cleveland JP, Anczykowski B, Schmid AE, Elings VB (1998) Energy dissipation in tapping-mode atomic force microscopy. Appl Phys Lett 72(20):2613–1615

Chapter 2

Two Ligand Binding Sites in Serotonin Transporter Revealed by Nanopharmacological Force Sensing

Rong Zhu, Hermann J. Gruber, and Peter Hinterdorfer

Abstract

The number of ligand binding sites in neurotransmitter-sodium symporters has been determined by crystal structure analysis and molecular pharmacology with controversial results. Here, we designed molecular tools to measure the interaction forces between the serotonin transporter (SERT) and S-citalopram on the single-molecule level by means of atomic force microscopy. Force spectroscopy allows for the extraction of dynamic information under physiological conditions which is inaccessible via X-ray crystallography. Two populations of distinctly different binding strength between S-citalopram and SERT were demonstrated in Na^+-containing buffer. In Li^+-containing buffer, SERT showed merely low-force interactions, whereas the vestibular mutant SERT-G402H only displayed the high force population. These observations provide physical evidence for the existence of two different binding sites in SERT when tested under near-physiological conditions.

Key words Nanopharmacology, Neurotransmitter-sodium symporter, Single-molecule force spectroscopy, Serotonin transporter, Atomic force microscopy, Citalopram, Binding site

1 Introduction

The serotonin transporter (SERT) reuptakes released serotonin into the presynaptic neuron, thereby modulating serotonergic neurotransmission and replenishing the neurotransmitter stores [1]. Besides its substrate serotonin, SERT is also a common target of addictive drugs (e.g., cocaine) and of therapeutic medicines (e.g., S-citalopram (S-CIT) for treatment of depression). SERT is a member of the neurotransmitter-sodium symporter (NSS) family, in which the number of ligand binding sites is controversial [2–8]. Some studies provided evidence for two sites, i.e., the central S1-site occupied by the substrate in the occluded state and the second S2-site, located within the extracellular vestibule [9–15]. However, some other studies found that the S1 site alone accounted for high-affinity binding of the inhibitors and of the substrate [4, 5, 16, 17].

Yuri L. Lyubchenko (ed.), *Nanoscale Imaging: Methods and Protocols*, Methods in Molecular Biology, vol. 1814,
https://doi.org/10.1007/978-1-4939-8591-3_2,

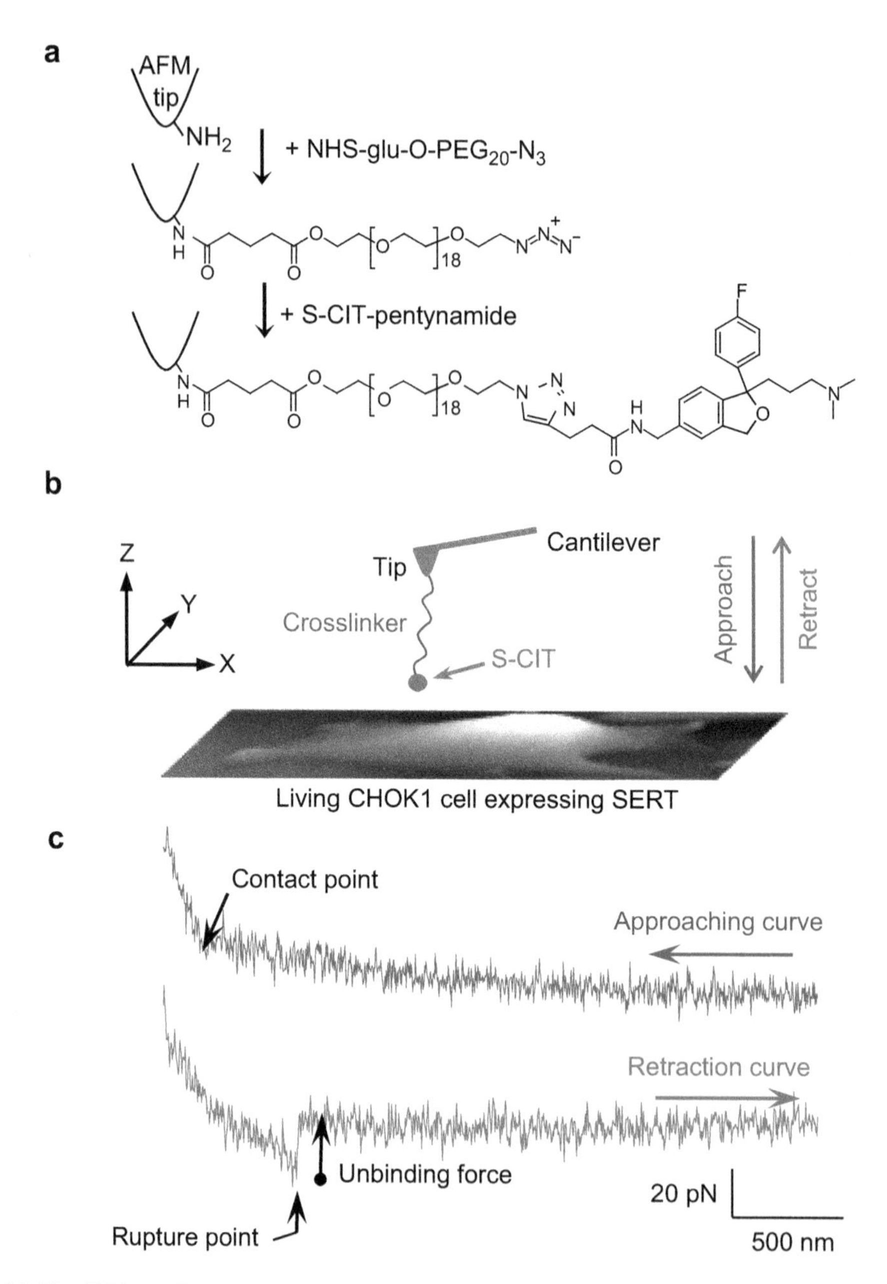

Fig. 1 (**a**) The AFM cantilever tip was functionalized with PEG linker and S-CIT. (**b**) The S-CIT-adorned cantilever tip was used to measure the force curve (**c**) on living CHOK1 cell expressing human SERT

To elucidate the number of ligand binding site(s) in SERT and their binding properties, we used a nanopharmacological sensing approach to measure the interaction forces between SERT and S-CIT by single-molecule force spectroscopy (SMFS) [18–20]. For this purpose, AFM tips were functionalized with an azide-terminated flexible PEG linker to which S-CIT (Fig. 1a) was coupled by click chemistry [21]. SMFS measurements were applied by performing force-distance cycles (Fig. 1b, c) on living CHOK1 cells expressing human SERT, and the unbinding forces were determined from the rupture events monitored during tip retraction (Fig. 1c). The distribution of unbinding forces from a large number of unbinding forces was generated in a continuous form by calculating the experimental probability density function (PDF) [22], as shown in Fig. 2f. Two distinct populations of unbinding forces were observed for S-CIT-modified tips (Fig. 2). Further force spectroscopy experiments revealed additional important characteristics of SERT. While two populations with distinctly different binding strengths of citalopram for SERT were found in Na^+-containing buffer, in Li^+-containing buffer, SERT only showed a single peak in the PDF of the unbinding forces [23]. The absence of the second peak indicated that the central S1 site is Na^+-dependent. Moreover, the vestibular mutant SERT-G402H merely displayed the higher force population, which confirmed that the weaker unbinding events of the wild-type SERT originate from the vestibule S2 site. Furthermore, the unbinding experiments were performed at different retraction speed, i.e., at different rates of force increase, also termed "loading rates" (Fig. 3). The loading rate dependence of the low and the high unbinding force was both fitted with Evan's theory [24], yielding the kinetic off rate k_{off} and the width of the energy barrier x_B for S1 and S2 site. Additional force measurements revealed that these two sites are allosterically coupled and exert reciprocal modulation [23].

2 Materials

All aqueous solutions were prepared from ultrapure water (with a resistivity of 18 MΩ-cm). All reagents were analytical grade or better. All preparations of solution, reactions, and measurements of pH value, weight, and forces, etc., were at 20–26 °C (unless indicated otherwise). The cleaning and functionalization of the cantilevers were performed in a well-ventilated hood. Strictly follow all regulations when handling dangerous and toxic materials. Diligently follow all waste disposal regulations when disposing waste materials.

2.1 AFM Tips with S-CIT

1. AFM cantilevers: Si_3N_4. Nominal spring constant: 10 pN/nm. Tips: Si_3N_4 or Si.
2. Argon gas 5.0 in a cylinder with pressure reducer for adjustment of low pressure and low to moderate flow.
3. Ultrapure water: prepared by purifying deionized water, to attain a resistivity of 18 MΩ-cm.
4. Parafilm™.

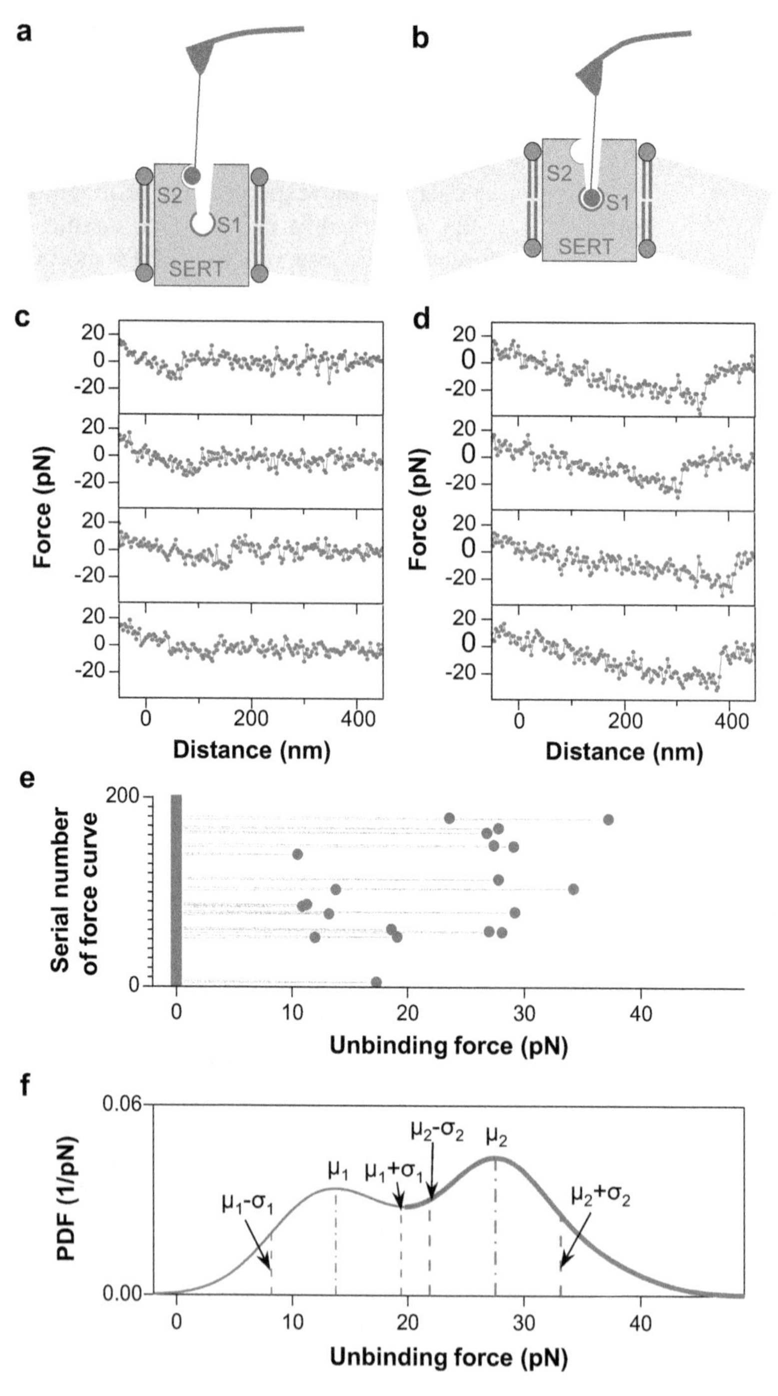

Fig. 2 S-CIT can bind at the vestibular S2 site (**a**) or central S1 site (**b**), resulting in weaker (**c**) or stronger (**d**) unbinding, respectively. From continuously repeated measurements (**e**), the probability density function (PDF) of the unbinding force was constructed (**f**). For further analysis of each individual binding site, the unbinding events were separated by using the range $\mu - \sigma$ to $\mu + \sigma$, where μ is the center of the peak and σ is the standard deviation of the Gaussian fitting

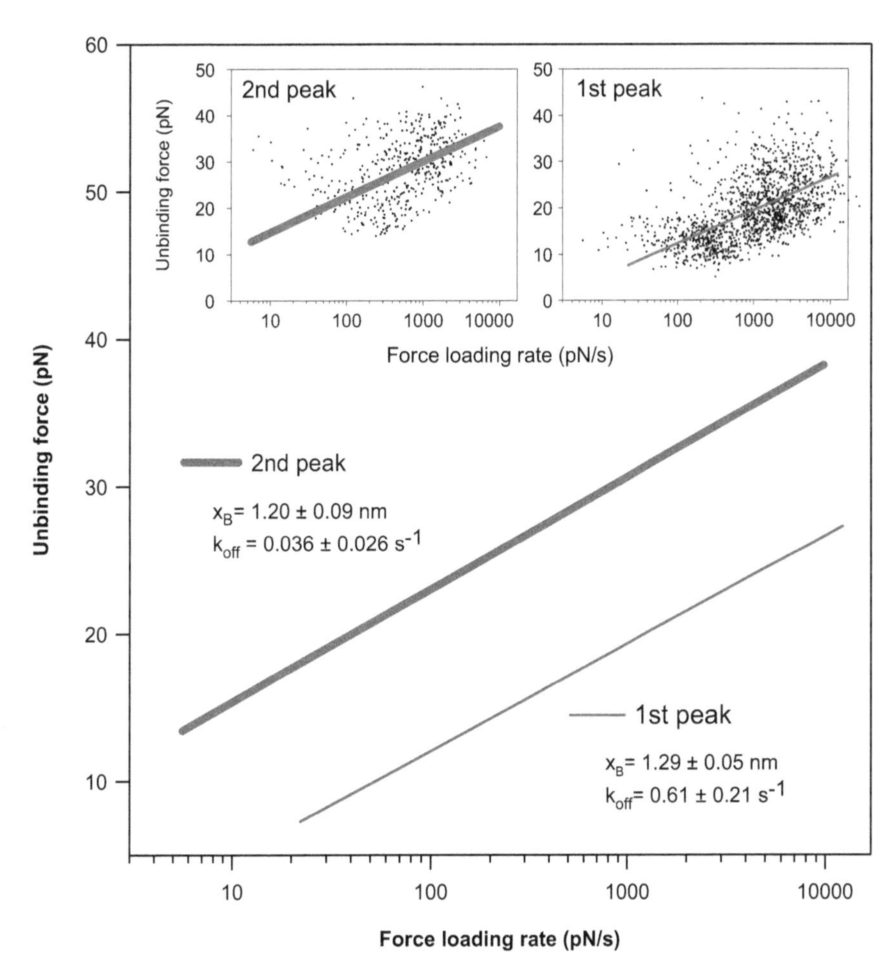

Fig. 3 The data points of unbinding forces for the S1 and S2 binding sites were plotted against the logarithm of the force loading rate (insets). They were fitted with Evans' single-energy barrier model, from which the kinetic off rate k_{off} and the width of the energy barrier x_B were extracted

5. Crystallizing dishes from glass (diameter: 50 mm, with a spout) for cantilever washing.
6. Small plastic Petri dishes (diameter: 35 mm) for cantilever storage, etc.
7. Two lids cut from the top of two 1.5 mL Eppendorf™ reaction vials.
8. Glass desiccator (5 L) with a large silicone O-ring as seal between cover and jar. No grease is allowed.
9. Chloroform.
10. Acidic piranha solution: 3 mL H_2O_2 (30%) is first added to a glass crystallizing dish, and then 7 mL H_2SO_4 (98%) is slowly added with cautious mixing (*see* **Note 1**).
11. 3-Aminopropyl-triethoxysilane (APTES): 30–60 μL (*see* **Note 2**).

12. Triethylamine (TEA): 10–20 μL.
13. Cross-linker: NHS-glu-O-PEG_{20}-N_3 (1 mg per vial, available as Azide-PEG_{18}-NHS from http://www.jku.at/biophysics/content).
14. Reaction chamber for coupling S-CIT to cantilevers (*see* **Note 3**).
15. 0.5 M Tris-(hydroxymethyl)-aminomethane (Tris) in water: pH 8.5 adjusted with HCl.
16. Dimethyl sulfoxide (DMSO).
17. 100 mM $CuSO_4$ in water.
18. 10 mM bathophenanthroline disulfonic acid disodium salt trihydrate in water.
19. 20 mM S-CIT-pentynamide in DMSO.
20. 1 M ascorbic acid in water.
21. 2 M NaOH.
22. Phosphate-buffered saline (PBS).

2.2 Cells

1. Wild-type CHOK1 cells.
2. CHOK1 cells transfected with human SERT fused with YFP: CHOK1-YFP-hSERT.
3. CHOK1 cells transfected with human SERT mutant (G402H) fused with YFP: CHOK1-YFP-hSERT-G402H.
4. Medium: mix DMEM (Dulbecco's Modified Eagle's Medium with high glucose) and HAM's F12 with the ratio of 1:1. Add 10% FBS (Fetal bovine serum) and 0.1% gentamycin (*see* **Note 4**).
5. Selective agent: 50 mg/mL Geneticin (G418) stock solution.
6. Trypsin solution: 0.05% trypsin with 0.02% EDTA in PBS.
7. 80% Isopropanol.

2.3 Force Curve Measurement

1. Plastic Petri dish: diameter 35 mm.
2. HEPES-Na buffer: 140 mM NaCl, 5 mM KCl, 1 mM $MgCl_2$, 1 mM $CaCl_2$, and 10 mM HEPES (pH 7.4 with NaOH).
3. HEPES-Li buffer: 70 mM Li_2CO_3, 5 mM KCl, 1 mM $MgCl_2$, 1 mM $CaCl_2$, and 10 mM HEPES (pH 7.4 with HCl).

3 Methods

3.1 Functionalization of AFM Tips

3.1.1 Cleaning of Tips

AFM cantilever tips were washed in chloroform three times (5 min each), dried, washed in acidic piranha solution (*see* **Note 5**) for 30 min in a well-ventilated hood using utmost care to avoid skin contact, washed in about 20 mL ultrapure water for three times, and dried by heating to 160 °C for 10 min.

3.1.2 Amino-Functionalization

The detailed protocol can be found in [25]. A short description is given as follows:

1. The pre-cleaned AFM tips were placed in a desiccator which was filled with argon.
2. 30 μL monomeric APTES (*see* **Note 2**) and 10 μL TEA were separately pipetted into the lids cut from the top of two 1.5 mL Eppendorf™ reaction vials. They were placed inside this desiccator, whereupon the cover of the desiccator was closed again.
3. After 2 h, the lids were removed, the desiccator was flushed with argon, and the tips were left in the closed desiccator for another 2 days to allow for "curing" of the APTES layer.

3.1.3 Before Coupling of the Cross-Linker

The cantilever chips were washed one more time in chloroform.

3.1.4 Coupling of Cross-Linker

1. One portion of cross-linker NHS-glu-O-PEG_{20}-N_3 (1 mg in a 1 mL crimp vial) was taken out of the −25 °C freezer and was put immediately into a box with dry gel. After 15 min, the vial was taken out of the dry box. The cross-linker can be seen on the bottom and the wall close to the bottom of the vial, which looks like colorless honey.
2. The lid of the vial can be removed with a special uncrimping tool or with plyers. Chloroform (0.5 mL) was transferred into the vial by using a Hamilton syringe. The cross-linker was quickly dissolved in chloroform. The solution was transferred into a cylindrical PTFE reaction chamber with a perfectly flat bottom which had an inner diameter and height of 12 mm.
3. About 7.5 μL TEA was added into the cross-linker solution by using a Hamilton syringe. The solution was mixed by pulling the solution into the Hamilton syringe (50 μL) and pushing out quickly for several times.
4. The APTES-treated cantilever chips were put into the solution, and the reaction chamber was covered with a PTFE lid to prevent the evaporation of the chloroform.
5. After 2 h, the cantilever chips were washed three times each in about 10 mL chloroform and dried in the air.

3.1.5 Functionalization with S-CIT

The alkyne-modified S-CIT analogue was coupled to the azido-terminated PEGs via co-catalyst-accelerated copper(I)-catalyzed azide-alkyne cycloaddition. The highly water-soluble bathophenanthroline disulfonic acid was chosen as co-catalyst [21]. It works best at a pH of 8.5 and may be used at concentrations equal to or up to twice the concentration of Cu(I). Co-catalyst-accelerated copper(I)-catalyzed azide-alkyne cycloaddition is a particularly favorable reaction mechanism that allows small molecule

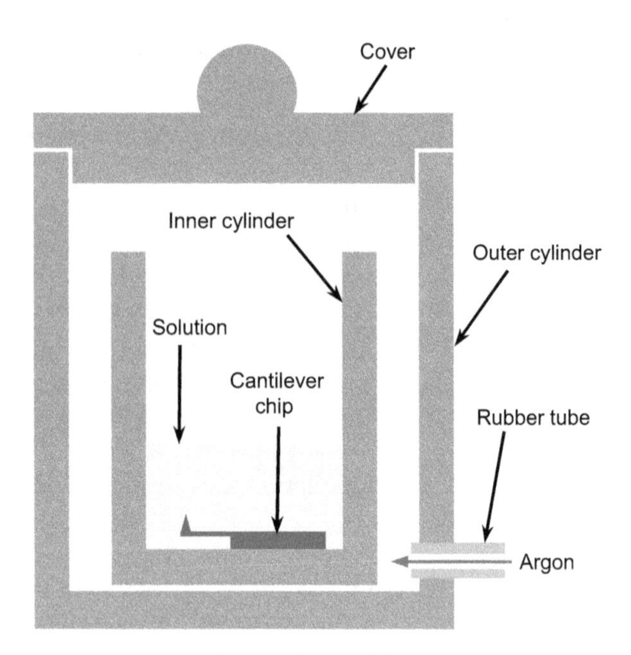

Fig. 4 The reaction chamber for coupling S-CIT to AFM tips. Both cylinders and the cover were made of PTFE. A small hole was drilled on the wall near the bottom of the outer cylinder. A rubber tube was connected with this hole, so that the argon gas was continuously transferred into the cylinder. Reaction solution and the cantilever chips were put in the inner cylinder. The argon flow was adjusted until the PTFE cover was lifted slightly

tip-coupling at a concentration of 1 mM of the component which is in excess (as used here) and even below [21]. The reaction was carried out in an argon-flooded PTFE reaction chamber (*see* Fig. 4 and **Note** 3) as follows:

1. 600 μL 0.5 M Tris (in water, pH 8.5 adjusted with HCl), 293 μL DMSO, 2.5 μL 100 mM $CuSO_4$ (in water), 25 μL 10 mM bathophenanthroline disulfonic acid disodium salt trihydrate (in water), and 50 μL 20 mM S-CIT-pentynamide (in DMSO) were mixed and gently bubbled with argon from a Pasteur pipette for 1–2 min, whereupon 20 μL 1 M ascorbic acid (in water) was added and argon bubbling was repeated. After adding the ascorbic acid, the color of the solution turned yellow to brown.
2. 10 μL 2 M NaOH was added to readjust the pH to 8.5, and argon bubbling was repeated. Thus, the final concentrations of Tris, $CuSO_4$, co-catalyst, S-CIT-pentynamide, ascorbic acid, and NaOH were 0.3 M, 0.25 mM, 0.25 mM, 1 mM, 20 mM, and 20 mM, respectively.
3. The cantilever chips were incubated in this solution for ~17 h with continuous protection from oxygen by gentle perfusion of the chamber with argon gas (Fig. 4).

4. The cantilevers were washed with about 3 mL PBS three times, placed in argon-treated PBS in small plastic Petri dish, sealed with Parafilm™, and stored at 4 °C.

3.2 Cell Culture

3.2.1 Preparation of New Passage

1. Remove the old medium after the cells were confluent on the bottom of the 50 mL flask (bottom area: 25 cm^2).
2. Wash the cells with 1–3 mL PBS for three times. Remove the PBS.
3. Add 1 mL 1× trypsin (0.05% trypsin with 0.02% EDTA in PBS) into the flask. Incubate for about 1 min at 20–26 °C.
4. Remove the cells from the bottom of the flask by flushing the cells using 1 mL pipette with the 1 mL trypsin solution for 25–30 times. The color of the solution became white, indicating a lot of cells in the solution.
5. Transfer about 10–200 μL (depending on the frequency of preparation of new passages: 1–3 time/week) cell solution into the new flask with 5 mL medium and 100 μL 50 mg/mL G418 (*see* **Note 6**).
6. Shake the flask gently so that the cells are distributed homogeneously in the solution.
7. Put the flask into the cell culture incubator (37 °C, 5% CO_2).

3.2.2 Preparation of Cells for Force Measurements

1. Transfer about 30, 12, or 5 μL (depending on the number of days of culture: 1, 2, or 3 days) cell solution from **step 4** (in Sect. 3.2.1) into the plastic Petri dish (diameter 35 mm) with 2 mL medium and 40 μL 50 mg/mL G418 (*see* **Note 6**).
2. Shake the dish gently so that the cells are distributed homogeneously in the solution.
3. Put the dish into the cell culture incubator (37 °C, 5% CO_2).

3.3 Force Curve Measurement

3.3.1 Cells

Living cells (CHOK1-YFP-hSERT or CHOK1-YFP-hSERT-G402H) on plastic Petri dish. The density of the cells should be 10–30% coverage of the dish surface. Remove the medium and wash the cells with 1 mL HEPES-Na buffer for three times. Apply 2 mL HEPES-Na or HEPES-Li buffer in the dish for measurements.

3.3.2 Selection of Cells

1. At first, the cells were examined by fluorescence microscope. An objective lens with magnification of 10× was used. Both GFP and YFP filter sets can be used to detect the expression level of YFP-hSERT. Cells with very low expression were excluded from force measurements. Optimal cells for force measurements are those with homogeneous distribution of SERT.
2. Besides the expression level, the shape of the cell is also important. Flat, smooth, and middle-sized cells can be selected.

3. At last, the position of the cell should be considered. Usually, we selected cells at the edge of a cluster of cells to ensure that there was no other cell underneath the cantilever arm(s).
4. Before transferring the dish from the fluorescence microscope to the AFM, make a mark on the outside wall of the dish with a pen, so that the dish can be mounted in the AFM with the same orientation. Normally, with the optical microscope of the AFM, it is not difficult to find the cells selected in the fluorescence microscope.

3.3.3 Force Distance Cycle

1. The sweep range was 3000 nm. The sweep rate was 0.25–2 Hz. For each cellular position, 100–200 force curves with 2000 data points per curve were recorded.
2. The S-CIT-functionalized cantilever was moved downward to the cell surface and moved upward after the tip touched it (*see* **Note** 7). The deflection (z) of the cantilever was monitored by a laser beam on the cantilever surface and plotted versus the z-position of the scanner.
3. When the tip-tethered CIT bound to a SERT on the cell surface, a pulling force developed during the upward movement of the cantilever, causing the cantilever to bend downward. At a critical force, i.e., the unbinding force, the tip-tethered CIT detached from SERT, and the cantilever jumped back to its neutral position. The force, F, can be determined according to Hook's law, $F = kz$, where k is the spring constant and z is the deflection of the cantilever.

3.3.4 Sensitivity

The deflection signal in **step 3** is the voltage signal from the position sensitive photo detector. The voltage signal can be converted to deflection in nanometer (nm) by multiplying the voltage signal with the sensitivity which can be obtained by recording force curves on the surface of the hard material such as plastic dish instead of cell. After the contact between the tip and the dish, the deflection of the cantilever equals to the movement of the scanner in z-direction. Therefore, the slope of the force curve on plastic surface provided the reciprocal of the sensitivity of the deflection signal.

3.3.5 Spring Constant

The spring constant of the cantilever, k, was determined from the thermal noise by using the equation [26] $k = k_B T / \langle q^2 \rangle$, where k_B is Boltzmann constant, T is the absolute temperature, q is the thermal displacement of the cantilever, and $\langle q^2 \rangle$ is the mean of the square of q (*see* **Note 8**).

3.4 Data Analysis

1. Unbinding event: the unbinding event was identified as upward jump in retraction curve by local maximum analysis using a signal-to-noise threshold of two.

2. Binding activity: the binding activity was calculated as the fraction of curves showing unbinding events. For example, if 150 curves from 1000 measured curves showed unbinding events, the binding activity is 15% or 0.15.
3. Unbinding force: the unbinding force of an unbinding event was determined as the force difference between the force after the rupture (mean value of the force data points on the right side of the rupture point, Fig. 1c) and the force before the rupture.
4. Force loading rate: the force loading rate of each unbinding event was determined by multiplying the pulling speed with the effective spring constant. The effective spring constant (k_{eff}) was obtained from the spring constant of the cantilever (k_c) and the spring constant of the pulled material (k_L), according to $k_{eff} = 1/(1/k_c + 1/k_L)$, where k_L was quantified by the slope of the force-distance curve before the rupture.
5. Probability density function (PDF): the PDF of unbinding force was constructed [22] from every unbinding event on the same cell at the same pulling speed. For each unbinding force value, a Gaussian of unitary area with its center representing the unbinding force and the width (standard deviation) reflecting its measuring uncertainty (square root of the variance of the noise in the force curve) was computed. All Gaussians from one experimental setting were accordingly summed up and normalized with its binding activity to yield the experimental PDF of unbinding force.
6. Separating unbinding events for each peak of the force PDF: to analyze the data for each binding site individually, we need to separate the unbinding events for each peak of the force PDF (Fig. 2f). To do so, the force PDF was at first fitted with multiple Gauss functions so that the standard deviation (σ) of the peaks was obtained. For each peak, the range was set as $\mu - \sigma$ to $\mu + \sigma$ (Fig. 2f), where μ is the center position of the peak. The data points of unbinding events within this range were used for further analysis of each binding site.
7. Extraction of kinetic off rate (k_{off}) and the width of energy barrier (x_B): the unbinding forces were plotted against the logarithm of the force loading rate for each binding site, respectively, (insets of Fig. 3) using the data points selected with the method described in the last step. A maximum likelihood approach [27] was employed to fit x_B and k_{off} for the obtained data by using Evans' single-energy barrier model [24]. The Bayesian information criterion (BIC) was designed to select the best fitting. The formulation for the BIC is given by

$$\mathrm{BIC} = -2(l(\psi; y) - l(\psi^*; y)) + 2p\log(\sqrt{n})$$

with $l(\psi;y)$ being the log-likelihood of the model under consideration, $l(\psi^*;y)$ being the log-likelihood of the most likely model in the subset of models considered, p being the number of parameters fit in the model, and n being the number of observations.

4 Notes

1. Wear protective goggles, lab coat, and gloves and pull the glass shield of the hood down as much as possible because the solution can explode.
2. Portions of freshly purchased or freshly vacuum-distilled APTES (30–60 μL) must immediately be transferred into crimp-vials, crimp-sealed with a PTFE-lined silicon septum under argon gas, and stored for up to 1 year at −25 °C. Only then APTES stays monomeric. All steps must be performed in a well-ventilated hood, preferably with additional protection by gas mask (filter type A against organic vapors); APTES is mutagenic and irritant.
3. PTFE reaction chamber for coupling S-CIT to AFM tip. The schematic illustration of the chamber is shown in Fig. 4. It consists of two PTFE cylinders. The reaction solution and the cantilever chips were put into the inner cylinder. The argon gas was perfused into the outer cylinder via an inlet in the wall close to the bottom. The outer cylinder was covered with a PTFE lid during the reaction. The lid and the outer cylinder were not sealed but loosely shut. The flow of the argon was adjusted until the PTFE cover was lifted slightly to make sure that there was continuous argon supply for the protection of the reaction.
4. As an alternative, FBS can be substituted with newborn calf serum (NBCS). However, the cells used here grow more slowly in medium with NBCS than that with FBS. 0.1% Gentamycin can be replaced with 1× penicillin/streptomycin (100 units penicillin and 100 μg streptomycin per mL).
5. During the treatment with piranha solution, small air bubbles can accumulate on the surface of the cantilever chip, which can turn the cantilever chip upside-down. To avoid this, one needs to keep watching the chip, shaking the chip with tweezers to get rid of the air bubbles, and pressing the chip onto the bottom of the glass ware. The tweezers should be made of PTFE or coated with PTFE or other material which can resist the piranha. Alternatively, the treatment with piranha can be replaced by treating the cantilever chips with ozone plasma for 15 min, followed by three washing steps in chloroform for 5 min.

6. The activity of G418 varies for different producers, different batches, and different cell lines. Therefore, the concentration of G418 should be tested for new batches and new cell lines. A suitable concentration must be found at which the transfected cells survive, while the untransfected cells do not.
7. After the tip has touched the cell surface, the cantilever should be moved upward immediately. The indentation should be as small as possible to minimize the contact area and nonspecific interaction. However, to make sure that the tip has actually touched the cell surface, we should see the increase of the force (bending of cantilever upward) after the contact. Optimally, the magnitude of the force increase upon cell contact should be about twice as high as the noise level (peak-peak value). This is normally difficult to achieve by automatic scanning with a given force limit, due to the movement of the cell and other fluctuations. The practical way to achieve this is to manually control the z-scanning range for each force curve to avoid too much indentation.
8. For soft cantilevers, it is not necessary to measure the frequency spectrum for determination of the spring constant. For instance, the resonance frequency of the cantilevers in buffer solution used in this study is much lower than 100 kHz. According to Nyquist-Shannon sampling theorem [28, 29], a sampling rate of 200 kHz for recording the thermal noise is high enough to capture all the information for the calculation of the spring constant. To do so, the tip was withdrawn from the dish surface by 100 μm. The sweep range was set as 0 nm. The sweep time for one force curve cycle was set as 0.01 s. The number of data points for one force curve cycle was set as 2000. Hundred curves of raw deflection data (thermal noise without filtration) were recorded. The product of the recorded raw deflection data and the sensitivity provided the thermal fluctuation in nm. The variance of these thermal fluctuation data equals to the mean of the square of the thermal displacement of the cantilever $<q^2>$ for calculation of the spring constant.

Acknowledgments

This work was supported by Austrian Science Fund Grant F35.

References

1. Kristensen AS, Andersen J, Jørgensen TN, Sørensen L, Eriksen J, Loland CJ, Strømgaard K, Gether U (2011) SLC6 neurotransmitter transporters: structure, function, and regulation. Pharmacol Rev 63(3):585–640
2. Yamashita A, Singh SK, Kawate T, Jin Y, Gouaux E (2005) Crystal structure of a bacterial homologue of Na^+/Cl^--dependent neurotransmitter transporters. Nature 437 (7056):215

3. Penmatsa A, Wang KH, Gouaux E (2013) X-ray structure of dopamine transporter elucidates antidepressant mechanism. Nature 503 (7474):85–90
4. Piscitelli CL, Krishnamurthy H, Gouaux E (2010) Neurotransmitter/sodium symporter orthologue LeuT has a single high–affinity substrate site. Nature 468(7327):1129
5. Wang H, Elferich J, Gouaux E (2012) Structures of LeuT in bicelles define conformation and substrate binding in a membrane-like context. Nat Struct Mol Biol 19(2):212–219
6. Shi L, Quick M, Zhao Y, Weinstein H, Javitch JA (2008) The mechanism of a neurotransmitter: sodium symporter—inward release of Na+ and substrate is triggered by substrate in a second binding site. Mol Cell 30(6):667–677
7. Zhao Y, Terry D, Shi L, Quick M, Weinstein H, Blanchard SC, Javitch JA (2011) Substrate-modulated gating dynamics in a Na+-coupled neurotransmitter transporter homolog. Nature 474(7349):109
8. Quick M, Shi L, Zehnpfennig B, Weinstein H, Javitch JA (2012) Experimental conditions can obscure the second high-affinity site in LeuT. Nat Struct Mol Biol 19(2):207–211
9. Plenge P, Mellerup ET (1997) An affinity-modulating site on neuronal monoamine transport proteins. Basic Clin Pharmacol Toxicol 80 (4):197–201
10. Chen F, Larsen MB, Neubauer HA, Sánchez C, Plenge P, Wiborg O (2005) Characterization of an allosteric citalopram-binding site at the serotonin transporter. J Neurochem 92(1):21–28
11. Plenge P, Gether U, Rasmussen SG (2007) Allosteric effects of R-and S-citalopram on the human 5-HT transporter: evidence for distinct high-and low-affinity binding sites. Eur J Pharmacol 567:1):1–1):9
12. Zhong H, Hansen KB, Boyle NJ, Han K, Muske G, Huang X, Egebjerg J, Sánchez C (2009) An allosteric binding site at the human serotonin transporter mediates the inhibition of escitalopram by R-citalopram: kinetic binding studies with the ALI/VFL–SI/TT mutant. Neurosci Lett 462(3):207–212
13. Sarker S, Weissensteiner R, Steiner I, Sitte HH, Ecker GF, Freissmuth M, Sucic S (2010) The high-affinity binding site for tricyclic antidepressants resides in the outer vestibule of the serotonin transporter. Mol Pharmacol 78 (6):1026–1035
14. Schmitt KC, Mamidyala S, Biswas S, Dutta AK, Reith ME (2010) Bivalent phenethylamines as novel dopamine transporter inhibitors: evidence for multiple substrate-binding sites in a single transporter. J Neurochem 112 (6):1605–1618
15. Plenge P, Shi L, Beuming T, Te J, Newman AH, Weinstein H, Gether U, Loland CJ (2012) Steric hindrance mutagenesis in the conserved extracellular vestibule impedes allosteric binding of antidepressants to the serotonin transporter. J Biol Chem 287 (47):39316–39326
16. Sinning S, Musgaard M, Jensen M, Severinsen K, Celik L, Koldsø H, Meyer T, Bols M, Jensen HH, Schiøtt B (2010) Binding and orientation of tricyclic antidepressants within the central substrate site of the human serotonin transporter. J Biol Chem 285 (11):8363–8374
17. Andersen J, Stuhr-Hansen N, Zachariassen L, Toubro S, Hansen SM, Eildal JN, Bond AD, Bøgesø KP, Bang-Andersen B, Kristensen AS (2011) Molecular determinants for selective recognition of antidepressants in the human serotonin and norepinephrine transporters. Proc Natl Acad Sci 108(29):12137–12142
18. Florin E-L, Moy VT, Gaub HE (1994) Adhesion forces between individual ligand-receptor pairs. Science 264(5157):415–417
19. Lee GU, Chrisey LA, Colton RJ (1994) Direct measurement of the forces between complementary strands of DNA. Science 266:771–771
20. Hinterdorfer P, Baumgartner W, Gruber HJ, Schilcher K, Schindler H (1996) Detection and localization of individual antibody-antigen recognition events by atomic force microscopy. Proc Natl Acad Sci U S A 93(8):3477–3481
21. Lewis WG, Magallon FG, Fokin VV, Finn M (2004) Discovery and characterization of catalysts for azide– alkyne cycloaddition by fluorescence quenching. J Am Chem Soc 126 (30):9152–9153
22. Baumgartner W, Hinterdorfer P, Schindler H (2000) Data analysis of interaction forces measured with the atomic force microscope. Ultramicroscopy 82(1):85–95
23. Zhu R, Sinwel D, Hasenhuetl PS, Saha K, Kumar V, Zhang P, Rankl C, Holy M, Sucic S, Kudlacek O et al (2016) Nanopharmacological force sensing to reveal allosteric coupling in transporter binding sites. Angew Chem Int Ed 55(5):1719–1722
24. Evans E, Ritchie K (1997) Dynamic strength of molecular adhesion bonds. Biophys J 72 (4):1541–1555
25. Riener CK, Stroh CM, Ebner A, Klampfl C, Gall AA, Romanin C, Lyubchenko YL, Hinterdorfer P, Gruber HJ (2003) Simple test

system for single molecule recognition force microscopy. Anal Chim Acta 479(1):59–75
26. Hutter JL, Bechhoefer J (1993) Calibration of atomic-force microscope tips. Rev Sci Instrum 64(7):1868–1873
27. Ebner A, Nevo R, Ranki C, Preiner J, Gruber H, Kapon R, Reich Z, Hinterdorfer P (2009) Probing the energy landscape of protein-binding reactions by dynamic force spectroscopy. In: Hinterdorfer P, van Oijen A (eds) Handbook of single-molecule biophysics. Springer, New York, pp 407–447
28. Nyquist H (1928) Certain topics in telegraph transmission theory. Trans Am Inst Electr Eng 47(2):617–644
29. Shannon CE (1949) Communication in the presence of noise. Proc IRE 37(1):10–21

Chapter 3

AFM-Based Single-Molecule Force Spectroscopy of Proteins

Zackary N. Scholl and Piotr E. Marszalek

Abstract

Single-molecule force spectroscopy by AFM (AFM-SMFS) is an experimental methodology that allows unequivocal sensitivity and control for investigating and manipulating the mechanical properties of single molecules. The past 20 years of AFM-SMFS has provided numerous breakthroughs in the understanding of the mechanical properties and force-induced structural rearrangements of sugars, DNA, and proteins. Here, we focus on the application of AFM-SMFS to study proteins, since AFM-SMFS has succeeded in providing abundant information about protein folding pathways, kinetics, interactions, and misfolding. In this chapter we describe the experimental procedures for conducting a SMFS-AFM experiment—including purification of protein samples, setup and calibration of the AFM instrumentation, and the thorough and unbiased analysis of resulting AFM data.

Key words Atomic force spectroscopy, Single molecule, Protein mechanics, Protein folding, Protein misfolding

1 Introduction

Single-molecule force spectroscopy by AFM (AFM-SMFS) has emerged in the last 20 years as one of the foremost methods for capturing the dynamic properties of proteins using mechanical perturbations. At its onset, AFM-SMFS brought new information to light about protein mechanics [1] and ligand binding [2] and has immediately found a foothold as tool for understanding protein-protein interactions [3], protein folding kinetics [4], protein unfolding pathways [5], and protein misfolding [6]. The applications and benefits of AFM-SMFS is well described elsewhere [7–13], whereas here we will describe how to efficiently perform AFM-SMFS experiments on a protein of interest.

An AFM instrument for performing SMFS consists of a piezoelectric element which suspends a sample containing the proteins immersed in buffer (*see* Fig. 1a). The AFM cantilever is suspended above the sample, and the deflection of the cantilever is continuously measured from a laser reflecting into a detector. The piezoelectric element controls the distance between the sample and a

Yuri L. Lyubchenko (ed.), *Nanoscale Imaging: Methods and Protocols*, Methods in Molecular Biology, vol. 1814,
https://doi.org/10.1007/978-1-4939-8591-3_3, © Springer Science+Business Media, LLC, part of Springer Nature 2018

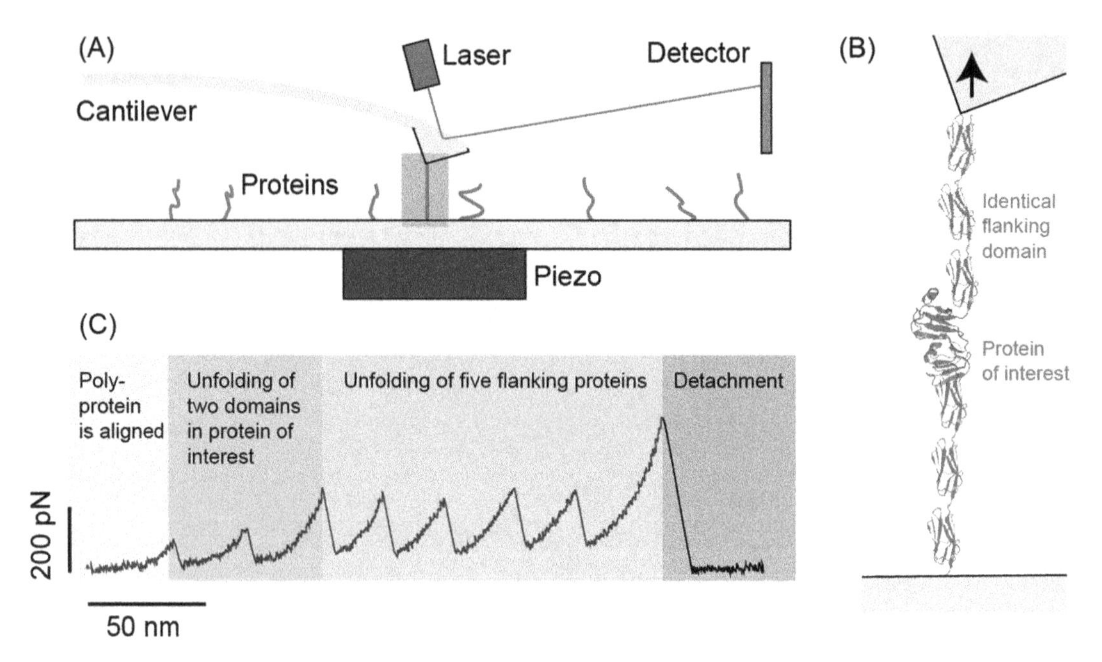

Fig. 1 A typical AFM-SMFS experiment. The schematic of an AFM experiment is shown in (**a**). The cantilever deflection is monitored by reflected laser light while pulling on a tethered protein that is moved away through the control of a piezoelectric element. A schematic of the grayed region in (**a**) is shown in (**b**), where AFM cantilever is in the process of pulling the polyprotein (illustrated in the cartoon form) that contains six identical flanking domains (blue) and a two-domain protein of interest (red). The force-extension trace shown in (**c**) is an example of pulling the protein shown in (**b**) which shows the individual events that occur during the pulling process. First, the polyprotein is initially aligned along the pulling direction (yellow shaded region), which is followed by the unfolding of the two domains of the protein of interest (red shaded region), followed by unfolding of five identical domains proteins (blue shaded region) and the detachment (gray shaded region)

suspended cantilever, and a measurement is initiated by bringing together the cantilever and the sample. When the sample surface and the AFM tip are in contact, the proteins on the surface will nonspecifically adsorb to the cantilever tip. A measurement is then performed by moving the cantilever away from the surface, pulling on the adsorbed protein. The pulling force that is exerted on the cantilever from the tethered protein is determined by Hooke's law through the deflection of the cantilever as measured by the reflected position of the laser in the photodiode. In a typical experiment, the protein is designed to be polyprotein composed of identical protein domains flanking a protein of interest (Fig. 1b) of which would give a characteristic saw-tooth pattern (Fig. 1c) where each "rip" corresponds to the unfolding of one of the domains in the polyprotein.

AFM-SMFS is performed in four main steps: protein purification, AFM calibration and setup, AFM experimentation, and analysis. A protein of interest is typically constructed and purified as synthetic polyprotein that contains a secondary flanking protein for the positive control in SMFS measurements. This step requires basic molecular biology reagents, tools, and techniques. The next steps require an AFM setup (commercial or homemade), and the preparation and operation is described in detail. The final step, the analysis of AFM-SMFS data, should be suited to the specific question of interest, and here we describe the basic analysis to determine the mechanical stability and verify the unfolding pathway of a protein of interest. Overall, this chapter should serve as a detailed guide to conduct a prototypical AFM-SMFS experiment in a systematic and reproducible way.

2 Materials

Prepare all solutions using ultrapure water, prepared by purifying deionized water to a specific resistance of 18 MΩ cm at room temperature. Materials listed here are from specific suppliers, though can be replaced by analogous items from a different supplier.

2.1 Protein Preparation

1. Chemically competent overexpression cells: e.g., C41 (DE3) pLysS cells.
2. Chemically competent high efficiency cells: e.g., Turbo cells.
3. LB broth: 20 g of LB powder diluted into 1 L of water and autoclaved for 15 min at 121 °C to sterilize.
4. Ampicillin stock solution: ampicillin diluted to 100 μg/mL in water.
5. Isopropyl β-D-1-thiogalactopyranoside (IPTG) diluted to 1 M in water.
6. Lysis buffer: 36 mL water, 10 μL DNase I, 10 mg lysozyme, 1 mM $CaCl_2$, 1 mM EDTA, 1 mM PMSF, 1 mM TCEP, and 5% glycerol.
7. Polyprotein vector: a DNA vector containing a polyprotein (e.g., pEMI91 Addgene #74888).
8. Gravity flow column for protein purification (Strep-tag or His-tag depending on application).
9. Spin filter columns: 0.5 mL spin filter column.

2.2 Material for Preparing Samples

1. Round glass cover slips: radius 7.5 mm, 0.13–0.16 mm thickness.
2. Rectangular microscope slides: 25 × 75 mm.
3. Double-sided adhesive tape.
4. Piranha solution: add 30 mL of 30% hydrogen peroxide into a glass beaker. Inside a chemical hood, add 10 mL sulfuric acid. Exercise caution when performing this step as these are highly volatile chemicals.

5. Tip tweezer.
6. Iron disks.
7. Adhesive tabs.
8. Ethanol and Acetone.

2.3 AFM Setup and Calibration

1. AFM cantilevers (Bruker #OBL-10 for mechanically weak proteins or Bruker #MLCT for mechanically stronger proteins).
2. AFM probe holder (Bruker #MTFML-V2).
3. AFM head (Bruker MMAFM Head) or commercial AFM setup.

3 Methods

Carry out all procedures at room temperature unless otherwise specified.

3.1 Protein Preparation

1. First clone the DNA of the protein of interest into a polyprotein vector that contains secondary flanking proteins for a positive control (*see* **Note 1**). First you identify the amino acid sequence for the protein of interest. Synthesize the DNA of the protein of interest with flanking restriction sites to be cloned into a polyprotein vector [14]. Use restriction enzymes to cut both the polyprotein vector and the linear DNA containing the protein of interest and use standard molecular biology techniques to ligate and transform the plasmid into *E. coli* cells. Grow up cells and purify the plasmid from the cell culture using ethanol precipitation. Verify the resulting plasmid by sequencing the DNA and checking to make sure the plasmid is free from insertions, deletions, and mutations. This should result in a stable, viable plasmid containing the DNA of a polyprotein-flanked protein of interest.
2. Prepare *E. coli* cells for production of synthetic protein. Transform the DNA of a polyprotein-flanked protein of interest into C41 (DE3) pLysS cells for protein production. Inoculate a freshly grown bacteria colony into 15 mL LB broth with appropriate antibiotic (100 μg/mL ampicillin if using plasmid #74888) and incubate at 37 °C overnight (12–14 h). Make a fresh 1 L of LB broth containing the appropriate antibiotic and add the overnight culture and shake for 4 h (or until OD 600 is greater than 0.8) at 37 °C. Then, add isopropyl β-D-1-thiogalactopyranoside (IPTG) to a final concentration of 1 mM and continue shaking overnight at room temperature. To harvest, centrifuge the cells at 4100 rpm (3300 × *g*) for 40 min. Then decant the broth, and freeze at −80 °C for at least 12 h.

3. Lyse cells containing protein. Take cells frozen at −80 °C containing the protein, and pour the lysis buffer over the top of them. Leave this to shake on ice for 30 min. Then freeze again at −80 °C until ready for purification.
4. Release protein from cells. Thaw an aliquot of the lysed cells at −80 °C by putting them in room temperature water for 15 min. Optionally, sonicate the cells six times at 1 min intervals at medium strength. Spin down the lysates at 13,100 rpm (10,700 × *g*) for 30 min at 4 °C. Optionally, clarify further by pressing the supernatant through a 0.2 μm filter.
5. Purify protein from lysed cells. Run the clarified protein extract through a gravity flow column (Strep-tag or His-tag column in the case of plasmid #74888). Wash several times according to the gravity flow column instructions, and then elute into 500 μL fractions.
6. Identify the fractions which contain the purified protein by running all the elution fractions on a SDS gel.
7. Dialyze the protein into a neutral buffer (e.g., 100 mM phosphate-buffered saline pH 7.4), and then use a spin filter to concentrate the protein. Store protein in its nominal buffer with 10% glycerol at −20 °C. Protein can be stored for up to 2 months.

3.2 AFM Sample Preparation

1. First prepare clean glass slides. Place 10–30 round glass cover slips into a 40 mL glass beaker. Inside a chemical hood, add piranha solution (make sure to wear PPE). Heat the piranha mixture for 10–30 min, and then carefully decant the piranha solution into a separate waste container. The excess piranha solution can be reused later or neutralized with baking soda and water. Rinse the slides with deionized water to remove excess piranha, and then suspend the slides in 40 mL of acetone. Decant acetone and resuspend the slides in 40 mL of ethanol. Use clean forceps to carefully extract one slide at a time, and dry with argon or purified air (*see* **Note 2**). Place the cleaned slides under a vacuum or argon-filled container for storage.
2. (Optional) Make gold-coated glass slides. Cut microscope slides in half using a glass cutter and apply double-sided adhesive tape in the middle of each glass slide. Take a sticky note, and place it so that the sticky side is face up. Press the cleaned glass slides (from Subheading 3.2, **step 1**) to each of the four corners so that they are held in place by the sticky side of the sticky note. Assemble these pieces into the mounting plate of an e-beam metal evaporator (e.g., CHA Industries Solution E-Beam, or similar). Follow the protocol of your evaporator in order to apply first 70 nm of chromium, and then apply

300 nm of gold to the surface of the cleaned glass slides. Store the resulting gold-covered slides under argon until use.

3. Prepare a sample to measure on the AFM. Select a clean iron disk and attach an adhesive tab to it. Select a piece of clean glass (from Subheading 3.2, **step 1**) or gold-coated glass (from Subheading 3.2, **step 2**), and place firmly to attach it to the adhesive side of the iron disk.
4. Dilute the purified protein. Take 5–10 μL of the purified protein in the −20 °C, and change the buffer using a spin filter column. Determine the concentration by measuring the absorbance at 280 nm, and dilute the purified protein to 10–100 μg/mL in a final volume of 100 μL.
5. Apply all 100 μL of diluted protein to the center of the slide. The drop of protein will form a spherical droplet on gold-coated slides (as this surface is more hydrophobic), or the drop of protein will spread across the surface on glass slides (as this surface is more hydrophilic).
6. Let this sample sit at room temperature for 10–60 min, which allows gold-thiol bonds to form (if using -Cys terminate polyprotein vector and the gold-coated slides).

3.3 AFM Experimentation

1. Carefully load the cantilever of your choice (*see* **Note 3**) onto the cantilever probe holder and load into the AFM head. Align the laser in the AFM head by mounting the head onto an inverted microscope with an attached camera and camera monitor. Attach a battery pack to the AFM head for powering the laser, and adjust the position of the laser so that the cantilever tip is positioned at the focus of the beam. Once aligned, flush the ports of the probe holder with 10–50 μL of the experimental buffer.
2. Place the sample slide onto the piezo stage of the AFM instrument (*see* **Note 4**), and decant 60 μL from the 100 μL original drop. Carefully place the AFM head overhead the sample, and adjust the height of the stage so that the protein sample on the slide comes into contact with buffer suspended on the bottom of the cantilever probe holder (careful not to go too far as it will break the cantilever tip). Adjust the laser position again by monitoring the laser light on a small piece of paper held between the photodetector and the path of the optical lever to make sure the light is concentrated to a small point and not cut off from any part of the photodetector.
3. Now, using the AFM software (*see* **Note 5**), maximize the total signal (A + B + C + D of the quadrant photodetector), and reduce the difference signal ((A + B − C − D)/(A + B + C + D) of the quadrant photodetector) so that it is zero.

4. Determine the spring constant of the AFM cantilever using the equipartition method [15]. While the equipartition method is the most straightforward to implement, there are other methods like using forced oscillations and static loading that can reduce the uncertainty in the determined spring stiffness [16]. Turn the photodiode signal filter settings on the AFM to measure at the full bandwidth, and make sure that the piezo is turned off. Take a power spectrum of approximately 1024 data points averaged over 512 calculations, and integrate across the first peak in the power spectral density (the main mode of vibration of the cantilever). Turn the piezo on and change the filter settings to low-pass filter with a desired cutoff of 100–1000 Hz (typical). Slowly bring the cantilever in contact with the surface by repeatedly moving the stage down a few hundred micrometers at a time. Take a constant velocity measurement so that the cantilever presses against the surface for about 100–500 nm and then moves through the buffer for another 500 nm. Measure the slope of the linear part of the resulting curve between the contact point and the detachment point. Calculate the spring constant as $k_c = {}^{k_b T}/_{(\mathrm{PSD} \times \mathrm{slope}^2)}$ where PSD is the integrated power spectral density determined earlier. A $k_b T$ at room temperature is about 4.1 pN nm.
5. The filter settings should be set so that the sampling frequency is at least twice the bandwidth (Nyquist criterion). The scan size should set equal to the theoretical length of the unfolded protein (number of amino acids × 0.365 nm) plus about 30–50% length to allow for pressing against the surface.
6. Position the cantilever so that it can touch the surface during each constant velocity ramp. Automation can be easily employed by adjusting the Z-position of the piezo, so it continues to press against the surface for about 30% of the length. Meanwhile, the raster scanning the X and Y directions of the piezo should be modulated as well to perform probe scans at different locations on the surface. Perform scans that start away from the sample, and then move with constant velocity toward the surface before retraction (*see* **Note 6**).
7. Collect data that have a maximum force above a small force threshold, in order to triage any recording that is not empty (*see* **Note 7**).

3.4 Data Analysis

1. Normalize the force-displacement data to force-extension curves, as shown in Fig. 2. Normalization takes place in three steps. First, the force interpreted from the spring constant is relative to the baseline, so the data must be baseline corrected. All force-displacement curves should be shifted by the mean value at high displacement (in which all molecules have detached), so that displacement is aligned on the zero of the

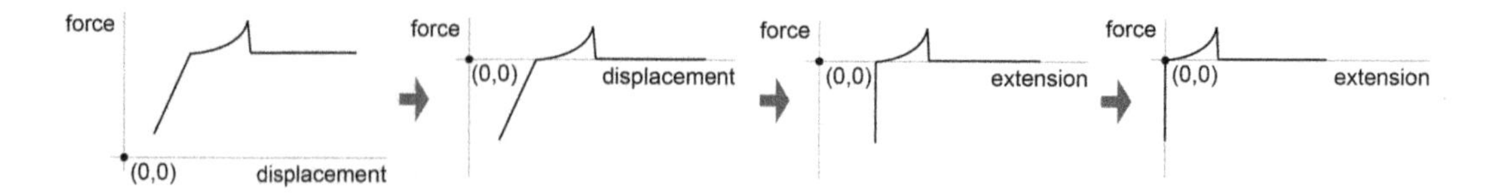

Fig. 2 A visualization of the data transformation for normalization of force-displacement data to force-extension curves. Data transformation occurs in three steps: normalizing the baseline (*y*-axis), converting displacement to extension, and then normalizing the approach (*x*-axis)

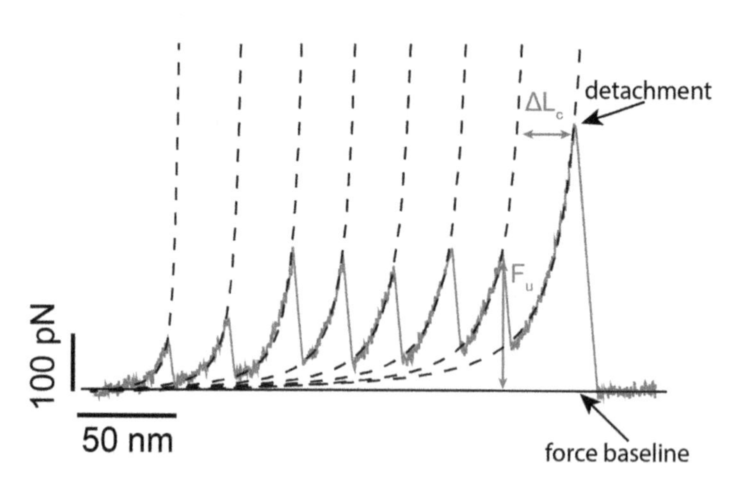

Fig. 3 An example of a force-extension curve (light gray) showing the unfolding of a protein of interest (peaks 2–3) and the unfolding of the flanking protein (peaks 3–7) until it detaches. The dashed line indicates the worm-like chain fit to that peak. Each peak, minus the detachment peak, has an unfolding force (F_u) and a contour-length increment (ΔL_c) which should be extracted for analysis

y-axis. Second, the displacement should be converted to molecular extension. The extension represents the molecular extension of the probed molecule which can be related to analytical models like the worm-like chain. The extension, e, is computed from the displacement, d, using the force at that displacement, $F(d)$, and the measured spring constant k_c: $e = d - F(d)/k_c$. Finally, the recordings should then be aligned so that the vertical position begins at zero extension (*see* **Note 8**).

2. Select only data which conforms to criteria that is indicative of a single-molecule event. Data selection is crucial to avoid interpreting artifacts of the AFM data as actual events (*see* **Note 9**).
3. Fit each peak in each force-extension curve with a worm-like chain model [17] (*see* **Note 10**) and determine the unfolding force and the contour-length increment of each peak in the force-extension curve (*see* Fig. 3).

4. Use the total contour-length increment of the protein of interest to further verify that the single-molecule events correspond to the protein of interest (*see* **Note 11**).
5. While the forces directly indicate the mechanical stability, they can also be used to determine the force-dependent unfolding kinetics of the protein of interest [18]. The series of contour-length increments can be mapped back to the protein structure to determine the unfolding pathway of the protein [19]. Utilize these particular methods or others as they pertain to rejecting or verifying your hypothesis.

4 Notes

1. The protein of interest should nominally have two properties: First, it should not have disulfide bridges. Disulfide bridges are covalent bonds that cannot be broken because these bonds are generally stronger than the strength of the adhesion between the protein and the surface, so these proteins will not experience a full mechanical denaturation. Second, the protein of interest should have an X-ray or NMR structural model available which will enable analysis of the protein unfolding pathway. The flanking proteins should also accommodate two major properties. First, the flanking protein should be well-characterized, so that they have a known unfolding behavior (unfolding force and contour-length increment). Second, the flanking proteins should be mechanically strong so that they will unfold after the protein of interest (typically I91 or GB1 domains are used, which have unfolding forces greater than 100 pN). In most cases the *E. coli* expression system can be used for the polyprotein; however if there is known glycosylation or other posttranslational effects, then the expression system that will produce those modifications should be used.
2. Purified air can be accomplished by simply attaching a filter pipette tip to the nozzle of the in-lab air hose.
3. The specific cantilever to use will depend on the application. The cantilevers with lower stiffness can be better suited for weak proteins as they will provide a better sensitivity. Recently, Edwards et al. have improved on the stability and temporal resolution of a commercial cantilever by stripping the gold coating and using a focused ion beam to modify the shorter cantilever [20].
4. An AFM for force spectroscopy can be purchased or be constructed in-house from its components. For building an AFM in-house, from scratch, *see* Rabbi et al. [21]. For building the

stage and assembling the piezo electronics in-house with a commercial AFM head, *see* Scholl [22] .

5. AFM software usually comes with a commercial instrument. In the case of using a homemade instrument, open-source software has been published by Pawlak and Strzelecki which can be used for free [23].
6. Typically, a speed of about 300 nm/s is used for the ramp. The pickup rate of a molecule is <1% [24], so over 10,000 pulls will need to be made to collect enough data to analyze. A typical scan size is about 500 nm, which means that at least 5 h of pulling is necessary to start acquiring datasets. Make sure to rehydrate the cell by applying more buffer every hour or so, in order to keep the sample from drying out. If the sample does not dry out, pulling may be done on a single experiment over the course of 8–10 h.
7. A typical force threshold is ~25 pN which is about twice the typical RMS of the noise in the force. If you notice that all recording events are empty, it could be that there are too few molecules deposited on the surface. In this case, start again but increase the concentration of the protein sample by tenfold. On the other hand, if you notice that there are too many events (attachment every pull), then you should decrease the concentration by about tenfold. Nominally there should be an event every ten pulls and a single molecule usually every hundred.
8. If you see that the force-extension curve has an initial region that does not align well vertically, it may be that the spring constant was not estimated correctly during the calibration. In this case, the calibration should be redone, or the cantilever could be changed.
9. Selection of the data should be done carefully as the contact location between the tip and the protein and the surface is not controlled. There are recent methods for acquiring specific attachment that can help mitigate this issue; *see* [11, 25]. The first criterion for a single-molecule event is that it is at least half of the events corresponding to the unfolding of the flanking proteins. For example, in Fig. 4, you can see that there are four flanking proteins, two on either side. In order to guarantee that unfolding of the middle protein took place, then you would require having at least three events (more than half of the flanking protein domains; *see* left on Fig. 4). Any other number of events would not necessarily ensure that the protein of interest has been unfolded (right on Fig. 4). Another criterion for a single-molecule event is that the full extension of the molecule is not any longer than what is expected for the full length of the molecule. Once you have established around 100 recordings which fulfill these criteria, then you can use

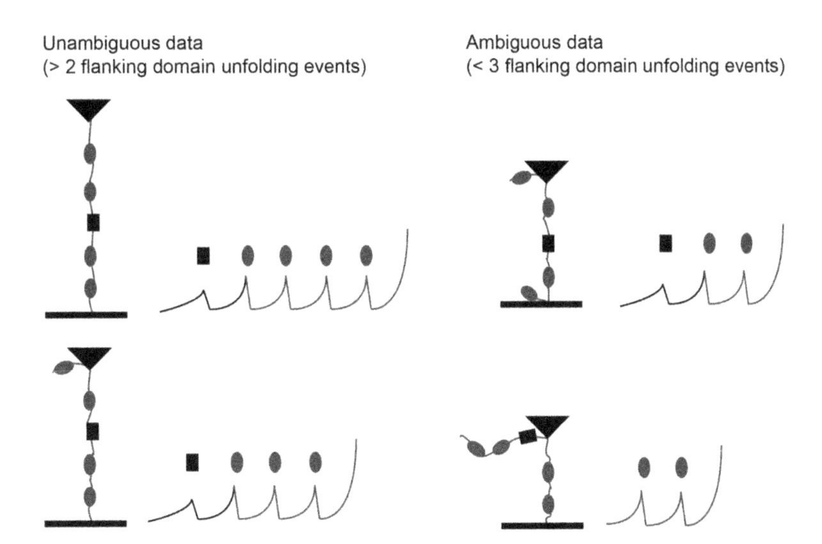

Fig. 4 Shown are schematics of tip and surface attachment with the polyprotein with their corresponding possible unfolding force-extension curve. Above each force-extension curve, the domain whose unfolding produces that force peak is shown. Each case illustrates pulling a protein of interest (black box) with four flanking domains (gray circles) for a positive control but with various examples of the random configuration of tip protein and protein surface attachment. In the two cases on the left, it is unambiguous that the protein of interest has unfolded because more than two of the flanking domains have unfolded. In the cases on the right, it is possibly ambiguous because the unfolding of only two domains does not necessitate that the protein of interest unfolds (black box in schematic and black line in force-extension schematic). Unless a fingerprint is available from unambiguous data, then the data with fewer than half of the flanking domains cannot be considered

this as a fingerprint. This fingerprint can be compared against data which has not enough flanking domain events but may contain the mechanical fingerprint of the protein of interest.

10. The worm-like chain model is a model that approximates the molecular extension of a semiflexible chain (like a protein). By fitting each peak in the force-extension curve, you can calculate the difference between the contour-length of subsequent curves to determine the number amino acids that were unfolded during that unfolding event (*see* **Note 11**). The worm-like chain model that approximates the exact solution of the Hamiltonian of elastic polymer under external force is given as [26, 27]

$$\frac{FP}{k_b T} = \frac{x}{L} + \frac{1}{4\left(1 - \frac{x}{L}\right)^2} - \frac{1}{4}.$$

In this equation, the force (F) is a function of the extension (x), persistence length (P), and contour-length (L). The persistence length is usually set to 0.4–0.6 nm and can be globally fit across all curves. This persistence length corresponds roughly to end-to-end distance of a single amino acid (0.365 nm); however the true persistence length can include contributions from other factors in the polypeptide—like residual secondary structure—which may act to increase the persistence length. In practice, this equation is used to fit the peak of a force rip, where the force, extension, and persistence are known variables. Since this equation is not a function of contour-length, it must be solved for the contour-length using a numerical solver. In practice, also, it is preferred to use the numerically approximated solution to the worm-like chain presented by Bouchiat et al. [17] which gives a seventh order correction to the above formula to decrease the deviation from the exact solution from 10% to less than 0.01%.

11. The total contour-length increment of the protein of interest (the sum of all contour-length increments that are contributed from the flanking protein) should be the length of all the residues minus the initial length of the protein, as the protein was initially pre-stretched and aligned along the pulling axis before unfolding. The length of all residues should be the number of residues in the protein times 0.365 nm per residue [28]. The initial length of the protein can be computed as the distance between the N- and C-termini from a NMR or X-ray structural model of the protein of interest.

Acknowledgments

This work was supported by NSF grant MCB-1517245 to P.E.M.

References

1. Rief M, Gautel M, Oesterhelt F, Fernandez JM, Gaub HE (1997) Reversible unfolding of individual titin immunoglobulin domains by AFM. Science 276(5315):1109–1112
2. Florin E-L, Moy VT, Gaub HE (1994) Adhesion forces between individual ligand-receptor pairs. Science 264(5157):415–417
3. Hinterdorfer P, Baumgartner W, Gruber HJ, Schilcher K, Schindler H (1996) Detection and localization of individual antibody-antigen recognition events by atomic force microscopy. Proc Natl Acad Sci U S A 93(8):3477–3481
4. Carrion-Vazquez M, Oberhauser AF, Fowler SB, Marszalek PE, Broedel SE, Clarke J, Fernandez JM (1999) Mechanical and chemical unfolding of a single protein: a comparison. Proc Natl Acad Sci U S A 96(7):3694–3699
5. Marszalek PE, Lu H, Li H, Carrion-Vazquez M (1999) Mechanical unfolding intermediates in titin modules. Nature 402(6757):100
6. Oberhauser AF, Marszalek PE, Carrion-Vazquez M, Fernandez JM (1999) Single protein misfolding events captured by atomic force microscopy. Nat Struct Mol Biol 6 (11):1025–1028
7. Hoffmann T, Dougan L (2012) Single molecule force spectroscopy using polyproteins. Chem Soc Rev 41(14):4781–4796
8. Žoldák G, Rief M (2013) Force as a single molecule probe of multidimensional protein energy landscapes. Curr Opin Struct Biol 23 (1):48–57. https://doi.org/10.1016/j.sbi.2012.11.007

9. Müller DJ, Dufrene YF (2008) Atomic force microscopy as a multifunctional molecular toolbox in nanobiotechnology. Nat Nanotechnol 3(5):261–269
10. Javadi Y, Fernandez JM, Perez-Jimenez R (2013) Protein folding under mechanical forces: a physiological view. Physiology 28 (1):9–17
11. Ott W, Jobst MA, Schoeler C, Gaub HE, Nash MA (2017) Single-molecule force spectroscopy on polyproteins and receptor–ligand complexes: the current toolbox. J Struct Biol 197 (1):3–12
12. Rico F, Rigato A, Picas L, Scheuring S (2013) Mechanics of proteins with a focus on atomic force microscopy. J Nanobiotechnol 11(1):S3
13. Schönfelder J, De Sancho D, Perez-Jimenez R (2016) The power of force: insights into the protein folding process using single-molecule force spectroscopy. J Mol Biol 428 (21):4245–4257. https://doi.org/10.1016/j.jmb.2016.09.006
14. Scholl ZN, Josephs EA, Marszalek PE (2016) Modular, nondegenerate polyprotein scaffolds for atomic force spectroscopy. Biomacromolecules 17(7):2502–2505
15. Florin E-L, Rief M, Lehmann H, Ludwig M, Dornmair C, Moy VT, Gaub HE (1995) Sensing specific molecular interactions with the atomic force microscope. Biosens Bioelectron 10(9):895–901
16. Burnham N, Chen X, Hodges C, Matei G, Thoreson E, Roberts C, Davies M, Tendler S (2002) Comparison of calibration methods for atomic-force microscopy cantilevers. Nanotechnology 14(1):1
17. Bouchiat C, Wang M, Allemand J-F, Strick T, Block S, Croquette V (1999) Estimating the persistence length of a worm-like chain molecule from force-extension measurements. Biophys J 76(1):409–413
18. Zhang Y, Dudko OK (2013) A transformation for the mechanical fingerprints of complex biomolecular interactions. Proc Natl Acad Sci U S A 110(41):16432–16437. https://doi.org/10.1073/pnas.1309101110
19. Scholl ZN, Yang W, Marszalek PE (2014) Chaperones rescue luciferase folding by separating its domains. J Biol Chem 289 (41):28607–28618
20. Edwards DT, Perkins TT (2017) Optimizing force spectroscopy by modifying commercial cantilevers: improved stability, precision, and temporal resolution. J Struct Biol 197 (1):13–25
21. Rabbi M, Marszalek PE (2007) Construction of a single-axis molecular puller for measuring polysaccharide and protein mechanics by atomic force microscopy. Cold Spring Harb Protoc 2007(12.) pdb. prot4899
22. Scholl ZN (2016) The (un) folding of multidomain proteins through the lens of single-molecule force-spectroscopy and computer simulation. Dissertation, Duke University
23. Pawlak K, Strzelecki J (2016) Nanopuller-open data acquisition platform for AFM force spectroscopy experiments. Ultramicroscopy 164:17–23
24. Scholl ZN, Marszalek PE (2014) Improving single molecule force spectroscopy through automated real-time data collection and quantification of experimental conditions. Ultramicroscopy 136:7–14
25. Popa I, Rivas-Pardo JA, Eckels EC, Echelman DJ, Badilla CL, Valle-Orero J, Fernández JM (2016) A halotag anchored ruler for week-long studies of protein dynamics. J Am Chem Soc 138(33):10546
26. Marko JF, Siggia ED (1995) Stretching DNA. Macromolecules 28(26):8759–8770. https://doi.org/10.1021/ma00130a008
27. Bustamante C, Marko JF, Siggia ED, Smith S (1994) Entropic elasticity of lambda-phage DNA. Science 265(5178):1599–1600
28. Scholl ZN, Li Q, Marszalek PE (2014) Single molecule mechanical manipulation for studying biological properties of proteins, DNA, and sugars. Wiley Interdiscip Rev Nanomed Nanobiotechnol 6(3):211–229

Chapter 4

High-Resolution AFM-Based Force Spectroscopy

Krishna P. Sigdel, Anna E. Pittman, Tina R. Matin, and Gavin M. King

Abstract

Atomic force microscopy (AFM)-based force spectroscopy is a powerful technique which has seen significant enhancements in both force and time resolution in recent years. This chapter details two AFM cantilever modification procedures that yield high force precision over different temporal bandwidths. Specifically, it explains a fairly straightforward method to achieve sub-pN force precision and stability at low frequencies (<50 Hz) by removing the metal coatings from a commercially available cantilever. A more involved procedure utilizing a focused ion beam milling machine is required to maintain high force precision at enhanced bandwidths. Both modification methods allow site-specific attachment of biomolecules onto the apex area of the tips for force spectroscopy. The chapter concludes with a comparative demonstration using the two cantilever modification methods to study a lipid-protein interaction.

Key words Atomic force microscopy, Cantilever, Tips, Functionalization, FIB, Peptide, Lipid bilayer, Interaction

1 Introduction

Atomic force microscope (AFM)-based single-molecule force spectroscopy (SMFS) is a powerful established technique used to probe physical properties of biological macromolecules and their interactions [1]. This technique can be used to extract unfolding pathways of individual domains of soluble proteins [2–4]. The method has also been used to study folding/unfolding mechanisms of integral membrane proteins such as bacteriorhodopsin [5–8] as well as to probe the interaction between polypeptide chains and fluid lipid bilayers [9–13].

AFM-based force spectroscopy experiments are often performed by attaching biomolecules such as a polypeptide chain or DNA on the apex area of an AFM tip via non-specific interactions. Unfortunately, this strategy often yields low throughput of high-quality data. Moreover, it is common to discard rupture events that occur very close to the sample surface and to consider such events non-specific [2, 5]. This can limit a number of studies, including those involving peptide-lipid bilayer interactions.

Yuri L. Lyubchenko (ed.), *Nanoscale Imaging: Methods and Protocols*, Methods in Molecular Biology, vol. 1814, https://doi.org/10.1007/978-1-4939-8591-3_4, © Springer Science+Business Media, LLC, part of Springer Nature 2018

Recent developments in AFM technology have made significant improvements in both force precision and temporal resolution. For example, modifications to commercially available cantilevers can routinely yield sub-pN force precision and achieve microsecond temporal response times [14–18].

Here, we detail the process of attaching a biomolecule onto customized AFM cantilevers to perform single-molecule force spectroscopy with high spatial-temporal precision. Two modification procedures of AFM cantilevers are presented, both of which were adopted from recent work [14–18]. The first method (discussed in Subheading 3.1), which involves removal of all metal coatings from a commercially available cantilever, can achieve sub-pN force precision. This method is straightforward to implement; however, it is restricted in its bandwidth. The second method (*see* Subheading 3.2) involves nanometer-precise milling of commercial cantilevers using a dual-beam focused ion beam/scanning electron microscope (FIB/SEM). FIB-modified cantilevers exhibit higher force precision over a wider bandwidth, but require significant time and resources to fabricate.

Subheadings 3.3–3.6 outline the methods used for functionalizing modified tips with specific biological molecules and for preparing supported lipid bilayers using common microscope cover glass [19, 20]. Finally, in Subheading 3.7, we compare peptide-lipid interaction studies using the two aforementioned cantilever modification procedures and highlight the tradeoff that exists between force precision and measurement bandwidth.

In this chapter, we detail methods to perform high-precision (sub-pN) AFM-based single-molecule force spectroscopy experiments with differing levels of temporal resolution. The ability to achieve such high precision in a biologically relevant setting is relatively new for AFM-based studies and opens the door for a variety of interesting future investigations. Modified tips are poised to reveal a host of details related to the energy landscape of many biological interactions of interest.

2 Materials

1. Atomic force microscope (homebuilt or commercial).
2. Ultrasonic bath (Branson).
3. MQ water: MilliQ Water (Millipore).
4. Anhydrous ethanol: absolute ethanol, 200 proof (Fisher Scientific).
5. Gold etch: Gold etch-type TFA (Transene).
6. Chromium etch: Chromium etchant 1020 (Transene).

7. Silane: 3-(ethoxydimethylsilyl)propyl-amine (Sigma-Aldrich) *or* (3-aminopropyl)dimethylethoxysilane (Oakwood Chemicals).
8. Toluene (Sigma-Aldrich).
9. PEG: NHS-PEG24-maleimide (Thermo Scientific).
10. DMSO: dimethyl sulfoxide (Thermo Scientific).
11. Glass coverslips (Corning).
12. Long working distance microscope (Leica).
13. Oxygen plasma cleaner (Harrick).
14. Oven (Napco).
15. Cantilevers (Olympus, BioLever [BL-RC-150VC] and BioLever mini [BL-AC40TS-C2]).
16. Kimwipes (Kimberly-Clark).
17. Nitrile gloves (Kimberly-Clark).
18. Filter paper (Fisher Scientific).
19. Focused ion beam/scanning electron microscope (FEI Scios Dual Beam FIB/SEM).
20. SEM pin stub (Ted Pella).
21. Carbon tape (Ted Pella).
22. POPC: (1-palmitoyl-2-oleoyl-sn-glycero-3-phosphocholine) (Avanti Polar Lipids).
23. Peptides (GenScript).
24. Disposable culture tubes 15 × 85 mm (Fisherbrand).
25. Liposome extruder (T & T Scientific).
26. Parafilm (Bemis).

3 Methods

We used a custom-built ultra-stable atomic force microscope [21] and a commercial AFM (Cypher, Asylum Research) to perform the force spectroscopy experiments presented here. It is expected that one can achieve similar performance on many commercial AFMs.

3.1 Procedure for Removing the Metal Off of a Cantilever

Removing the metal coatings (gold and chromium) off of Olympus BioLever long cantilevers is a straightforward procedure for achieving sub-pN force precision [14]. For applications where high temporal resolution is not required, a procedure for obtaining high force precision is as follows:

1. Take three small-sized (10 mL) and three medium-sized (100 mL) beakers. Fill the three small beakers with ethanol, gold etch solution, and chromium etch solution and the larger ones with MQ water. Arrange them in a single row according to

the following sequence: ethanol, MQ water, gold etch, MQ water, chromium etch, and MQ water (*see* **Notes 1** and **2**).

2. Hold the cantilever chip to be etched with the help of a pair of tweezers. Make sure that the tips do not bump into the container wall in any of the following steps. Also, make sure that the tweezer material is compatible with both etching solutions. The compliant cantilevers can bend drastically upon contact with fluid meniscus, preventing complete etching of the metal. To minimize this (especially for the gold etchant which is highly viscous), pick up the chip sideways by tweezers and proceed with dipping into the various fluids listed below.
3. Submerge the whole AFM chip into the beaker with ethanol for approximately 10 s.
4. Immerse into MQ water for 10 s.
5. Immerse into gold etchant for approximately 40 s.
6. Dip into MQ water again for about 10 s.
7. Dry the tip by pressing the chip (not the tip!) onto filter paper. Kimwipes can also be used here, but filter paper is more rigid and thus reduces the chances of damage to the tip in case of poor handling. It is safe to press the side of the chip (that does not have any tips) into the absorbing material.
8. Dip into chromium etchant for 30 s.
9. Dip into MQ water one last time.

   (*Note: Move the chip up-down and side to side to ensure proper mixing.*)
10. Wick dry tips by pressing the chip onto filter paper.
11. Store the cantilevers immersed in ethanol (in a clean glass petri dish with a cover) (*see* **Note 3**).

3.2 Dual-Beam FIB/SEM Modification of Cantilevers

Focused ion beam milling of an Olympus BioLever mini cantilever can be used to achieve higher temporal resolution force spectroscopy data without compromising force precision [17]. A procedure for obtaining higher temporal resolution is as follows:

1. Load AFM cantilevers (with metal coatings intact) onto a SEM/FIB sample holder using double-sided carbon tape (*see* **Note 4**). A photograph of the mounting block we use is shown in Fig. 1.

   Place the cantilever so that roughly half of it overhangs the carbon tape. Position the cantilever with the tip pointed away from the beam, so that the tip is not inadvertently irradiated during imaging.
2. Place the holder in the specimen chamber and initiate the vacuum pump.
3. Use the SEM to image the cantilever. The standard electron beam parameters are 5 kV, 400 pA, 300 ns/pixel, 1× frame

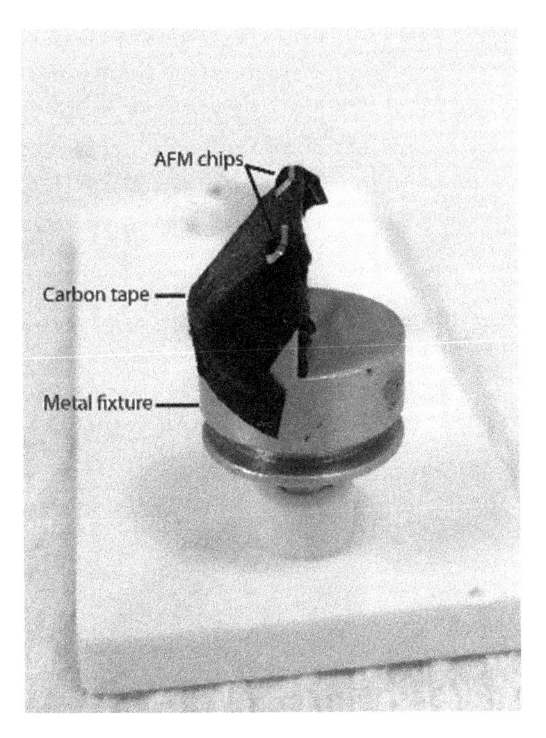

Fig. 1 Photograph of SEM/FIB sample fixture with cantilevers mounted

averaging, and 1024 × 884 pixels. Move the stage until the cantilever is in view and optimize imaging with a horizontal field width (HFW) less than 100 μm.

4. Rotate the stage so that the cantilever is horizontal.
5. Next, move the stage so that it is at the correct eucentric and coincident height. The eucentric height ensures that the specimen-aperture distance does not change when the sample is tilted, and the coincidence position ensures that the SEM and the FIB beams will impinge on the same location.
6. Now, imaging with the FIB as little as possible, find the cantilever with the FIB beam. The standard parameters are 1.5 pA, 30 kV, 300 ns/pixel, and 1024 × 884 pixels. The more the FIB beam is used to image the cantilever, the more stress and bending will be induced in the cantilever.
7. Use a reduced size area (RSA) to image a smaller part of the cantilever and fix the focus. When using BioLever mini tips (Olympus), the "wings" (as shown in Fig. 5a) about the perimeter of the cantilever are a good feature to focus on, as they are not necessary to functionality and thus can be imaged many times with the FIB to little ill effect.
8. Take a single image of the cantilever with the FIB using the lowest magnification and center the cantilever, zooming in to 100 μm HFW.

9. Once the cantilever is in view and properly focused—using the RSA to image the wings is the best way to optimize the parameters—set the ion beam current to ~90 pA.
10. Zoom out to roughly 50 μm HFW and use the RSA to center and focus the cantilever again.
11. Image the cantilever with the SEM. If the cantilever needs to be centered, do not move the stage but rather use *x*-,*y*-beam-shifts, as moving the stage will require you to re-center the cantilever with respect to the FIB.
12. In order to prepare for milling, take a single FIB image of the entire cantilever. Typical milling parameters for our FIB are 30 keV, 93 pA, 200 ns/pixel.
13. Draw a rectangular pattern, with a width of 6 μm and a length of 30 μm, on one side of the cantilever extending from the base of the cantilever, near the chip. When this pattern is executed on both sides of the cantilever, one leg with a width of 4 μm is left to support the end of the cantilever (*see* **Note 5**).
14. Create an exclusion zone by drawing another rectangle, smaller than the first, and placing it inside the first rectangle. This will have the effect of milling out the edge of the pattern as opposed to milling out the entire rectangular area, which speeds up the process. Extend the end of the pattern up onto the chip, so that only two sides of the rectangle will be milled—the inner and upper side. Extend the exclusion zone so that the outer edge of the cantilever will not be milled, as there is no need to. This pattern will mill out two sides of a rectangle and leave the extra flap of material attached to the chip. This is because the part of the cantilever that is getting milled off has a tendency to flip up over the last edge that is being milled. If the last side of the rectangle that has yet to detach is the inner side, then the flap will flip up overhanging the cantilever and becomes very difficult to detach. A way around this problem is to choose which side of the rectangle will be the last to detach—in this case we chose the side nearest the chip (Fig. 2) (*see* **Note 6**).

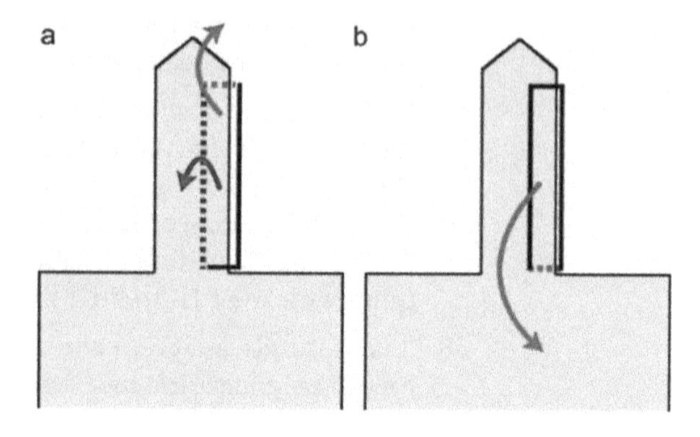

Fig. 2 Cartoon demonstrating (**a**) improper and (**b**) proper milling sequences

15. After two sides of the rectangle have been milled, the flap will only be attached at one point near the chip. Begin milling a thin rectangle on the flap close to the chip. As the milling begins, the flap will flip over toward the chip. In our experience, this is adequate, and there is no need to further attempt to remove the extra material as it does not interfere with the cantilever itself.
16. Create the same pattern on the other side of the cantilever, starting by milling out the two sides of the rectangle, and once they are finished, begin to mill the final edge. Once again, the flap of material should flip up toward the chip and out of the way. You should be left with one leg, ~1–5 μm wide, supporting the end of the cantilever.
17. Next, the cantilever bending should be assessed. Rotate the stage so that the SEM can view the cantilever in profile. Typically, the leg is bent above horizontal. If the leg is bent, then once in the AFM, the photodiode may not be able to properly zero the signal.
18. If the leg of the cantilever is bent, tilt the stage back to perpendicular to the FIB. Draw a rectangular pattern over the entire leg, but avoid overlapping the pattern on the edge of the cantilever where the tip is, as that would remove the gold coating.
19. While the FIB is milling, take SEM images. When the leg appears to relax down to horizontal, stop the milling. If necessary, tilt the stage, view the tip in profile, and decide if further milling is needed. This step removes the gold coating off of the leg by thinning it. If the leg is thinned too much, it will bend under the horizontal and become more fragile, so watch the cantilever with the SEM while the FIB is milling in order to not thin the leg too much.
20. Due to sputtering, some of the reflective gold coating will be removed from the end of the cantilever. However, in our experience, there is enough gold coating left to produce a signal strong enough to be picked up by the AFM without undue addition of noise.

3.3 Tip Functionalization Procedure

After cantilever modifications are complete, the next step in a force spectroscopy experiment is to affix a molecule of interest to the tip. This process is described below:

1. Take cantilevers without metallic coatings (or FIB modified), and place them in a glass petri dish facing all tips upward.
2. Plasma clean them for ~10 min in oxygen at 250 mTorr using ~20 W power.
3. Turn oven "ON" and set the temperature to 90 °C.

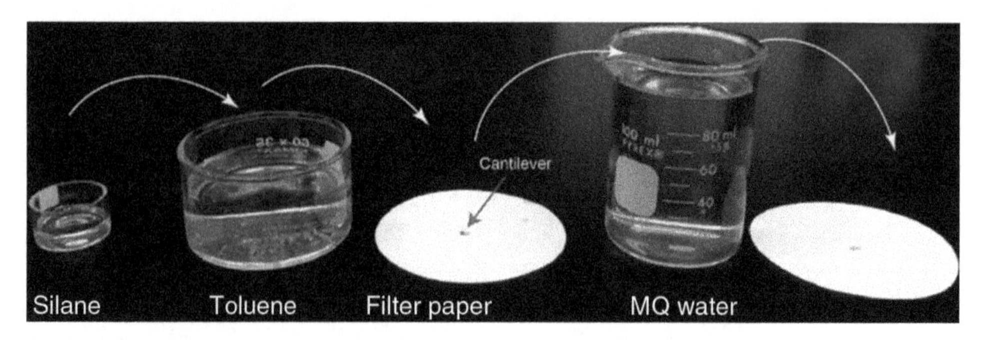

Fig. 3 Process of tip silanization

4. Pour contents of 3-aminopropyldimethylethoxysilane (Silane) (1 g) bottle to a modified low-profile beaker (made by cutting a 10 mL beaker in half, shown at the far left of Fig. 3).
5. Fill a tall glass petri dish with toluene (*see* **Note 7**).
6. Fill a 100 mL beaker with MQ water.
7. Arrange filter paper with the other contents in a row as shown in Fig. 3.
8. Submerge the cantilever chips in Silane for ~60 s. The chip should be gently agitated up and down and side to side.
9. Immerse the chips into toluene for 2 min.
10. Wick the tips; dry them by pressing the chips onto filter paper or Kimwipes.
11. Immerse the tips into MQ water for about 2 min.
12. Deposit the cantilevers in a clean glass petri dish.
13. Repeat the same process for other cantilevers.
14. Bake the silanized cantilevers in an oven at 90 °C for 30 min.
15. Prepare water-saturated environment using several water-filled small beakers under an inverted trough. Alternately, this environment can be created by partially filling the bottom of a 4 L desiccator with water and covering it with its lid.
16. Cool the cantilevers down to room temperature for about 1–2 min.
17. For each cantilever chip, pipette 50 μL borate buffer (50 mM sodium borate, pH 8.5) in a disposable plastic petri dish, forming a single well-contained droplet.
18. Insert the cantilever into the droplet, and place the petri dish in a water-saturated environment at room temperature.
19. Incubate the contents for 1 h.
20. Prepare PEG stock solution (250 mM PEG in DMSO) by dissolving 100 mg of PEG with 187 μL of DMSO (one may have to adjust the amount of DMSO if a different length linker

is used). Remove DMSO out from the bottle using a Hamilton syringe. Fill the bottle with dry argon at the end.

21. Prepare 25 μL of 50 mM PEG solution in borate buffer for each chip (mix 5 μL of PEG stock with 20 μL of borate buffer).
22. Immerse each chip into a 25 μL droplet of PEG solution prepared above, and incubate it in water-saturated environment for 1 h at room temperature.
23. Wash the cantilevers in 100 mL MQ water, and store them in clean petri dish containing MQ water.
24. Prepare peptide stock solution at a concentration of ~1 mM (or less, depending upon the solubility).
25. Dilute stock solution down to 1–100 μM peptide in buffer: 50 mM sodium phosphate pH 7.2, 50 mM NaCl, and 10 mM EDTA, and add 1 mM TCEP (tris(2-carboxyethyl)phosphine). Depending on the physical properties of the peptide under study, the ideal concentration is variable. We have found charged peptides are easier to work with due to their self-limiting densities. Using concentrations <1 μM is challenging, due to the low probability of affixing a peptide to the apex area of the tip, which is vanishingly small.
26. Immerse each cantilever into 25 μL of peptide solution prepared above.
27. Incubate for 2–4 h at 4 °C (in refrigerator).
28. Wash the cantilevers by immersing in them ~50 mL buffer (75 mM sodium phosphate buffer of pH 7.2) for ~2 min.
29. Store cantilevers in buffer in a clean petri dish at 4 °C until use (about 1 week maximum).

3.4 Glass Cleaning Procedure

Here, we detail the procedure for the preparation of clean glass coverslips onto which one can prepare a supported lipid bilayer:

1. Put a large stir bar, 350 mL anhydrous ethanol, and ~100 g potassium hydroxide pellets into a 1 L autoclaved beaker. For safety, cover the beaker to prevent splashing.
2. Turn on stirrer. Set at 400 rpm. Solution should be saturated and should turn dark orange in color; if not, add more KOH pellets. It might take 2–4 h to achieve the desired color change.
3. Fill Branson 5200 ultrasonic bath ¼ full with water.
4. Place 2, 1 L autoclaved beakers ½ full of MQ water in the bath.
5. Place the beaker of KOH-ethanol solution in the bath. Turn the bath power “ON.” Heating is not required.
6. Load the Teflon coverslip baskets (Fig. 4) with coverslips, and attach a handle.

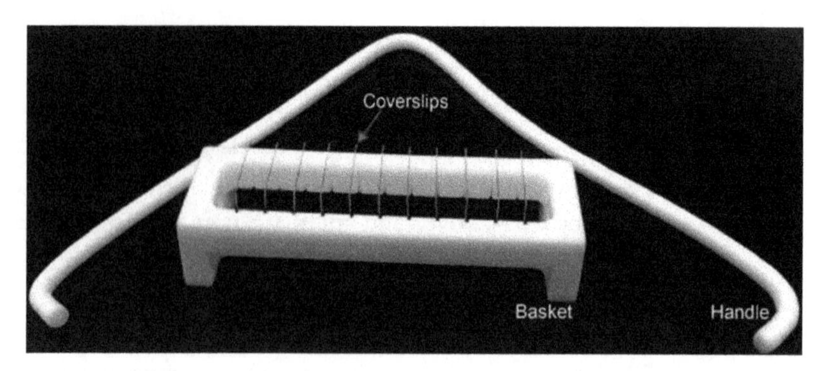

Fig. 4 Teflon coverslip basket

7. Immerse the basket with coverslips in KOH solution with ultrasound for 3 min, and move the basket up and down and side to side to enhance the reaction of coverslips with the KOH solution.
8. Rinse coverslips with MQ water using squirt bottle.
9. Immerse the basket into the first beaker of MQ water in the ultrasonic bath for 3 min.
10. Rinse the basket with MQ water from a squirt bottle again.
11. Immerse the basket into the second beaker of MQ water. Keep in the bath for 3 min.
12. Finally, rinse coverslips with ethanol using a squirt bottle.
13. Blow-dry with ultrapure N_2.
14. Place coverslips in an auxiliary container.
15. Repeat **steps 7–14** for the rest of the baskets.
    Dispose MQ water in sink and solution of KOH and alcohol in an approved hazardous material bottle.

3.5 Liposome Preparation

Liposomes are prepared from POPC lipids and allowed to rupture onto clean glass coverslips to form supported lipid bilayers. The detailed procedure is as described below:

1. Distribute a bottle of POPC lipids in chloroform (Avanti Polar Lipids # 850457C-100 mg) into eight different culture tubes so that each one contains 12.5 mg of lipids.
2. Prepare a thin homogeneous lipid film around the tube walls by evaporating the chloroform using a flow of argon gas while rotating the tube (should be done under the hood).
3. Place the tube in a vacuum chamber (~10 mTorr) overnight using a dry vacuum pump (*see* **Note 8**).
4. For long-term storage, fill the tubes with dry argon. Seal the tubes using parafilm and tape, and store them at −20 °C.

5. Take a tube of lipid film prepared above (or out from −20 °C freezer).
6. Add 0.55 mL of buffer to the tube (50 mM sodium phosphate, 50 mM sodium chloride, 10 mM EDTA pH 7.2).
7. Let the tube sit for about an hour to swell the lipids.
8. Add 2 mL of buffer and manually agitate, for example, by gently flicking the tube using fingertips.
9. Use a pipette to rigorously rinse the walls of the tube with the lipid solution itself.
10. Read the pH of the solution using pH paper, and adjust it if needed so that it has a pH around 7.
11. Open the package containing a clean NanoSizer extruder.
12. Open two disposable syringes that come with the extruder package.
13. Fill one of the syringes with lipid suspension.
14. Mark one of the sides of the extruder "R."
15. Load the filled syringe on the side opposite to "R."
16. Pass the solution *25* times through the extruder (or until the solution is clear enough to see through).
17. Remove the contents from the "R" side into Eppendorf tube (s).
18. Make aliquots of 5 μL.
19. Store at −80 °C.

3.6 Preparing Supported Bilayers

1. Take one aliquot from the −80 °C freezer.
2. Mix it with 495 μL buffer (75 mM sodium phosphate pH 8.0).
3. Apply 100–200 μL of this solution to freshly cleaned glass coverslips.
4. Incubate about 30–45 min at room temperature (elevated temperatures of ~60 °C can also be used).
5. Rinse surface three times (100 μL) with buffer.
6. The bilayer is ready for force spectroscopy experiments.

3.7 Force Spectroscopy Comparison

Here we compare the two high-resolution force spectroscopy tip modification methods discussed above in a peptide-lipid bilayer force spectroscopy assay:

Figure 5 demonstrates a significant improvement in response to time between a FIB-modified BioLever mini tip and a BioLever long tip which has had its metal removed, but was not FIB modified. Figure 5b shows data from a single-molecule force spectroscopy experiment using a FIB-modified BioLever mini tip similar to that shown in the scanning electron micrograph (Fig. 5a). Figure 5c

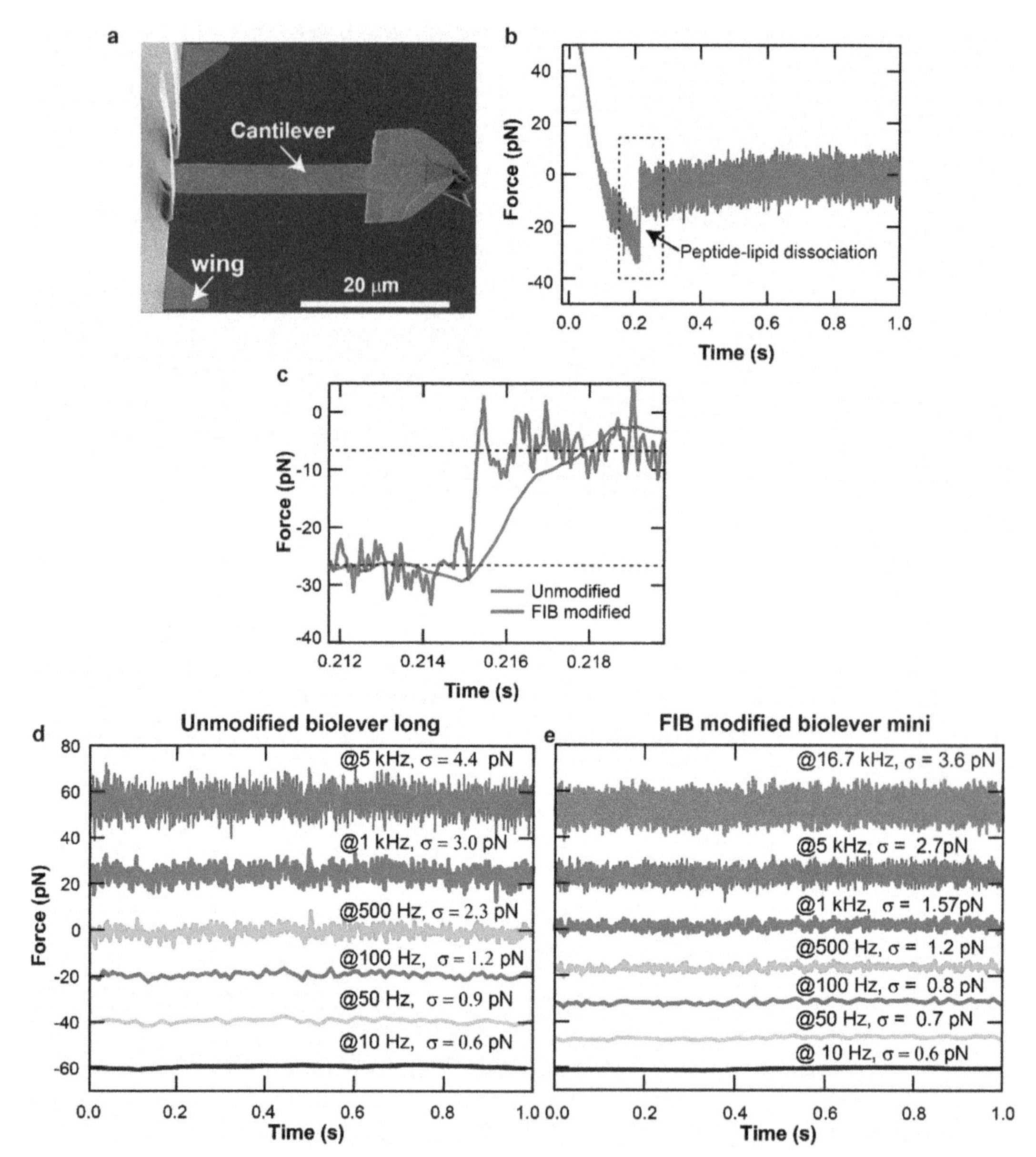

Fig. 5 FIB-modified tips can achieve higher time resolution without sacrificing force precision. (**a**) SEM image of a FIB-modified BioLever mini cantilever (as described in Subheading 3.2). (**b**) A representative retraction force curve using a FIB-modified tip. Here, the tip was functionalized with a peptide (Cys-Trp-Lue-Lue-Lue-Arg) via a NHS-PEG24-maleimide linker and was interacted with a 1-palmitoyl-2-oleoyl-sn-glycero-3-phosphocholine (POPC) bilayer. (**c**) Zoom in of the curve during dissociation indicates the enhanced time resolution of the FIB-modified tip. The blue curve in (**c**) shows similar data obtained with a nonmetalized and non-FIB-modified BioLever long tip (as described in Subheading 3.1). Force precision at various bandwidths for (**d**) the BioLever long cantilever and (**e**) the FIB-modified BioLever mini cantilever. Note that both tips exhibit similar sub-pN force precision at low frequency: 0.6 pN at 10 Hz; however, the FIB-modified tip maintains sub-pN precision at frequencies above 100 Hz

zooms into the rupture event corresponding to dissociation of a peptide from a model lipid bilayer surface. The data shows the significantly faster response time of the FIB-modified BioLever mini tip (red trace) as compared to that of the BioLever long (blue trace) which has had its metal removed, but was not FIB modified (as described in Subheading 3.1). Figure 5d, e compares the force precision of these two tip types at different bandwidths and shows the compromise between force precision and time resolution.

4 Notes

1. The cantilever etching process, tip functionalization steps (e.g., salinization), and lipid film preparation process should be performed in a fume hood.
2. Wear nitrile gloves while etching and functionalizing the cantilevers.
3. Soft cantilevers have the tendency to fold back upon themselves in which case the gold and chromium might not be removed completely during the etching process. It is recommended to check them under an inspection microscope before functionalizing them. If some gold or chromium exists, etch them again using the abovementioned protocol. If the cantilevers remain folded, one may entice them to unfold by sequentially dipping them into the anhydrous ethanol and MQ water.
4. For the FIB milling process, in order to check the final tilt of the cantilever, it is beneficial to load the cantilevers on a tilted holder so that with only a little extra tilting, you can get a view of the cantilever in profile.
5. Because the cantilever will likely bend during patterning, it is advisable to underestimate the length of your pattern. As it bends, the FIB will drill out a longer pattern than the one you created on the horizontal cantilever. In order to preserve the gold coating on the end of the cantilever, be careful when deciding how long to make the pattern.
6. The cantilever may drift during the milling process. It is helpful to periodically stop milling, take an image with the FIB, and re-center the pattern. If drift becomes problematic, allow the samples to sit in the vacuum chamber for an extended period prior to milling.
7. Never use plastic petri dishes for toluene. It reacts with plastic.
8. A dry vacuum pump (i.e., no oil) is used during the lipid film drying process to prevent the back streaming of oil, which is a potential contaminate.

Acknowledgments

This work was supported by National Science Foundation CAREER Award 1054832 (G.M.K.), a Burroughs Wellcome Fund Career Award at the Scientific Interface (G.M.K.), and the MU Research Board.

References

1. Muller DJ (2008) AFM: a nanotool in membrane biology. Biochemistry 47 (31):7986–7998
2. Rief M et al (1997) Reversible unfolding of individual titin immunoglobulin domains by AFM. Science 276(5315):1109–1112
3. Borgia A, Williams PM, Clarke J (2008) Single-molecule studies of protein folding. Annu Rev Biochem 77:101–125
4. Kim BH, Lyubchenko YL (2014) Nanoprobing of misfolding and interactions of amyloid beta 42 protein. Nanomedicine 10 (4):871–878
5. Oesterhelt F et al (2000) Unfolding pathways of individual bacteriorhodopsins. Science 288 (5463):143–146
6. Bippes C, Müller D (2011) High-resolution atomic force microscopy and spectroscopy of native membrane proteins. Rep Prog Phys 74:086601
7. Petrosyan R et al (2015) Single-molecule force spectroscopy of membrane proteins from membranes freely spanning across nanoscopic pores. Nano Lett 15(5):3624–3633
8. Yu H et al (2017) Hidden dynamics in the unfolding of individual bacteriorhodopsin proteins. Science 355(6328):945–950
9. Matin TR et al (2017) Single-molecule peptide-lipid affinity assay reveals interplay between solution structure and partitioning. Langmuir 33(16):4057–4065
10. Ganchev DN et al (2004) Strength of integration of transmembrane alpha-helical peptides in lipid bilayers as determined by atomic force spectroscopy. Biochemistry 43 (47):14987–14993
11. Desmeules P et al (2002) Measurement of membrane binding between recoverin, a calcium-myristoyl switch protein, and lipid bilayers by AFM-based force spectroscopy. Biophys J 82(6):3343–3350
12. Andre G, Brasseur R, Dufrene YF (2007) Probing the interaction forces between hydrophobic peptides and supported lipid bilayers using AFM. J Mol Recognit 20(6):538–545
13. Schwierz N et al (2016) Mechanism of reversible peptide-bilayer attachment: combined simulation and experimental single-molecule study. Langmuir 32(3):810–821
14. Churnside AB et al (2012) Routine and timely sub-picoNewton force stability and precision for biological applications of atomic force microscopy. Nano Lett 12(7):3557–3561
15. Edwards DT et al (2015) Optimizing 1-mus--resolution single-molecule force spectroscopy on a commercial atomic force microscope. Nano Lett 15(10):7091–7098
16. Walder R et al (2017) Rapid characterization of a mechanically labile alpha-helical protein enabled by efficient site-specific bioconjugation. J Am Chem Soc 139(29):9867–9875
17. Faulk JK et al (2017) Improved force spectroscopy using focused-ion-beam-modified cantilevers. Methods Enzymol 582:321
18. Edwards DT, Perkins TT (2017) Optimizing force spectroscopy by modifying commercial cantilevers: improved stability, precision, and temporal resolution. J Struct Biol 197 (1):13–25
19. Zimmermann JL et al (2010) Thiol-based, site-specific and covalent immobilization of biomolecules for single-molecule experiments. Nat Protoc 5(6):975–985
20. Mingeot-Leclercq MP et al (2008) Atomic force microscopy of supported lipid bilayers. Nat Protoc 3(10):1654–1659
21. Sigdel KP, Grayer JS, King GM (2013) Three-dimensional atomic force microscopy: interaction force vector by direct observation of tip trajectory. Nano Lett 13(11):5106–5111

Chapter 5

Polymer Nanoarray Approach for the Characterization of Biomolecular Interactions

Sibaprasad Maity, Ekaterina Viazovkina, Alexander Gall, and Yuri L. Lyubchenko

Abstract

Pair-wise interactions at the single-molecule level can be done with nanoprobing techniques, such as AFM force spectroscopy, optical tweezers, and magnetic tweezers. These techniques can be used to probe interactions between well-characterized assemblies of biomolecules, such as monomer-dimer, dimer-dimer, and trimer-monomer. An important step of these techniques is the proper assembly of dimers, trimers, and higher oligomers to enable the interactions to be probed. We have developed a novel approach in which a defined number of peptides are assembled along a flexible polymeric molecule that serves as a linear matrix, termed as *flexible nanoarray (FNA)*. The construct is synthesized with the use of phosphoramidite chemistry (PA), in which non-nucleoside PA spacers and standard oligonucleotide synthesis are used to grow the polymeric chain with the desired length. The reactive sites are incorporated during FNA synthesis. As a result, the FNA polymer contains a set of predesigned reactive sites to which the peptides are covalently conjugated. We describe the protocol for the synthesis of FNA and the application of this methodology to measure the molecular interactions between amyloid peptides of monomer-monomer, monomer-dimer, and dimer-dimer.

Key words Protein-protein interactions, Nanoprobing, AFM, Force spectroscopy, Amyloid oligomers, Click chemistry

1 Introduction

All biological dynamics are controlled by intermolecular interactions. Molecular biophysicists measure these interactions to increase our understanding of the molecular mechanisms of biological processes. A breakthrough in the techniques used to study these interactions was made in the last two decades with the invention of single-molecule methods that enable intermolecular interactions between numerous biological molecules to be probed, as described in recent reviews [1–4]. These techniques primarily include optical trapping or tweezers (OT), force spectroscopy studies with atomic force microscopy (AFM), and magnetic tweezers

Yuri L. Lyubchenko (ed.), *Nanoscale Imaging: Methods and Protocols*, Methods in Molecular Biology, vol. 1814, https://doi.org/10.1007/978-1-4939-8591-3_5,

(MT). These techniques enable one to measure interactions in the range of hydrogen bonds (primarily OT and MT methods) up to interactions approaching the strength of a covalent bond (AFM force spectroscopy) [5]. Many biological systems consist of several biological molecules; therefore, nanoprobing approaches need to be able to measure the interactions among these molecules within the systems. One example is the characterization of amyloid protein assembly from monomers into oligomers. The self-assembly of amyloid protein into nano-aggregates is a hallmark of diseases such as Alzheimer's and Parkinson's diseases. Characterizing this self-assembly process will improve our understanding of the molecular mechanisms leading to the development of these devastating diseases ([6–9]). We have recently introduced an approach that allowed us to characterize dimers formed by different amyloidogenic polypeptides [10]. The use of a flexible polymeric platform that contained the assembled oligomers allowed us to probe oligomers larger than dimers [11]. The flexible polymer is synthesized from the non-nucleoside PA spacers using a commercial DNA synthesizer. An important feature of this approach is that phosphoramidites with reactive groups can be incorporated at defined stages during the synthesis process, leading to the incorporation of polymers at the reactive sites at predetermined positions. We termed this linear polymeric template flexible nanoarray (FNA, [12, 13]).

This chapter describes protocols for the synthesis of FNA, with an emphasis on the incorporation of reactive groups inside the FNA, and modifications of the FNA ends to immobilize the tether onto the AFM tip and surface. This approach is illustrated by experimental data from AFM probing of dimers, trimers, and tetramers of Aβ(14-23) peptides, as described in [11, 13].

2 Materials

2.1 Reagents

1. Biotin-conjugated controlled pore glass (*3'-Protected* Biotin Serinol CPG; Glen Research, Sterling, VA).
2. Oxidizer: 0.02 M I_2 in THF/pyridine/H_2O (Glen Research, Sterling, VA).
3. Capping reagent A: THF/2,6-lutidine/Ac_2O (Glen Research, Sterling, VA).
4. Capping reagent B: 16% 1-MeIm in THF (Glen Research, Sterling, VA).
5. Deblocking reagent: 3% dichloroacetic acid in dichloromethane (Glen Research, Sterling, VA).
6. Activator: 0.25 M 5-ethylthio-1H-tetrazole in acetonitrile (Glen Research, Sterling, VA).

7. Spacer 18 phosphoramidite (18-*O*-dimethoxytritylhexaethyleneglycol, 1-[(2-cyanoethyl)-(*N,N*-diisopropyl)]-phosphoramidite) (Glen Research, Sterling, VA).
8. DBCO-dT-CE phosphoramidite (5′-dimethoxytrityl-5-[(6-oxo-6-(dibenzo[b,f]azacyclooct-4-yn-1-yl)-capramido-*N*-hex-6-yl)-3-acrylimido]-2’-deoxyuridine, 3′-[(2-cyanoethyl)-(*N,N*-diisopropyl)]-phosphoramidite) (Glen Research, Sterling, VA).
9. Thiol modifier C6 S-S (1-*O*-dimethoxytrityl-hexyl-disulfide,1-′-[(2-cyanoethyl)-(*N,N*-diisopropyl)]-phosphoramidite) (Glen Research, Sterling, VA).
10. Azide- terminated Aβ(14-23) peptides (e.g., Peptide 2.0 Inc, VA, USA).
11. Tris(2-carboxyethyl) phosphine (TCEP) hydrochloride (Hampton Research, Aliso Viejo, CA).
12. Heterobifunctional MAL-PEG-SVA (M.wt 3400 g/mol) (Laysan Bio, Arab, AL).
13. *N*-(g-Maleimidobutyryloxysuccinimide ester) GMBS (Pierce Biotechnology, Grand Island, NY).
14. Streptavidin thiol (SAVT) (Protein Mods, Madison, WI).
15. 1-(3-Aminopropyl) silatrane (APS), use the previously described protocol [14] for synthesis.
16. Mica sheet (Asheville-Schoonmaker Mica Co., Newport News, VA).
17. MSNL-10 AFM probe (Bruker, Camarillo, CA).
18. Epoxy glue (EPOTEC 301, Epoxy Technology, Inc., Billerica, MA).
19. 0.22 μm syringe filter (Millipore express PES membrane, Sigma-Aldrich, USA).
20. Triethylammonium bicarbonate buffer, pH 8.0 (Sigma-Aldrich, St. Louis, MO).

2.2 Instruments

1. MerMade 12 (Bioautomation, Irving, Texas) DNA synthesizer.
2. UV chamber with UV lamp (366 nm wavelength) (Homemade).
3. Eppendorf Vacufuge, model 5301 (Hamburg, Germany).
4. Water purification system (aquaMAX-Ultra, YL Instruments Co., Ltd., Gyeonggi-do, Republic of Korea).
5. MFP-3D AFM instrument (Asylum Research, Santa Barbara, CA).
6. HPLC system (Agilent HPLC 1100, Agilent Technologies, Waldbronn, Germany).

7. Phenomenex Gemini C18, 5μ, 250 × 4.6 mm (Phenomenex, Torrance, CA).
8. Nanodrop UV-visible spectrometer (ND-1000, Thermo Scientific, Grand Island, NY).
9. Electrospray ionization (ESI) mass spectrometer (Thermo Scientific LTQ LCMS system, Grand Island, NY).

2.3 Buffers and Solutions

1. 0.1 M spacer 18 phosphoramidite solution: Dissolve 0.25 g of the reagent in 3.18 mL of anhydrous acetonitrile (*see* **Note 1**).
2. 0.1 M DBCO-dT-CE phosphoramidite solution: Dissolve 0.25 g of reagent in 2.06 mL of anhydrous acetonitrile/dichloromethane solvent (*see* **Note 1**).
3. 0.1 M thiol modifier C6 S-S solution: Dissolve 0.25 g of reagent in 3.25 mL of anhydrous acetonitrile solvent (*see* **Note 1**).
4. 10 mM sodium phosphate buffer, pH 7.0: Dissolve 0.142 g of sodium phosphate monobasic (NaH_2PO_4) in 90 mL DI water. The pH should be adjusted to 7 with 1 M sodium hydroxide or 10% HCl, and the volume should be made up to 100 mL.
5. 50 mM GMBS cross-linker solution: 50 mM of stock solution of GMBS is prepared by dissolving 1.4 mg of GMBS (M.wt 280.23 g/mol) in 100 μL of DMSO. Dilute the solution to the required concentration with DMSO.
6. 10 mM β-mercaptoethanol: 7 μL of β-mercaptoethanol (M.wt 78.13 g/mol, density 1.114 g/cm^3) is dissolved in 10 mL of 10 mM phosphate buffer (pH 7.0) (*see* **Note 2**).
7. 0.1 M TCEP solution: 140 mg of TCEP-HCl (M.wt 286.65 g/mol) is dissolved in 5 mL 10 mM phosphate buffer (pH 7.0), and then pH is adjusted to 7.0 by strong sodium hydroxide solution. The solution is diluted as required in the experiments (*see* **Note 3**).
8. 167 μM APS solution: 44 mg of solid APS powder is dissolved in 3.78 mL of DI water to make 50 mM of stock solution. The solution is then filtered with a 0.22 μm syringe filter to remove any undissolved substances. To make 167 μM solution, 1 μL of stock solution is diluted with 299 μL of DI water.
9. HPLC solvents: Solvent A (100% acetonitrile HPLC grade) and Solvent B: 0.1 M triethylammonium bicarbonate (pH 7.5), which is prepared from commercially available 1 M triethylammonium bicarbonate by diluting 10 times with DI water.

3 Methods

The schematics for the assembly of FNA with the conjugated peptide Aβ(14-23) are shown in Fig. 1. The FNA with two DBCO groups separated with six PA units is synthesized in the DNA synthesizer (Fig. 1a). The product is reacted with azide-terminated Aβ(14-23) to obtain the final product (Fig. 1b). The assembly of the final product is verified by ESI mass spectroscopy.

3.1 FNA Synthesis

1. Load 20 mg of biotin CPG in the reaction column.
2. Connect all the reagent and solvent bottles to the proper channels in the DNA synthesizer.
3. Program the sequence in Mermade 12 software.
4. Each phosphoramidite conjugation requires five key steps (DMT deprotection-coupling-capping-oxidation-capping). After each step, wash two times with acetonitrile.
5. Deprotect the DMT group from biotin-connected linkers to form free –OH groups with 3% dichloroacetic acid in dichloromethane.
6. Couple the 18 phosphoramidite (S18) spacer using 0.25 M 5-ethylthio-1H-tetrazole as an activator. The coupling time should be extended to 6 min to maximize the yield.

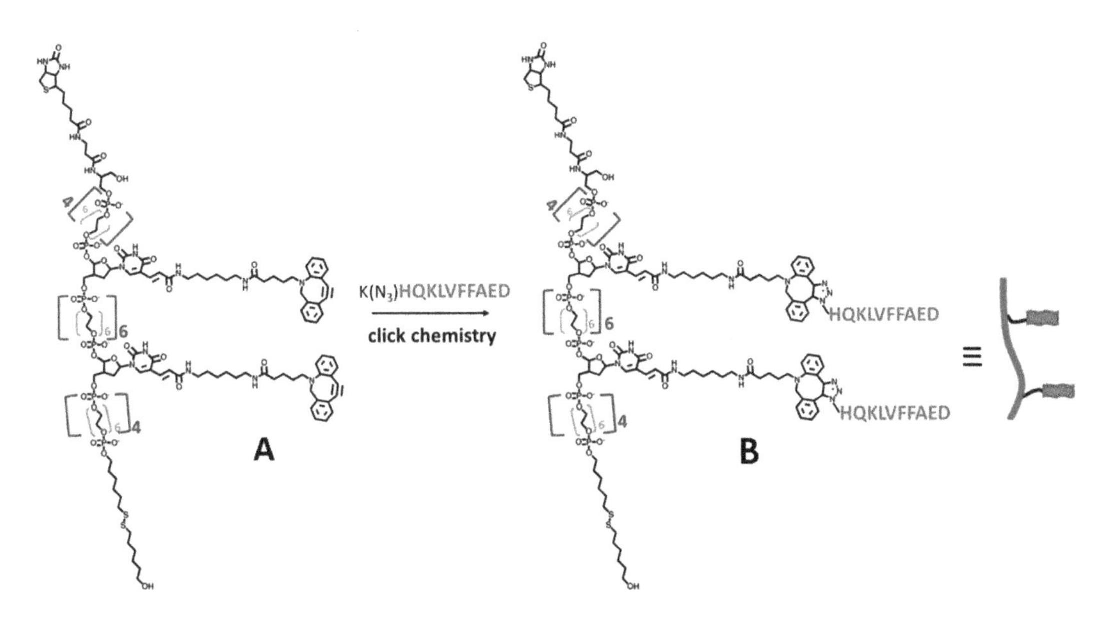

Fig. 1 Schematic presentation of the FNA structure. (**a**) The FNA polymer is synthesized in a DNA synthesizer, in which two DBCO groups are incorporated into the polymer chain at specified positions. (**b**) The FNA construct with two Aβ(14-23) peptides conjugated to DBCO groups is further synthesized using a metal-free "click reaction"

7. Cap the unreacted OH groups with a mixture of cap A and B reagent.
8. Oxidation is performed with 0.02 M iodine in water/pyridine/THF (1:2:7) to obtain a stable phosphate linkage.
9. Conjugate phosphoramidite monomers to assemble FNA in the following order: (1) four units of S18 spacer, (2) DBCO phosphoramidite, (3) six units of S18 spacer, (4) second DBCO phosphoramidite, and (5) four units of S18 spacer (*see* **Note 4**).
10. Add a thiol modifier at the end of synthesis, and keep the DMT group ON.
11. Cleave the product from CPG by treating with 30% ammonium hydroxide for 16 h at room temperature.
12. Add Trizma base to prevent DMT removal and evaporate the ammonia with the vacuum concentrator.
13. Filter the solution through a 0.22 μm syringe filter and discard the precipitate.
14. Collect the filtrate part.
15. Purify the product using reversed-phase high-performance liquid chromatography (RP-HPLC) (acetonitrile gradient in 0.1 M triethylammonium bicarbonate, pH 7.5 buffer, gradient 20–45% in 40 min, Phenomenex Gemini C18 column, 5μ, 250 × 4.6 mm).
16. The final structure of the linker (Fig. 1a) is as follows: DMT-O-$(CH_2)_6$-S-S-$(CH_2)_6$-$(S18)_4$-(DBCO-dT)-$(S18)_6$-(DBCO-dT)-$(S18)_4$-biotin.
17. Measure the concentration of the product using the Nanodrop instrument. The extinction coefficient of DBCO at 308 nm is 12,000 M^{-1} cm^{-1}.

3.2 Conjugation of the Azide-Modified Peptide Using Click Chemistry (Fig. 1)

1. Dilute the FNA construct to 50 μM with 10 mM sodium phosphate buffer.
2. Prepare a 1 M stock solution of azide-labeled Aβ(14-23) in DMSO, and dilute the solution to 200 μM with 10 mM sodium phosphate buffer.
3. Mix 100 μL of FNA solution (at 50 μM concentration) with 100 μL peptide solution (at 200 μM concentration), and vortex.
4. Incubate this mixture for 24 h at room temperature.
5. Evaporate the solvent in a vacuum.
6. Purify the product (Fig. 1b) with RP-HPLC (acetonitrile gradient in 0.1 M triethylammonium bicarbonate, pH 7.5 buffer,

gradient 20–45% in 40 min, Phenomenex Gemini C18 column, 5μ, 250 × 4.6 mm).

7. Confirm the final product with mass spectroscopy.

3.3 Probing of Monomer-Monomer Interactions

3.3.1 Mica Surface Functionalization

1. Cut a piece of mica (~200 μm thickness) into a ~1.5 × 1.5 cm square, and glue it to a glass slide using epoxy glue (*see* **Note 5**).
2. Cleave the mica with scotch tape (*see* **Note 6**).
3. Place 150 μL of 167 μM APS solution on the mica, allow it to react for 30 min, and then wash the surface with DI water.
4. Add 150 μL of 167 μM MAL-PEG-SVA (M.wt. 3400 g/mol) in DMSO and incubate for 3 h, and then wash the surface with DMSO and DI water.
5. Prepare 10 nM streptavidin thiol (SAVT) in 10 μM TCEP in sodium phosphate buffer, and incubate for 30 min.
6. Add the protein solution onto the surface, and incubate overnight at 4 °C.
7. Wash the surface with DI water, add 100 μL of a 10 mM solution of β-mercaptoethanol, and keep for 15 min.
8. Wash the surface with DI water and store at 4 °C.

3.3.2 AFM Tip Functionalization

1. Clean the AFM tip by immerging it into ethanol for 15 min, rinsing with water, and drying with gentle argon flow (*see* **Note 7**).
2. Place the AFM tip in the UV chamber, and expose it with UV light (366 nm wavelength) for 45 min.
3. Immerse the tip in 200 μL of a 1 μM APS solution for 30 min, and rinse it with DI water.
4. Incubate the tip in 200 μL of 1 μM *N*-(g-maleimidobutyryloxysuccinimide ester) (GMBS) solution for 1 h, and rinse it with water.
5. Prepare a 10 nM solution of FNA Aβ(14-23) dimer (Fig. 2a) in 10 mM sodium phosphate buffer (pH 7.0). Activate the thiol group by treating the FNA construct with 10 μM Tris (2-carboxyethyl) phosphine (TCEP), and incubate it for 30 min.
6. Place a droplet of FNA solution (50 μL) onto a piece of parafilm, and incubate the tip in that solution overnight at 4 °C.
7. Wash the tip with water and quench the unreacted maleimide groups by treating the tips with a 10 mM solution of β-mercaptoethanol for 10 minutes (*see* **Note 2**).
8. Wash the tip in water and store at 4 °C.

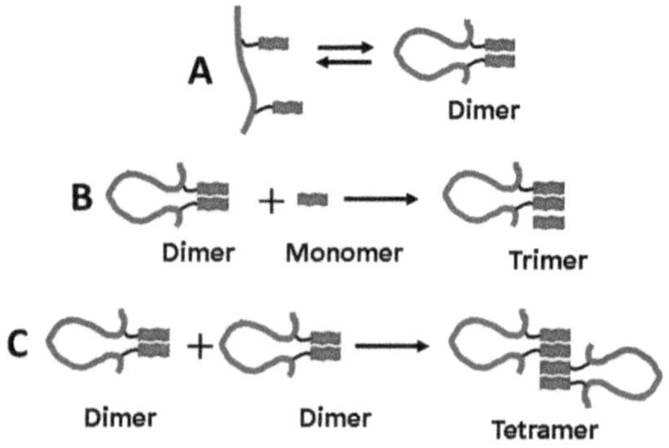

A Dimer
B Dimer + Monomer → Trimer
C Dimer + Dimer → Tetramer

Fig. 2 Schematic for AFM force spectroscopy studies of FNA-Aβ(14-23) dimer. (**a**) The spontaneous assembly of Aβ(14-23) dimers within the FNA construct. (**b**) The interaction between the FNA-Aβ(14-23) dimer and an Aβ(14-23) monomer leading to a trimer. (**c**) The interaction of two FNA-Aβ(14-23) dimers leading to a tetramer

3.3.3 Force Measurements

Force-distance (F-D) measurements were performed on the Asylum MFP 3D AFM instrument. Aβ(14-23) oligomers were probed in 10 mM sodium phosphate buffer (7.0) at room temperature.

1. Mount the functionalized AFM tip in the tip holder, and adjust the laser and photodiode accordingly.
2. Place the functionalized mica surface on the AFM stage; add 200 μL of probing buffer. Manually move the tip toward the surface so that the tip remains immersed in the water layer. Adjust the laser deflection.
3. Approach the AFM tip to the surface.
4. Calculate the actual spring constant of the cantilever by the thermal noise analysis method as recommended by the manufacturer. The calculated spring constant for MSNL-10 D probes is in the range of 20–30 pN/nm.
5. During measurement, set the trigger force at 100 pN for 0.5 s, and perform the approach-retract cycle at a speed of 500 nm/s.
6. Collect several thousand force curves to obtain sufficient statistics.

3.4 Probing of Monomer-Dimer Interactions

The FNA approach for the characterization of Aβ(14-23) trimers [11] (Fig. 2b).

1. Modify the mica surface using the same protocol as described in Subheading 3.3.1, except at **step 5**; inject 40 nM Aβ(14-23) pretreated with TCEP.
2. Functionalize the AFM tip using the same method as discussed in Subheading 3.3.2, except using 167 μM of Mal-PEG-SVA (M.wt 3400 g/mol) instead of GMBS.
3. Perform force measurements with the functionalized mica surface and AFM tip as described in Subheading 3.3.3.

3.5 Probing of Dimer-Dimer Interactions (Fig. 2c)

1. Modify the mica surface using the same protocol as described in Subheading 3.3.1, except at **step 5**; inject 40 nM FNA Aβ (14-23) dimer (pretreated with TCEP).
2. Functionalize the AFM tip using the same method as discussed in Subheading 3.3.2, except using 167 μM of Mal-PEG-SVA (M.wt. 3400 g/mol) instead of GMBS.
3. Perform force measurements with the functionalized mica surface and AFM tip as described in Subheading 3.3.3.

We illustrate below how the FNA-based methodology was used to characterize the interactions within self-assembled amyloid dimers. The FNA construct with two Aβ(14-23) peptides terminated with biotin is immobilized onto the AFM tip (Fig. 3). Streptavidin is tethered to the mica surface via a flexible PEG tether. As the tip approaches the surface to make biotin-streptavidin contact, the peptides spontaneously assemble into a dimer, as shown in Fig. 3a (i). Upon retracting the tip from the surface, the PEG tether stretches, as shown in Fig. 3a (ii). Further retraction of the tip can lead to dimer rupture and stretching of the FNA tether (Fig. 3a (iii)). Finally, at a highly applied force, the biotin-streptavidin link ruptures, as shown in Fig. 3a (iv). All of these events are identified in the force-distance curve shown in Fig. 3b. The arrows between frames (**a**) and (**b**) illustrate the appearance of each event on the force-distance curve, and the rupture forces for each event are measured (F_1 = 50 pN and F_2 = 62 pN). Each event is validated by measuring the contour lengths Lc_1 and Lc_2, based on the known lengths of the tethers. The details for this type of analysis can be found in reference [13]. Specifically, the distance between the two peaks in Fig. 3b is ~12 nm, which is consistent with the distance between the DBCO units [13]. Additional analysis of the rupture process, as performed in [12] for Aβ(14-23) dimer, can reveal interesting properties of dimer assembly. The FNA approach was critical in probing interactions between Aβ(14-23) dimer and monomer and dimer-dimer interactions [11]. This suggests that FNA can be used to probe higher order oligomers with three or more conjugated monomer units. Although the examples above described probing interactions between peptides, any other interacting partners can be probed after conjugating to an appropriately designed FNA template.

4 Notes

1. Phosphoramidite solutions should be prepared as fresh as possible with the use of anhydrous solvents because added moisture can lower the yield. The phosphoramidite solution should be used within 2–3 days because it degrades in solution rapidly.

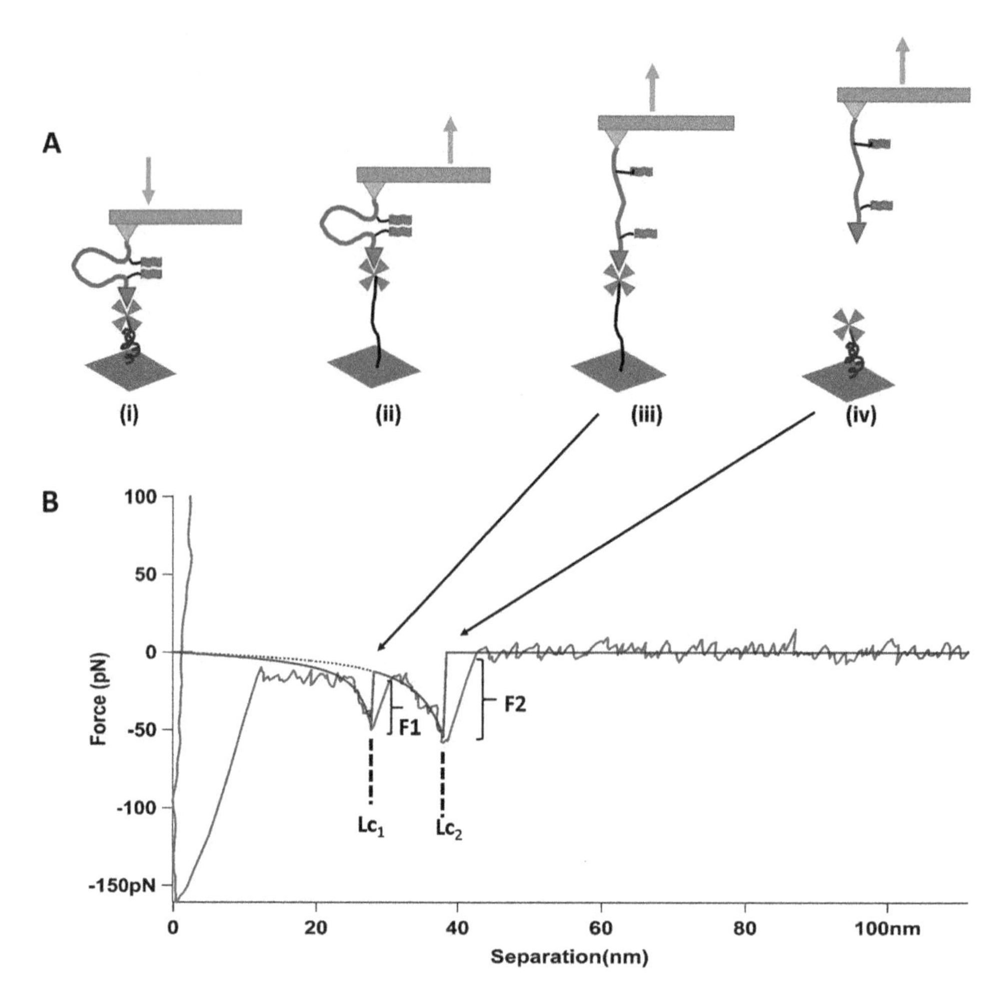

Fig. 3 AFM force spectroscopy experiment for probing dimers assembled within the FNA-Aβ(14-23) dimer construct. (**a**) Cartoon for selected stages of the AFM probing experiment: (i) Approaching the FNA-Aβ(14-23) dimer to the surface yields the complex between terminal biotin (red triangle) and the surface-immobilized streptavidin (green four-leaf construct illustrating the tetramer structure of the protein), (ii) retracting the tip and stretching the PEG polymer tethered to streptavidin, (iii) rupture of the Aβ(14-23) dimer, and (iv) rupture of the biotin-streptavidin link. (**b**) A typical force vs. distance (*F*–*D*) retraction curve for the event in (**a**). F1 and F2 are rupture forces for the dissociation of the Aβ(14-23) dimer and biotin-streptavidin links, respectively. Lc_1 and Lc_2 are the contour lengths corresponding to these two events. The red lines on the graph show the approximation of the extension parts of the force curves according to the wormlike chain model [15]. The arrows point to the events in the force curves that correspond to the cartoons in frame (**a**)

2. β-Mercaptoethanol is a sulfur-containing compound with an unpleasant smell. This solution should be prepared inside a safety hood. The handler should wear proper protecting eyeglasses, hand gloves, and a lab coat.
3. When TCEP is dissolved in 10 mM sodium phosphate buffer (pH 7.0), the solution becomes acidic. The pH should be adjusted to 7.0 by adding a few microliters of a 1 M NaOH

solution. TCEP should be prepared fresh to maintain its reactivity.

4. According to http://www.glenresearch.com/GlenReports/GR27-17.html, DBCO undergoes oxidation by I_2 oxidizer if the number of oxidation steps is extensive. As an alternative, CSO ((1S)-(+)-(10-camphorsulfonyl)-oxaziridine) in acetonitrile can be used instead of iodine. In our case, we had only ten oxidation steps from the first DBCO and four steps from the second DBCO, and we were able to use a regular iodine oxidizer.
5. It is important to pay attention while gluing the mica sheet to a glass slide using EPOTEC 301 glue. To work properly, at least 4 g part A of the glue should be mixed with 1 g part B, followed by a thorough mixing. A little amount of this mixture should be applied as a thin layer onto the glass surface where the mica sheet will be placed. Then, the whole system should be cured at 70 °C for 2 h inside a vacuum chamber. A vacuum would help to remove air bubbles between the mica and glass. The surface should be cooled before being used.
6. To get a uniform mica surface, perform cleaving with scotch tape until a uniform layer of cleaved mica is clearly seen on the tape.
7. The AFM tip should be handled very carefully to avoid breaking the cantilevers. During tip washing, the tip should be immerged into ethanol for 10 min, and then transferred to pure water for another 10 min. Before putting the tip into the UV chamber, the tip should be dried with a gentle flow of argon gas.

Acknowledgments

The work was supported by grants GM096039 and GM118006 to Y.L.L. from the National Institutes of Health.

References

1. Neuman KC, Nagy A (2008) Single-molecule force spectroscopy: optical tweezers, magnetic tweezers and atomic force microscopy. Nat Methods 5(6):491–505
2. Lyubchenko YL, Kim BH, Krasnoslobodtsev AV, Yu J (2010) Nanoimaging for protein misfolding diseases. Wiley Interdiscip Rev Nanomed Nanobiotechnol 2(5):526–543. https://doi.org/10.1002/wnan.102
3. Dulin D, Berghuis BA, Depken M, Dekker NH (2015) Untangling reaction pathways through modern approaches to high-throughput single-molecule force-spectroscopy experiments. Curr Opin Struct Biol 34:116–122. https://doi.org/10.1016/j.sbi.2015.08.007
4. Scholl ZN, Li Q, Marszalek PE (2014) Single molecule mechanical manipulation for studying biological properties of proteins, DNA, and sugars. Wiley Interdiscip Rev Nanomed Nanobiotechnol 6(3):211–229. https://doi.org/10.1002/wnan.1253

5. Lyubchenko YL (2017) An introduction to single molecule biophysics. CRC Press, Taylor & Francis Group, Boca Raton
6. Lyubchenko YL (2015) Amyloid misfolding, aggregation, and the early onset of protein deposition diseases: insights from AFM experiments and computational analyses. AIMS Mol Sci 2(3):190–210. https://doi.org/10.3934/molsci.2015.3.190
7. Krasnoslobodtsev AV, Volkov IL, Asiago JM, Hindupur J, Rochet JC, Lyubchenko YL (2013) Alpha-synuclein misfolding assessed with single molecule AFM force spectroscopy: effect of pathogenic mutations. Biochemistry 52(42):7377–7386. https://doi.org/10.1021/bi401037z
8. Maity S, Lyubchenko YL (2015) Probing of amyloid A beta (14-23) trimers by single-molecule force spectroscopy. J Mol Transl Med 1(1):004
9. Maity S, Hashemi M, Lyubchenko YL (2017) Nano-assembly of amyloid beta peptide: role of the hairpin fold. Sci Rep 7(1):2344. https://doi.org/10.1038/s41598-017-02454-0
10. Tong Z, Mikheikin A, Krasnoslobodtsev A, Lv Z, Lyubchenko YL (2013) Novel polymer linkers for single molecule AFM force spectroscopy. Methods 60(2):161–168. https://doi.org/10.1016/j.ymeth.2013.02.019
11. Maity S, Viazovkina E, Gall A, Lyubchenko YL (2017) Single-molecule probing of amyloid nano-ensembles using the polymer nanoarray approach. Phys Chem Chem Phys 19 (25):16387–16394. https://doi.org/10.1039/C7CP02691A
12. Krasnoslobodtsev AV, Zhang Y, Viazovkina E, Gall A, Bertagni C, Lyubchenko YL (2015) A flexible nanoarray approach for the assembly and probing of molecular complexes. Biophys J 108(9):2333–2339. https://doi.org/10.1016/j.bpj.2015.03.040
13. Maity S, Viazovkina E, Gall A, Lyubchenko Y (2016) A metal-free click chemistry approach for the assembly and probing of biomolecules. J Nat Sci 2(4):e187
14. Shlyakhtenko LS, Gall AA, Lyubchenko YL (2013) Mica functionalization for imaging of DNA and protein-DNA complexes with atomic force microscopy. Methods Mol Biol 931:295–312. https://doi.org/10.1007/978-1-62703-056-4_14
15. Bouchiat C, Wang MD, Allemand J, Strick T, Block SM, Croquette V (1999) Estimating the persistence length of a worm-like chain molecule from force-extension measurements. Biophys J 76(1 Pt 1):409–413

Chapter 6

Measuring Single-Molecule Twist and Torque in Multiplexed Magnetic Tweezers

Franziska Kriegel, Willem Vanderlinden, Thomas Nicolaus, Angelika Kardinal, and Jan Lipfert

Abstract

Magnetic tweezers permit application of precisely calibrated stretching forces to nucleic acid molecules tethered between a surface and superparamagnetic beads. In addition, magnetic tweezers can control the tethers' twist. Here, we focus on recent extensions of the technique that expand the capabilities of conventional magnetic tweezers by enabling direct measurements of single-molecule torque and twist. Magnetic torque tweezers (MTT) still control the DNA or RNA tether's twist, but directly measure molecular torque by monitoring changes in the equilibrium rotation angle upon overwinding and underwinding of the tether. In freely orbiting magnetic tweezers (FOMT), one end of the tether is allowed to rotate freely, while still applying stretching forces and monitoring rotation angle. Both MTT and FOMT have provided unique insights into the mechanical properties, structural transitions, and interactions of DNA and RNA. Here, we provide step-by-step protocols to carry out FOMT and MTT measurements. In particular, we focus on multiplexed measurements, i.e., measurements that record data for multiple nucleic acid tethers at the same time, to improve statistics and to facilitate the observation of rare events.

Key words Single-molecule techniques, Magnetic tweezers, Multiplexing, Freely orbiting magnetic tweezers, Magnetic torque tweezers, DNA, Torque, Twist

1 Introduction

1.1 Magnetic Tweezers Apply Forces and Torques to Biological Macromolecules

Magnetic tweezers (MT) are a powerful technique to study the mechanical properties, dynamics, and conformational transitions of nucleic acids and to probe their interactions with proteins and other ligands [1–5]. In MT, molecules of interest are tethered between a flow cell surface and superparamagnetic beads (Fig. 1). Using permanent magnets or electromagnets, controlled stretching forces can be applied to the molecular tethers. In addition, rotation of the external magnetic field permits to rotate the magnetic particles in a controlled fashion and, therefore, systematically twist the molecular tethers, as first introduced by Strick et al. [5]. Conventional MT assays track the (x, y, z)-position of the magnetic beads

Yuri L. Lyubchenko (ed.), *Nanoscale Imaging: Methods and Protocols*, Methods in Molecular Biology, vol. 1814,
https://doi.org/10.1007/978-1-4939-8591-3_6, © Springer Science+Business Media, LLC, part of Springer Nature 2018

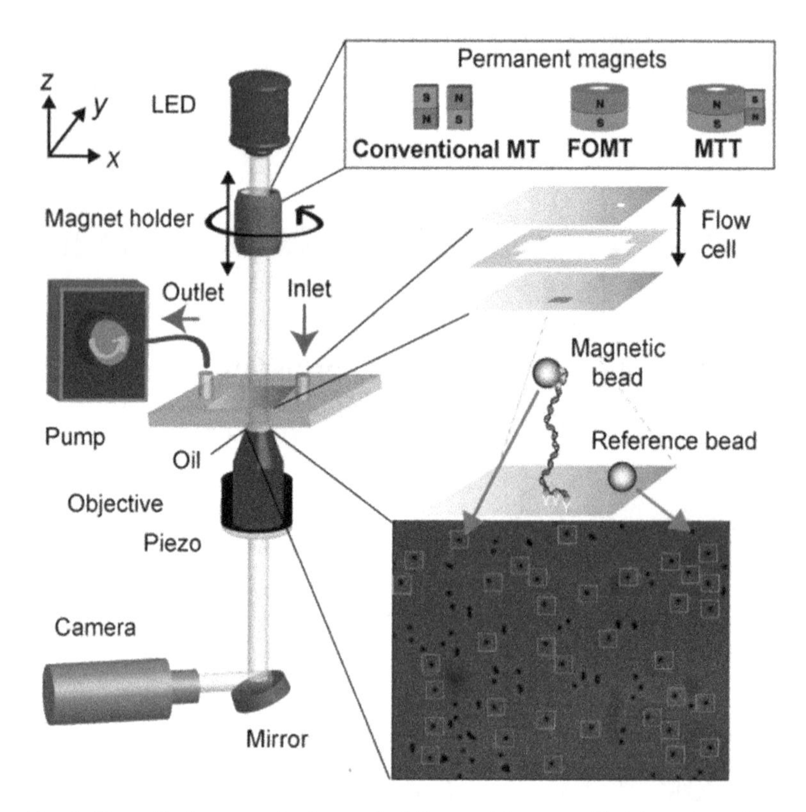

Fig. 1 Magnetic tweezers instrument. A flow cell is assembled from two glass cover slides (light blue on the right) that are separated by a Parafilm (light grey) layer that provides a measurement chamber. The flow cell is installed in a flow cell holder (light brown color) that is connected to a pump for fluid handling, and mounted on a stage to establish direct contact between the flow cell and an objective. The objective is placed on a piezo stage to control the focal plane, used in particular to sensitively calibrate the bead's *z*-position via their diffraction pattern. A LED monochromatically illuminates the sample through the magnet assembly and is directed via a mirror and a tube lens (not shown) to a camera, which is connected to a computer for read-out and bead tracking. A magnet holder, whose position in *z* and rotation around the vertical axis can be controlled by translation and rotation motors, is placed on top of the flow cell. Changing the magnet assembly changes the overall magnetic field, which in turn alters the force and the rotational trap stiffness applied to the magnetic beads (*see* also Fig. 2)

and apply stretching forces that are calibrated from transverse fluctuations [6–9]. MT have been applied to bare DNA (and RNA) to study nucleic acids upon overtwisting and undertwisting, or to probe mechanical properties, such as the bending persistence length or the stretch modulus [10–12]. In addition, MT have been used to study interactions of small molecules with DNA [13–15] and provided a platform to probe DNA– and RNA–protein interactions [4, 16–21] and processing. While conventional MT control the twist of the molecular tether, they do not track rotation directly and do not measure torque. To overcome these limitations, recently new types of torque and twist measuring MT have been developed that are introduced in the next sections and that are the focus of this protocol paper.

1.2 Magnetic Tweezers for Twist and Torque Measurements: FOMT and MTT

Several variants of MT have been developed to enable direct measurements of torque and twist at the single-molecule level [2] (Fig. 2). Two main schemes can be distinguished. In freely orbiting magnetic tweezers (FOMT) and related approaches, one end of the molecular tether (typically DNA) can rotate freely, thus enabling measurements of angular changes within the DNA [22–25]. FOMT have enabled, for example, studying the assembly of nucleosomes on DNA [26] and resolved changes in the tether extension and the chiral winding of DNA in tetrasomes. Similarly, FOMT have been used to detect the assembly and helicity of nucleoprotein filaments involved in DNA repair formed by RecA [27] or Rad51 [28]. In a related assay, transcription of DNA by RNA polymerase [29] or the activity of DNA gyrase [23, 30] and the ensuing rotation of the DNA tether have been directly observed.

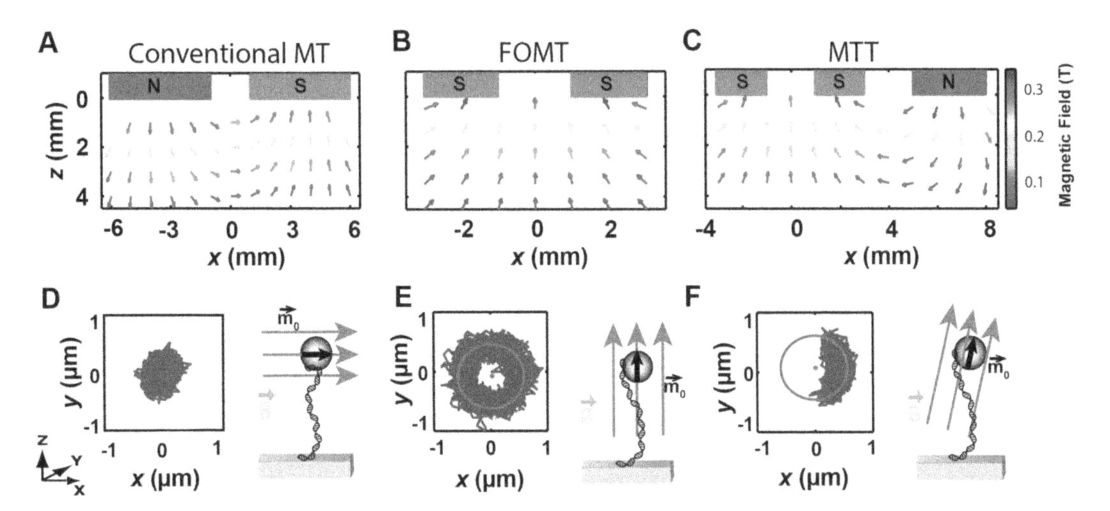

Fig. 2 Magnetic fields and bead fluctuations in conventional tweezers, FOMT, and MTT. Magnetic field calculations for conventional MT (**a**), freely orbiting magnetic tweezers (**b**), and magnetic torque tweezers (**c**). In conventional MT the bead's fluctuations in the (*x*, *y*)-plane are confined and trace out a small ellipse (**d**). The bead is free to rotate about the tether axis in FOMT, tracing out a doughnut-like pattern as illustrated in (**e**). The small side magnet in MTT adds a small horizontal field component such that the beads fluctuations are confined to an arc-like shape (**f**), i.e., in the MTT the bead does not trace out a full circle, but a segment of the doughnut, due to the rotational trap exerted in MTT by the magnets. Note that the bead aligns with the direction of the magnetic field $\vec{B}$ due to its preferred magnetization axis $\vec{m}_0$. In a typical FOMT or MTT experiment, initial assessment of tethered beads is performed using conventional MT magnets. Consequently, the geometry of the DNA–bead system in the flow cell is altered on changing to the FOMT or MTT magnet geometry (*see* also Fig. 4). Figure adapted from Ref. [54] with the permission from Elsevier Inc. Copyright© 2017 Elsevier Inc.

In magnetic torque tweezers (MTT) the twist of the molecular tether is still controlled by the external magnets, but the angular trap k_{ROT}, which is calibrated from the variance of the angular thermal fluctuations by $k_{ROT} = k_B \cdot T/\mathrm{Var}(\Theta)$ (where Θ is the rotation angle about the tether axis, k_B the Boltzmann constant, and T the absolute temperature), is weak enough to permit measurements of molecular torque from changes in the equilibrium angle $\langle\Theta\rangle$ [31–37] by $\tau = -k_{ROT} \cdot (\langle\Theta_N\rangle - \langle\Theta_0\rangle)$, where the subscript denotes the number of applied turns. MTT measurements have revealed the torsional stiffness of DNA and RNA as a function of applied stretching force [33, 38]. In addition, they have been used to probe, for example, the torsional response of DNA--protein filaments [28, 33], nucleosomes [26, 31], and self-assembled DNA origami structures [37]. Basic aspects of how to perform MTT and FOMT measurements have been published in a video-based format previously [39].

Recently, torque measuring MT assays have been extended beyond simple molecular tethers to for example probe the torque generated by the bacterial flagellar motors in live cells [40–42]; however, here we focus on applications to DNA tethers in vitro.

1.3 Multiplexing Magnetic Tweezers

MT assays are used to address increasingly complex questions, e.g., involving several interaction partners in protein–nucleic acid interactions or nonprocessive reactions that are more demanding than "DNA only" measurements. The more complex the biological question, the more important is a reliable and robustly working instrument to perform precise measurements and to capture a maximum amount of data per measurement run. In addition, due to the relatively low torsional stiffness of ~kbp DNA tethers, molecular torque measurements even on bare DNA require measuring small angular deviations against thermal fluctuation noise, making it desirable to improve statistics beyond "one-molecule-at-a-time" measurements, again highlighting the need for parallelized measurements. Tracking of many beads in parallel in MT is possible (Fig. 1) and has been presented in earlier studies for conventional MT [43–45]. Recently, multiplexed measurements have also been demonstrated for MTT [46]. Multiplexed MTT have helped to increase statistics and have revealed the subtle changes in torsional stiffness of DNA upon variations in stretching force and salt concentration [46]. The multiplexed MTT measurements have resolved a small decrease (<10%) in the effective torsional stiffness of DNA for stretching forces smaller 2 pN at high salt concentrations (>500 mM NaCl or 10 mM $MgCl_2$), but found that the intrinsic torsional stiffness (at 6.5 pN stretching force) is independent of salt concentration. In addition, high-resolution multiplexed DNA torque measurements have enabled precise comparisons to coarse-grained simulations of DNA to test two mechanical models, the isotropic and the anisotropic rod model [46, 47].

Combining high-resolution torque measurements with simulations, the best fitting value for the intrinsic twist-bend coupling G was determined to be 40 ± 10 nm.

In this Chapter, we give detailed protocols on how to perform multiplexed magnetic torque tweezers measurements similar to the ones reported in Ref. [46, 47]. In addition, we provide guidelines for FOMT assays and demonstrate that FOMT measurements, too, can be carried out in a parallelized fashion.

2 Materials

2.1 Magnetic Tweezers Microscope

1. Torque measuring MT are currently only available as custom-built instruments. However, conventional MT instruments can be converted in a straightforward fashion to torque and twist measuring instruments by replacing the magnets and updating measurement and data analysis procedures. Here, we describe our implementation of a custom-made parallelized instrument.
2. A light emitting diode (Osram Oslon SSL, red, 612 nm) is used to illuminate the sample monochromatically from above.
3. An oil immersion objective (typically 60×, Plan Fluorite with correction collar, NA 0.9 or 40×, Plan Fluorite, NA 0.75, Olympus) is placed on a piezo (Pifoc, P-726.1CD and controller, E-753.1CD, Physik Instrumente) stage underneath the flow cell holder.
4. The flow cell holder is an important element used in MT to position and mount the flow cell onto the objective. It is (custom-) made of an aluminum (or steal) bottom holder and a top part made of PEEK (polyetheretherketon). The top part provides an inlet and outlet for fluid handling. The outlet is connected via an adapter (TECHLAB, VBM 101.538) to the tubing of the pumping system (*see* Subheading 2.1, **item 8**). The fluid system is sealed with rubber O-rings between the glass flow cell and flow cell holder.
5. The flow cell is imaged on the chip of a camera (Falcon PT-41-4M60, Dalsa) via a mirror (20D20ER.1, Newport) and tube lens (G322304000, Newport).
6. In conventional MT (Fig. 1), two cubic magnets ($5 \times 5 \times 5$ mm^3; W-05-N50-G, Supermagnete), separated by a small gap (typically 1 mm) are placed on a motorized arm and can be moved vertically using a translational motor (C-863.11-Mercury controller and M-126.PD2 motor, Physik Instrumente) and rotated with a rotation motor (C-863.11-Mercury controller and C-150.PD motor, Physik Instrumente).
7. The magnet arm itself is placed on two orthogonal translational micrometer-stages (M-UMR8.25, Newport) to precisely

control and align the relative position of the magnets main axis with the objectives main axis (*see* Subheading 3.5).

8. The flow cell outlet is connected to a peristaltic pump (ISM832C, Ismatec) for fluid handling [48].
9. The setup is controlled using a computer (DELL Precision T3600) equipped with a frame grabber (PCIe-1433, National Instruments) and using software written in LabVIEW (National Instruments) described by Cnossen et al. [45].
10. Measurements are generally performed at 60 Hz, but the acquisition frequency can be increased up to >1 kHz if a reduced field of view is used.

2.2 Magnetic Tweezers Software

1. In MT, the motion of the bead in the (x, y)- and z-position is tracked using the diffraction pattern of the bead, as described previously [6, 49].
2. Tracking is possible for many (>100) beads in parallel and carried out using the optimized routines described and made available by Cnossen et al. [45]. If necessary, the software allows tracking on GPUs for enhanced speed.

2.3 Flow Cells

Flow cells are constructed from a pair of glass microscope coverslips, separated by a parafilm spacer that is cut to form the flow channel with a scalpel. The top cover slide has two holes drilled with a laser cutter (*see* Subheading 3.2) to enable liquid exchange via an inlet and an outlet. The bottom cover slide is functionalized to bind digoxigenin-labeled DNA segments (*see* Fig. 1).

1. Coverslips: Glass cover slips 24 × 60 mm, No. 1, 130 μm thickness (Carl Roth).
2. Custom-made teflon holder for the preparation of clean glass cover slides.
3. (3-glycidoxypropyl)trimethoxysilane (abcr GmbH).
4. Parafilm M, 130 μm thickness (Carl Roth).
5. Anti-digoxigenin (Roche).
6. Bovine serum albumin (BSA) or BlockAid (BA) blocking solution (*see* Subheading 2.4, **item 3**).

2.4 Buffer Solutions

1. Phosphate-buffered saline (PBS) buffer to bind the DNA constructs to streptavidin-coated beads.
2. For nucleic acid measurements, we frequently employ Tris-EDTA (TE) buffer (1 mM EDTA and 10 mM Tris at pH 7.4) supplemented with varying concentrations of salt, e.g., NaCl or $MgCl_2$.
3. Solution of bovine serum albumin (BSA, ~25 mg/mL) and Tween (~0.1% v/v) or commercial available BlockAid blocking solution (ThermoFisher) to passivate the flow cell.

2.5 Superparamagnetic Beads

1. To couple DNA to magnetic beads we use streptavidin-coated, commercially available Dynabeads already functionalized with multiple streptavidin binding sides.
2. We generally use two types of beads (*see* **Note 1**): streptavidin-coated 1.0 μm diameter MyOne or 2.8 μm diameter M270 beads (Life Technologies).

2.6 DNA Constructs for Tweezers Measurements

1. pBluescript II SK plasmid (Stratagene).
2. Miniprep Kit (Qiagen).
3. Restriction enzymes (Xho1 and Pci1, NEB).
4. 10x Tango buffer solution (Thermo Fisher).
5. Monarch, Gel extraction kit (NEB).
6. λ-Phage DNA (NEB).
7. Primer sequence (for λ-phage DNA):

   Xho1-forward (5′AGTGGCTACGGCTCAGTTTG′3);
   Xho1-reverse (5′AACATTCGCTTATGCGGATTATTGC′3);
   Pci1-forward (5′CCGGCAATACTCGTAAACCATATCAA′3);
   Pci1-reverse (5′CCGCAGAGTGGATGTTTGAC3′).
8. dNTP mix (10 mM, ThermoScientific).
9. DMSO.
10. Taq DNA polymerase (NEB).
11. Biotin-16-dUTP (Sigma-Aldrich).
12. Digoxigenin-11-dUTP (Sigma-Aldrich).
13. 0.8 to 1%-high resolution agarose gels.
14. Ethidium bromide (EtBr; Sigma).
15. PCR purification kit (Qiagen).
16. Gel extraction Kit (Qiagen).
17. T4 DNA ligase (ThermoScientific).
18. Proteinase K (20 mg/mL, Biolabs).

3 Methods

3.1 Preparation of DNA Constructs

The aim is to construct DNA molecules with a specific length and sequence (the "middle part") and with multiple labels at each end ("handles") for MT measurements. Typically, one end is labeled with multiple biotin linkages to bind to streptavidin-coated magnetic beads and the other end is labeled with multiple digoxigenin moieties to bind specifically to anti-digoxigenin at the surface of the flow cell. We construct the three parts of DNA independently and then ligate them together to assemble: handle with biotin–middle part–handle with digoxigenin.

3.1.1 Middle Part

1. The plasmid containing the DNA sequence (pBluescript II SK) for the middle part is amplified in *E. coli* using standard protocols (*see* for example Ref. [50]). The plasmid DNA is purified using a Miniprep kit.
2. To generate linear, double-stranded DNA molecules we start with the plasmid (500 ng/μL) from which we generate a 7927 base pair long DNA strand by cutting with two restriction enzymes (XhoI and PciI) in Tango buffer solution simultaneously.
3. Run the restriction products on a 0.8% agarose gel in the absence of any DNA stain. To load large sample volumes (>50 μL), we create a large well by joining three protrusions of the gel comb using scotch tape. After electrophoresis, the gel is cut using a clean scalpel along the electrophoresis direction, and through the broad lane (at 0.5 cm from one edge) containing the sample. This procedure separates the gel along the broad sample lane into a smaller and a larger portion.
4. The smaller gel portion is stained by incubation in an aqueous EtBr solution. On a UV trans-illuminator, the product bands are visualized, and compared to a 1 kb DNA ladder run in an adjacent lane. The band containing the product at the expected length is excised and removed from the gel.
5. The smaller gel portion with excised product band is aligned to the larger, unstained gel portion. The position of the excised gel band in the small gel portion indicates the position of the desired restriction fragment in the larger (unstained) gel fraction. Excise the corresponding gel slice in the unstained larger gel portion with a clean scalpel. Use the gel extraction kit to extract DNA from the gel slice.
6. The overall yield of the restriction digest and subsequent purification steps, is ~40% (DNA concentration ~200 ng/μL).

3.1.2 Handles

We generate handles that have multiple bindings sides, either biotin or digoxigenin, by PCR from a λ-phage DNA template. We choose the biotin handle to bind to the PciI restriction side and the digoxigenin handle to bind to the XhoI restriction side.

1. We use λ-phage DNA to (PCR-) generate DNA strands of approximately 700 base pairs with multiple biotin or digoxigenin-labels incorporated to bind to the magnetic beads (via the biotin:streptavidin linkages) and to the surface of the flow cell (via digoxigenin–anti-digoxigenin interactions).
2. For PCR we use 5 μL of reverse and forward primers (10 μM), 2 μL of dNTPs (10 mM), 5 μL of DMSO, 5 μL of λ-phage DNA (stock solution), 1 μL of Taq DNA polymerase, and 7.5 μL of labeled nucleotides at 1 mM concentration in a total volume of 100 μL.

3. We combine the PciI restriction side with the digoxigenin labels to construct DNA with 712 base pairs and XhoI with biotin labels to construct DNA with 778 base pairs.
4. The restriction site can be a part of the λ-phage DNA sequence. If so, make sure that the primers produce a DNA strand with only one restriction site at the correct position.
5. If the desired restriction side is not part of the λ-phage DNA sequence, the sequence of the restriction enzyme can be included into the primer sequence (in this protocol the restriction sides are parts of the plasmid sequence).
6. To control the PCR products we run agarose gels using ~3 μL of the reaction volume.
7. The remaining volume of the PCR products is treated with a PCR cleanup kit and result in concentrations around ~40–70 ng/μL DNA.
8. Both of the labeled DNA fragments are cut using the same restriction enzymes as in Subheading 3.1.1, **step 2**. to generate single stranded DNA overhangs complementary to the 7927-base-pair-long middle part.
9. We use 30 μL of handle-DNA and 3 μL of PciI (or XhoI) in a total volume of 40 μL.
10. Incubate the restriction reaction overnight at 37 °C.
11. Afterward perform a "heat shock" at 65 °C for 10 min to degrade the restriction enzymes.
12. Use the PCR cleanup kit for each DNA construct (biotin labeled and digoxigenin labeled).
13. Final DNA concentrations are around 40 ng/μL.

3.1.3 Ligation

1. We ligate the middle construct with both handles in a ratio of 1:4:4 and try to maximize the concentration of DNA.
2. We use 8 μL of ligase in 139 μL overnight at 22 °C.
3. Heat-shock at 75 °C for 10 min to inactivate the ligase.
4. Use the whole amount of the ligated DNA construct and add Proteinase K (3 μL in a total volume of 157 μL) for 3 h at 37 °C.
5. Heat shock at 75 °C for 10 min to inactivate the Proteinase K.
6. The DNA construct is ready to use.
7. Aliquot the DNA preparation and store aliquots at −20 °C. After thawing, store the aliquot at +4 °C to avoid repeated freeze–thaw cycles. DNA remains useable for several months even at +4 °C.

3.2 Flow Cell Assembly

3.2.1 Top Glass Coverslips

1. We use a laser-cutter (Speedy 100, Trotec) to create holes of ~1 mm diameter on both sides of a glass coverslip at a distance of ~5 mm from the short edge for the top glass coverslips to serve as fluid inlets and outlets that match the dimensions of the flow cell holder.
2. To create the holes we use a power of 4.5 W, a laser with wavelength 1060 nm and a frequency of 1000 Hz.
3. The instrument provides software to create a layout from the glass coverslips to direct the laser to the desired positions.
4. It is best to probe several glass coverslips and optimize the layout file, as glass coverslips are cheap and the creation of two holes takes less than 1 min.
5. Top glass coverslips are sonicated in a 1:1 mixture of H_2Odd and 2-Propanol for 10 min before use.
6. Do not touch the flat surface of the coverslip. Coverslips can be touched at the edges.

3.2.2 Bottom Glass Coverslips

1. Install glass cover slides in teflon holder that is placed in a glass container, and sonicate in a 1:1 mixture of H_2Odd and 2-propanol for 10 min.
2. Next, the cover slides are treated with a 1:1 mixture of H_2SO_4 (concentrated) and H_2O_2 (30%) for 20 min.
3. Extensively rinse with H_2Odd.
4. Rinse cover slides with ethanol *p.a.*
5. Place cover slides in silane-solution (88% ethanol *p.a.*, 10% H_2O, 2% Silane ((3-glycidoxypropyl)-trimethoxy-silane) and incubate for 1 h.
6. Rinse with ethanol *p.a.*
7. Rinse with H_2Odd.
8. Dry in N_2 stream.
9. Heat for 45 min at 80 °C.
10. Store bottom glass cover slides under argon (ready to use, do not touch the binding side!).

3.2.3 Sealing the Flow Cell

Top and bottom glass cover slides are assembled with a single layer parafilm spacer (that is cut manually with a scalpel, like in our case, or using the laser cutter) on a heating plate for 1 min at 80 °C (*see* also Fig. 1 for the approximate pattern of the parafilm layer, light grey on the right).

3.2.4 Installing the Flow Cell to the MT Setup

Next, the assembled flow cell is placed into a flow cell holder and the flow cell assembly is mounted onto the oil objective of the instrument (*see* Fig. 1).

3.2.5 Flow Cell Surface Functionalization

Anti-digoxigenin at 100 μg/mL in PBS buffer is incubated for at least 1 h, followed by flushing the flow cell with PBS buffer (~600 μL, corresponding to >10 cell volumes).

3.2.6 Flow Cell Surface Passivation

To passivate the flow cell surface the chamber is filled with BlockAid or BSA-Tween solution (*see* Subheading 2.4). Before introducing the DNA–bead solution (*see* Subheading 3.3) the flow cell is rinsed with ~500 μL PBS buffer.

3.3 Preparation of DNA–Bead Solution and DNA Tethering

1. Measurements, shown within this work, were performed using the 7.9 kbp DNA construct described previously [33] and prepared as outlined in Subheading 3.1.
2. The DNA construct is first coupled to the streptavidin-coated beads (*see* **Note 2**) by incubating ~1 ng of the DNA construct with 2 μL of MyOne (or, alternatively, 5 μL M270) beads in a final volume of 20 μL PBS buffer for ~12 min (~5 min for M270 beads).
3. The DNA–bead solution is subsequently diluted into 100 μL PBS. The dilution needs to be adjusted for different DNA molecules or beads, such that the final surface coverage in the MT is optimized, i.e., to have enough useful beads to track, but not too many beads such that the diffraction pattern of the beads overlap.
4. Finally, 60 μL of the diluted DNA–bead solution, which corresponds to approximately one flow cell volume, are flushed into the flow cell at 170 μL/min.
5. After ~7 min (~40 s for M270 beads) unbound beads are removed by flushing with PBS buffer at 400 μL/min for at least 2 min (1 min for M270 beads).
6. Finally, the magnet holder with magnets in conventional MT geometry is mounted on the setup to apply stretching forces (~2 pN).
7. The DNA and bead concentrations and the bead-to-DNA ratio should be adjusted such that DNA-tethered beads are almost exclusively attached via single double-stranded DNA molecules. Lower the DNA concentration if many beads are attached via multiple DNA molecules. *See* the next sections on how to discriminate single from multiple tethers.
8. Typically, ~50–80% of the single DNA tethers in a flow cell can be supercoiled. The exact number depends on the quality of the DNA preparation and the length of the construct.

3.4 Tether Selection with Conventional MT

For initial selection of good DNA tethers, we use conventional MT magnets in a vertical configuration (Fig. 1), which exert a horizontally aligned magnetic field, as explained in Subheading 2.1, **item 6**.

To select for magnetic beads tethered to a single, torsionally constrained double-stranded DNA molecule, we perform three tests:

1. As a first test, we select DNA tethers that have (close to) the expected contour length. We check the tether length by varying the magnet distance to the flow cell from 0.5 mm to 12 mm, which corresponds to 4.4 pN and <0.1 pN, for MyOne beads, respectively and 55 pN to <0.1 pN for M270 beads, respectively. At a magnet height of 0.5 mm the DNA tethers are stretched to >95% of their contour length whereas at 12 mm (close to zero force) the beads drop the bottom surface, corresponding to zero tether extension. The expected tether length is 0.34 nm per base pair times ~7900 base pairs, equal to ~2700 nm. Beads with a tether length close to the expected contour length are further considered.
2. To select beads that are tethered by a single double-stranded DNA molecule, we take advantage of the asymmetric response of DNA upon underwinding and overwinding at stretching forces >1 pN (*see* also Fig. 3). We exert a pulling force of ~1.5 pN and turn the magnets 20 times against the helical orientation of DNA (clockwise). If the tether length decreases, we assume multiple molecules to be bound to that bead and discard them from further analysis.
3. Finally, we select DNA tethers that are fully torsionally constrained, which requires multiple attachment points at both ends and no nicks along the DNA molecule. We move the magnets to exert a stretching force of ~0.4 pN and rotate the magnets by ±20 turns. If there is a single, coilable DNA molecule attached to the bead the molecule buckles resulting in a decrease of the tether length (*see* also Fig. 3). Note, that the number of turns required will vary for longer (or shorter) DNA constructs. Only beads that pass all three tests are used for further analysis.

3.5 FOMT Alignment and Measurements

Having selected beads that are tethered by single, torsionally constrained DNA constructs, we replace the conventional magnet configuration by a stack of cylindrical magnets with a central aperture (R-06-02-02-G, N-45, Supermagnete) for FOMT measurements (Figs. 1 and 2). This stack consists of three ring magnets with an outer diameter of 6 mm, an inner diameter of 2 mm and a height of 2 mm per magnet (*see* **Note 3**). Magnets are not brought closer to the flow cell than 3 mm (*see* **Note 4**). For the force calibration and the analysis of the (x, y)-fluctuations in FOMT, measurements were performed at 3, 3.5, 4, 4.5, and 5 mm magnet height for 600 s for MyOne beads and 1200 s for M270 beads (*see* **Note 5**). FOMT measurements require precise alignment of the magnets relative to the flow cell, to achieve a situation where the

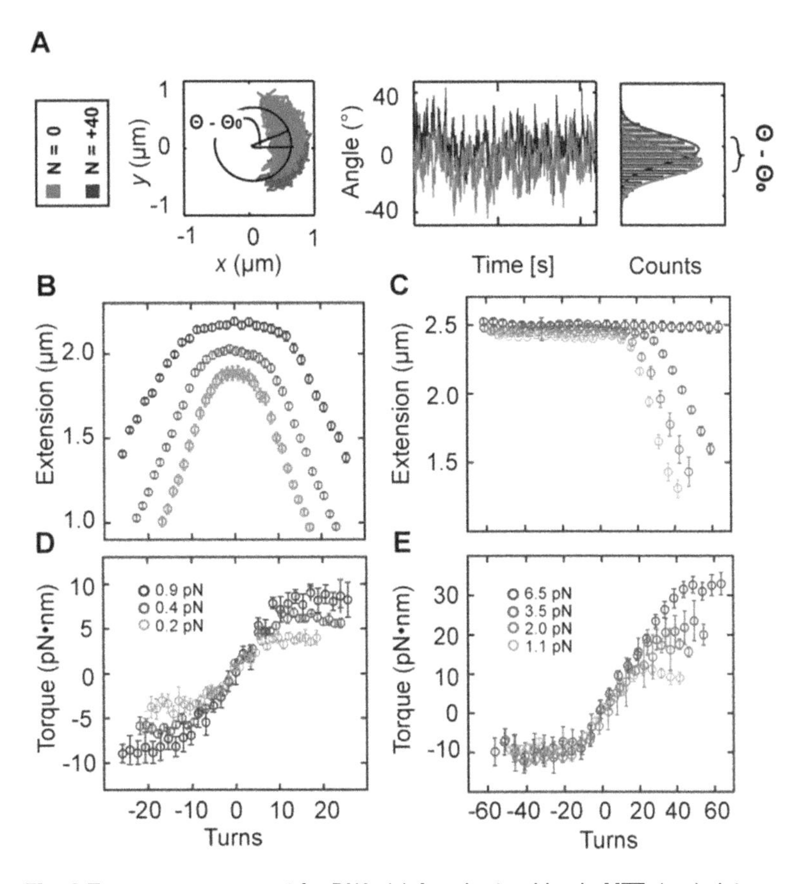

Fig. 3 Torque measurement for DNA. (**a**) Angular tracking in MTT. (*x*, *y*)-data are transformed to polar coordinates. The bead's position has an equilibrium angle Θ_0 for a torsionally relaxed tether (pink). The fluctuations around Θ_0 are a measure of the rotational trap stiffness k_{ROT}. The shift in the equilibrium angle Θ_0 after *N* turns to Θ (grey) can be converted to molecular torque. In MTT two signals are recorded at the same time: the tethers' extensions (**b**, **c**) and their molecular torque (**d**, **e**) upon the application of twist. The response of DNA to induced twist depends on the stretching force. (**b**, **d**) For forces smaller 1 pN, the response of DNA to twist is symmetric around zero turns. Upon overwinding and underwinding of DNA, its extension initially stays nearly constant, while its molecular torque increases linearly. At a critical supercoil density the molecule buckles and forms plectonemes, resulting in a decrease in the molecules extension and a saturation of the molecular torque. (**c**, **e**) For forces larger 1 pN the response of DNA is asymmetric: no buckling occurs for a negative number of turns (against the helical nature of DNA) and the corresponding torque reaches a plateau at −10 pN·nm [22, 55, 56]. Here, additional torsional stress is released by torque-induced melting of base pairs. At the highest stretching force (here 6.5 pN) no buckling occurs at all, since a transition of B-DNA to P-DNA occurs at ~35 pN·nm. Data are for 7.9 kbp DNA measured with multiplexed magnetic torque tweezers with MyOne beads for forces <1 pN and M270 beads for forces >1 pN in TE-buffer (10 mM Tris and 1 mM EDTA) with 100 mM NaCl [46]. Figure adapted from Ref. [46].

beads' rotation about the vertical tether axis is unconstrained by the magnets. If precise alignment is achieved, beads will trace out a full doughnut shaped pattern in the (x, y)-plane, corresponding to a situation where the residual effect of the magnets on the rotational energy landscape is less than $\sim k_B T$, the thermal energy. The alignment is carried out as follows:

3.5.1 Coarse Alignment of the FOMT

1. From the geometry of the magnets used in FOMT, with an inner diameter of 2 mm, we expect the perfect aligned situation (only vertical field components, no radial field components, corresponding to a situation where the beads rotation about the vertical tether axis is unconstrained by the magnets) if the objective's main axis is in line with the magnets main axis.
2. To evaluate and improve alignment, we make use of the beads characteristic transverse fluctuations in the (x, y)-plane that trace out a doughnut-like shape if perfectly aligned (Fig. 2e).
3. In our instrument, we change the relative position of magnet and objective by moving the arm, which carries the magnet holder and the light emitting diode, in x and y direction using micrometer screws ($\sim$100 μm accuracy).
4. As a starting point we place the magnets such that we record maximal light intensity (on average) over the field of view ("light-based alignment").
5. At a fixed magnet height (typically 3 mm) we then record the fluctuations of beads and observe the fluctuation pattern in the (x, y)-plane for several times of the characteristic time of the system (*see* **Note 5**).
6. Focusing on the fluctuation pattern, one can maximize, i.e., optimize the number of aligned beads within a field of view by changing the relative position of magnet and objective using the micrometer screws ("coarse alignment").
7. For MyOne beads we find that this coarse grained method is often sufficient to have multiple aligned beads at the same time within the field of view (Fig. 4a).
8. In contrast, for M270 beads this coarse alignment is typically insufficient: We find no aligned bead, i.e., a bead that traces out a full doughnut in the (x, y)-fluctuations, in the field of view. Instead beads trace out only a segment of the doughnut. We achieve alignment of individual M270 beads by varying the x and y position of the magnet, independently, relative to the objective by steps of $\sim$100 μm.

3.5.2 Fine Alignment of the FOMT

1. To achieve a "perfect" aligned bead that fulfills the criteria to rotate about the tether axis unconstrained by the external magnets, it is best to focus on a single bead instead of many in parallel.

2. When plotting the recorded (x, y)-trace of a particular bead as a heat map, it is apparent that even a complete "doughnut" trace not necessarily suffices to have equal probability for all positions on the doughnut, indicating that there are still residual effects on the rotational energy landscape from the magnets (Fig. 4c shows a bead that is almost perfectly aligned).
3. To optimize the relative position of the bead and the magnet, one can translocate the flow cell itself using the micrometer screws of the base plates underneath the flow cell holder (or using a nano-positioning stage) with <10 μm accuracy ("fine tuning").
4. This finetuning is particularly relevant for larger beads, here M270 beads.
5. Note that sometimes coarse alignment is sufficient to align a single M270 bead.

3.6 DNA Torque Measurements

FOMT do not allow rotating the magnetic beads and, therefore, the DNA tethers cannot be supercoiled. MTT can be thought of as an extension of FOMT. To switch from FOMT to MTT a side magnet is added to the magnet holder that carries the stack of cylindrical magnets, as indicated in Fig. 1. For the side magnet, we use a cylindrical magnet with a diameter of 3 mm and a height of 6 mm (S-03-06-N, N-48, Supermagnete).

3.6.1 Optimization of the Instrument for Multiplexed Torque Measurements

1. The effect of the side magnet on the beads fluctuations in the (x, y)-plane is large: Displayed in Fig. 4b are the same beads as in Fig. 4a with the side magnet added to the FOMT configuration. All beads trace out a segment of the doughnut in the (x, y)-plane and are oriented toward the added side magnet.
2. MTT exert stretching forces on the molecular bead and allow tracking of the rotation angle of the bead, similar to FOMT, while additionally enable control of the twist of the DNA tether (Fig. 3a).
3. The magnitude of bead fluctuations is used to determine the stiffness of the rotational trap k_{ROT} (*see* Subheading 3.7, **step 10**).
4. k_{ROT} varies by 20% from bead-to-bead (*see* Supplementary Fig. S10 of Ref. [46]). This is understood from the anisotropy of the magnetic content within the particles [40].
5. To determine the trap stiffness of each individual bead in MTT, the (x, y)-fluctuations were measured for 600 s at each magnet height (3, 3.5, 4, 4.5 and 5 mm).
6. One can tune the effect of the side magnet on the beads by moving the magnets toward or away from the direction of the added side magnet [46]. This is of interest when performing multiplexed torque measurements, where the actual strength of

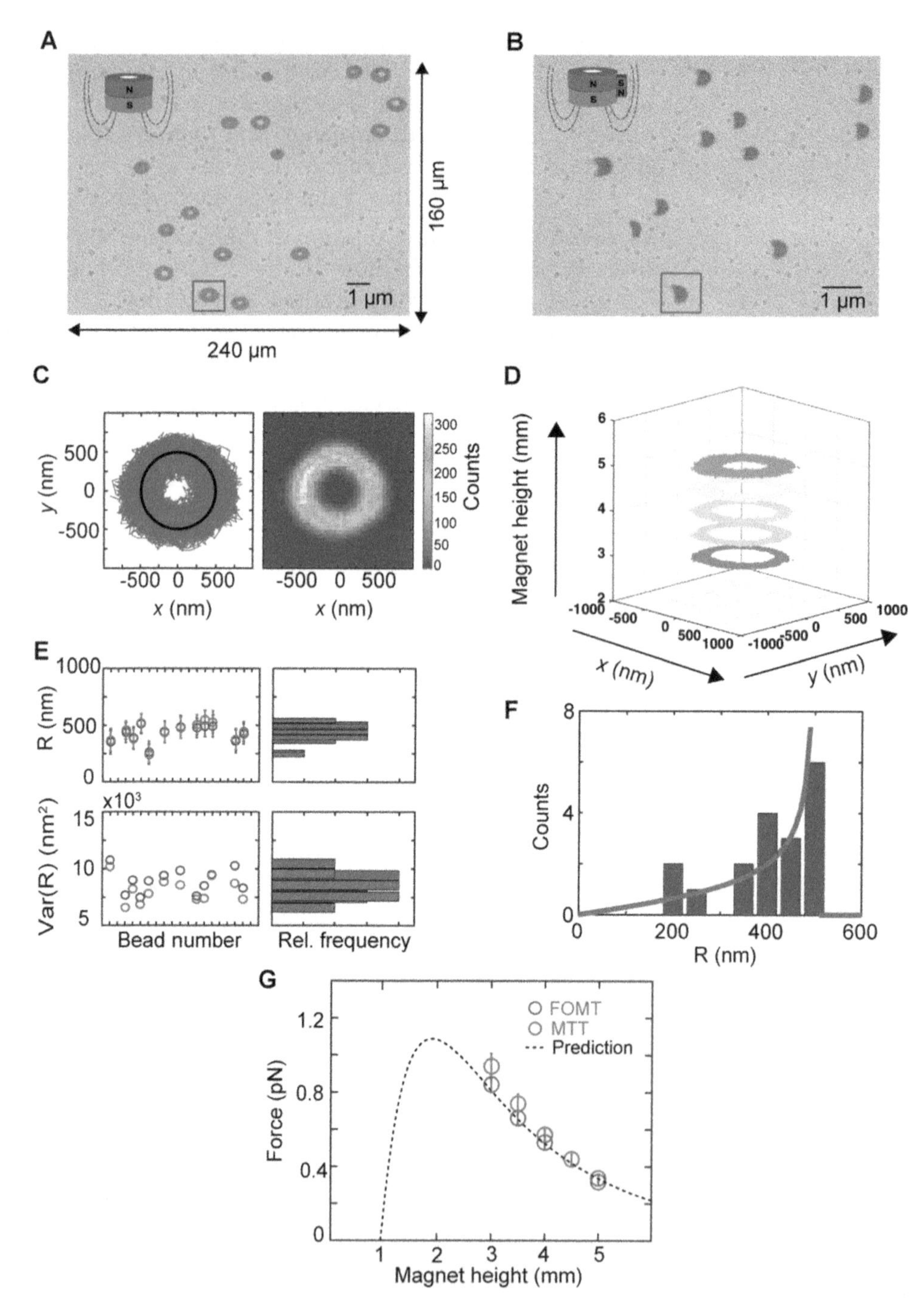

Fig. 4 Multiplexed FOMT and MTT and force calibration. (**a**) Multiplexed field of view in the FOMT geometry using 60× magnification and 7.9 kbp DNA attached to MyOne beads in PBS at 3 mm magnet height. Shown in blue are fluctuations of different beads in the (*x*, *y*)-plane. Fluctuations are all from the same 600 s recording and shown magnified compared to the camera image (1 μm scale bar gives the scale for the fluctuations; the size of the field of view is indicated on the sides). (**b**) Same field of view and same beads as in (a) (the number of beads was slightly reduced as some beads became unspecifically stuck to the surface). Upon adding a side magnet in the MTT configuration to the cylindrical magnet, the beads' (*x*, *y*)-fluctuations trace out arc-like

the trap stiffness plays a crucial role in determining molecular torque. There is an upper and a lower limit for k_{ROT}: on the one hand the trap needs to be strong enough such that the bead follows the rotation of the magnet (typically $k_{ROT} > 35$ pN·nm/rad), on the other hand the trap needs to be soft enough such that small changes in the shift in the equilibrium angle can be resolved (typically $k_{ROT} < 1000$ pN·nm/rad).

7. For multiplexed MTT measurements, the position of the magnet relative to the field of view can be varied (similar to "coarse alignment") such that the rotational trap stiffness that the beads experience are within these limits. Note that there is no need to perform a FOMT alignment prior to MTT measurements.

3.6.2 Protocol for Multiplexed Torque Measurements

1. During MTT measurements magnets are placed at a specific magnet height, i.e., stretching force, while the magnets are rotated.
2. If several forces should be measured, it is recommended to start with the highest stretching force to reduce unspecific interactions of the beads with the flow cell surface at lower stretching forces.
3. Typically starting at zero turns, the magnets are rotated against the helical nature of DNA, then back to zero, to a positive number of turns and back to zero.
4. Depending on tether length and the stretching force, the magnets are rotated several times in positive and negative direction of rotation (*see* Fig. 3 as an example for 7.9 kbp DNA attached with MyOne beads Fig. 3b, d, and attached to M270 beads, Fig. 3c, e).

Fig. 4 (continued) patterns shown in red. (**c**) (*x, y*)-fluctuations for one selected bead from the field of view in panel (a), tracked for 600 s. A histogram of the data, illustrated as a heat map (right panel), shows that the fluctuations of the bead are mostly uniformly distributed around the tether axis. (**d**) (*x, y*)-fluctuations of an aligned M270 bead attached to 7.9 kbp DNA in PBS at 3, 3.5, 4, 4.5, and 5 mm magnet height (blue to orange). The variance of the fluctuations of the bead around the doughnut radius *R* become larger with greater magnet distance, corresponding to lower stretching forces, while the mean of *R* stays constant (*see* also **Note 7**). (**e**) Mean doughnut radii *R* for the tethers shown in panel (a) and (b) in FOMT (blue circles) and MTT (red circles) geometry with the standard deviation as error bars. The close agreement between red and blue data points indicates that fitting the doughnut radius to time traces of thermal fluctuation in the FOMT or to traces during active magnet rotation in the MTT (*see* Subheading 3.7) gives the same result, within error. Corresponding variances of the radius in FOMT vs. MTT (same color code as before). (**f**) Histogram of doughnut radii *R* for the beads shown in panel A. The data are reasonably well described by a simple model based on spherical geometry (*see* **Note 8**; magenta line). (**e**) Forces computed from radial fluctuations in FOMT (blue circles) and MTT (red circles) for MyOne beads. Co-plotted is the force prediction in FOMT (dashed line) [27, 51]

5. Rotations are performed at 0.1 turn per second (typically in steps of two or five) to ensure that the bead follows the magnet rotation.
6. Data are recorded at a specific number of turns for several multiples of the characteristic time of the system at each measurement step, typically >50–100 s (*see* **Note 5**).
7. Data are typically recorded at 60 Hz.

3.7 Analysis of DNA Torque and Twist Measurements

1. The MT software saves the time, the (x, y, z)-position data, and the positions of the piezo and of both motors (translational and rotational) as text files.
2. The position of the translational motor and thus the magnet height above the flow cell can be used to determine the stretching force from a calibration measurement. For MyOne and M270 beads the bead-to-bead variation in stretching force is ~10% for a given setting of the magnets [51, 52]. Therefore, it is often sufficient to record a force-magnet height calibration (Fig. 4g) for a given setup configuration (*see* Subheading 3.8) and to then use the calibrated forces. We note, however, that other types of magnetic beads can be less homogeneous, requiring force calibration for every bead.
3. The position of the rotational motor gives the number of applied rotations and can be read out to identify the segments of the torque measurements where the magnets are rotating, *see* **step 5** below.
4. To calibrate the extension offset, i.e., to determine z-position where the beads touch the surface of the flow cell, it is recommended to record traces of each individual bead under condition where it contacts the surface (e.g., by tracking close to zero force) and to determine the corresponding z-value for subtraction in the following analysis of z.
5. In MTT, the segments of the measurements while rotating the magnets (*see* Subheading 3.6.2, **step 5** above) are used to (i) count and control the number of turns the bead rotated; (ii) fit a circle to the recorded (x, y)-positions in order to determine a radius and center position.
6. In FOMT, aligned beads trace out full circles, such that the circle fit can be performed on the whole recorded data set.
7. Using the fitted center position and radius of the doughnut in FOMT and MTT, one can transform the (x, y)-position of the bead to polar (r, Θ)-coordinates (Fig. 3a). *See* Eqs. 9–11 of Ref. [27] for details. We note that fitting a circle to the segments of traces where the bead is actively rotated in the MTT (*see* Subheading 3.6.2, **step 5** above) or to data where the bead traces out

a full "doughnut" by thermal fluctuation alone in the FOMT gives identical results, within experimental error (Fig. 4e).

8. In MTT, the segments of the measurements at a fixed number of rotations are used to determine changes in the equilibrium angle $\langle \Theta_N \rangle$ (Fig. 3a) and to monitor the DNA tether extension via the z-position of the bead.
9. The mean of the tracked angle positions is defined as equilibrium angle $\langle \Theta_N \rangle$.

   The beads fluctuations around the equilibrium angle $\langle \Theta \rangle$ are used to determine k_{ROT}:

$$k_{ROT} = k_B T / \mathrm{Var}(\Theta)$$

   where k_B is the Boltzmann constant and T the absolute temperature.
10. The shift in equilibrium angle after N turns is a direct measure for molecular torque (Fig. 3a, d, e):

$$\tau = -k_{ROT} \cdot (\langle \Theta_N \rangle - \langle \Theta_0 \rangle)$$

11. z is a direct measure of the extension of the molecule (Fig. 3b, c), after correcting for an z-offset (*see* Subheading 3.7, **step 4** above), i.e., after ensuring that the bead touching the surface is taking as the zero position.

3.8 Force Calibrations

While force calibration in conventional MT typically relies on analysis of the transverse (x, y)-fluctuations, in both FOMT and MTT the force calibration is based on analysis of the radial fluctuations after coordinate transformation from (x, y) to (r,Θ)-coordinates (*see* Subheading 3.7). Here, we compare force calibration in the FOMT with force calibration in the MTT.

1. In FOMT, we fit a circle to the radial thermal fluctuations of the bead, while in MTT we fit a circle to the recorded trace of the bead when turning the magnet to determine a center point and such transfer the data to polar coordinates (here: six rotations at 0.1 turn per second).
2. We performed force calibrations (F) similar to Ref. [27] for different magnet heights z_{mag}, where l is the measured tether length and Var(R) the variance of radial fluctuations:

$$F = k_B T \cdot l / \mathrm{Var}(R)$$

3. We use the recorded transverse fluctuations in x and y to determine the radius R and the recorded z trace to calculate the tether length l.
4. Both methods are in good agreement (Fig. 4g) allowing us to directly perform force calibrations in MTT or to use existing

FOMT force calibrations for torque measurements (*see* **Notes 1** and **6**). The force calibrations based on the radial fluctuations are, in addition, in good agreement with prediction from magnetic field and bead magnetization calculations [27, 51] (Fig. 4g, dashed line).

4 Notes

1. The stretching force is dependent on the beads' magnetization (in turn determined by the size and type of the beads) and on the magnetic field and its gradient [51]. For both MyOne and M270 beads, the bead-to-bead variability in the force is ~10%. The inhomogeneity in force across the field of view is <10% in the MT configurations described here [46]. The inhomogeneity in the stiffness of the rotational trap in MTT measurements is about 20% [40, 46].
2. In particular if the biological system is sensitive to buffer conditions, it is recommended to perform a buffer exchange for the magnetic beads prior to incubation with the DNA, by using a magnet pipette aid to pull the magnetic beads to one side of the Eppendorf tube so that one can easily remove the solution with a pipette and add the desired measurement buffer. Here, we typically use PBS for the buffer exchange.
3. In conventional MT, FOMT and MTT, the set of magnets are assembled in a magnet holder mount made from aluminum that is itself not magnetic. The magnet holder is fixed to the instrument via a screw. After bringing the translational motor to its maximal height, which is usually 2 cm above the flow cell surface, it is easy to manually replace the magnet holder with another magnet configuration. For fast exchange, it is convenient to place each magnet configuration in its own magnet holder.
4. In the FOMT and MTT configurations field gradients change signs, such that at a certain distance (<1 mm for the magnets described here that feature a 2 mm diameter aperture) of the magnet to the flow cell the magnet exerts pushing instead of pulling forces (Fig. 4g) [27, 51].
5. For high-resolution measurements (e.g., of molecular torques in the MTT or of angular steps in the FOMT) the overall measurement time T_{meas} should be—ideally—significantly larger than the characteristic time of the system τ_c. The characteristic time of the system is generally given by the friction coefficient γ divided by the trap stiffness k. For the rotational degree of freedom, the friction coefficient is $\gamma_{\text{ROT}} \approx 140\pi\eta R^3$ (the exact formula is given as Eq. 18 in Ref. [27]) and

$k_{ROT} = k_B \cdot T/\mathrm{Var}(\Theta)$, where η is the viscosity of the solution ($\eta \approx 0.001$ Pa·s for aqueous buffers). In the FOMT, $k_{ROT} = k_B \cdot TC/L_C$, where C is the torsional persistence length ($\approx$50–100 nm, depending on the stretching force) and L_C the contour length of the employed DNA molecule [27]. In particular for long DNA molecules and large beads, characteristic times in the FOMT can be very long (e.g., $\tau_c \sim 1000$ s for M270 beads and the 7.9 kbp DNA), which severely limits the ability to detect steps in the angular coordinate, since steps can only be reliably detected if they occur more slowly than τ_c. For small beads ($R_{bead} \leq 0.5$ μm) and short DNA tethers ($L_C \leq 0.5$ μm) characteristic times in the FOMT are reduced to $\leq$15 s. The MTT characteristic times are generally much shorter than in the FOMT, due to the (much) higher rotational trap stiffness.

6. While the fitted radius in FOMT and MTT is similar, the beads fluctuations, i.e., the variance of the radius is slightly smaller in MTT, which leads to an overestimation of the forces in MTT compared to FOMT. This is understood as the fluctuations while rotation the magnet (as used in MTT) are smaller than fluctuations in FOMT.

7. Figure 4d: the position of the magnet was adjusted when changing the magnet height in order to measure full circle fluctuations (“*fine tuning*”). In general, the magnet alignment in FOMT is so sensitive that significant changes in the magnet height require realignment of the (x, y)-position of the magnets.

8. The paramagnetic beads used in MT experiments have a preferred magnetization axis [40, 53], $\vec{m}_0$, that aligns with the external field (Fig. 2). A DNA attachment at the bottom of the bead in conventional MT (where the magnetic field is horizontal) corresponds to an attachment at the side of the bead in FOMT or MTT (where the field is (nearly) vertical); in both cases the DNA is attached at the “equator” of the bead relative to the preferred magnetization axis that defines the “poles” (this is the situation shown schematically in Fig. 2). Similarly, for an attachment of the DNA at one of the poles, the DNA would be attached to the bead at the side in conventional MT and at the bottom (or top) in FOMT or MTT. Since the beads are typically uniformly functionalized and DNA–bead coupling is carried out in free solution, we do not control the attachment point of the DNA to the bead. However, attachment at or near the equator is much more likely than attachment at or near one of the poles due to spherical geometry. The area per polar angle θ segment $dA = 4\pi R_{bead}^2 \cdot \sin(\theta) d\theta$ is much larger at the equator (where $\theta \sim \pi/2$ such that $\sin(\theta) \sim 1$) compared to the poles (with $\theta \sim 0$ or π such that $\sin(\theta) \sim 0$). From this argument

follows that, assuming random attachment anywhere on the bead's surface, the probability to observe a given doughnut radius R is proportional to $P \sim R/R_{\mathrm{bead}}[1-(R/R_{\mathrm{bead}})^2]^{-1/2}$, in good agreement with the experimental observations (Fig. 4f).

Acknowledgments

We thank Jelle van der Does for help with instrument development, Susanne Hage for help with development of the DNA construct, Philipp Walker for instrument construction and useful discussions, the Rief chair at the TU Munich for use of the laser cutter, and the German Research Foundation (DFG) via Sonderforschungbereich SFB 863 "Forces in Biomolecular Systems" for funding.

References

1. Bryant Z, Oberstrass FC, Basu A (2012) Recent developments in single-molecule DNA mechanics. Curr Opin Struct Biol 22:304–312
2. Lipfert J, van Oene MM, Lee M, Pedaci F, Dekker NH (2015) Torque spectroscopy for the study of rotary motion in biological systems. Chem Rev 115:1449–1474
3. Neuman KC, Nagy A (2008) Single-molecule force spectroscopy: optical tweezers, magnetic tweezers and atomic force microscopy. Nat Methods 5:491–505
4. Manosas M, Meglio A, Spiering MM, Ding F, Benkovic SJ, Barre F-X et al (2010) Magnetic tweezers for the study of DNA tracking motors. Methods Enzymol 475:297–320
5. Strick TR, Allemand JF, Bensimon D, Bensimon A, Croquette V (1996) The elasticity of a single supercoiled DNA molecule. Science 271:1835–1837
6. Gosse C, Croquette V (2002) Magnetic tweezers: micromanipulation and force measurement at the molecular level. Biophys J 82:3314–3329
7. te Velthuis AJW, Kerssemakers JWJ, Lipfert J, Dekker NH (2010) Quantitative guidelines for force calibration through spectral analysis of magnetic tweezers data. Biophys J 99:1292–1302
8. Lansdorp BM, Saleh OA (2012) Power spectrum and Allan variance methods for calibrating single-molecule video-tracking instruments. Rev Sci Instrum 83:025115
9. Daldrop P, Brutzer H, Huhle A, Kauert DJ, Seidel R (2015) Extending the range for force calibration in magnetic tweezers. Biophys J 108:2550–2561
10. Strick TR, Allemand JF, Bensimon D, Croquette V (2000) Stress-induced structural transitions in DNA and proteins. Annu Rev Biophys Biomol Struct 29:523–543
11. Abels JA, Moreno-Herrero F, Van Der Heijden T, Dekker C, Dekker NH (2005) Single-molecule measurements of the persistence length of double-stranded RNA. Biophys J 88:2737–2744
12. Herrero-Galán E, Fuentes-Perez ME, Carrasco C, Valpuesta JM, Carrascosa JL, Moreno-Herrero F et al (2013) Mechanical identities of RNA and DNA double helices unveiled at the single-molecule level. J Am Chem Soc 135:122–131
13. Lipfert J, Klijnhout S, Dekker NH (2010) Torsional sensing of small-molecule binding using magnetic tweezers. Nucleic Acids Res 38:7122–7132
14. Salerno D, Brogioli D, Cassina V, Turchi D, Beretta GL, Seruggia D et al (2010) Magnetic tweezers measurements of the nanomechanical properties of DNA in the presence of drugs. Nucleic Acids Res 38:7089–7099
15. Wang Y, Schellenberg H, Walhorn V, Toensing K, Anselmetti D (2017) Binding mechanism of fluorescent dyes to DNA characterized by magnetic tweezers. Mater Today Proc 4:S218–S225
16. Vilfan ID, Lipfert J, Koster DA, Lemay SG, Dekker NH (2009) Magnetic tweezers for single molecule measurements. Springer, New York, NY

17. Carrasco C, Dillingham MS, Moreno-Herrero F (2014) Single molecule approaches to monitor the recognition and resection of double-stranded DNA breaks during homologous recombination. DNA Repair (Amst) 20:119–129
18. Dulin D, Berghuis BA, Depken M, Dekker NH (2015) Untangling reaction pathways through modern approaches to high-throughput single-molecule force-spectroscopy experiments. Curr Opin Struct Biol 34:116–122
19. Berghuis BA, Köber M, van Laar T, Dekker NH (2016) High-throughput, high-force probing of DNA-protein interactions with magnetic tweezers. Methods 105:90–98
20. Ordu O, Lusser A, Dekker NH (2016) Recent insights from in vitro single-molecule studies into nucleosome structure and dynamics. Biophys Rev 8:33–49
21. Hodeib S, Raj S, Manosas M, Zhang W, Bagchi D, Ducos B et al (2016) Single molecule studies of helicases with magnetic tweezers. Methods 105:3–15
22. Bryant Z, Stone MD, Gore J, Smith SB, Cozzarelli NR, Bustamante C (2003) Structural transitions and elasticity from torque measurements on DNA. Nature 424:338–341
23. Gore J, Bryant Z, Stone MD, Nöllmann M, Cozzarelli NR, Bustamante C (2006) Mechanochemical analysis of DNA gyrase using rotor bead tracking. Nature 439:100–104
24. Lebel P, Basu A, Oberstrass FC, Tretter EM, Bryant Z (2014) Gold rotor bead tracking for high-speed measurements of DNA twist, torque and extension. Nat Methods 11:456–462
25. Oberstrass FC, Fernandes LE, Bryant Z (2012) Torque measurements reveal sequence-specific cooperative transitions in supercoiled DNA. Proc Natl Acad Sci U S A 109:6106–6111
26. Vlijm R, Lee M, Lipfert J, Lusser A, Dekker C, Dekker NH (2015) Nucleosome assembly dynamics involve spontaneous fluctuations in the handedness of Tetrasomes. Cell Rep 10:216–225
27. Lipfert J, Wiggin M, Kerssemakers JWJ, Pedaci F, Dekker NH (2011) Freely orbiting magnetic tweezers to directly monitor changes in the twist of nucleic acids. Nat Commun 2:439–439
28. Lee M, Lipfert J, Sanchez H, Wyman C, Dekker NH (2013) Structural and torsional properties of the RAD51-dsDNA nucleoprotein filament. Nucleic Acids Res 41:7023–7030
29. Harada Y, Ohara O, Takatsuki A, Itoh H, Shimamoto N, Kinosita K (2001) Direct observation of DNA rotation during transcription by *Escherichia coli* RNA polymerase. Nature 409:113–115
30. Basu A, Schoeffler AJ, Berger JM, Bryant Z (2012) ATP binding controls distinct structural transitions of *Escherichia coli* DNA gyrase in complex with DNA. Nat Struct Mol Biol 19:538–546
31. Celedon A, Nodelman IM, Wildt B, Dewan R, Searson P, Wirtz D et al (2009) Magnetic tweezers measurement of single molecule torque. Nano Lett 9:1720–1725
32. Celedon A, Wirtz D, Sun S (2010) Torsional mechanics of DNA are regulated by small-molecule intercalation. J Phys Chem B 114:16929–16935
33. Lipfert J, Kerssemakers JWJ, Jager T, Dekker NH (2010) Magnetic torque tweezers: measuring torsional stiffness in DNA and RecA-DNA filaments. Nat Methods 7:977–980
34. Lipfert J, Kerssemakers JJW, Rojer M, Dekker NH (2011) A method to track rotational motion for use in single-molecule biophysics. Rev Sci Instrum 82:103707
35. Janssen XJA, Lipfert J, Jager T, Daudey R, Beekman J, Dekker NH (2012) Electromagnetic torque tweezers: a versatile approach for measurement of single-molecule twist and torque. Nano Lett 12:3634–3639
36. Mosconi F, Allemand JF, Croquette V (2011) Soft magnetic tweezers: a proof of principle. Rev Sci Instrum 82:034302
37. Kauert DJ, Kurth T, Liedl T, Seidel R (2011) Direct mechanical measurements reveal the material properties of three-dimensional DNA origami. Nano Lett 11:5558–5563
38. Lipfert J, Skinner GM, Keegstra JM, Hensgens T, Jager T, Dulin D et al (2014) Double-stranded RNA under force and torque: similarities to and striking differences from double-stranded DNA. Proc Natl Acad Sci U S A 111:15408–15413
39. Lipfert J, Lee M, Ordu O, Kerssemakers JWJ, Dekker NH (2014) Magnetic tweezers for the measurement of twist and torque. J Vis Exp 19 (87). https://doi.org/10.3791/51503
40. van Oene MM, Dickinson LE, Pedaci F, Köber M, Dulin D, Lipfert J et al (2015) Biological magnetometry: torque on superparamagnetic beads in magnetic fields. Phys Rev Lett 114:218301. https://doi.org/10.1103/PhysRevLett.114.218301
41. van Oene MM, Dickinson LE, Cross B, Pedaci F, Lipfert J, Dekker NH (2017) Applying torque to the *Escherichia coli* flagellar motor using magnetic tweezers. Sci Rep 7:43285

42. Nord AL, Gachon E, Perez-Carrasco R, Nirody JA, Barducci A, Berry RM et al (2017) Catch bond drives stator mechanosensitivity in the bacterial flagellar motor. Proc Natl Acad Sci U S A 114:12952–12957
43. Ribeck N, Saleh OA (2008) Multiplexed single-molecule measurements with magnetic tweezers. Rev Sci Instrum 79:094301
44. De Vlaminck I, Henighan T, van Loenhout MTJ, Pfeiffer I, Huijts J, Kerssemakers JWJ et al (2011) Highly parallel magnetic tweezers by targeted DNA tethering. Nano Lett 11:5489–5493
45. Cnossen JP, Dulin D, Dekker NH (2014) An optimized software framework for real-time, high-throughput tracking of spherical beads. Rev Sci Instrum 85:103712
46. Kriegel F, Ermann N, Forbes R, Dulin D, Dekker NH, Lipfert J (2017) Probing the salt dependence of the torsional stiffness of DNA by multiplexed magnetic torque tweezers. Nucleic Acids Res 45:5920–5929
47. Nomidis SK, Kriegel F, Vanderlinden W, Lipfert J, Carlon E (2017) Twist-bend coupling and the torsional response of double-stranded DNA. Phys Rev Lett 118:217801
48. Dulin D, Vilfan ID, Berghuis BA, Hage S, Bamford DH, Poranen MM et al (2015) Elongation-competent pauses govern the Fidelity of a viral RNA-dependent RNA polymerase. Cell Rep 10:983–992
49. van Loenhout MTJ, Kerssemakers JWJ, De Vlaminck I, Dekker C (2012) Non-bias-limited tracking of spherical particles, enabling Nanometer resolution at low magnification. Biophys J 102:2362–2371
50. Lipfert J, Koster DA, Vilfan ID, Hage S, Dekker NH (2009) Single-molecule magnetic tweezers studies of type IB topoisomerases. Methods Mol Biol 582:71–89
51. Lipfert J, Hao X, Dekker NH (2009) Quantitative modeling and optimization of magnetic tweezers. Biophys J 96:5040–5049
52. De Vlaminck I, Henighan T, van Loenhout MT, Burnham DR, Dekker C (2012) Magnetic forces and DNA mechanics in multiplexed magnetic tweezers. PLoS One 7:e41432
53. Klaue D, Seidel R (2009) Torsional stiffness of single Superparamagnetic microspheres in an external magnetic field. Phys Rev Lett 102:028302
54. Kriegel F, Ermann N, Lipfert J (2017) Probing the mechanical properties, conformational changes, and interactions of nucleic acids with magnetic tweezers. J Struct Biol 197:26–36
55. Mosconi F, Allemand JF, Bensimon D, Croquette V (2009) Measurement of the torque on a single stretched and twisted DNA using magnetic tweezers. Phys Rev Lett 102:078301
56. Sheinin MY, Forth S, Marko JF, Wang MD (2011) Underwound DNA under tension: structure, elasticity, and sequence-dependent Behaviors. Phys Rev Lett 107:108102

Chapter 7

AFM-Based Characterization of Electrical Properties of Materials

John Alexander, Sergey Belikov, and Sergei Magonov

Abstract

Capabilities of atomic force microscopy (AFM) for characterization of local electrical properties of materials are presented in this chapter. At the beginning the probe–sample force interactions, which are employed for detection of surface topography and materials properties, are described theoretically in their application in different AFM modes and electrical techniques. The electrical techniques, which are based on detection of electrostatic probe–sample forces, are outlined in AFM contact and oscillatory resonant modes. The basic features of the detection of surface potential and capacitance gradients are explained. The applications of these techniques are illustrated on metals, surfactant compounds, semiconductors, and different polymers. Practical recommendations on use of the AFM-based electrical methods and the related challenges are given in the last section.

Key words Atomic force microscopy (AFM), Electrical force microscopy (EFM), Kelvin force microscopy (KFM), Piezoresponse force microscopy (PFM), Surface potential, Dielectric permittivity

1 Introduction

The increasing importance of functional nanoscale structures has spread across majority of research areas: semiconductors, data storage, biomaterials, health care, and others. This trend has demanded a development of analytical methods for materials characterization at small scales. Atomic force microscopy (AFM) was introduced in practice in the 1980s [1], and at present it is a leading microscopic method for visualization of nanoscale surface structures and exploring their local properties. Another advantage of AFM compared to other microscopic techniques is the applicability of this method in different environments (air, vacuum, liquid, vapors, etc.). The high-resolution profiling of surface features with subnanometer accuracy, which was verified in studies of numerous samples, has been expanded to quantitative measurements of mechanical and electric properties at the nanometer scale. The primary source of information is force interaction between the sample surface and

Yuri L. Lyubchenko (ed.), *Nanoscale Imaging: Methods and Protocols*, Methods in Molecular Biology, vol. 1814, https://doi.org/10.1007/978-1-4939-8591-3_7,

AFM probe, which is a microfabricated cantilever with the sharp tip at its end. The cantilever bends or changes its vibration in response to the tip–sample force, and this makes it a highly sensitive detector of mechanical and electromagnetic interactions. The contributions of different forces to the probe response can be used for extraction of quantitative mechanical properties (elastic modulus, work of adhesion, etc.) and electrical properties (surface charge, surface potential, dielectric permittivity, doping type and level, etc.). The maps of local properties are invaluable for discerning the dissimilar sample locations or constituents of heterogeneous and multicomponent materials. In the following we will address the electrostatic tip–sample interactions and their use in AFM electrical techniques, which are applied to study local electrical properties. In this section we outline the backgrounds of AFM and electrical modes.

1.1 AFM Modes and Tip–Sample Interactions

The ongoing developments of AFM electronics have led to expansion of operational modes and their capabilities. Contemporary AFM modes can be classified into quasi-static contact and oscillatory techniques, Table 1.

Among the oscillatory modes there are resonant and nonresonant techniques, in which the probe is brought in periodical motion at frequency close to or far from its main flexural resonance, respectively. Primary AFM function is a gentle surface profiling that is achieved in these modes by keeping the tip–sample force constant while the probe is rastering over sample surface. In the contact mode and in nonresonant oscillatory modes (Pulsed Force, PeakForce, Hybrid, QuickSense, etc. [2, 3]) for topography imaging the

Table 1
AFM-based electric techniques

AFM modes and techniques	Contact mode	Oscillatory nonresonant mode	Oscillatory resonant modes (AM-PI, AM-FI, FM-AI)
Current sensing (conducting AFM)	✓	✓	
Scanning capacitance microscopy	✓	✓	
Scanning microwave microscopy	✓	✓	
Scanning Seebeck microscopy	✓	✓	
Electric force microscopy	✓	✓	✓
Piezoresponse force microscopy	✓	✓	
Kelvin force microscopy	*p*	*p*	✓
Capacitance gradient microscopy	*p*	*p*	✓
Maxwell stress microscopy			✓

Check mark means that this operation was practically realized. Mark "*p*" means that this operation is possible

probe deflection is kept at set-point level (D_{sp}) by z-servo control. In contact mode, D_{sp} is directly related to the tip–sample interaction force, and it is described by a simple expression:

$$F(Z_c) = kD_{sp}, \tag{1}$$

where Z_c is the rest position of the tip; k—spring constant of the probe; D_{sp}—set-point deflection; and F—is the z-projection of the tip–sample force.

In the nonresonant modes the periodic tip–sample contact proceeds at frequencies (1–3 kHz) well below the resonant frequency of the applied probe. As the probe deflection cyclically changes, its peak value is chosen for the feedback of topography tracking. Therefore, a force equation for nonresonant oscillatory modes is expressed as follows:

$$\max_{t\in[0,T]} F(Z_c(t)) = kD_{sp}, \tag{2}$$

where T is the period of oscillation, and t is the time from the beginning of the latest cycle.

The direct correspondence between the probe deflection and applied force has been utilized for quantitative mechanical measurements in these modes by recording deflection-vs.-(vertical) distance curves at individual surface locations. Such curves can be converted to force vs. sample deformation dependencies, from which elastic modulus, work of adhesion and other mechanical properties can be extracted using different solid state deformation models. In nonresonant modes these steps are performed at higher rates than in contact mode, and quantitative maps of local mechanical properties are displayed practically simultaneously with topography images.

The situation is more complicated in the resonant modes, where the probe frequency (f), amplitude (A), and phase (θ) are influenced by the tip–sample interactions, and each parameter can be used as a relative measure of the acting force for the feedback control. The AFM probe oscillations, which are influenced by the tip–sample forces, can be modeled by Euler–Bernoulli equation [4, Eq. 5.82]. The Krylov–Bogoliubov–Mitropolsky approach has been applied to derive asymptotic amplitude-phase dynamics [5]. In particular, steady state equations for the first eigenmode relate the characteristics of the oscillating probe (f, A, and θ) with sample topography described by Z_c and the tip–sample forces (F_a and F_r) when the probe approaches a sample and retracts from it, respectively. In conservative cases, which include electromagnetic forces, $F_a = F_r$, and two steady-state equations describing the force interactions in resonant modes are the following [6, 7]:

$$\begin{cases} \sqrt{1+G^2}\sin\theta = A/A_0 \\ \sqrt{1+G^2}\cos\theta = \dfrac{2Q_1}{\pi A_0 k}\displaystyle\int_0^{\pi} F_z(Z_c + A\cos y)\cos y\,dy - G\dfrac{A}{A_0}, \end{cases} \tag{3}$$

where

$$G = 2Q_1\frac{f - f_1}{f_1}$$

is the magnified relative frequency shift of the generated frequency f from f_1 (first resonant frequency of free oscillation).

These two equations contain four variables (A, f, θ, and Z_c), and their solutions can be obtained for two variables when the other two are fixed. Surface topography, Z_c cannot be fixed because it is the variable a researcher always wants to measure. There are three remained cases, in which the frequency and amplitude (case 1), the phase and amplitude (case 2), or the phase and frequency (case 3) held constant. (For full classification please check [7, table 1].) Essentially, these cases are implemented in different oscillatory AFM modes when the operation proceeds near or at the probe resonant frequency.

In the most applicable mode—amplitude modulation with phase imaging (AM-PI) [7], a particular implementation of which is known as tapping mode [8], the probe is brought into oscillation near or at its resonance frequency. The damped amplitude A_{sp} of prior-to-interaction amplitude A_0 is used as a set-point for z-servo control in surface profiling. In this mode, in addition to the amplitude, the frequency is fixed to its free resonance value, whereas the topography and phase changes are recorded. At high forces, the phase contrast distinctively reveals dissimilar surface features of heterogeneous samples. Therefore, this mode offers compositional mapping of multicomponent materials. The evaluation of the tip–sample force in AM-PI is based on Eq. (3) with $G = 0$. Phase θ can be excluded and two equations will be transformed to the following one:

$$\left(\frac{2Q_1}{\pi A_0 k}\int_0^{\pi} F_z(Z_c + A\cos y)\cos y\,dy\right)^2 + \left(\frac{A}{A_0}\right)^2 = 1. \tag{4}$$

We have applied Eq. (5) to calculate the maximal forces and deflections as functions of A_0 and A_{sp} for materials with different elastic moduli [6]. The results have showed that the force (deformation) in tapping mode is higher (lower) for samples with higher modulus, and it reaches the peak values at $A_{sp} \approx 0.5A_0$. These findings are consistent with earlier experimental data [9].

The pathway for calculations of elastic modulus in AM-PI mode was also pointed out [3, 6].

For our consideration of the electrostatic forces we need to use the equation that relates the cosine phase and probe force (from Eq. (3) with $G = 0$):

$$\cos\theta = \frac{2Q_1}{\pi A_0 k}\int_0^{\pi} F_z(Z_c + A\cos y)\cos y\,dy \tag{5}$$

There is another implementation of AM mode (Case 2), in which in addition to z-servo for keeping A_{sp}, phase-locked-loop (PLL) is employed for keeping the probe phase at 90° [7, 10]. In this mode, which is undervalued so far, the frequency changes can be used for recognition of different components of heterogeneous samples. The abbreviation AM-FI (FI—frequency imaging) fits this operation mode. In AM-FI, the probe drive is kept at the effective probe resonance (i.e., phase $\theta = 90°$), and this helps avoiding the fake height contrast often happened on smooth surfaces of heterogeneous samples in tapping mode [11].

Frequency Modulation (FM) mode is one more AFM resonant technique [12], in which a frequency shift caused by tip interactions with a sample, is employed for tracking sample topography (Case 3). In this mode the phase θ is kept constant at 90° by using PLL, frequency shift is the set-point for z-servo control, and amplitude changes, which are related to energy dissipation, are recorded as a separate image [7]. Therefore, this mode can be connoted as FM-AI (AI—amplitude imaging). Originally, this mode was introduced for experiments in ultra-high vacuum, where AFM probe has extremely high Q that restricts the use of amplitude damping for feedback control. Later, this mode was applied in different media (liquid, air, etc. [13]) where it offers a number of advantages compared to AM-PI and AM-FI. Particularly, FM mode allows expansion of AFM imaging to a broader force range with an emphasis on low-force imaging of soft materials [14] and weakly bonded adsorbates.

For FM-AI and AM-FI ($\theta = 90°$), the probe frequency shift relates to the force interaction by the following equation derived from (Eq. 3) for $\theta = 90°$:

$$\frac{\Delta f}{f_1} = \frac{f - f_1}{f_1} = \frac{1}{\pi k A}\int_0^{\pi} F_z(Z_c + A\cos y)\cos y\,dy \tag{6}$$

It is worth noting that Eqs. (6) and (7), which relate the probe phase and frequency changes with the tip-force, contain the same integral that in some cases can be analytically calculated to facilitate the analysis of AFM data.

In approximation of small amplitudes, the asymptotic relation between the integral and the force gradient can be obtained.

$$\lim_{A\to 0}\frac{1}{A}\int_0^{\pi} F_z(Z_c + A\cos y)\cos y\mathrm{d}y = \int_0^{\pi}\frac{\partial F_z}{\partial z}\bigg|_{z=Z_c}\cos^2 y\mathrm{d}y = \frac{\pi}{2}\frac{\partial F_z}{\partial z}\bigg|_{z=Z_c} \quad (7)$$

This gradient approximation is often used by AFM researchers dealing with applications of frequency modulation techniques [15, 16]. Below for brevity we will use force gradient always meaning that it is either approximation of the integral (Eq. 7) being accurate for small amplitude, or the integral itself.

1.2 AFM-Based Electrical Modes

AFM was introduced for high-resolution imaging and further expanded to probing and mapping of local mechanical and electromagnetic properties. Studies of these properties can be performed practically in all AFM modes described above. AFM-based electrical techniques can be classified into two groups, Table 1. First group includes techniques that employ additional sensors (current detector, microwave gauge, capacitors, thermocouple, etc.) for measurements of electrical properties. These techniques are conducting AFM [17], scanning capacitance microscopy [18], scanning microwave microscopy [19], and scanning Seebeck microscopy [20]). The recordings of these sensors are synchronized with the probe profiling of sample surface and are displayed simultaneously with sample topography. Optimization of such measurements on soft materials requires a lowering of the tip force that can be a challenge in the contact mode.

In this document we will be primarily focused on techniques of second group consisting of methods, which utilize the probe sensitivity to electrostatic tip–sample forces. The response of the AFM probe to the electrostatic tip–sample forces was examined shortly after the introduction of this technique [21]. Nowadays the AFM electrical modes, which are based on detection of tip–sample electrostatic forces, embrace electrical force microscopy (EFM), piezoresponse force microscopy (PFM) [22], Kelvin force microscopy (KFM) [23], and probing of capacitance gradients—$\mathrm{d}C/\mathrm{d}Z$ and $\mathrm{d}C/\mathrm{d}V$ (i.e., capacitance gradient microscopy) as well as Maxwell stress microscopy [24] and others. Detection and high-resolution mapping of variations of electrostatic force, surface potential and dielectric permittivity are the main functions of these modes. For theoretical considerations, the tip–sample junction can be rationally modeled by considering a probe tip as an electrode in a tiny capacitor it forms with a typically grounded sample that acts as the second electrode. When the conducting probe operates in the contact or oscillatory modes, its electrostatic force interactions can be promoted by an external DC and AC voltage: $U_{DC} + U_{AC}\cos\omega_{elec}t$ (Fig. 1a).

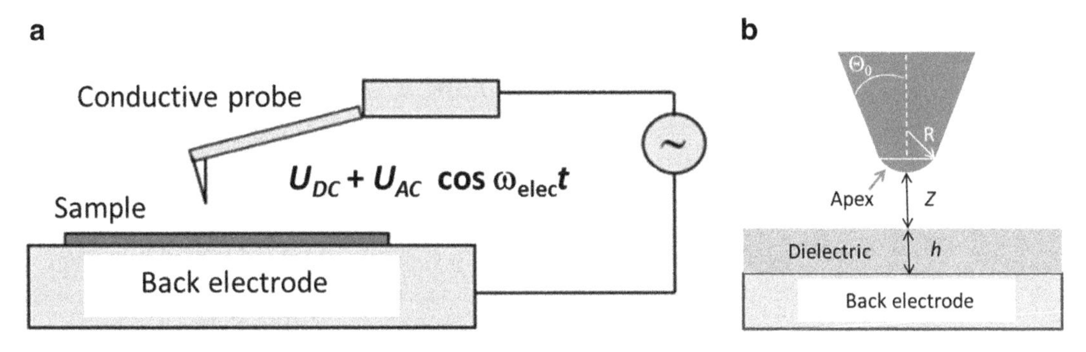

Fig. 1 (**a**, **b**) Sketch describing the AFM setup for electrostatic force measurements (**a**) and a model of tip–sample junction for local dielectric measurements (**b**)

The equations, which relate the overall electrostatic force $F_{\text{elec}}(Z)$ and its DC and AC constituents: $F_{\text{DC}}(Z)$, $F_{\omega_{\text{elec}}}(Z)$, and $F_{2\omega_{\text{elec}}}(Z)$ with tip–sample capacitance—C and surface potential difference—ϕ, are given below.

$$F_{\text{elec}}(Z) = -\frac{1}{2}\frac{\partial C}{\partial Z}[(\phi - U_{\text{DC}} + U_{\text{AC}}\cos(\omega_{\text{elec}}t))]^2 = F_{\text{DC}}(Z) + F_{\omega_{\text{elec}}}(Z) + F_{2\omega_{\text{elec}}}(Z) \quad (8)$$

$$F_{\text{DC}}(Z) = -\frac{1}{2}\frac{\partial C}{\partial Z}\left[(\phi - U_{\text{DC}})^2 + \frac{1}{2}{U_{\text{AC}}}^2\right] \quad (9)$$

$$F_{\omega_{\text{elec}}}(Z) = -\frac{\partial C}{\partial Z}(\phi - U_{\text{DC}})U_{\text{AC}}\cos(\omega_{\text{elec}}t) \quad (10)$$

$$F_{2\omega_{\text{elec}}}(Z) = -\frac{1}{4}\frac{\partial C}{\partial Z}{U_{\text{AC}}}^2\cos(2\omega_{\text{elec}}t) \quad (11)$$

In formulas (8)–(11) is assumed that Z changes slowly and, therefore, it can be considered constant during several periods of ω_{elec}.

The force components (Eqs. 10 and 11) are the most useful for practical AFM-based electrical measurements. By adjusting DC voltage offset (U_{DC}) to surface potential difference one can nullify ω_{elec}—force component. This procedure is employed in KFM for finding surface potential of sample locations. Finding the probe response at $2\omega_{\text{elec}}$ provides essential data for extraction of tip–sample capacitance and local dielectric permittivity. In this consideration we can use the formula for nanoscale capacitance for thin dielectric film [25] for geometry shown in Fig. 1b:

$$C_{\text{apex}} = 2\pi\varepsilon_0 R\left\{1 + \frac{R(1 - \sin\theta_0)}{z + h/\varepsilon_{\text{r}}}\right\} + C_0(R, \theta_0) \quad (12)$$

where geometrical parameters are shown in this figure; ε_0 and ε_{r} are the vacuum dielectric constant and relative dielectric permittivity of the film respectively. Term C_0 does not depend on Z and has no influence on the force that is proportional to the capacitance gradient. In practical measurements at $2\omega_{\mathrm{elec}}$ we can get amplitude of mechanical phase cosine $\cos\,\theta$, which is related to integral $\int_0^{\pi} F_z(Z_c + A\cos y)\cos y dy$ according to Eq. (5). At small amplitudes, this integral is approximated by force gradient (Eq. 8). In general case the integral can be expressed in close form for Eq. (12) as shown below. In both cases, expressions contain dielectric permittivity. Therefore, by measurements of the probe response at $2\omega_{\mathrm{elec}}$ one can extract the local dielectric permittivity and the quantitative permittivity values were obtained in studies of thin polymer films with known thickness [26].

In addition to the data treatment with analytical methods, finite element analysis (FEA) can bring into interplay a detailed description of the probe and the sample. We illustrate this point by considering the role of the probe as one of the major factors determining the sensitivity, spatial resolution and reliability of detection of local electrical properties. The electric field maps, which were calculated for conducting probes of two geometries that were positioned at different distances above a surface, are shown in Fig. 2. These calculations were performed with FEA software of Field Precision LLC [27].

The analysis of these maps shows that electric field at surface location under the probe apex is of the same value for probes with different geometry separated from the sample by 10 nm, Fig. 2a, b. Naturally, the electric field from the larger tip is more delocalized. The strength of electric field increases by approximately one order

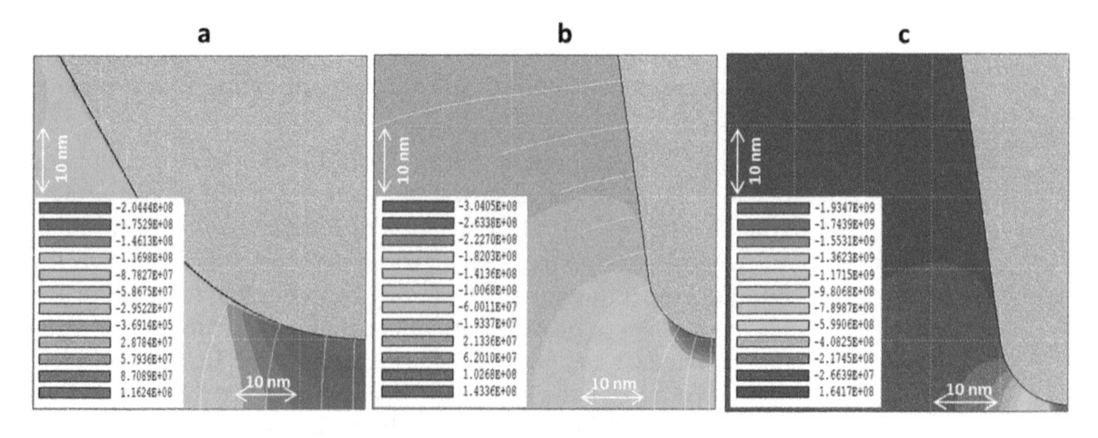

Fig. 2 The maps of electric field Z-component – $E_z(N/C)$ for the tip–metal configuration with grounded metal and tip with 2 V bias. The calculations were performed for probe with tip radius of 40 nm and its opening angle—30 degrees in (**a**); for probe with tip radius of 10 nm and opening angle of 10 degrees in (**b**) and (**c**). Tip–sample distance $Z = 10$ nm for (**a**) and (**b**), and $Z = 1$ nm for (**c**). White lines represent the electric field lines. In (**c**) electric field outside the probe apex is weak and the field lines are not shown. The color regions represent the different field levels and the color codes are shown in the inserts

of magnitude and focused into a couple of nm, when the sharper probe has approached the surface to a separation of 1 nm, Fig. 2c. This finding hints on spatial resolution of few nanometers in mapping of electrical properties with the probe having tip radius of 10 nm or more. As a better spatial resolution for imaging is expected with the sharper probe, the sensitivity of the sharp probe to the electrostatic force is most likely weaker because for the dull probe the force is integrated over the larger area.

Another question that has been discussed concerns the contributions of the tip apex, tip body and cantilever to the measured probe force and integral in (Eqs. 3–7). (The integral is asymptotically converges to force gradient according to (Eq. 8)). In earlier FEA application the contributions of the objects mimicking the probe apex, a tip extended body and a whole probe to electrostatic force and force gradient (integral) were calculated [28]. In this procedure the objects were divided in a large number of elements and their interactions were examined by the surface charge method. It was found that the force acting on the tip body and cantilever increases the total bending of the probe yet the force gradient and the integral are less susceptible to the influence of these objects. This conclusion had been later confirmed by other researchers [29], who found out that at very small tip–sample distances the total force is dominated by contributions from the tip apex. Therefore, the use of the integral or force gradient for small amplitude was suggested for measurements of electric field changes and surface potential.

After bringing theoretical backgrounds we will explain main AFM-based electric modes and outline their protocols. The capabilities of these modes will be illustrated by a number of practical examples.

2 Materials and Instrumentation

The capabilities of AFM-based electrical modes will be demonstrated on several materials.

1. Polymer blend of commercially available syndiotactic polystyrene—sPS and poly(vinyl difluoride)—PVDF.
2. Polymer blend of commercially available atactic polystyrene—PS and poly(vinyl acetate)—PVAC.
3. Thermoplastic vulcanizate—TPV, which is a composition of isotactic polypropylene, EPDM rubber and carbon black.
4. Semifluorinated alkane $CH_3(CH_2)_{20}(CF_3)_{14}CF_3$—F14H20.
5. Macroscopic crystal of lithium niobate—$LiNbO_3$, which was parallel polarized.
6. Crystal of ferroelectric triglycine sulphate—TGS.

7. Incomplete alloy of Bi and Sn metals with 40/60 composition.
8. The described below AFM-based electrical methods are realized with scanning probe microscopes whose controllers incorporating three dual lock-in amplifiers (LIA) and software that enable single-pass and double-pass measurements. For the experiments the electrical contacts to the probe and sample are needed for excitation of electric field in the probe–sample junction.
9. AFM microscopes made by Bruker, Asylum, Keysight Technologies, NT-MDT, and other manufacturers can be applied for electric measurements. The described data were recorded with the 5500 microscope (Keysight Technologies) and NEXT microscope (NT-MDT).
10. Common probes for AFM-based electrical modes are made of Si with Pt coating, and their spring constants (k) are typically in the 2–5 N/m range, rarely in the 30–40 N/m range.

3 Methods

Main AFM-based electrical modes, which are based on detection of electrostatic tip–sample forces, are collected in this section.

3.1 Studies in the Contact Mode: Electrical Force Microscopy (EFM)

Electrical force microscopy (EFM) and Piezoresponse force microscopy (PFM) in contact mode are realized when the conducting probe stays in contact with electrically active sample and AC electric field excitation at ω_{elec} is provided between these electrodes, Fig. 3.

The vertical and horizontal components of the probe deflection are analyzed at ω_{elec} and its second harmonic ($2\omega_{elec}$) with two dual LIAs. Below are several steps of the experimental procedure:

1. Prior to the measurements in any AFM-based electrical techniques, for a control purpose, set some values of AC and/or DC bias voltage in software in either configuration of biased probe or sample depending on a particular microscope.
2. Check that the applied AC or DC voltage between the probe and sample stage is sensed by a voltmeter or multimeter.
3. Place a sample directly on a microscope stage or on a conducting substrate, which should be placed on a microscope stage. The conducting glue such as colloidal silver paste is needed to make the sample immobile and for better electrical contact, which also must be verified with the multimeter (ohm-meter) after glue is dried.
4. In the contact mode, engage the probe to the sample at smallest possible force (set-point cantilever deflection) and keep the force low during scanning to avoid a possible wear of

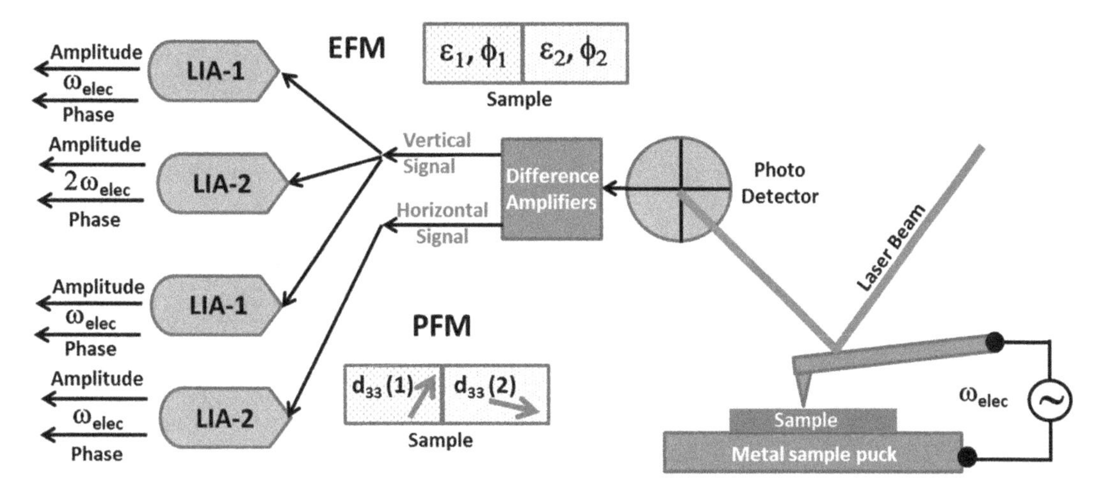

Fig. 3 Diagram of the electrical measurements in AFM contact mode. In EFM technique deflection signal is sensed by LIA-1 and LIA-2, which are tuned to ω_{elec} and $2\omega_{\text{elec}}$. The amplitude and phase signals at these frequencies are mapped to reflect differences in low electric properties (surface potential—φ and dielectric permittivity ε). In PFM, the vertical and horizontal signals are analyzed by two LIAs, which are tuned to ω_{elec}. The related maps reflect the magnitude (piezoelectric constant d_{33}) and orientation of local piezoresponse

conducting coating. The scanning with the probe deflection corresponding to overall attractive force suites this purpose.

5. Electrical excitation (3–5 V) should be provided at ω_{elec} below the contact resonance of the probe (*see* **Note 1**). The ω_{elec} range from few to 300 kHz is suitable for probes with $k = 2$–5 N/m as their contact resonances are above 350 kHz. For stiffer probes with $k = 30$–40 N/m the contact resonance is approaching 1 MHz that makes ω_{elec} range broader.
6. Select EFM operation by directing the vertical deflection signal to LIA 1 at ω_{elec}.
7. Choose Amplitude and Phase at ω_{elec} and/or $2\omega_{\text{elec}}$ as imaging channels depending on a number of LIA available in a particular microscope.
8. Perform an optimization of EFM signal by observing line trace of amplitude or phase channel while adjusting such parameters as AC drive voltage, ω_{elec}, and tip–sample mechanical force (set-point deflection). AC drive voltages can be increased to ~5 V yet one should avoid a possible charging of the sample. In choosing a frequency of electric field one should avoid low frequency range below 3 kHz that can be contaminated by $1/f$ noise of the optical detection system. If sample stiffness allows the set-point deflection should be gradually increased to enhance the electric field in the tip–sample junction.

3.2 Studies in the Contact Mode: Piezoresponse Force Microscopy (PFM)

1. Follow the **steps 1–3** of Subheading 3.1.
2. In PFM a good tip–sample contact is required for reliable sensing of sample displacement caused by a piezo-effect. Therefore the probe contact can be improved by increasing the set-point deflection/force and by using the conducting probes with larger tip apex (*see* **Note 2**).
3. Select PFM operation by directing vertical and lateral deflections to LIA 1 and LIA 2 both at ω_{elec}.
4. Choose excitation frequency ω_{elec} and AC voltage applied to the tip–sample junction. For ω_{elec} use frequency in the 10–100 kHz range that is below the contact probe resonance of the recommended probe with $k = 2–4$ N/m (*see* **Note 1**). Besides this probe stiff conducting probes with $k = 30–40$ N/m can be also used and they are more preferable compared to soft conducting probes with $k < 1$ N/m (*see* **Note 3**). Select 10 V for AC voltage and this value can be adjusted and in some commercial instruments voltages as high as 150 V are available for PFM excitation.
5. Select Amplitude and Phase at ω_{elec} of LIA 1 and/or LIA 2 as imaging channels depending on a number of LIA available in a particular microscope.
6. In optimization of the PFM experiment by adjusting ω_{elec} and AC voltage watch for line scan of phase signal that should show good signal-to-noise character, and for many samples the phase change of 180° is common for differently polarized domains (*see* **Note 4**).
7. When PFM contrast shows the different domains, record individual surface points the piezoresponse amplitude and phase as function of applied DC voltage sweeping between negative and positive bias values, e.g., ~10 V (or higher voltage if the microscope is capable).

3.3 Double-Pass Electrical Measurements in AFM Resonant Modes

As oscillatory resonant modes overcame the contact mode in studies of soft materials, we introduce the electrical studies in these modes in lift or double-pass procedure [30], which allows for a separation of mechanical and electrical responses, Fig. 4.

In the first pass, surface topography is measured with AM-PI mode at the resonant frequency of the probe (ω_{mech}). In the second pass, the conductive probe follows the learned profile being lifted 10–40 nm above the sample (the feedback is off) and it is driven into oscillation at ω_{mech}. The probe–sample DC bias, which is applied in the second pass, causes long-distance electrostatic force and the related probe amplitude and phase variations. The latter will be detected and presented in combination with the height image recorded in first pass. This EFM operation can be extended to KFM in the second pass with an additional servo that finds DC

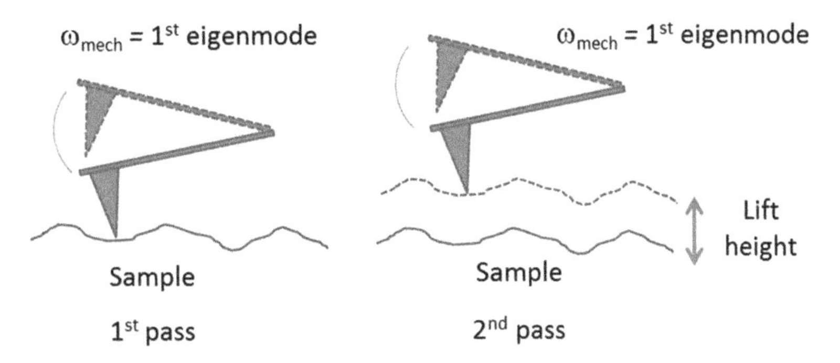

Fig. 4 Sketch illustrating two-pass operation in EFM. In first pass the mechanical tip–sample forces are employed for a detection of surface profile. In second pass the probe is pivoted above the sample mimicking the surface profile. At the same time the oscillating probe is subjected to tip–sample electrostatic interactions

voltage eliminating the probe response caused by a difference of surface potentials (φ) of the probe and sample. The experimental protocol is as follows:

1. Follow the **steps 1–3** of Subheading 3.1.
2. For double-pass or lift operation in AM-PI/tapping mode use a conducting probe ($k = 2$–5 N/m) with the cantilever length of 225 μm and resonant frequency is in the 60–90 kHz range.
3. After engagement in AM-PI/tapping mode, for further operations choose the free amplitude A_0 in the 30–40 nm and set-point amplitude $A_{sp} = 0.6$–$0.8A_0$. The calibration of the probe amplitude is important for optimizing the lift operation (*see* **Note 5**).
4. Scan a chosen sample area in order to verify that sample topography is relatively smooth (corrugations below 50 nm are desirable). This helps avoiding artifacts in the lift operation that are caused by nonwanted tip touching the surface in second pass.
5. Prior to activation of the double-pass experiment choose the parameters for second pass: lift height, drive voltage in the second pass and DC bias voltage, which can be applied to the probe or to the sample (*see* **Note 6**). The initial lift height and DC bias can be set to 100 nm and 3 V, respectively. Verify that deflection signal in second pass is featureless, i.e., the probe is flying over the surface features.
6. In next step the second pass parameters need to be optimized by monitoring real-time traces of height, deflection and phase signals. By minimizing the lift height the electrostatic force interactions will increase, therefore, the lift height should be decreased to a value just prior to a disturbance by tip touching detected by the second pass deflection signal.

7. When the drive signal in second pass is the same as in the first pass (*see* **Note 7**), adjust the sample or tip DC bias voltage to confirm that phase contrast in the second pass is due to electrical force interactions. The phase contrast should increase or decrease as the tip–sample voltage is changed up and down. Choose the voltage value for high quality phase signal but have in mind that high voltage might lead to a nondesirable charging of the sample.
8. The experiment should continue with scanning a desirable sample area and recording sets of first pass height and second pass and phase images.
9. For surface potential operation—KFM, which requires AC excitation, an additional servo needs to be activated in software to find at what probe voltage the EFM signal disappear.
10. As the surface potential signal feedback loop can be unstable (the oscillating potential signal), the feedback gains should be minimized, lift height increased, and/or drive amplitude reduced.

3.4 Single-Pass Electrical Measurements in AFM Resonant Modes

A remote position of the probe in the two-pass procedure limits sensitivity and spatial resolution of the lift operation. This limitation has been overcome is single-pass operation, in which separation of mechanical and electrostatic forces by simultaneous use, respectively, flexural resonant frequency (ω_{mech}) and a lower frequency (ω_{elec}) is realized by using 2 LIAs, Fig. 5.

The probe–sample electrostatic interactions, which are stimulated at ω_{elec}, are sensed at ω_{mech}, ω_{elec}, and $2\omega_{elec}$ by LIA-1 and LIA-2 in the parallel or in-series fashions. LIA-1 provides the

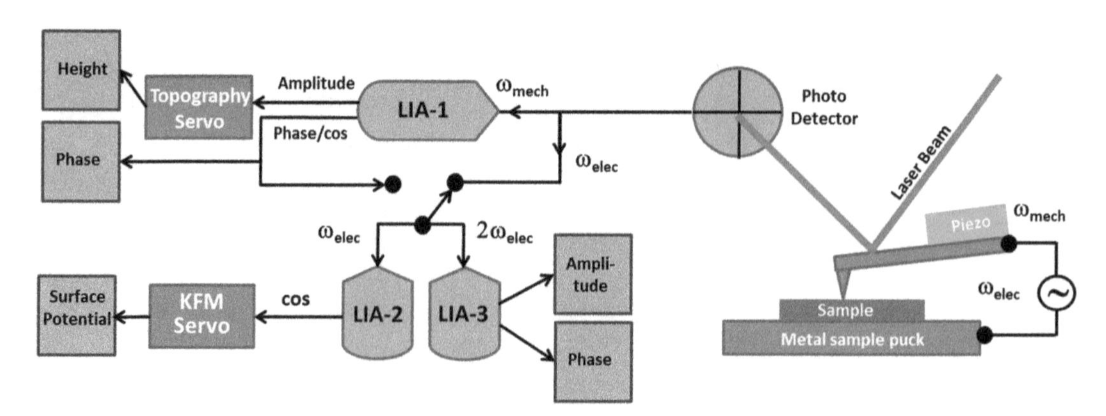

Fig. 5 Diagram of instrumental setup for single-pass electric measurements in the oscillatory resonant modes. Sample topography is detected in AM-PI mode at ω_{mech}. A switch between the parallel and in-series setups of LIAs allows for a choice between amplitude modulation (electrostatic force) and phase modulation (electrostatic force gradient) detection of surface potential using LIA-2 and KFM feedback. Simultaneously with topography and surface potential the capacitance gradients (dC/dZ and d^2C/dZ^2) are measured with LIA-3 and displayed as amplitude and phase images

components of ω_{mech} signal with the amplitude being used in topography feedback leading to height image, and the phase changes - to phase image. In the setup of parallel LIAs, the components of the deflection signal at ω_{elec}, which are selected by LIA-2, are adjusted to maximize the magnitude of cosine that comes to the input of KFM feedback. The DC bias voltage of this feedback, which nullifies the input signal, is reflected in the surface potential image. The amplitude and phase of $2\omega_{elec}$ signal from LIA-3 are presented in the images that reflect the local variations of dC/dZ. As the measurements in this operation are performed far from the probe mechanical resonance, the amplitude changes are directly related to *electrostatic force*.

When LIAs are connected in series, the electrostatic interactions are excited at $\omega_{elec} = 2$–5 kHz, which is within the bandwidth of LIA-1 tuned to ω_{mech}. The electrostatic responses at the heterodyne frequencies $\omega_{mech} \pm \omega_{elec}$ and $\omega_{mech} \pm 2\omega_{elec}$ are confined to ω_{elec} and $2\omega_{elec}$ components by LIA-2 and LIA-3. One of the component outputs of LIA-1 (typically, cosine component or phase) at ω_{mech} is minimized to near zero by adjusting reference phase in this LIA, and it is used as the input to LIA-2, LIA-3. The cosine component from LIA-2 is maximized by adjusting reference phase of this LIA, while a DC bias voltage is applied between the probe and the sample, and used as input to KFM servo to provide surface potential image. In this case, the electrical signal to be nullified is related to phase cosine ($\cos\omega$), which is proportional to the integral in Eq. (6) simplified to *force gradient* for small amplitudes according to Eq. (8). As it was discussed above, such KFM measurements are more sensitive (less error sources) and they provide higher spatial resolution compared to those based on the electrostatic force [29]. The components at $2\omega_{elec}$ are presented as amplitude and phase images of d^2C/dZ^2. Both capacitance gradients are related to local dielectric permittivity and being recorded as amplitude and phase components they can be applied for extraction of complex dielectric response. In the microscope software the LIA connections are typically preset for force or force gradient detections that simplify the experimental steps presented below.

1. Follow the **steps 1–3** of Subheading 3.1.
2. As a conducting probe (k = 2–5 N/m) with the cantilever length of 225 μm and ω_{mech} in the 60–90 kHz range is usually applied in these modes for both (force and force gradient) measurements, the stiffer and softer conducting probes can be also used depending on mechanical and adhesive properties of a sample.
3. After engagement in AM-PI/tapping mode, for further operations choose the free amplitude A_0 in the 30–40 nm and setpoint amplitude $A_{sp} = 0.6$–$0.8A_0$ to ensure a gentle tracking of sample surface.

4. For single-pass operation with electrostatic force detection, ω_{elec} can be chosen in the 2–20 kHz range well below ω_{mech} (*see* **Note 8**) LIAs will be set to ω_{elec} and $2\omega_{elec}$. With this excitation frequency choice the probe deflection will be measured at ω_{elec} and $2\omega_{elec}$.
5. For single-pass operation with electrostatic force gradient detection ω_{elec} can be chosen in a narrow range of 2–5 kHz that this frequency will be in the bandwidth of applied LIAs.
6. AC voltage for electrical field excitation at ω_{elec} should be chosen below 3 V to avoid a possible surface charging and artefact in surface tracking in AM-PI mode.
7. For optimization of parameters for KFM measurements in both modes set the probe sample bias to a known DC potential greater than expected surface potential (3 V usually will be sufficient). Then the probe response should be maximized in one of the components (cos or sin) by adjusting the LIA reference phase. The maximized component is used as the input to the servo controlling the DC bias voltage between the tip and sample.
8. Prior to scanning, select height, surface potential and other channels for imaging and optimize gains for these signals. This procedure can be assisted by real-time observation of the signals and adjusting the gains for their best quality (*see* **Note 9**).

3.5 Mapping of Capacitance Gradient and Quantitative Measurements of Dielectric Permittivity of Thin Polymer Films

An extraction of quantitative dielectric permittivity from capacitance gradient data requires knowledge of sample thickness because according to Eq. (12) capacitance depends on the ratio of sample thickness to permittivity: $p = h/\varepsilon_r$. In applying Eq. (6) that relates the amplitude of cosine at $2\omega_{elec}$—$G_{2\omega_e}^{\cos\theta}$ to electrostatic force and substituting the force using capacitance gradient we came to Eq. (13).

$$\frac{A_0 k\tilde{R}}{Q_1 {U_{ac}}^2 \varepsilon_0 R}\left[G_{2\omega_e}^{\cos\theta}\right] = \int_0^{\pi} C'(Z_c + A\cos y)\cos y dy = \frac{\pi}{\bar{A}}\left[\frac{1}{\sqrt{1-\bar{A}^2x^2}} - \frac{1+x}{\sqrt{(1+x)^2-\bar{A}^2x^2}}\right], \tag{13}$$

where

$$\tilde{R} = R(1-\sin\theta_0);\, C = \frac{C_{apex}\tilde{R}}{2\pi\varepsilon_0 R};\, \bar{A} = A/\tilde{R};\;\; x = \frac{\tilde{R}}{Z_c + p};\;\; p = \frac{h}{\varepsilon_r} \tag{14}$$

This integral equation was solved analytically and applied for calculations of ε_r for thin PS and PVAC films on ITO substrate [26].

Practically, quantitative measurements of dielectric permittivity of thin polymer films on conducting substrates can be performed in the following steps:

1. Prior to performing experiment of polymer film on a conducting substrate (i.e., ITO glass) a scratch should be made on the film with a sharpened wood stick for creating a step that helps measuring the film thickness in AFM height image.
2. Follow the **steps 1–3** of Subheading 3.1 and **steps 2, 3, 5, 6** of Subheading 3.4.
3. Do imaging of the surface area with the scratch in a way that the polymer material and substrate are both seen in the images. Record height image and map of amplitude of phase cosine cos $G_{2\omega_e}^{\cos\theta}$ and change voltage of electric field excitation at $\omega_{elec} = 3$ kHz (*see* **Note 10**) in 1 V steps, Fig. 6a–d. The finding that the cosine amplitude for PVAC film increases with AC bias in quadratic fashion (Fig. 6b, d), justifies that the

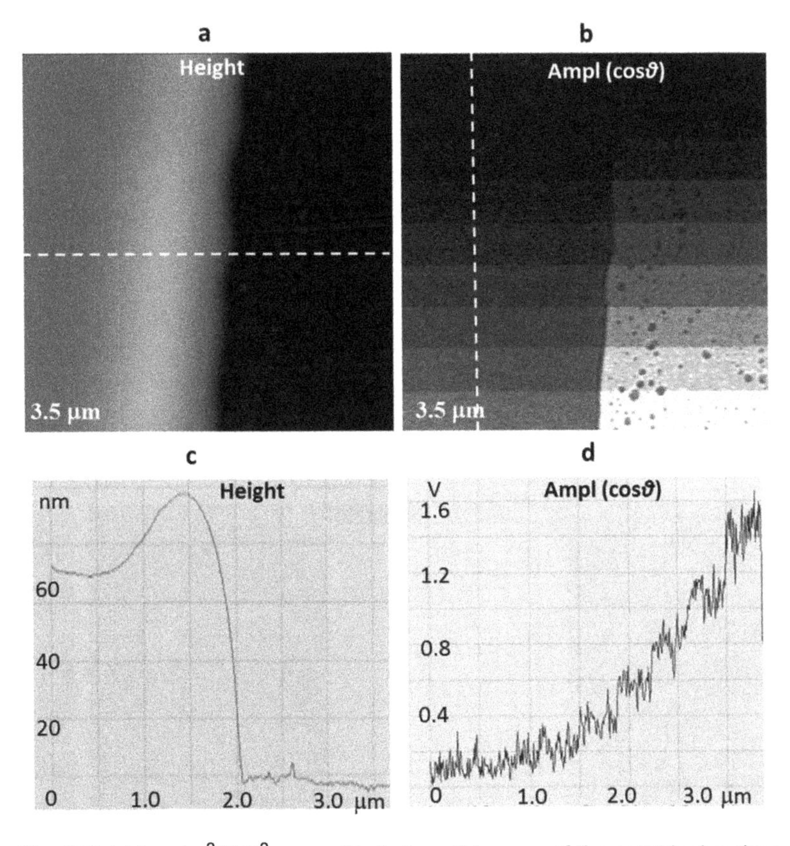

Fig. 6 Height and d^2C/dZ^2 or amplitude (cos θ) images of the scratched regions of PVAC film on ITO glass. During scanning the AC voltage at 3 kHz has increased up to 1.6 V. The height profile along the dashed white line in (**a**) is shown in (**c**). The changes of amplitude (cos θ) across the dashed white line in (**b**) are shown in (**d**)

measurements were conducted correctly. Measure the film thickness from the height image ($h = 65$ nm, Fig. 6c) and use the amplitude value of the AC bias (say 1 V) for ε_r calculations.

4. After finding the amplitude of $\cos\theta$ and using the probe dynamic (k and Q_1) and tip shape (R, θ_0) parameters one can find x from Eq. 13 and ε_r from the following equation:

$$\varepsilon_r = \frac{h}{\tilde{R}/x - Z_c}, \tag{15}$$

where Z_c equals to A_{sp} (*see* **Note 11**).

3.6 Practical Examples of AFM—Electrical Studies

1. The example of EFM in contact mode is demonstrated by observing the electrical-dipole rich locations of PVDF in PVDF/sPS blend (Fig. 7).

   Morphology of the blend film on Si substrate (Fig. 7a) is characterized by circular domains embedded into raised flat matrix. The circular domains with lamellar nanostructure can be assigned to PVDF—the polymer with high electrical dipole content compared to sPS. This explains the different electrical response of the probe on the matrix and domains, Fig. 7b, c.

2. The example of PFM is given in Fig. 8 in which images of parallel polarized $LiNbO_3$ are shown.

   A polished surface is smooth and without traces of the domain morphology, Fig. 8a. The image of vertical PFM amplitude reveals the domain pattern with two 5 μm wide "fingers', Fig. 8b, and the amplitude variations show only domain walls, Fig. 8c. The cross-section amplitude trace in Fig. 8d indicates that the depressions representing the domain walls have a

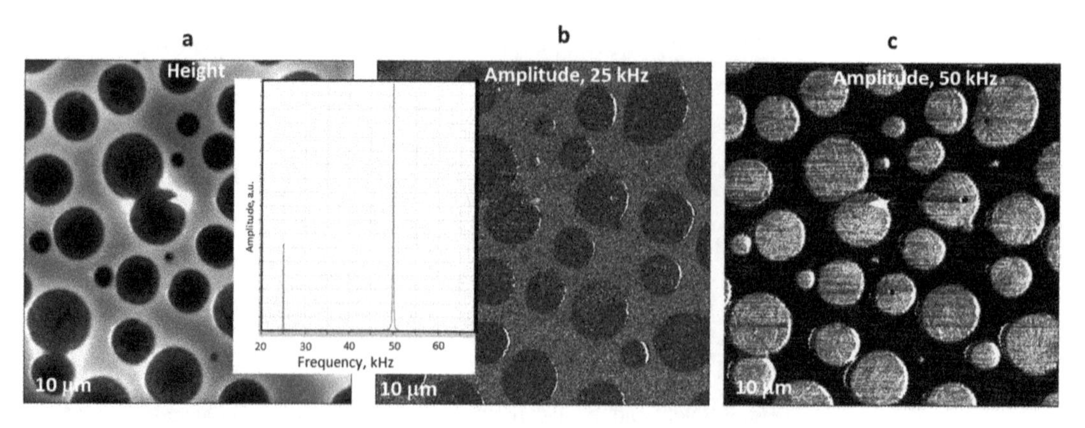

Fig. 7 (**a–c**) Height and amplitude images, which were obtained in electric force microscopy in contact mode on film of sPS/PVDF blend on Si substrate. The electric excitation of 5 V was applied to the conducting probe at frequency of 25 kHz. The amplitude responses at 25 kHz and 50 kHz, which were collected with LIAs, are shown in the insert

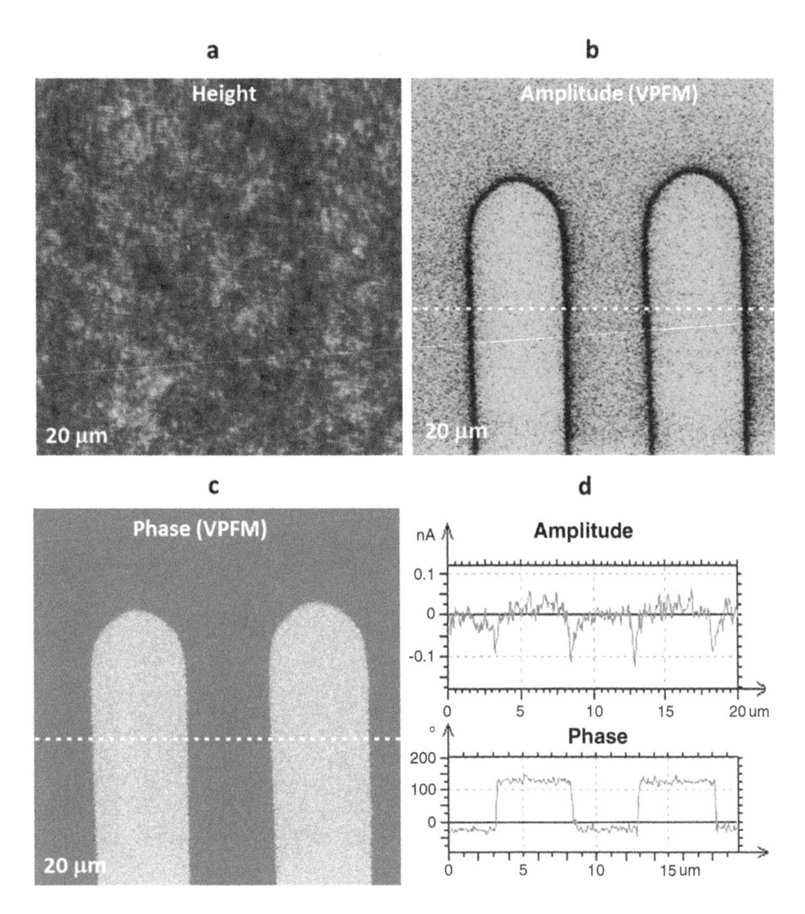

Fig. 8 Height (**a**) and vertical PFM amplitude (**b**) and phase (**c**) images of parallel polarized $LiNbO_3$ crystal. The surface corrugations in (**a**) are in the 0–2.2 nm range. The cross-section profiles along the white-dashed lines in the amplitude and phase images are shown in (**d**). The measurements were performed with the electric excitation 10 V at 30 kHz

sub-100 nm width. The pronounced phase changes of ~180 degrees in Fig. 8c and in the cross-section in Fig. 8d point on the opposite polarization of neighboring domains.

3. A common challenge in electrical measurements in contact mode, which is related to a possible combined effect of the electrostatic force and piezo response, is illustrated by images of ferroelectric material—crystal of triglycine sulphate (TGS) [31]. This crystal undergoes spontaneous polarization on cooling below Curie point (49 °C) with domains oriented vertically to the cleavage crystallographic ac plane, Fig. 9a–c. Multiple steps of 1.26 nN in height, which correspond to the crystallographic b-constant, are seen in height image (Fig. 9a).

   The amplitude and phase maps (Fig. 9b, c), which were recorded at $\omega_{elec} = 25$ kHz, reveal two uniaxial ferroelectric domains with a lens-like shape. The amplitude map shows

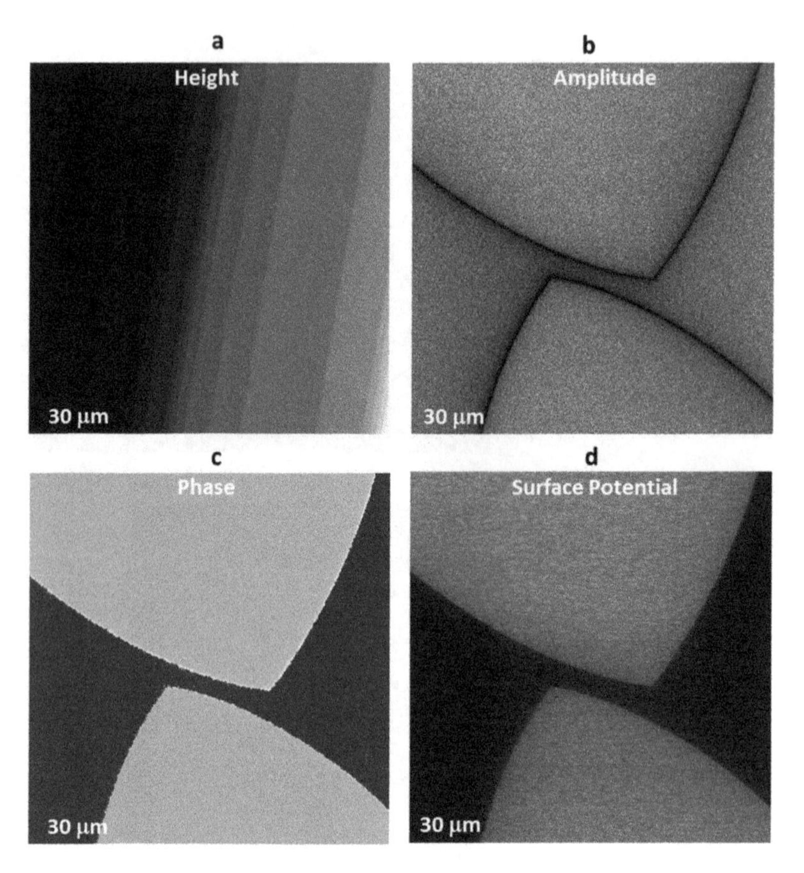

Fig. 9 (**a–c**) AFM height, amplitude and phase images of surface of TGC crystal, which were recorded in the contact mode. The tip–sample electric field was excited with 8 V at 25 kHz and the amplitude and phase signals were detected at the same frequency. The surface potential image in (**d**) was recorded in single-pass KFM mode with force gradient detection

primarily the domains' walls, and the phase map exhibits 180°steps between the lens-shaped domains and their surroundings. This observation can be explained either by the polarization charge density or thickness variations caused by piezoelectric effect. The first explanation is supported by finding surface potential variations, Fig. 9d. In contrast to TGS, KFM images of $LiNbO_3$ did not exhibit surface potential changes of the domains that serve as indication of pure piezoresponse nature of PFM images.

4. The double-pass study of TPV sample (Fig. 10) shows that the amplitude and phase images distinctively reveal the locations of carbon black particles, which experience attractive electrostatic force interactions with a conducting tip.

   Groups of carbon particles, which are emphasized with darker (brighter) contrast in amplitude (phase) images, do not necessarily have their counterparts in height image, which shows surface structure, Fig. 10a–e. This hints on signals of

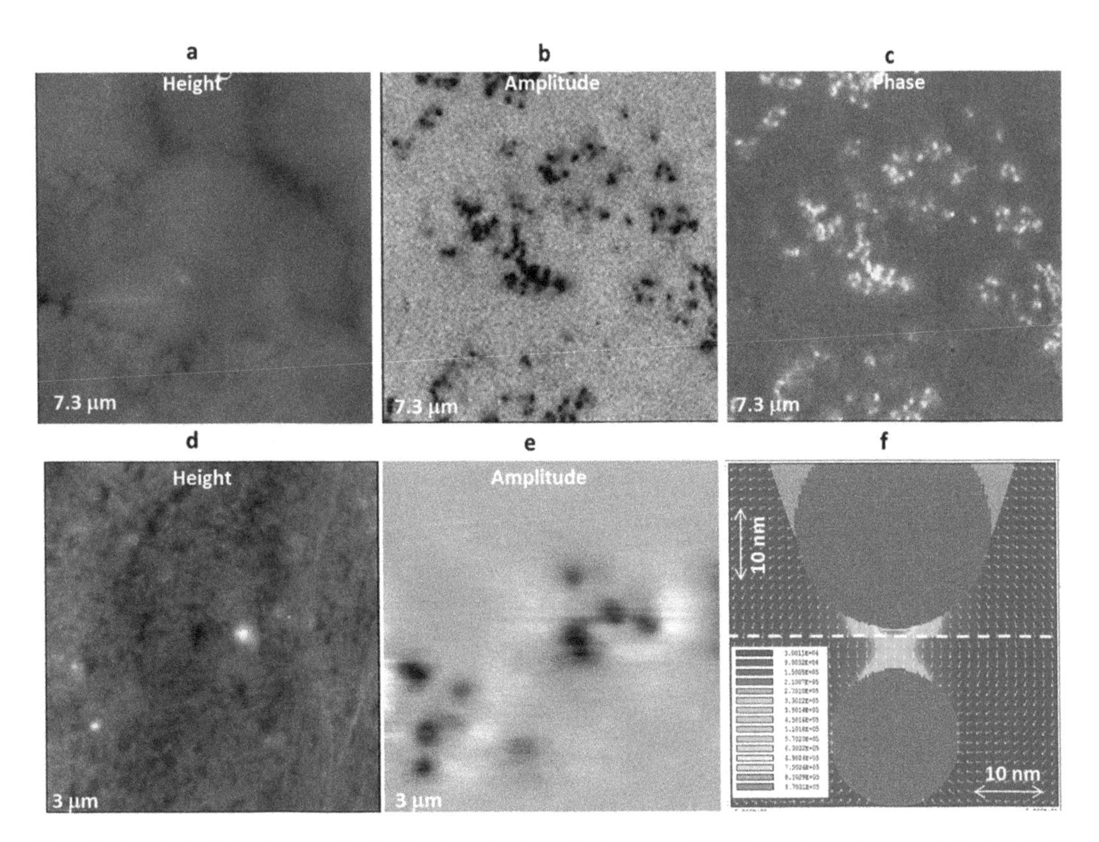

Fig. 10 The height (**a**, **d**), amplitude (**b**, **e**), and phase (**c**) images of TPV sample and map of electric field between the conducting probe and particle embedded in dielectric media (**f**). The height images were obtained in the first pass and the amplitude and phase images—in the second pass

subsurface particles in amplitude and phase images. Indeed, FEA calculation of electric field between the conducting probe and subsurface particle showed an enhancement of electric field at surface location above the particle, Fig. 10f.

5. Single-pass studies with low-force imaging in intermittent contact mode as well as sensitive detection of surface potential and capacitance gradient with nanometer-scale spatial resolution are illustrated by images of F14H20 self-assemblies, whose molecules exhibit strong dipole moment of 3.1D oriented along the chains consisting of fluorinated and hydrogenated parts. On Si surface F14H20 molecules form the curved and donut-shaped self-assemblies of 4 nm in height, Fig. 11a.

   These domains exhibit surface potential of ~ 0.8 V, Fig. 11b close to one determined in macroscopic studies [32]. In Fig. 11c, the d^2C/dZ^2 contrast, which is proportional to force gradient, differentiates electrically active donuts from two particles with disordered structure, which are located at the bottom of the images. F14H20 adsorbate on HOPG (Fig. 11d) is formed of flat-lying lamellar structures (left) and nanoscale ribbons (right). The surface potential of these

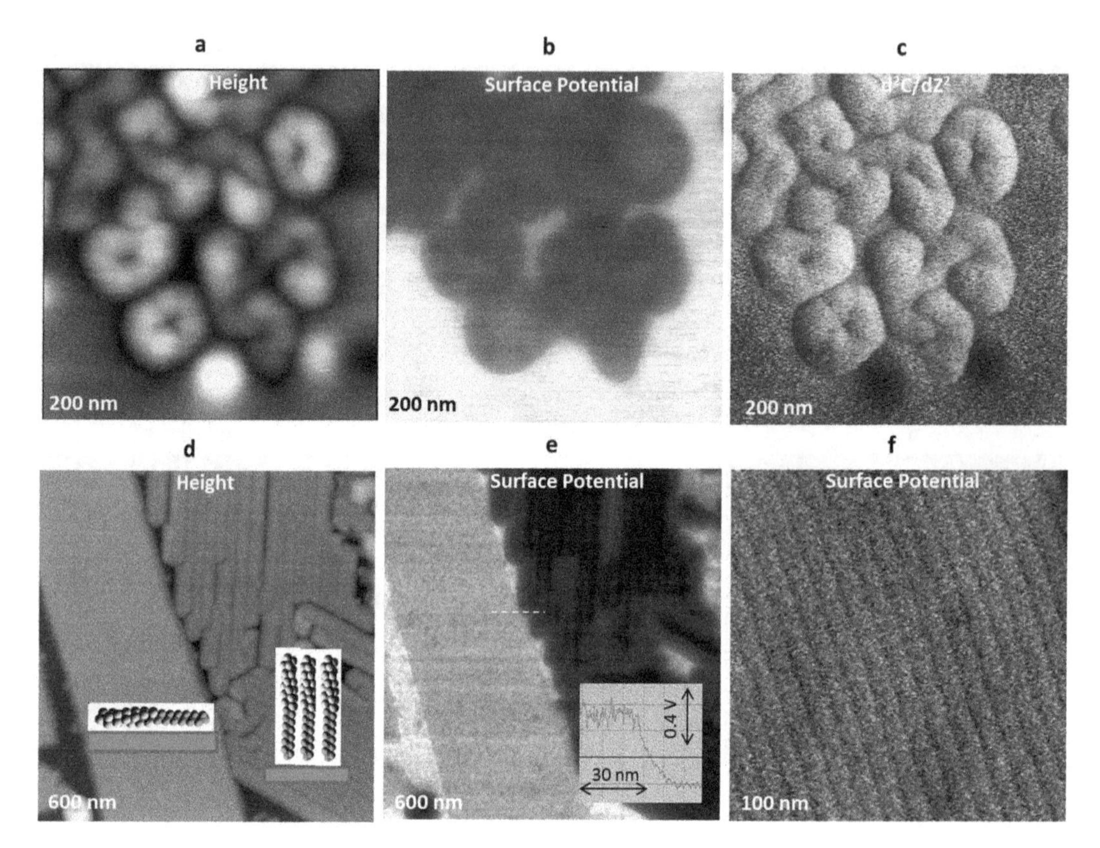

Fig. 11 Height (**a**), surface potential (**b**) and d^2C/dZ^2 (**c**) images of F14H20 self-assemblies on Si substrate. The images were obtained in single-pass operation with force gradient detection. Height (**d**) and surface potential (**e**, **f**) images of F14H20 self-assemblies on HOPG substrate. The molecular models of F14H20 show differences in their orientation in the left and right parts of the examined location. The image in (**f**) was recorded on a small area of the flat surface region in the left part of (**d**, **e**)

structures is different due to parallel and perpendicular orientation of molecular dipoles in the lamellae and ribbons, Fig. 11e. High-resolution mapping of surface potential image, in which the contrast on the edges of 6 nm lamellae is due to dipoles of $-CF_3$, groups, is shown in Fig. 11f [33].

6. KFM studies in single-pass (force gradient detection) and double-pass modes, which were recorded on F14H20 ribbons on mica, are shown in Fig. 12.

   In double-pass technique, several self-assemblies, which are seen in height image Fig. 12a, exhibit the surface potential that substantially varies at different lift heights. The amplitude of probe oscillation in first pass was around 50 nm, and second pass was performed not only at positive but also at negative lift heights up to 40 nm without touching a sample. The surface potential contrast is the largest when the probe was closest to the sample (negative lift 40 nm), Fig. 12b. In the single-pass

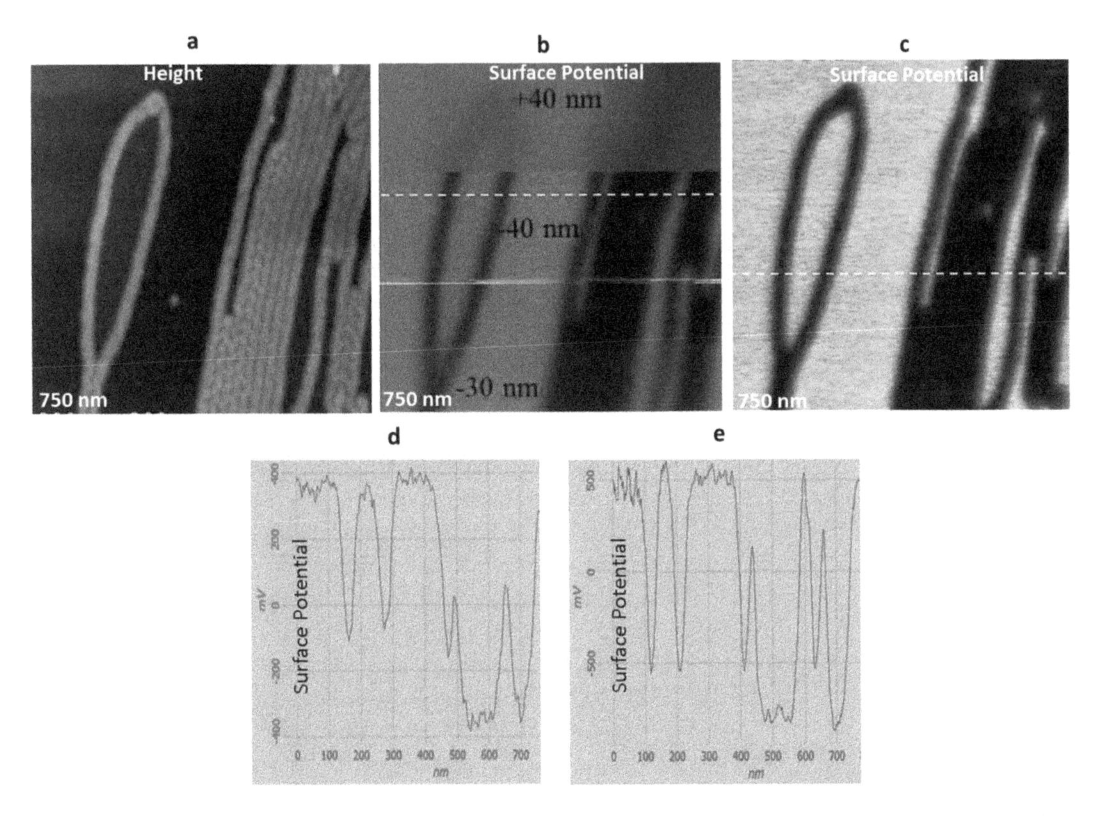

Fig. 12 (**a–c**) Height and surface potential images, which were recorded on F14H20 self-assemblies on mica. The surface potential image in (**b**) was recorded in double-pass operation with a positive lift height of 40 nm in the top part of the image and with the negative lift of −40 nm in the center and − 30 nm at the bottom of the image. The surface potential image in (**c**) was recorded in single-pass KFM more with force gradient detection. (**d**, **e**) The surface potential profiles, which were taken along the white-dashed lines in (**b**) and (**c**)

KFM study the surface potential at the same sample location (Fig. 12c) exhibits 1.5 times higher sensitivity. This is evident from the cross-section profiles along the same sample locations, which are shown in Fig. 12d, e.

7. Important question of KFM studies is related with quantitative values of surface potential. We have applied a sample of Bi/Sn alloy (Fig. 13) to verify that KFM measurements provide the surface potential difference (~0.18 V), which is close to one between the tabulated values of their work functions (Bi—4.22 eV; Sn—4.42 eV).

These images were obtained on a freshly prepared surface as its oxidation in air leads to decrease of the surface potential contrast. The oxidation and contamination of surfaces in air do not allow precise absolute measurements of surface potential compared to UHV conditions, where quantitative KFM data match to the results of ultraviolet photoelectron spectroscopy [34].

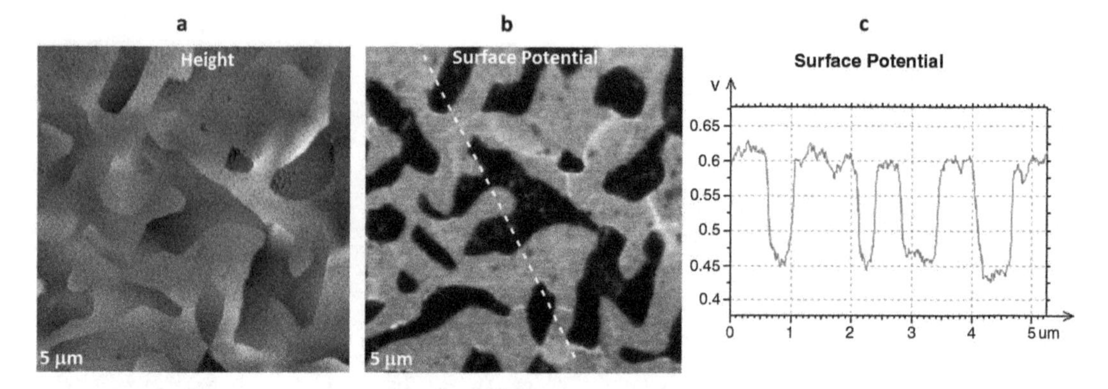

Fig. 13 The height (**a**) and surface potential (**b**) images obtained on surface of incomplete alloy Bi/Sn in single-pass KFM mode with force gradient detection. Surface corrugations are in the 0–25 nm range in (**a**). The potential profile along the dashed-white line in (**b**) is shown in (**c**)

8. A detection of capacitance gradient at $2\omega_{elec}$ is often shown for PS/PVAC blend [35–37], in which hydrophobic PS has glass transition T_g at 100 °C and hydrophilic PVAC has $T_g = 45$ °C. The height and d^2C/dZ^2 images of PS/PVAC film on ITO, which were recorded in single-pass operation with force gradient detection at 25 °C and 75 °C, are shown in Fig. 14.

   The topography of the blend is characterized by circular domains of different height and depression compared to a flat matrix, Fig. 14a. The domains are separated from the matrix by the interface rims due to immiscibility of the components. The circular domains, which can be assigned to PVAC due to their selective swelling in humid air and methanol vapor [36], show slightly brighter contrast in the d^2C/dZ^2 map (Fig. 14b). At $T = 75$ °C (above T_g of PVAC), the circular domains have raised and above the matrix level and the surrounding rims have disappeared, Fig. 14c. The d^2C/dZ^2 contrast has intensified (Fig. 14d) making PVAC domains well distinguished. This illustrates the contrast use for compositional imaging.

9. The other comment is related to use of AM-PI mode in single-pass KFM studies, where high AC voltage might deteriorate the contrast of height image as the electric field shifts the probe resonant frequency. This effect is seen in the bottom part of the height image in Fig. 15a when the AC voltage at 3 kHz was raised to 4 V.

   The surface potential contrast in Fig. 15b was the same through the entire image. The deterioration of the height contrast can be avoided if the surface topography is measured with AM-FI mode. In this mode the measurements of the free and set-point amplitudes are always made at the effective resonant frequency, which is tracked with a PLL. Therefore, the height image in Fig. 15c and surface potential map in Fig. 15d accurately represent the changes of topography and surface potential.

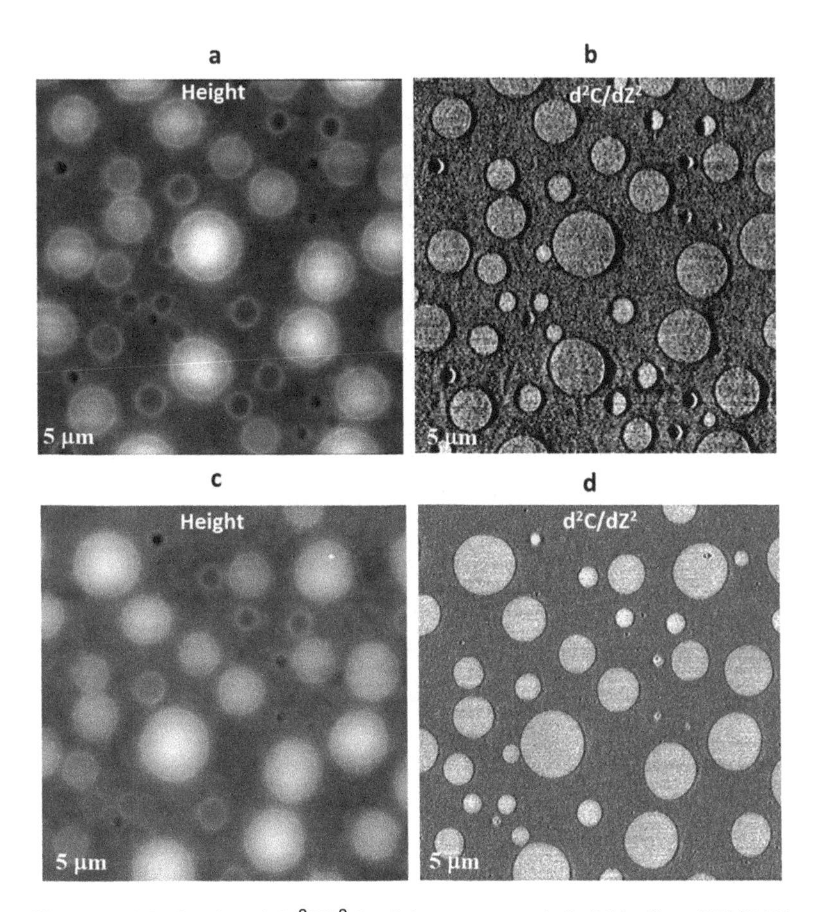

Fig. 14 Height (**a**, **c**) and dC^2/dZ^2 (**b**, **d**) images recorded of thin film of PS/PVAC on ITO glass at 25 and 75 °C. The images were obtained in single-pass operation with force gradient detection

4 Notes

1. In choosing ω_{elec} frequency for excitation of EFM or PFM signals one should have in mind that the response at subcontact resonance frequency is directly related to electrostatic or piezo-force compared to a choice of $\omega_{\mathrm{elec}} = \omega_{\mathrm{contact}}$. The latter choice can be more suitable for a better contrast between locations with different electrical or piezoelectric properties due to the resonance character of such frequency.
2. The probes with special coatings such as diamond or silicide, which are used for conducting AFM, have typically a larger tip apex and can be tried in PFM experiments.
3. In case of hard samples the use of stiffer probes can be beneficial as in PFM they are less subjective to electrostatic force that might "contaminate" the piezoresponse.

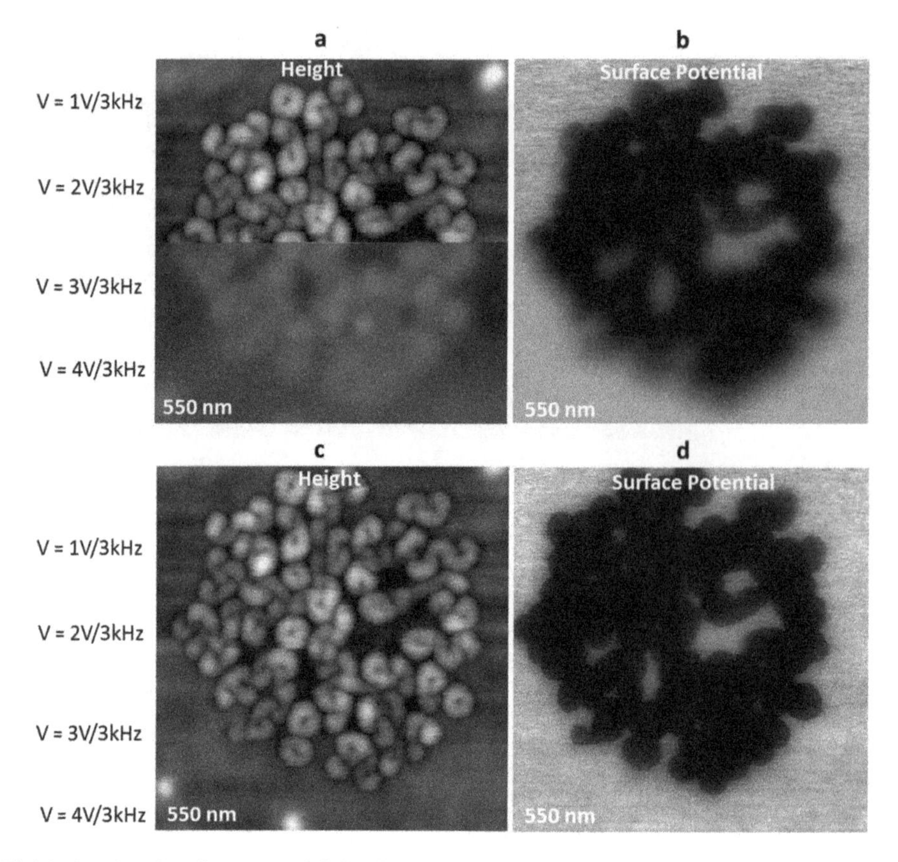

Fig. 15 Height (**a**, **c**) and surface potential (**b**, **d**) images, in topography tracking was made with AM-PI mode for in (**a**) and with AM-FI mode in (**c**). In the top half of the images the AC excitation at 3 kHz mode was performed with 1 V and 2 V. In the bottom half the AC excitation was made with 3 V and 4 V

4. An accurate detection of weak AC electrical signals and, particularly, their phase information demands a proper isolation of the related wires. Our experience shows that this problem is most common for detection of PFM response, where a presence of high-voltage excitation and probe deflection signals in the same cable leads to the unwanted cross talk. This results in incorrect phase changes, for example, on $LiNbO_3$ domains, Fig. 9d. A separation of the related wires, i.e., by applying excitation signal through a separate shielded cable to a sample, can solve this problem.
5. A knowledge of the probe amplitude in oscillatory modes is important in many respects with two listed below. (a) As set-point amplitude in first pass determines a rest position of the probe in the second pass, one can use negative lift of a size just smaller than the set-point amplitude to bring the probe closer to a sample and thus increase the electrostatic tip–sample interactions. (b) Numerical values of the amplitudes help a rational comparison of experimental results with FEA calculations.

6. A given description of lift mode operation refers to introduction of this mode at the time when LIAs were not common for AFM microscope and only one LIA or its substitute were used. Now this mode can be enhanced by adding AC excitations at different frequencies. The double-pass technique can be further diversified by changing a requirement of topography reproduction in the second pass.
7. At a particular DC bias, the phase contrast in second pass is higher for larger amplitudes and for smaller probe–sample separation, which is limited, however, by the probe amplitude. In principle, the optimal amplitude and separation can be found theoretically, but practical trial-and-error method can be good enough.
8. Be aware that the use of frequencies in the 2–5 kHz might be restricted due to the controller noise in this region.
9. Single-pass KFM mode with force gradient detection provides most accurate data with high spatial resolution but it does not completely eliminate the use of double-pass mode. In some cases, the measurements in single-pass mode, in which the probe intermittently contacts the sample, might induce surface charging. Therefore, the application of double-pass mode can be beneficial in such situations, particularly, for control measurements for possible charging.
10. The reported measurements and mapping of capacitance gradient were conducted at low frequency 2–5 kHz below frequency of first eigenmode of conducting probe. In general, the detection of $2\omega_{elec}$ response can be conducted in a broader frequency range extended to MHz [38] and even GHz. A substantial enhancement of the gradient contrast can be achieved by choosing ω_{elec} at half of second eigenmode that enlarge the probe response at $2\omega_{elec}$.
11. We described the extraction of quantitative dielectric permittivity for thin dielectric films on conducting substrate, for more complicated samples the use of FEA might be needed [39].

References

1. Binnig G, Quate CF, Gerber C (1986) Atomic force microscope. Phys Rev Lett 56:930–933
2. Krotil H-J, Stifter T, Waschipky H et al (1999) Pulsed force mode: a new method for the investigation of surface properties. Surf Interface Anal 27:336–340
3. Belikov S, Alexander J, Wall C (2013) Tip-sample forces in atomic force microscopy: Interplay between theory and experiment. MRS Proc 1527. mrsf12-1527-uu02–04
4. Timoshenko S, Young DH, Weaver W Jr (1974) Vibration problems in engineering. Willey, New York
5. Belikov S, Magonov S (2009) Classification of dynamic atomic force microscopy control modes based on asymptotic nonlinear mechanics Proceedings American Control Conference, p 979–985
6. Belikov S, Yermolenko I, Magonov S (2015) Modeling and measurements in atomic force

microscopy resonance modes. Proceedings American Control Conference, p 3484–3489
7. Belikov S, Alexander J, Surtchev M et al (2016) Implementation of atomic force microscopy resonance modes based on asymptotic dynamics using Costas loop. Proceedings American Control Conference, p 6201–6208
8. Zhong Q, Innis D, Kjoller K et al (1993) Fractured polymer/silica fiber surface studied by tapping mode atomic force microscopy. Surf Sci Lett 290:L688–L692
9. Magonov S (2000) AFM in analysis of polymers. In: Meyers RA (ed) Encyclopedia of analytical chemistry. Wiley, Chichester, pp 7432–7491
10. Hölscher H, Schmutz JE, Schwarz UD (2011) Dynamic force microscopy and spectroscopy in ambient conditions: theory and applications. In: Kalinin SV, Gruverman A (eds) Scanning probe microscopy of functional materials: nanoscale imaging and spectroscopy. Springer, New York, pp 71–94
11. Bai M, Trogisch S, Magonov S et al (2008) Explanation and correction of false step heights in amplitude modulation atomic force microscopy measurements on alkane films. Ultramicroscopy 108:946–952
12. Albrecht TP, Gruetter P, Horne D et al (1991) Frequency modulation detection using high *Q* cantilevers for enhanced force microscope sensitivity. J Appl Phys 61:668–673
13. Fukuma T, Ichii T, Kobayashi K et al (1995) True-molecular resolution imaging by frequency modulation atomic force microscopy in various environments. Appl Phys Lett 86:034103–034105
14. Magonov S, Alexander J, Surtchev M (2017) Compositional mapping of bitumen using local electrostatic force interactions in atomic force microscopy. J Microsc 265:196–206
15. Sader JE, Jarvis SP (2004) Interpretation of frequency modulation atomic force microscopy in terms of fractional calculus. Phys Rev B 70:012303–012305
16. Sader JE, Jarvis SP (2004) Accurate formulas for interaction force and energy in frequency modulation force spectroscopy. Appl Phys Lett 84:1801–1803
17. Reid OG, Munechika K, Ginger DS (2008) Space charge limited current measurements on conjugated polymer films using conductive atomic force microscopy. Nano Lett 8:1602–1609
18. Kopanski JJ (2007) Scanning capacitance microscopy for electrical characterization of semiconductors and dielectrics. In: Kalinin S, Gruverman A (eds) Scanning probe microscopy. Springer, New York, NY, pp 88–112
19. Lai K, Ji MB, Leindecker N, Kelly MA et al (2007) Atomic-force-microscope-compatible near-field scanning microwave microscope with separated excitation and sensing probes. Rev Sci Instrum 78:063702. https://doi.org/10.1063/1.2746768
20. Cho S, Kang SD, Kim W et al (2013) Thermoelectric imaging of structural disorder in epitaxial graphene. Nat Mater 12:913–918
21. Martin Y, Abraham DA, Wickramasinghe HK et al (1988) High-resolution capacitance measurement and potentiometry by force microscopy. Appl Phys Lett 52:1103–10005
22. Kholkin AL, Kalinin SV, Roelofs A et al (2007) Review of ferroelectric domain imaging by piezoresponse force microscopy. In: Kalinin S, Gruverman A (eds) Scanning probe microscopy, vol 1. Springer, New York, pp 173–214
23. Nonnenmacher M, O'Boyle MP, Wickramasinghe HK (1991) Kelvin probe force microscopy. Appl Phys Lett 58:2921–2923
24. Inoue T, Yokoyama H (1994) Imaging of surface electrostatic features in phase-separated phospholipid monolayers by scanning Maxwell stress microscopy. J Vac Sci Technol B 12:1569–1571
25. Gomila G, Toset J, Fumagalli L (2008) Nanoscale capacitance microscopy of thin dielectric films. J Appl Phys 104(024315):1–8
26. Belikov S, Alexander J, Magonov S et al (2012) Atomic force microscopy control system for electrostatic measurements based on mechanical and electrical modulation. Proceedings American Control Conference, p 3228–3233
27. Humphries S Jr (1998) Field solutions on computers. CRC Press, Boca Raton, FL
28. Watanabe S, Hane K, Ohye T et al (1993) Electrostatic force microscope imaging analyzed by the surface charge method. J Vac Sci Technol B11:1774–1781
29. Colchero J, Gil A, Baro AM (2001) Resolution enhancement and improved data interpretation in electrostatic force microscopy. Phys Rev B 64(245403):1–11
30. Elings V B, Gurley J A (1994) Scanning probe microscope using stored data for vertical probe positioning. US Patent 5,308,974, 5 Mar 1994
31. Hong JW, Noh KH, Park S et al (1998) Surface charge density and evolution of domain structure in triglycine sulfate determined by electrostatic-force microscopy. Phys Rev B 58:5078–5084
32. Abed AE, Faure M-C, Pouzet E et al (2002) Experimental evidence for an original

two-dimensional phase structure: an antiparallel semifluorinated monolayer at the air-water interface. Phys Rev E 5:051603–051604

33. Magonov S, Alexander J, Wu S (2011) Advancing characterization of materials with atomic force microscopy - based electrical techniques (Chapter 9). In: Kalinin SV, Gruverman A (eds) Scanning probe microscopy of functional materials: nanoscale imaging and spectroscopy. Springer, New York, pp 233–300
34. Zerweck U, Loppacher C, Otto T et al (2005) Accuracy and resolution limits of kelvin probe force microscopy. Phys Rev B 71:125424
35. Riedel C, Sweeney R, Israeloff NE et al (2010) Imaging dielectric relaxation in nanostructured polymers by frequency modulation electrostatic force microscopy. Appl Phys Lett 96 (213110):1–3
36. Magonov S, Alexander J (2012) Single-pass kelvin force microscopy and d*C*/d*Z* measurements in the intermittent contact: applications to polymer materials. Beilstein J Nanotechnol 2:15–27
37. Magonov S, Alexander J, Belikov S (2012) Exploring surfaces of materials with atomic force microscopy(Chapter 7). In: Korkin A, Lockwood DJ (eds) Nanoscale applications for information and energy systems, Nanostructure Science and Technology. Springer, New York, pp 203–253
38. Fumagalli L, Esteban-Ferrer D, Cuervo A et al (2012) Label-free identification of single dielectric nanoparticles and viruses with ultraweak polarization forces. Nat Mater 11:808–816
39. Gramse G, Dols-Peres A, Edwards MA et al (2013) Nanoscale measurements of the dielectric constant of supported lipid bilayers in aqueous solutions with electrostatic force microscopy. Biophys J 104:1257–1262

Chapter 8

Supported Lipid Bilayers for Atomic Force Microscopy Studies

Zhengjian Lv, Siddhartha Banerjee, Karen Zagorski, and Yuri L. Lyubchenko

Abstract

Nanoimaging methods, atomic force microscopy (AFM) in particular, are widely used to study the interaction of biological molecules with the supported lipid bilayer (SLB), which itself is a traditional model for cellular membranes. Success in these studies is based on the availability of a stable SLB for the required observation period, which can extend several hours. The application of AFM requires that the SLB have a smooth morphology, thus enabling visualization of proteins and other molecules on its surface. Herein, we describe protocols for SLB assembly by using 1-palmitoyl-2-oleoyl-sn-glycero-3-phosphocholine (POPC) and 1-palmitoyl-2-oleoyl-sn-glycero-3-phospho-L-serine (POPS) on a mica support. Our methodology enables us to assemble defect-free POPC and POPS SLBs that remain stable for at least 8 h. The application of such smooth and stable surfaces is illustrated by monitoring of the on-surface aggregation of amyloid proteins with the use of time-lapse AFM.

Key words Supported lipid bilayer, Nanoimaging, Atomic force microscope, Time-lapse imaging, Amyloid aggregation

1 Introduction

Cell membranes are essential components where various cellular processes take place and essentially act as the site of communication between the intra- and extracellular environments [1–4]. To understand cellular events that occur on the surfaces, researchers have developed many in vitro model systems to mimic the milieu of the cell, like bicelles [5], micelles [6], vesicles or liposomes [7], supported lipid bilayers (SLB), and planar lipid bilayers [8–10], among which SLBs have gained interest due to their relatively simple preparation and compatibility with a variety of surface-based techniques [10, 11].

Nanoimaging with atomic force microscope (AFM) allows researchers to directly monitor on-membrane events in a near-physiological environment. However, obtaining a stable,

Yuri L. Lyubchenko (ed.), *Nanoscale Imaging: Methods and Protocols*, Methods in Molecular Biology, vol. 1814,
https://doi.org/10.1007/978-1-4939-8591-3_8,

homogeneous, and relatively defect-free SLB surface is essential for any reliable experiments that mimic events on cell membranes. A smooth, relatively vesicle-free surface is also critical for probing the interaction of proteins or small molecules with the surface of the SLB. Herein, we describe methods for the preparation and characterization of SLB surfaces that meet these criteria by using the following lipids on a mica surface: 1-palmitoyl-2-oleoyl-sn-glycero-3-phosphocholine (POPC) and 1-palmitoyl-2-oleoyl-sn-glycero-3-phospho-L-serine (POPS). Notably, the tapping mode of AFM in buffer medium, where the tip gently scans the biological sample [12, 13], is used to directly characterize the topography and stability of the SLB surfaces. The results show that the homogenous, defect-free SLB surfaces constructed by using this technique remain stable for as long as 3 days, whereas SLB surfaces with small defects can be stable for at least 8 h. Time-lapse imaging was employed to visualize the in situ interaction of amyloid proteins with these SLB surfaces in real time.

2 Materials

2.1 Reagents for SLB Preparation

1. POPC and POPS: Both lipids were purchased from Avanti Polar Lipids Inc. (Alabaster, Alabama, USA). POPC and POPS samples come in sealed glass ampules (25 mg), which should be stored in −20 °C. All the methods and results presented herein were obtained using these samples. Notably, there are a few other companies that also provide lipid molecules, including Matreya, LLC (College, Pennsylvania, USA), and Lipoid LLC (Newark, New Jersey, USA).
2. Chloroform. HPLC grade (amylene and ethanol stabilized).
3. Sodium phosphate buffer (10×): To prepare 100 mM sodium phosphate buffer (pH 7.4), dissolve 3.1 g of $NaH_2PO_4 \cdot H_2O$ and 10.9 g of Na_2HPO_4 in water. Next, filter the solution filtered through a disposable Millex-GP syringe filter unit (0.22 μm) before use. All buffer solutions should be prepared with deionized (DI) water (18.2 MΩ at room temperature).

2.2 Instruments and Related Accessories

1. AFM instruments. The following AFM instruments were used due to their availability: NanoScope MultiMode 8 system (Bruker, Santa Barbara, California, USA) and MFP-3D AFM (Oxford Instrument Asylum Research, Santa Barbara, California, USA).
2. MSNL AFM cantilevers E and F (Bruker, Santa Barbara, California, USA).
3. AFM specimen discs (Ted Pella, Inc., Redding, California, USA).

4. 2 mL glass vial and glass pipette (Fisher Scientific, Waltham, Massachusetts, USA).
5. Micro glass slides (Allegiance Healthcare Corporation, McGaw Park, Illinois, USA).
6. Muscovite mica (Asheville Schoonmaker Mica Co., Newport News, Virginia, USA).
7. Texwipe TX604 wipers (Texwipe, Kernersville, North Carolina, USA).
8. Ruban Invisible Tape (Staples Office Superstore, Framingham, Massachusetts, USA).
9. Double-sided sticky black, conductive carbon tape (Ted Pella, Inc., Redding, California, USA).
10. Aron Alpha Industrial Krazy Glue™ (Toagosei America, West Jefferson, Ohio, USA).
11. Ultrasonicator—Branson 1210 (Branson Ultrasonics, Danbury, Connecticut, USA).
12. Vacuum oven—VWR Sheldon, Model 1400E (J&M Scientific, Woburn, Massachusetts).
13. ImmEdge Hydrophobic Barrier Pen (Vector Laboratories, Inc., Burlingame, California, USA).
14. Dry bath incubator (Fisher Scientific, Waltham, Massachusetts, USA).

3 Methods

3.1 Preparation of POPC SLB

1. Open the mouth of the glass ampule containing 25 mg of POPC and then add 1 mL chloroform. This is the stock solution. Always use a glass pipette (*see* **Note 1**).
2. Pipette 20 μL of the above solution into a glass vial and evaporate the chloroform. First, gently purge the vial by using argon flow, and then place the vial in a vacuum chamber overnight, or for 8 h, to ensure complete drying.
3. To prepare a 0.5 mg/mL solution of POPC, add 1 mL of 10 mM sodium phosphate buffer (pH 7.4), and then sonicate the solution until it becomes clear (*see* **Note 2**).
4. Using Krazy™ glue, paste a piece of mica (1.0 × 1.0 cm) onto a glass slide and allow it to dry for 30 min. After drying, draw a line along edges of the mica with a hydrophobic pen and again allow it to dry for 10 min (*see* **Note 3**).
5. Once sonication is complete from **step 3**, cleave the mica by using scotch tape until a clear layer of mica is visible on the tape.

6. Add ~200 μL of the 0.5 mg/mL POPC solution onto the mica, and place it in a dry heating bath at 60 °C for 1 h (*see* **Note 4**).
7. Add buffer periodically to prevent sample drying.
8. After incubating for 1 h in the heating bath, remove the sample from the heating plate, allow it to reach room temperature, and then rinse the surface gently, but thoroughly, with the 10 mM sodium phosphate buffer (pH 7.4).
9. Add ~100 μL of 10 mM phosphate buffer (pH 7.4) to the newly prepared POPC SLB surface, and immediately place it on the AFM stage for imaging.

3.2 Preparation of POPS SLB

1. Add 1 mL of chloroform immediately after opening the mouth of the glass ampule containing 25 mg of POPS powder. Transfer the solution to the glass vial. This is the stock solution. Always use a glass pipette (*see* **Note 1**).
2. To prepare the working solution, use a glass pipette to extract 20 μL of the stock solution and place it into a new glass vial. Completely evaporate the chloroform from the working solution by first applying an argon stream and then leaving the sample in vacuum chamber overnight.
3. Add 1 mL of the 10 mM sodium phosphate buffer (pH 7.4) to prepare the 0.5 mg/mL POPS solution, which will be used for SLB preparation. Then sonicate the POPS solution for 30–40 min at room temperature until the solution becomes clear.
4. Repeat the **steps 4** and **5** in Subheading 3.1.
5. Add ~150 μL of the 0.5 mg/mL POPS solution onto the freshly cleaved mica surface attached to a glass slide.
6. Incubate for 1 h at 60 °C in a dry heating bath.
7. Repeat **steps 7–9** described in Subheading 3.1.

3.3 Testing of the Surfaces

3.3.1 Ex Situ AFM Imaging of POPC SLB Prepared at Room Temperature

The formation of the SLB surface involves a few characteristic steps, including the deposition of vesicles on a solid support (Fig. 1a), followed by deformation and rupture of the vesicles (Fig. 1b), and then the transformation of vesicles into lipid patches (Fig. 1c), which coalesce and reorganize into larger patches (Fig. 1d) [8, 14]. These steps can be directly visualized by ex situ AFM imaging described in the following steps.

1. Perform **steps 1–3** in Subheading 3.1.
2. Deposit 30 μL of POPS solution onto a freshly cleaved mica surface, which has been mounted on a metal disc using a double-sided tape.

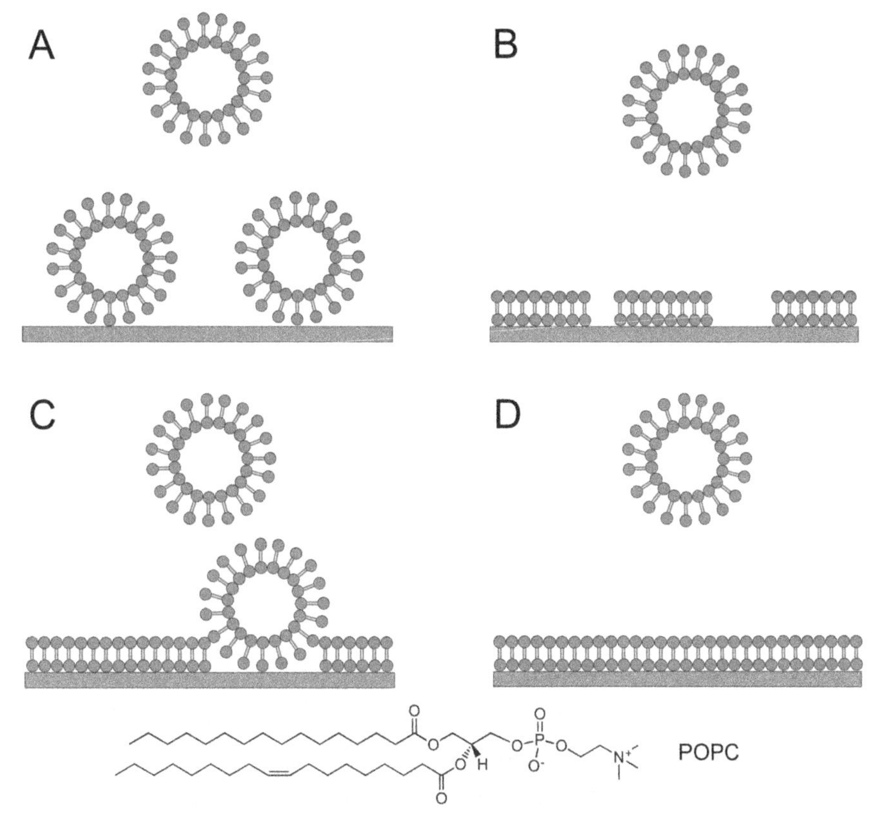

Fig. 1 Proposed formation process of phospholipids onto mica surface. (**a**) Deposition/adhesion, vesicles deposit onto mica over time; (**b**) rupture, vesicles rupture and form patches; (**c**) coalescence of lipid patches; adjacent lipid patches merge to form larger patches while some vesicles are trapped; (**d**) completion, trapped vesicles and vesicles from solution fix the SLB, resulting in homogenous SLB. Objects are out of scale. Water layer between SLB and mica surface is omitted for simplicity. The chemical structure of POPC is shown at the bottom

3. Incubate the POPC solution at room temperature in a humid environment, which is created by placing wet paper towels at the bottom of a petri dish and covering it with the lid.
4. After the desired incubation times are achieved, wash excess POPC by thoroughly, but gently, exchanging the POPC solution with the sodium phosphate buffer. Keep the subsequent POPC SLB in buffer in a humid chamber.
5. Perform imaging of the samples in liquid by using a MultiMode 8 AFM equipped with the PeakForce-HR (High Rate) mode (*see* **Note 5**). Use the MSNL cantilever E, with a nominal spring constant of 0.1 N/m.
6. Initially (i.e., after 1 min incubation), only the vesicles are observed on the surface (Fig. 2a). After 3 min of incubation, the number of vesicles deposited on the surface is increased (Fig. 2b). Vesicles start forming small patches after 6 min

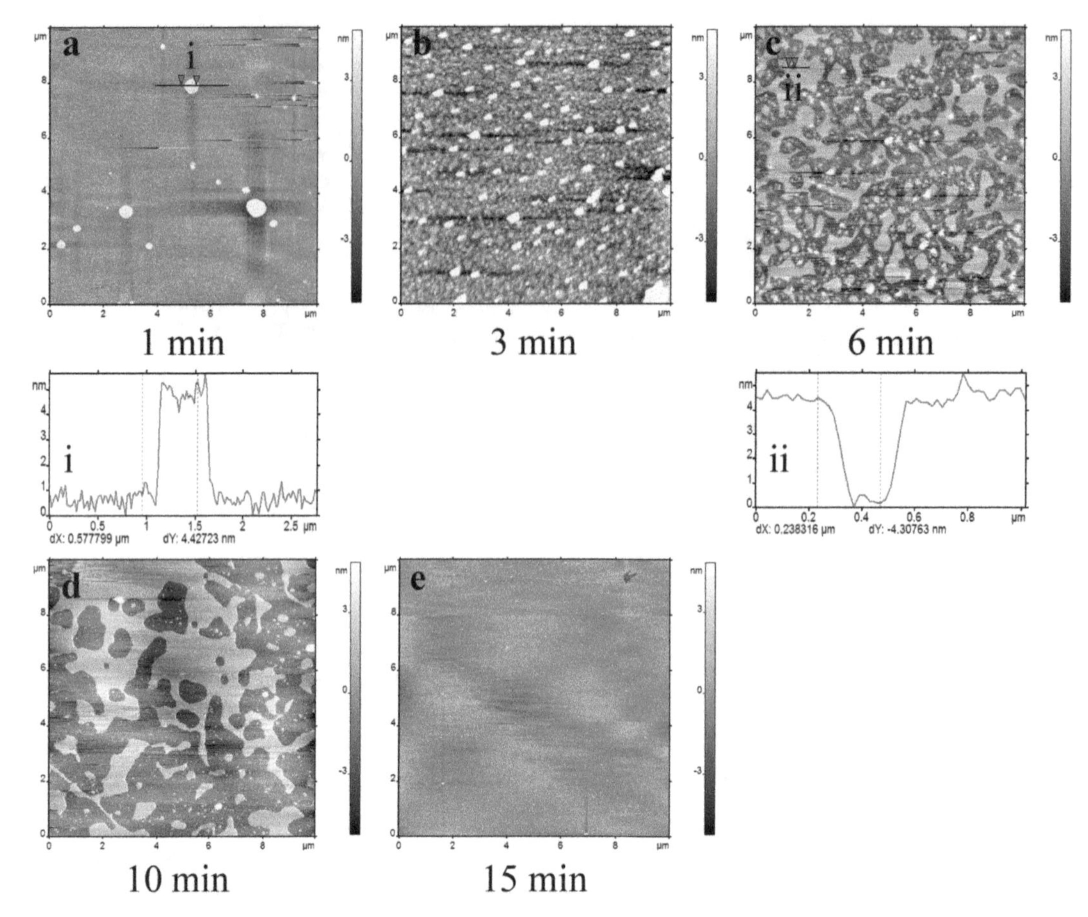

Fig. 2 Ex situ time-lapse of the formation of POPC SLB. (**a–e**) are AFM images taken at different time periods. At 1 min (**a**), there are a few vesicles on the substrate, evidenced by the cross section (i) of the round-shaped feature indicated a black straight line. At 3 min (**b**), more vesicles deposit on the substrate. Most vesicles transformed into lipid patches/islands (6 min, plate (**c**)). A cross section shows the height of the lipid islands are about 4.3 nm (ii). The vesicles then coalesce into large patches (**d**, 10 min). After incubated for 15 min (**e**), almost full coverage of SLB has been achieved. Images are 10 $\times$ 10 μm. The concentration of POPC was 0.5 mg/mL

(Fig. 2c). These patches coalesce to form bigger patches (Fig. 2d), and after 15 min of incubation, large lipid patches are obtained (Fig. 2e).

3.3.2 In Situ Time-Lapse AFM Imaging of the Formation of the POPC SLB at Room Temperature

In situ time-lapse AFM imaging provides the unique opportunity to monitor the same surface area for a considerable period of time. Further, this technique enables one to directly visualize the formation of the POPC SLB on the mica surface in real time. Detailed methods for this procedure are described below.

1. For the MultiMode 8 AFM, a metal disc can be used to mount the mica substrate onto the sample stage. The MSNL cantilever F, with a nominal spring constant of 0.6 N/m, can

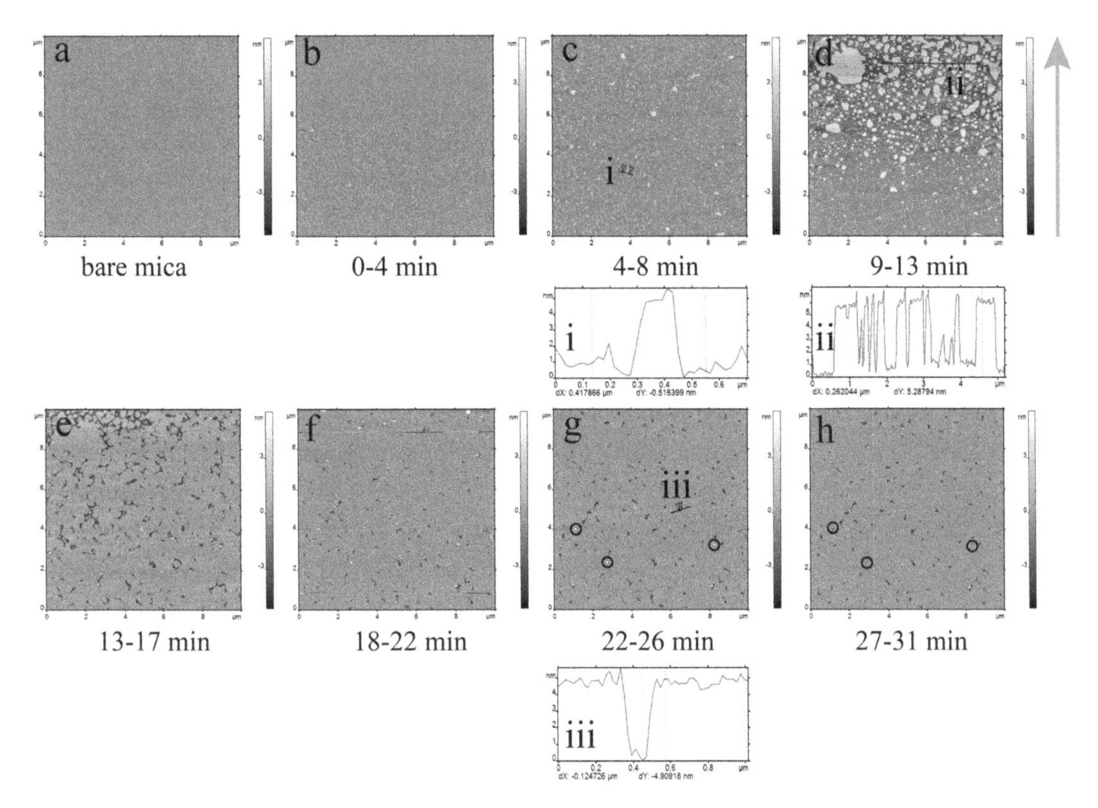

Fig. 3 In situ time-lapse AFM imaging of 5 μg/mL on freshly cleaved bare mica. (**a**) Prior to adding of POPC, bare mica surface is imaged which is very smooth. (**b–h**) are in situ time-lapse images showing the process of the formation of a POPC SLB. At 0–4 min, there are only a few small vesicles, if there are any (**b**). More and more vesicles deposit onto the substrate (**c**). Vesicles rupture into patches (**d**). Lipid patches coalesce into uniform SLB with small defects (**e**). Trapped vesicles or vesicles from the solution fix packing defects (**f–h**). i, ii, and iii are cross sections indicated with black solid lines at **c**, **d**, and **g**, respectively. Green arrow to the right of frame "**d**" shows the scanning direction

be used to image in liquid. The sample stage is surrounded by rubber to prevent the leakage of liquid. On top of the sample stage, a fluid cell is mounted in an inverted fashion. This fluid cell has an inlet and an outlet, both of which allow for exchange of buffer in situ and are each connected to a disposable 1 mL syringes. Buffer exchange is achieved by streaming fresh buffer into the inlet while evacuating the existing buffer through the outlet.

2. First, an image of the mica support alone should be acquired to ensure that the substrate is clean (Fig. 3a).
3. Add 5 μg/mL of POPC solution (*see* **Note 6**); images are to be taken immediately after injecting the POPC solution (Fig. 3b). The number of vesicles is increased after 8 min of imaging (Fig. 3c) and starts forming patches after 13 min of imaging

(Fig. 3d). Patches coalesce to form the large patch with small defects (Fig. 3e, f). The circled features in Fig. 3g, h indicate how the vesicle, which is trapped in the bilayer or appears from the solution, fixes the defects. The height of the patches (4–5 nm) obtained from the cross-section profiles indicates the formation of a single bilayer (Fig. 3i–iii).

4. Use PeakForce mode to image (*see* **Note 5**). The imaging parameters will be automatically adjusted by the NanoScope 8 software to ensure the best optimal imaging quality.

3.3.3 Heating in Order to Improve Homogeneity of POPC SLB Surface

1. Notably, preparation of bilayers at room temperature usually results in defects. In turn, SLB surfaces prepared at elevated temperature (60 °C) provide homogeneous bilayers with complete coverage.
2. Follow the steps mentioned in Subheading 3.1 (*see* **Note 7**).
3. In the case of preparing the POPC SLB at an elevated temperature (60 °C), a cooling step is recommended in order to create more homogeneous SLB surfaces. Extensive exchange of the lipid solution with buffer is suggested to remove any excess vesicles. It is also important to note that the POPC SLB should not be harshly rinsed; this force can potentially damage the bilayer surface.
4. After gently rinsing with the sodium phosphate buffer, add 100 μL of the same buffer on top of the prepared SLB, and keep the sample in a humid chamber until imaging (*see* **Note 8**). Samples should be imaged as soon as possible after the rinsing. AFM can be used to monitor the homogeneity of the POPC SLB by scanning a relatively large area of the surface. Figure 4a demonstrates a defect-free, large patch of a POPC SLB. A larger area (80 × 80 μm) is scanned to locate any defects (Fig. 4b). A hole/defect and a second layer of bilayer is found, whose depth and height are shown in the cross-section profiles (Fig. 4i and ii).
5. This increased homogeneity due to the incubation at elevated temperature is also true for POPS SLB formation.

3.4 Validation of Homogeneity of the POPC SLB Surface

1. Nano-scratching or nano-shaving is a suitable method to characterize the homogeneity and the number of layers formed during the SLB preparation. These experiments are performed using a MultiMode 8 AFM. Prior to nano-scratching, the sample is scanned to acquire a reference image using the Peak-Force Tapping mode in liquid. The AFM tip is then retracted and engaged again with the contact mode.
2. Scratching is performed using the MSNL cantilever F, with a nominal spring constant of 0.6 N/m. The applied force is 8–10 nN (*see* **Note 9**). A small surface area of 500 × 500 nm

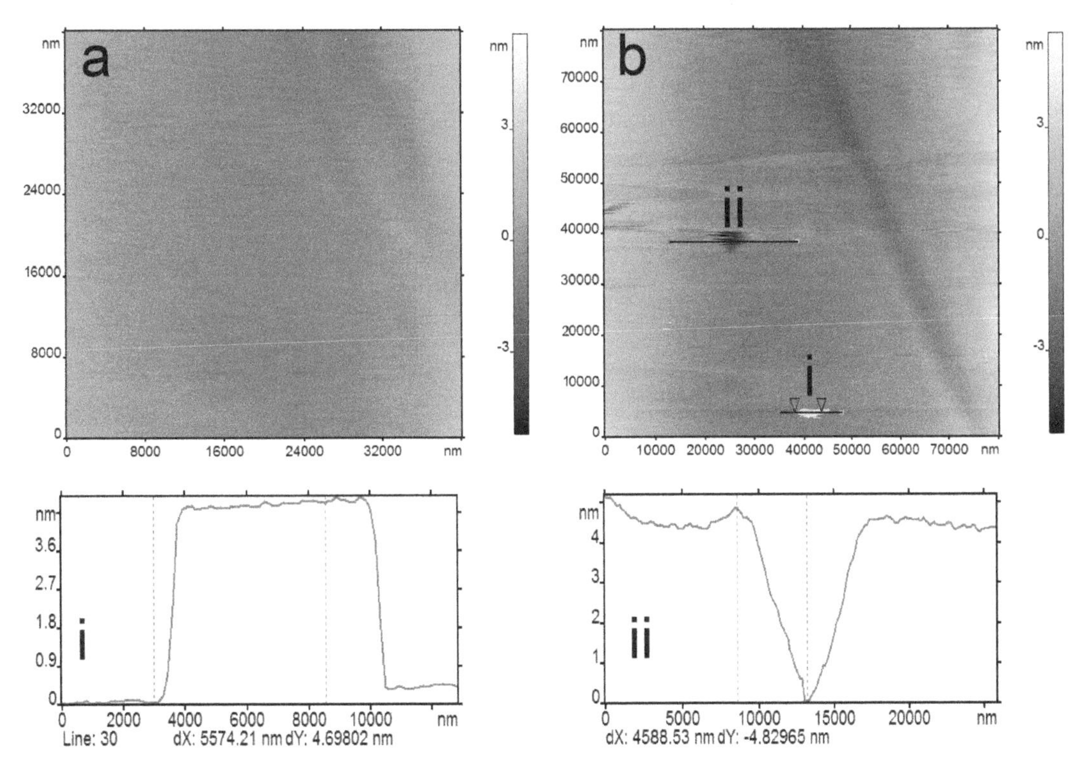

Fig. 4 AFM images of a POPC SLB prepared by incubating 0.5 mg/mL at 60 °C for 1 h. (**a**) A 20 × 20 μm image shows uniform coverage with no packing defects. (**b**) shows a zoomed-out view of the same area of 80 × 80 μm, which shows a large packing defect and a second layer on top of the POPC SLB. Cross sections show the height (i) and the depth (ii) of the packing defect and the second layer, respectively

is selected on the bilayer, and then the scratching is performed at scanning rate of 44 Hz for 3–5 min, depending on the sharpness of the tip.

3. After completion of scratching, the tip is retracted, and the imaging mode is switched back to PeakForce imaging. The same area is imaged again to inspect if there is a scratched hole with an expected size and height or depth (Fig. 5a). Figure 5b shows the 3D projection of the scratched area squared on the SLB (Fig. 5b). In this case, the height of the shaved area (~4.0 nm) indicates the formation of one bilayer (Fig. 5i). The scratching experiment can also be performed after 48 hours of formation of the SLB (Fig. 5c, d). The depth of the scratched area is ~5.0 nm, which is typical for a SLB (Fig. 5ii).

3.5 Stabilities of the POPC SLB Surface

1. To examine the stability of the obtained POPC SLBs, one can use the MFP3D AFM in the tapping mode and the MSNL cantilever E, with a nominal spring constant of 0.1 N/m, in liquid. The driving frequency will vary between 7 and 9 kHz. The drive amplitude is set to 1.5 V, and the setpoint is ~0.6–0.8 V (*see* **Note 10**). 10 mM sodium phosphate buffer

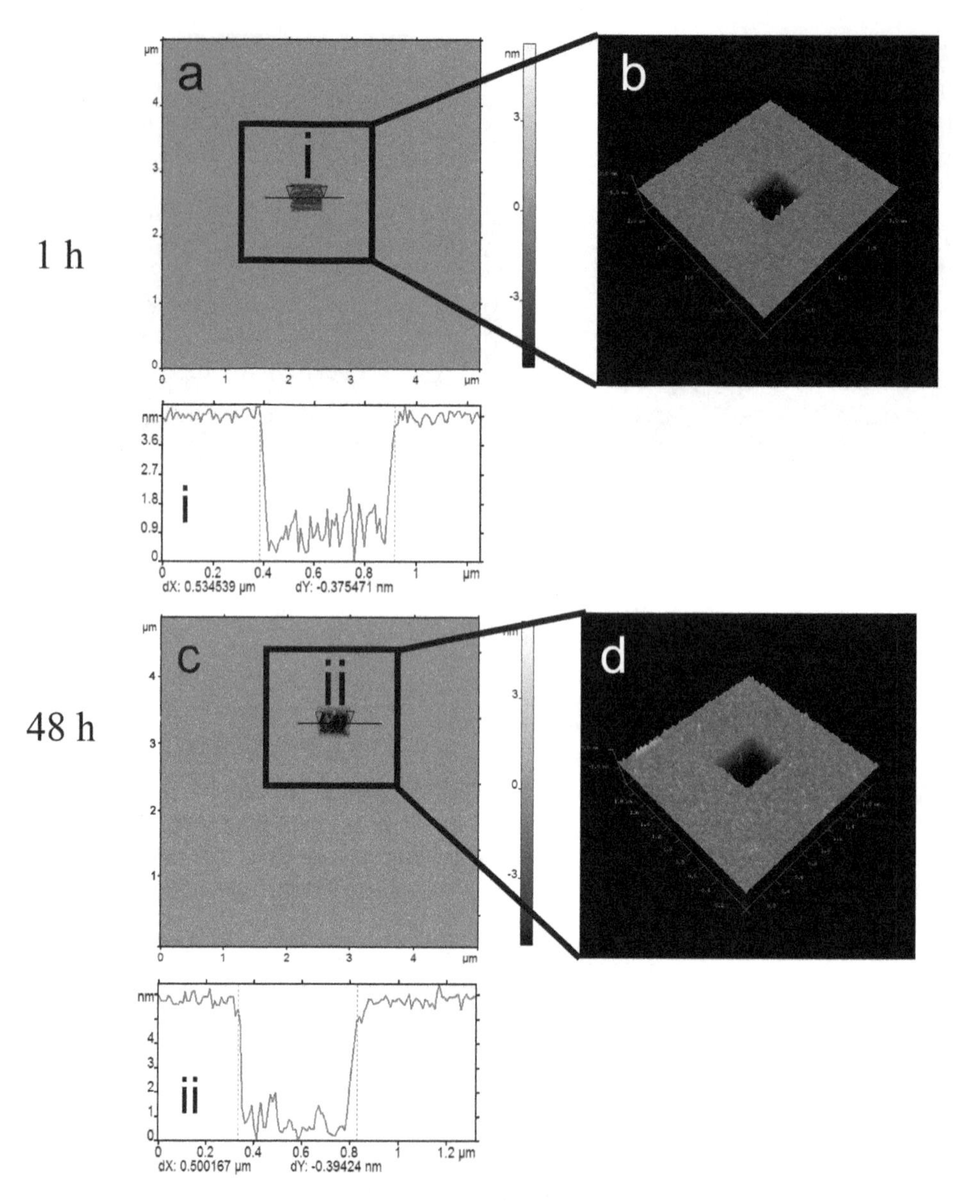

Fig. 5 Characterization of POPC SLB morphology and homogeneity. Nano-lithography was conducted on freshly prepared homogenous sample (**a**). A square with size of 500 × 500 nm can be clearly seen. The zoom-in 3D display shows the shape and depth of the hole (**b**). The cross section shows the hole is ~500 nm wide and ~4 nm deep (i). Nano-lithography was also performed on the homogenous sample after 48 h (**c**). A square sized in 500 × 500 nm can be clearly seen. The zoom-in 3D display shows the shape and depth of the hole (**d**). The cross section shows the hole is ~500 nm in width and ~5 nm in depth (ii)

is to be injected periodically to keep the sample wet and the instrument stable.

2. The morphology of defective POPC SLB surfaces can be imaged by using AFM (Fig. 6). The stability, in terms of disrupted area, has been successfully captured (*see* **Note 11**). Figure 6a shows the topographic image of a POPC SLB with

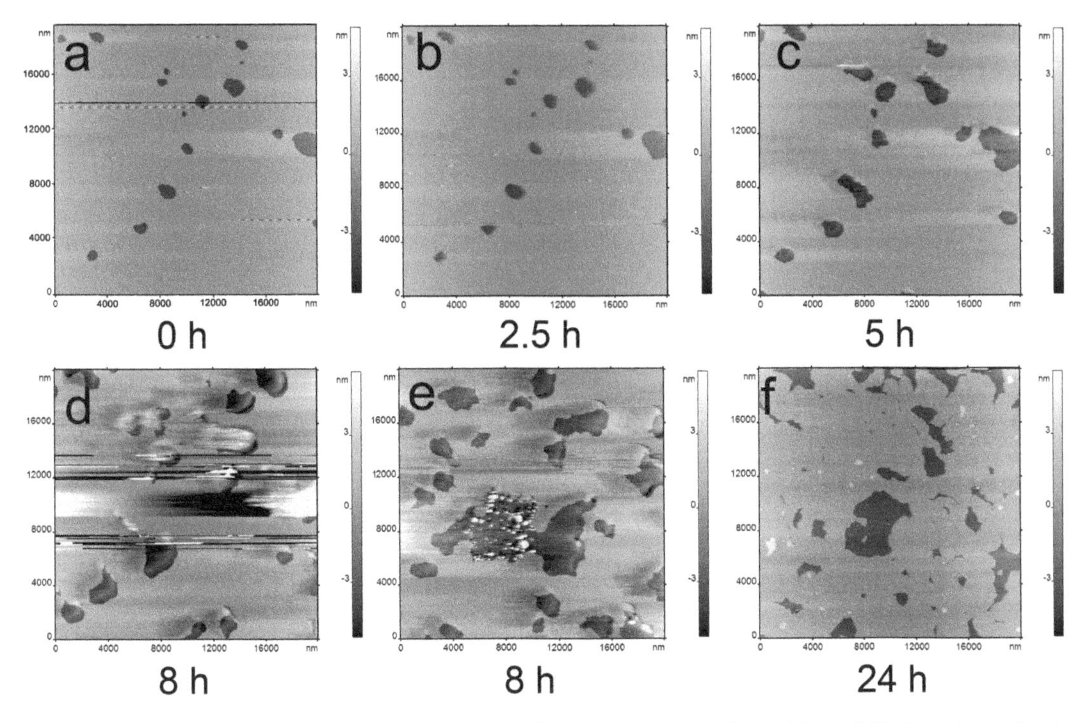

Fig. 6 In situ time-lapse AFM images (**a–d**) of a POPC SLB with packing defects. The stability test shows that it is stable for 2.5 h (**b**). After 2.5 h, the defects start to grow (**c**) and result in complete damage of the bilayer (**d**). An image taken from a different spot after 8 h is shown in (**e**). The bilayer was imaged again after 24 h (**f**). Images are 20×20 μm

defects. The bilayer surface remains stable up to 2.5 h (Fig. 6b). After 2.5 h, the defects begin to grow (Fig. 6c) and finally disrupt the bilayer with large defects (Fig. 6d, e). The same bilayer sample can be imaged again after 24 h, which also shows large holes/defects (Fig. 6f).

3. In contrast to SLB surfaces formed at room temperature, these same types of bilayers formed under elevated temperatures (*see* Subheading 3.3.3) not only form homogeneous and relatively defect-free bilayers but are also more stable (Fig. 7). Figure 7a shows a large area (i.e., 20×20 μm) of a defect-free SLB surface. This same area was monitored using time-lapse imaging, which shows that the SLB remained stable up to 11 h (Fig. 7b–d). AFM images have been captured from several other areas after 24, 48, and 72 h (Fig. 7e–g), which did not show any indication of disruption of the SLB surface. After 96 h, the surface started forming some defects, which are shown by black circles (Fig. 7h).

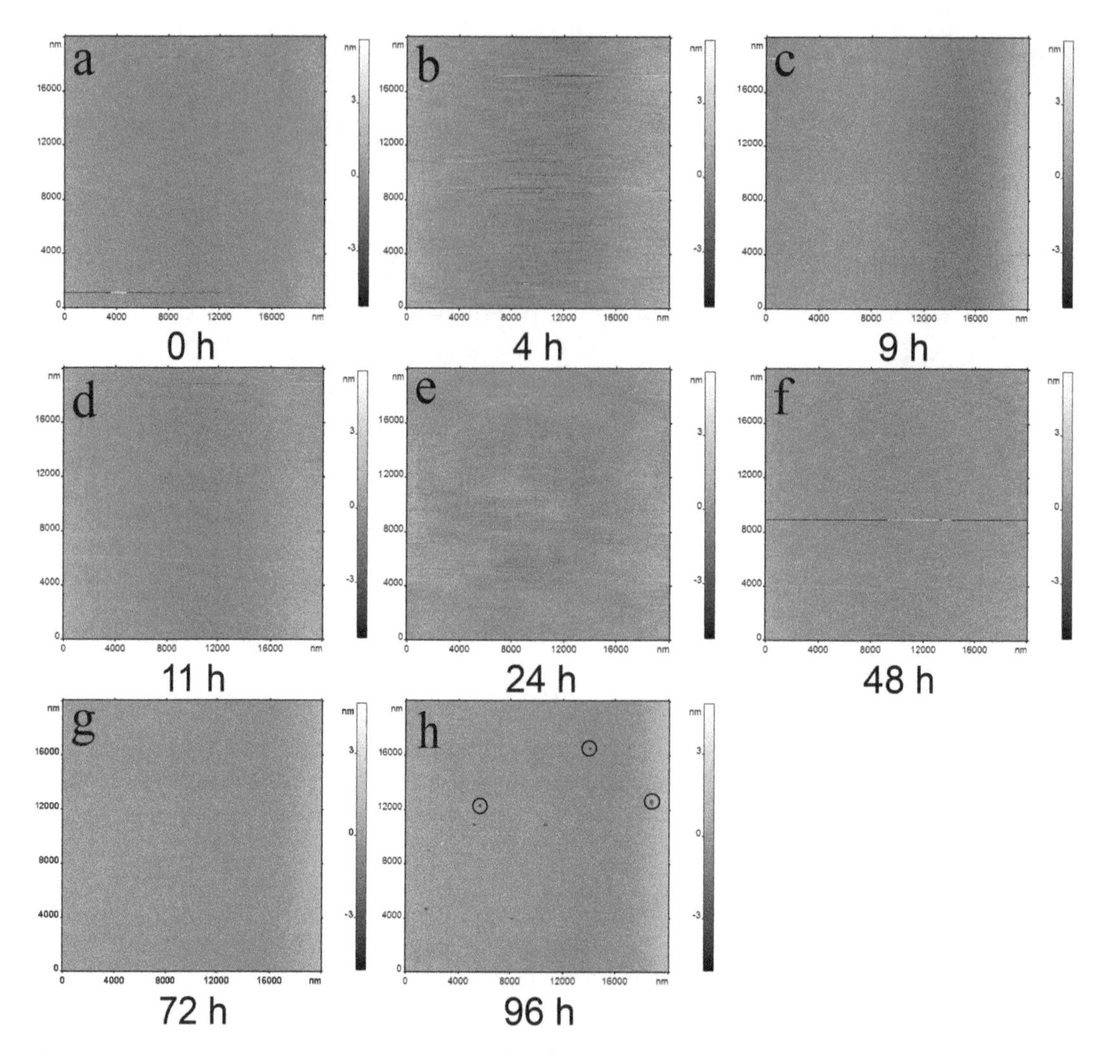

Fig. 7 In situ time-lapse AFM characterization of POPC SLB prepared by incubating 0.5 mg/mL at 60 °C for 1 h. (**a–c**) Images are taken during first 9 h. The POPC shows no packing defects. The stability test shows that it is stable for 11 h (**d**). Representative images from different spots after 11 h demonstrate the lipid is stable for 72 h (**e–g**). The bilayer starts to disintegrate after 96 h (**h**). Defects are highlighted with black circles. Images are 20 × 20 μm

3.6 Observation of Interaction of α-Syn with POPS SLB Surface

These smooth, stable, and defect-free POPS SLB surfaces were used as model membrane systems to probe the interaction of amyloid proteins with bilayers. The POPS SLB surface was prepared as described in Subheading 3.2, and a solution of 10 nM α-synuclein (α-syn) protein was placed above the bilayer. Figure 8a shows AFM images taken just after the tip approached the surface. Here, a smooth, homogeneous, relatively vesicle-free surface is essential because otherwise the vesicles could be mistaken for protein aggregates. The α-syn aggregates appear as clear, small,

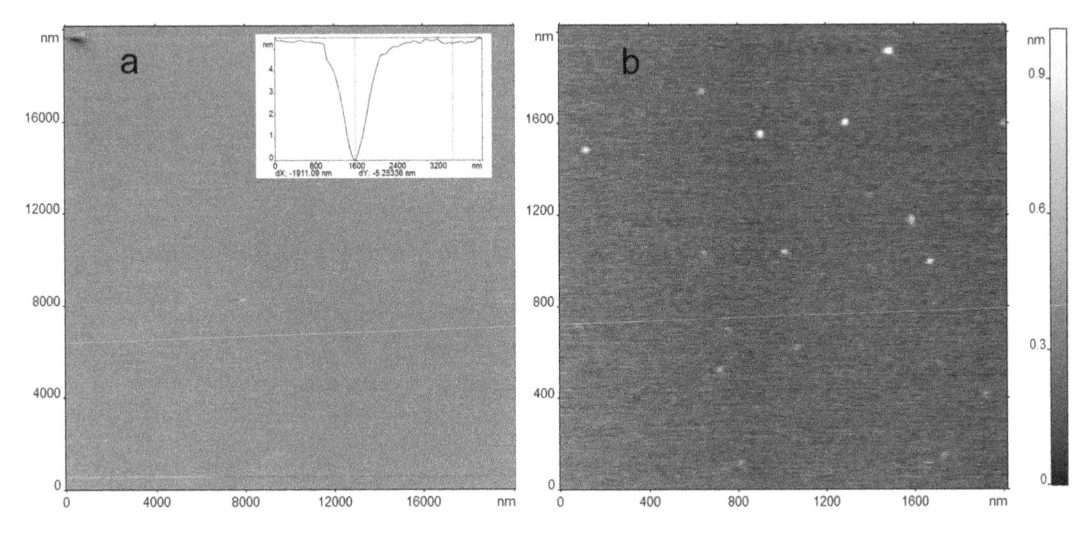

Fig. 8 (**a**) AFM image of a 20 × 20 μm area of POPS SLB prepared by incubating 0.5 mg/mL lipid in phosphate buffer at 60 °C for 1 h. A single packing defect is visible in the upper left-hand corner of the scanned image. Cross-section profile in the inset demonstrates the expected height for a packing defect. (**b**) α-Syn aggregates on POPS bilayer after 1 h

globular features after 1 h of incubation of protein on the POPS SLB (Fig. 8b). Thus, the POPS SLB assembled, as described above, can be used to monitor the aggregation of the amyloid proteins in situ using time-lapse imaging.

4 Notes

1. Lipids solution should always be stored in a glass container with a Teflon closure under −20 °C. Glass pipette/tips are preferred for handling lipid solutions because chloroform solutions leech contaminants from plastics.
2. The appropriate time to stop ultrasonication is when the solution becomes clear, and there are no visible lipid flakes. To reduce hydrolysis, it is necessary to maintain temperature not more than 5–10 degrees above transition temperature of the lipid.
3. For MFP 3D heat- and humidity-resistant double-sided tape can be used instead of glue, for example, "Atemto PET Acrylic Double Sided Adhesive Sticker Tape." This may also be useful, if mica needs to be attached post SLB formation, when volatile compounds from glue are unacceptable.
4. During heating stage of sample preparation, it can be placed into a humidified chamber (e.g., petri dish with wet tissue) to minimize buffer evaporation; however, liquid meniscus must

still be monitored, since some evaporation will happen anyway, and if meniscus was too small it may dry out.

5. The AFM cantilever operated under the regular tapping mode may also induce the rupture of vesicles. The PeakForce module is an excellent option here because of its low invasiveness and fast scan rate.
6. The reasoning behind using 5 μg/mL instead of 0.5 mg/mL is that increased concentration results in faster rupture of vesicles, thus causing difficulty in capturing the initial stages of bilayer formation. The deposition of vesicles on the mica support is diffusion-dependent. In this case, a reduced concentration is preferred.
7. A transparent glue is very useful for this experimental setup. In particular, when aligning the laser on an asylum MFP 3D AFM, this type of glue gives one a clear bottom view of the cantilever. Nontransparent glue also works but requires additional adjustments, in which a transparent glass slide needs to be used to align the laser before mounting the sample.
8. While a defect-free membrane can be stable for as long as 72 h, the membrane "aging" and degradation process starts regardless of the start of scanning. Lipids that form the bilayer start hydrolyzing and forming lysolipids as soon as they are exposed to water; therefore, it is best to start imaging the SLB immediately after preparation.
9. In our case while using NanoScope MultiMode 8 system, the drive amplitude is 3 V for a brand-new sharp tip and 8 V for a worn wear tip. The voltage can be converted to force if the spring constant and deflection sensitivity of the cantilever are both known.
10. The scanning rate can greatly vary depending on multiple factors: variable tip quality and frequency, gain, drive amplitude, surface smoothness, and adhesion. Typically, a 90 × 90 μm SLB with no other components can be scanned at 0.7 Hz with MSNL-E or MSNL-A and at 1 Hz with MSNL-F. Smaller areas, like 15 × 15 μm and below, can be scanned at 2–4 Hz.
11. The AFM tip is placed on idle (electronically retract once, about 20 μm above the sample) to let it exert minimum influence on the sample. When the sample needs to survive overnight, it can be taken out of the sample stage and occluded with a cap that contains a wet Texwipe wiper. This treatment prevents the sample from drying and is able to preserve the sample in aqueous solution for prolonged time.

Acknowledgments

The work at University of Nebraska Medical Center was supported by grants from the National Institutes of Health (NIH) GM096039, GM118006, and NS101504 to Y.L.L. We thank Jean-Christophe Rochet (Purdue University) for providing us with the α-syn protein.

References

1. Coskun Ü, Simons K (2011) Cell membranes: the lipid perspective. Structure 19 (11):1543–1548. https://doi.org/10.1016/j.str.2011.10.010
2. Ingólfsson HI, Melo MN, van Eerden FJ, Arnarez C, Lopez CA, Wassenaar TA, Periole X, de Vries AH, Tieleman DP, Marrink SJ (2014) Lipid organization of the plasma membrane. J Am Chem Soc 136 (41):14554–14559. https://doi.org/10.1021/ja507832e
3. Nicolson GL (2014) The fluid-mosaic model of membrane structure: still relevant to understanding the structure, function and dynamics of biological membranes after more than 40 years. Biochim Biophys Acta 1838 (6):1451–1466. https://doi.org/10.1016/j.bbamem.2013.10.019
4. Simons K, Toomre D (2000) Lipid rafts and signal transduction. Nat Rev Mol Cell Biol 1 (1):31–39. https://doi.org/10.1038/35036052
5. Durr UH, Gildenberg M, Ramamoorthy A (2012) The magic of bicelles lights up membrane protein structure. Chem Rev 112 (11):6054–6074. https://doi.org/10.1021/cr300061w
6. Ahmad Z, Shah A, Siddiq M, Kraatz H-B (2014) Polymeric micelles as drug delivery vehicles. RSC Adv 4(33):17028–17038. https://doi.org/10.1039/C3RA47370H
7. Akbarzadeh A, Rezaei-Sadabady R, Davaran S, Joo SW, Zarghami N, Hanifehpour Y, Samiei M, Kouhi M, Nejati-Koshki K (2013) Liposome: classification, preparation, and applications. Nanoscale Res Lett 8(1):102. https://doi.org/10.1186/1556-276x-8-102
8. Richter RP, Berat R, Brisson AR (2006) Formation of solid-supported lipid bilayers: an integrated view. Langmuir 22(8):3497–3505. https://doi.org/10.1021/la052687c
9. Pfefferkorn CM, Jiang Z, Lee JC (2012) Biophysics of alpha-synuclein membrane interactions. Biochim Biophys Acta 1818 (2):162–171. https://doi.org/10.1016/j.bbamem.2011.07.032
10. Castellana ET, Cremer PS (2006) Solid supported lipid bilayers: from biophysical studies to sensor design. Surf Sci Rep 61 (10):429–444. https://doi.org/10.1016/j.surfrep.2006.06.001
11. Plant AL, Brigham-Burke M, Petrella EC, O'Shannessy DJ (1995) Phospholipid/alkanethiol bilayers for cell-surface receptor studies by surface plasmon resonance. Anal Biochem 226(2):342–348
12. Shlyakhtenko LS, Gall AA, Lyubchenko YL (2013) Mica functionalization for imaging of DNA and protein-DNA complexes with atomic force microscopy. Methods Mol Biol 931:295–312. https://doi.org/10.1007/978-1-62703-056-4_14
13. Lyubchenko YL, Shlyakhtenko LS (2009) AFM for analysis of structure and dynamics of DNA and protein-DNA complexes. Methods 47(3):206–213. https://doi.org/10.1016/j.ymeth.2008.09.002
14. Andrecka J, Spillane KM, Ortega-Arroyo J, Kukura P (2013) Direct observation and control of supported lipid bilayer formation with interferometric scattering microscopy. ACS Nano 7(12):10662–10670. https://doi.org/10.1021/nn403367c

Chapter 9

Quantifying Small Molecule Binding Interactions with DNA Nanostructures

Xuye Lang, Yingning Gao, and Ian Wheeldon

Abstract

DNA nanostructures and hybrid DNA–protein materials are attractive solutions to many applications in biotechnology and material science because of their controllable molecule-level features. Critical to a complete description and characterization of these technologies is the quantification of binding affinity between DNA nanostructures and small molecules relevant to the application at hand. This protocols chapter described a series of experimental and in silico analyses that can be used to described and quantify ligand binding interactions between DNA nanostructures (DNA DX tiles), short double stranded DNA fragments, and arbitrary small molecules. The described methods include microscale thermophoresis, ligand completion assays, circular dichroism spectroscopy, and AutoDock simulations. The protocols use organophosphates and model chemical nerve agents as examples, but the methods described here are broadly applicable.

Key words DNA-sequence dependent binding, Enzyme–DNA, Microscale thermophoresis, Methylene blue displacement assay, Protein conjugation

1 Introduction

The past decade has seen a rapid rise in DNA-nanotechnology. At the root of these technologies are two critical properties, Watson--Crick base pairing and the rigid structure of DNA's double helix [1]. The specificity of Watson–Crick base pairing allows for the design of structures with arbitrary configuration, while the rigidity of the double helix provides stability at the nanometer scale. Building on these properties, researchers have created a vast array of two- and three-dimensional nanoscale structures with molecule-level control of the chemical and physical features. Such DNA nanostructures have been used as carriers for gene therapies [2], as scaffolds for the spatial organization of enzymes and proteins [3–7], and as drug delivery vehicles [8, 9], among many other applications [1, 10, 11].

Yuri L. Lyubchenko (ed.), *Nanoscale Imaging: Methods and Protocols*, Methods in Molecular Biology, vol. 1814, https://doi.org/10.1007/978-1-4939-8591-3_9,

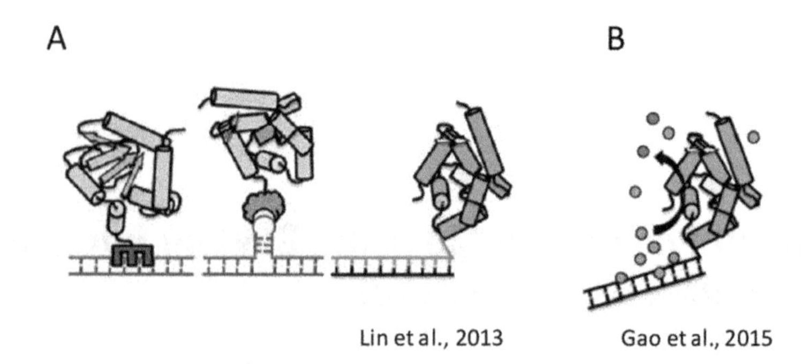

Fig. 1 The application of DNA in the scaffold engineering (**a**) and biocatalysis field (**b**). Reprinted with permission from Lin et al. (2014) ACS Catalysis 4(2): 505–511 [6] and Gao et al. (2015) ACS Catalysis, 5(4): 2149–2153 [15]. Copyright 2017 American Chemical Society

One of the challenges (or opportunities) facing DNA nanostructures is that the same properties that enable the novel structures are also beneficial to ligand binding (i.e., the double helix is a good target for small molecule binding). The interactions between small molecules and DNA are well known and have been exploited in traditional and new biotechnologies. For example, electrophoretic gel analysis of DNA is accomplished using intercalating dyes, many anticancer drugs are also intercalators [12, 13], and DNA templating of polymer precursors has been well-studied [14]. In our research, we have used DNA nanostructures and simple double stranded DNA fragments to modified enzymes (Fig. 1) [6, 15, 16]. The purpose of the enzyme modifications was to introduce new ligand binding interactions between the modified enzyme and the enzyme's reactant (or substrate). The end result is enhanced enzyme kinetics: micromolar binding interactions between the DNA appended to the enzyme and the enzyme substrates increases the local substrate concentration, thus increasing catalysis in solutions with low bulk concentrations [17]. With an enzyme engineering application in mind, we have also studied the binding interactions of organophosphates and model chemical nerve agents to double stranded DNA [18].

The quantification of ligand binding interactions is an active field of study and there are many different computational and experimental techniques that can be applied to the problem. We have focused on a select set of analytical methods to study and quantify small molecule interactions with double stranded DNA. These methods include small molecule docking simulations that can be used to predict the interaction between known DNA structures and small molecules, as well as experimental techniques such as microscale thermophoresis (MST) and binding competition assays. We have also used circular dichroism (CD) spectroscopy to investigate ligand binding modes—intercalation and groove binding [18].

MST is a relatively new technique that measures the movement of molecules in a temperature gradient created by a focused infrared laser. Molecular movement is observed by fluorescence detection (samples often require modification with a suitable dye). The technique is sensitive to changes in ligand–receptor binding because the movement of each molecule within the induced temperature gradient is proportional to molecule size, charge, and hydration, as well as other chemical factors [19]. In our experiments, MST has proven valuable because it can be accomplished with small solution volumes, requires only small amounts of ligand and receptor, and can be adapted for compounds that do not have intrinsic fluorescence.

As a separated measure of ligand binding we have used a methylene blue displacement assay. This method was selected because, in comparison with MST, it does not require modification of the ligand or receptor with a fluorescent dye or marker. We have also used CD spectroscopy to analysis binding mode. In this protocols chapter, we describe in detail the experimental procedures of MST, methylene blue competition assay, and CD spectroscopy analysis to study the binding interaction of a series of organophosphate and DNA double stranded fragment and DNA tile nanostructures. We also describe ligand–receptor docking simulation that can be used to provide initial estimates of the binding strength and potential for DNA sequence-dependent binding.

2 Materials

2.1 Docking Simulation

1. AutoDock 4.2 simulation software (http://autodock.scripps.edu).
2. Lamarckian 4 genetic algorithm, a search function option within AutoDock 4.2.
3. AutoDock tools (http://mgltools.scripps.edu/downloads).
4. 3D DART web server (http://haddock.chem.uu.nl/enmr/services/3DDART/).
5. Online ChemSpider database (www. Chemspider.com).
6. VEGA ZZ (http://nova.disfarm.unimi.it/cms/index.php?Software_projects:VEGA_ZZ:Download#VEGA_Z).

2.2 MicroScale Thermophoresis (MST)

1. NanoTemper Monolith NT.115 instrument.
2. Cy5-labeled single stranded DNA molecules and the nonlabeled complementary strand (Integrated DNA Technologies (IDT)). DNA strands DNA 1, 2, and 3 used in this protocol are shown in Table 1. All the DNA fragments from IDT were purchased with standard desalting purification and lyophilizing treatment.

Table 1
Cy5-labeled single stranded DNA used in MST binding experiments

Cy5′-labeled DNA	Sequence
Single-stranded DNA 1	Cy5′-CAGGTTGCAGCAGGTTGCAG-3′
Single-stranded DNA 2	Cy5′-GAATCTTCGGGAATCTTCGG-3′
Single-stranded DNA 3	Cy5′-CCTAAAAGAGCCTAAAAGAG-3′

3. Hydrochloric acid (Fisher Bioreagents, LOT: 111440).
4. Dimethyl sulfoxide (Fisher Bioreagents, LOT: 160239).
5. Standard treatment capillaries for MST analysis (NanoTemper).
6. Methyl parathion (Fluka, LOT: SZBA221XV).
7. Paraoxon (Fluka, LOT: SZBD172XV).
8. *p*-nitrophenol (Acros, LOT: B0130126D).
9. MST buffer: 50 mM Tris–HCl, 150 mM NaCl, 10 mM $MgCl_2$, 0.05% Tween 20.
10. Water: Milli-Q integral water purification system (Quantum, Q-Gard 2, and 0.22 μm filter were installed) was employed for water purification.

2.3 Methylene Blue Displacement Assay

1. BioTek Synergy 4 hybrid microplate reader, or similar microplate reader equipped with absorbance and fluorescence measurement capabilities.
2. Methylene blue (Sigma-Aldrich, LOT: M9140).
3. Single stranded DNA molecules and the nonlabeled complementary strand (IDT). DNA strands DNA 1, 2, and 3 used in this protocol are provided in Table 2. All the DNA fragments from IDT were purchased with standard desalting purification and lyophilizing treatment.
4. Sodium phosphate buffer (50 mM Tris–HCl, pH 7.4).
5. Methyl parathion (Fluka, LOT: SZBA221XV).
6. Paraoxon (Fluka, LOT: SZBD172XV).
7. *p*-nitrophenol (Acros, LOT: B0130126D).
8. Water: Milli-Q integral water purification system (Quantum, Q-Gard 2, and 0.22 μm filter were installed) was employed for water purification.

2.4 DNA Tile Assembly

1. Single stranded DNA molecules used to create the DNA DX tiles used in this protocol are provided in Table 3. A description of DNA DX tile nanostructures is provided in reference [20]. All the DNA fragments from IDT were purchased with standard desalting purification and lyophilizing treatment.

Table 2
Unmodified single-stranded DNA used in methylene blue displacement assay

DNA	Sequence
Single-stranded DNA 1	5′-CAGGTTGCAGCAGGTTGCAG-3′
Single-stranded DNA 2	5′-GAATCTTCGGGAATCTTCGG-3′
Single-stranded DNA 3	5′-CCTAAAAGAGCCTAAAAGAG-3′

Table 3
DNA DX tile strands

Tile	Single-stranded DNA sequence
Tile 1	5′-CGCAGACATCCTGCCGTAGCCTGAGGCACACG-3′ 5′-CGTGTGCCTCACCGACCAATGC-3′ 5′-GCATTGGTCGGACTGAACAGGACTACGCTGGC-3′ 5′-GCCAGCGTAGTGGATGTCTGCG-3′ 5′-TCAGTGGCTACGGCACCTGT-3′
Tile 2	5′-CAAGTGTGTTCTGCCGTAGCCTGAGGCACACG-3′ 5′-CGTGTGCCTCACCGACCAATGC-3′ 5′-GCATTGGTCGGACTGAACAGGACTACGCTGGC-3′ 5′-GCCAGCGTAGTGAACACACTTG-3′ 5′-TCAGTGGCTACGGCACCTGT-3′

2. 95, 65, 50 and 37 °C Water bath.
3. TAE-Mg^{2+} buffer: 40 mM Tris–acetic acid, pH 8.0, 2 mM EDTA-Na2, and 12.5 mM magnesium acetate.
4. Zymoclean gel DNA recovery kit (Zymo Research).
5. Water: Milli-Q integral water purification system (Quantum, Q-Gard 2, and 0.22 μm filter were installed) was employed for water purification.

2.5 Circular Dichroism Spectroscopy

1. Jasco J-815 circular dichroism spectrometer (JASCO Corporation).
2. 1 mm path length quartz cuvette (JASCO Corporation).
3. Sodium phosphate buffer (50 mM Tris–HCl, pH 7.4).
4. Single stranded DNA molecules and the nonlabeled complementary strands (IDT). DNA 1, 2, and 3 used in these protocols are provided in Table 2. All the DNA fragments from IDT were purchased with standard desalting purification and lyophilizing treatment.
5. DNA Tile 1 and DNA Tile 2 (Table 3; IDT).

6. Methyl parathion (Fluka, LOT: SZBA221XV).
7. Paraoxon (Fluka, LOT: SZBD172XV).
8. *p*-nitrophenol (Acros, LOT: B0130126D).
9. Diethy hydrogen phosphate (Sigma-Aldrich, LOT: 32449).
10. Water: Milli-Q integral water purification system (Quantum, Q-Gard 2, and 0.22 μm filter were installed) was employed for water purification.

3 Methods

3.1 Docking Simulation

This protocol was used for predicting the binding affinity (K_d) between double stranded DNA and small molecules of interest. In our work, we have used such simulations to inform experimental designs and provide initial estimates of DNA sequence-dependent binding of organophosphates and phenolic substrates of the enzyme horseradish peroxidase [16, 18]. The protocols described here focus on the analysis of organophosphates methyl parathion, paraoxon, and their hydrolysis products *p*-nitrophenol and diethyl hydrogen phosphate.

1. Obtain the structure of methyl parathion, paraoxon, *p*-nitrophenol, and diethyl hydrogen phosphate, or ligands of interest, from the online ChemSpider database (www.chemspider.com) [21].
2. Assign hydrogen and Gasteiger charges by VEGA ZZ.
3. Create 3D structure files of double-stranded DNA using the 3D DART Web server [22].
4. Assign polar hydrogen atoms and Kollman charges using Autodock 4.2 tools (*see* **Note 1**).
5. Select Lamarckian 4 genetic algorithm (LGA) as the search method for Autodock 4.2 simulations.
6. Allow ligand torsional bands to rotate within a 60 × 60 × 100 grid with point separation of 0.375 Å.
7. Run each LGA project 30 (or more) times, with the following conditions: initial popular of 150 individuals; energy evaluation of 2.5×10^6; maximum iterations of 2.7×10^4; mutation rate of 0.02; and crossover rate of 0.80.
8. Group docking poses after completion of the simulations for easy visual inspection.

3.2 MicroScale Thermophoresis (MST)

The protocol described below is an MST method for rapidly and accurately determining the binding affinity (K_d) between DNA nanostructures (including DNA DX tiles and double stranded

DNA fragments 20–80 bp in length) and small molecules of interest. This protocol uses organophosphate ligands.

1. Prepare MST buffer (50 mM Tris–HCl, 150 mM NaCl, 10 mM MgCl2, 0.05% Tween 20).
2. Separately, incubate 16 nM Cy5′-labeled single stranded DNA 1, DNA 2, and DNA 3 (Table 1) and 16 nM nonlabeled complementary strand (same volume) at 42 °C for 5 min (*see* **Note 2**).
3. Mix Cy5′-labeled single stranded DNA with nonlabeled complementary strand.
4. Anneal double stranded DNA by incubating the mixed solution at room temperature for 30 min.
5. Determine the predicted K_d for each double stranded DNA and ligand combination from the docking simulation methods described above in Subheading 3.1 (can be accomplished beforehand).
6. Open NanoTemper software.
7. Set excitation wavelength to 650 nm and emissions filter set to 680 nm.
8. Choose "Binding Affinity" mode.
9. Enter DNA molecule and ligand information (e.g., expected K_d and ligand stock solution concentration) in the planning page of "Binding Affinity" mode.
10. Set up the experimental temperature as needed.
11. Dilute ligand stock solution according to the instruction outline on the "Instruction Page" of the Binding Affinity mode module.
12. Mix Cy5′-labeled double stranded DNA fragments with ligand molecules in PCR, or other suitable tube.
13. Transfer mixed Cy5′-labeled double stranded DNA–ligand solution into standard treatment MST capillaries.
14. Start the MST measurement.
15. Obtain results and measurement details from the results page.
16. Repeat each experiment and DNA/ligand combination three times.

3.3 Methylene Blue Displacement Assay

The interactions between DNA nanostructures and small molecules also can be identified by the analytical chemistry method described here. In our work, this assay was used as a second, independent measure of ligand–DNA binding to complement the MST experiments described above in Subheading 3.2.

1. Separately, incubate single stranded DNA 1, 2, and 3 (Table 2) and the complementary strands at 42 °C for 5 min.
2. Mix, equal volumes of single stranded DNA 1, 2, and 3 with their corresponding complementary strand.
3. Anneal double stranded DNA by incubating the mixed solutions at room temperature for 30 min.
4. Prepare double stranded DNA/methylene blue complexes by incubating solutions from **step 3** above with equal molar concentrations of methylene blue in sodium phosphate buffer (50 mM Tris–HCl, pH 7.4) (*see* **Note 3**).
5. Allow solution to equilibrate for at least 50 min.
6. Add ligand of interest to create a series of solutions with final ligand concentrations of 0, 2.5, 5, 7.5, 10, 25, 50, 100, and 250 μM. Allow the mixed solutions to equilibrate for at least 10 min.
7. For each solution, acquire fluorescence spectra from 650 to 750 nm with an excitation wavelength of 615 nm.
8. Determine K_d from the acquired data by plotting $1/(F-F_0)$ vs. $1/[\text{Ligand}]_0$, where F_0 and F are the fluorescence signal intensity of DNA–methylene blue with and without ligands, and $[\text{Ligand}]_0$ is the initial concentrations of the ligand [18, 23].
9. Repeat **steps 1** through **8** three times to acquire triplicate measures of the K_d.

3.4 DNA Tile Assembly

The above protocols outlined in Subheadings 3.2 and 3.3 are described for double stranded DNA fragments. These protocols can be repeated with DNA DX tile nanostructures. The assembly of those structures is described here. Representative images of the DNA tiles used in this work are shown in Fig. 2.

1. Prepare TAE-Mg^{2+} buffer: 40 mM Tris–acetic acid, pH 8.0, 2 mM EDTA-Na2, and 12.5 mM magnesium acetate.
2. Prepare 200 μM solutions of each of the single stranded DNA strands listed in Table 3. Solutions should be prepared in TAE-Mg^{2+} buffer.
3. Mix equal volumes of each of the five single stranded DNA solutions into a single tile assembly mixture. Total volume of the mixed solution can range from 50 to 150 μL.
4. Incubate the tile assembly solution from **step 3** for 5 min. At 95 °C.
5. To assemble tiles use a thermocycler or controlled temperature incubator with the following temperature ramp: 65 °C for

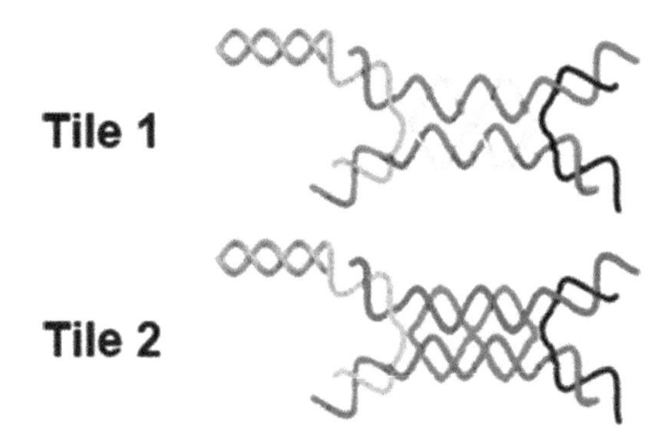

Fig. 2 Schematic representations of the structure of DNA DX tiles, Tile 1 and Tile 2. Reprinted with permission from Gao et al. (2016) ChemBioChem 17(15): 1430–1436 [3]. Copyright 2017 ChemBioChem

20 min, 50 °C for 20 min, 37 °C for 20 min and room temperature for 30 min.

6. Tiles can be used directly from the assembly solution or purified by electrophoresis in a 1% agarose gel (*see* **Note 4**).

3.5 Circular Dichroism (CD) Spectroscopy

The molecular interactions between DNA nanostructures and the small molecules of interest can also be investigated by CD spectroscopy. This analysis can provide information on the mode of binding and changes in secondary and tertiary structure of the DNA upon ligand binding [18].

1. Create double stranded DNA 1, 2, and 3 as described in **steps 1**, **2**, and **3** of Subheading 3.3.
2. Create DNA tiles as described in Subheading 3.4.
3. Create DNA–ligand solutions with a range of ligand concentrations as described in Subheading 3.3 (*see* **step 6**).
4. Acquire CD spectra from 200 to 300 nm for each solution using the following parameters: 1 mm path length cuvette; 0.2 nm data pitch, 50 nm min^{-1} scanning speed, and three spectra accumulations for sample averaging.
5. Repeat **steps 1–4** to acquire triplicate data.

4 Notes

1. Docking Simulation
   The AutoDock software suite is available from the Scripps Institute. The software is designed to predict how small molecules interact with known 3D structures. For a detailed description of the software and development notes see the Material and Methods provided by the Scripps Institute (autodock.scipps.edu).

2. MicroScale Thermophoresis (MST)
   For optimal MST results, the manufacturer provided the following guidance: (1) the concentration of fluorescent molecules should be equivalent or lower than the K_d; (2) The highest ligand concentration should be at least 20 times that of the expected K_d; (3) Final sample volume for each titration point should be approximately 20 μL; (4) Use small tubes (e.g., PCR tubes) for serial ligand dilutions; (5) The serial dilution buffer and sample buffer should contain equivalent salt and analyte concentrations; (6) Accurate pipetting is critical for obtaining repeatable data; (7) Solutions should be mixed by pipetting and not by vortexing; (8) Handle capillaries by the ends and avoid touching the length of the tubes; and (9) In the case of methyl parathion, 2% DMSO was added to the buffer to ensure that all ligand was soluble.
3. Methylene Blue Displacement Assay
   In our experience we have achieved repeatable results by: (1) slowly adding stock methylene blue to the sample solutions; (2) mixing the DNA–methylene blue complex and ligand solutions by pipetting prior to fluorescence measurements; and (3) monitoring the fluorescence intensity signal (F) over a ~5 min time period to identify potential changes in intensity. In addition, when comparing the binding strength of various ligands, it is critical to use the same DNA–methylene blue solution.
4. DNA Tile Assembly
   Tile assembly can be checked by running a sample of the putative DNA nanostructure on a 1% agarose gel. Assembled tiles should result in a single band, while partially assembled tiles produce multiple bands or a large smeared band. The DNA Tiles cut from the agarose gel and recovered by Zymoclean gel DNA recovery kit (use protocols as described in the kit manual).

Acknowledgments

This work was supported by HDTRA1-14-1-0045. The authors declare no conflict of interest.

References

1. Jones MR, Seeman NC, Mirkin CA (2015) Nanomaterials. Programmable materials and the nature of the DNA bond. Science 347 (6224):1260901
2. Mansouri S, Lavigne P, Corsi K, Benderdour M, Beaumont E, Fernandes JC (2004) Chitosan-DNA nanoparticles as non-viral vectors in gene therapy: strategies to improve transfection efficacy. Eur J Pharm Biopharm 57(1):1–8
3. Fu JL, Yang YR, Johnson-Buck A, Liu MH, Liu Y, Walter NG, Woodbury NW, Yan H

(2014) Multi-enzyme complexes on DNA scaffolds capable of substrate channelling with an artificial swinging arm. Nat Nanotechnol 9 (7):531–536

4. Fu J, Liu M, Liu Y, Yan H (2012) Spatially-interactive biomolecular networks organized by nucleic acid nanostructures. Acc Chem Res 45(8):1215–1226
5. Fu JL, Liu MH, Liu Y, Woodbury NW, Yan H (2012) Interenzyme substrate diffusion for an enzyme cascade organized on spatially addressable DNA nanostructures. J Am Chem Soc 134 (12):5516–5519
6. Lin J-L, Palomec L, Wheeldon I (2014) Design and analysis of enhanced catalysis in scaffolded multienzyme cascade reactions. ACS Catal 4(2):505–511
7. Wheeldon I, Minteer SD, Banta S, Barton SC, Atanassov P, Sigman M (2016) Substrate channelling as an approach to cascade reactions. Nat Chem 8(4):299–309
8. Jiang Q, Song C, Nangreave J, Liu XW, Lin L, Qiu DL, Wang ZG, Zou GZ, Liang XJ, Yan H, Ding BQ (2012) DNA origami as a carrier for circumvention of drug resistance. J Am Chem Soc 134(32):13396–13403
9. Douglas SM, Bachelet I, Church GM (2012) A logic-gated nanorobot for targeted transport of molecular payloads. Science 335 (6070):831–834
10. Kuzyk A, Schreiber R, Fan ZY, Pardatscher G, Roller EM, Hogele A, Simmel FC, Govorov AO, Liedl T (2012) DNA-based self-assembly of chiral plasmonic nanostructures with tailored optical response. Nature 483 (7389):311–314
11. Qi H, Ghodousi M, Du Y, Grun C, Bae H, Yin P, Khademhosseini A (2013) DNA-directed self-assembly of shape-controlled hydrogels. Nat Commun 4:2275
12. Neidle S (2011) DNA minor-groove recognition by small molecules. Nat Prod Rep 18 (3):291–309
13. Palchaudhuri R, Hergenrother PJ (2007) DNA as a target for anticancer compounds: methods to determine the mode of binding and the mechanism of action. Curr Opin Biotechnol 18(6):497–503
14. Ma YF, Zhang JM, Zhang GJ, He HX (2004) Polyaniline nanowires on Si surfaces fabricated with DNA templates. J Am Chem Soc 126 (22):7097–7101
15. Gao YN, Roberts CC, Zhu J, Lin JL, Chang CEA, Wheeldon I (2015) Tuning enzyme kinetics through designed intermolecular interactions far from the active site. ACS Catal 5 (4):2149–2153
16. Lin J-L, Wheeldon I (2013) Kinetic enhancements in DNA-enzyme nanostructures mimic the Sabatier principle. ACS Catal 3 (4):560–564
17. Gao Y, Roberts CC, Toop A, Chang CEA, Wheeldon I (2016) Mechanisms of enhanced catalysis in enzyme–DNA nanostructures revealed through molecular simulations and experimental analysis. Chembiochem 17 (15):1430–1436
18. Gao YN, Or S, Toop A, Wheeldon I (2017) DNA nanostructure sequence-dependent binding of organophosphates. Langmuir 33 (8):2033–2040
19. Jerabek-Willemsen M, Andre T, Wanner R, Roth HM, Duhr S, Baaske P, Breitsprecher D (2014) MicroScale thermophoresis: interaction analysis and beyond. J Mol Struct 1077:101–113
20. Li XJ, Yang XP, Qi J, Seeman NC (1996) Antiparallel DNA double crossover molecules as components for nanoconstruction. J Am Chem Soc 118(26):6131–6140
21. Pence HE, Williams A (2010) ChemSpider: an online chemical information resource. J Chem Educ 87(11):1123–1124
22. van Dijk M, Bonvin AMJJ (2009) 3D-DART: a DNA structure modelling server. Nucleic Acids Res 37:W235–W239
23. Shen HY, Shao XL, Xu H, Li J, Pan SD (2011) In vitro study of DNA interaction with trichlorobenzenes by spectroscopic and voltammetric techniques. Int J Electrochem Sci 6 (3):532–547

Part II

High-Speed AFM of Biomolecules

Chapter 10

Optimum Substrates for Imaging Biological Molecules with High-Speed Atomic Force Microscopy

Takayuki Uchihashi, Hiroki Watanabe, and Noriyuki Kodera

Abstract

Recent progresses in high-speed atomic force microscopy (HS-AFM) have enabled us to directly visualize dynamic processes of various proteins in liquid conditions. One of the key factors leading to successful HS-AFM observations is the selection of an appropriate substrate depending on molecules to be observed. For the HS-AFM imaging, a target molecule must be absorbed on a substrate by controlling its orientation without impairing the dynamics or physiological function of the molecule. In this chapter, we describe protocols for preparation of substrates that have been used for HS-AFM and then introduce observation examples on dynamic processes of biological molecules.

Key words High-speed AFM, Substrate, Biological molecules, Imaging

1 Introduction

Since the atomic force microscopy (AFM) enables observation of surface structure of biomolecules under a solution environment with high spatial resolution, it has been used for analyzing the structure or physical properties of biological samples ranging from nucleic acids and proteins to living cells. Although its application range in biological research with conventional AFM has been limited because of its slow imaging speed, recent technological advances for both fast and low-invasive imaging of AFM have allowed us to directly visualize biological molecules at work with subsecond time resolution. The fastest imaging rate by the latest HS-AFM has reached 80 ms/frame under the condition of a scan range of ~250 nm and ~100 scan lines. The HS-AFM can now capture various dynamic events on individual molecules such as conformational change, molecular interactions (association/dissociation), assembly and diffusion.

AFM observations require solid substrates on which the target specimens are adsorbed because the AFM probes the mechanical interaction between the sharp needle and the sample through the

Yuri L. Lyubchenko (ed.), *Nanoscale Imaging: Methods and Protocols*, Methods in Molecular Biology, vol. 1814,
https://doi.org/10.1007/978-1-4939-8591-3_10,

deflection of the cantilever. In the dynamics observation by HS-AFM, dynamics relevant to the physiological function of the molecule must be retained on the solid substrate. Furthermore, the molecules should be specifically anchored with controlled orientation on the substrate in order to facilitate the interpretation of the obtained images. Therefore, one of the crucial factors for successful HS-AFM observation is development of optimum substrates suitable for the molecules to be observed. This chapter focuses on describing preparation of several substrates according to the biological samples and their typical applications.

2 Materials

1. Cantilever (BL-AC10FS-A2 and BL-AC10DS-A2, Olympus, Tokyo, Japan).
2. Mica disk (Clear Mica, Fuuchi Chemical, Tokyo, Japan).
3. Acryl adhesive (Super X, CEMEDINE, Tokyo, Japan).
4. Milli-Q water (resistivity 18.2 MΩ·cm).
5. α-synuclein (provided by Prof. Koichi Kato, Okazaki Institute for Integrative Bioscience).
6. His_6-tagged UGGT (provided by Prof. Koichi Kato, Okazaki Institute for Integrative Bioscience).
7. 10 mM Tris–HCl (pH 7.5), 150 mM NaCl, and 2 mM $CaCl_2$.
8. (3-aminopropyl)triethoxysilane (APTES).
9. 50 mM Tris–HCl (pH 7.5), 50 mM KCl.
10. Glutaraldehyde.
11. Bacteriorhodopsin.
12. 10 mM phosphate (pH 7.0) and 300 mM KCl.
13. 1,2-dipalmitoyl-*sn*-glycero-3-phosphocholine (DPPC).
14. 1,2-dipalmitoyl-3-trimethylammonium-propane (DPTAP)
15. 1,2-dipalmitoyl-*sn*-glycero-3-phosphoethanolamine-*N*-(cap biotinyl) (biotin-CAP-DPPE).
16. Cas9 from *Streptococcus pyogenes* (provided by Prof. Osamu Nureki, The University of Tokyo).
17. Double-stranded DNA (provided by Prof. Osamu Nureki, The University of Tokyo).
18. 20 mM Tris–HCl, pH 8.8.
19. 20 mM Tris–HCl, pH 8.0, 30 mM KCl, 0.01 mM EDTA.
20. HOPG (highly oriented pyrolytic graphite) (SPI Supplies, PA, USA).

21. Cellulose (provided by Prof. Kiyohiko Igarashi, The University of Tokyo).
22. Cel7A from *Tricoderma reesei* (provided by Prof. Kiyohiko Igarashi, The University of Tokyo).
23. 20 mM sodium acetate buffer, pH 5.0.
24. Thin gold surface on mica (Phasis, Geneva, Switzerland).
25. Cellobiose dehydrogenases from *Phanerochaete chrysosporium* (*Pc*CDHs) (provided by Prof. Akira Onoda and Prof. Takashi Hayashi, Osaka University).
26. 6-hydroxy-1- hexanethiol/11-amino-1-undecanethiol.
27. 1-ethyl-3-(3- dimethylaminopropyl)carbodiimide (EDC).
28. Dimethyl sulfoxide (DMSO).
29. 50 mM NaOAc buffer (pH 4.5).
30. Polydimethylsiloxane (PDMS) (SYLGARD 184, Dow Corning, MI, USA).
31. Polystyrene bead (Polyscience, PA, USA).
32. Silica bead (Micromod, Germany).
33. Biopsy punch (BPP-15F, Kai industries, Japan).

3 Methods

3.1 Outline of High-Speed AFM

We here briefly outline the laboratory-built HS-AFM operated in the tapping mode. A similar type of the HS-AFM is commercially available from Research Institute of Biomolecule Metrology Co., Ltd., Tsukuba, Japan. In tapping mode, the AFM probe is oscillated at near the eigenmode resonant frequency and hence intermittently contacts the sample surface, which can reduce the lateral force between the probe and the sample surface during the raster scanning and therefore the tapping mode is frequently used for imaging of soft biological molecules [1]. The basic configuration of the HS-AFM instrument is similar to that of conventional tapping-mode AFM consisting of cantilever, deflection sensor, amplitude detector, feedback controller and scanner (Fig. 1) [2].

1. Cantilever

   For construction of an AFM image in tapping mode, the AFM probe should contact to the surface at least once for individual image pixels, indicating that the cantilever should oscillate with a high frequency for the fast imaging. Mechanical properties of the cantilever are defined by material and dimension of the cantilever. In general, a stiffer and smaller cantilever has a high resonant frequency. However, the stiff cantilever often damages soft biological molecule. To satisfy both the high-resonant frequency and the soft spring constant, size of the

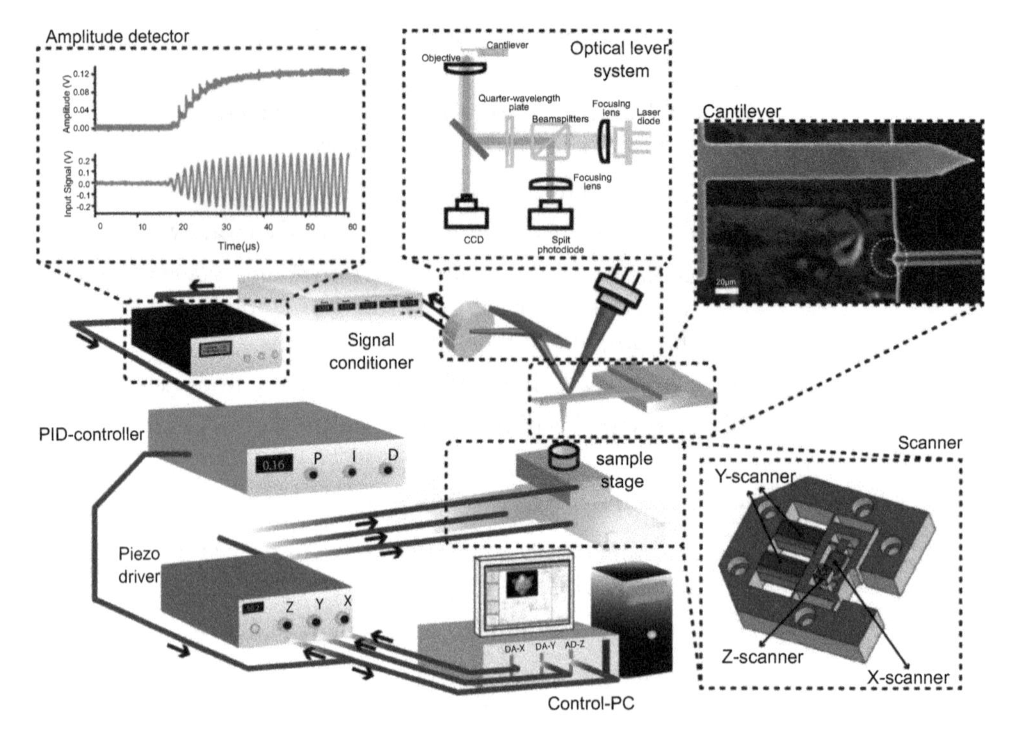

Fig. 1 Schematic diagram of HS-AFM. Essential components (cantilever, optical beam deflection system, amplitude detector, and scanner) are highlighted by squares

cantilever is reduced down to 10 μm in length, 2 μm wide, and 90 nm thick, which is less than one-tenth of size of a conventional cantilever. The resonant frequency, the spring constant and the quality factor of the small cantilever in liquid are 400–600 kHz, ~0.1 N/m and 2–3, respectively (*see* **Note 1**).

2. Amplitude detector and PID feedback

   The oscillation amplitude should be quickly detected by minimizing delays. The fast amplitude detector which can measure the amplitude at each oscillation cycle was developed [3]. The probe-surface distance is controlled by using an ordinary PID (Proportional-Integral-Derivative) circuit with the dynamic PID control, which dynamically adjusts the feedback gain depending on the input signal level [4].

3. Scanner

   Fast actuation of the sample in three dimensions requires a mechanical rigidity of the scanner to suppress unwanted mechanical vibrations leading inaccuracy in the feedback control. The HS-AFM scanner is designed using flexure stages made of blade springs for the x- and y-scanners which are fabricated by monolithic processing to have a mechanical rigidity [2]. Mechanical excitation of the x- and y-scanning are further suppressed by the feedforward control using inverse compensation method [5]. Fast response of the z-piezoactuator is crucial for the low-invasive and precise imaging. Hence an extension of the resonant frequency of the z-actuator has been endeavored. Furthermore, the mechanical excitations of the z-scanner is damped by the active *Q*-control technique [6].

Thanks to all essential components and control techniques, the HS-AFM can now capture dynamic behaviors of individual proteins at the imaging rate less than 100 ms/frame [7]. More technical details about the instrumentation can be found in the comprehensive review regarding the HS-AFM [8].

3.2 Solid Substrate

The properties required for the substrate are highly dependent on the dynamic phenomena of biological molecules to be captured. Two major important issues are controllability of the orientation and maintaining physiological activity of the molecule adsorbed on the substrate. For an observation of conformational change of proteins, the protein must be firmly fixed on the substrate with well-controlled orientation [9–14]. On the other hand, when observing molecular interaction such as association and dissociation, or self-assembly of proteins, the molecules must two-dimensionally diffuse on the substrate with a speed enough slower than the imaging speed [15–18]. Furthermore, when observing interaction between two different proteins such as a motility of a linear motor protein along the rail protein, the rail protein must be immobilized on the substrate, while the other is needed to move almost freely without nonspecifically binding to the substrate [19, 20].

In this section, we describe several substrates that have been ever used so far; bare mica, chemically treated mica, supported planar lipid bilayers, graphite, gold and patterned elastomer.

3.3 Bare Mica

Mica is the most commonly used substrate in AFM observations because one can easily obtain atomically flat surface by cleavage with a scotch tape. Bare mica has net negative chargers (area/charge ~0.48 nm^2) on the surface and hence hydrophilic. Various proteins on the bare mica surface have been observed with HS-AFM [7, 9, 13, 14]. Binding affinity of proteins to the bare mica can be controlled to some extent by solution conditions such as pH and salt concentrations.

Figure 2a, b show an example of HS-AFM images showing affinity change of α-synuclein on the bare mica by changing ion concentrations. α-synuclein is a natively disordered presynaptic protein with a molecular weight of 14 kDa, which is a key player in the pathogenesis of Parkinson's disease [21]. This protein has a tendency to aggregate due to its hydrophobic non-amyloid-β regions and strongly immobilized onto the bare mica in Milli-Q water as shown in the images at 16.2 and 91.0 s of Fig. 2a. After the

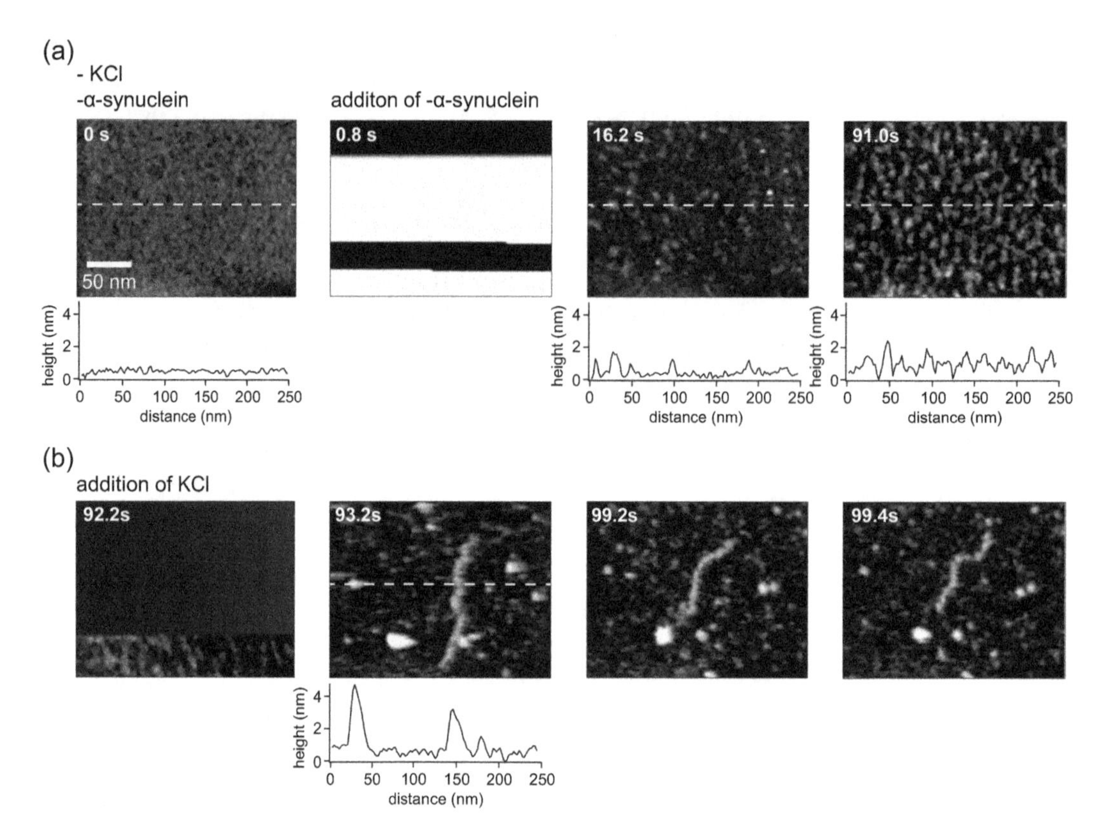

Fig. 2 HS-AFM images of α-synuclein on bare mica surface. The solution of α-synuclein was added into the water pool at 0.8 s during the imaging (final concentration: 1.1 μM). The globular molecules adsorb onto the bare mica with time (16.2 and 91.0 s) and eventually the surface is almost fully covered by the molecules. Then a KCl solution was added into the water pool during the imaging at 92.2 s (final concentration: 60 mM). Most of α-synuclein desorbed from the surface and further the remained molecules showed flexible string-like structure. Panels below the images at 0, 16.2, 91.0, and 93.2 s show the cross-sectional profiles along the broken lines on the corresponding images. Imaging rate, 0.2 s/frame

addition of KCl solution (the final concentration of 60 mM) to the water at the time of $t = 92.2$ s of Fig. 2b, most of aggregated proteins were quickly desorbed and eventually disentangled fibrils of α-synuclein were observed.

3.3.1 Preparation of Bare Mica

1. Prepare mica disk with a diameter 1.5 or 2 mm.
2. Glue the mica disk to a cylindrical glass stage (2 mm in height and 1.5 or 2 mm in diameter) using an acryl adhesive and dry it at least 1 h (*see* **Note 2**).
3. Press Scotch tape evenly onto the mica disk and remove the tape from the mica. Observe the cleaved mica sheet on the Scotch tape and check whether the mica surface is evenly cleaved.
4. Place a sample solution on the freshly cleaved surface and incubate the sample droplet appropriate duration (*see* **Note 3**), and then wash with an observation buffer (*see* **Note 4**).

3.4 Ni^{2+}-Coated Mica

By treating the bare mica surface with Ni^{2+}, histidine-tagged (His-tag) proteins can be easily immobilized through metal–chelate affinity. Figure 3 shows an example of the HS-AFM observation of a multidomain enzyme where specific domain with His-tag is tethered on the Ni-coated mica. The folding-sensor enzyme of endoplasmic reticulum (ER), glycoprotein glucosyltransferase (UGGT) is a large enzyme with a molecular mass of around 160 kDa consisting of two main parts: a 125-kDa N-terminal region, which is thought to act as a folding-sensor region, and a C-terminal 35-kDa catalytic domain [22]. The smaller catalytic C-terminal domain is tethered to the larger N-terminal folding-sensor region. When the

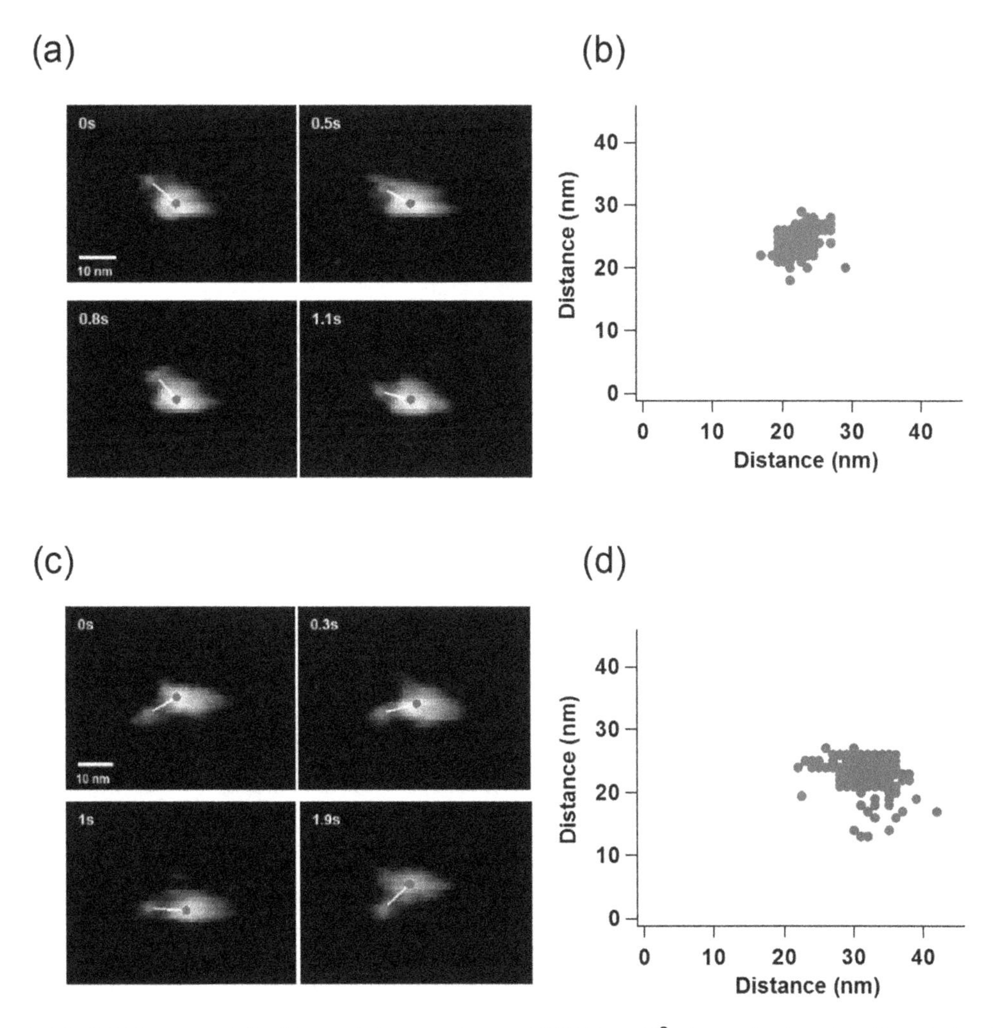

Fig. 3 HS-AFM images of flexible modular structures of UGGT on Ni^{2+}-treated mica. (**a**) Clipped HS-AFM images of N-terminally His_6-tagged UGGT. The N-terminal and the C-terminal domains are marked blue and red dots, respectively. (**b**) Distribution of center positions of the C-lobes (red circles) relative to the center of the immobilized N-lobes (blue circles). (**c**) Clipped HS-AFM images of C-terminally His_6-tagged UGGT. (**d**) Distribution of center positions of the N-lobes (red circles) relative to the center of the immobilized C-lobes (blue circles)

N-terminally His_6-tagged UGGT was immobilized onto the Ni^{2+}-treated mica surface, the smaller lobe dynamically fluctuated around the larger lobe (Fig. 3a, b) [23]. In contrast, when the C-terminally His_6-tagged UGGT was used, the position of the larger lobe dynamically moved (Fig. 3c, d). These results clearly demonstrate that Ni^{2+}-treated mica is useful for specific immobilization of His-tagged proteins on solid substrates.

3.4.1 Preparation of Ni^{2+}-Covered Mica

1. Prepare freshly cleaved mica disk according to the **steps 1–3** in the previous section.
2. Deposit a solution of 1 mM $NiCl_2$ or $NiSO_4$ and leave it for 3 min (*see* **Note 5**). Then wash the solution thoroughly by Milli-Q water.
3. Place a sample solution on the Ni^{2+}-coated mica substrate and incubate the sample droplet appropriate duration, and then wash with an observation buffer (*see* **Note 6**). In the case of observing UGGT, the incubation duration is 3 min and the observation buffer is 10 mM Tris–HCl (pH 7.5), 150 mM NaCl, and 2 mM $CaCl_2$.

3.5 Aminosilane-Functionalized Mica

The affinity between some proteins and bare mica is too weak to be imaged depending on charge distributions of proteins and buffer conditions. In such cases, aminosilane-functionalized mica is available to immobilize proteins because the functionalization with aminosilane of mica produces a positively charged surface [24]. It has been used for HS-AFM imaging of DNA and DNA–protein complexes [25, 26].

Figure 4 shows comparison of the affinity of monoclonal antibodies of immunoglobulin G (IgG, molecular weight of 150 kDa) purified from mice to the bare mica and the APTES-mica. The HS-AFM images of the IgG on the bare mica surface showed obscure features and fast diffusion due to the weak binding affinity between the IgG and the substrate (Fig. 4a). On the other hand, the treatment of the mica with 0.02% APTES (volume ratio) suppressed the fast diffusion and thus provided clear features of IgG with a typical Y-shaped structure (Fig. 4b), composed of two antigen-binding fragments (Fab region) and one crystallizable fragment (Fc region) [27].

3.5.1 Preparation of APTES/Mica

1. Dissolve the APTES in Milli-Q water (volume ratio: 0.01–0.1%) (*see* **Note 7**).
2. Deposit the APTES solution on the freshly cleaved mica disk and leave it for 3 min.
3. Wash the surface thoroughly with Milli-Q water.

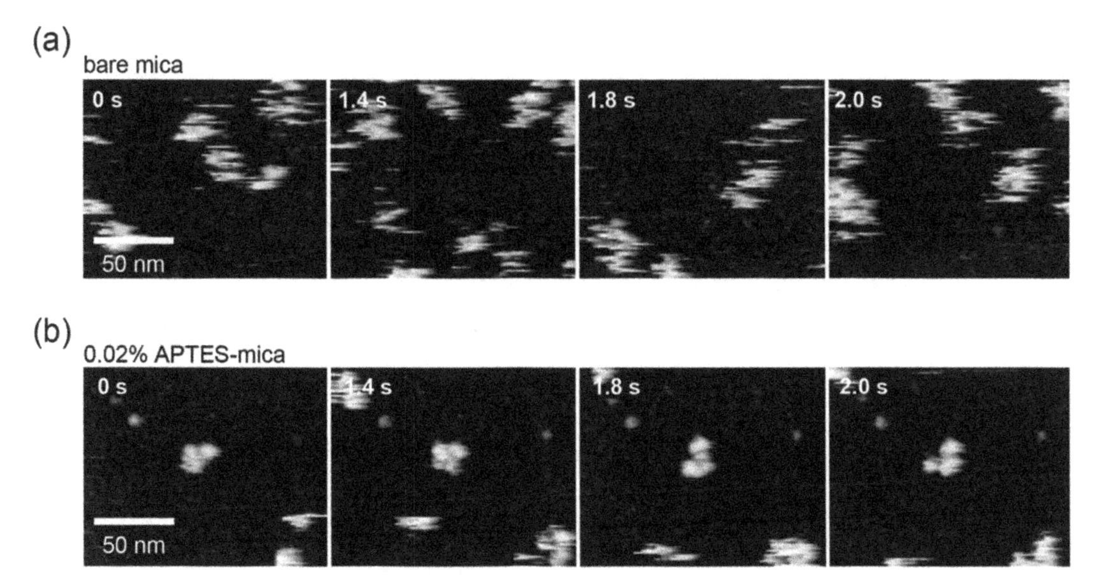

Fig. 4 HS-AFM images of IgG on APTES-mica. Clipped HS-AFM images of IgG on (**a**) bare mica and (**b**) 0.02% APTES-mica. Imaging rate, 0.2 s/frame

4. Place a sample solution on the APTES/mica substrate. Incubate the sample droplet appropriate duration, and then wash with an appropriate observation buffer (*see* **Note 8**). For observing IgG antibody, the incubation time is 3 min and the buffer solution is 50 mM Tris–HCl (pH 7.5), 50 mM KCl.

3.6 Glutaraldehyde/Aminosilane-Functionalized Mica

Some proteins do not have sufficient affinity to be observed even after the treatment of mica with aminosilane. In such case, glutaraldehyde (GA) is useful because GA chemically crosslinks amino groups of APTES/mica and proteins. Figure 5a, b show an example of the immobilization of proteins on the GA-treated mica, isolated bacteriorhodopsin in a lipid bilayer. Bacteriorhodopsin (bR) usually forms two-dimensional crystal in the membrane [28] and is naturally immobilized even on bare mica [29]. Disassembly of bR crystal introduced by the point mutation at the interaction site (W12I) between bR trimers produced isolated trimers [11]. The isolated bR trimers translationally and rotationally diffuse in the membrane on the bare mica (Fig. 5a) and even on the APTES/mica. On the other hand, the GA-treated APTES mica effectively repressed the diffusion of the isolated trimers, suggesting that the GA treatment chemically crosslinked the primary amine on the APTES/mica and residues with amines contained in bR (Fig. 5b).

3.6.1 Preparation of GA/APTES/Mica

1. Dissolve the GA in Milli-Q water (volume ratio: 0.1–0.2%) and deposit the GA solution on the APTES/mica.
2. Incubate it for 3 min. Then wash the GA solution with Milli-Q water (*see* **Note 9**).

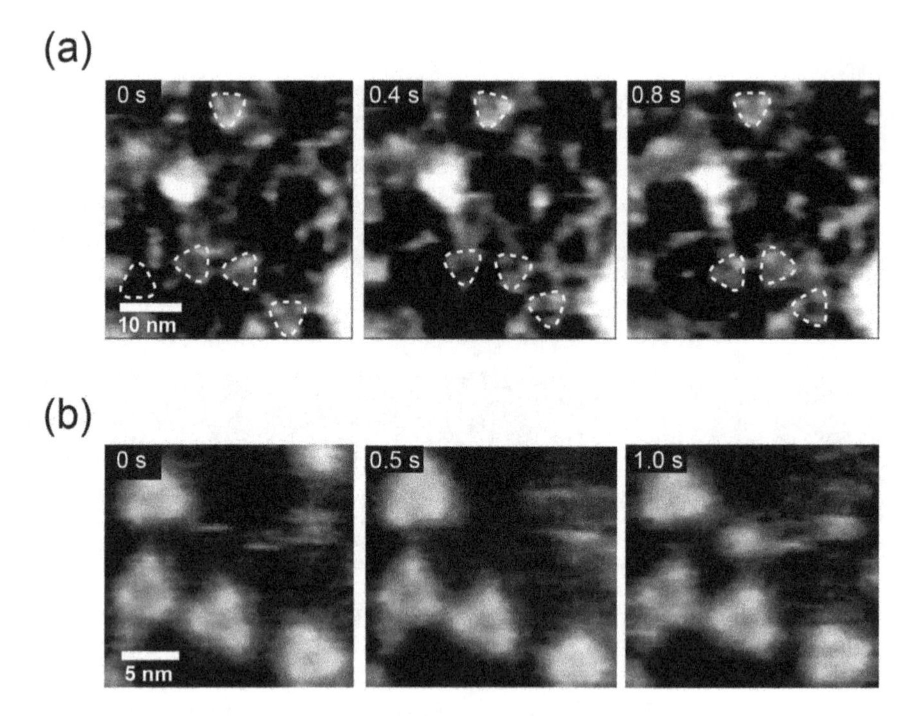

Fig. 5 HS-AFM images of bacteriorhodopsin. (**a**) Bacteriorhodopsin trimers which the two dimensional assembly was disrupted by introduction of the point mutation (W12I) on bare mica. The trimers quickly diffuse in the membrane. Imaging rate, 0.2 s/frame. (**b**) The isolated trimers in the membrane adsorbed on the GA/APTES-mica are firmly fixed and the diffusion was markedly suppressed. Imaging rate, 0.5 s/frame

3. Place a sample solution on the substrate. Incubate the sample droplet appropriate duration (usually 3–5 min), and then wash with an observation buffer (*see* **Note 10**). In the case of observing disassembled bacteriorhodopsin, the incubation time is 3 min and the observation buffer is 10 mM phosphate (pH 7.0) and 300 mM KCl.

3.7 Supported Lipid Bilayer

Lipid bilayers supported by mica substrate are useful substrates for HS-AFM imaging because they provide an ideal substrate with orientation controllability and selective affinity of molecules to be observed [20]. Applications of the lipid bilayer as a substrate for HS-AFM imaging are described in other literatures [30, 31]. Here we focus on the usability of partially positively charged lipid bilayers, which is in the gel phase at room temperature, for the substrate. This substrate is particularly useful for observing DNA--protein complexes and their formation processes although these observations on mica are generally difficult. This is because both the surfaces of DNA and mica are negatively charged and the DNA binding sites of DNA binding proteins are generally positively charged, resulting in the preferential binding of DNA binding proteins to bare mica surface but not to DNA. Slight additions of the lipid molecules with the bulky head group such as biotin or

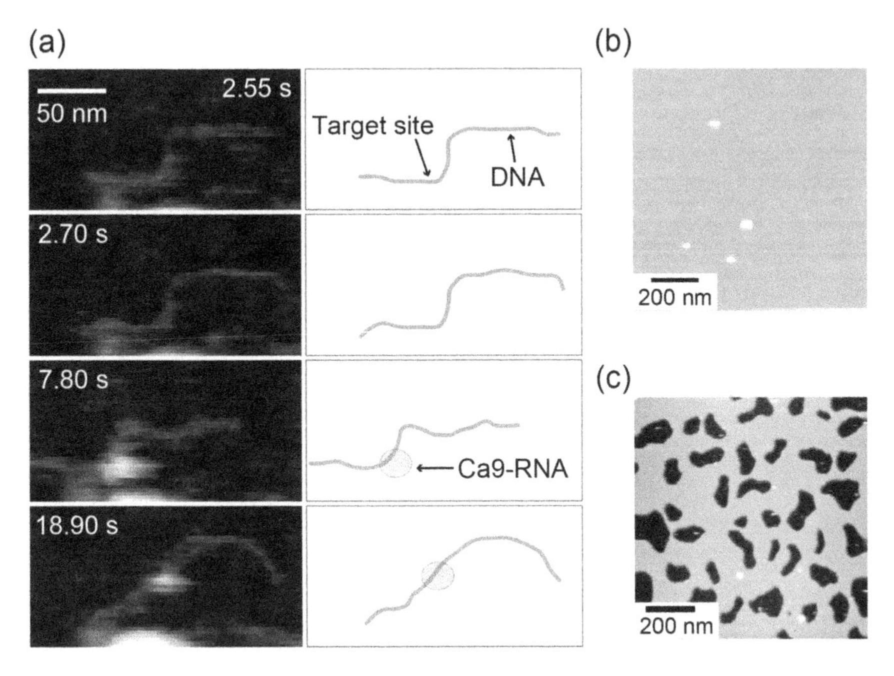

Fig. 6 HS-AFM images of Cas9-RNA complex binding to the target site of dsDNA. (**a**) Clipped HS-AFM images and schematics of target-specific binding of Cas9-RNA to the dsDNA with 600 bp adsorbed on the mixed lipid (DPPC:DPTAP:biotin-cap-DPPE = 0.90:0.05:0.05). Imaging rate, 0.15 s/frame. (**b**) A typical AFM image of a lipid bilayer composed of DPPC, DPTAP, and biotin-cap-DPPE. (**c**) An AFM image of the lipid bilayer after the drying during the incubation

PEG to the lipid bilayers make a small gap between the lipid bilayers and DNA, assisting the binding of DNA binding proteins to DNA in some cases.

Figure 6a demonstrates an application of the positively charged lipid membrane, showing the formation of DNA–protein complex. Cas9 is a CRISPR (Clustered Regularly Interspaced Short Palindromic Repeats) associated endonuclease and the complex with a guide RNA specifically binds to double-stranded DNA with a sequence complementary to the RNA [32]. The dsDNA of 600 bp with the target sequence of 20 nucleotides was adsorbed on the lipid composed of DPPC, DPTAP, and biotin-cap-DPPE in a weight ratio of 0.90:0.05:0.05 (lipid composition A). The HS-AFM images showed direct binding of the Cas9-RNA to the target site (Fig. 6a) without nonspecific binding of the Cas9-RNA to the lipid or nontarget sites on DNA.

3.7.1 Preparation of Lipid Bilayer on Mica

1. Prepare liposome solution [30, 33] of multilamellar vesicle form (1 mg/mL), make small aliquots (20 μL) and store at −80 °C before use. In the case of observing the formation process of Cas9-RNA and DNA complexes, a typical lipid composition is the lipid composition A (*see* **Note 11**).

2. Thaw the liposome solution stored at −80 °C. Dilute the solution to ~0.2 mg/mL by adding a buffer solution. In the case of lipid composition A, 10 mM $MgCl_2$ solution is added.
3. To obtain small unilamellar vesicles (SUV), sonicate the diluted liposome solution with a bath sonicator (35 W type) for 1 min.
4. Deposit the SUV solution of 2 μL on fleshly cleaved mica for 3 min, followed by 1 μL addition of 20 mM Tris–HCl, pH 8.8 to promote vesicle ruptures and surface coverage by lipid bilayers. Usually, we prepare ~10 vesicle-deposited sample stages at once and stored at room temperature in a sealed container that can maintain high humidity to avoid surface drying. The stored stages with lipid bilayer can be used for up to ~2 weeks unless the surface dries (*see* **Note 12**).
5. After waiting for more than 3 h, take one of the stored stages from the sealed container and glue on the z-scanner.
6. Rinse the sample surface with Milli-Q water dropwise (20 μL × 5) and replace the solution with a buffer solution suitable for sample dilution and deposition (*see* **Note 13**). In the case of observing DNA, the buffer solution is 20 mM Tris–HCl, pH 8.0, 30 mM KCl, 0.01 mM EDTA (*see* **Note 14**).
7. Place a sample solution for an appropriate incubation time, wash the sample solution with the observation buffer and perform AFM imaging. The incubation time for DNA adsorption is 3 min.

3.8 Graphite

Graphite is also frequently used as a substrate for conventional AFM observations as with mica because atomically flat surface is easily acquired. Among several types of graphite, highly ordered pyrolytic graphite (HOPG) is usually used because of low density of defects and steps. The HOPG surface is highly hydrophobic and therefore has been used for immobilization of samples with hydrophobic region. For example, crystalline cellulose and chitin are not immobilized on bare mica surface because hydrophobic fibril is hard to adsorb to hydrophilic mica surface in liquid environments [19, 34]. Figure 7a shows crystalline celluloses immobilized on the HOPG surface due to the hydrophobic interaction. Further, the stable fixation of the cellulose allows us to observe high-resolution observation of degradation process of cellulose by processive movements of cellulase, Cel7A from *T. reesei* (Fig. 7b), which is an enzyme that hydrolyzes β-1,4-glucosidic linkage of cellulose-producing soluble cellooligosaccharides.

3.8.1 Preparation of HOPG Substrate

1. Cut out an HOPG plate with thickness of ~0.1 mm from an HOPG block (10 × 10 × 2 mm^3) using a cutter blade.

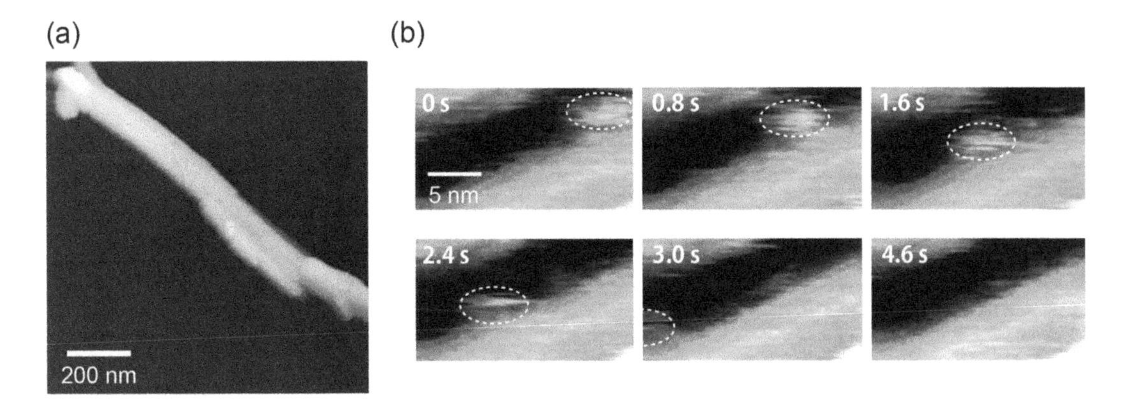

Fig. 7 Cellulose fixed on HOPG surface and cellulase moving along the cellulose. (**a**) Bundled cellulose fibrils are immobilized on hydrophobic HOPG surface. (**b**) Clipped HS-AFM images of cellulase encircled by broken circles moving along the cellulose due to hydrolysis of a polysaccharide chain. Imaging rate, 0.2 s/frame

2. Punch the HOPG disk (φ1.5 mm) by a puncher.
3. Glue the HOPG disk onto a cylindrical glass stage with epoxy glue. After gluing, the sample is left for more than 1 h to be tightly fixed on the stage.
4. Press a Scotch tape onto the HOPG surface evenly by wiping a cotton bud. Then peel off the top layers of the HOPG surface by removing the tape (*see* **Note 15**).
5. Place a sample solution for an appropriate incubation time, wash the sample solution with the observation buffer and perform AFM imaging. For the adsorption of cellulose, first the cellulose solution is placed on the HOPG and incubated for 5 min. Then the residual cellulose is washed with the observation buffer of 20 mM sodium acetate buffer, pH 5.0.

3.9 Gold-Coated Mica

Gold is often used as a substrate for formation of self-assembled monolayers (SAMs) of alkanethiols. Since the end of alkanethiol can be modified by various head groups, the SAMs on gold is useful to anchor proteins in the well-defined orientation. One example is site-specific immobilization of cellobiose dehydrogenases from *Phanerochaete chrysosporium* (*Pc*CDHs) (Fig. 8a), which are extracellular flavocytochromes [35]. By using technique for reconstitution of hemoproteins on a heme-immobilized surface via the heme–heme pocket interaction, oriented *Pc*CDHs are anchored on the SAM/gold surface (Fig. 8b). Interdomain dynamics between the closed and open configuration depending on concentrations of the cellobiose substrate were observed (Fig. 8c) [35]. Furthermore, HS-AFM combined with electrochemical measurement allows us to image adsorption dynamics of cytochrome c to the gold surface with simultaneous measurement of cyclic

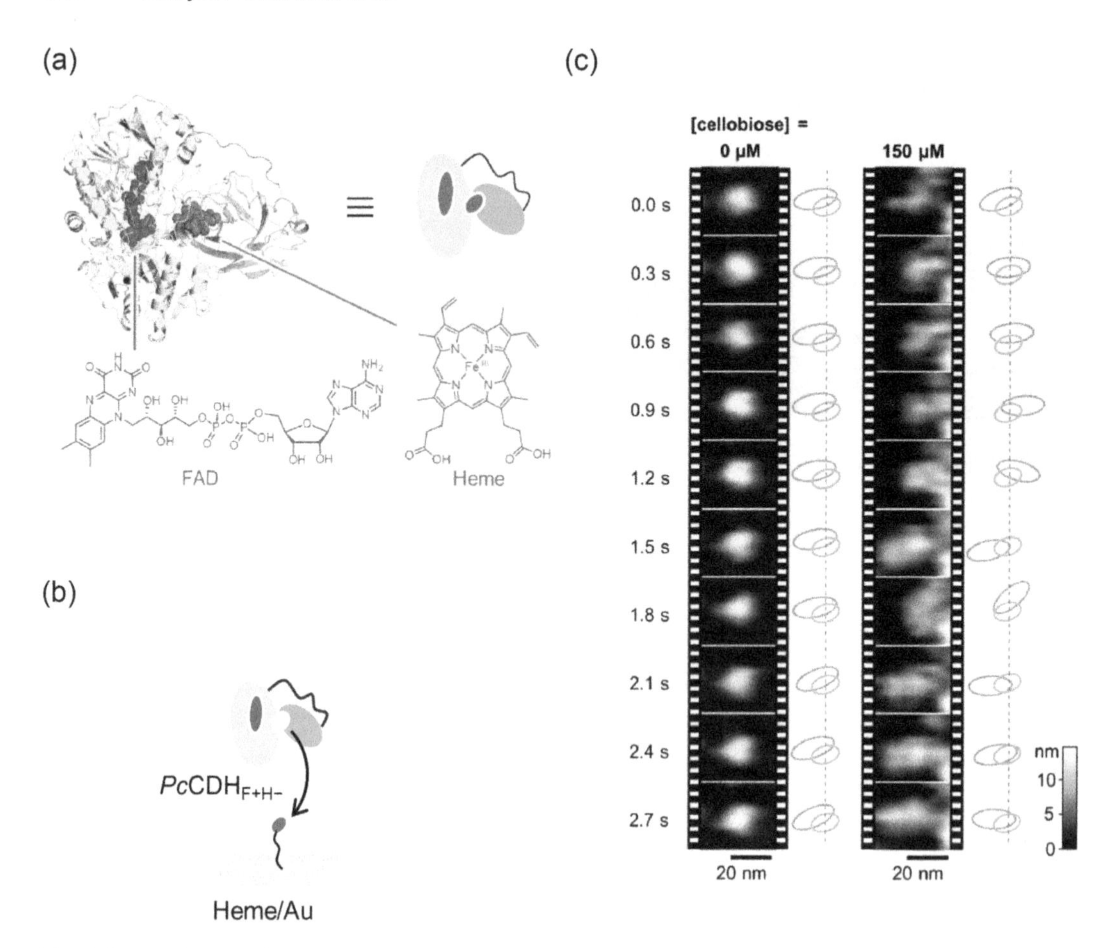

Fig. 8 Immobilization *PcCHD* via the heme–heme pocket interaction on SAM/gold surface and interdomain flip-flop motion imaged by HS-AFM. (**a**) PcCDH and consists of the FAD-binding DH domain in yellow (PDB: 1NAA) and the heme-binding CYT domain in green (PDB: 1D7C): the linker region is illustrated as a gray line. Chemical structures of oxidized FAD and heme are shown. (**b**) Strategy for the immobilization of catalytically active *PcCDH* on the heme-modified flat gold surface (heme/Au). (**c**) Successive HS-AFM images of the domain motion of *PcCDH* in a different concentration of cellobiose (0 and 150 μM). Estimated relative positions of the DH domain (orange) and the CYT domain (green) are illustrated on the right side of each set of image frames

voltammetry due to the electron transfer from the protein to the substrate [36].

3.9.1 Preparation of Gold Substrate

1. Anneal a gold /mica sheet in a flame of a gas burner for a few seconds (*see* **Note 16**).
2. For observing *Pc*CDH, the gold surface is first modified with heme/SAM. The substrate is immersed in the solution of 1 mM mixed thiol (6-hydroxy-1- hexanethiol/11-amino-1-undecanethiol = 10,000/1) at 42 °C for 12 h. Then it is washed with ethanol and dried under Ar flow (99.99%) to afford a SAM-modified gold surface (SAM/Au). SAM/Au is

immersed in 2.5 mM hemin and 7.5 mM 1-ethyl-3-(3-dimethylaminopropyl)carbodiimide (EDC), and 3 mM triethylamine in DMSO (dimethylsulfoxide) (10 mL) and incubated in dark at 42 °C for 12 h. The surface is thoroughly washed with DMSO and ethanol, then dried under Ar flow [35].

3. Punch the gold/mica disk (φ1.5 mm) by a puncher. (*see* **Note 17**)
4. Glue the gold/mica disk onto a cylindrical glass stage with adhesive.
5. Place a sample solution for an appropriate incubation time, wash the sample solution with the observation buffer and perform AFM imaging. In the case of observing *Pc*CDH, the incubation time is 10 min and the observation buffer is 50 mM NaOAc buffer (pH 4.5)

3.10 PDMS

One of next challenges of application studies on AFM in biology is to observe biological phenomena occurring at nonflat surfaces. PDMS (polydimethylsiloxane) is a thermosetting biocompatible elastomer with chemical resistance and useful to fabricate submicrometer-sized shapes without using expensive and hazardous materials [37]. AFM substrates with small concave and/or convex shapes can be formed on PDMS substrate by the nanosphere imprinting method [38]. Since such substrates may deform biological membranes or biomolecules, one can investigate the phenomena depending on the curvature of biological membranes or the function modulations induced by the mechanical deformation of such objects placed on the substrates. Moreover, by sealing tiny holes created on a substrate with biological membrane continuing membrane proteins, the functional and structural relationships of the membrane proteins can be analyzed [39]. Figure 9a, b show convex and concave structures formed on PDMS surfaces, respectively. The substrate with the concave pattern can be applied to form a suspended membrane, for example purple membrane (Fig. 9c), which will enable imaging membrane proteins with separated conditions at outer and inner membrane surfaces. Here we describe the protocol for preparation of PDMS substrate with submicrometer-sized convex or concave shapes using the nanosphere imprinting method.

3.10.1 Preparation Patterned PDMS Substrate

1. Deposit a solution containing nanospheres (0.01–0.1 mg/mL) on a glass slide for overnight. The incubation is done in a sealed container that can maintain high humidity to avoid surface drying. Polystyrene and silica beads are used for making concave and convex shapes, respectively.

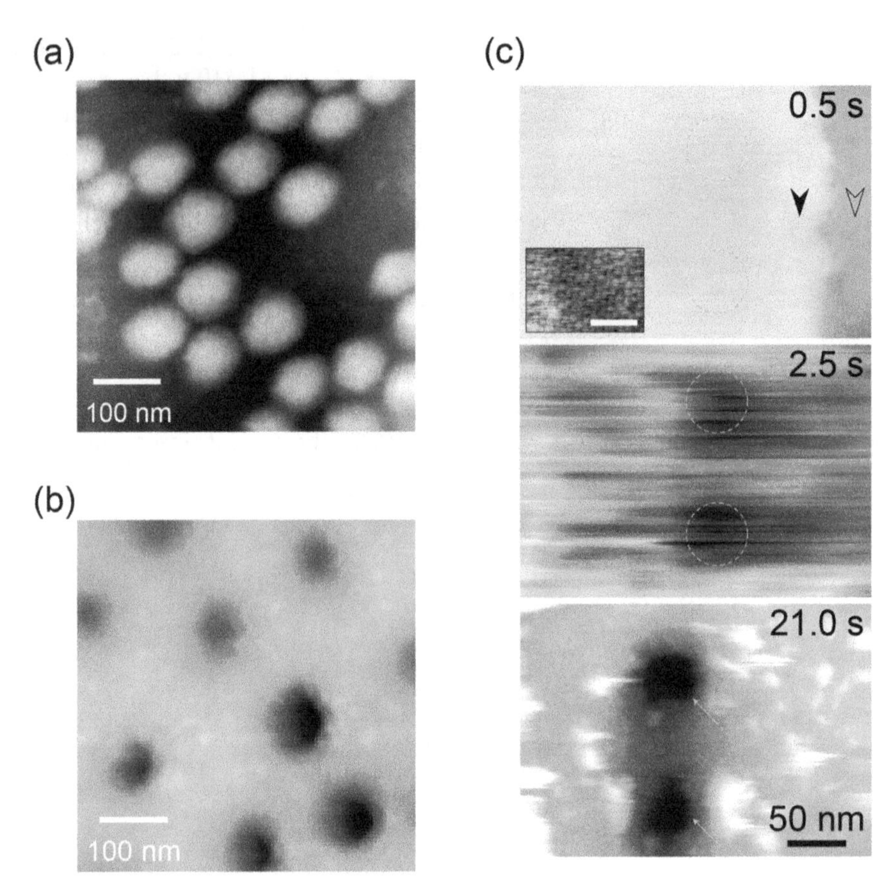

Fig. 9 Patterned PDMS substrate and application to formation of suspended membrane (**a**, **b**). AFM substrates with convex (**a**) and concave (**b**) shapes formed on PDMS surface. Convex and concave shapes were made by silica and polystyrene beads (100 nm in diameter), respectively. (**c**) Successive AFM images showing purple membrane placed on PDMS substrate with concave shapes made by 100 nm polystyrene beads. Closed and open arrow heads indicate purple membrane and PDMS surfaces, respectively. Inset, magnified image showing the lattice structure of bacteriorhodopsin formed in the purple membrane. Scale bar, 20 nm. From 2.5 s, the purple membrane was removed by the AFM tip, and the holes indicated by green arrows appeared under the purple membrane at 21.0 s. The positions where the holes appeared are indicated by green dashed lines on the images of 0.5 and 2.5 s. Z-scale, 40 nm

2. Wash out the nanosphere solution with large amount of Milli-Q water (approximately 50 ml or more) without drying the glass surface.
3. Blow away the water remained on the glass surface by N_2 gas vigorously. (*see* **Note 18**)
4. Deposit PDMS on the glass surface and cure PDMS at 80 °C for more than 3 h (*see* **Note 19**). After cooling down, cut out PDMS disk (1.5 mm in diameter) from the cured PDMS sheet using a biopsy punch.
5. Glue the back side of the PDMS disk, the surface not facing with nanospheres, onto an AFM glass stage.

6. For the applications using the convex shapes, one can directly use the AFM substrate made in the **step 5**. For the applications using the concave shapes, immerse the PDMS disk in acetone, and remove and dissolve polyethylene beads by sonication for 5 min followed by rinsing with sonication in Milli-Q water for 5 min.
7. Place a sample solution for an appropriate incubation time, wash the sample solution with the observation buffer and perform AFM imaging. For the observation of purple membrane sealing the concave structure, the incubation time is 10 min and the observation buffer is 10 mM phosphate (pH 7.0) and 300 mM KCl.

3.11 Conclusions

Here, we present solid substrates for HS-AFM to observe the dynamic behavior of biomolecules. We describe usability and preparation methods of various substrates: bare mica, Ni/mica, aminosilane/mica, glutaraldehyde/aminosilane/mica, lipid bilayer, graphite, gold, and patterned PDMS. The choice of substrate is one of the key factors for successful experiments of HS-AFM. However, the variety of available substrates is still limited for further applications of HS-AFM to diverse samples. Development of more versatile substrates would be required in the near future.

4 Notes

1. The small cantilevers are produced by Olympus Corp., Tokyo, Japan. Two types of cantilevers, BL-AC10FS-A2 and BL-AC10DS-A2, are available. The BL-AC10FS-A2 has a carbon-nanofiber tip with the end diameter less than 10 nm at the end of the cantilever, while the BL-AC10DS-A2 has no sharp tip. A sharp tip can be formed by electron-beam-induced carbon deposition using electron microscope [40].
2. A whole surface of the mica disk should be firmly fixed on the glass stage. If a part of the mica disk protrudes from the glass stage and the AFM probe approaches the protruded part, the imaging stability markedly deteriorates. Therefore one should carefully inspect the adhesive condition between the mica and the glass before use.
3. The incubation duration highly depends on protein samples. For a small protein, it is usually 3–5 min, while a larger protein complex such microtuble and actin filament needs a longer incubation time over 10 min. It should be noted that the sample solution is not dried during the incubation of sample. Drying sample sometimes causes not only denature of proteins but also aggregations and contaminations of the surface. Avoid drying by covering the stage with a small cap with wet Texwipe.

4. Choice of an observation buffer highly depends on protein samples to be observed. In general, a high salt concentration in the buffer weakens the affinity between the proteins and the samples on the bare mica surface and K^+ influences the affinity more than Na^+.
5. During the 3-min incubation, avoid drying the $NiCl_2$ or $NiSO_4$ solution by covering the stage with a small cap with wet Texwipe.
6. pH of the observation buffer is critical for the Ni^{2+}-chelating ability of His-tag because the binding occurs at neutral or slightly alkali pH (6.5–8.0). Therefore, one should avoid using the observation buffer with a pH less than 6.0 and with chelating agents such as EDTA, EGTA and imidazole.
7. The Si-OR bonds of APTES hydrolyze with water to form silanol Si-OH groups, which can condense with each other to form polymeric structures with very stable siloxane Si–O–Si bonds. The oligomerization causes less reactive to the mica substrate. Therefore the APTES should be dissolved in water just before use.
8. An appropriate buffer solution for the APTES/mica also depends on samples to be observed. But, in general, it would be better to avoid using the sample solution with a high pH because an alkaline pH reduces positive charges of the amino-silane. 3-min incubation is usually enough to adsorb proteins onto the APTES/mica. But a longer incubation time may be required for relatively large biomolecules such as microtuble, actin filament and DNA.
9. During the 3-min incubation, avoid drying the GA solution by covering the stage with a small cap with wet Texwipe.
10. Since GA reacts with amine groups, treatment with crosslinkers should be conducted in buffers free from amines. Tris, glycine, or imidazole should be avoided to use a buffer for diluting proteins. After the proteins were immobilized on GA-treated APTES/mica, it would be better to terminate the reaction by washing with a 1 M Tris–HCl, pH 8.0.
11. When one could not see more than 70% surface coverage by lipid bilayers after 5 min incubation, the solution used for liposome dilution is not appropriate.
12. Avoid slow temperature changes across the transition temperature of lipid molecules. This may induce phase separation of lipid bilayers, resulting in nonuniform surface properties.
13. Do not dry the surface, otherwise most of the bilayers will certainly be damaged. Typically, one can see the hole-like appearance (~100 nm in diameter) in the lipid bilayers, when the surface is dried once (Fig. 6b, c).

14. Avoid the observation buffer including detergent higher than the critical micelle concentration. Otherwise, the lipid bilayer is disrupted.
15. Complete cleavage of HOPG surface is bit harder compared to mica because a thin HOPG layer is flexible and the cleavage layers are easily fragmented. Also the edge of the HOPG disk tends to be serrate. Serrated edge prevents the approach of the AFM prove to the surface and produces fatal vibrations on the image. Therefore one should carefully inspect whether the peripheral parts of the disk show no burrs.
16. After the flame annealing of the gold/mica substrate, it should be used as soon as possible to avoid contamination and possible changes in surface roughness.
17. To keep the gold surface clean, preparation of the disk-shaped substrate by a puncher should be carried out without the puncher head touching the gold surface. Also after making the disk shape, one should be careful not to touch the gold surface by tips of tweezers as handling.
18. Slow blowing induces aggregation of nanosphere.
19. Degas PDMS using an aspirator for 30 min before deposition, or bubbles would appear in the products.

Acknowledgments

We thank Dr. Christian Ganser for critical reading and for improving the language of the manuscript. This work was supported by JSPS KAKENHI Grant Numbers 15H04360 (N.K.), 24227005 and 16H00830, 16H00758 and 15H03540 (T.U.), and JST/PRESTO (JPMJPR13L4) (N.K.).

References

1. Zhong Q, Inniss D, Kjoller K, Elings VB (1993) Fractured polymer silica fiber surface studied by tapping mode atomic-force microscopy. Surf Sci 290(1–2):L688–L692. https://doi.org/10.1016/0167-2584(93)90906-y
2. Ando T, Kodera N, Takai E, Maruyama D, Saito K, Toda A (2001) A high-speed atomic force microscope for studying biological macromolecules. Proc Natl Acad Sci U S A 98 (22):12468–12472. https://doi.org/10.1073/pnas.211400898
3. Kokavecz J, Toth Z, Horvath ZL, Heszler P, Mechler A (2006) Novel amplitude and frequency demodulation algorithm for a virtual dynamic atomic force microscope. Nanotechnology 17(7):S173–S177. https://doi.org/10.1088/0957-4484/17/7/s12
4. Kodera N, Sakashita M, Ando T (2006) Dynamic proportional-integral-differential controller for high-speed atomic force microscopy. Rev Sci Instrum 77(8):6. https://doi.org/10.1063/1.2336113
5. Schitter G, Allgower F, Stemmer A (2004) A new control strategy for high-speed atomic force miciroscopy. Nanotechnology 15 (1):108–114. https://doi.org/10.1088/0957-4484/15/1/021
6. Kodera N, Yamashita H, Ando T (2005) Active damping of the scanner for high-speed atomic force microscopy. Rev Sci Instrum 76(5):5. https://doi.org/10.1063/1.1903123

7. Ando T, Uchihashi T, Scheuring S (2014) Filming biomolecular processes by high-speed atomic force microscopy. Chem Rev 114 (6):3120–3188. https://doi.org/10.1021/cr4003837
8. Ando T, Uchihashi T, Fukuma T (2008) High-speed atomic force microscopy for nano-visualization of dynamic biomolecular processes. Prog Surf Sci 83(7–9):337–437. https://doi.org/10.1016/j.progsurf.2008.09.001
9. Shibata M, Yamashita H, Uchihashi T, Kandori H, Ando T (2010) High-speed atomic force microscopy shows dynamic molecular processes in photoactivated bacteriorhodopsin. Nat Nanotechnol 5(3):208–212. https://doi.org/10.1038/nnano.2010.7
10. Uchihashi T, Iino R, Ando T, Noji H (2011) High-speed atomic force microscopy reveals rotary catalysis of rotorless F-1-ATPase. Science 333(6043):755–758. https://doi.org/10.1126/Science.1205510
11. Yamashita H, Inoue K, Shibata M, Uchihashi T, Sasaki J, Kandori H, Ando T (2013) Role of trimer-trimer interaction of bacteriorhodopsin studied by optical spectroscopy and high-speed atomic force microscopy. J Struct Biol 184(1):2–11. https://doi.org/10.1016/j.jsb.2013.02.011
12. Yokokawa M, Takeyasu K (2011) Motion of the Ca2+−pump captured. FEBS J 278 (17):3025–3031. https://doi.org/10.1111/j.1742-4658.2011.08222.x
13. Rangl M, Miyagi A, Kowal J, Stahlberg H, Nimigean CM, Scheuring S (2016) Real-time visualization of conformational changes within single MloK1 cyclic nucleotide-modulated channels. Nat Commun 7:12789. https://doi.org/10.1038/ncomms12789
14. Kowal J, Chami M, Baumgartner P, Arheit M, Chiu PL, Rangl M, Scheuring S, Schroder GF, Nimigean CM, Stahlberg H (2014) Ligand-induced structural changes in the cyclic nucleotide-modulated potassium channel MloK1. Nat Commun 5:3106. https://doi.org/10.1038/ncomms4106
15. Yamashita H, Voitchovsky K, Uchihashi T, Contera S, Ryan J, Ando T (2009) Dynamics of bacteriorhodopsin 2D crystal observed by high-speed atomic force microscopy. J Struct Biol 167(2):153–158. https://doi.org/10.1016/j.jsb.2009.04.011
16. Casuso I, Khao J, Chami M, Paul-Gilloteaux P, Husain M, Duneau JP, Stahlberg H, Sturgis JN, Scheuring S (2012) Characterization of the motion of membrane proteins using high-speed atomic force microscopy. Nat Nanotechnol 7(8):525–529. https://doi.org/10.1038/Nnano.2012.109
17. Gregoire C, Marco S, Thimonier J, Duplan L, Laurine E, Chauvin JP, Michel B, Peyrot V, Verdier JM (2001) Three-dimensional structure of the lithostathine protofibril, a protein involved in Alzheimer's disease. EMBO J 20 (13):3313–3321. https://doi.org/10.1093/emboj/20.13.3313
18. Watanabe-Nakayama T, Ono K, Itami M, Takahashi R, Teplow DB, Yamada M (2016) High-speed atomic force microscopy reveals structural dynamics of amyloid beta(1-42) aggregates. Proc Natl Acad Sci U S A 113 (21):5835–5840. https://doi.org/10.1073/pnas.1524807113
19. Igarashi K, Uchihashi T, Koivula A, Wada M, Kimura S, Okamoto T, Penttila M, Ando T, Samejima M (2011) Traffic jams reduce hydrolytic efficiency of cellulase on cellulose surface. Science 333(6047):1279–1282. https://doi.org/10.1126/Science.1208386
20. Kodera N, Yamamoto D, Ishikawa R, Ando T (2010) Video imaging of walking myosin V by high-speed atomic force microscopy. Nature 468(7320):72–76. https://doi.org/10.1038/nature09450
21. Abeliovich A, Schmitz Y, Farinas I, Choi-Lundberg D, Ho WH, Castillo PE, Shinsky N, Verdugo JM, Armanini M, Ryan A, Hynes M, Phillips H, Sulzer D, Rosenthal A (2000) Mice lacking alpha-synuclein display functional deficits in the nigrostriatal dopamine system. Neuron 25 (1):239–252
22. Zhu T, Satoh T, Kato K (2014) Structural insight into substrate recognition by the endoplasmic reticulum folding-sensor enzyme: crystal structure of third thioredoxin-like domain of UDP-glucose: glycoprotein glucosyltransferase. Sci Rep 4:7322. https://doi.org/10.1038/srep07322
23. Satoh T, Song C, Zhu T, Toshimori T, Murata K, Hayashi Y, Kamikubo H, Uchihashi T, Kato K (2017) Visualisation of a flexible modular structure of the ER folding-sensor enzyme UGGT. Sci Rep 7(1):12142. https://doi.org/10.1038/s41598-017-12283-w
24. Shlyakhtenko LS, Gall AA, Filonov A, Cerovac Z, Lushnikov A, Lyubchenko YL (2003) Silatrane-based surface chemistry for immobilization of DNA, protein-DNA complexes and other biological materials. Ultramicroscopy 97(1–4):279–287. https://doi.org/10.1016/S0304-3991(03)00053-6
25. Lyubchenko YL, Shlyakhtenko LS, Ando T (2011) Imaging of nucleic acids with atomic

force microscopy. Methods 54(2):274–283. https://doi.org/10.1016/j.ymeth.2011.02.001

26. Miyagi A, Ando T, Lyubchenko YL (2011) Dynamics of nucleosomes assessed with time-lapse high-speed atomic force microscopy. Biochemistry 50(37):7901–7908. https://doi.org/10.1021/bi200946z
27. Feinstein A, Rowe AJ (1965) Molecular mechanism of formation of an antigen-antibody complex. Nature 205:147–149
28. Lanyi JK (2004) Bacteriorhodopsin. Annu Rev Physiol 66:665–688. https://doi.org/10.1146/annurev.physiol.66.032102.150049
29. Butt HJ, Downing KH, Hansma PK (1990) Imaging the membrane-protein bacteriorhodopsin with the atomic force microscope. Biophys J 58(6):1473–1480
30. Yamamoto D, Uchihashi T, Kodera N, Yamashita H, Nishikori S, Ogura T, Shibata M, Ando T (2010) High-speed atomic force microscopy techniques for observing dynamic biomolecular processes. Methods Enzymol 475:541–564. https://doi.org/10.1016/S0076-6879(10)75020-5
31. Colom A, Redondo-Morata L, Chiaruttini N, Roux A, Scheuring S (2017) Dynamic remodeling of the dynamin helix during membrane constriction. Proc Natl Acad Sci U S A 114 (21):5449–5454. https://doi.org/10.1073/pnas.1619578114
32. Garneau JE, Dupuis ME, Villion M, Romero DA, Barrangou R, Boyaval P, Fremaux C, Horvath P, Magadan AH, Moineau S (2010) The CRISPR/Cas bacterial immune system cleaves bacteriophage and plasmid DNA. Nature 468(7320):67–71. https://doi.org/10.1038/nature09523
33. Uchihashi T, Kodera N, Ando T (2012) Guide to video recording of structure dynamics and dynamic processes of proteins by high-speed atomic force microscopy. Nat Protoc 7 (6):1193–1206. https://doi.org/10.1038/Nprot.2012.047
34. Igarashi K, Uchihashi T, Uchiyama T, Sugimoto H, Wada M, Suzuki K, Sakuda S, Ando T, Watanabe T, Samejima M (2014) Two-way traffic of glycoside hydrolase family 18 processive chitinases on crystalline chitin. Nat Commun 5:3975. https://doi.org/10.1038/ncomms4975
35. Harada H, Onoda A, Uchihashi T, Watanabe H, Sunagawa N, Samejima M, Igarashi K, Hayashi T (2017) Interdomain flip-flop motion visualized in flavocytochrome cellobiose dehydrogenase using high-speed atomic force microscopy during catalysis. Chem Sci 8(9):6561–6565. https://doi.org/10.1039/c7sc01672g
36. Takeda K, Uchihashi T, Watanabe H, Ishida T, Igarashi K, Nakamura N, Ohno H (2015) Real-time dynamic adsorption processes of cytochrome c on an electrode observed through electrochemical high-speed atomic force microscopy. PLoS One 10(2): e0116685. https://doi.org/10.1371/journal.pone.0116685
37. Jo BH, Van Lerberghe LM, Motsegood KM, Beebe DJ (2000) Three-dimensional microchannel fabrication in polydimethylsiloxane (PDMS) elastomer. J Microelectromech Syst 9 (1):76–81. https://doi.org/10.1109/84.825780
38. Chen JY, Chang WL, Huang CK, Sun KW (2011) Biomimetic nanostructured antireflection coating and its application on crystalline silicon solar cells. Opt Express 19 (15):14411–14419. https://doi.org/10.1364/Oe.19.014411
39. Goncalves RP, Agnus G, Sens P, Houssin C, Bartenlian B, Scheuring S (2006) Two-chamber AFM: probing membrane proteins separating two aqueous compartments. Nat Methods 3(12):1007–1012. https://doi.org/10.1038/nmeth965
40. Wendel M, Lorenz H, Kotthaus JP (1995) Sharpened electron beam deposited tips for high resolution atomic force microscope lithography and imaging. Appl Phys Lett 67 (25):3732–3734. https://doi.org/10.1063/1.115365

Chapter 11

High-Resolution and High-Speed Atomic Force Microscope Imaging

Francesca Zuttion, Lorena Redondo-Morata, Arin Marchesi, and Ignacio Casuso

Abstract

The advent of high-speed atomic force microscopy (HS-AFM) over the recent years has opened up new horizons for the study of structure, function and dynamics of biological molecules. HS-AFM is capable of 1000 times faster imaging than conventional AFM. This circumstance uniquely enables the observation of the dynamics of all the molecules present in the imaging area. Over the last 10 years, the HS-AFM has gone from a prototype-state technology that only a few labs in the world had access to (including ours) to an established commercialized technology that is present in tens of labs around the world. In this protocol chapter we share with the readers our practical know-how on high resolution HS-AFM imaging.

Key words High speed atomic force microscope, HS-AFM, Cantilever, Electron beam deposited tip, Plasma EBD tip sharpening, Tip–sample approach, Oscillation amplitude, PID, Feedback optimization, Image treatment

1 Introduction

The atomic force microscopy (AFM) [1] is a powerful tool for direct visualization of biological samples in an aqueous solution with submolecular resolution. It allows noninvasive imaging in physiological conditions of unlabeled biological samples, such as nucleic acids [2–4] and cell membrane proteins [5, 6]. AFM has rapidly emerged as an effective structural analysis tool, complementing atomic structure data acquired by other techniques such as X-ray crystallography, NMR, and electron microscopy. In contrast to the latter methods, which heavily rely on ensemble averaging, AFM exhibits a superior signal-to-noise ratio, enabling molecules to be observed individually under an aqueous environment. To acquire a topographic image, a sharp tip attached to the free end of a flexible cantilever is brought into contact and scanned over a sample and its profile height is recorded over the selected scan area. The feedback control allows for adjusting the force

Yuri L. Lyubchenko (ed.), *Nanoscale Imaging: Methods and Protocols*, Methods in Molecular Biology, vol. 1814, https://doi.org/10.1007/978-1-4939-8591-3_11,

applied to the sample to avoid damaging fragile biological structures. As a result of the imaging process, during which the distance between the sample stage and the AFM probe varied following the sample topography, a 3D profilometric image of the sample surface is obtained. With the advantage of operating in nearly physiological conditions, AFM is an ideal platform to study biological dynamic processes at the molecular scale. Nevertheless, the framerate of one image every few minutes of conventional AFM is not sufficient to visualize most dynamic processes of biological molecules that take place in its majority at subsecond time scales. The introduction of the high-speed atomic force microscopy (HS-AFM) in 2001 opened the door to the visualization of biological molecule dynamics [7]. The technical development of the HS-AFM was performed by the laboratory of Toshio Ando in the University of Kanazawa, Japan.

HS-AFM allows imaging at maximum rates of five to ten frames per second [8–11]. The operating principle of the HS-AFM is based on the miniaturization of the moving components of the AFM (cantilever and scanner) to increase their velocity by 1000 times, in the interest of achieving reaction speeds of tens of microseconds. For the HS-AFM probe, the gain in speed does not suppose a lost in performance. In the case of the scanner, the *X*, *Y*, *Z* displacement range is decreased ~1:10 over conventional AFMs. With this technique, centrosome [12, 13], DNA origami [14], micelles [15], and many other biological systems have been imaged at high speed and high resolution.

In this book chapter, we describe in detail our HS-AFM imaging procedure for high resolution and high speed imaging, applicable to any version of the Ando-type HS-AFM system "SS-NEX". Such procedure is based on our experience imaging individual molecules [10, 16] or big macromolecular assemblies [17, 18].

2 Materials

2.1 High-Speed AFM

The protocol here described is adapted for any Ando-type HS-AFM setup "SS-NEX", commercialized by the Research Institute of Biomolecule Metrology (RIBM), Japan. A picture of the HS-AFM hardware of our "SS-NEX" HS-AFM is presented in Fig. 1, where the cantilever is mounted up-facing in a cantilever holder that comes with a pool of 110 μL while the sample is mounted on top of the scanner facing down. In general, imaging experiments are performed in tapping mode where the tip is oscillated at constant amplitude. For that, the cantilever is excited at its resonance frequency with the help of a miniaturized piezoelectric actuator. The sample can move in the *XY*-plane and also in the *Z* vertical direction thanks to three small multilayer piezoelectric actuators. The *Z*-piezo the one displaced at the highest frequency

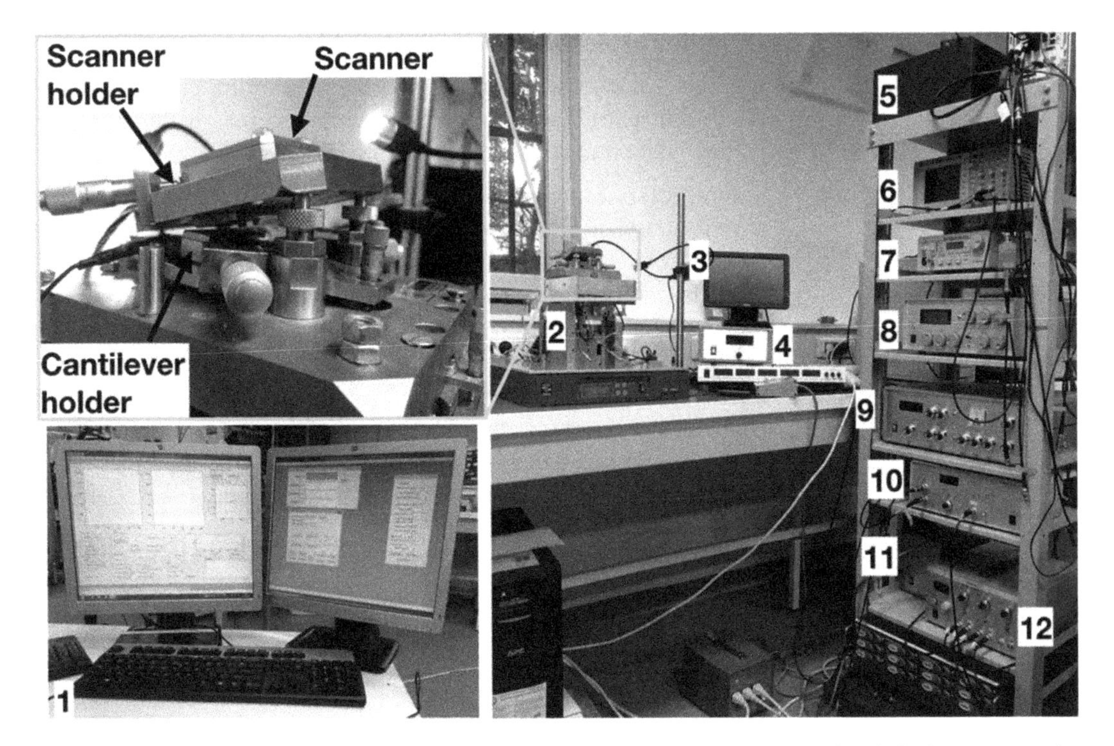

Fig. 1 HS-AFM hardware. Image of the HS-AFM hardware. Relevant parts are signaled: the scanner, the scanner holder and the cantilever holder are shown. Right, zoom-in of the HS-AFM hardware, where the sample stage is plunged in the pool, above the cantilever

(nominal resonance frequency ~600 kHz and ~150 kHz once glued on the scanner). The deflection of the cantilever is monitored by an optical system. A laser beam with a 20× optical microscope objective is focused on the backside of the cantilever and the reflected beam tracked with a 15 MHz bandwidth photodiode which sends the signal to a Lock-In amplifier (Fourier Analyzer). The signal goes to the Proportional Integral Derivative (PID) controller which compares the amplitude readout with the value settled by the operator (set point amplitude, A) to feed the adjustment of the clearance distance between the sample stage and the HS-AFM probe by the feedback loop.

2.2 Sample Preparation and Mounting

1. Glass-rod (small glass cylinder of 1.5 mm in diameter and 2 mm in height purchased from RIBM, Japan or SCHOTT, Switzerland), colorless nail polish (nitrocellulose dissolved in butyl acetate or ethyl acetate), mica sheet, puncher purchased from RIBM, cyanoacrylate glue (Super Glue®) or two components epoxy adhesive (Araldite®), adhesive tape (Scotch®), cellulose wipers (Kimwipes®), humid hood, microfuge tube of 1 mL.
2. HS-AFM probe: AC10 from Olympus ($f_c = 1.5$ MHz in air and $k_c = 0.1$ N/m) or USC-F1.5-k0.6 Nanoworld ($f_c = 1.5$ MHz in air and $k_c = 0.6$ N/m).

3. Optical microscope, tweezers, screwdrivers.
4. Rinsing solutions: anionic precision detergent cleaner (Alconox® dilute to 2% in water), ultrapure water (Milli-Q®), tap water, 70% ethanol, acetone.
5. Sample case: Supported Lipid Bilayer (SLB). Lipids: 1-palmitoyl-2-oleoyl-*sn*-glycero-3-phospho-(1′-rac-glycerol) (POPG), 1-palmitoyl-2-oleoyl-*sn*-glycero-3-phosphoethanolamine (POPE), chloroform, plastic tube (15 mL Falcon®), microfuge tube of 2 mL, Hamilton syringe, fume hood, vacuum desiccator, vacuum pump, potassium chloride, calcium chloride, working buffers: "liposome hydration buffer" 150 mM KCl, 10 mM Tris (tris(hydroxymethyl)aminomethane), pH 7.4 and "incubation buffer" 10 mM Hepes (4-(2-hydroxyethyl)-1-piperazineethanesulfonic acid), 100 mM NaCl, 5 mM $CaCl_2$, pH 7.4.

2.3 Plasma Sharpening of the Electron Beam Deposited Tips

1. A plasma system for surface treatment. We use a 40 kHz Diener® Zepto system.
2. Piece of $2 \times 1 \times 1$ cm^3 of glass or Teflon.
3. Double sided carbon conductive tape.
4. Bottles of oxygen and helium gas.

2.4 Image Treatment

1. Image processing software supplied with your AFM instrument (*see* **Note 1**).
2. MATLAB (MathWorks, USA). This is a proprietary programming language specifically designed for simple and efficient matrix manipulation, data plotting and algorithm implementation (*see* **Note 2**). A 30-day trial version is available for free download (https://www.mathworks.com). All the procedures described in Subheading 3.5 can be carried out in MATLAB software environment. Routines written in this language for batch image processing described in Subheading 3.5 are available at https://sites.google.com/view/fm4b-lab/group-members/dr-arin-marchesi.
3. ImageJ (NIH). This is a free, Java-based image processing platform (https://imagej.nih.gov/ij/). Although not specifically designed for AFM image treatment, a number of user-written plugins exists to efficiently tackle several of the procedures described in Subheading 3.5. A routine written in MATLAB to convert the *.asd* HS-AFM file format in multipage 32-bit *.tiff* files supported by ImageJ is available at https://sites.google.com/view/fm4b-lab/group-members/dr-arin-marchesi.

3 Methods

3.1 Sample Preparation

1. Glue a cleaned glass-rod on the *Z*-piezo by using nail polish and let it dry for at least 20 min. Glass-rods can be cleaned by sonication ($f \sim 35$ kHz) in acetone and stored in a plastic box until use.
2. Glue a mica disk on the glass-rod with cyanoacrylate glue and let it dry for 10 min (*see* **Note 3**).
3. Cleave the mica disk with adhesive tape to obtain a clean, atomically flat surface. To facilitate the cleavage, mica can be previously cleaned with acetone (*see* **Note 4**).
4. Deposit a 1.2–2 μL drop of the sample solution on the mica surface immediately after cleavage. Cover the sample with a humid hood to avoid drying (Fig. 2, *see* **Note 5**).
5. Fill a cap of a microfuge tube of 1 mL with the working buffer solution (~300 μL). Immerse the sample to wash it; this will remove the excess of molecules that have not adsorbed on the surface. Repeat the procedure at least five times without the need to replace the buffer between rinses.
6. Keep the sample wet with the working buffer solution under the humid hood until it is mounted on the HS-AFM.

3.2 Sample Mounting

1. Wash abundantly the cantilever holder with 2% anionic detergent, tap water, acetone, 70% ethanol, and ultrapure water. Dry it gently with air or nitrogen flow if possible (*see* **Note 6**).

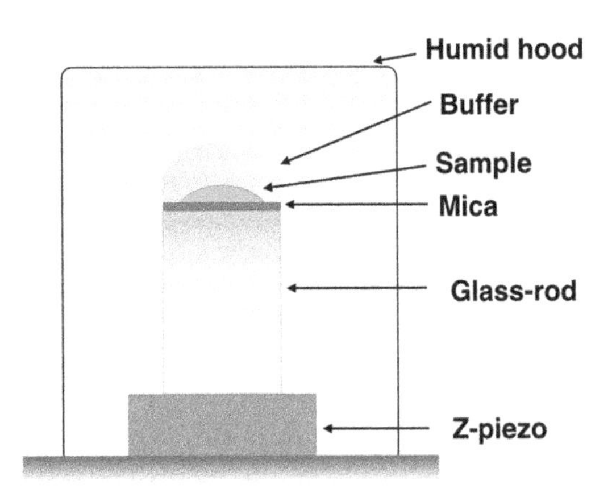

Fig. 2 Sample incubation. Image of the humid hood used for sample incubation on a sample stage (mica disk glued on the glass rod) attached to the *Z*-piezo of the scanner. The humid ambience is maintained by a wet cellulose wipe inserted in the hood

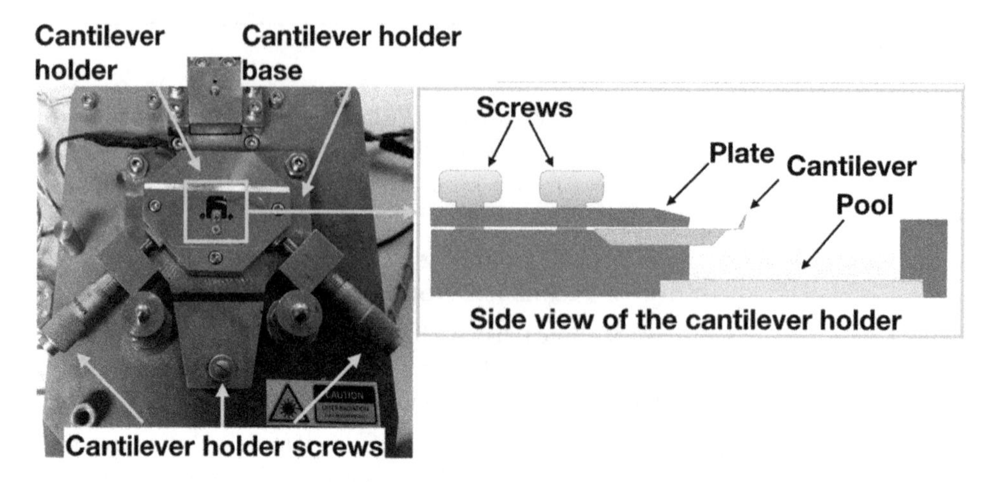

Fig. 3 HS-AFM cantilever holder. Left, top view of the high-speed cantilever holder, showing how the cantilever holder is positioned on the cantilever holder base and the micrometer screws. Right, sketch of the side view of the chip cantilever holder, showing the cantilever holding screws, the plate, a side view of the chip with its cantilever and the pool filled with aqueous solution

2. Set a small cantilever on the cantilever holder: position the cantilever on the trench of the cantilever holder and fix it with the cantilever holding screws (Fig. 3, *see* **Note 7**).
3. Mount the cantilever holder on its base by placing the two magnets of the cantilever holder in contact with the micrometer screws of the base.
4. Wash the pool with 110 μL ultrapure water for ten times.
5. Fill the pool with 110 μL of the working buffer.
6. Switch on the laser controller, select the "*I mode*" and adjust the intensity taking into account the SUM signal and the cantilever thermal fluctuation. This means that the chosen value should allow reaching a high SUM signal (in the range of 0.5–1.2 V) when the intensity of the reflected laser spot from the cantilever on the photodetector is maximized, while the thermal fluctuation of the cantilever is kept lower (in the range of 10–20 mV) (*see* **Note 8**).
7. Focus the cantilever on the camera screen and find the infrared laser spot ($\lambda = 808$ nm) (*see* **Note 9**).
8. Align the laser on the cantilever and adjust the photodiode parameters: use the micrometer screws of the cantilever holder base to adjust both the lateral and vertical positions of the holder until the intensity of the reflected laser spot from the cantilever on the photodetector is maximized. This corresponds to a maximum value of the SUM parameter on the photodetector controller. Set the DIFF parameter of the photodetector controller as close as possible to zero by using the *Y*-

photodetector micrometer screw. In general, the SUM value should be around 0.5–1.2 V for new cantilevers.

9. Troubleshooting the laser alignment: If the SUM is too low, then check the position of the cantilever and try to adjust it. If the cantilever is not in focus point of the laser beam, modify its vertical position by using the *Z*-cantilever holder micrometer screw. The SUM will increase as the cantilever reaches the focus point of the laser beam. If the laser spot is not well centered, use the XY-cantilever holder micrometer screws to adjust the lateral position of the cantilever. If the back of the cantilever is dirty or is an already-used cantilever then the SUM signal could be lower, in that case replace the cantilever.
10. Place the scanner holder. The scanner holder presents on the backside one linear and one circular grooves. The grooves have to be placed in correspondence with the support screws of the stage that determine the distance between the cantilever holder and the scanner. Ensure that the scanner holder is well positioned and does not move. Check the position of the screws of the scanner holder and ensure that they are retracted.
11. Carefully mount the sample scanner (Fig. 4) on its holder by first accommodating the back (cable) side, putting the magnets in contact with the screws of the scanner holder and then placing the front side of the scanner. During this step, the sample will be plunged in the pool. Pay constant attention to the distance between the scanner and the cantilever while plunging in the sample, always looking at the video monitor. If they are too close to each other, then use the handy

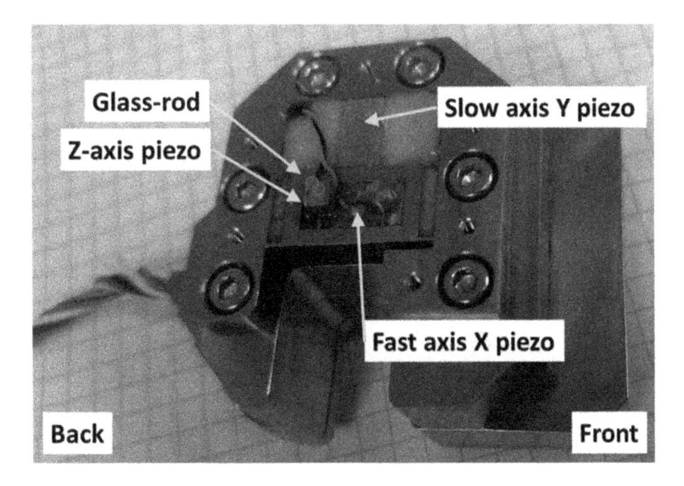

Fig. 4 HS-AFM sample scanner. Image of the standard HS-AFM sample scanner. Relevant parts are signaled: slow axis *Y* piezoelectric actuator, fast axis *X* piezoelectric actuator, *Z*-axis piezoelectric actuator, and the glass-rod where the mica disk substrate is glued

controller ("*UP*") to move the step motors and lift up the scanner (*see* **Note 10**).

12. Check the alignment of the glass-rod and the cantilever by looking if the glass-rod is in the same vertical plane of the cantilever.
13. Connect the scanner to the break-out box (*see* **Note 11**).

3.3 Tip-Sample Approach in Amplitude Modulation Mode

1. Before approaching the sample surface and the tip, check in the video monitor (video camera or CCD) that there are no bubbles around the surface or the AFM chip (that would cause abnormal cantilever deflection). Likewise, ensure that in the camera display there are no particles or dust diffusing around in the liquid, due to contamination (*see* **Note 12**). The edges of both the sample support (mica, glass, etc.) and the cantilever should look clean.
2. The tip is placed relative to the sample stage around a millimeter away in the *Z*-axis, as shown in Fig. 5a. In the camera display, the surface edge must be blurred or unfocused, ensuring that there is enough distance between the substrate surface and the cantilever (Fig. 5b). Bring the sample close to the cantilever using the micrometer screws in *XY*-plane, as indicated by the arrow in Fig. 5a.

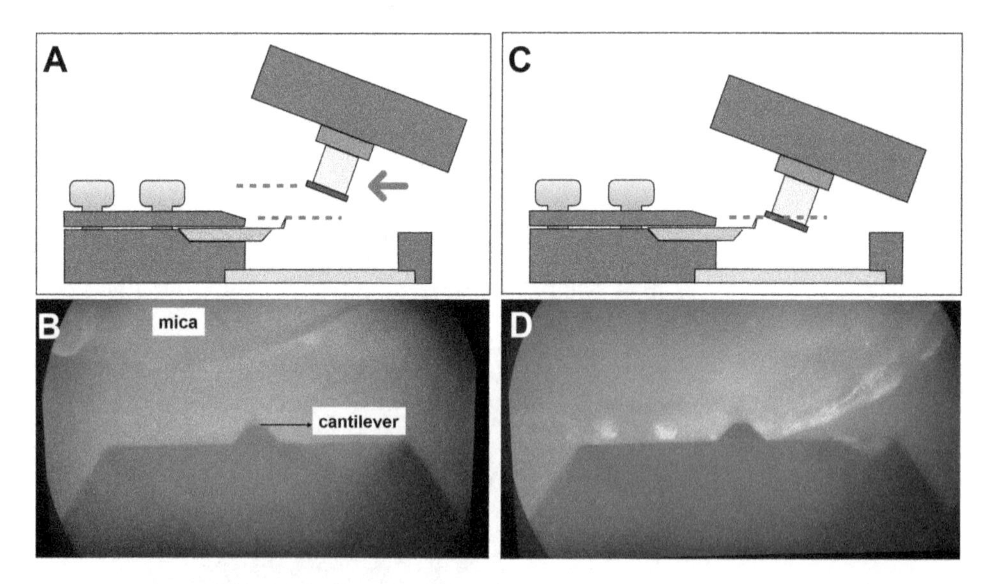

Fig. 5 Cantilever–sample positioning for approaching. (**a**) Scheme of the cantilever holder and the sample stage before approaching (side view). The mica surface is placed ~1 mm away from the cantilever in the *Z*-axis. (**b**) Screen capture of the video camera for the situation sketched in A; the mica surface is unfocused. (**c**) Scheme of the cantilever holder and the sample stage for approaching (side view). The *X*- and *Y*-micrometer positions are moved towards the cantilever; the cantilever should be finally placed in the vicinity of the mica edge. (**d**) Screen capture of the video camera for the situation sketched in **c**; the proximity to the cantilever brings mica to the same focal plane

3. In the *Z*-axis, bring the sample close to the cantilever performing a coarse approach with the *Z*-step motor (*see* **Note 13**). For approaching, the tip is exactly positioned in the vicinity of the edge of the surface (Fig. 5c, d, *see* **Note 14**).
4. To find the first (or fundamental) resonance frequency of the cantilever, measure the thermal fluctuation of the HS-AFM probe cantilever performing a power spectrum of the photodiode signal (*see* **Note 15**).
5. Using the Fourier Analyzer, apply an AC sinusoidal voltage to the excitation piezo located on the cantilever holder, to excite the cantilever at the measured resonance frequency in point 3.4 (*see* **Note 16**).
6. Select the cantilever free amplitude between 0.1 and 0.5 V (*see* **Note 17**). This typically corresponds to amplitudes between 1 and 5 nm (*see* **Note 18**).
7. The amplitude set point of the cantilever should be lower than the free amplitude. In general, for the approach the amplitude set point is set around 20–40% lower than the free amplitude.
8. Ensure that everything is plugged, *Z*-scanner, isolation table (if used), X-, *Y*- and *Z*-drivers and PID are on (*see* **Note 19**).
9. Set parameters to start approaching: center the offset for the output voltage of the *X*-, *Y*-, and *Z*-drivers (*see* **Note 20**). In the PID controller: the integral gain is set typically at 20–40% of the maximum, the proportional and derivative gains are set to minimum. The tilt for the *X*- and *Y*-axes is also centered. If there are two time constants in the controller (slow and fast), switch to slow.
10. Approach automatically, selecting the velocity of the step motor and the percentage of amplitude reduction that will halt the automatic approach (*see* **Note 21**).
11. When the tip is very close to the surface, the free oscillation amplitude may vary significantly—usually increasing—due to unspecific interactions. If the oscillation amplitude varies, stop the automatic approach and readjust the set point amplitude 20–40% lower than the free amplitude. Proceed again to approach automatically.
12. If the software detects a tip–sample contact (when the set point amplitude approximates the free amplitude, detected thanks to the readout of the amplitude and the PID output), the automatic approach stops automatically. In case of a manual approach, stop approaching when the *Z*-piezo starts to retract.
13. After the automatic approach stopped, offset the voltage of the piezo-driver until the output voltage in the PID is close to 0 V (*see* **Note 22**).

14. Troubleshooting the approach step: several situations may lead to an anomalous engage or a complete approach failure. It can be observed that the *Z*-piezo does not hold the voltage value (it moves slowly to fully expanded or contracted position, or back and forth):
    (a) Mica is not tightly attached to the glass-rod or the glass-rod is not firmly attached to the piezo stage. Sometimes mica partially exfoliates allowing some fluid leakage between the different layers. All these situations lead to oscillations of the surface while scanning that may prevent a successful approach.
    (b) Due to a bad cleavage, or an excess of glue between the mica and the glass, the substrate plane is too inclined respect to the tip, hence the contact is not stable.
    (c) The cantilever is tip-less (*see* **Note 23**). In this case, decreasing the amplitude set point also changes the SUM value in the photodetector (negative deflection values).
    (d) There are big aggregates loosely adsorbed on the surface with strong and unspecific tip–sample interactions that vary the oscillation amplitude. Try to approach with lower amplitude set point, higher free oscillation amplitude or a different area; eventually reconsider the sample preparation.
    (e) The solution is dusty or there is material adsorbing on the cantilever, which may lead to false engaging in the surface. Exchange the buffer solution in the pool or eventually increase the free oscillation amplitude. Eventually reconsider the sample preparation.

3.4 HS-AFM Imaging

At this point the clearance distance between the sample stage and the HS-AFM probe is sufficiently small for the *Z*-piezo range to bring sample and the tip in and out of contact.

1. Out of contact, turn off the cantilever excitation, with the AFM probe, observe in the oscilloscope the width of the thermal vibration of the cantilever A_{Thermal}.
2. Turn on the excitation of the cantilever oscillation and set an out of contact oscillation $A_0 = 10 \cdot A_{\mathrm{Thermal}}$.
3. Set a limited pixel number (~100 × 100 pixels), a set point amplitude $A > A_0$ so the tip is separated from sample, a small scan size (~50 × 50 nm), and a scan rate of 1 frame/s (*see* **Note 24**).
4. Start scanning, then slowly decrease the A; when it is below A_0, the highest features on the sample are visualized. In the case of observing very high objects of hundreds of nanometer height

that could damage the probe, move away to another region. If the area is appropriate for the HS-AFM tip to scan, decrease A a little more, so most of the area scanned is under slight tip contact.

5. Increase of the feedback speed: First, set the proportional (P) and the derivative (D) gains at their slowest possible reaction speed, next localize the integral (I) gain feedback speed saturation, for this increase the I gain slowly until strips of feedback resonance appear in the image (*see* **Note 25**), next reduce the I gain to a close to saturation position. If needed, the A can be slightly further decreased below A_0 to obtain an optimal surface profile. Typical A values range from $0.90 \cdot A_0$ to $0.5 \cdot A_0$. In the case of having several time constants in the controller it is possible that you may have to switch to the fastest reaction speed one to achieve the I gain saturation. Specific $P\ I\ D$ values are not given as they will depend on the gain adjustments of the rest of the feedback loop.
6. Localization of "area of interest": Increase the scan area progressively to several hundred nanometers and increase the number of pixels accordingly. It is recommended to have between 1 and 2 pixels per nanometer. Look for the area of interest where to perform the imaging, displacing on the sample using the sample stage. Use always minimal force (maximal possible set point amplitude A) for imaging to reduce the risk of tip damage. In the case that it is not found an area of sufficient interest in the available scan area, withdraw the sample from the tip, displace macroscopically using the screws and approach on other area (*see* **Note 26**).
7. In the case you localize an area of interest, zoom in to frame your experiment, the frame should show clearly all the areas that are important for posterior data analysis.
8. Optimization of oscillation amplitude: Reduce the amplitude excitation signal, the probe will go out of contact, reduce the set point amplitude A to regain tip–sample contact, further reduce the excitation signal, further reduce A... repeat this procedure until you arrive to a cantilever oscillation that is A_0 $\sim 5 \cdot A_{Thermal}$.
9. Check again the feedback saturation limit position by increasing the feedback speed of the I gain, next slow down slightly the I gain to avoid feedback resonance, and next increase the feedback speed of the P gain parameter to increase contrast on fine topographic details.
10. Speed increase: progressively increase the imaging speed, evaluate the loss of quality on the image relative to speed increase. Select the speed that provides a fair compromise between image quality and maximum possible speed.

11. If the sample presents height jumps over a few nanometers, the standard PID feedback speed of reaction may be too slow to correctly contour the sample during downhill scanning (so-called parachuting), in this case the HS-AFM dynamic PID should be turned on [19].
12. Dynamic PID tuning: set all values to 0 (the upper threshold, the lower threshold, and their corresponding gains). Slow down the imaging rate to its lowest speed where parachuting is observed. Next, increase the gain of the upper threshold until the parachuting disappears. Next, increase the imaging speed until parachuting reappears and again increase the gain of the upper threshold until the parachuting disappears. Repeat this action until resonances start appearing in the imaging, then reduce slightly. The gain of the lower threshold is usually set to 0 unless the sample shows very sharp uphill slopes (for example, trenches).
13. The shape of the tip apex is the most important parameter regarding the resolution. If the resolution obtained is not satisfactory, tip sharpening of the amorphous carbon tip deposited by electron beam deposited (EBD) on the end of the small cantilever is advised, for this use the gas plasma chamber.
    (a) In a piece of clean glass or Teflon, place on the border a strip of double sided carbon tape. Position the HS-AFM probe in a way that only the cantilever chip is in contact with the carbon tape, hence the cantilever and its tip are poked out in the air (*see* **Note 27**).
    (b) Introduce the HS-AFM cantilever probe with its support in the plasma chamber of the plasma. Select the gas to fill in the chamber (*see* **Note 28**). Run the plasma chamber for a few tens of seconds, inert gases require longer times. Typical parameters we use are pressure ~1 mbar, power ~20 W.
    (c) Image the sample with the sharpened tip. If resolution is not good enough further repeat **steps** (**a**)–(**c**). Figure 6 shows the image quality improvement on Outer Membrane Porin F (OmpF) following consecutive sharpening cycles (*see* **Note 29**).

3.5 HS-AFM Video Processing and Displaying

HS-AFM produces large data sets of image sequences that are affected by a number of instrument-specific peculiarities (such as drift of the piezoelectric elements, contrast discontinuities along the fast- and slow-scan axes, feedback parameters, etc.) that need to be accounted for before any analysis is performed.

Hereafter we provide some clues and suggestions in order to fix these instrument-specific hitches in what we believe is an accurate and time effective manner.

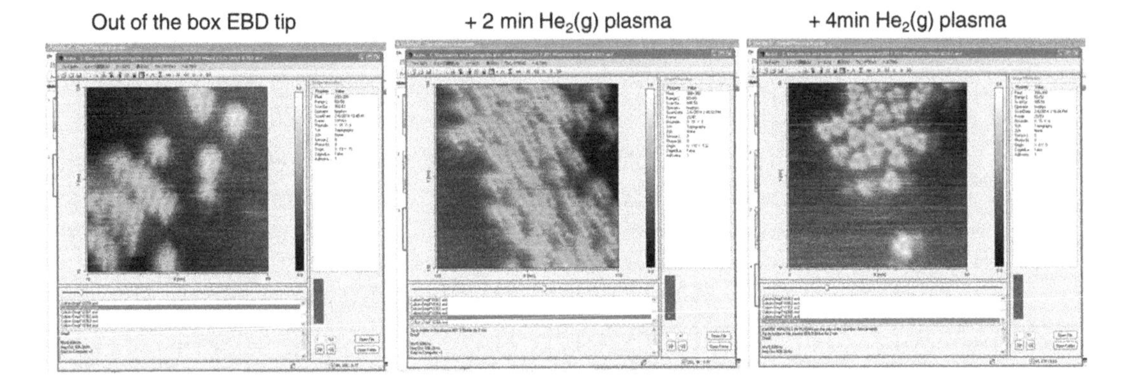

Fig. 6 Effect of helium plasma etching on the EBD tip imaging of OmpF. Left, Image using out of the box Nanoworld USC-F1.5-k0.6 HS-AFM probe. Central, image using the same probe after 2 min in helium plasma. Right, image using the sample probe after 4 min in total in helium plasma. Images are not processed

1. Open the HS-AFM data file with your favorite AFM software or programming language. Our own routine to read *.asd* HS-AFM file formats in MATLAB is available at https://sites.google.com/view/fm4b-lab/group-members/dr-arin-marchesi. To normalize the brightness across all the image stack subtract from each frame the frame average intensity (*see* **Note 30**).
2. Visually inspect your image stack and remove all the corrupted frames eventually present (changes in imaging conditions might result in insufficient forces or even the tip being for a certain amount of time out-of-contact). You might also consider removing frames skewed or dilated by sudden and fast motions of the scanner or sample.
3. Align all the frames in the stack to a reference image (*see* **Note 31**) in order to compensate for *X*- and *Y*-axis drift (*see* **Note 32**). This can be first tried in an automated way using normalized cross-correlation (or other image registration algorithms) and reiterated throughout all the frames using control flow statements (*see* **Note 33**). Visually inspect the aligned image stack and delete -or manually overlay- badly registered frames to the reference image.
4. Remove *X*- and *Y*-axis tilting by means of three-point plane fitting (*see* **Note 34**). To attain this, first filter each frame with a 2D Gaussian blur filter to remove high frequency noise. Afterwards, identify three points that sit on the substrate/background of the image. Make sure these very same points are on the substrate only throughout all the frames. Fit each triad with a plane and subtract it from the aligned video.
5. Perform a horizontal line-by-line leveling (*see* **Note 35**). As the background has been already subtracted, a median (0 order) leveling will suffice (*see* **Note 36**). To avoid leveling artifacts make sure to exclude features not belonging to the substrate/

background (usually by means of their height) from the computation. Define a region of interest over the substrate and fit the histogram with a Gaussian function to verify that the variance of the background has decreased after the procedure.

6. Fine align all the frames in the stack to a reference image using subpixel precision. For this purpose a software package that performs template matching by means of zero mean normalized cross-correlation was developed by our group some time ago [20]. Alternatively, use the ImageJ plugin Template Matching and Slice Alignment (*see* **Note 32**).
7. (Optional) Filter frame by frame the image stack with a matrix Gaussian filter to remove unwanted high-frequency noise. Be aware that overapplication of low-pass filters or use of large matrices will tend to blur the data.

4 Notes

1. Although AFM image processing and analysis is in general best done with the software supplied by the instrument manufacturer, software packages provided with HS-AFM instruments might be rather limited, and more powerful third party packages are often needed.
2. Alternatively, a free software compatible with many MATLAB scripts is GNU Octave (https://www.gnu.org/software/octave/). All the procedures described in Subheading 3.5 can be implemented in Octave.
3. Mica disks with a diameter of 1.5 mm are used as sample surfaces. To obtain them, ~0.05 mm thick mica sheets are perforated with a puncher. Careful attention must be paid in the selection of the mica disk. The disk should present a regular and flat border since only areas near the edges will be used for imaging. For gluing the mica disk on the glass-rod it can be used whether cyanoacrylate adhesive or a two-component epoxy adhesive.
4. To facilitate the cleaving of the mica disk glued on the glass-rod, you can clean the surface with a cellulose wipe soaked with acetone before performing the cleavage. Acetone can eliminate possible residues of organic material from previous sample preparation and facilitate the attachment of the adhesive tape on the mica sheet that will be removed. The final surface will not be affected by the previous use of acetone.
5. Use a wet cellulose wipe to create a humid environment inside the hood. Incubation conditions (time, temperature, etc.) will depend on the sample type. *Sample case: Preparation of supported lipid bilayers (SLBs) on mica*. POPG and POPE lipid

stocks are kept in solution in chloroform. The stock solutions are mixed in a microfuge tube of 2 mL to obtain the desired molar ratio and diluted in chloroform:methanol (3:1, v:v). The organic solvent is evaporated under a gentle nitrogen flow. After the sample is kept in a desiccator under reduced pressure overnight. The "liposome hydration buffer" is heated up at 65 °C and it is used to rehydrate the sample. Then, the sample is dropped in a thermal bath at 65 °C for 45 s, following by 45 s vortexing; this heating-vortexing cycle is repeated five times. Finally, the lipid suspension is sonicated for 30 min in an ultrasonic bath ($f \sim$ 35 kHz) to obtain large unilamellar vesicles. 1 μL of the "incubation buffer" and 1 μL liposomes suspension are incubated at room temperature on a freshly cleaved mica under a humid hood for 20 min. The sample is then rinsed ten times with imaging buffer and then imaged by HS-AFM.

6. Pay attention not to wet the backside of the cantilever holder.
7. To settle the cantilever, completely remove the plate and position the cantilever, then first secure the screw at the bottom of the plate and afterwards the one in the front (*see* Fig. 3). Do not screw to strong, screws can break easily, but enough to ensure the tighten fix of the cantilever and avoid possible vibration during imaging. Alternatively, slide the cantilever into place with the help of a tweezers to rise up the plate and secure the screws as explained before. Pay attention not to damage the cantilever during this operation. The cantilever must be parallel to the holder otherwise it would be easier to break it when mounted on the setup. For imaging experiment, the small cantilever listed in the "*Material*" section should be used.
8. Imaging experiments can be performed with the laser diode driver set in constant intensity "*I mode*" or constant power "*P mode*", depending on the type of sample and conditions in which to be imaged. For high-resolution imaging, we recommend using the laser diode driver in the "*I mode*". The "*P mode*" operates a feedback system that may introduce noise to the image. It can be used to avoid local heating when performing 1-day long measurements.

   Characteristic values we use are: in the "*I mode*", set the output between 21.8 and 23 mA, while in the "*P mode*" set the output between 0.13 and 0.18 mW for the laser diodes used in HS-AFM, although this number may vary with the ageing of the laser diode.
9. To facilitate this task, set the video camera in high light sensitivity ("*night shot*" mode) so the laser spot is intensified and easily visible on the screen.

10. In the case the wide XY range scanner (scan range: XY = 4 × 4 μm, Z = 0.7 μm) is used to perform imaging experiments, the screws of the scanner holder are in the front side of the element. Also in this case, to accommodate the scanner, the backside is first positioned and afterwards the front side is placed by engaging the magnets in contact with the screws.
11. Before connecting the scanner, check the liquid in the pool neither to form a pronounced meniscus nor wet the Z-scanner, because while applying a voltage during imaging may cause an electric arc and consequently damage the Z-scanner.
12. If the fluid in the pool does not look clear or there are particles, dust visible in the video monitor, renew your buffer solution exchanging the fluid several times in the chamber. If the problem persists, consider these situations: (i) the adsorption of the sample to the surface is not strong enough, (ii) the sample concentration is too high, (iii) there is a possible buffer solution–sample interaction (for example, magnesium salts precipitate in phosphate buffer).
13. To prevent the tip crashing on the surface, it is critically to be very attentive to the video camera while performing a coarse approach. Stop the coarse approach when the surface edge starts to define but is still blurred or unfocused. If the tip crashes brutally on the surface it will be most likely damaged.
14. The cantilever position on the surface is important for high-speed imaging. If the cantilever is not placed in the edge but closer to the middle area of the surface, during imaging, the Z-scanner exerts a viscous drag or a hydrodynamic pressure that drags the base or chip of the cantilever, which leads to an increase in the response time and imprecise cantilever deflection [8, 21, 22].
15. Usually, AFM operating software programs include this function.
16. The fundamental resonant frequency is convoluted when using piezo acoustic excitation of the cantilever with a myriad of peaks due to "echoes" in the liquid cell pool, the so-called *the forest of peaks*. HS-AFM imaging is performed selecting a frequency that "echoes" in the fluid cell in the range of frequencies of the fundamental resonance of the cantilever [23].
17. Typically, the free oscillation amplitude is set in concordance with the roughness of the sample; the flatter the sample, the lower the amplitude.
18. To convert the amplitude value from V to nm, it should be calibrated once approached using the optical beam detection sensitivity, which requires performing force curves (approach

and retract cycles between the tip and the sample). In general, avoid calibrating the sensitivity before imaging—or before the experiment—to prevent tip damage or tip wearing. The calibration can be performed at the end of the experiment, after imaging. At this step—to approach—use an approximate value: the sensitivity does not change significantly for the same type of cantilevers.

19. When the PID controller is active and the amplitude set point is set lower than the free oscillation amplitude, the *Z*-piezo fully expands (to positive voltage).
20. If the full voltage range of the *Z*-piezo driver is 50 V, center it to 25 V. If the full voltage range of the *X*- and *Y*-piezo drivers are 100 V, center them to 50 V.
21. A typical percentage of amplitude reduction for halting the automatic approach is ~30%. Alternatively, manually approach slowly moving the step-motor to bring close the cantilever to the sample surface. This is particularly useful when the surface is rough and/or heterogeneous.
22. Alternatively, slowly up or down with the step-motor.
23. In general for high-resolution imaging, it is very common to use amorphous carbon or carbon nanofibers tips. These materials are brittle and therefore they break easily. Always scan with very gentle tapping.
24. During the first images the risk of tip damage is high, here it is intended to limit it.
25. If the integral *I* gain is increased too much, saturation of the feedback loop bandwidth occurs; the *Z*-displacement cannot be done any faster even if the PID controller is requiring it. Then when *I* saturation is reached the feedback loop breaks, and error accumulates in the integrator-circuit, an *I* signal overshoot is created. Next, the circuit will create to compensate an overshoot of the *I* signal in the other direction. This oscillatory sequence of overshoots creates strips in the topographic image of the HS-AFM. Importantly, the *I* saturation depends on the topography of the sample under the tip. *I* saturates on the sample topographies that demand faster motions of the z-displacement, the uphill and downhill motions.
26. To avoid tip damage during the withdraw, first retract the tip from the sample using the *Z*-piezo displacement; for this set point $A > A_0$ before using the motor.
27. Otherwise, as surfaces get charged with plasma, the cantilever can break due to electrostatic repulsions.
28. The nature of the gas injected in the plasma chamber affects the etching rate of the amorphous carbon EBD tip. For example,

oxygen oxidizes the carbon or the EBD tip, hence is more aggressive than inert gases such as helium, whose atoms just physically bombard the carbon. As a general rule, the less is the imaging resolution, the more etching power should be applied on the tip apex.

29. The final shape of the sharpened apex depends on its initial shape, high resolution apexes are not achievable always by plasma etching, a minimum of sharpness of the initial apex is required. So in the case that after a few cycles of sharpening the required resolution is not achieved, the HS-AFM probe should be replaced.
30. This operation can be implemented in few simple steps using a *for* loop. Routines to perform in MATLAB this and other described processing steps are freely available at https://sites.google.com/view/fm4b-lab/group-members/dr-arin-marchesi.
31. The performance of image registration also depends upon the choice of the reference image (template). To increase the signal-to-noise ratio you shall build your reference image by averaging over several frames (usually 10–20). For this purpose, we recommend to use the median, which is more robust than the mean. Cropping is also commonly used to remove unwanted features from the edges of the averaged image.
32. Mechanical and piezo-scanner drifts may result into movies in which the sequential frames are not perfect real-space superposition. Therefore, before any data can be extracted from a HS-AFM image stack, all images must be perfectly aligned with respect to a stable coordinate origin.
33. For this purpose, an excellent option is the ImageJ plugins Template Matching and Slice Alignment (available at https://sites.google.com/site/qingzongtseng/template-matching-ij-plugin). Please note that this routine requires your video to be converted in 16-bit format, which might result in an unwanted loss of information. If this is the case, discard the obtained aligned video and apply the computed X and Y shifts to your original image stack to create a new registered image stack.
34. Nonlinearities of the piezoelectric scanners to the applied voltage can result in curved trajectories of the probe or sample stage (depending on the configuration of your system). If scanner bow artifacts are present, you shall consider fitting your image stack with a second order polynomial surface through five points.
35. As the horizontal axis is usually the fast scan axis, any change in imaging conditions over the time of the scan will lead to

horizontal discontinuities in the topographs. It will thus be well accounted for by a horizontal line-by-line leveling.

36. In this routine, for each row of the image, the row median is subtracted from every element in the row.

Acknowledgments

This work was supported by French National Institute of Health and Medical Research (Inserm); the program "Investissements d'Avenir" by the French National Research Agency, ANR-10-LABX-0083 (Labex EFL); Grant ANR-16-CE15-0023 (ANR SalmoTubes); Grant ANR-15-CE11-0020 (ANR MoBaRhE). A.M. acknowledges support from the Long Term EMBO Fellowship (ALTF 1427-2014) and the Marie Curie Action (MSCA-IF-2014-EF-655157).

References

1. Binnig G, Quate CF (1986) Atomic force microscopy. Phys Rev Lett 56:930–933
2. Ido S, Kimura K, Oyabu N, Kobayashi K, Tsukada M, Matsushige K, Yamada H (2013) Beyond the helix pitch: direct visualization of native DNA in aqueous solution. ACS Nano 7:1817–1822
3. Ares P, Fuentes-Perez MEE, Herrero-Galan E, Valpuesta JMM, Gil A, Gomez-Herrero J, Moreno-Herrero F (2016) High resolution atomic force microscopy of double-stranded RNA. Nanoscale 8:11818–11826
4. Hansma HG, Vesenka J, Siegerist C, Kelderman G, Morrett H, Sinsheimer RL, Elings V, Bustamante C, Hansma PK (1992) Reproducible imaging and dissection of plasmid DNA under liquid with the atomic force microscope. Science 256:1180–1184
5. Henderson E, Haydon PG, Sakaguchi DS (1992) Actin filament dynamics in living glial cells imaged by atomic force microscopy. Science 257:1944–1946
6. Schabert F, Henn C, Engel A (1995) Native Escherichia coli OmpF porin surfaces probed by atomic force microscopy. Science 268:92–94
7. Ando T, Kodera N, Takai E, Maruyama D, Saito K, Toda A (2001) A high-speed atomic force microscope for studying biological macromolecules. Proc Natl Acad Sci U S A 98:12468–12472
8. Ando T, Kodera N, Maruyama D, Takai E, Saito K, Toda A (2002) A high-speed atomic force microscope for studying biological macromolecules in action. Jpn J Appl Phys 41:4851–4856
9. Ando T, Uchihashi T, Kodera N (2013) High-speed AFM and applications to biomolecular systems. Annu Rev Biophys 42:393–414
10. Casuso I, Khao J, Chami M, Paul-Gilloteaux P, Husain M, Duneau J-P, Stahlberg H, Sturgis JN, Scheuring S (2012) Characterization of the motion of membrane proteins using high-speed atomic force microscopy. Nat Nanotechnol 7:525–529
11. Ando T (2012) High-speed atomic force microscopy coming of age. Nanotechnology 23:62001
12. Gorle S, Pan Y, Sun Z, Shlyakhtenko LS, Harris RS, Lyubchenko YL, Vukovic L (2017) Computational model and dynamics of monomeric full-length APOBEC3G. ACS Cent Sci 3:1180–1188
13. Pan Y, Sun Z, Maiti A, Kanai T, Matsuo H, Li M, Harris RS, Shlyakhtenko LS, Lyubchenko YL (2017) Nanoscale characterization of interaction of APOBEC3G with RNA. Biochemistry 56:1473–1481
14. Rajendran A, Endo M, Katsuda Y, Hidaka K, Sugiyama H (2011) Programmed two-dimensional self-assembly of multiple DNA origami jigsaw pieces. ACS Nano 5:665–671
15. Inoue S, Uchihashi T, Yamamoto D, Ando T (2011) Direct observation of surfactant aggregate behavior on a mica surface using high-speed atomic force microscopy. Chem Commun 47:4974

16. Munguira I, Casuso I, Takahashi H, Rico F, Miyagi A, Chami M, Scheuring S (2016) Glass-like membrane protein diffusion in a crowded membrane. ACS Nano 10:2584–2590
17. Chiaruttini N, Redondo-Morata L, Colom A, Humbert F, Lenz M, Scheuring S, Roux A (2015) Relaxation of loaded ESCRT-III spiral springs drives membrane deformation. Cell 163:866–879
18. Mierzwa BE, Chiaruttini N, Redondo-Morata-L, Moser von Filseck J, König J, Larios J, Poser I, Müller-Reichert T, Scheuring S, Roux A, Gerlich DW (2017) Dynamic subunit turnover in ESCRT-III assemblies is regulated by Vps4 to mediate membrane remodelling during cytokinesis. Nat Cell Biol 19:787–798
19. Kodera N, Sakashita M, Ando T (2006) Dynamic proportional-integral-differential controller for high-speed atomic force microscopy. Rev Sci Instrum 77:83704
20. Husain M, Boudier T, Paul-Gilloteaux P, Casuso I, Scheuring S (2012) Software for drift compensation, particle tracking and particle analysis of high-speed atomic force microscopy image series. J Mol Recognit 25:292–298
21. Ando T, Uchihashi T, Scheuring S (2014) Filming biomolecular processes by high-speed atomic force microscopy. Chem Rev 114:3120–3188
22. Uchihashi T, Kodera N, Ando T (2012) Guide to video recording of structure dynamics and dynamic processes of proteins by high-speed atomic force microscopy. Nat Protoc 7:1193–1206
23. Carrasco C, Ares P, De Pablo PJ, Gómez-Herrero J (2008) Cutting down the forest of peaks in acoustic dynamic atomic force microscopy in liquid. Rev Sci Instrum 79:111–113

Chapter 12

High-Speed Atomic Force Microscopy of Individual Amyloidogenic Protein Assemblies

Takahiro Watanabe-Nakayama and Kenjiro Ono

Abstract

High-speed atomic force microscopy (HS-AFM) with high spatiotemporal resolution allows for the video imaging of the conformational changes of individual molecules in an observation area in liquid at nanometer-scale spatial resolution. This method verifies the molecular mechanism and reveals the structural dynamics of relevant biomolecules for various biological phenomena. Here, we describe the methods for HS-AFM observation and the analysis of the structural dynamics of individual amyloidogenic protein assemblies using amyloid β 1–42 as an example.

Key words Atomic force microscopy, Video recording, Single-molecule imaging, Amyloid, Amyloidogenic proteins, Amyloidosis

1 Introduction

High-speed atomic force microscopy, which can take movies of the structural dynamics of individual biomolecules, has had a great impact on the studies of molecular processes of various biological phenomena [1–3]. Biomolecules on a stage are rapidly scanned by a sharp tip on a small soft cantilever that oscillates at a resonance frequency with a setting amplitude. This oscillating amplitude is changed when the tip comes across ups and down caused by biomolecules on the stage. This change is rapidly recovered and maintained to the set point by a high-speed feedback system that can move the cantilever position (for the tip-scanning AFM) or the stage (for the stage-scanning AFM) in the *z*-direction in response to changes in the oscillating amplitude of the cantilever. The obtained feedback signals that correspond to the information for the sample structure are combined in two dimensions to form an AFM image. Finally, the acquisition of a sequence of AFM images at a high-speed frame rate makes a HS-AFM movie.

To achieve successful HS-AFM, we must optimize the sample scanning rate, the sample–stage interaction, and the procedure for

Yuri L. Lyubchenko (ed.), *Nanoscale Imaging: Methods and Protocols*, Methods in Molecular Biology, vol. 1814,
https://doi.org/10.1007/978-1-4939-8591-3_12,

HS-AFM movie processing. The scanning rate for which the configuration procedure is described in refs. [1–5] should be used to take a clear image at low tapping force and at high speed. The sample–stage interaction should be optimized so that biomolecules can work without detachment from the HS-AFM sample stage. The tight interaction between the sample-stage helps AFM image acquisition but can prevent an accurate movement of the structural dynamics and function of biomolecules [6]. In contrast, the loose sample–stage interaction reduces the possibility of the inhibition of sample functions but promotes the detachment of samples from the stage surface. The detached molecules cannot be imaged in HS-AFM movies. For the regulation of the sample–stage interaction, the electrolytes and their concentration in a sample solution was optimized to modulate the electrostatic interaction between the biomolecules and mica surface [6–9], and lipid planar bilayers [10] or two-dimensional crystal of streptavidin [11, 12] were used to reduce the nonspecific binding of samples to the stage surface. The processing of HS-AFM movies is needed to remove thermal drift noise from the *x*, *y*, and *z* scanners and to obtain information for the structural dynamics of individual biomolecules and their assemblies.

Here, we describe the procedures for HS-AFM observations and the HS-AFM movie analysis of amyloid fibril growth using amyloid β 1–42 (Aβ1–42) fibrils.

2 Materials

Prepare all buffer solutions using ultrapure water (18 MΩ-cm at room temperature) and the JIS special grade, guaranteed or higher-grade reagents from manufactures. For gel filtration high-performance liquid chromatography, the buffer solution should be filtrated using a 0.22-μm filter and degassed.

2.1 HS-AFM Observation

1. Instrument: The high-speed atomic force microscope (HS-AFM) we use is laboratory built [4]. The scanners with the maximum scan ranges of the type 1: $x = {\sim}1.8$ μm, $y = {\sim}3.6$ μm, $z = {\sim}0.2$ μm; the type 2: $x = {\sim}6.1$ μm, $y = {\sim}5.8$ μm, $z = {\sim}0.9$ μm are used with the Q-control circuits [13] and with the feedback system with the dynamic proportional-integral-derivative (PID) controller [14]. The same type of the HS-AFM (NanoExplorer®) is commercially available from Research Institute of Biomolecule Metrology Co., Ltd. (RIBM) of Japan. The preparation and operation of the HS-AFM are according to ref. [4]. The sharp probe is also prepared on the probeless small cantilever (BL-AC10DS-A2, Olympus) by electron beam deposition (EBD) according to ref. [4] with modifications (*see* **Note 1**). The EBD probe is

longer than the carbon-nanotube probe of commercially available cantilever (BL-AC10FS-A2, Olympus) and can be removed as described in **Note 1**, so that we can reuse the cantilever several times.

2. Low molecular weight (LMW) fraction of Aβ1–42 stock solution: Purify LMW Aβ1–42 in 10 mM sodium phosphate, pH 7.4 (imaging buffer) by using gel filtration chromatography as reported ref. [15] (*see* **Note 2**). Quantify the peptide concentration using the Bradford method. It should be more than 2.5 μM (typically ~25 μM) and stored at −80 °C before use.
3. 10 mM sodium phosphate, pH 7.4 (imaging buffer): Store at room temperature.
4. 2.5 M NaCl: Store at room temperature.
5. 2.5 M KCl: Store at room temperature.

2.2 Processing and Analysis of HS-AFM Movies

1. Image processing software: Install the latest version (1.51r on Oct. 15, 2017) of ImageJ, which is a free software developed for image processing by the National Institutes of Health (NIH).
2. Additional plugins for ImageJ: Install the plugins for template matching to the "Plugins" folder from https://sites.google.com/site/qingzongtseng/template-matching-ij-plugin according to the instructions on the website.

3 Methods

Carry out all procedures at room temperature unless noted.

3.1 HS-AFM Observation

1. Mount the HS-AFM liquid cell with a cantilever, 60 μL of imaging buffer, and the HS-AFM scanner with a glass rod stage with a mica disc according to ref. [4] (Fig. 1). For

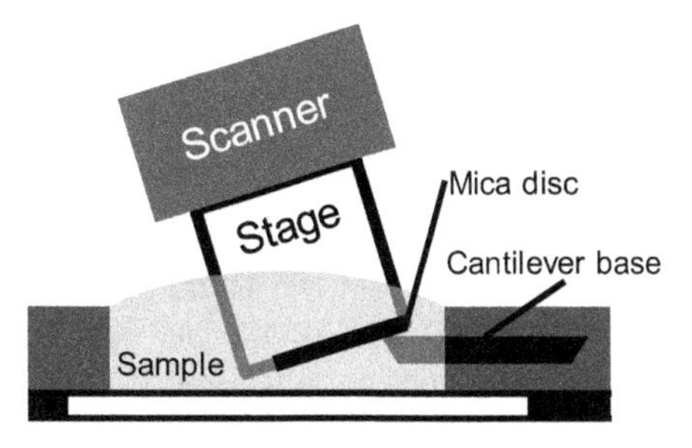

Fig. 1 A schematic side view of the HS-AFM assembly. The mica disc on the glass rod, which is immobilized on the tilted scanner, contacts the cantilever tip in sample chamber

preparation of the mica stage, glue the mica disc with 1.2 mm in diameter and 0.05 mm in thickness on the glass rod with 2 mm in diameter and 2 mm in height by epoxy under a stereoscopic microscope. Then, glue the reverse side of the glass rod on the *z*-piezo by a colorless touch-up paint for an automobile. After drying the glue, cleave the mica with scotch tape.

2. Approach the scanner to the stage until the cantilever tip reaches the mica surface.
3. Take HS-AFM images of the mica surface (*see* **Note 3**).
4. Retract the scanner from the cantilever a little (about 0.5 μm) (*see* **Note 4**).
5. Remove the solution in the liquid cell using a pipette connected to a peristatic pump (*see* **Note 5**).
6. Just before the next step, prepare 60 μL of 2.5 μM Aβ1–42 by diluting the stock with 10 mM sodium phosphate, pH 7.4 and add 2.5 μL of 2.5 M NaCl or KCl to the diluted Aβ1–42 solution.
7. Immediately introduce the Aβ1–42 solution in **step 5** (*see* **Note 5**).
8. Approach the stage with the scanner to the cantilever again, and take and save the HS-AFM images (Fig. 2).

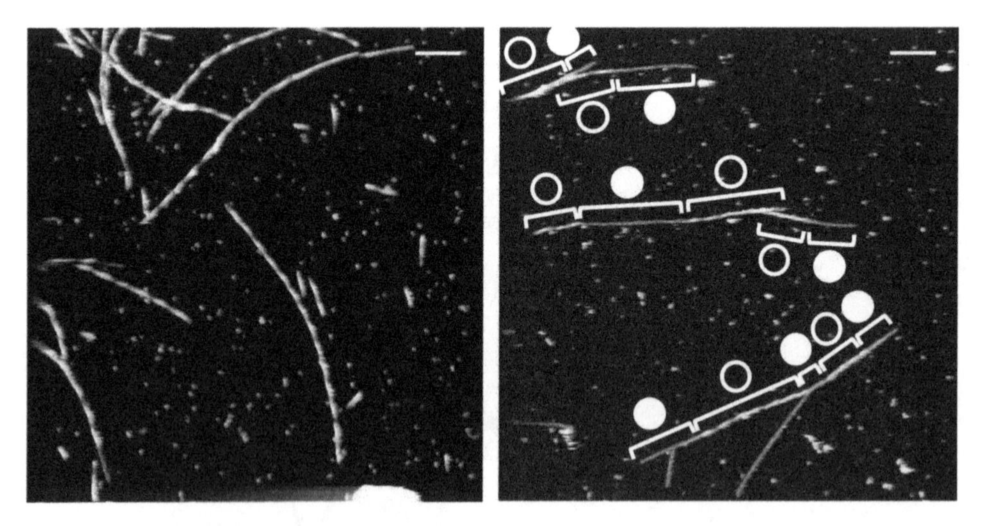

Fig. 2 HS-AFM images of Aβ1–42 amyloid fibrils grown on mica in 10 mM sodium phosphate, pH 7.4 supplemented with 100 mM NaCl (*left*) or 100 mM KCl (*right*) (reproduced from ref. [9]). Bars: 200 nm. Different electrolytes in buffer solution introduced different structural dynamics in Aβ1–42 fibril growth to produce fibrils with different structures. Sodium ions constraint Aβ1–42 fibrils to grow as a spiral, while potassium ions increase the switch of growth modes between spiral (*open circles*) and straight (*closed circles*)

3.2 Processing and Analysis of HS-AFM Movies

3.2.1 Processing

1. Open ImageJ. Import the HS-AFM movies (*see* **Note 6**).
2. Subtract the background, which is mainly due to the drift of the HS-AFM scanner in the *z*-direction (Fig. 3a, b). The methods for subtraction of the background depend on the scanning time necessary for one single frame. For low scanning rate images in which the background along *y*-direction is complicated and cannot be easily corrected by polynomial fitting, apply line-by-line leveling with polynomial fitting or the sliding paraboloid algorithm (*see* **Notes 7** and **8**) (Fig. 3c, d). For high-speed scanned images, simply apply polynomial fitting in both the *x*- and *y*-directions (Plugins/Filters/Fit polynomial with *x* and *y* without *xy*) or rolling ball to the two-dimensional images (Process/Subtract Background). Optionally, adjust the background height level to zero after background leveling. In many cases, the background level can be checked as the modal value. The modal value is displayed on the pixel histogram (Analyze/

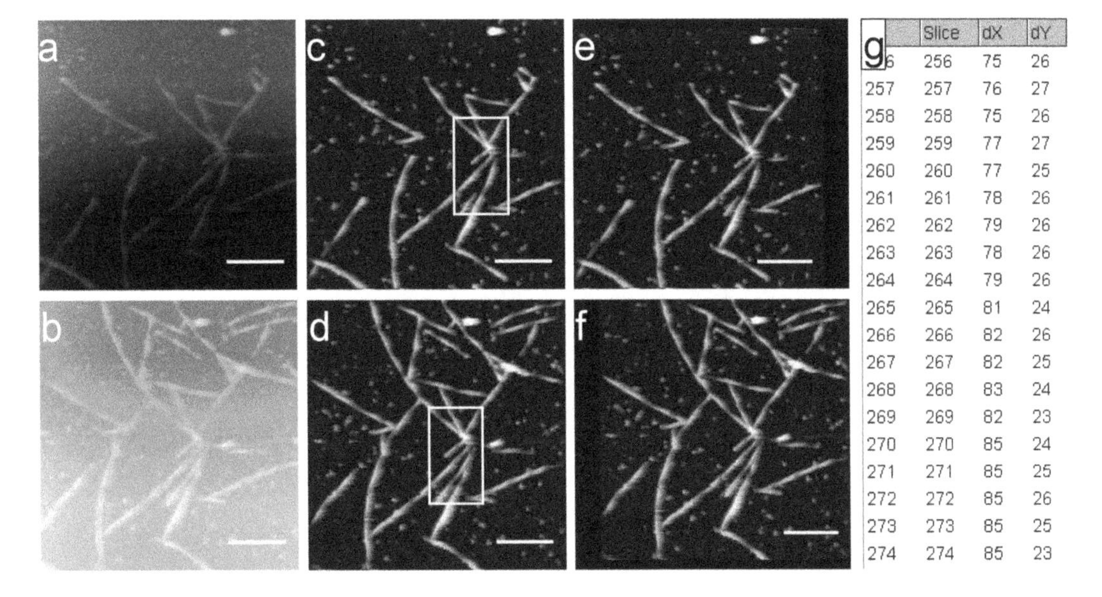

Fig. 3 The processing of HS-AFM movies. (**a**–**d**) The subtraction of background by line-by-line leveling. (**a**, **b**) Two different frames in a HS-AFM movie scanned at 2 s/frame. 900 × 900 nm (300 pixels × 300 pixels). Bars: 200 nm. *Z*-scale: 60 nm. Background level in the *y*-direction shows complicated ups and downs within a single frame and even between the two frames, mainly because of the drift of the scanner in the *z*-direction and the scan speed in the *y*-direction is lower than in *x*-direction. (**c**, **d**) The two images after the subtraction of the background by line-by-line leveling with the sliding paraboloid algorithm (**a** and **b** corresponds to **c** and **d**) (reproduced from ref. [9]). Bars: 200 nm. *Z*-scale: 15 nm. The changes in background level were removed. (**e**–**g**) The subtraction of the position gaps between image frames. (**e**, **f**) Two different HS-AFM frames after the alignment of their *x*- and *y*-positions (**e** and **f** corresponds to **c** and **d**). The box in (**c**) is the area used as the template for the template matching algorithm. The canvas size in images (**e**, **f**) is enlarged so that each frame can cover the entire region of the original image, even after translation (*see* **Note 9**). (**g**) The output text file from the "Align slices in stack" command. The file contains the number of pixels for the translation of each frame

Histogram) or can be measured ("Analyze/Measure" with "Modal gray value" in the "Set Measurements" dialog).

3. Align the image frames to fix the drift of the HS-AFM scanner in the *x*- and *y*-directions (Fig. 3e, f). Because the "Align slices in stack" plugin does not work on 32-bit images, this step contains the following three steps: (1) make a copy movie with 8-bit or 16-bit (the original is 32-bit float), (2) apply the "Align slices in stack" plugin (Plugins/Template Matching/ Align slices in stack) to the copy movie to obtain the file containing the number of pixels for individual frame movements (Fig. 3g), and (3) translating each frame of the original movie using this file (*see* **Note 9**). The "Duplicate" plugin (Image/Duplicate) with "duplicate stack" option makes a copy of the original movie. The "8-bit" (Image/Type/8-bit) and "16-bit" (Image/Type/16-bit) commands change the type of copy images to 8-bit and 16-bit. For the "Align slices in stack" plugin, we usually set the matching method to be "Normalized correlation coefficient" without subpixel resolution in the option dialog. The "Translate" plugin (Image/ Transform/Translate) can translate each frame of the 32-bit float type original movie but cannot automatically move all frames in the movie by the appropriate number of pixels. We made a plugin that can load the file containing the number of pixels for individual frame movements and apply it to the corresponding frames in an original movie to move them by the appropriate amounts.
4. Extract the regions of individual amyloid fibrils into individual movie files. This step uses "Segmented Line" selection with spline fit in the ImageJ main window and the "Straighten" plugin (Edit/Selection/Straighten) (*see* **Note 10**) (Fig. 4).

3.2.2 Kymograph

1. Before making the kymograph for fibril growth, process the individual movies according to the previous steps (Subheading 3.2.1). Align image frames for single fibrils in the horizontal direction. This step uses the command (Image/Stacks/Make Montage) with "row = 1 and column = the number of image frames" (Fig. 5).

3.2.3 Time Course of Fibril Growth

1. Reconstruct the orthogonal image frames (*x* frames of *t* pixels × *y* pixels) from the movie (*t* frames of *x* pixels × *y* pixels) in **step 4** of Subheading 3.2.1. This step uses the "Reslice" plugin (Image/Stacks/Reslice with "Start at Top or Left (along fibril axis)" and "Avoid interpolation") (Fig. 6a, b).
2. Make an image from the orthogonal image frames. This step uses the "*Z* Project" plugin with "Average Intensity," "Maximum Intensity," or "Sum Slices" (Fig. 6c).

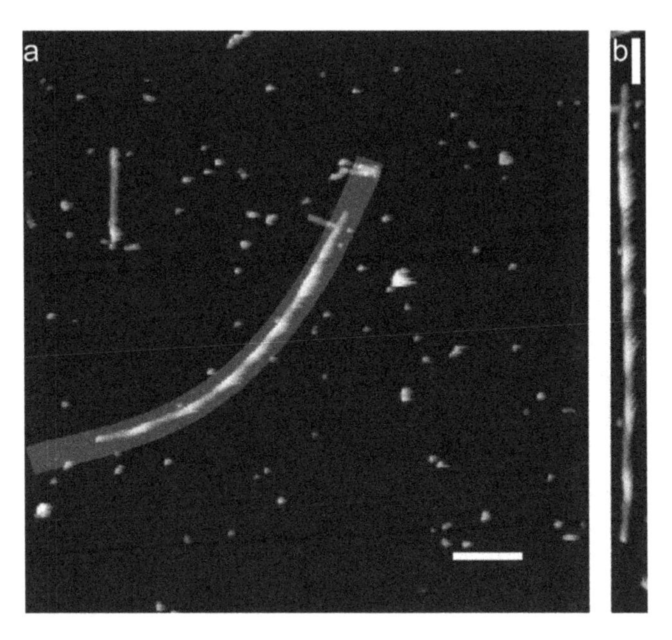

Fig. 4 The selection and extraction of individual amyloid fibrils. (**a**) The processed original HS-AFM image with a curve (segmented line with spline fit, *yellow*) for the selection of an individual amyloid fibril. Bar: 200 nm. (**b**) The extracted and straightened image for the fibril. Bar: 100 nm. *Z*-scale: 15 nm

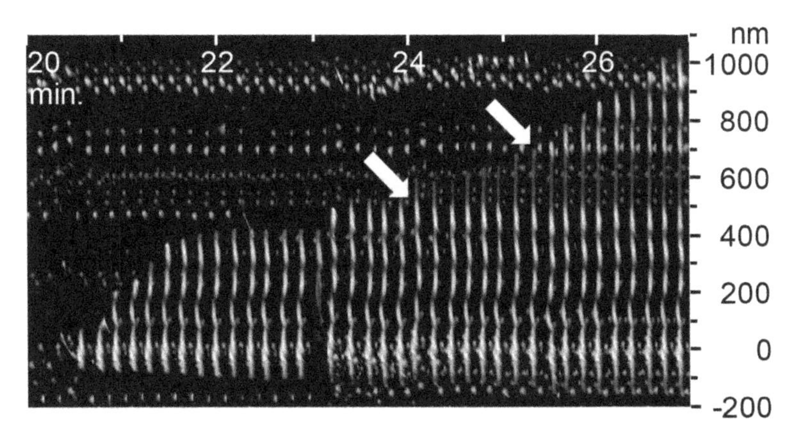

Fig. 5 The kymograph for an individual amyloid fibril showing difference in growth rate between the two ends of the fibril and the switching of growth mode (spiral to straight at 24′00″ or straight to spiral at 25′20″ after the start of aggregation) (reproduced from ref. [9])

3. Adjust the threshold and make a binary image to differentiate between background and the amyloid fibril. This step uses the "Threshold" plugin (Image/Adjust/Threshold) (*see* **Note 11**) (Fig. 6d).

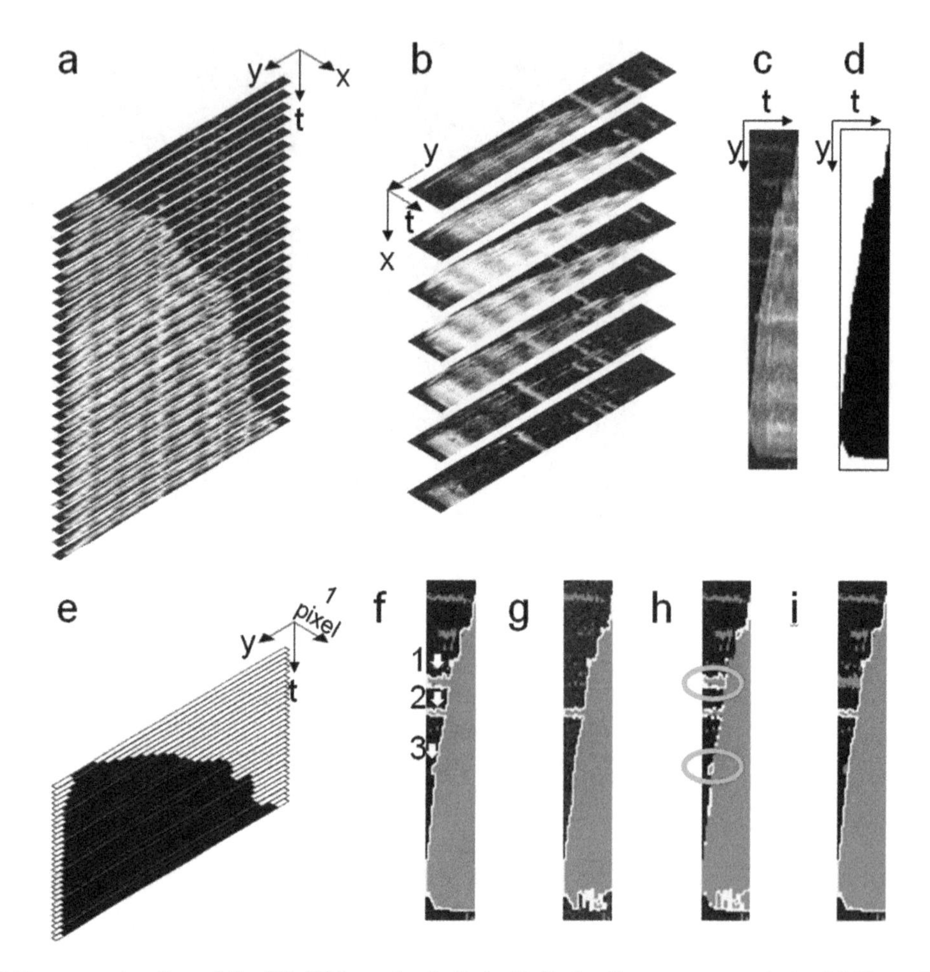

Fig. 6 The reconstruction of the HS-AFM movie of single fibrils for time course analysis. (**a**) The extracted movie (*t* frames of *x* pixels × *y* pixels) for a single fibril. (**b**) The reconstructed orthogonal image frames (*x* frames of *t* pixels × *y* pixels) from (**a**). (**c**) The averaged image of (**b**). (**d**) The binary image of (**c**). (**e**) The reconstructed orthogonal image frames (*t* frames of 1 pixel × *y* pixels) from (**d**). (**f–i**) Schematic images for exclusion of the particle regions from the fibril on the reconstructed images. (**f**) The averaged image (**c**) with selection of the region (*yellow line*) covering the fibril with the particles 1, 2 and 3 (*arrows*) using the threshold (*red*). (**g**) The third frame from the top of the reconstructed images (**b**) with selection of the region (*yellow line*) covering the fibril without the particles 1 and 3 using the threshold (*red*). This ROI also lacks a part of the fibril (*bottom*). (**h**) The averaged image (**c**) with the difference in the selection between (**f**) and (**g**) (*green circles*). The other ROIs (*yellow* out of *green circles*) can be removed by the "Brush" tool in the menu window. (**i**) The averaged image (**c**) with the ROI (*yellow*) which is subtraction of the particle regions 1 and 3 (*green circles* in (**h**)) from the ROI in (**f**). The region of particle 2 can be removed as the same way using the second frame from the top of the reconstructed images (**b**)

4. Reconstruct the orthogonal image frames (*t* frames of 1 pixel × *y* pixels) from the binary image. This step uses the "Reslice" plugin with (Image/Stacks/Reslice with "Start at Top or Left (along fibril axis)" and "Avoid interpolation"). Each frame has a single bar showing fibril length and position (Fig. 6e).

Fig. 7 The structural periodicity of spiral fibrils is detected by finding local maxima (Process/Find Maxima). This fibril is the same fibril as Fig. 4b. Bar: 100 nm. Red corresponds to the background region lower than the threshold. Crosses with dots (*yellow*) indicate local maxima on a spiral fibril

5. Obtain coordinate data for both fibril ends. This step uses the "Analyze Particles" plugin (Analyze/Analyze Particles). Before the use of the "Analyze Particles," select "Bounding rectangle" in the "Set measurements" dialog box (Analyze/Set Measurements). The coordinates for both fibril ends and the fibril length in each frame are calculated from the obtained upper left coordinate ("BX" and "BY") and width (or height) of the rectangle which corresponds to the single bar in each frame.

3.2.4 Analysis of the Structural Periodicity

1. Adjust the threshold to hide the background on the movie in **step 4** in Subheading 3.2.1. This step uses the "Threshold" plugin (Image/Adjust/Threshold) (Fig. 7).
2. Obtain the coordinate data for the local maxima of a spiral fibril. This step uses the "Find Maxima" plugin (Process/Find Maxima) with "Point selection" (for displaying a multi-point selection in which each point corresponds to each maximum) and "above lower threshold", and then applies the "Measure" plugin (Analyze/Measure) (Fig. 7). The spiral periodicity is analyzed from the distribution of the distance between local maxima. The spiral diameter of each fibril corresponds to the pixel value at the local maximum.

4 Notes

1. We use a commercially available tipless HS-AFM cantilever (BioLever Fast; BL-AC10DS-A2, Olympus, Japan). Before tip preparation, the cantilever surface is cleaned using UV/O-zone irradiation for 30 min in an UV/ozone cleaner (UV/O-zone ProCleaner plus PC450, BioForce Nanosciences, Inc., USA) followed by surface inactivation where cantilevers are immersed in ultrapure water (for ELS-7500 described below, acetone is used instead of ultrapure water) in a glass dish for more than 1 h at room temperature. Just before electron beam deposition (EBD), the cantilevers are immersed in acetone in a glass dish for more than 15 min. After the removal of acetone from each cantilever using paper, the cantilevers are put in the sample chamber of field-emission scanning electron microscope (FE-SEM). We use FE-SEM, ERA-8000FE (Elionix, Japan) or electron beam lithography with ELS-7500 (Elionix, Japan) in which an oil diffusion pump or a turbo molecular

pump is used for evacuation of sample chamber. For the carbon source of EBD, contaminant gas from the oil diffusion pump for ERA-8000FE or oil for the diffusion pump put on the stage for ELS-7500 is used. The tip is grown on the top of the cantilever by the irradiation of the electron beam for 2 min at 30 kV with a probe current as low as possible. The used tips after HS-AFM observation are removed by UV/ozone as described above.

2. For comparison of the buffer electrolyte effects, the same lot of the LMW fraction should be used.
3. Check surface flatness and that there are not any particles on the surface.
4. This step helps avoid damages on the cantilever tip and shorten the time for approaching the scanner in **step 8**.
5. Do not move the liquid cell using the pipettes.
6. The ImageJ command (Import/Raw) can open the HS-AFM movie files (-.asd) which are produced by the HS-AFM operation program (Eagle™). The positions of information necessary for loading the ASD file (file header, frame header separating frames, body of image, calibration value for scanner, number of pixels in *x*- and *y*-directions, scanning ranges in *x*- and *y*-directions, etc.) are written in the procedure file, "Simple Viewer.ipf", of the Eagle™ 2.5.1. Input the fields for the import command in the ImageJ as follows: Image type = 16-bit signed; Width and Height = the number of pixels in width and height of the scanning range; Offset to first image = the number of bytes in the file header and the frame header: Number of images: the number of frames in a HS-AFM movie: Gap between images: the number of bytes in the frame header: Little-endian byte order. After loading, flip the movie upside down because the direction of *y*-axis for the ImageJ is opposite to the Eagle™. Then, change the file type from 16-bit to 32-bit (Image/Type/32-bit) and convert the pixel values to nanometer scale using the "Process/Math/Macro" with the equation described in the "Simple Viewer.ipf" and with "stack." At last, set *x*- and *y*-scales by using the command (Analyze/Set Scale) with the number of pixels and the scanning ranges in *x*- and *y*-directions.
7. For line-by-line leveling, select either polynomial fitting or sliding paraboloid algorithm, considering artifact(s). Line-by-line leveling with polynomial fitting can result in artifacts where the background level around samples on the fast scan axis appeared to be lower than other area. Line-by-line leveling with the sliding paraboloid algorithm can subtract samples across the edge of the images.

8. Line-by-line leveling can be applied to each frame of the movie using the following steps: (1) copy a single frame of the HS-AFM movie (x pixels in horizontal direction (fast scan axis), y pixels in vertical direction, t frames) (Image/Duplicate without "duplicate stack"), (2) reconstruct orthogonal images (y frames of x pixels × 1 pixel) from the copy (Image/Stacks/Reslice with "start at Top" and "Avoid interpolation"), (3) apply polynomial fitting (Plugins/Filters/Fit polynomial) or sliding paraboloid filtering (Process/Subtract Background with "Sliding paraboloid") to the orthogonal images, (4) reconstruct the corrected image from the orthogonal images (Image/Stacks/Reslice with "start at Top" and "Avoid interpolation"), and (5) put the corrected image on the original frame (Edit/Copy and Paste).
9. Translating loses the translated region outside of the canvas of the original movie. To avoid it, before the translation of the frames in a 32-bit original movie, enlarge the canvas size of the movie by using the "Canvas Size" plugin (Image/Adjust/Canvas Size) with the "center" option. After translation, the extra region can be checked by the "*Z* project" plugin (Image/Stacks/*Z* project) and can be removed by area selection and the "Crop" plugin (Image/Crop).
10. Information for the lines for selection can be saved as ROI (.roi) files using the "ROI manager" (Analyze/Tools/ROI manager). The ROI files can be loaded on the original movie by the ROI manager.
11. If the background contains some particles, not fibrils, they should be removed. For exclusion of particle regions, check each frame of the reconstructed movie, and adjust the threshold. Proceed as follows: (1) make a selection of the region, including the particles and fibril, using the "Wand" tool in the menu window and record it on the "ROI Manager" (Analyze/Tools/ROI Manager) (Fig. 6f); (2) select the fibril region which completely excluded the particle region(s) and lacks a part of fibril, using a frame of the reconstructed images (Fig. 6g); (3) make the ROIs for the particles from the difference between (1) and (2) ("XOR" in the "More" in the "ROI manager" (Analyze/Tools/ROI Manager): "Brush" in the menu window for removal of the non-particle regions) (Fig. 6h); and (4) make the complete fibril region from the difference between (1) and (3) (Fig. 6i). After selection, fill the region and remove the outside using the "Clear Outside" and "Fill" commands (Edit/Clear Outside and Edit/Fill).

Acknowledgments

The authors would like to thank Prof. T. Ando, Prof. T. Uchihashi, Dr. N. Kodera, and Dr. H. Konno for the technical advice concerning our HS-AFM observations.

References

1. Ando T (2017) Directly watching biomolecules in action by high-speed atomic force microscopy. Biophys Rev 9:421–429. https://doi.org/10.1007/s12551-017-0281-7
2. Ando T, Uchihashi T, Scheuring S (2014) Filming biomolecular processes by high-speed atomic force microscopy. Chem Rev 114:3120–3188. https://doi.org/10.1021/cr4003837
3. Ando T, Uchihashi T, Kodera N (2013) High-speed AFM and applications to biomolecular systems. Annu Rev Biophys 42:393–414. https://doi.org/10.1146/annurev-biophys-083012-130324
4. Uchihashi T, Kodera N, Ando T (2012) Guide to video recording of structure dynamics and dynamic processes of proteins by high-speed atomic force microscopy. Nat Protoc 7:1193–1206. https://doi.org/10.1038/nprot.2012.047
5. Ando T, Uchihashi T, Fukuma T (2008) High-speed atomic force microscopy for nano-visualization of dynamic biomolecular processes. Prog Surf Sci 83:337–437. https://doi.org/10.1016/j.progsurf.2008.09.001
6. Watanabe-Nakayama T, Itami M, Kodera N et al (2016) High-speed atomic force microscopy reveals strongly polarized movement of clostridial collagenase along collagen fibrils. Sci Rep 6:28975. https://doi.org/10.1038/srep28975
7. Shao Z, Czajkowsky DM (2003) Inhibition of protein adsorption to muscovite mica by monovalent cations. J Microsc 211:1–7. https://doi.org/10.1046/j.1365-2818.2003.01208.x
8. Yamamoto D, Uchihashi T, Kodera N et al (2010) High-speed atomic force microscopy techniques for observing dynamic biomolecular processes. Methods Enzymol 475:541–564. https://doi.org/10.1016/S0076-6879(10)75020-5
9. Watanabe-Nakayama T, Ono K, Itami M et al (2016) High-speed atomic force microscopy reveals structural dynamics of amyloid β1-42 aggregates. Proc Natl Acad Sci U S A 113:5835–5840. https://doi.org/10.1073/pnas.1524807113
10. Kodera N, Yamamoto D, Ishikawa R, Ando T (2010) Video imaging of walking myosin V by high-speed atomic force microscopy. Nature 468:72–76. https://doi.org/10.1038/nature09450
11. Yamamoto D, Nagura N, Omote S et al (2009) Streptavidin 2D crystal substrates for visualizing biomolecular processes by atomic force microscopy. Biophys J 97:2358–2367. https://doi.org/10.1016/j.bpj.2009.07.046
12. Yamamoto D, Uchihashi T, Kodera N, Ando T (2008) Anisotropic diffusion of point defects in a two-dimensional crystal of streptavidin observed by high-speed atomic force microscopy. Nanotechnology 19:384009. https://doi.org/10.1088/0957-4484/19/38/384009
13. Kodera N, Yamashita H, Ando T (2005) Active damping of the scanner for high-speed atomic force microscopy. Rev Sci Instrum 76:53708. https://doi.org/10.1063/1.1903123
14. Kodera N, Sakashita M, Ando T (2006) Dynamic proportional-integral-differential controller for high-speed atomic force microscopy. Rev Sci Instrum 77:83704. https://doi.org/10.1063/1.2336113
15. Teplow DB (2006) Preparation of amyloid β-protein for structural and functional studies. Methods Enzymol 413:20–33. https://doi.org/10.1016/S0076-6879(06)13002-5

Chapter 13

Direct Observation of Dynamic Movement of DNA Molecules in DNA Origami Imaged Using High-Speed AFM

Masayuki Endo and Hiroshi Sugiyama

Abstract

The visualization of biomolecules is a straightforward way to elucidate the physical properties of molecules and their reaction processes. Atomic force microscopy (AFM) enables the direct imaging of biomolecules under physiological conditions at nanometer-scale spatial resolution. Because AFM visualizes all molecules in a scanning area, an observation scaffold is required for the target-specific imaging of molecules in the dynamic state. The DNA origami technology allows the precise placement of target molecules in a designed nanostructure, and the detection of the molecules at the single-molecule level. DNA origami is applied for visualizing the detailed motions of molecules using high-speed AFM (HS-AFM), which enables the analysis of the dynamic movement of biomolecules in a subsecond time resolution. Here, we describe the combination of the DNA origami system with HS-AFM for the imaging of DNA structural changes controlled by photoresponsive molecules. The hybridization and dehybridization of photoresponsive oligonucleotides were visualized directly using this observation system. These target-oriented observation systems should contribute to the detailed analysis of biomolecules in real time with molecular resolution.

Key words DNA nanotechnology, DNA origami, Single molecule observation, High-speed atomic force microscopy, Photoreaction

1 Introduction

Direct imaging of target biomolecules is a commonly used way to study the physical properties of molecules in various phenomena involved in living systems. Single-molecule imaging using a probe microscope is a practical approach for investigating the motions of biomolecules during reactions. Atomic force microscopy (AFM) enables the direct observation of biomolecules with nanoscale spatial resolution, and the imaging can be performed under physiological conditions. Because AFM visualizes all molecules in a scanning area, an observation scaffold is required to facilitate the target-specific imaging of molecules in the dynamic state [1–3]. A novel DNA self-assembly system, DNA origami, has recently been developed for the construction of various two-dimensional and three-

Yuri L. Lyubchenko (ed.), *Nanoscale Imaging: Methods and Protocols*, Methods in Molecular Biology, vol. 1814, https://doi.org/10.1007/978-1-4939-8591-3_13,

dimensional nanostructures [4]. DNA origami nanostructures are used as scaffolds for the incorporation of various functional molecules and nanoparticles at desired positions. The DNA origami system is also used for the single-molecule detection of target molecules and for the analysis of single chemical reactions, which are imaged by AFM [1, 2]. Furthermore, the detailed dynamics of the molecules could be visualized if the single-molecule imaging is performed on the DNA origami nanostructure.

In the past decade, the visualization of molecular movement during biological reactions in a subsecond time scale has been achieved using high-speed AFM (HS-AFM) [5–8]. For the analysis of the enzyme reaction that occurs on the DNA strand, the direct extraction of the target reaction process using AFM is challenging because flexible long dsDNA strands do not keep a uniform structure. By combining the DNA origami technology with HS-AFM analysis, the dynamic movement of biomolecules has been imaged in a robust origami structure (Fig. 1). The DNA origami system can be expanded to visualize the dynamic movement of various biochemical reactions, including enzymatic reactions, DNA structural changes, DNA photoreactions, DNA catalytic reactions, and RNA interactions at the single-molecule level [2, 9]. By using the designed DNA origami scaffolds and improving the HS-AFM imaging technique, these observation systems can be used

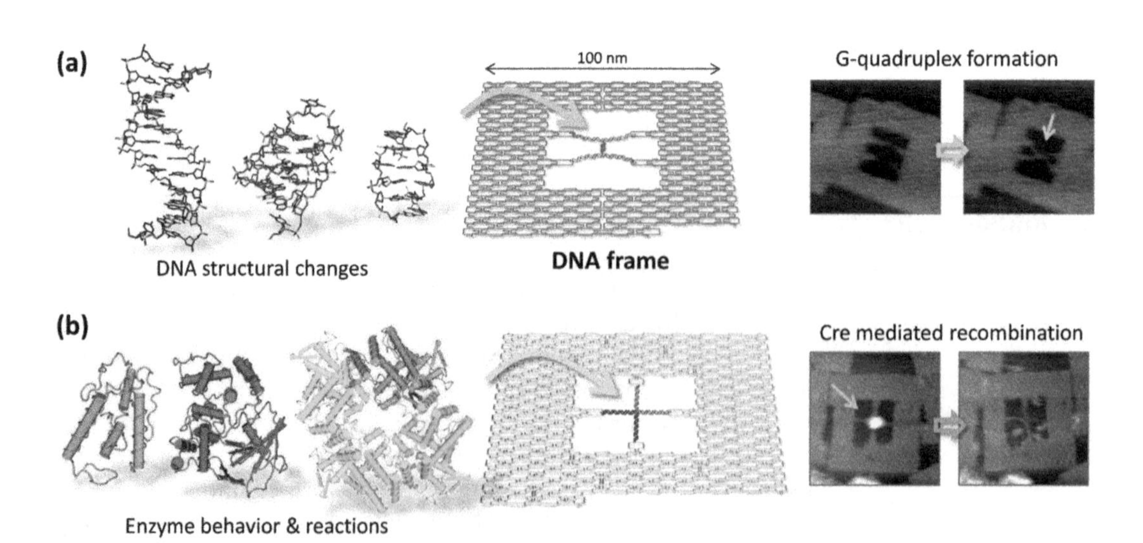

Fig. 1 Single-molecule observation of the DNA structural changes and enzymatic reactions in the DNA origami frame using high-speed atomic force microscope (HS-AFM). (**a**) A target DNA structure of interest is attached to the DNA origami scaffold called "DNA frame" and the structural change of the target DNA is observed directly using HS-AFM. The AFM images show the formation of G-quadruplex (green arrow) during AFM scanning. (**b**) A DNA substrate of the target enzyme and protein is attached in the DNA frame, and the interaction with the target enzyme and protein, as well as the behavior and reaction, are observed directly using HS-AFM. The AFM images show the recombination reaction mediated by Cre recombinase (orange arrow)

extensively to elucidate the physical properties of individual molecules involved in various biological and nonbiological reactions.

In this chapter, we describe the direct observation of the hybridization and dissociation of dsDNA with molecular resolution. AFM-based single-molecule imaging can visualize whole nanostructures by directly monitoring the shape of DNA strands. We originally designed a DNA origami frame and prepared single-stranded DNA and complementary strands called staple strands by annealing of M13mp18 (Fig. 2a) [4, 9]. Photoswitching DNAs containing azobenzene molecules were employed to observe the hybridization and dissociation of DNA strands (Fig. 2b) [10, 11]. Photoswitching DNA strands can hybridize in the *trans*-form of the azobenzene moiety and dissociate in the *cis*-form by photoisomerization using UV irradiation. The dissociated DNA strands in the *cis*-form hybridize again upon visible light (Vis) irradiation. We prepared photoswitching DNA strands for incorporation into the DNA frame (Fig. 3). We incorporated a pair of photoswitching DNA strands connected to individual supporting dsDNA chains in the cavity of the DNA frame, to visualize the photoinduced hybridization and dissociation at the single-molecule level (Fig. 2c) [12]. The hybridized photoswitching dsDNA in the center of the supporting dsDNA chains in the DNA frame was visualized clearly (Fig. 2d). Using this origami system, the hybridization and dissociation of the DNA strands can be clearly visualized by observing the global structural change of the two supporting dsDNA chains. The hybridization and dissociation were directly identified as either an X-shape or a separated shape in the DNA frame, respectively. The dissociation of the two dsDNA chains that were in contact at the center (X-shaped) was imaged during UV irradiation. The contact of the two separated dsDNA chains in the center was imaged again during Vis irradiation (Fig. 4a). The alternating dissociation and hybridization of the photoswitching DNA strands were visualized in the DNA frame at the single-molecule level.

The reversible activities of a pair of photoswitching DNA strands in a DNA frame were successfully imaged under a UV–Vis–UV light irradiation series on the AFM stage (Fig. 4b). The distance between the two centers of the dsDNAs was monitored, and the alternating dissociation and hybridization of the photoresponsive domains were observed by HS-AFM. These results show that the reversible dissociation and hybridization of the photoswitching DNA strands proceeded in the DNA frame under the control of UV–Vis irradiation, and their detailed behaviors were visualized. These findings indicate that the DNA observation system developed here is a versatile tool for the observation of the reversible switching behavior of DNA strands at the single-molecule level.

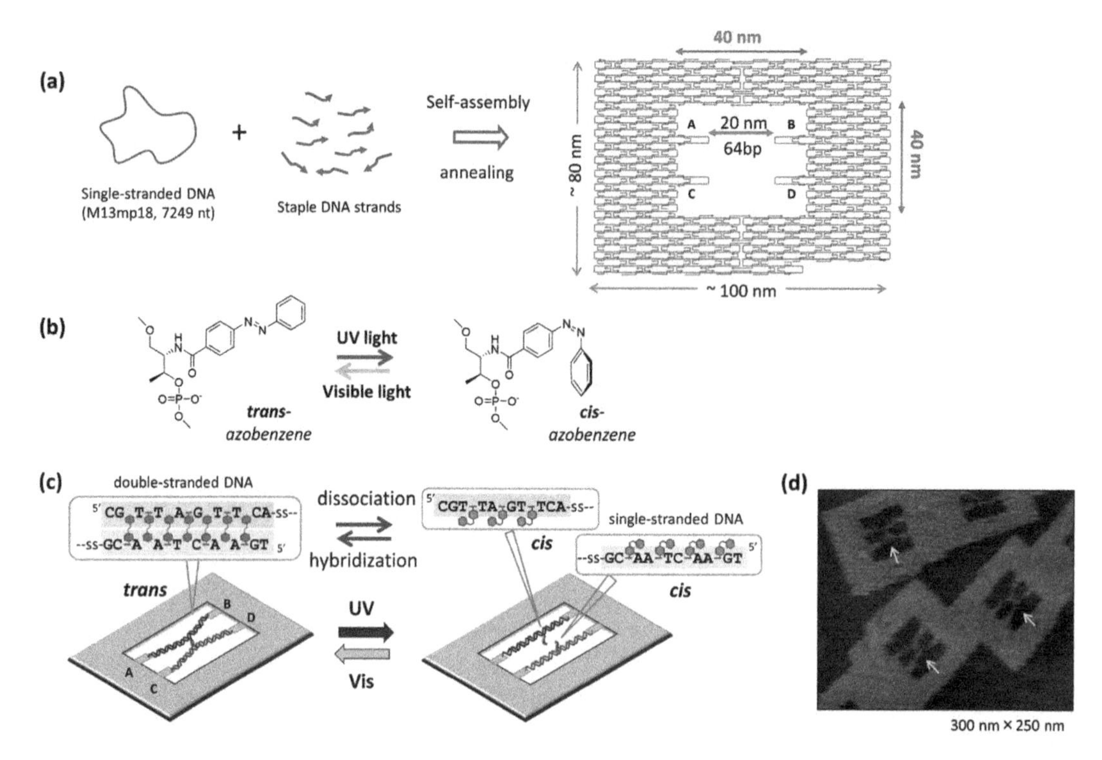

Fig. 2 (**a**) Preparation of a DNA origami frame. The single-stranded scaffold DNA and the complementary staple DNA strands were self-assembled by annealing to form the predesigned structure. Direct observation of the hybridization and dissociation of a pair of photoswitching DNA strands. (**b**) Structure of an azobenzene moiety in the DNA strand and the *trans–cis* photoisomerization by UV–Vis irradiation. (**c**) Single-molecule observation system. A pair of pseudocomplementary photoswitching strands was used in this experiment. In the *trans*-form of azobenzene, two photoswitching domains hybridize to form a duplex. In the *cis*-form induced by UV irradiation, the two DNA strands dissociate. Two dsDNA chains containing photoswitching DNA strands were placed in the DNA frame to observe the dissociation and hybridization by UV and Vis irradiation, respectively. Two dsDNA chains containing photoswitching DNA strands were connected between the specific sites in the DNA frame via the four corresponding connectors. (**d**) AFM image of the photoswitching DNA duplex supported by two dsDNA chains in the DNA frame (orange arrow)

2 Materials

2.1 Design and Preparation of DNA Origami

1. M13mp18 single-stranded DNA (New England Biolabs, USA).
2. Staple DNA strands (Sigma Genosys, Tokyo, Japan).
3. Thermal cycler.
4. Electrophoresis apparatus.
5. Electrophoresis buffer: 0.5 × TBE (Tris–Borate–EDTA buffer).
6. Agarose gel: 1% Agarose, 5 mM $MgCl_2$, 0.5× TBE.

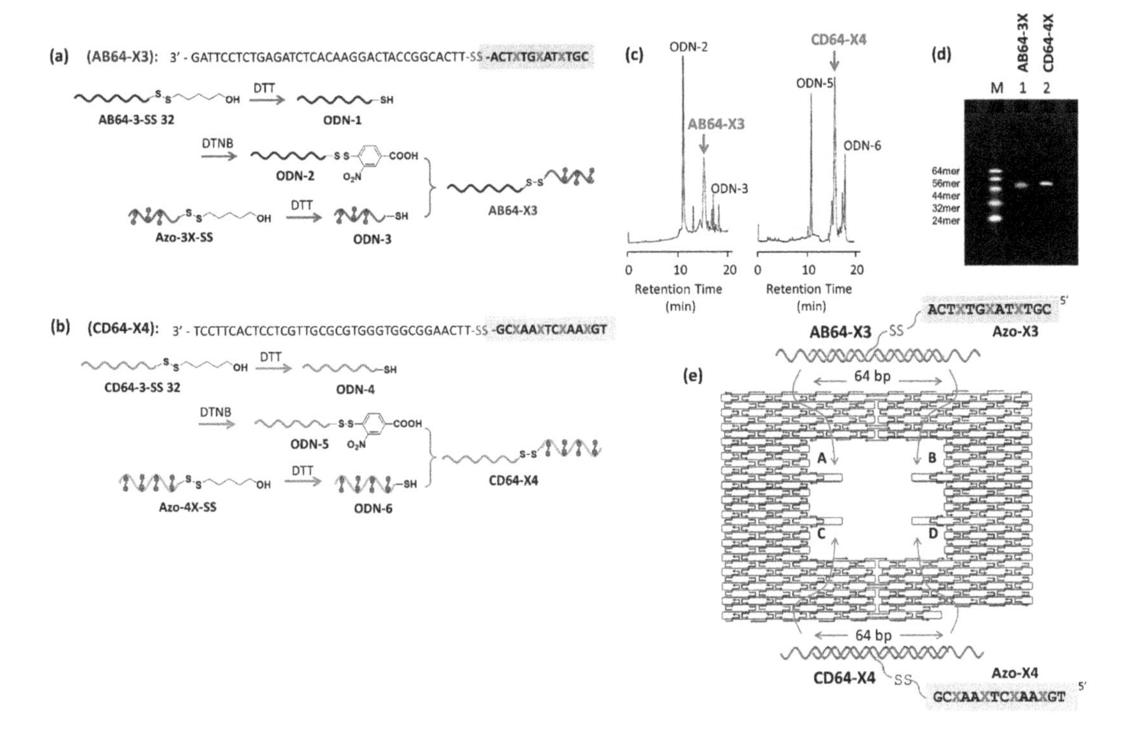

Fig. 3 Schematic representation of the synthesis of oligonucleotides (ODNs) containing the photoswitching DNA strands *X3* (three azobenzene moieties) and *X4* (four azobenzene moieties): (**a**) *AB64-X3* and (**b**) *CD64-X4*. The red X represents the azobenzene molecules incorporated in the ODNs. HPLC profiles and denaturing PAGE analysis. (**c**) Purification of *AB64-X3* and *CD64-X4*. (**d**) Denaturing PAGE analysis of *AB64-X3* and *CD64-X4*. PAGE conditions: 12% polyacrylamide, 1× TBE buffer, 8 M urea, 120 V, and ethidium bromide as a staining dye. Lane M: markers; lane 1: *AB64-X3*; lane 2: *CD64-X4*. (**e**) Incorporation of two preassembled dsDNAs containing photoswitching strands into the DNA frame (*see* Fig. 2c)

7. 1 M Tris–HCl buffer pH 7.6 (Nacalai Tesque, Kyoto, Japan).
8. $MgCl_2$ (Nacalai Tesque, Kyoto, Japan).
9. EDTA (Nacalai Tesque, Kyoto, Japan).
10. Milli-Q water (Millipore, USA).

2.2 Preparation of Photoswitching DNA Strands for Incorporation into the DNA Origami Frame

1. Synthetic oligonucleotides (ODNs) containing azobenzene molecules and a thiol group (purchased from Japan Bioservice, Saitama, Japan).
2. Synthetic ODNs containing a thiol group (purchased from Japan Bioservice, Saitama, Japan).
3. Dithiothreitol (DTT) (Takara Bio Inc., Kyoto, Japan).
4. 5, 5′-Dithiobis (2-nitrobenzoic acid) (DTNB) (Nacalai Tesque Inc., Kyoto, Japan).
5. High performance liquid chromatography (HPLC) instrument LC-2000 plus (JASCO, Tokyo, Japan).

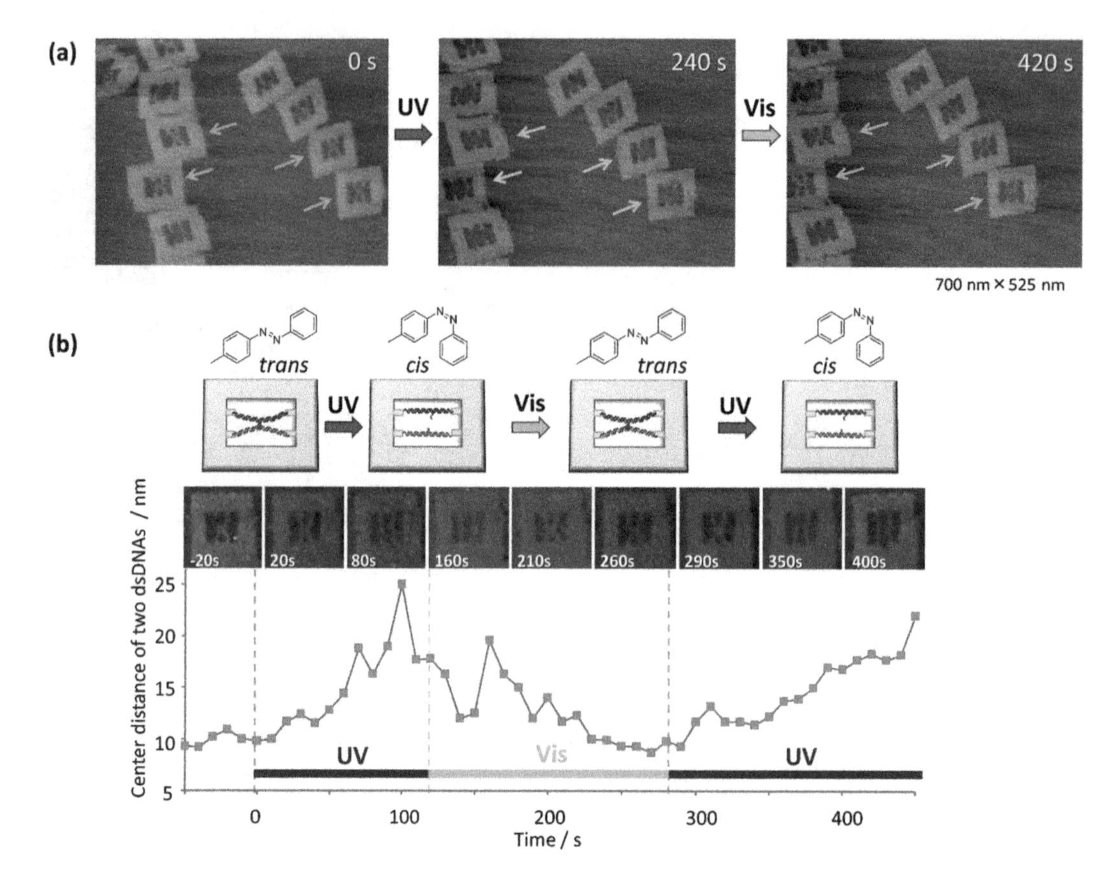

Fig. 4 Live imaging of the hybridization and dissociation of a pair of photoswitching DNA strands using HS-AFM. (**a**) HS-AFM images under the successive UV (240 s) and Vis (180 s) irradiation during AFM scanning. The red and blue arrows represent hybridized (X-shape) and dissociated (separated) photoswitching strands in the DNA frame, respectively. The samples were imaged at a scanning rate of 0.1 frame/s. (**b**) Single-molecule switching motions of photoswitching DNA strands in the DNA frame. Repeating dissociation and hybridization were visualized by HS-AFM during alternating UV/Vis photoirradiation. The distance between the centers of two dsDNA chains was plotted. The appearance of the X-shape and separated structures is shown as orange and blue rectangles in the graph, respectively. The samples were imaged at a scanning rate of 0.1 frame/s

6. Nacalai Cosmosil C18 reversed-phase column (7.5 × 150 mm) (Nacalai Tesque Inc., Kyoto, Japan).
7. Acetonitrile (HPLC grade, Wako Pure Chemical, Japan).
8. Ammonium formate (Nacalai Tesque Inc., Kyoto, Japan).
9. Electrophoresis apparatus.
10. Electrophoresis buffer: 1 × TBE (Tris/Borate/EDTA buffer) (Nacalai Tesque, Kyoto, Japan).
11. Denatured polyacrylamide gel: 1% Agarose, 5 mM $MgCl_2$, 0.5× TBE.
12. 1 M Tris–HCl buffer pH 8.0 (Nacalai Tesque, Kyoto, Japan).
13. Milli-Q water (Millipore, USA).

2.3 Preparation of a DNA Origami Frame with Photoswitching DNA Strands

1. Synthetic dsDNAs with azobenzene-containing ODNs prepared using the procedure described in Subheading 3.2.
2. DNA origami frame prepared using the procedure described in Subheading 3.1.
3. Thermal cycler.
4. 1 M Tris–HCl buffer pH 7.6 (Nacalai Tesque, Kyoto, Japan).
5. $MgCl_2$ (Nacalai Tesque, Kyoto, Japan).
6. EDTA (Nacalai Tesque, Kyoto, Japan).
7. Milli-Q water (Millipore, USA).

2.4 Photoirradiation of the Photoswitching DNA Strands in the DNA Frame

1. Xe-lamp (300 W, Ashahi-spectra MAX-303, Japan) with light guide (optical fiber cable).
2. Band-pass filter (10 nm full width at half maximum); LX0350 (350 nm for UV-light) and LX0450 (450 nm for visible-light).
3. Dry thermobath (Eyela MG-1200, Tokyo, Japan).

2.5 High-Speed Atomic Force Microscopy (HS-AFM)

1. Mica disks (diameter of 1.5 mm) (Furuuchi Chemical Corporation, Tokyo, Japan).
2. Solution of 10 nM DNA nanostructure prepared using the procedure described in Subheading 3.3.
3. Observation buffer: 20 mM Tris–HCl (pH 7.6), 10 mM $MgCl_2$.
4. High-speed atomic force microscopy (Nano Live Vision, RIBM, Tsukuba, Japan) integrated in the Olympus IX-71 microscope platform.
5. Hg lamp light source (Olympus U-RFL-T).
6. Cantilever (BL-AC10EGS, Olympus, Tokyo, Japan).

3 Methods

3.1 Design and Preparation of a DNA Origami Frame

The DNA frame structure was designed according to the rules of the DNA origami method (*see* **Note 1**) [4]. The M13mp18 single-stranded DNA was used for the scaffold DNA strand, and the complementary staple strands were assigned according to the design of the origami DNA frame (Fig. 2a) [9] . The sequences of the staple DNA strands were extracted from the design of the DNA frame. These sequences are listed in Ref. [13]. The DNA frame lacks the right-bottom corner for the identification of the orientation of the DNA frame structure (orientation marker). The DNA origami frame was prepared as follows.

1. Mix all staple DNA strands (226 strands, 100 μM each) to a final concentration of 0.2 μM.

2. Prepare a 20 μL solution containing the M13mp18 single-stranded DNA (10 nM), premixed staple strands (50 nM; 5 equiv.), Tris buffer (20 mM; pH 7.6), EDTA (1 mM), and $MgCl_2$ (10 mM).
3. Anneal the mixture from 85 to 15 °C at a rate of −1.0 °C/min using a thermal cycler.
4. Observe the formation of the structure by AFM using the procedure described in Subheading 3.5.

3.2 Preparation of Oligonucleotides Containing Photoswitching DNA Strands

The preparation of *AB64-X3* and *CD64-X4* was achieved using thiolated azobenzene-containing ODNs and thiolated ODNs via selective disulfide bond formation as shown in Fig. 3a, b (*see* **Note 2**).

1. Add 5 mM DTT and 0.1 M Tris–HCl buffer (pH 8.0) to a single DNA strand "*AB64-3-SS*" (32mer, 100 μM, 20 μL) tethered with a disulfide bond at the 5′ terminus.
2. Incubate at 37 °C for 2 h for reduction.
3. Purify the thiolated ODN (*ODN-1*) by HPLC [linear gradient using 2–30% acetonitrile–water (20 min) containing 20 mM ammonium formate, Nacalai Cosmosil C18 reversed-phase column (7.5 × 150 mm), 2.0 mL/min, 260 nm].
4. Treat the reduced thiolated ODN with DTNB (1 μL, 50 mM DMF solution) directly after the purification.
5. Incubate at 40 °C for 1 h during the evaporation of the HPLC solvent.
6. Purify the ODN-nitrobenzoic acid conjugate (*ODN-2*) using HPLC [the same conditions as those described above].
7. Treat the azobenzene-tethered ODN ("*Azo-3X- SS*", 40 μM, 20 μL), which also had the disulfide protection at the 3′ terminus, with 5 mM DTT and 0.1 M Tris–HCl buffer (pH 8.0) at the same time.
8. Incubate at 37 °C for 2 h for reduction.
9. Purify the thiolated ODN (*ODN-3*) using HPLC [the same conditions as those described above].
10. Mix equivalent amounts of *ODN-1* and *ODN-3*.
11. Incubate at 40 °C for 1 h under evaporation of the HPLC solvent.
12. Analyze and purify the concentrated mixture using HPLC [the same conditions as those described above] (Fig. 3c).
13. Lyophilize the purified azobenzene-modified ODN conjugate (*AB64-X3*).

14. Check the final product using denaturing polyacrylamide gel electrophoresis (12%, 1 × TBE buffer, 8 M urea, 120 V, ethidium bromide as a staining dye) (Fig. 3d).

The method used for the preparation of *CD64-X4* was the same as that used to prepare *AB64-X3*. The only difference was that the "*Azo-4X- SS*" ODN was used as the photoswitching DNA strand. Moreover, the final ODN product was purified using HPLC (Fig. 3c) and confirmed by denaturing polyacrylamide gel electrophoresis (same conditions as those described above) (Fig. 3d).

3.3 Incorporation of Two dsDNAs Containing Photoswitching ODNs into the DNA Frame

The preassembled dsDNAs containing a photoswitching DNA strand were incorporated into the DNA frame at a lower annealing temperature (Fig. 3e). The sample was purified by gel filtration, and the assembled structures were confirmed by AFM imaging.

1. Assemble the bridging strand photoswitching ODNs by annealing from 85 to 15 °C at a rate of −1.0 °C/min using a thermal cycler.
2. Mix the two assembled dsDNAs with photoswitching ODNs (20 nM; 2 equiv.) described above with the preassembled DNA frame (10 nM).
3. Assemble the dsDNAs in the DNA frame by heating at 40 °C and then cooling to 15 °C at a rate of −0.5 °C/min using a thermal cycler.
4. Purify the sample by gel filtration chromatography (GE sephacryl-300).
5. Observe the assembled structures using HS-AFM.
6. Calculate the yield of incorporation of the strands to the DNA frame manually from the AFM images, in which >200 DNA frames are counted.

3.4 Photoirradiation to Photoswitching DNA Strands Attached to the Supporting dsDNAs in the DNA Frame

Photoirradiation of the sample was performed using a Xe lamp with a band-pass filter of 350 nm and 450 nm for UV and Vis, respectively.

1. Place a microtube containing the sample prepared in Subheading 3.3 on a dry thermobath to maintain the temperature constant during photoirradiation.
2. Keep the temperature at 25, 30, and 35 °C.
3. Irradiate the sample with UV light for 15 min or visible light for 10 min using the Xe-lamp through the optical fiber cable.
4. Image the irradiated samples by AFM using the procedure described in Subheading 3.5 to characterize the state of the photoswitching DNA strands.
5. Count the number of connected (X-shape) and separated (parallel shape) structure in the AFM images (>200 DNA frames).

3.5 High-Speed AFM Imaging of the Behavior of the Photoswitching DNA Strands in the DNA Frame

AFM images were obtained via HS-AFM (Nano Live Vision, RIBM, Tsukuba, Japan) (*see* **Note 3**) using a silicon nitride cantilever (Olympus BLAC10EGS) (*see* **Note 4**). Samples for AFM imaging were prepared, and the HS-AFM operation was performed.

1. Attach the mica disc onto the glass scaffold.
2. Cleave the mica disc to obtain a fresh surface.
3. Dilute the DNA nanostructures sample to ~5 nM by adding observation buffer.
4. Place ~2 μL of the sample solution onto the mica surface for 5 min.
5. Rinse the surface with imaging buffer (~10 μL) three times to remove unbound molecules.
6. Place the cantilever on the cantilever holder.
7. Fill the liquid cell with ~120 μL of observation buffer.
8. Place the mica plate with a glass scaffold on onto the scanner stage.
9. Set up the scanner over the liquid cell in which the cantilever is immersed in the observation buffer.
10. Align the laser focusing position and the photodetector position to maximize the intensity of the laser light reflected back from the cantilever.
11. Find the resonant frequency of the cantilever using a fast Fourier transform (FFT) analyzer.
12. Excite the cantilever at the resonant frequency by applying sinusoidal AC voltage.
13. Execute the approach until the software stops the motor automatically.
14. Adjust the set point voltage to ~75–95% of the free oscillation amplitude.
15. Gradually decrease the set point voltage until the sample is clearly imaged
16. Image the samples in the observation buffer with a scanning rate of 0.1–1.0 frame/s (fps).

3.6 Photoirradiation to the Sample on the AFM Stage

Photoirradiation of the sample during AFM scanning, we carried out on the AFM stage (using an Olympus IX71 microscope as a base unit) using a Hg-lamp light source with band-pass filters (330–380 nm for UV irradiation and 440–470 nm for Vis irradiation). The shutter on the microscope was manually operated during AFM scanning.

1. Image the sample in the same procedure in described in Subheading 3.5.

2. Open the shutter manually to introduce UV light through the band-pass filter to the sample stage during AFM scanning.
3. Stop the UV light irradiation by closing the shutter.
4. Change the revolver position and open the shutter to introduce visible light through the band-pass filter.
5. Stop the visible light irradiation by closing the shutter.

4 Notes

1. The design of DNA origami structures is currently carried out using the caDNAno software (http://cadnano.org/) [14].
2. The photoswitching DNA strands were prepared using the sequences reported in the literature [11] .
3. Details on the information of the instrument are available at the home page of RIBM (http://www.ribm.co.jp).
4. For HS-AFM imaging, small cantilevers are used. Small cantilevers (9 μm long, 2 μm wide and 130 nm thick; BL-AC10DS, Olympus, Tokyo, Japan) made of silicon nitride with a spring constant ~0.1 N/m, and a resonant frequency of ~300–600 kHz in water are commercially available from Olympus.

Acknowledgments

This work was supported by JSPS KAKENHI (grant numbers 15H03837, 16K14033, 16H06356) and a Grant-in-Aid for Scientific Research on Innovative Areas "Molecular Robotics" (No. 24104002) from MEXT, Japan.

References

1. Torring T, Voigt NV, Nangreave J, Yan H, Gothelf KV (2011) DNA origami: a quantum leap for self-assembly of complex structures. Chem Soc Rev 40:5636–5646
2. Rajendran A, Endo M, Sugiyama H (2012) Single-molecule analysis using DNA origami. Angew Chem Int Ed 51:874–890
3. Endo M, Yang Y, Sugiyama H (2013) DNA origami technology for biomaterials applications. Biomater Sci 1:347–360
4. Rothemund PW (2006) Folding DNA to create nanoscale shapes and patterns. Nature 440:297–302
5. Ando T, Kodera N, Takai E, Maruyama D, Saito K, Toda A (2001) A high-speed atomic force microscope for studying biological macromolecules. Proc Natl Acad Sci U S A 98:12468–12472
6. Ando T, Kodera N (2012) Visualization of mobility by atomic force microscopy. Methods Mol Biol 896:57–69
7. Uchihashi T, Kodera N, Ando T (2012) Guide to video recording of structure dynamics and dynamic processes of proteins by high-speed atomic force microscopy. Nat Protoc 7:1193–1206
8. Rajendran A, Endo M, Sugiyama H (2014) State-of-the-art high-speed atomic force microscopy for investigation of single-molecular dynamics of proteins. Chem Rev 114:1493–1520

9. Endo M, Katsuda Y, Hidaka K, Sugiyama H (2010) Regulation of DNA methylation using different tensions of double strands constructed in a defined DNA nanostructure. J Am Chem Soc 132:1592–1597
10. Asanuma H, Liang X, Nishioka H, Matsunaga D, Liu M, Komiyama M (2007) Synthesis of azobenzene-tethered DNA for reversible photo-regulation of DNA functions: hybridization and transcription. Nat Protoc 2:203–212
11. Liang X, Mochizuki T, Asanuma H (2009) A supra-photoswitch involving sandwiched DNA base pairs and azobenzenes for light-driven nanostructures and nanodevices. Small 5:1761–1768
12. Endo M, Yang Y, Suzuki Y, Hidaka K, Sugiyama H (2012) Single-molecule visualization of the hybridization and dissociation of photoresponsive oligonucleotides and their reversible switching behavior in a DNA nanostructure. Angew Chem Int Ed 51:10518–10522
13. Endo M, Katsuda Y, Hidaka K, Sugiyama H (2010) A versatile DNA nanochip for direct analysis of DNA base-excision repair. Angew Chem Int Ed 49:9412–9416
14. Douglas SM, Marblestone AH, Teerapittayanon S, Vazquez A, Church GM, Shih WM (2009) Rapid prototyping of 3D DNA-origami shapes with caDNAno. Nucleic Acids Res 37:5001–5006

Chapter 14

Assembly of Centromere Chromatin for Characterization by High-Speed Time-Lapse Atomic Force Microscopy

Micah P. Stumme-Diers, Siddhartha Banerjee, Zhiqiang Sun, and Yuri L. Lyubchenko

Abstract

Atomic force microscopy (AFM) is an imaging technique that enables single molecule characterization of biological systems at nanometer resolution. Imaging in ambient conditions can provide details of the conformational states and interactions of a population of molecules which is well complemented by single-molecule imaging of the systems dynamics using time-lapse AFM imaging, in which images are capture at rates of 10–15 frames per second in an aqueous buffer. Here we describe the assembly and preparation of nucleosomes containing centromere protein A (CENP-A) for AFM imaging in both static and time-lapse modes. The AFM imaging and data analysis techniques described enable characterization of the extent of DNA wrapping around the histone core and time-resolved visualization of the systems intrinsic dynamic behaviors.

Key words High-speed atomic force microscopy, Chromatin dynamics, Nucleosomes, Centromere, CENP-A, Nanoimaging

1 Introduction

The presence of aberrantly functioning centromeres results in aneuploidy, where chromosomes are improperly distributed to daughter cells, a hallmark of many cancers [1]. A closer look at the centromere reveals the most distinct difference of centromere chromatin at the nucleosome level: centromere nucleosome contain histone protein A (CENP-A) in place of histone H3 of bulk chromatin, assembling with histones H2A, H2B and H4 to form a histone octamer [2, 3]. Characterization of the intrinsic conformational and dynamic properties of this centromeric chromatin is critical to our understanding of centromere function. Atomic force microscopy (AFM) is an imaging technique capable of probing dynamic properties of biological systems, such as nucleosomes, with nanometer resolution and little to no interaction with the sample [4–7]. The cutting-edge development of AFM enabling the

Yuri L. Lyubchenko (ed.), *Nanoscale Imaging: Methods and Protocols*, Methods in Molecular Biology, vol. 1814,
https://doi.org/10.1007/978-1-4939-8591-3_14,

instrument to acquire images of fully hydrated molecules and molecular assemblies with video rate speed (high-speed AFM (HS AFM)) made it possible to characterize nanoscale dynamics of various biological molecules and their complexes [8, 9] including nucleosomes [4, 10, 11]. We have recently applied this instrumentation and the methodology to study nanoscale dynamics of centromere nucleosomes that revealed a number of novel properties of them that distinct dramatically centromere chromatin from the bulk one [12].

Here we describe the methodology for the preparation of centromere nucleosomes starting with the assembly process of CENP-A nucleosomes by a salt gradient dilution method, followed by characterization of the prepared nucleosomes by SDS-PAGE [13]. Following this, we will describe the preparation of the nucleosome sample enabling AFM imaging in several conditions. We will first describe imaging of dry AFM specimens at ambient conditions in which a snapshot of bulk nucleosome conformation is assessed. Next, we will show how to use standard time-lapse AFM imaging to probe the overall dynamics of nucleosome particles through continuous imaging in aqueous buffers at a rate of several minutes per frame. Finally, high-speed time-lapse AFM (HS AFM) will be described in which image acquisition rates as fast as 10–15 frames per second can be used thereby capturing dynamic behaviors beyond the spatiotemporal resolution of standard time-lapse imaging [14]. The chapter ends with a description of analysis that can be employed to characterize the nucleosome samples by these three AFM imaging techniques.

2 Materials

2.1 Preparation of Nucleosomal DNA Substrate

1. Plasmid pGEM3Z-601 (Addgene, Cambridge, MA).
2. PCR primers: (FP) 5′CAGTGAATTGTAATACGACTC3′; (RP) 5′ACAGCTATGACCATGATTAC3′.
3. Each reagent of the PCR Master Mix has the following final concentration and is added in the order listed on ice. 1× DreamTaq polymerase buffer (contains 2 mM $MgCl_2$), 0.2 mM each dNTP, 0.2 μM FP, 0.2 μM RP, 1.25 U/50 μL reaction DreamTaq polymerase.
4. PCR Purification Kit.

2.2 Continuous Dilution Assembly of CENP-A Nucleosomes

1. 10× TE (pH 7.5): Add 100 mL of Tris–HCl (pH 7.5) and 20 mL of 0.5 M EDTA to 880 mL of 18.2 MΩ•cm H_2O (ddI).
2. $(CENP\text{-}A/H4)_2$: This can be purified and reconstituted [15] or purchased from EpiCypher (Durham, North Carolina).

3. H2A/H2B: This can be purified and reconstituted [15] or purchased from EpiCypher (Durham, North Carolina).
4. H3 octamers: This can be purified and reconstituted [15] or purchased from EpiCypher (Durham, North Carolina).
5. Slide-A-Lyzer Mini Dialysis Unit 10 K MWCO (ThermoFisher Scientific).
6. Low salt buffer: 10 mM Tris–HCl (pH 7.5), 0.25 mM EDTA and 2.5 mM NaCl.
7. Centrifugal filter 10 K MWCO.

2.3 Histone Composition of Nucleosome Assembly by 15% SDS-PAGE

1. Separating buffer (1.5 M, 50 mL): 1.5 M Tris base (9.09 g) in ddI H_2O. Bring to pH 8.9 with HCl.
2. Stacking buffer (1.0 M, 50 mL): 1.0 M Tris base (6.06 g) in ddI H_2O. Bring to pH 6.8 with HCl.
3. Cathode running buffer (10×, 50 mL): 1.0 M Tris base (6.06 g), 1 M tricine (8.96 g), 1% SDS (0.5 g). Dilute to 1× immediately before use. No need to adjust pH as it should already be ~8.25.
4. Anode running buffer (10×, 50 mL): 2.0 M Tris base (12.11 g). Bring to pH 8.9 with HCl and dilute to 1× immediately before use.
5. 10% ammonium persulfate (AmmPS): 20 mg of AmmPS in 200 μL ddI H_2O. Make fresh as needed.
6. 30% acrylamide–bis solution (37.5:1)
7. 15% separating gel mixture: 2.5 mL 1.5 M separating buffer, 5 mL 30% acrylamide–bis (37.5:1) solution, 2.5 mL ddI H_2O, and 40 μL 10% AmmPS.
8. 6% stacking gel mixture: 650 μL 1.0 M stacking buffer, 1 mL 30% acrylamide–bis (37.5:1) solution, 2.5 mL ddI H_2O, 50 μL 10% AmmPS.
9. Tetramethylethylenediamine (TEMED).
10. 4× Laemmli Sample Buffer: 277.8 mM Tris–HCl, pH 6.8, 44% (v/v) glycerol, 4.4% LDS, 0.02% bromophenol blue. Add 100 μL of 2-mercaptoethanol (2-ME) to 900 μL of 4× Laemmli buffer immediately before use.
11. Protein ladder: PageRuler Prestained Protein Ladder from Thermo Scientific works well but many similar options are available from various suppliers.
12. Bio-Safe Coomassie G-250 Stain (Bio-Rad, Hercules, CA).

2.4 Preparation of Aminopropyl Silatrane Functionalized Mica

1. Nonwoven cleanroom wipes: TX609 TechniCloth Cellulose and Polyester Cleanroom Wipers made by TexWipe

(Kernersville, North Carolina) are recommended but others are available as well.

2. Clean argon gas.
3. Sheet of high grade mica.
4. 50 mM aminopropyl silatrane (APS). APS was synthesized using the protocol described in [16].

2.5 Preparation of Nucleosome Samples for Imaging in Air

1. Nucleosome dilution buffer: 10 mM HEPES, pH 7.5 and 4 mM $MgCl_2$ passed through a 0.22 μm filter.
2. Magnetic AFM sample puck (1 cm diameter).
3. Double-faced adhesive tape.

2.6 AFM Imaging in Air

An AFM tip with a spring constant of approximately 40 N/m and a resonant frequency between 300 and 340 kHz. A tip with these properties that works reliably for air imaging by tapping/oscillating mode is the TESPA-V2 (Bruker AFM Probes, Camarillo, CA) which has a spring constant of 37 N/m and a resonance frequency of 320 kHz.

2.7 Time-Lapse Imaging in Aqueous Solutions

1. Tips used for liquid imaging should have a spring constant of around 0.06 N/m and a resonance frequency of ~7–10 kHz. A probe that fits these criteria and works reliably for the described application is cantilever "E" on MSNL-10 (Bruker AFM Probes, Camarillo, CA).
2. Aron Alpha Industrial Krazy Glue (Toagosei America, West Jefferson, Ohio).
3. Reprorubber Thin Pour (Reprorubber, Islandia, NY).
4. Glass microscope slide (75 × 25 mm)PMMA ring with an inner diameter of at least 1.5 cm and a height of 0.5 cm.
5. Nucleosome imaging buffer: 10 mM HEPES, pH 7.5 and 4 mM $MgCl_2$ passed through a 0.22 μm filter.

2.8 Time-Lapse Imaging in Aqueous Solutions with High-Speed AFM

1. For HS AFM probes, we recommend BL-AC10DS-A2 cantlievers (Olympus, Japan) which are modified using electron beam deposition (EBD) and sharped by plasma etching [9].
2. Compound FG-3020C-20 (FluoroTechnology Co., Ltd., Kagiya, Kasugai, Aichi, Japan).
3. Compound FS-1010S135-0.5 (FluoroTechnology Co., Ltd., Kagiya, Kasugai, Aichi, Japan).
4. Nucleosome imaging buffer: 10 mM HEPES, pH 7.5 and 4 mM $MgCl_2$ passed through a 0.22 μm filter.

2.9 AFM Imaging in Air

1. AFM tip with a spring constant of approximately 40 N/m and a resonant frequency between 300 and 340 kHz on the AFM tip

holder. A tip with these properties that works reliably for air imaging by tapping/oscillating mode is the TESPA-V2 (Bruker AFM Probes, Camarillo, CA) which has a spring constant of 37 N/m and a resonance frequency of 320 kHz.

2. Compound FG-3020C-20 (FluoroTechnology Co., Ltd., Kagiya, Kasugai, Aichi, Japan).
3. Compound FS-1010S135-0.5 (FluoroTechnology Co., Ltd., Kagiya, Kasugai, Aichi, Japan).
4. Nucleosome imaging buffer: 10 mM HEPES, pH 7.5 and 4 mM $MgCl_2$ passed through a 0.22 μm filter.

3 Methods

3.1 Preparation of Nucleosomal DNA

1. Substrate DNA with a slightly off-center 601-positioning singlet is generated by PCR using plasmid pGEM3Z-601 and the forward and reverse primers listed above (*see* **Note 1**). Preheat the thermal cycler to 95 °C and add tubes containing the reaction mixture. Run the following program for 33 cycles following an initial denaturation at 95 °C for 5 min: Denaturation at 95 °C for 30 s, annealing at 49 °C for 30 s, extension at 72 °C for 35 s. Following the 33 cycles is a final extension at 72 °C for 10 min.
2. Purify DNA from the PCR mixture using a PCR Purification Kit (*see* **Note 2**).
3. Combine the purified substrate DNA and collect an absorbance at 260 nm.
4. Aliquot 25 pmol of purified DNA into a 0.6 mL microfuge tube and remove water using a vacuum centrifuge (*see* **Note 3**).

3.2 Continuous Dilution Assembly of CENP-A Nucleosomes

Salt gradient deposition of histones onto a DNA substrate is a well-established method that utilizes the binding of $(CENP\text{-}A/H4)_2$ tetramers to DNA at higher salt concentrations than H2A/H2B dimers [13, 15]. To assemble nucleosomes by this technique, DNA and histones are first mixed with 2 M NaCl which is then lowered to 0.2 M NaCl or less either at a continuous or stepwise gradient. When working with microliter volumes as described here it is beneficial to use a technique in which minimal volumes of sample can be easily recovered. Dilution assembly with a syringe pump provides a means by which a continuous salt gradient is achieved while ensuring maximum recovery once assembled.

1. Determine the required volume of each assembly reagent by referring to Table 1 (*see* **Note 4**).
2. Prepare the reaction assembly mixture on ice in the order listed in Table 1 by adding each assembly component to the

Table 1
Components needed for a 25 pmol mononucleosome assembly

	Volume added
DNA (25 pmol)	Dried in tube
10× TE (pH 7.5)	1.0 μL
DDI H_2O	2.5 μL
5 M NaCl	2.75 μL
20 μM H2A/H2B dimer	2.5 μL
20 μM $(CENP\text{-}A/H4)_2$ tetramer	1.25 μL
Total	10 μL

Starting assembly volume should total 10 μL. The H2A/H2B and $(CENP\text{-}A/H4)_2$ stocks used in this example contain 2 M NaCl and 1 M NaCl, respectively

Fig. 1 Syringe pump configuration used for microscale CENP-A nucleosome assembly. The assembly mixture is positioned to be in contact with the end the syringe needle. As the dilution buffer is delivered by the syringe pump to the assembly mixture the concentration of NaCl is decreased, promoting nucleosome assembly

microfuge tube containing the 25 pmol of DNA prepared in **step 4** of Subheading 3.1. Incubate this mixture at room temperature for 30 min.

3. Following the 30 min incubation at room temperature the assembly mixture will be continuously diluted from the initial 2 M of NaCl to 0.25 M of NaCl using a syringe pump (Fig. 1).
4. Fill the syringe with 100 μL of room temperature dilution buffer and place on the pump (*see* **Note 5**). Insert the needle of the syringe into the assembly mixture tube.

5. Set the pump to deliver 90 μL of the dilution buffer at a rate of 0.75 μL/min. This will dilute the assembly mixture from 2.0 M to 200 mM NaCl with a final DNA concentration of 250 nM.
6. Transfer the diluted assembly mixture to a 10 K MWCO Slide-A-Lyzer MINI dialysis button and dialyze against 200 mL of the low salt buffer for 1 h.
7. Transfer the dialyzed solution to a clean microfuge tube and store at 4 °C until use. Nucleosomes remain stable for at least 2 months when stored in these conditions.

3.3 SDS-PAGE for Assessment of Histone Stoichiometry

In addition to characterization of nucleosomes by AFM as described below, the histone stoichiometry content of the assembly should be checked by discontinuous SDS-PAGE with a 15% separating and 6% stacking gel. Mini-gel setups provide sufficient separation of all histones and require a low volume of gel reagents. The volumes listed are sufficient to cast two mini-gel cassettes with a gel thickness of 0.75 mm.

1. Wash glass plates and combs with a laboratory detergent followed by thorough rinsing with distilled water.
2. Assemble the cassette on a casting stand and add distilled water or ethanol to check for leaks. Remove the water or ethanol from the cassette using filter paper. If leaking from the bottom, a 2% agarose solution can be used as a seal. Insert the comb and make a mark 1 cm below the teeth to indicate the level to which the separating gel should be poured.
3. Degas the 15% separating gel mixture.
4. Add 8 μL TEMED to one of the tubes and invert several times to mix. Quickly add this solution to each cassette until it reaches the mark indicating 1 cm below comb teeth. Immediately (but gently!) add ddI H_2O above the separating layer. Allow polymerization to proceed for 1 h.
5. Degas the 6% stacking gel mixture.
6. Remove water from above the separating gel using filter paper being careful not to disrupt the separating gel. Add 6 μL TEMED to the degassed stacking gel solution and invert several times to mix. Apply this solution to the top of each cassette and insert combs avoiding trapped bubbles.
7. Allow polymerization to proceed for 1 h. When done, transfer the cassette to the electrophoresis unit and fill the lower chamber with 1× anode buffer and the middle/upper chamber with 1× cathode buffer. Remove the comb and rinse each well with cathode buffer to remove any polymerization debris.
8. Run the following samples in adjacent lanes on the prepared gel: 1–2 μg of H3 histone octamer, 1–2 μg of CENP-A assembly mixture of histones and ~10–20 μL of the assembled

nucleosome samples (*see* **Note 6**). Add 4× Laemmli Sample Buffer (with 2-ME freshly added) to a working concentration of 1–2×. Heat the samples at 95 °C for ~5 min.

9. Add each sample to adjacent lanes in the gel. Add protein ladder to one or two lanes adjacent to the samples and fill the remaining lanes with sample buffer to promote even band migration. Run at 65 V until the dye front moves through the stacking gel and then increase to 150 V and run until the dye front has completely migrated out of the gel.
10. Upon migration of the dye front from the bottom of the gel, disassemble the electrophoresis unit, carefully separate the glass plates of the gel cassette and transfer the gel to a container filled with ddI H_2O (*see* **Note 7**). Wash three times for 5 min each in a total of ~300 mL of ddI H_2O. After the third wash, remove the water from the container and add 50–100 mL of Bio-Safe Coomassie Stain (Bio-Rad, Hercules, CA). Gently shake/agitate the container for at least 1 h. Remove the stain from container and rinse the gel with ddI H_2O followed by an additional 30 min soak in fresh ddI H_2O. The gel should look like that shown in Fig. 2 with the assembled nucleosome samples appearing as a faint copy of the assembly mixture control.

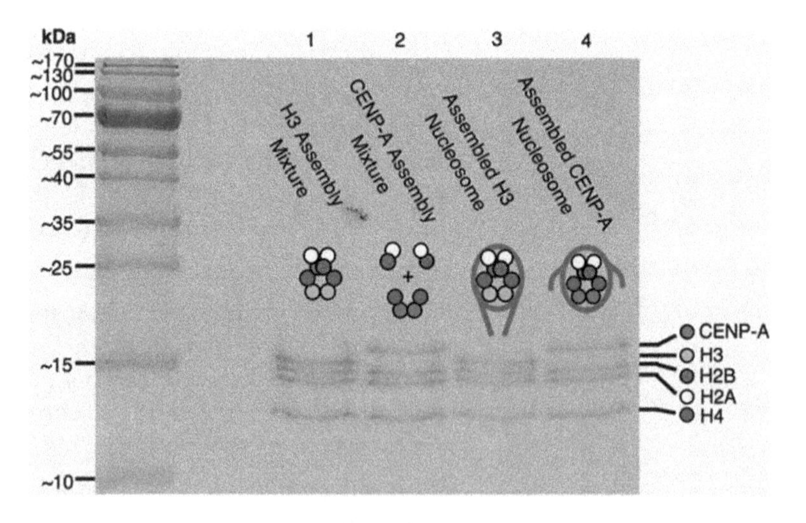

Fig. 2 Assembled nucleosome samples are first checked by SDS-PAGE. Lanes 1 and 2 contain the H3 octamer and the CENP-A assembly of histones, respectively. Lanes 3 and 4 contain the assembled H3 nucleosomes and the assembled CENP-A nucleosomes, respectively. Comparison of the assembled nucleosomes to the histone only controls in lanes, confirm that nucleosomes were properly assembled. The cartoon schematic above each lane indicates which histone components are present

3.4 Preparation of APS Functionalized Mica

1. Before handling the mica surface, thoroughly clean all tweezers and scissors that will be used (*see* **Note 8**).
2. Cut a 1 × 4 cm piece of mica using a paper trimmer or sharp scissors.
3. Cleave layers from each side of the mica piece using a razor blade and a sharp pair of tweezers until you have a piece as thin as ~0.1 mm with both sides freshly cleaved (*see* **Note 8**).
4. Promptly place the mica piece into a cuvette that was prerinsed with ddI H_2O and subsequently filled with APS solution diluted 300× with ddI H_2O. Let the surface functionalize for 30 min.
5. Rinse the functionalized mica piece by dipping it into a cuvette filled with ddI H_2O several times followed by thorough drying of both sides under a light flow of clean argon gas.
6. The substrate is now ready for sample deposition but may also be stored in a vacuum or desiccator for up to several weeks before any loss of DNA binding activity will occur.

3.5 Preparation of Nucleosome Samples for Imaging in Air

1. Cut the APS functionalized mica with a sharp and clean scissors to a size to fit the imaging platform of the AFM system being used. A ~1 × 1 cm square piece is optimal if mounting samples on the 1 cm diameter magnetic puck.
2. Dilute the nucleosome sample to 0.5–1.0 nM in nucleosome dilution buffer. Deposit 5–10 μL of this freshly diluted sample on the middle of the functionalized mica piece and let it sit for 2 min.
3. After the 2 min, Grab the mica substrate at a corner with a tweezer and gently rinse the substrate with 1–3 mL of ddI H_2O (*see* **Note 9**).
4. Dry the sample under a light flow of clean argon gas (*see* **Note 10**). Fix the dried substrate to a magnetic puck using a double-faced adhesive tape. Dry the sample for at least 2 h under vacuum before imaging to avoid tip adhesion.

3.6 AFM Imaging in Air

1. Mount the correct tip on the AFM tip holder.
2. If the sample was not mounted on magnetic puck in **step 4** of Subheading 3.5, use double-faced adhesive tape to mount the sample on the imaging platform.
3. Tune the cantilever to find the expected resonance frequency for the probe being used. If using the recommended tip, a target amplitude of 0.5 V will typically produce a drive amplitude in the range of 6–10 mV. Change image size to 100 × 100 nm and then begin the approach.
4. Once approached, increase the scan size to at least 1 × 1 μm and reduce the set point until the surface is being tracked well.

An image resolution of 4 nm per pixel is sufficient for quickly checking a sample by eye; however, the resolution should be increased to 2 nm per pixel when capturing images intended for analysis. For example, when capturing a 2 × 2 μm image, a pixel density of 1024 points per line should be used.

3.7 Time-Lapse Imaging in Aqueous Solutions

Time-lapse imaging in liquid provides a window to the general dynamic properties of the nucleosome sample at a spatiotemporal resolution on the order of minutes per frame. The following protocol assumes the use of the NanoWizard 4a (JPK, Berlin, Germany) but is easily adapted to any AFM system.

1. Cut and glue a 1 × 1 cm piece of un-functionalized mica to a glass microscope slide using Aron Alpha Industrial Krazy Glue (Toagosei America, West Jefferson, Ohio). A pipette tip works well for application of glue and only a very thin layer is needed. Let the glue dry for a minimum of 5 min.
2. Prepare a working solution of APS by diluting 1 μL of 50 mM APS stock in 299 μL of ddI H_2O. Cleave layers of mica using scotch tape until the top layer of mica appears to be unbroken as evidenced by the cleaved layer stuck on the tape. Add working APS solution to the freshly cleaved mica surface until a large droplet covers the surface and then let it functionalize for 30 min. After the 30 min, rinse the surface with several milliliters of ddI H_2O. Leave a droplet of water on the surface to prevent drying of the surface before sample deposition.
3. Mix equal parts of Reprorubber Thin Pour (Reprorubber, Islandia, NY) until a light green final color is produced. Apply an even layer of this mixture around the bottom of the PMMA ring and quickly fix the ring to the microscope slide around the functionalized mica being careful not to touch the surface. Rotate the ring to evenly distribute the glue and let dry for at least 10 min.
4. Mount a tip on the glass block tip holder (*see* **Note 11**).
5. Prepare the nucleosome sample as described in **step 2** of Subheading 3.5. Remove the droplet of water from the mica surface and deposit ~10–20 μL of the diluted nucleosome sample. After letting it sit for 2 min, rinse the surface with ~1 mL of nucleosome imaging buffer. Following the rinse, fill the PMMA ring with 1.5–2.0 mL of nucleosome imaging buffer.
6. Mount the AFM head so that the glass block tip holder is well centered over the liquid chamber. Using the step motors, lower the tip holder until the probe side of the glass block is fully submerged in liquid. Use the topview camera to focus on the cantilever; if bubbles appear anywhere around the cantilever they must be removed prior to imaging (*see* **Note 11**).

7. Using the laser adjustment knobs, position the laser over the cantilever in the position that provides the maximum sum. If no sum can be found or if it is lower than expected, change the mirror position. Adjust the lateral deflection to ~0 V and the vertical deflection to ~4–5 V (*see* **Note 12**).
8. Calibrate and auto tune the cantilever as described in the system manual and then begin the approach (*see* **Note 13**).
9. Adjust the drive amplitude and relative set point to improve image quality. An initial scan size of 2 × 2 μm is helpful in finding particles of interest, at which time the scan size can be decreased to ~500 nm for acquisition of nucleosome dynamics. At a scan size of 500 nm, an image acquisition rate of 1–2 Hz typically provides the best results.

3.8 Time-Lapse Imaging in Aqueous Solutions with High-Speed AFM

High-speed AFM (HS AFM) imaging, as designed by T. Ando [9], provides the unique opportunity to visualize nucleosome dynamics at a millisecond image acquisition rate. While standard time-lapse imaging provides the general dynamic trends of a system, HS AFM can probe the various pathways that make up these general trends.

1. Glue the glass rod to the scanner stage using compound FG-3020C-20. Let this dry for at least 10 min. Punch a 2 mm diameter circle of ~100 μm thick mica and attach it to the glass rod with compound FS-1010S135-0.5. Let this dry for 10 min.
2. Cleave layers from the glued mica disk using scotch tape until a well cleaved layer is seen on the scotch tape. Functionalize the freshly cleaved surface with 2.5 μL of 500 μM APS solution for 30 min. To prevent drying of the surface during functionalization, place a cap fit with a damp piece of filter paper over the scanner. When functionalization is complete, rinse the mica with 20 μL of ddI H_2O by applying ~3 μL per rinse. Water should be removed from each rinse by placing an absorbent wipe at the edge of the mica piece. Following these rinses, place a droplet of ddI H_2O on the surface and let it sit for at least 5 min to remove any nonspecifically bound APS.
3. Place the probe in the holder using a Teflon-coated tweezer to avoid damage to tip via static discharge. Place the holder on the AFM stage with the tip facing up. Rinse the holder with ~100 μL of ddI H_2O followed by two ~100 μL rinses of nucleosome imaging buffer. Fill the chamber with ~100 μL of nucleosome imaging buffer so that the tip is submerged. With buffer in chamber, adjust the cantilever position so the laser hits it.
4. Rinse the APS functionalized mica with 20 μL of nucleosome imaging by applying ~4 μL for each rinse. Deposit 2.5 μL of nucleosome sample to the surface and let it sit for 2 min. Rinse

with ~4 μL of imaging buffer two times and leave the sample covered in buffer.

5. Position the scanner with the sample on top of the tip holder so the deposited sample is face down.
6. To approach use the auto-approach function with the set point, A_s close to the free oscillation amplitude, A_0. Ideally, $A_s = 0.95\ A_0$, however, operating at 82% of the free amplitude will work if careful. Adjust the set point while imaging until the surface is being well tracked. The cantilever amplitude should be kept small to minimize energy transfer from the tip to the sample. Stable operation of the instrument is possible at amplitudes as low as 1 nm.
7. An imaging area of 150 × 150 nm to 200 × 200 nm typically contains several nucleosome particles. An imaging rate of ~300 ms per frame is sufficient for observing nucleosome dynamics such as looping and sliding of the core (*see* Subheading 3.10 below).

3.9 Analysis of NCPs Imaged in Air

1. Analyze the contour length of all DNA free of a histone core (free DNA) using an analysis software capable of contour measurement (Fig. 3), such as FemtoScan Online (Advanced Technologies Center, Moscow, Russia).
2. Generate a histogram from all free DNA contour length measurements and fit it with a normal (Gaussian) distribution. Divide the resulting peak center (x_c) by the total base pairs of

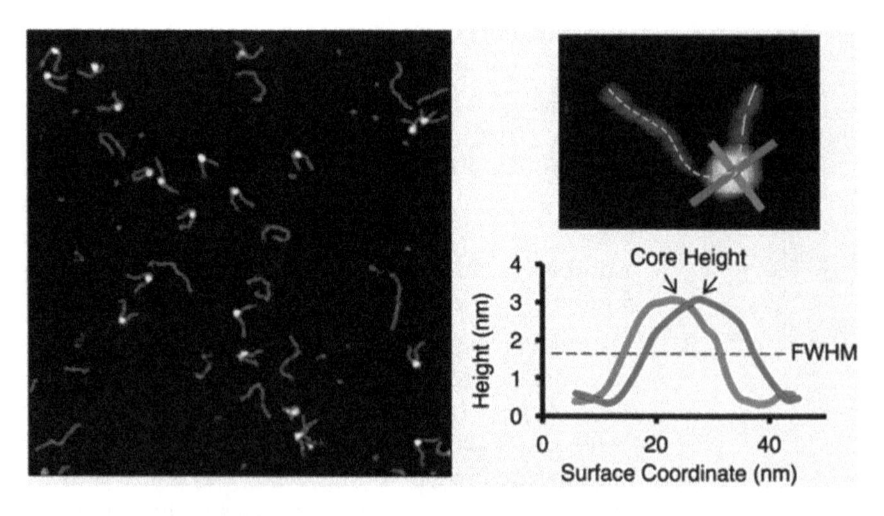

Fig. 3 Air imaging of CENP-A nucleosomes provides conformational details of the nucleosome population. (Left) A 1 × 1 μm scan of CENP-A nucleosomes deposited at a concentration of 1 nM (right, top). Each nucleosome core is bisected with two cross sections (blue and red lines) and contour lengths of the two DNA "arms" of the nucleosome are measured (long and short green dashes) to determine the extent of DNA wrapping around each nucleosome core. (Right, bottom) The two cross sections measured for each nucleosome can be plotted and the height and FWHM calculated

the DNA substrate to generate a conversion factor from nanometers to base pairs of DNA.

3. Measure the arm lengths of each nucleosome by measuring from the end of each arm to the center of the nucleosome core as shown in Fig. 3 (*see* **Note 14**).
4. Measure two cross sections of the core and calculate the full width at half maximum for each (Fig. 3). Subtract one-half the averaged FWHM from each nucleosome arm (*see* **Note 15**). Convert the corrected arm lengths to base pairs by dividing each arm by the conversion factor calculated in **step 2**.
5. Calculate the extent of DNA wrapping by subtracting the sum of both nucleosome arms from the total base pairs of the DNA substrate. Plot the wrapped base pairs of all nucleosomes as a histogram and fit the peak(s) with a Gaussian to get the mean wrapped base pairs of DNA.

3.10 Analysis of Time-Lapse (and HS) AFM Images of NCP's

The ability to track an individual nucleosome while undergoing dynamic behaviors is the major benefit of both stand and HS time-lapse AFM imaging. Since both techniques use time-based imaging, the analysis of images is largely the same, with HS AFM only requiring analysis of hundreds if not thousands of images more for a given particle. The general parameters to measure are the same as those for dry images: length of each nucleosome "arm," wrapped base pairs of DNA, height, FWHM and volume. However, dynamic events such as looping that are not possible to image by conventional AFM require that a different type of measurement is made. The software used for HS AFM movie generation is FalconViewer kindly provided by T. Ando. Cross sections are obtained using Gwyddion and DNA contour lengths using FemtoScan Online (Advanced Technologies Center, Moscow, Russia).

1. To begin analysis, all images must be converted from the HS AFM data type to .tiff images. To do this, first flattening the images in the HS AFM software using either plane or line flattening until the image background is uniform in contrast. Set the image contrast to automatic for analysis; this can be changed later for images used as a visualization.
2. With the movie images flattened, save the selected range of frames as a .mov file. It is critical that scale bars are removed from the file before making the movie as these will skew the size of the images created. With the .mov file created, convert to . tiff images using a software such as FFmpeg (www.ffmpeg. org). with a frame rate for conversion the same as that used for creation of the .mov file.
3. Before measuring particles in FemtoScan Online, correct the image size to match that at which it was captured. Measure the length of each nucleosome arm from the end of the arm to the center of the nucleosome core. Record this data in a

spreadsheet along with the frame in which it is from. Upon unwrapping of the nucleosome, measure the length of free DNA for several frames as this is used in determining the extend of DNA wrapping around the core.

4. Open each frame in Gwyddion and enter the *x*, *y* and *z* scale of the image when prompted; these values can be found in the HS AFM analysis software. In each frame, measure three cross sections of the nucleosome core and three cross sections of DNA.
5. With cross sections collected for all frames of interest, calculate the height and FWHM of each cross section and average the three for the nucleosome core and the three for free DNA. For each frame, subtract ½ the FWHM value from each arm length value. Plot these arm values as well as the sum of the arms as a scatter plot as a function of frame number. Divide the mean contour length of the dissociated DNA and divide by the total base pairs of DNA. Use this value to determine wrapped DNA in each frame as described in the air image analysis section. An example of the wrapped DNA analysis for an unwrapping CENP-A nucleosome is shown (Fig. 4, top).

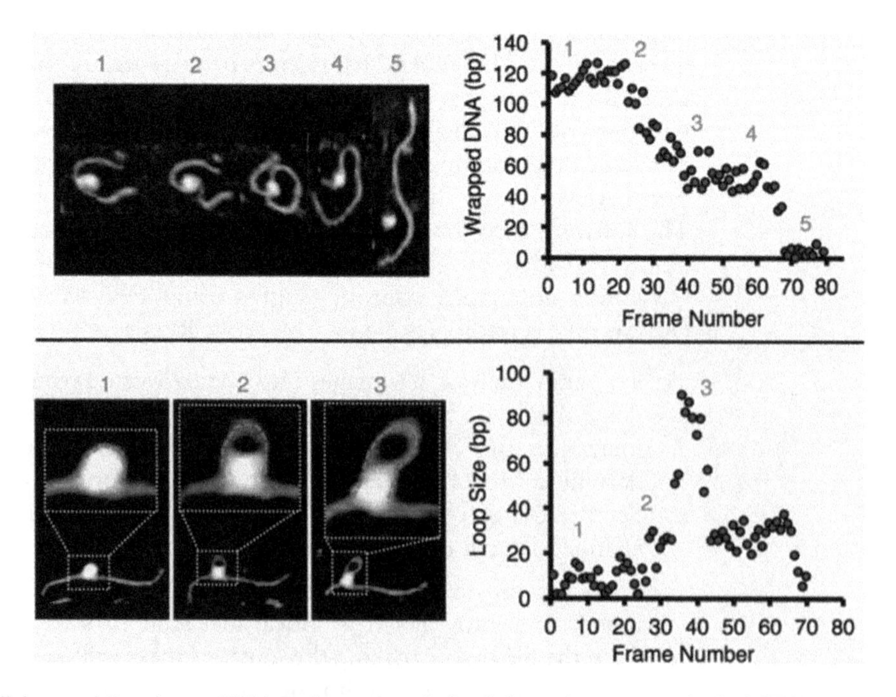

Fig. 4 High speed time-lapse AFM imaging and analysis of dynamic events typical of CENP-A nucleosomes. (Top) DNA unwraps from around a CENP-A nucleosome core (frames 1–4) until fully dissociating from the stable core (frame 5). The base pairs of wrapped DNA were measured as they were for air images and plotted for the time-trajectory of the particles dynamics. The unwrapping dynamics seen in the images match the graph in that less base pairs of DNA are wrapped as the unwrapping event progresses. (Bottom) Looping of DNA from the CENP-A nucleosome is a feature unique to this centromere derived chromatin. The contour lengths described in **step 6** of Subheading 3.10 are shown in red over the zoomed portions of the three frames

6. To determine the base pairs of DNA that make up a DNA loop, measure the contour length from one DNA exit/entry site to the other for frames before, during and after the looping event. Convert these values to base pairs and subtract the lowest value (before the loop) from all lengths. This will set the lengths before and after looping to ~0 bp as shown in the provided example (Fig. 4, bottom).

4 Notes

1. To achieve a high yield of well positioned nucleosomes it is best to use a DNA substrate that contains the Widom 601 positioning sequence flanked on both sides by nonspecific DNA. To prevent the formation of dinucleosomes, each of the DNA segments flanking the 601 motif should not exceed ~150 bp. If using a less specific nucleosome positioning sequence than 601, the total substrate length should be no greater than ~250 bp. Other sequences can be used in place of 601; however, the low yield can complicate time-lapse AFM experiments.
2. Elution of DNA from the spin column should be done using ddI H_2O Eluting DNA in the provided elution buffer can cause difficulties when measuring DNA concentration in later steps as it contains EDTA which absorbs strongly at ~230 nm. If a suboptimal 260/230 ratio (<2.0) is produced and DNA was eluted in ddI H_2O, purify the DNA once more being extra careful not to carry residual buffer over between steps.
3. Allowing the DNA to dry for too long in vacuum can make resuspension difficult in later steps, so be sure to check the sample frequently and remove it from the vacuum as soon as no liquid is visible.
4. It is important to account for the NaCl contribution from the histone stock solution. A second assembly of canonical nucleosomes (containing H3 octamers are a useful control when assembling CENP-A nucleosomes for the first time. The gel in Fig. 2 includes this H3 nucleosome control.
5. When mounting the syringe on the pump, expel a few microliters of buffer while aligning the pusher block to the syringe plunger to make sure solid contact is being made and that there will be no lag in buffer delivery. Any such lag will result in an incomplete dilution gradient and most likely a failed nucleosome assembly.
6. Histone H3 octamer is included as a reference for CENP-A migration while also serving as a nice control for the gel. The CENP-A assembly mixture of histones is run to confirm that

the correct histone stoichiometry was used for assembly while also serving to check histone quality.

7. When working with thin gels extra care should be taken to keep it from breaking. This is most likely to happen when transferring the gel from the glass plates to the staining container or from the staining container to an imaging platform. When transferring from the glass plates to a container, first add ddI H_2O to the container and dip one end of the glass plate, gel side down, into the water. By moving the glass plate around in this position, water should begin to creep under the gel until it detaches from the glass surface and slides to the container.
8. To clean tweezers, soak in acetic acid, methanol and isopropanol, in that order for 5 min each. Following the isopropanol soak, dry the tweezers using an argon flow and a nonwoven cleanroom wipe to remove excess liquid. Store the tweezers in a container cleaned thoroughly with isopropanol and well-sealed until use. In addition, once the surface has been cleaved, any contact with the surface will result in contamination and/or damage to the surface and it must be recleaved.
9. A 1 mL syringe works well for rinsing samples with since it provides a gentile and controllable flow. Others have used 1 mL pipettes with success as well.
10. Check the strength of the argon flow before drying sample by directing it toward bare skin. The flow should just barely be felt from a few millimeters away. To strong of a flow and DNA will become oriented all in one direction and disrupt the natural position of the particles.
11. Position the tip in such a way that the cantilever is completely within the optical window without much of the silicon chip showing. This helps to prevent air bubbles from becoming trapped when immersing the tip and holder in liquid. If bubbles do become trapped under the cantilever, remove the AFM head from the stage and wick the liquid away from the glass block with an absorbent wipe being careful not to touch the probe itself. Remount the AFM head and check again for bubbles with the topview camera.
12. The vertical deflection will always drift from positive to negative when imaging in liquid, so this positive starting value provides longer imaging time before adjustment is needed. The signal is linear from +5 V to −5 V so adjustment is only needed when it is not within this range.
13. The softness of the cantilevers used for liquid imaging can often give rise to failed approaches. To overcome this, lower the relative set point to between 70% and 80% and restart the approach. When the surface is reached, retract 20 μm and retune.

14. Before starting analysis of the nucleosome population, number all NCPs in each image using software with numbering capabilities such as FemtoScan Online (Advanced Technologies Center, Moscow, Russia). This will be necessary when matching the arm length measurements of each particle with the line profile measurements.
15. Measuring each nucleosome arm to the center of the core provides a consistent reference point for these measurements to be made. However, in doing this the measured length of each arm is artificially increased so one-half the FWHM of the nucleosome core is subtracted from the arm length to correct for this measurement.

Acknowledgments

This work was supported by the National Science Foundation [grant MCB 1515346 to Y.L.L.].

References

1. Gordon DJ, Resio B, Pellman D (2012) Causes and consequences of aneuploidy in cancer. Nat Rev Genet 13(3):189–203
2. Black BE, Jansen LET, Maddox PS, Foltz DR, Desai AB, Shah JV, Cleveland DW (2007) Centromere identity maintained by nucleosomes assembled with histone H3 containing the CENP-A targeting domain. Mol Cell 25 (2):309–322. https://doi.org/10.1016/j.molcel.2006.12.018
3. McKinley KL, Cheeseman IM (2016) The molecular basis for centromere identity and function. Nat Rev Mol Cell Biol 17(1):16–29. https://doi.org/10.1038/nrm.2015.5
4. Lyubchenko YL (2014) Nanoscale nucleosome dynamics assessed with time-lapse AFM. Biophys Rev 6(2):181–190. https://doi.org/10.1007/s12551-013-0121-3
5. Lyubchenko YL, Shlyakhtenko LS (2015) Chromatin imaging with time-lapse atomic force microscopy. In: Chellappan SP (ed) Chromatin protocols. Springer New York, New York, NY, pp 27–42. https://doi.org/10.1007/978-1-4939-2474-5_3
6. Menshikova I, Menshikov E, Filenko N, Lyubchenko YL (2011) Nucleosomes structure and dynamics: effect of CHAPS. Int J Biochem Mol Biol 2:129–137
7. Miyagi A, Ando T, Lyubchenko YL (2011) Dynamics of nucleosomes assessed with time-lapse high-speed atomic force microscopy. Biochemistry 50(37):7901–7908
8. Lyubchenko YL, Shlyakhtenko LS, Ando T (2011) Imaging of nucleic acids with atomic force microscopy. Methods 54(2):274–283. https://doi.org/10.1016/j.ymeth.2011.02.001
9. Uchihashi T, Ando T (2011) High-speed atomic force microscopy and biomolecular processes. In: Braga PC, Ricci D (eds) Atomic force microscopy in biomedical research: methods and protocols. Humana Press, Totowa, NJ, pp 285–300. https://doi.org/10.1007/978-1-61779-105-5_18
10. Lyubchenko YL (2006) AFM methods for DNA analysis. In: Encyclopedia of analytical chemistry. John Wiley & Sons, Ltd, Chichester. https://doi.org/10.1002/9780470027318.a9258
11. Lyubchenko YL (2014) Centromere chromatin: a loose grip on the nucleosome? Nat Struct Mol Biol 21(1):8–8. https://doi.org/10.1038/nsmb.2745
12. Stumme-Diers MP, Banerjee S, Hashemi M, Sun Z, Lyubchenko YL (2018) Nanoscale dynamics of centromere nucleosomes and the critical roles of CENP-A. Nucleic Acids Res 46:94–103
13. Luger K, Rechsteiner TJ, Richmond TJ (1999) Preparation of nucleosome core particle from

recombinant histones. Methods Enzymol 304:3–19

14. Uchihashi T, Ando T (2011) High-speed atomic force microscopy and biomolecular processes. Atomic Force Microscopy Biomed Res Methods Protoc 736:285–300
15. Guse A, Fuller CJ, Straight AF (2012) A cell free system for functional centromere and kinetochore assembly authors. Nat Protoc 7 (10):1847–1869. https://doi.org/10.1038/nprot.2012.112
16. Shlyakhtenko LS, Gall AA, Lyubchenko YL (2013) Mica functionalization for imaging of DNA and protein-DNA complexes with atomic force microscopy. In: Taatjes DJ, Roth J (eds) Cell imaging techniques: methods and protocols. Humana Press, Totowa, NJ, pp 295–312. https://doi.org/10.1007/978-1-62703-056-4_14

Chapter 15

High-Speed Force Spectroscopy for Single Protein Unfolding

Fidan Sumbul, Arin Marchesi, Hirohide Takahashi, Simon Scheuring, and Felix Rico

Abstract

Single-molecule force spectroscopy (SMFS) measurements allow for quantification of the molecular forces required to unfold individual protein domains. Atomic force microscopy (AFM) is one of the long-established techniques for force spectroscopy (FS). Although FS at conventional AFM pulling rates provides valuable information on protein unfolding, in order to get a more complete picture of the mechanism, explore new regimes, and combine and compare experiments with simulations, we need higher pulling rates and μs-time resolution, now accessible via high-speed force spectroscopy (HS-FS). In this chapter, we provide a step-by-step protocol of HS-FS including sample preparation, measurements and analysis of the acquired data using HS-AFM with an illustrative example on unfolding of a well-studied concatamer made of eight repeats of the titin I91 domain.

Key words High-speed force spectroscopy, Atomic force microscopy, Protein unfolding, Titin, Dockerin, Cohesion

1 Introduction

As the last step of the central dogma in molecular biology, understanding protein folding is still a fundamental challenge in biology, but also from a physical point of view [1–3]. The folding of proteins into their native conformations to perform function is one of the most essential processes within the living cell, as proteins are responsible of many biological processes in living organisms, such as catalysis of metabolic processes, gene expression, transport of molecules/solutes between and across the cell, cellular communication and molecular recognition. The unfolding of proteins by mechanical force is also biologically relevant as, many proteins, like muscle protein titin in muscle cells, spectrin in red blood cells, integrin in cell adhesion cascade, and cytoskeleton linker talin, are subjected to mechanical cues and often unfold to function [4–8]. Force spectroscopy (FS) using nanotools, such as magnetic

Yuri L. Lyubchenko (ed.), *Nanoscale Imaging: Methods and Protocols*, Methods in Molecular Biology, vol. 1814,
https://doi.org/10.1007/978-1-4939-8591-3_15,

tweezers, optical tweezers, and AFM, allows for the manipulation of individual proteins and monitoring of the forces required to unfold individual protein domains or subdomains [9–14]. This method provides information about the mechanisms of folding and unfolding under applied force, allowing the characterization of the (un)folding energy landscape along the axis of applied force.

FS using AFM is a well-established technique, however, the time resolution is limited to a few hundreds of microseconds [15]. The recent development of high speed AFM (HS-AFM) using micrometer sized cantilevers [16, 17] and its application for force measurements (what we termed high-speed force spectroscopy, HS-FS) provides access to the μs timescale and mm/s pulling velocities thanks to their low viscous drag coefficient [18–21]. HS-AFM cantilevers, with common dimensions of about $8 \times 2 \times 0.1$ μm, are considerable smaller than conventional AFM cantilevers. These reduced dimensions give access to the velocities of all atom molecular dynamics (MD) simulations and allow, thus, direct comparison of unfolding forces. Force spectroscopy using HS-AFM cantilevers allowed revisiting well-studied protein systems with better force and time resolution, revealing unfolding intermediate states and accessing new dynamic regimes predicted by theory [18, 19].

The use of micrometer-sized cantilevers for force spectroscopy measurements is emerging and, thus, still presents important bottlenecks. The optical lever detection method involves the focalization of the laser into a spot of a few μm wide cantilever, which leads to substantial optical interference artifacts [19, 22]. This effect is particularly important when using highly reflective samples, like gold-coated surfaces, a common practice for AFM unfolding measurements using cysteine residues to immobilize the proteins [5, 11, 23]. Although tilting the sample support by 45° and using low coherence sources, such as superluminescent diodes [18], minimize optical interference, it still represents a major issue to be solved in AFM in general, and in HS-AFM in particular. Apart from potential optical interference artifacts, only few companies have the required technology and know-how to fabricate such small and thin cantilevers, limiting the availability of alternative cantilever types and tip geometries. Indeed, conventional AFM cantilevers with blunt tips, which provide a larger contact area between the tip and the sample, frequently used for force spectroscopy as they enhance the probability of picking up molecules, are commercially available. This is not the case for HS-AFM cantilevers, which feature sharp tips appropriate for imaging. While nickel sputtering has been used in the past to enlarge the tip radius and favor binding with histidine tags, the resulting binding probability remained extremely low, of ~1 per 1000 or less. Finally, in some HS-AFM setups [16, 24], the small piezoelectric elements that allow high pulling rates limit considerably the size of the sample support, limiting the type of surface used to immobilize the protein. Thus, more robust techniques for protein immobilization are important to improve binding efficiency and reproducibility in HS-FS measurements. The recent discovery of mechanically ultrastable receptor/ligand complexes now allows protein unfolding experiments by grabbing the molecule from specific sites, with precise knowledge of the pulling direction and with higher efficiency than previous methods based on unspecific attachment [25–29]. One of the most versatile methods uses the ultrastable complex formed by dockerin/cohesin III. In practice, a DNA construct is engineered concatenating the protein to be studied and dockerin III. This construct is covalently immobilized to the sample with the free end exposing dockerin III to the bulk, while cohesin III is covalently attached to the tip [26, 27]. The dockerin/cohesin III complex dissociates at forces above 300 pN at conventional pulling rates which is higher than the unfolding forces of most proteins, it also provides an unfolding fingerprint that further allows identification of specific unfolding events. This assures specificity of unfolding of the desired protein, proper orientation, reversibility and reproducibility, while avoiding the use of highly reflecting surfaces, such as gold. Therefore, using this ultrastable molecular complex turns out to be an excellent option for HS-FS measurements.

In this chapter, we describe the use of HS-FS unfolding measurements on single molecules, addressing the specificities of using small cantilevers and sample supports and the construction of chimera proteins featuring the dockerin III domain and their covalent immobilization on the tip and sample support. Although the chapter is written having in mind the Ando-type HS-AFM (commercialized by RIBM, Japan), the adaptation to other AFM systems is straightforward and we provide some notes in that sense.

2 Materials

2.1 Protein Expression and Purification

1. BL21(DE3) competent *E. coli* cells.
2. LB-Agar solution prepared according to the provided instructions with kanamycin monosulfate (50 μg/mL).
3. LB-Broth solution prepared according to the provided instructions with kanamycin monosulfate (50 μg/mL).
4. 2 mM $CaCl_2$ solution.
5. Lysis Buffer: 100 mM NaCl, 20 mM Hepes–NaOH, pH 7.6, 2 mM $CaCl_2$.
6. PMSF (phenylmethylsulfonyl fluoride) stock solution: Prepare a 0.1 M stock solution of PMSF (MW 174.2) in isopropanol.

Aliquot and store at −20 °C. Note that the half-life is short in aqueous solutions (110 min at pH = 7 and 35 min at pH = 8).

7. IPTG (Isopropyl β-D-1-thiogalactopyranoside).
8. DNase I, Triton X-100.
9. Imidiazole stock solution: Prepare a 5 M imidiazole (MW 68.1) stock solution and adjust the pH to 7.6 with HCl. Store at −20 °C and protect from light.
10. Wash Buffer: 100 mM NaCl, 20 mM Hepes–NaOH, pH 7.6, 50 mM imidiazole, 2 mM $CaCl_2$ (*see* **Notes 1** and **2**).
11. Elution Buffer: 100 mM NaCl, 20 mM Hepes–NaOH, pH 7.6, 2 mM $CaCl_2$, 250 mM imidiazole.
12. Dialysis Buffer: 100 mM NaCl, 20 mM Hepes–NaOH, pH 7.6, 2 mM $CaCl_2$.
13. Ni-NTA affinity resin.
14. Dialysis membranes with a 10 K molecular weight cutoff (MWCO).
15. Orbital shaker incubator to grow bacteria.
16. OD_{600} (optical density at a wavelength of 600 nm) spectrophotometer to measure bacterial growth.
17. Centrifuge and appropriate rotors to harvest bacteria and pellet the bacterial lysate.

2.2 Surface Functionalization

2.2.1 Equipment

1. Silicon nitride cantilevers (AC10DS, Olympus).
2. 1.5 mm diameter glass surface (glass rods or coverslips).
3. Ozone or plasma cleaner for cleaning cantilevers.
4. Oven to bake cantilevers and glass surfaces.
5. Fine stainless steel tweezers for handling cantilevers.
6. Pyrex Petri dishes/similar inert vessel for treating cantilevers with acetone, acids, and other reactive reagents.
7. 24-well tissue culture plate for rinsing and treating cantilevers.
8. Parafilm.
9. Plastic petri dishes.

2.2.2 Chemicals

1. Nanopure MilliQ water.
2. Phosphate Buffered Saline (PBS): 10 mM Na_2HPO_4, 1.76 mM KH_2PO_4, 137 mM NaCl, 2.7 mM KCl pH:9 and pH:7.2.
3. HPLC grade or >95% purity acetone.
4. Analytical grade or >99.9% purity ethanol (EtOH).
5. Piranha solution (75% sulfuric acid (H_2SO_4) and 25% hydrogen peroxide (H_2O_2)).

6. 5% (3-aminopropyl)-dimethyl-ethoxysilane (3-APDMES) in EtOH.
7. 5 mM maleimidopropionyl-PEG(27)-NHS Ester in PBS pH:7.2.
8. 20 mM coenzyme A trilithium salt (93%) in 50 mM Na_2PO_4, 50 mM NaCl, 10 mM EDTA pH 7.2.
9. 1 μM Sfp *phosphopantetheinyl transferase in* 50 mM Hepes--NaOH, pH 7.5.
10. O_2, Argon and N_2 gas.

2.3 High-Speed Atomic Force Microscopy (HS-AFM)

1. Ando-type HS-AFM (SS-NEX HS-AFM, commercialized by RIBM, Japan). HS-AFM has some particularities: small piezo-electric elements, an optical microscope objective to focus the laser into a μm-sized spot and fast acquisition boards (*see* **Note 3**) (Fig. 1).
2. Micrometre sized AFM cantilevers (*see* **Note 4**, Fig. 2).

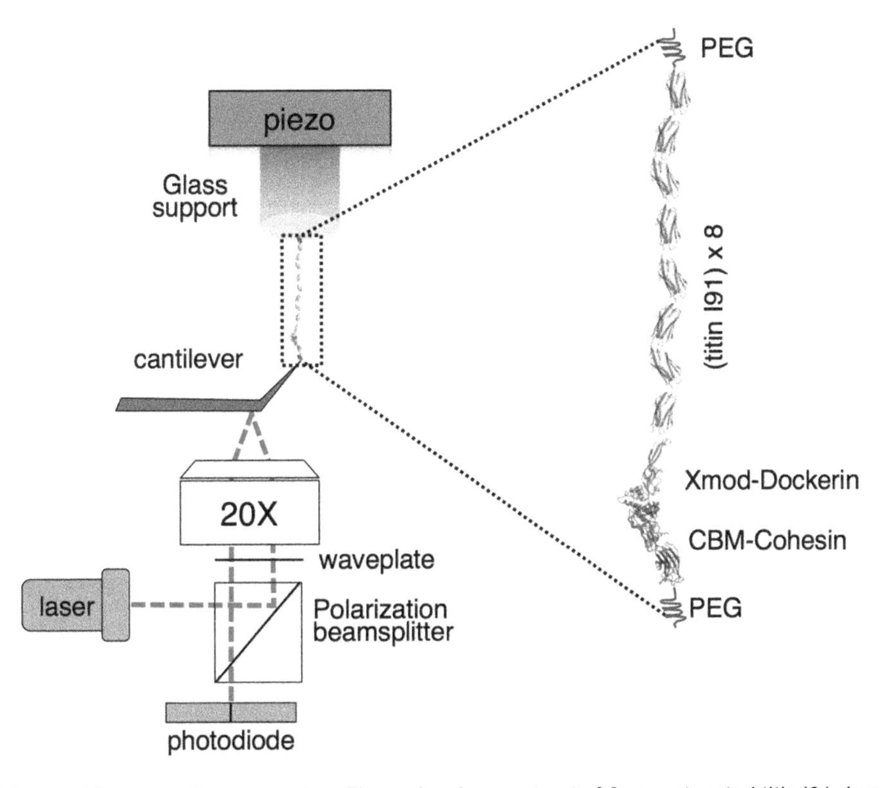

Fig. 1 High speed force spectroscopy setup. The molecular construct of 8 concatanated titin I91 domains and the Xmod-dockerin III complex covalently attached to the support and the CBM-Cohesin III complex covalently attached to the tip (shown in the inset). NMR structure of titin monomer (PDB:1TIT), crystal structure of Xmod-dockerin/Cohesin complex (PDB: 4IU3) and crystal structure of CBM (PDB:4B9F) are used to show the full complex used in the experiments

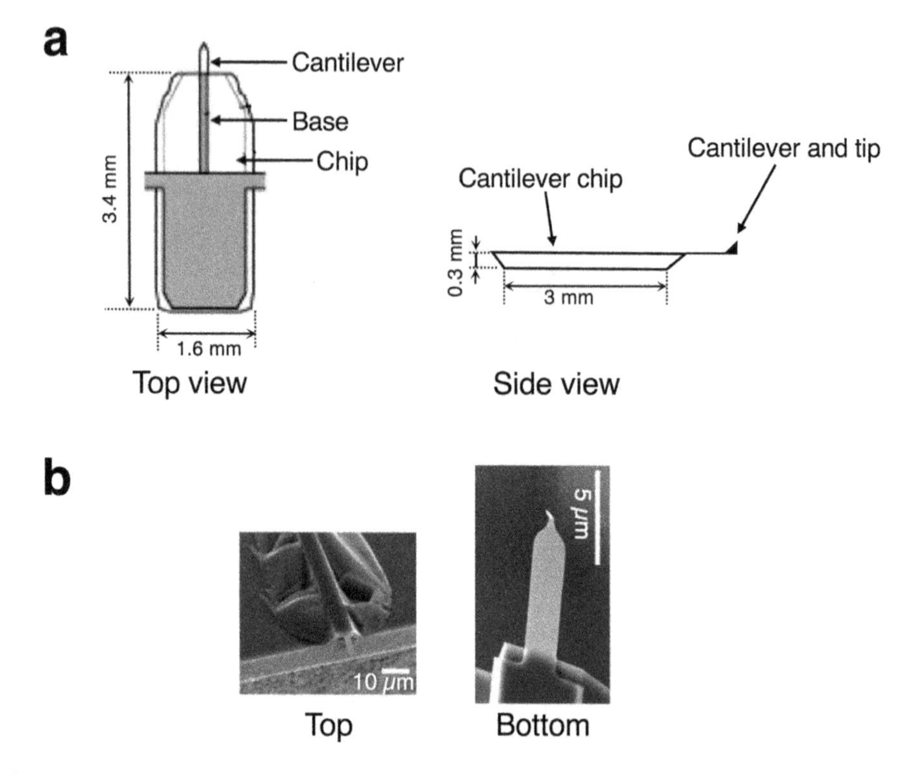

Fig. 2 Short AFM cantilever. (**a**) Olympus AC10DS chip from top and side view with its geometry. (**b**) The cantilever from different angles with the shape on the geometry of the tip

3 Methods

3.1 Protein Purification and Sample Preparation

As a well-studied system in the literature, here we use a concatamer made of 8 repeats of titin I91 domain with Xmod-dockerin III complex covalently immobilized on the sample support via ybbR peptide (DSLEFIASKLA) as an exemplary system to describe unfolding measurements with HS-FS. The 8 concatenated repeats of titin I91 domains were cloned with ybbR peptide tag at their N-terminus and Xmod-dockerin III at the C-terminus. The covalent attachment of ybbR peptide to coenzyme A (CoA, attached to the free end of the PEG linker) is mediated by the catalytic protein, Sfp phosphopantetheinyl transferase. On the cantilever side, cohesin III-CBM-ybbR complex was used and covalently attached to the cantilever from the ybbR-tag using the same procedure. The dockerin/cohesin III complex provides a robust and mechanically stable system to unfold protein domains pulling from a controlled location [26, 27, 30–33]. The ybbR-8x(titin I91)-Xmod-dockerin III concatamer was recombinantly expressed in *Escherichia coli* and purified according to the following protocol.

3.1.1 Affinity Purification of His-Tagged Titin Chimera

1. Transform the BL21 competent cells with the chimeric pET28a plasmid (*see* **Note 5**, Fig. 3a) according to the protocol provided by the competent cell manufacturer.
2. Using proper aseptic technique spread 50-100 μL of transfected cells onto an LB/agar selection plate containing kanamycin (50 μg/mL) and incubate overnight at 37 °C.
3. Select a bacterial colony from the agar plate and grow it up in a 20 mL LB broth + kanamycin (50 μg/mL) for 4-6 h in a shaking incubator at 37 °C (250 rpm) or until optical density at 600 nm (O.D.600) reaches 1.0.
4. Expand the culture by growing 10 mL of starting culture in 1 L LB broth + kanamycin + 2 mM $CaCl_2$ in 2.8 L Fernbach culture flask (*see* **Note 1**). Grow in a shaking incubator at 37 °C (200 rpm) until O.D.600 reaches 0.6.
5. Induce protein expression by adding IPTG to a final concentration of 0.3 mM. Induction should be carried out overnight at 20 °C in a shaking incubator (200 rpm).
6. After induction harvest the cells by centrifugation (i.e., 3500 × *g* for 15 min) and gently resuspend the cells in 5 mL Lysis Buffer.

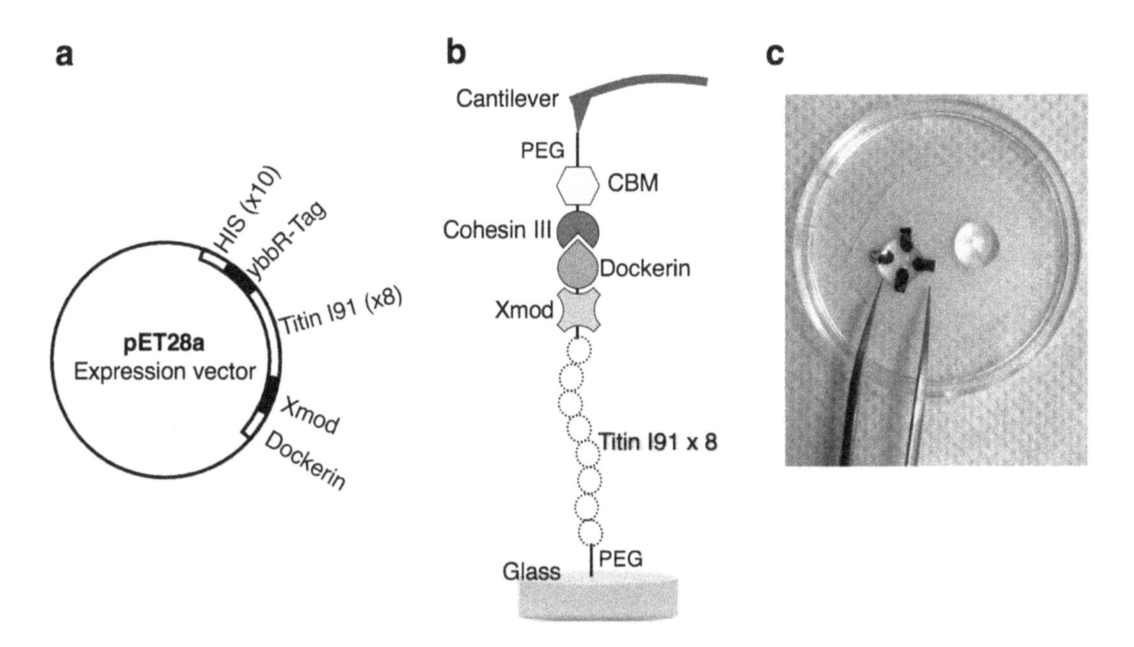

Fig. 3 DNA construct and cantilever and surface coating. (**a**) DNA construct for the ybbR-titin8-XMod-dockerin chimera. Diagram showing the pET28a expression vector after insertion of the titin chimera. Histidine and ybbR tags were added upstream the titin 8x repeat. Xmod and dockerin III domain were added downstream. (**b**) Handling the AFM cantilevers during coating. (**c**) Schematic illustration of surface and cantilever coating with relevant molecules

7. Add PMSF to a final concentration of 1 mM (*see* **Note 6**) and lyse the cells with a probe-type sonicator (place the sample in an ice bath, and use three cycles of 30 s bursts at 75% output amplitude and 30 s pauses to allow for heat dissipation and cooling).
8. Add DNase and $MgCl_2$ at a final concentration of 10 μg/mL and 2 mM, respectively (*see* **Note 7**). Add Triton X-100 at a final concentration 0.1% (*see* **Note 8**). Solubilize the protein by shaking the suspension for 30 min at 4 °C.
9. Centrifuge the crude cell extract at 10,000 × *g* at 4 °C for 45 min to remove cell debris.
10. Collect the supernatant (without touching the pellet) and mix it with 1 mL of Ni^{2+}-NTA affinity resin that has been already equilibrated with the Lysis Buffer (*see* **Note 9**).
11. Incubate the mixture on an end-ever-end shaker for 90 min at 4 °C to allow binding of the His-tagged protein (*see* **Note 8**).
12. Wash the resin with 5 mL Wash Buffer in order to decrease the *unspecific* adsorption of contaminating proteins. Repeat the washing step at least three times (*see* **Note 9**).
13. Elute the His-tagged protein three times with 0.5 mL Elution Buffer (*see* **Note 9**).
14. To remove the imidazole, dialyze the purified protein solution against 0.5 L of dialysis buffer for 1 h (use 10 K MWCO membranes) at room temperature to speed up diffusion. Change the dialysis buffer and dialyze overnight at 4 °C. Analyze all fractions by SDS-PAGE.

3.1.2 Surface Functionalization

1. Rinse the glass surfaces and silicon nitride cantilevers with acetone for 10 min (*see* **Notes 10–12**).
2. If necessary (in case they are used) immerse the glass surfaces in a mixture of sulfuric acid (H_2SO_4) (75%) and hydrogen peroxide (H_2O_2) (25%) (so-called piranha solution) for 30 min (*see* **Notes 13–15**).
3. Rinse the glass surfaces by dipping into ~1 mL milli-Q water in a 24-well tissue culture plate (five times).
4. Dry the glass surfaces and cantilevers with a gentle flow of N_2.
5. Clean glass surfaces and silicon nitride cantilevers with plasma cleaner 80 W power under oxygen for 5 min at 0.6 mbar.
6. Silanize the glass surfaces and cantilevers with 5% (3-amino-propyl)-dimethyl-ethoxysilane (3-APDMES) in EtOH for 10 min at room temperature (*see* **Notes 16–18**).
7. Rinse the glass surfaces and cantilevers in analytical grade ethanol (more than 99.9% purity),

8. Bake the glass surfaces and cantilevers at 80 °C for ~1 h for curing (*see* **Note 19**).
9. Immerse immediately the glass surfaces and cantilevers in PBS pH:9 and incubate overnight at 4 °C. This process is necessary to deprotonate the amino groups on the surface of coverslips and cantilevers to help amide-bond formation with the NHS-ester group of the linker (*see* **Notes 20** and **21**).
10. Rinse the glass surfaces and cantilevers by dipping into ~1 mL milli-Q water in a 24-well tissue culture plate (five times).
11. Incubate the glass surfaces and cantilevers in ~5 mM NHS-PEG-Maleimide in PBS pH:7.2 for 1 h at room temperature to PEGylate the amino functionalized sides of the glass surfaces. In order to coat more than one surfaces and/or cantilevers with limited amount of solutions, we recommend to place them as illustrated in Fig. 3b.
12. Rinse the glass surfaces and cantilevers by dipping into ~1 mL milli-Q water in a 24-well tissue culture plate (five times).
13. Incubate the PEGylated glass surfaces and cantilevers in 20 mM coenzyme A in coupling buffer 50 mM Na_2PO_4, 50 mM NaCl 10 mM EDTA pH 7.2 for ~1 h at room temperature.
14. Rinse the glass surfaces and cantilevers by dipping into ~1 mL milli-Q water in a 24-well tissue culture plate (five times) to remove unbound CoA.
15. Incubate the glass surfaces with ~100 μg/mL ybbR-8x(titin)-XMod-dockerin in the reaction buffer 50 mM Hepes–NaOH, pH 7.5, 20 mM $MgCl_2$, and 1 mM $CaCl_2$ in the presence 1 μM Sfp for 1 h at room temperature. This process enables covalent immobilization of protein via Sfp-catalyzed ligation of coenzyme A and the ybbR-tags (*see* **Notes 22** and **23**).
16. Incubate the cantilevers with ~100 μg/mL ybbR-Cohesin-III in the reaction buffer 50 mM Hepes–NaOH, pH 7.5, 20 mM $MgCl_2$ in the presence 1 μM Sfp for 1 h at room temperature.
17. Rinse the functionalized glass surfaces and cantilevers with measurement buffer (PBS pH 7.2) and store in measurement buffer at 4 °C until the measurements (*see* **Note 24**). The final experimental design of the molecules is illustrated in Fig. 3c.

3.2 High-Speed Force Spectroscopy (HS-FS)

FS measurements compose of three major steps: calibration of AFM cantilevers, FS measurements, and data processing and analysis. The detailed procedures of each steps are explained in Subheadings 3.2.1–3.2.3.

3.2.1 Calibration of AFM Cantilevers

In force spectroscopy, as the AFM cantilever deflects due to the interaction forces between the tip and the sample surface, the angle of the deflected laser beam changes, which translates into a difference between currents in the quadrants of the photodiode. The photodiode signal is in volts. In order to convert this signal into units of force, the spring constant of the cantilever (k) and the deflection sensitivity (also known as invOLS, inverse of the optical lever sensitivity, in nm/V) of the detection system must be calibrated before measurements [34]. The available techniques used in AFM to calibrate the cantilevers can be classified into two classes: contact-based and contact-free methods. In the contact based method the invOLS is determined by acquiring a force–distance curve on a stiff surface in liquid and then the cantilever spring constant is calculated by thermal fluctuations of the cantilever in liquid using the equipartition theorem [35, 36]. In the noncontact method, the invOLS is determined using the thermal fluctuations in liquid together with a priori knowledge of the spring constant. Noncontact methods have various advantages: (1) the spring constant calibration is independent of the determination of detector sensitivity (invOLS) minimizing propagation of errors, and (2) the invOLS determination does not require the acquisition of force curves on a hard substrate, preventing damaging the cantilever tip or the coating. Thus, we recommend using non-contact methods. Sader method [37, 38] is one of the non-contact methods to calibrate the spring constant of the cantilevers in air, which can be easily implemented in any AFM system, without any additional equipment. The method uses the measured plan view dimensions of the cantilever, its resonance frequency and quality factor (Q) in air (*see* **Notes 25** and **26**), and the physical properties of ambient fluid (density and viscosity). Then the invOLS value can be calculated by using this spring constant and the power spectral density of the thermal fluctuations in liquid [39, 40]. The procedure for spring constant and invOLS calibration using the noncontact method is as follows.

1. Mount the cantilever on the cantilever holder.
2. Focus the laser beam at the very end of the cantilever where the tip is positioned trying to maximize the sum signal on the photodiode.
3. Adjust the position of the reflected beam to the center of the bi-segmented photodiode (zero horizontal deflection).
4. Acquire and save the thermal fluctuation spectrum of the cantilever (*see* **Note 27**).
5. Extract the resonance frequency and the quality factor from the power spectral density (PSD) of the thermal fluctuation response of the cantilever (*see* Fig. 4a).

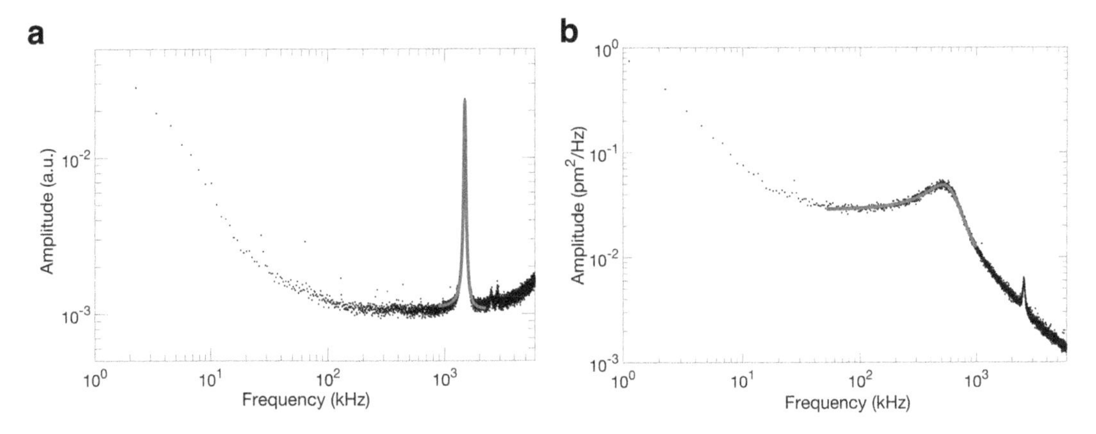

Fig. 4 Contact free calibration of the spring constant and invOLS. (**a**) Power spectral density (PSD) of the thermal fluctuations of a cantilever (AC10DS) in air with the respective fit to cantilever's first mode. The spring constant determined using the Sader method was 0.11 N/m; the fitted resonance frequency and Q-factor were 1487 kHz and 35, respectively. (**b**) PSD of the thermal fluctuations of the same cantilever (AC10DS) in liquid with the respective fit to cantilever's first mode. The invOLS value was 76.5 nm/V, the fitted resonance frequency and Q-factor were 609 kHz and 1.3, respectively

6. Calculate the spring constant of the cantilever using Sader formula by using measured width and length of the cantilever (using optical or electron microscopy) (*see* **Notes 25** and **26**).
7. After functionalization of the cantilever, prior to engaging the cantilever on the surface, repeat **steps 2–5** and record and save the thermal fluctuation spectrum in liquid.
8. Using the calculated spring constant at **step 6** and the PSD in liquid (*see* Fig. 4b), determine the invOLS value.

3.2.2 Force Spectroscopy Measurements

After calibration and functionalization of the AFM cantilever, HS-FS measurements can be performed according to the detailed procedure described below.

1. Clean the cantilever holder by rinsing with Alconox, de-ionized water, acetone, de-ionized water, ethanol, de-ionized water in the given order.
2. Dry the cantilever holder head with dustless tissue and if possible blow some air or an inert gas to further clean the chamber.
3. Mount the cantilever on the cantilever holder under microscope with the help of clean stainless-steel tweezers.
4. Immediately fill the chamber with measurement buffer to prevent the functionalized cantilever to dry up, as this may damage the linked biomolecule. We recommend filling the chamber with the measurement buffer prior to mounting the cantilever where this will prevent any drying up possibilities. If the cantilever holder does not have a pool, add a drop of buffer.

5. Mount the cantilever holder head on the setup and adjust the connections
6. Focus the laser beam at the end of the cantilever where the tip is located trying to maximize the sum signal on the photodiode.
7. Clean the surface of the piezo located on the scanner of HS-AFM with a dustless tissue.
8. Immobilize the glass rod on the piezo using glue, nail polish or vacuum grease (we recommend vacuum grease for the sake of operational simplicity and to prevent exposure to glue or nail polish chemicals). In case you are using coverslips as a coated surface, place the coverslip on top of the glass rod with vacuum grease.
9. Immediately add a drop of measurement buffer (PBS) on the surface to prevent drying up.
10. Immediately place the sample holder (scanner) on the AFM setup by carefully dipping the sample surface into the chamber where the cantilever is mounted.
11. Align the coated surface on the cantilever, ensuring that the tip is sufficiently away from the surface (at least 20 μm).
12. If necessary, refocus the laser beam at the end of the cantilever.
13. Adjust the position of the reflected beam near the center of the segmented photodiode (zero horizontal deflection).
14. Acquire the thermal fluctuation spectrum of the cantilever in liquid away from the surface to calibrate the invOLS (*see* **Note 27**).
15. Determine the invOLS using the known spring constant as determined in Subheading 3.2.1.
16. Turn on the closed loop feedback on and adjust the set-point.
17. Engage the surface on the cantilever with the feedback on to make a slight contact or close enough to make contact manually (*see* **Note 28**).
18. Turn off the feedback control, as the HS-FS measurements will be performed without a feedback signal in order to be able to reach high velocities.
19. Start acquiring force–distance curves ensuring that the tip and the surface gets in contact; follow **steps 20–25**.
20. Approach the surface to the tip at ~1 μm/s until the required force is reached (<300 pN).
21. Maintain this force for the desired contact time (*see* **Note 29**).
22. Retract to initial position with constant velocity.
23. Move within the XY plane to probe fresh regions and prevent surface degradation.

24. Repeat the cycle (a–d) until a satisfactory amount of unbinding/unfolding force–distance curves are collected (50–100 successful events per velocity).
25. Repeat the force–distance data collection at different velocities logarithmically spaced, covering the widest possible range (usually from ~10 nm/s to ~10 mm/s) for dynamic force spectroscopy (DFS) analysis.

3.2.3 Data Processing and Analysis

During the HS-FS measurements using the described protocol, thousands of force–distance curves are collected with a success of 5–10% (sometimes up to 50%) rate of specific dockerin–cohesin III unbinding with single molecule unfolding events. The specific unfolding forces are extracted by inspecting the individual force–extension curves. However, considering the large number of data collected during the experiments and the low probability of specific single-molecule unfolding events, an automated or semiautomated data processing tool is required. Most of the commercially available AFM equipment provides their own software developed to process the force–distance curves in a semiautomated fashion. However, we recommend writing one's own tool especially if you are using an in-house developed HS-FS data acquisition software. Designing a molecular complex with a well-defined fingerprint of unfolding mechanism or using linkers with known unfolding or extension profile, such as PEG, assists to automated data processing.

The 8x(titin I91)-Xmod-Dock/Coh-CBM complex provides a characteristic unfolding of single domains followed by an unbinding of dockerin–cohesin III complex fingerprint shown in Fig. 5, facilitating the recognition of successful events. A HS-FS raw data processing should contain the following steps.

1. Take a segment (around 10–20%) of the retraction curves from the end where the cantilever moves freely from any interaction with the surface.
2. If necessary, correct the optical interference by extracting the periodic interference signal from the force–distance data (*see* **Note 30**).
3. Correct the force offset (baseline) using the mean force recorded in the segment defined in 1 by fitting a straight line to this part of the curve which will give the baseline (*see* **Note 31**).
4. Calculate the relative tip displacement (piezo movement minus deflection, z–d) (*see* **Note 32**).
5. Determine the instantaneous velocity of the tip at each point by computing numerically the first derivative of the relative tip displacement over time.

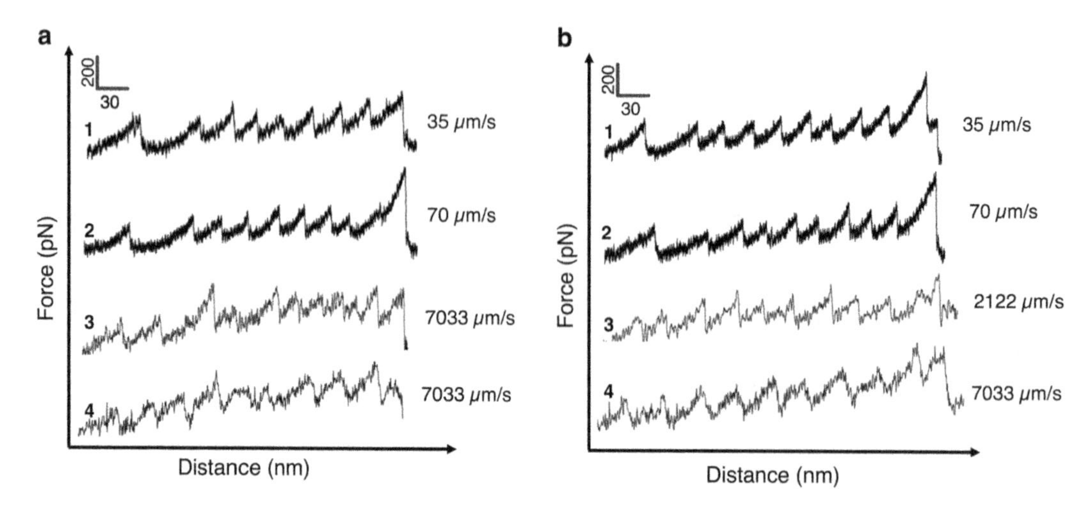

Fig. 5 Example force–distance curves. (**a**) Rejected unfolding traces during data processing. The unfolding curves which have unexpected unfolding events, such as, no dockerin–cohesin III unbinding at the last event (*curves 1 and 4*), less than eight titin unfolding peaks (*curve 2*), longer or shorter unfolded chain lengths (*curve 3*) should be rejected during data processing. (**b**) Accepted unfolding traces during data processing. In the *curve 1*, the Xmod domain unfolds before dockerin–cohesin III unbinding as also reported in [26]

6. If necessary, correct the recorded force trace at each point by multiplying the instantaneous velocity by the viscous drag coefficient (*see* **Note 33**).
7. Identify the point of contact (*see* **Note 34**).
8. Automatically detect the peaks in the retract profile (*see* **Note 35**).
9. Extract the specific unbinding/unfolding events by differentiating them from the nonspecific adhesions (*see* **Note 36**).
10. Determine the unfolding force. The unfolding force is the peak height, the difference between the peak point and the baseline.
11. Determine the loading rate by fitting a first order polynomial to a short time interval (corresponding to 1–3 nm) just before the rupture event in the force–time plot.
12. If force measurements span over various velocities, pool events by loading rate intervals, calculate histograms for each interval and determine the most probable unfolding force and median loading rate.
13. Extract the kinetic parameters, dissociation rate at zero force (k_{off}) and the distance to the transition state (x_β) of the unfolding energy landscape fitting a theoretical model describing the dependence of unfolding force with loading rate (*see* **Note 37**).

4 Notes

1. Buffers and bacterial culture media should include calcium, which is necessary for the proper folding of cellulosome ultra-stable complex proteins (cohesin and dockerin).
2. Imidazole concentration might need to be tuned for optimal washing.
3. The vertical motion of the sample stage is performed using a miniature multilayer piezoelectric actuator of dimensions $3 \times 2 \times 2\ mm^3$ (PL033.33, Physik Instrumente, Germany) with high resonance frequency (nominal ~600 kHz, actual after immobilization ~150 kHz) and 2.2 μm displacement range (nominal sensitivity 22 nm/V).

   The use of short AFM cantilevers (*see* Fig. 2) requires focusing the laser beam (808 nm, Newport) with a 20× optical microscope objective (ELWD, Nikon) with long working distance into a spot of ~3 μm diameter. The reflected beam is tracked using a 15 MHz bandwidth photodiode (MPR-1 AFM, Graviton).

   The HS-FS setup is controlled with in-house built software based on Labiew using a multichannel analog to digital converter with maximum acquisition rate of 100 megasamples per second and channel. This allows control of the drive piezo displacement and acquisition of the cantilever deflection signal (PXI-5122, National Instruments).

   In this system, the cantilever is mounted facing up in a holder featuring a liquid pool of ~150 μL. The scanner with the sample support is mounted on top, facing down (*see* Fig. 1).
4. Short cantilevers are the most essential part of HS-AFM as they allow for attaining μs-time resolution with a relatively low spring constant (0.1–0.6 N/m). There are mainly two companies that produce HS-AFM cantilevers Olympus (Japan) and Nanoworld (Switzerland). To allow functionalization with the protocol described in this chapter (*see* Subheading 3.1.2), a cantilever tip of silicon or silicon nitride is required. Nanoworld cantilevers (USC-F1.2-k0.15 or USC-F1.5-k0.6) are made of quartz and feature an electron beam deposition (EBD) tip, thus not convenient to be functionalized with the protocol described here. Olympus cantilevers (AC10DS, $2 \times 8 \times 0.1$ μm, Au backside coated, nominal spring constant 0.1 N/m) are made of silicon nitride and feature a bird beak shape tip (*see* Fig. 2) made also of silicon nitride and, thus, suitable for functionalization with the protocol described here. Note that Olympus commercializes a version of AC10 cantilevers featuring an electron beam deposited carbon tip

(AC10FS). This type is not suitable for functionalization using the protocol described here.

5. Titin concatamer construct was assembled by standard molecular biology methods using polymerase chain reaction (PCR). ybbR-tag and Xmod-dockerin III domains were added at the N- and C-terminus of an 8x titin I91 domain concatamer, respectively. The PCR product was purified and ligated to pET28a vector (*see* Fig. 3a) for transformation and protein expression in E. Coli BL21 strain. The expected weight of the encoded protein is about 109 kDa.
6. To protect the integrity of the protein from multiple proteases you may consider adding an EDTA-free protease inhibitor cocktail in addition to PMSF.
7. During sonication, nucleic acids are expected to undergo mechanical breakdown by hydrodynamic shearing. RNase treatment of RNA (which is chemically unstable compared to DNA) is therefore unnecessary.
8. DNase and Triton X -100 will reduce the lysate viscosity and prevent aggregate formation, respectively.
9. Affinity resin manufacturers usually provide their products with an optimized, step-by-step protocol. We recommend you to integrate our recommendations to this material.
10. Glass coverslips are recommended for better functionalization.
11. Since cantilevers are quite small it is a difficult task to handle them and it is common to lose some of them during the whole coating process. Thus, we recommend functionalizing more than one cantilever for each experiment. In addition, it is recommended to check whether the cantilevers on the chip are intact and undamaged before starting the functionalization protocol. If you coat more than one probe simultaneously, label the probes before calibration. A tungsten or diamond pen may be used for that.
12. When working with acetone or ethanol, pyrex/glass petri dishes and pipets should be used to prevent corrosion and should be performed under a well-ventilated hood.
13. The mixing process of H_2O_2 and H_2SO_4 is a highly exothermic reaction. In order to avoid any boiling or splashing, H_2O_2 must be mixed with H_2SO_4 SLOWLY. The solution itself is highly explosive and hazardous, and therefore the necessary safety precautions must be followed. Piranha solution cannot be stored for future use.
14. Since the piranha solution is highly corrosive, only glass or Teflon tools must be used while working with it.

15. The cantilevers should not be cleaned with piranha solution because they easily flip and break and it may damage the gold coating.
16. It is always good practice to put the larger volume component first and then the smaller volume component. Therefore, EtOH should be first put into the pyrex/glass petri dish and then 3-APDMES and mixed well by pipetting several times.
17. From now on, since only one side of the glass surfaces is treated, attention should be paid to use always the same side of the glass surfaces.
18. This 3-APDMES solution is very hygroscopic and can hydrolyze very quickly. In order to prevent air contact, Argon should be purged onto 3-APDMES bottle immediately and the bottle should be sealed properly with parafilm after purging Argon. We also recommend purchasing the solution in small volumes.
19. In order to allow evaporation but prevent contaminations the glass petri dish should be left semiopen in the oven.
20. Amino-functionalized surfaces can be stored for several days in the PBS buffer at alkaline pH solution at 4 °C. If the cantilevers need to be stored more than few days, we recommend storing them dried under Argon.
21. Handling small coverslips and/or glass rods for functionalization might be difficult. To reduce the amount of protein used and to prevent evaporation, sandwiching a protein drop between two coverslips or between the coverslip and a clean surface (like Parafilm, common in immunostaining) is a useful strategy.
22. The final reaction solution should contain 20 mM $MgCl_2$ for better functionalization. Thus, necessary amount of $MgCl_2$ should be added into Sfp solution before use.
23. Usually the stock solution of Sfp is 10 μM in 50 mM Hepes--NaOH, pH 7.5 and 10 mM $MgCl_2$. The necessary amount of Sfp + $MgCl_2$ mixture should be added onto the cantilevers after the protein solution to achieve 1 μM of Sfp in final incubation solution and mixed well via pipetting several times.
24. For better efficiency, it is recommended to use the coated surfaces within 2 days.
25. Sader method is valid only for high Q-factor cantilevers. This is the case of AC10DS cantilevers in air.
26. Sader and coworkers initiated a web-based platform for spring constant calibration, called the Global Calibration Initiative (GCI). Using this portal, any AFM user can upload the calibration parameters (spring constant, resonance frequency and Q-factor) of their own cantilevers, establishing a global

database [41]. The assessment of the uploaded data from individual users facilitates calculation of a universal coefficient, called *A*-coefficient, for each specific cantilever geometry which completes the functional relationship between the spring constant (k), the resonance frequency (f_R) and the quality factor (Q) measured in air. The portal also allows calibration of the spring constant using the globally calculated *A*-coefficient. The spring constant determination via the GCI becomes more and more accurate as users upload their own calibrations to the database.

27. The cantilever must be sufficiently away from any surface, at least 50 μm.
28. In this step, it is recommended to use a set-point ~10% below the zero-deflection value in order to minimize the applied force and time in contact. This step allows getting close enough to the cantilever, for then finish the actual engagement manually.
29. The adhesion frequency (fraction of successful unfolding events) has to be kept below 30%, preferable 10% in general, as low adhesion frequency ensures that the majority of the unfolding events are due to single molecule unfolding. However, given the specific fingerprint of the proposed protein construct, this success rate can be higher provided no multiple events are observed. Moreover, the contact time between adhering surfaces and/or the protein densities on the surfaces can be adjusted to control the adhesion frequency. A typical contact time using the proposed protocol is about 100 ms, but for higher velocities, longer contact time may be needed.
30. In order to correct the periodic interference signal, a sinusoidal relation to the force–distance data with the correct amplitude and period can be fitted to the data. More advanced functions have been proposed (*see*, for example, supplementary material of ref. [19]). In the curves reported in our work, no correction was necessary.
31. Ideally, after the last rupture event, the signal should be a straight, flat line with a certain noise level caused by thermal fluctuations until the end of the piezo movement. A signal presenting a slope is often observed on tip moving AFM systems and long z-ranges. This may be corrected by fitting and removing a straight line.
32. The actual tip displacement is different than the piezo movement because of the bending of the cantilever due to the viscous forces acting on the cantilever caused by the surrounding fluid movement.
33. Due to the viscous drag force exerted on the cantilever caused by movement of the surrounding medium, the baselines of the approach and retract traces may be shifted away from each

other. The viscous drag force depends on the retract velocity, the separation between the tip and the surface and the cantilever geometry. At very high velocities, the viscous drag effect should be corrected [42, 43]. The difference in force between approach and retract baselines where the cantilever moves freely will give the viscous drag force at the corresponding velocity. The viscous drag coefficient (b) can be extracted from the linear relationship between the viscous drag force and the relative tip velocity.

34. The intersection of the smoothed retract trace and the extrapolation of the corrected baseline is a way to determine the point of contact. Another method involves detecting the first data point that changes from positive to negative deflection (starting from the contact region).
35. Sharp jumps after a stretching regime in the force–distance profile are candidates for rupture events of protein-ligand complexes or single domain unfolding. In order to locate the sharp jumps, one can calculate the first derivative of the deflection within a certain interval which provides the slope of the curve within that interval and large slope values reflect sharp jumps. Defining a threshold for the slope is commonly used to locate inflection points and thereby the location of the peaks. Any other peak detection algorithms can also be applied.
36. Often, nonspecific adhesion occurs between the tip and the coated surface. Using linker molecules like PEG with a known length will help to prevent unspecific binding.
37. The Bell model [44] is the first and still the most conventional *phenomenological* model used to describe the rupture of molecular bonds under an external mechanical force. The Bell model was further developed for nonconstant force loading by Evans and Ritchie (so-called Bell–Evans model) [45]. This model describes a linear relation between the rupture force and the logarithm of the loading rate. There are other models that predict a nonlinear dependence of the most probable rupture force with the logarithm of the loading rate by taking into account the modulation of the distance between initial and transient state [46], the possibility of rebinding [47], the intrinsic properties of the molecule and the cantilever [48, 49] and that extend the applicable dynamic range [50, 51].

Acknowledgments

This work was supported by Agence National de la Recherche grants BioHSFS ANR-15-CE11-0007 and ANR-11-LABX-0054 (Labex INFORM). A.M. was supported by a Long Term EMBO Fellowship (ALTF 1427-2014) and a Marie Curie Action (MSCA-

IF-2014-EF-655157). We thank Michael Nash and Wolfgang Ott for sharing the XMod-dockerin-III and cohesin-III plasmids and for technical assistance.

References

1. Kubelka J, Hofrichter J, Eaton WA (2004) The protein folding 'speed limit'. Curr Opin Struct Biol 14(1):76–88
2. Alberts B, Bray D, Lewis J, Raff M, Roberts K, Watson JD (1994) Molecular biology of the cell, 3rd edn. Garland Publishing, New York
3. Bryngelson JD, Onuchic JN, Socci ND, Wolynes PG (1995) Funnels, pathways, and the energy landscape of protein folding: a synthesis. Proteins 21(3):167–195. https://doi.org/10.1002/prot.340210302
4. Kellermayer MS, Smith SB, Granzier HL, Bustamante C (1997) Folding-unfolding transitions in single titin molecules characterized with laser tweezers. Science 276 (5315):1112–1116. https://doi.org/10.1126/science.276.5315.1112
5. Rief M, Gautel M, Oesterhelt F, Fernandez JM, Gaub HE (1997) Reversible unfolding of individual titin immunoglobulin domains by AFM. Science 276:1109–1112. https://doi.org/10.1126/science.276.5315.1109
6. del Rio A, Perez-Jimenez R, Liu R, Roca-Cusachs P, Fernandez JM, Sheetz MP (2009) Stretching single talin rod molecules activates vinculin binding. Science 323 (5914):638–641. https://doi.org/10.1126/science.1162912
7. Johnson CP, Tang H-Y, Carag C, Speicher DW, Discher DE (2007) Forced unfolding of proteins within cells. Science 317 (5838):663–666. https://doi.org/10.1126/science.1139857
8. Dobson CM (2003) Protein folding and misfolding. Nature 426(6968):884–890
9. Schwaiger I, Kardinal A, Schleicher M, Noegel AA, Rief M (2004) A mechanical unfolding intermediate in an actin-crosslinking protein. Nat Struct Mol Biol 11(1):81–85
10. Marszalek PE, Lu H, Li H, Carrion-Vazquez-M, Oberhauser aF, Schulten K, Fernandez JM (1999) Mechanical unfolding intermediates in titin modules. Nature 402 (November):100–103. https://doi.org/10.1038/47083
11. Oberhauser AF, Hansma PK, Carrion-Vazquez M, Fernandez JM (2001) Stepwise unfolding of titin under force-clamp atomic force microscopy. Proc Natl Acad Sci U S A 98(2):468–472
12. Oesterhelt F, Oesterhelt D, Pfeiffer M, Engel A, Gaub HE, Müller DJ (2000) Unfolding pathways of individual bacteriorhodopsins. Science 288(5463):143–146. https://doi.org/10.1126/science.288.5463.143
13. Woodside MT, Block SM (2014) Reconstructing folding energy landscapes by single-molecule force spectroscopy. Annu Rev Biophys 43:19
14. Neupane K, Foster DA, Dee DR, Yu H, Wang F, Woodside MT (2016) Direct observation of transition paths during the folding of proteins and nucleic acids. Science 352 (6282):239–242
15. Hughes ML, Dougan L (2016) The physics of pulling polyproteins: a review of single molecule force spectroscopy using the AFM to study protein unfolding. Rep Prog Phys 79 (7):076601. https://doi.org/10.1088/0034-4885/79/7/076601
16. Ando T, Kodera N, Naito Y, Kinoshita T, Ky F, Toyoshima YY (2003) A high-speed atomic force microscope for studying biological macromolecules in action. ChemPhysChem 4 (11):1196–1202. https://doi.org/10.1002/cphc.200300795
17. Viani MB, Schaffer TE, Chand A, Rief M, Gaub HE, Hansma PK (1999) Small cantilevers for force spectroscopy of single molecules. J Appl Phys 86:2258–2262
18. Rico F, Gonzalez L, Casuso I, Puig-vidal M, Scheuring S (2013) High-speed force spectroscopy molecular dynamics simulations. Science 342:741–743. https://doi.org/10.1126/science.1239764
19. Yu H, Siewny MG, Edwards DT, Sanders AW, Perkins TT (2017) Hidden dynamics in the unfolding of individual bacteriorhodopsin proteins. Science 355(6328):945–950. https://doi.org/10.1126/science.aah7124
20. Rigato A, Miyagi A, Scheuring S, Rico F (2017) High-frequency microrheology reveals cytoskeleton dynamics in living cells. Nat Phys 13:771–775
21. Edwards DT, Faulk JK, Sanders AW, Bull MS, Walder R, LeBlanc MA, Sousa MC, Perkins TT (2015) Optimizing 1-mus-resolution single-molecule force spectroscopy on a commercial atomic force microscope. Nano Lett 15

(10):7091–7098. https://doi.org/10.1021/acs.nanolett.5b03166

22. Kassies R, van der Werf KO, Bennink ML, Otto C (2004) Removing interference and optical feedback artifacts in atomic force microscopy measurements by application of high frequency laser current modulation. Rev Sci Instrum 75 (3):689–693. https://doi.org/10.1063/1.1646767
23. Carrion-Vazquez M, Oberhauser AF, Fowler SB, Marszalek PE, Broedel SE, Clarke J, Fernandez JM (1999) Mechanical and chemical unfolding of a single protein: a comparison. Proc Natl Acad Sci 96(7):3694–3699. https://doi.org/10.1073/pnas.96.7.3694
24. Ando T, Kodera N, Takai E, Maruyama D, Saito K, Toda A (2001) A high-speed atomic force microscope for studying biological macromolecules. Proc Natl Acad Sci 98 (22):12468–12472. https://doi.org/10.1073/pnas.211400898
25. Zakeri B, Fierer JO, Celik E, Chittock EC, Schwarz-Linek U, Moy VT, Howarth M (2012) Peptide tag forming a rapid covalent bond to a protein, through engineering a bacterial adhesin. Proc Natl Acad Sci U S A 109 (12):E690–E697. https://doi.org/10.1073/pnas.1115485109
26. Otten M, Ott W, Jobst MA, Milles LF, Verdorfer T, Pippig DA, Nash MA, Gaub HE (2014) From genes to protein mechanics on a chip. Nat Methods 11(11):1127–1130. https://doi.org/10.1038/nmeth.3099
27. Schoeler C, Malinowska KH, Bernardi RC, Milles LF, Jobst MA, Durner E, Ott W, Fried DB, Bayer EA, Schulten K (2014) Ultrastable cellulosome-adhesion complex tightens under load. Nat Commun 5:5635
28. Pippig DA, Baumann F, Strackharn M, Aschenbrenner D, Gaub HE (2014) Protein-DNA chimeras for nano assembly. ACS Nano 8 (7):6551–6555. https://doi.org/10.1021/nn501644w
29. Popa I, Berkovich R, Alegre-Cebollada J, Badilla CL, Rivas-Pardo JA, Taniguchi Y, Kawakami M, Fernandez JM (2013) Nanomechanics of HaloTag tethers. J Am Chem Soc 135(34):12762–12771. https://doi.org/10.1021/ja4056382
30. Zimmermann JL, Nicolaus T, Neuert G, Blank K (2010) Thiol-based, site-specific and covalent immobilization of biomolecules for single-molecule experiments. Nat Protoc 5 (6):975–985. https://doi.org/10.1038/nprot.2010.49
31. Yin J, Lin AJ, Golan DE, Walsh CT (2006) Site-specific protein labeling by Sfp phosphopantetheinyl transferase. Nat Protoc 1(1):280–285. https://doi.org/10.1038/nprot.2006.43
32. Durner E, Ott W, Nash MA, Gaub HE (2017) Post-translational sortase-mediated attachment of high-strength force spectroscopy handles. ACS Omega 2(6):3064–3069. https://doi.org/10.1021/acsomega.7b00478
33. Sumbul F, Marchesi A, Rico F (2018) History, rare, and multiple events of mechanical unfolding of repeat proteins, The Journal of chemical physics, 148:12 https://doi.org/10.1063/1.5013259
34. Zhang X, Rico F, Xu AJ, Moy VT (2009) Atomic force microscopy of protein–protein interactions. Springer, New York, NY, pp 555–570. https://doi.org/10.1007/978-0-387-76497-9_19
35. Hutter JL, Bechhoefer J (1993) Calibration of atomic-force microscope tips. Rev Sci Instrum 64(7):1868–1873
36. Butt HJ, Jaschke M (1995) Calculation of thermal noise in atomic force microscopy. Nanotechnology 6:1–7
37. Sader JE, Chon JWM, Mulvaney P (1999) Calibration of rectangular atomic force microscope cantilevers. Rev Sci Instrum 70 (10):3967–3969. https://doi.org/10.1063/1.1150021
38. Sader JE, Larson I, Mulvaney P, White L (1995) Method for the calibration of atomic force microscope cantilevers. Rev Sci Instrum 66(7):3789–3798
39. Higgins MJ, Proksch R, Sader JE, Polcik M, Mc Endoo S, Cleveland JP, Jarvis SP (2006) Noninvasive determination of optical lever sensitivity in atomic force microscopy. Rev Sci Instrum 77(1):1–5. https://doi.org/10.1063/1.2162455
40. Schillers H, Rianna C, Schäpe J, Luque T, Doschke H, Wälte M, Uriarte JJ, Campillo N, Michanetzis GPA, Bobrowska J, Dumitru A, Herruzo ET, Bovio S, Parot P, Galluzzi M, Podestà A, Puricelli L, Scheuring S, Missirlis Y, Garcia R, Odorico M, Teulon J-M, Lafont F, Lekka M, Rico F, Rigato A, Pellequer J-L, Oberleithner H, Navajas D, Radmacher M (2017) Standardized nanomechanical atomic force microscopy procedure (SNAP) for measuring soft and biological samples. Sci Rep 7(1):5117. https://doi.org/10.1038/s41598-017-05383-0
41. Sader JE, Borgani R, Gibson CT, Haviland DB, Higgins MJ, Kilpatrick J, Lu, J, Mulvaney, P, Shearer, CJ, Slattery, AD, Thorén, P, Tran, J, Zhang H, Zheng T (2016) A virtual instrument to standardise the calibration of atomic

force microscope cantilevers. Review of Scientific Instruments, 87 093711-1-093711-14

42. Janovjak H, Struckmeier J, Müller DJ (2005) Hydrodynamic effects in fast AFM single-molecule force measurements. Eur Biophys J 34(1):91–96. https://doi.org/10.1007/s00249-004-0430-3
43. Alcaraz J, Buscemi L, Puig-de-Morales M, Colchero J, Baro A, Navajas D (2002) Correction of microrheological measurements of soft samples with atomic force microscopy for the hydrodynamic drag on the cantilever. Langmuir 18(3):716–721
44. Bell G (1978) Models for the specific adhesion of cells to cells. Science 200(4342):618–627. https://doi.org/10.1126/science.347575
45. Evans E, Ritchie K (1997) Dynamic strength of molecular adhesion bonds. Biophys J 72:1541–1555. https://doi.org/10.1016/S0006-3495(97)78802-7
46. Dudko OK, Hummer G, Szabo A (2006) Intrinsic rates and activation free energies from single-molecule pulling experiments. Phys Rev Lett 96(10):108101. https://doi.org/10.1103/PhysRevLett.96.108101
47. Friddle RW, Noy a DYJJ (2012) Interpreting the widespread nonlinear force spectra of intermolecular bonds. Proc Natl Acad Sci 109 (34):13573–13578. https://doi.org/10.1073/pnas.1202946109
48. Evans E, Ritchie K (1999) Strength of a weak bond connecting flexible polymer chains. Biophys J 76(5):2439–2447. https://doi.org/10.1016/S0006-3495(99)77399-6
49. Maitra A, Arya G (2010) Model accounting for the effects of pulling-device stiffness in the analyses of single-molecule force measurements. Phys Rev Lett 104(10):108301. https://doi.org/10.1103/PhysRevLett.104.108301
50. Bullerjahn JT, Sturm S, Kroy K (2014) Theory of rapid force spectroscopy. Nat Commun 5:4463. https://doi.org/10.1038/ncomms5463
51. Hummer G, Szabo A (2003) Kinetics from nonequilibrium single-molecule pulling experiments. Biophys J 85(1):5–15

Part III

Imaging and Probing of Biomolecular Complexes

Chapter 16

Probing RNA–Protein Interactions with Single-Molecule Pull-Down Assays

Mohamed Fareh and Chirlmin Joo

Abstract

Recent advances in single-molecule techniques allow for dynamic observations of the interactions between various protein assemblies and RNA molecules with high spatiotemporal resolution. However, it remains challenging to obtain functional eukaryotic protein complexes and cost-effective fluorescently labeled RNAs to study their interactions at the single-molecule level. Here, we describe protocols combining single-molecule fluorescence with various protein complex pull-down techniques to determine the function of RNA-interacting protein complexes of interest. We provide step-by-step guidance for using novel single-molecule techniques including RNA labeling, protein complexes purification, and single-molecule imaging. As a proof-of-concept of the utility of our single-molecule approaches, we show how human Dicer and its cofactor TRBP orchestrate the biogenesis of microRNA in real time. These single-molecule pull-down and fluorescence assays provide sub-second time resolution and can be applied to various ribonucleoprotein complexes that are essential for cellular processes.

Key words Single-molecule fluorescence, Single protein pull-down, Protein complex, RNA ligation, RNA labeling

1 Introduction

Interactions between protein complexes and different RNA molecules are essential elements of cellular processes, such as splicing, translation, and chromatin modification. A comprehensive analysis of RNA–protein complexes is a key step toward the understanding of cellular functions. Recent advances in analytical and biochemical methods have led to numerous breakthroughs in the characterization of multicomponent protein assemblies in complexes with nucleic acids. High-throughput approaches, including large-scale tandem affinity purification, the yeast two-hybrid system, and mass spectrometry analysis, have been used to identify thousands of new protein complexes in yeast [1], *Drosophila melanogaster* [2], and *Caenorhabditis elegans* [3]. Advanced computational methods allowed to predict the formation of various protein complexes

Yuri L. Lyubchenko (ed.), *Nanoscale Imaging: Methods and Protocols*, Methods in Molecular Biology, vol. 1814,
https://doi.org/10.1007/978-1-4939-8591-3_16,

within living cells [4]. Major advances in crystallography and cryo-electron microscopy have also enabled to determine the structure of large protein complexes interacting with RNA substrates at near-atomic resolution [5–7].

The analytical and structural data listed above paved the way for the understanding of the assembly and function of ribonucleoprotein complexes. Single-molecule approaches are great tools to achieve high spatiotemporal observations of the dynamics of RNA-interacting proteins. Recently, we and other groups developed single-molecule fluorescence methods to observe protein complexes performing their function with milliseconds resolution. Hoskins et al. revealed the order of spliceosome assembly during pre-mRNA maturation in cell extract via single-molecule multicolor fluorescence [8]. Lee et al. used a single-molecule coimmunoprecipitation approach to investigate weak interactions between different proteins [9, 10]. Jain et al. developed single-molecule pull-down techniques to determine the stoichiometry of protein complexes [11]. Our group also developed single-molecule pull-down assays to gain insight into the molecular mechanism of large nucleoprotein complexes involved in microRNA biogenesis [12–14].

Here, we describe various RNA preparation and single-molecule pull-down approaches and provide step-by-step protocols for the purification and immobilization of ribonucleoprotein complexes associated with their native cofactors. Our pull-down methods in combination with single-molecule fluorescence allow for real-time visualization of protein complexes and RNA interactions. As a proof-of-concept, we show an example of a protein complex involved in microRNA biogenesis called human Dicer and TRBP. We illustrate how to elucidate the molecular bases of their functions. These protocols will be helpful to make single-molecule approaches accessible for researchers to study RNA and protein complexes interactions in real time [13].

2 Materials

2.1 RNA Labeling, Ligation, and Purification

1. Synthetic RNA strands with various modifications can be purchased from several suppliers (e.g., iba, IDT, STPharm).
2. TE buffer (10 mM Tris (pH 8), 1 mM EDTA).
3. RNA elution buffer (0.3 M NaOAc, 2% SDS).
4. T4 RNA ligase (AM2140, Invitrogen).
5. T7 RNA polymerase kit (Takara, 2540A).
6. rNTP (Takara, 4041-4044).
7. Recombinant RNase inhibitor (Takara, 2313A).
8. DNA template with T7 promoter sequence for in vitro transcription.

9. BSA (AM2616, Ambion).
10. Acid-Phenol–Chloroform, pH 4.5 (Invitrogen, AM9722).
11. Ethanol (Sigma-Aldrich, 64-17-5).
12. Urea polyacrylamide gels.
13. NHS-ester form of cyanine dyes (GE Healthcare; Cy3, PA13101; Cy5, PA15101).
14. pCp-Cy5 (Jena Biosciences, NU-1706-CY5).
15. Dimethyl sulfoxide (DMSO, Sigma-Aldrich, D8418-100ML).
16. Sodium chloride (NaCl, 3 M, Fisher scientific, 7647-14-5).
17. Sodium tetraborate ($B_4Na_2O_7 \cdot 10\ H_2O$, Fisher Scientific, 1303-96-4).
18. Hydrochloride acid (HCL, 12.1 M, Fisher Scientific, 7647-01-0).
19. Glycogen (5 mg/mL, Invitrogen, AM9510).
20. SYBER Safe.
21. Spectrophotometer (NanoDrop).
22. Typhoon (GE Healthcare).

2.2 Cell Culture, Transfection and Lysis

1. Human embryonic kidney cells (HEK 293T).
2. Dulbecco's Modified Eagle's Medium (Gibco DMEM, 31885023).
3. Fetal bovine serum (FBS, Greiner Bio-One) decomplemented by 30 min incubation at 56 °C in a water bath.
4. Cell incubator connected to a CO_2 source (5% CO_2).
5. Calcium phosphate transfection kit [15].
6. Mammalian cell expression vectors (plasmids of interest) containing FLAG tag (for the immunoprecipitation) and an Acceptor Peptide (AP, amino-acid sequence: GLNDIFEAQ**K**IEWHE) tag for the in vivo biotinylation.
7. Biotin (B4639, Sigma).
8. Dulbecco's Phosphate-Buffered Saline (DPBS, Gibco, 14200).
9. Buffer D (20 mM Tris–HCl [pH 8.0], 200 mM KCl and 0.2 mM EDTA).
10. 1 mL syringes and needles (30½ gauge, BD).

2.3 Immunoprecipitation and Elution

1. Anti-FLAG antibody-conjugated agarose beads (50% slurry, anti-FLAG® M2 affinity gel, A2220, Sigma).
2. *Tobacco Etch Virus* TEV protease (0.05 U/μL) (Promega, V6101).
3. 3xFLAG® peptide (Sigma, F4799).

4. Buffer D (20 mM Tris–HCl [pH 8.0], 200 mM KCl and 0.2 mM EDTA) supplemented with 10% glycerol for long term storage at −80 °C.

2.4 Single-Molecule Pull-Down and Imaging

1. mPEG (Laysan Bio, MPEG-SVA-5000-1g) biotin-PEG (Laysan Bio, Biotin-PEG-SVA-5000-100 mg), and MS4-PEG (Pierce, 22341).
2. 5% Tween-20 (v/v in T50 buffer: 10 mM Tris [pH 8.0], 50 mM NaCl).
3. Piranha-etched quartz slides (Finkenbeiner, 1″ × 3″, 1 mm thick) and glass coverslips (VWR, 631-0136 631-0144 631-0147).
4. Double sticky tape (Scotch, 10 mm wide) and epoxy (Thorlabs, G14250).
5. Streptavidin (Invitrogen, S888).
6. Biotinylated antibodies (66 nM) that recognize specific tags attached to the protein of interest (e.g., anti-Flag, anti-MBP, anti-6xHis).
7. 0.8% glucose (v/v), 0.1 mg/mL glucose oxidase (Sigma, G2133), 17 μg/μL catalase (Roche, 11668153103), and 1 mM Trolox (Aldrich, 238813).

2.5 Single-Molecule Fluorescence Microscopy, Data Acquisition, and Analysis

1. Optical table.
2. Inverted microscope (IX71, Olympus).
3. 473-nm solid-state laser (OBIS LX 75 mW, Coherent).
4. 532-nm solid-state laser (Compass 215M-50, Coherent).
5. 632-nm solid-state laser (25 LHP 928, CVI Melles Griot).
6. 60× water immersion objective (UPlanSApo, Olympus).
7. 473-nm long-pass filter (Chroma), a 550-nm long-pass filter (Chroma) and a 633-nm notch filter (SemRock).
8. Dichroic mirror ($\lambda_{cutoff} = 645$ nm, Chroma).
9. Lenses (Thorlabs).
10. Electron-multiplying charge-coupled device (EMCCD camera, iXon 897, Andor Technology).
11. Lab-made software written in Visual C++ with a time resolution of 0.03–1 s.
12. IDL software (ITT Visual Information Solutions).
13. Matlab software (MathWorks).
14. Origin software (OriginLab Corporation).

3 Methods

3.1 RNA Labeling, Ligation, and Purification

The binding and processing of RNA molecules by RNA-interacting protein complexes can be observed in real time with our single-molecule assay. This requires labeling of RNA of interest with a fluorescent dye. Several companies offer the synthesis of RNA oligonucleotides that are fluorescently labeled with various dyes at their 5′ and/or 3′ end. Since many RNA-interacting proteins (e.g., Dicer, Ago) possess binding pockets that engage tight interactions with RNA ends (5′ and 3′), these labeling strategies might interfere with the function of those proteins of interest. Therefore, we recommend to use internal labeling of RNA molecules by incorporating an amine-modified uridine in a suited position during the synthesis process (Fig. 1a–c). Synthetic RNAs containing an amino-modified uridine are commercially available by several nucleic-acid suppliers (e.g., IDT, iba, STPharm).

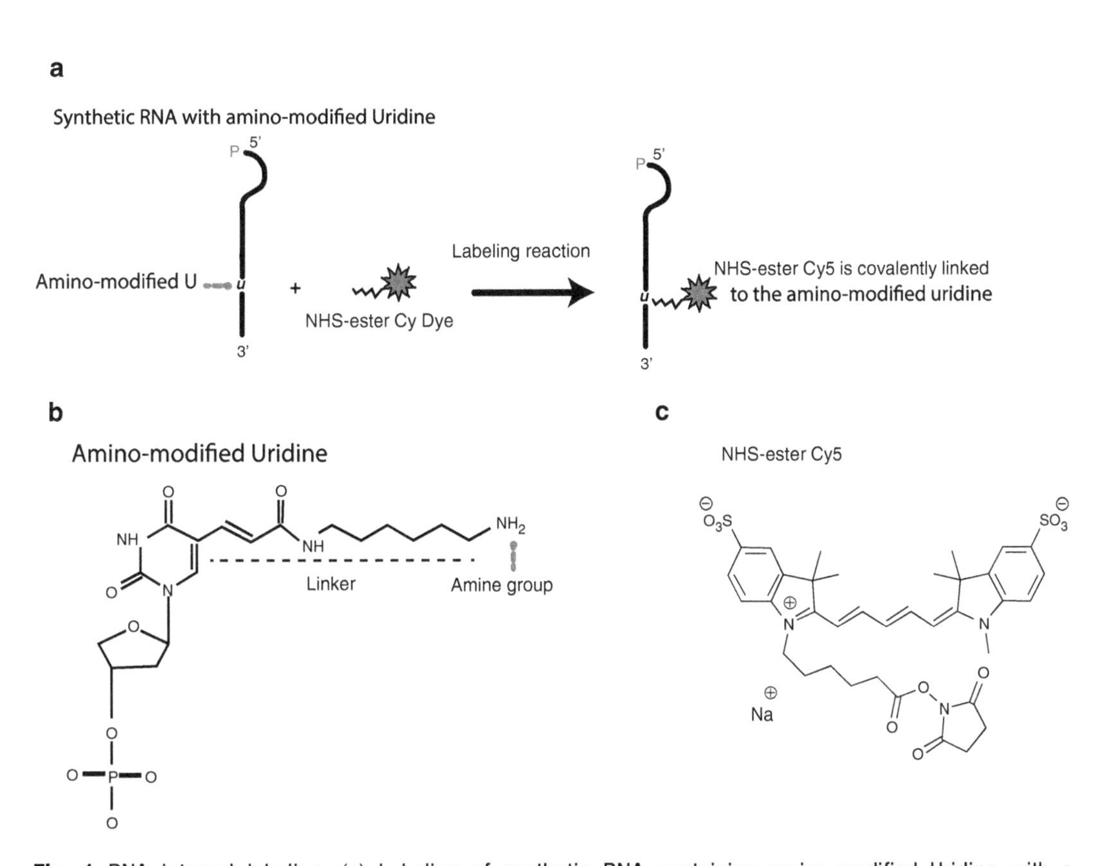

Fig. 1 RNA internal labeling. (**a**) Labeling of synthetic RNA containing amino-modified Uridine with a monofunctional NHS ester form of cyanine dye (Cy3, Cy5). (**b**) The molecular structure of an amino-modified uridine used for conjugation to the NHS-ester cyanine dye. A carbon linker connects the uridine to a amine group that reacts with NHS-ester (Glenn Research). (**c**) The molecular structure of NHS-ester form of the Cy5 dye

3.1.1 RNA Labeling

1. Dissolve RNA (containing an amine-modified uridine) in RNase-free milliQ water to obtain a concentration of 1 mM (*see* **Note 1**).
2. Prepare fresh labeling buffer (0.1 M) by dissolving 38 mg of sodium tetraborate in 1 mL of milliQ H_2O. Add 6.5 μL of 12.1 M HCl to adjust the pH to 8.5.
3. Dissolve 1 mg of cyanine dyes (Cy3 or Cy5) in 56 μL of DMSO to obtain 20 mM concentration solution. (Aliquot the rest of the dyes and store at −20 °C).
4. In a microcentrifuge tube, mix 2.5 μL of dye (Cy3 or Cy5), 2.5 of 1 mM RNA (containing an amino-modified U), and 12.5 μL labeling buffer (0.1 M) (*see* **Note 2**).
5. Incubate the mixture overnight at 4 °C with gentle mixing. Protect the sample from light sources.
6. Perform ethanol precipitation by adding 43.75 μL of cold ethanol, 1.75 μL of 3 M NaCl, and 1 μL of glycogen (5 mg/mL). Incubate the sample at least for 30 min at −20 °C.
7. Centrifuge the sample at 12,000 × *g* for 30 min at 4 °C. Discard the supernatant and rinse the pellet three times with 1 mL of 70% cold ethanol very gently.
8. Remove the supernatant and dry the pellet at 60 °C for 10 min.
9. Dissolve the pellet in a RNase-free H_2O or in an appropriate buffer.
10. Measure the labeling efficiency by comparing the absorption spectra of the RNA and conjugated dye (*see* **Note 3**).

3.1.2 RNA Ligation

Hairpin RNAs harboring stem and loop structure such as precursor microRNA (pre-miRNA) has a relatively long sequence (~70 nucleotides) and can be expensive to order from a supplier. Such a long RNA can be constructed by ligating two synthetic RNA strands as follow (Fig. 2a–c):

1. Mix 1 μL of donor single-stranded RNA (100 pM) containing a 5′ phosphate with 1 μL of acceptor single-stranded RNA (200 pM) lacking 5′ phosphate. The two strands possess sequence complementarity in the stem that allows for the annealing and subsequent ligation (*see* **Note 4**).
2. Add 18 μL of TE buffer supplemented with 100 mM NaCl and anneal the two RNA strands by heating it to 80 °C, followed by a slow cooling down to 4 °C (−1 °C/4 min in a thermal cycler).
3. Perform the ligation of the annealed substrate (20 μL) by adding 3 μL of T4 RNA ligase, 3 μL of 0.1% BSA, 5 μL of the 10× ligation buffer (included in the RNA ligase kit), and 19 μL of H_2O and incubate the mixture at 16 °C for 24 h.

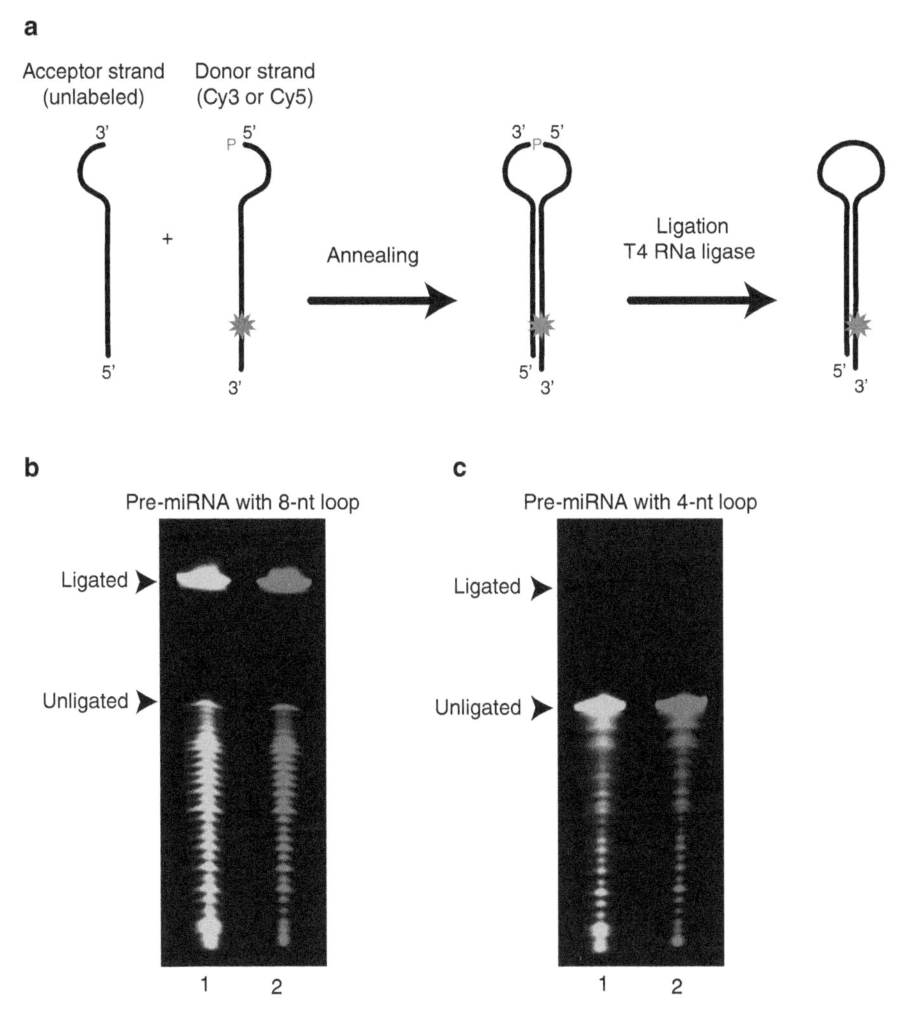

Fig. 2 Production of dye-labeled hairpin RNA by ligation of two RNA strands. (**a**) Two RNA strands (~35–38 nucleotides) with sequence complementarity are annealed together to form hairpin structure. Acceptor and donor (with 5′ phosphate) RNA strands are ligated together with RNA ligase 4 with high efficiency. (**b**) 10% PAGE gel displaying the ligation efficiency of acceptor and donor (with 5′ phosphate) RNA strands with T4 RNA ligase. After annealing, the two RNA strands form a eight nucleotides long single-stranded terminal loop. The donor strand was labeled with NHS-ester Cy3 (lane 1) or Cy5 (lane 2). The upper arrow indicates ligated products and the bottom arrow indicates unligated RNA strand. The gel was scanned with Typhoon using appropriate lasers and filters. (**c**) The ligation of RNA hairpin with four nucleotides long single-stranded terminal loop failed, likely because of the short length of the loop

4. Perform acid phenol–chloroform extraction and ethanol precipitation.
5. Run 12.5% urea polyacrylamide gel to purify the ligated product (upper band) from the unligated strands (lower bands).
6. Scan the gel with Typhoon using the appropriate lasers and filters (Fig. 2b, c).

7. Cut the ligated RNA (upper band) out of the gel and transfer it into an Eppendorf tube containing ~400 μL RNA-elution buffer.
8. Incubate at 40 °C overnight to elute the RNA out of the gel, then transfer the supernatant into a new Eppendorf tube.
9. Perform acidic phenol extraction followed by ethanol precipitation.
10. Dissolve the ligated RNA in TE buffer (10 mM Tris (pH 8), 1 mM EDTA) and measure the concentration using spectrophotometer (*see* **Note 5**).

3.1.3 In Vitro Transcription of Long RNAs

Certain RNAs such as mRNA and long noncoding RNA are very long and their synthesis would be very expensive. In this case, we suggest to perform in vitro synthesis using T7 RNA polymerase (Takara, 2540A) as following:

1. In an Eppendorf tube, mix 2 μL of 10× T7 RNA Polymerase buffer, 2 μL of 50 mM DTT, 2 mM of each rNTP (ATP, CTP, GTP, and UTP), 20 unites of RNase inhibitor, and 50 unites of T7 RNA polymerase.
2. Add 200–1000 ng of a DNA template (containing T7 promotor followed by the sequence of interest). Add milliQ water up to 20 μL final volume.
3. Incubate the mixture for 1 h at 37 °C.
4. Remove the DNA template by adding DNAse to the sample followed by 1 h of incubation at 37 °C.
5. Perform acidic phenol extraction followed by ethanol precipitation.
6. Dissolve the in vitro-transcribed RNA in TE buffer (10 mM Tris (pH 8), 1 mM EDTA) and measure the concentration using spectrophotometer.
7. Load a sample of the in vitro-transcribed RNA into denaturing polyacrylamide gel (PAGE) to check the quality of the RNA. The type of gel depends on the size of the transcript to be purified.
8. Visualize the RNA by SYBER safe staining.
9. In case the gel shows multiple bands, it is required to cut the upper band (full-length product) out of the gel, and carry out acidic phenol and ethanol precipitation to obtain homogeneous RNA population.

3.1.4 3′ Labeling of In Vitro-Transcribed RNAs

In vitro-transcribed RNAs can be labeled at their 3′ end by the ligation of a dye-labeled cytidine containing 5′-phosphate group (pCp-Cy5, Jena Biosciences, NU-1706-CY5) as follow:

1. In an Eppendorf tube, mix 0.5 μL of ~100 mM in vitro-transcribed RNA, 0.5 μL of 1 mM pCp-Cy5, 1 μL of 10× RNA ligase buffer, 0.5 μL of 100 mM ATP, 1 μL of T4 RNA ligase, and 6.5 μL of RNase-free milliQ water.
2. Incubate the mixture at 16 °C overnight.
3. Perform acidic phenol extraction followed by ethanol precipitation.
4. Dissolve the in vitro-transcribed RNA in TE buffer (10 mM Tris (pH 8), 1 mM EDTA) and measure the concentration using spectrophotometer.
5. Run a PAGE gel to check the purity of the in vitro-transcribed and labeled RNA (*see* **Note 6**).

3.2 Cell Culture, Transfection, and Lysis

Various eukaryotic protein complexes cannot be expressed in *E. coli* and reassembled in vitro. In this case, we advise to use human cells (e.g., 293 HEK) to express and purify functional protein complexes with the possibility of in vivo biotinylation of those proteins in a specific position (Fig. 3).

3.2.1 Cell Culture and Transfection

1. Grow human embryonic kidney cells (HEK-293T) in Dulbecco's Modified Eagle's Medium (DMEM) supplemented with 10% fetal bovine serum (FBS, heat-inactivated) at 37 °C and 5% CO_2.
2. Before transfection, split the cells into 10 cm cell culture dishes to a confluence of 25%. After 24 h of growth, transfect the cells with plasmids required to express protein complexes of interest using a $CaPO_4$ transfection method [15]. In case the protein of interest contains AP (Acceptor Peptide) tag for in vivo biotinylation, a plasmid coding for BirA enzyme should be co-transfected in the cells.
3. After 5 h, exchange the medium with fresh DMEM containing 10% FBS. In case of in vivo biotinylation, include free biotin (1 μg/mL) into the fresh medium.
4. Incubate transfected cells for another 48 h to enable protein expression and in vivo biotinylation.

3.2.2 Cell Lysis

1. Remove the medium and wash the cells with ice-cold Dulbecco's Phosphate-Buffered Saline (DPBS, 14200 Gibco®).
2. Add 10 mL of ice-cold DPBS and harvested the cells with scrapers.
3. Transfer the cells into a cold 15 mL tubes and centrifuge at 276 × *g* and 4 °C for 5 min to form cell pellet.
4. After the removal of the supernatant, freeze and store the cell pellet at −80 °C until further processing.

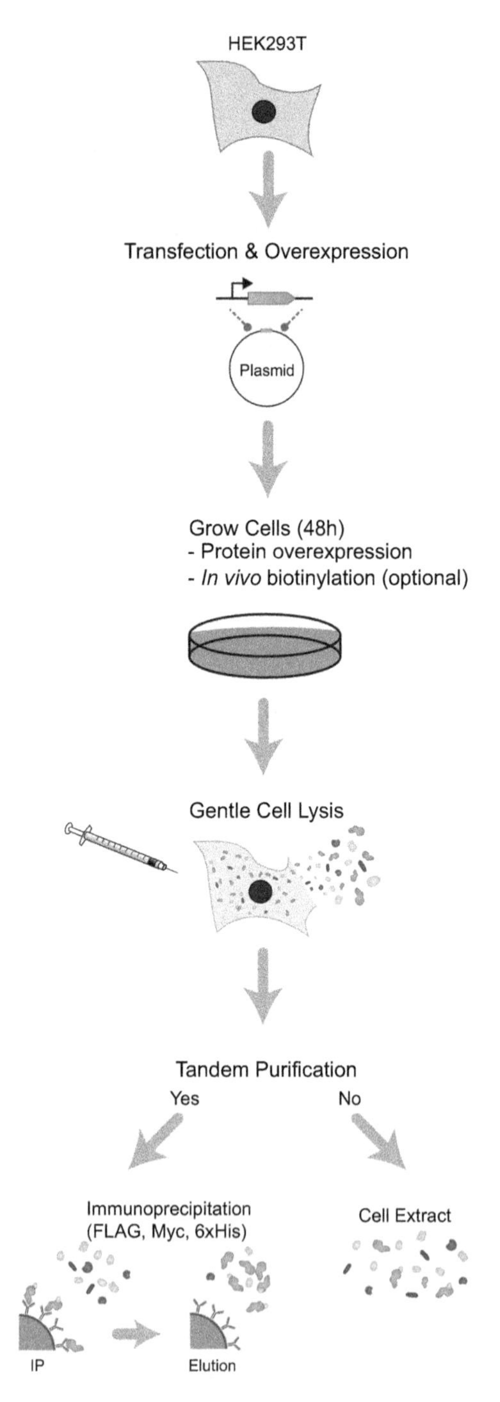

Fig. 3 Overview of the protein overexpression and immunoprecipitation approaches

5. Before lysis, thaw the cell pellet on ice for 30 min and resuspend it in cold buffer D (20 mM Tris–HCl [pH 8.0], 200 mM KCl and 0.2 mM EDTA).
6. Lyse the cells by carefully passing them ten times through a needle (30½ gauge, BD), while avoiding the formation of air bubbles.
7. Centrifuge the sample twice (16,100 × *g* at 4 °C, for 20 min) to remove cell debris (pellet) and transfer the supernatant into a new Eppendorf tube (*see* **Note 7**).

3.3 Immunoprecipitation and Elution

1. For immunoprecipitation of 1xFLAG-tagged proteins, incubate 1 mg of total protein in the cell extract with 2.5 μL of anti-FLAG antibody-conjugated agarose beads (50% slurry, anti-FLAG® M2 affinity gel, A2220, Sigma) under gentle agitation at 4 °C for 60 min (*see* **Note 8**).
2. After incubation, gently wash the beads five times with cold buffer D (~1 mL).
3. Resuspend the beads in 10 μL of buffer D.
4. Elute the proteins of interest from the beads by site-specific cleavage using *Tobacco Etch Virus* TEV protease (0.05 U/μL) at 30 °C for 90 min. Alternatively, 1xFlag tagged proteins can be eluted from the beads using 2 mM 3xFLAG peptide.
5. Add 15% glycerol (final concentration), aliquot and snap-freeze the eluted proteins using liquid nitrogen, and store them at −80 °C (*see* **Note 9**).

3.4 Single-Molecule Pull-Down

To increase purity of the IPs, an additional purification step is carried out directly on the surface of the imaging chamber using streptavidin and/or specific antibodies targeting the proteins of interest with nanomolar affinity range. This allows for an efficient immobilization of the protein of interest, while discarding unwanted contaminant proteins (Fig. 4).

To eliminate the nonspecific surface adsorption of proteins and nucleic acids to a quartz surface, piranha-etched slides (Finkenbeiner) are passivated with polyethylene glycol (PEG) over two rounds of PEGylation as we described previously [16].

3.4.1 Biotinylated Proteins (Fig. 4a)

1. Assemble microfluidic flow chambers using PEGylated quartz slide, PEGylated glass coverslip, double sticky tape, and epoxy as described previously [16].
2. Introduce in the flow chamber 50 μL of streptavidin (0.1 mg/mL) and incubate for 2 min, then wash the unbound streptavidin out of the chamber with 100 μL of buffer of interest (protein buffer).

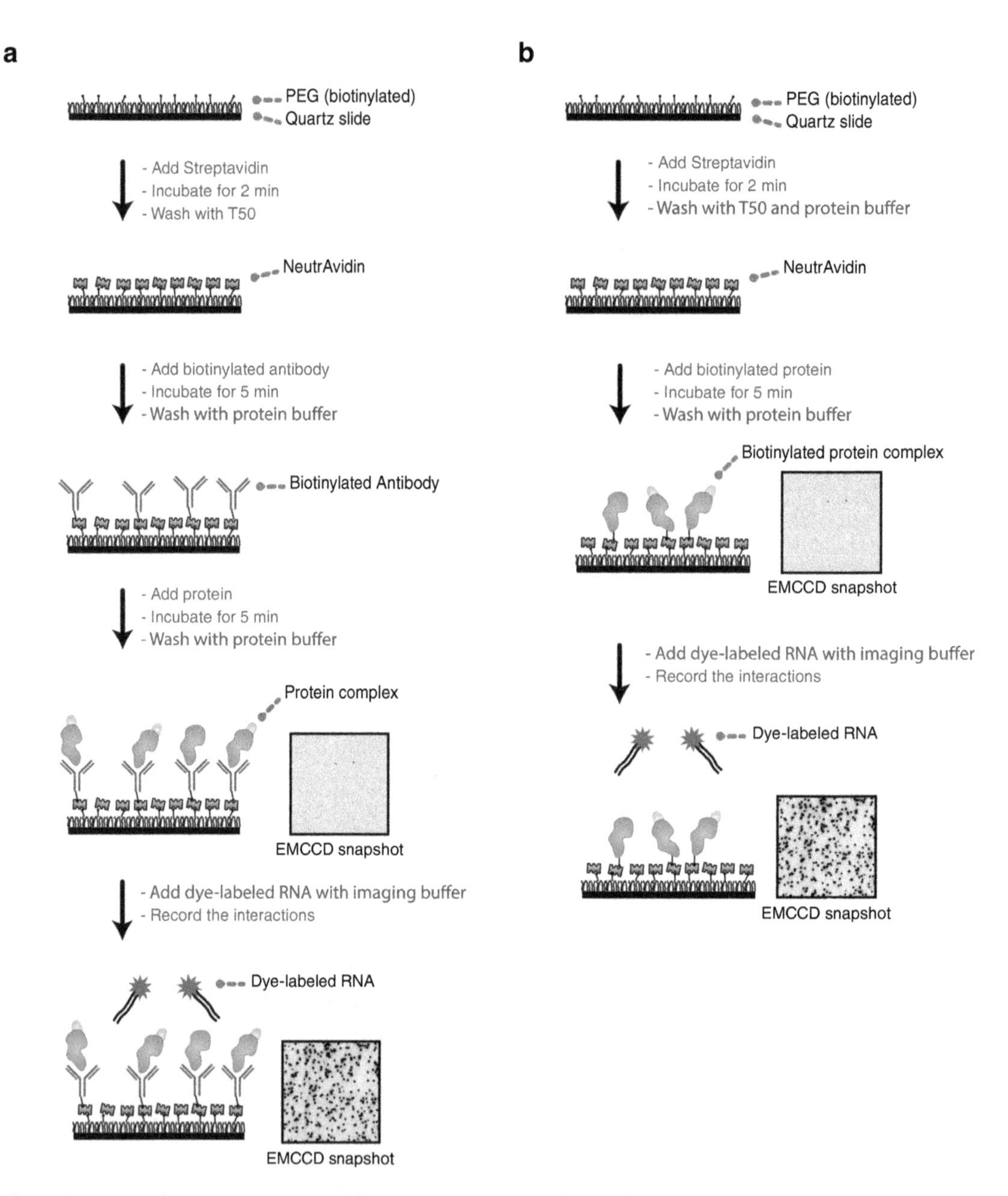

Fig. 4 Overview of single-molecule pulldown assays on the surface of the imaging chamber. (**a**) Biotinylated protein complexes are directly immobilized via streptavidin–biotin interaction. (**b**) Nonbiotinylated protein complexes are immobilized through a biotinylated antibody (e.g., antibody anti-MBP, anti-FLAG, anti-6xHis, anti-Myc) that recognize a tag attached to the C-terminus or N-terminus of a protein of interest (e.g., MBP, FLAG, 6xHis, Myc)

3. Introduce in the imaging chamber 20 μL of immunoprecipitate containing biotinylated protein of interest in order to immobilize them on the surface via streptavidin–biotin interaction (*see* **Note 10**).

4. Incubate for 5 min and wash out unbound proteins with 100 μL of protein buffer (*see* **Note 11**).
5. Dilute the dye-labeled RNA in 50 μL protein buffer containing oxygen scavenging system (0.8% glucose (v/v), 0.1 mg/mL glucose oxidase, 17 μg/μL catalase). Add the mixture into the imaging chamber while recording the interaction between the RNA and surface-immobilized protein complexes with EMCCD camera (*see* **Note 12**).

3.4.2 Nonbiotinylated Proteins (Fig. 4b)

1. Assemble microfluidic flow chambers using PEGylated quartz slide, PEGylated glass coverslip, double sticky tape, and epoxy as described previously [16].
2. Introduce in the flow chamber 50 μL of streptavidin (0.1 mg/mL) and incubate for 2 min, then wash the chamber with 100 μL of TE buffer.
3. Use a commercially available biotinylated antibody for specific immobilization. Incubate the imaging chamber with 50 μL of biotinylated antibody (66 nM) for 5 min, then wash away the remaining unbound antibodies with 100 μL of buffer of interest (*see* **Note 13**).
4. Introduce in the imaging chamber 20 μL of immunoprecipitate containing the protein of interest to immobilize on the surface via the interaction with biotinylated antibody.
5. Incubate for 5 min and wash out unbound proteins with 100 μL of protein buffer.
6. Dilute the dye-labeled RNA in 50 μL protein buffer containing oxygen scavenging system. Add the mixture into the imaging chamber while recording the interaction between the RNA and surface-immobilized protein complexes with EMCCD camera.

3.5 Single-Molecule Fluorescence Microscopy, Data Acquisition, and Analysis

A lab-made prism-type total internal reflection microscope (prism-TIRFM) was used for the single-molecule experiments as described elsewhere [17]. Briefly:

1. Use a 473-nm solid-state laser (OBIS LX 75 mW, Coherent) to excite eGFP tagged-proteins, a 532-nm solid-state laser (Compass 215M-50, Coherent) to excite Cy3 molecules, and a 632-nm solid-state laser (25 LHP 928, CVI Melles Griot) to excite Cy5 molecules. We excite eGFP, Cy3 and Cy5 molecules as weakly as possible to minimize their rapid photobleaching during the observation time.
2. Collect the fluorescence signals from single molecules through a 60x water immersion objective (UPlanSApo, Olympus) with an inverted microscope (IX71, Olympus).

3. To block 473, 532 and 632 nm laser scattering, use a 473-nm long-pass filter (Chroma), a 550-nm long-pass filter (Chroma) and a 633-nm notch filter (SemRock), respectively.
4. Collect data in either single or dual color mode. For dual color measurements, fluorescence signals should be spatially split with a dichroic mirror ($\lambda_{cutoff} = 645$ nm, Chroma) and imaged onto two halves of an EMCCD camera (iXon 897, Andor Technology).
5. Use a lab-made software written in Visual C++ for data acquisition with a time resolution of 0.03–1 s.
6. Extract fluorescence images and time traces with programs written in IDL (ITT Visual Information Solutions). To systematically select single-molecule fluorescence signals of eGFP, Cy3 or Cy5 from the acquired images, we employ an algorithm written in IDL that search for fluorescence spots with a defined Gaussian profile and with signals above a threshold.
7. Analyse the data with Matlab (MathWorks) and Origin (OriginLab Corporation).

3.6 Example of Single-Molecule Investigation Using Pulldown Assay: Observation of Pre-miRNA Processing by Human Dicer-TRBP Complex

MicroRNAs (miRNAs) are small noncoding RNAs that regulate the majority of physiological and physiopathological processes including cell fate determination and tumorigenesis [18, 19]. MiRNA biogenesis occurs through several enzymatic processing steps including the cleavage of precursor miRNA (or pre-miRNA) by Dicer proteins in the cytoplasm [19]. Human Dicer is a multidomain enzyme that consists of several RNA binding domains including the PAZ domain, tandem RNase III domains and the helicase domain in the N-terminus [19, 20]. Biochemical and structural data showed that the PAZ domain recognizes the 3′ overhang end of the pre-miRNA, and the region between the PAZ and RNase III domains acts as a molecular ruler that defines the microRNA size [21–24]. In addition to its own RNA interacting domains, Dicer often associates with a protein partner called TRBP (transactivation response element RNA binding protein) that assist in pre-miRNA processing and miRNA loading [25–28]. TRBP is an RNA binding cofactor that is tightly associated with Dicer in human cells [25, 26]. While the molecular bases of pre-miRNA cleavage were extensively investigated, it remained unclear how Dicer and TRBP coordinate the recognition of pre-miRNA from another competitor cellular RNAs.

We used our single-molecule pulldown assay and RNA preparation strategies listed above to probe the substrate recognition process in real time employed by Dicer-TRBP complex (Fig. 5). We purified in vivo biotinylated Dicer protein in a complex with its cofactor TRBP using 293 HEK human cells (Fig. 3). We immobilized this protein complex on the surface of imaging chamber and

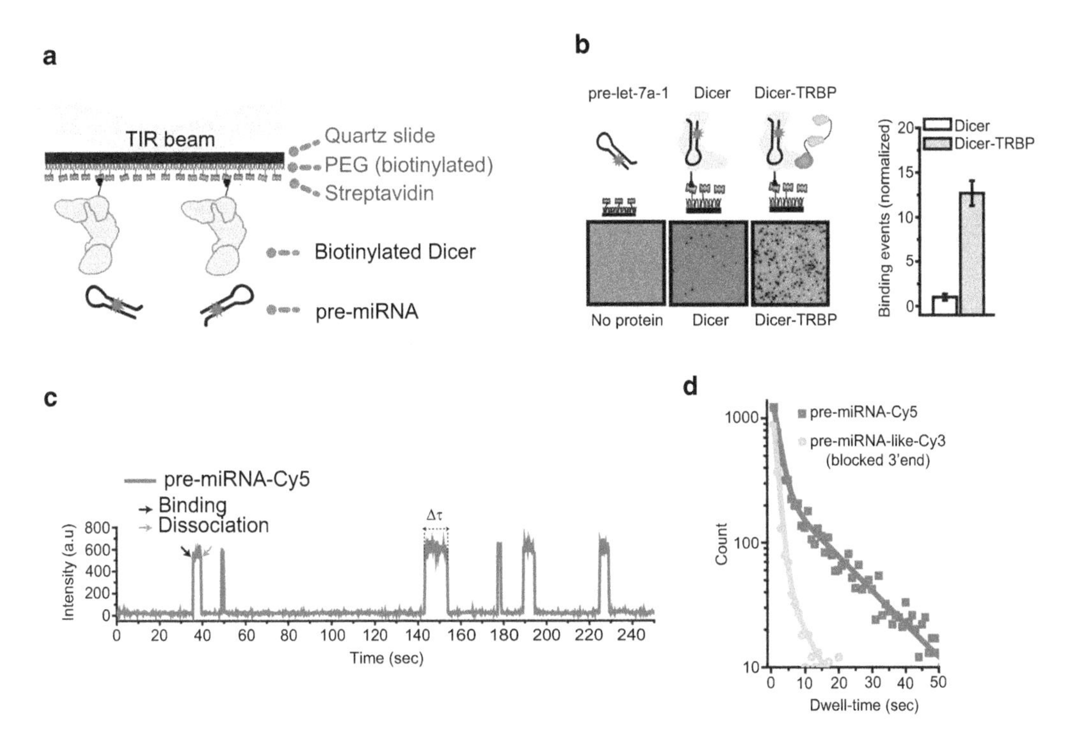

Fig. 5 Single-molecule measurement of pre-miRNA recognition by Dicer and its cofactor TRBP. (**a**) Schematic overview of a single-molecule pull-down assay for probing the recognition of pre-miRNA by Dicer and TRBP. Immunoprecipitated human Dicer-TRBP complexes were immobilized on a PEGylated surface using biotin–-streptavidin interaction. Cy5-labeled pre-miRNA was introduced into the imaging chamber by flow while recording the binding events in real time (300 ms). (**b**) CCD images display the RNA binding activity of Dicer alone (middle) and the Dicer-TRBP complex (right). Passivated surface without Dicer immobilized was used as negative control (left). The histogram on the right displays a normalized number of pre-miRNA bound to Dicer (white) or to Dicer-TRBP (blue). (**c**) Representative time trajectory (obtained with a time resolution 300 ms) displays six binding events of Cy5-labeled pre-miRNAs to a single Dicer-TRBP complex. The black arrow indicates the binding and the gray arrow indicates the dissociation of a pre-miRNA. Δt indicates the interaction life-time between surface immobilized Dicer-TRBP complex and pre-miRNA-Cy5. (**d**) Two-color competition assay to probe how Dicer-TRBP complex can discriminate pre-miRNA from other pre-miRNA-like RNAs in the same single-molecule assay. The standard pre-miRNA was labeled with Cy5. A biotin group was attached to the 3′-end of Cy3-labeled pre-miRNA to block its 3′-end (pre-miRNA-like with blocked 3′-end). The Dwell-time histogram is derived from binding of standard pre-miRNA (red) and pre-miRNA-like with blocked 3′-end (green). The distributions were fitted with a double-exponential decay. Reprinted and modified with permission from Fareh et al. (2016) *Methods*, **105**, 99–108 [13]. Copyright 2018 Elsevier

recorded the interactions with dye-labeled pre-miRNA and pre-miRNA-like substrates in real time (Fig. 5a). We found that TRBP greatly enhances the RNA binding activity of Dicer by one order of magnitude (Fig. 5b). Biochemical cleavage data suggest Dicer-TRBP might act as a processive multi-turnover enzyme that can cleave several pre-miRNA molecules. Indeed, our data allowed as to visualize the processing of six pre-miRNAs by Dicer-TRBP within less than 4 min of observation (Fig. 5c). We estimated that

Dicer-TRBP needs ~15 s to cleave a single pre-miRNA and release the products [12].

Next, we used our single-molecule technique to understand how Dicer-TRBP distinguishes pre-miRNA molecules from other cellular RNA species. We performed a two-color competition assay in which we labeled canonical pre-miRNA with Cy5 (red), while we labeled a pre-miRNA-like molecule harboring a blocked 3′-end with Cy3 (green). We introduced a mixture containing both Cy5 and Cy3 labeled RNAs into the imaging chamber and visualized how a single Dicer-TRBP probes those two RNA substrates in the same experiment. We found that Dicer-TRBP rejected noncanonical substrates (green) very quickly (>1 s) and engaged long and productive binding with canonical pre-miRNA (red) (Fig. 5d). Taking together, our data suggest that Dicer-TRBP engages weak interactions with RNA substrates containing dsRNA features. Dicer's PAZ domain and TRBP coordinate the substrate selection process and rapidly reject noncanonical pre-miRNAs after the initial probing. Only canonical pre-miRNA molecules can bypass PAZ-TRBP checkpoint and get fully transferred to the vicinity of Dicer's RNase III domains for subsequent cleavage of the terminal loop within ~15 s [12]. Our single-molecule approach allowed to uncover the molecular basis of substrate recognition employed by Dicer-TRBP complex. This technique that we describe in the protocol above can be applied to a variety of RNA-interacting protein complexes to gain a dynamic view of their functions.

4 Notes

1. Buffers containing Tris or reagents with amine groups should be avoided since it will interfere with the labeling reaction. Such buffer should be removed by ethanol precipitation prior the labeling.
2. The concentration of RNA, dye, and labeling buffer can be rescaled linearly in case of low concentration of RNA.
3. RNA strands are typically labeled with the NHS-ester form of Cy3 or Cy5 with near 100% efficiency
4. We noticed a correlation between the size of the loop and ligation efficiency. We successfully ligated RNA constructs harboring ≥8 nucleotides in the loop (Fig. 2b), while we failed to ligate RNA constructs with a loop containing only four nucleotides or less (Fig. 2c).
5. Strict work in RNase-free conditions is required to avoid RNA degradation.
6. Alternatively, a short dye-labeled oligonucleotide harboring a 5′ phosphate group can be ligated to the in vitro-transcribed

RNA using the conventional RNA ligation protocol with T4 RNA ligase

7. The recovered cell extract (supernatant) can be either directly used for single-molecule experiments, or alternatively, tandem purification steps can be carried out to obtain higher purity samples. To prevent disturbing the protein complexes, it is important to perform the cell lysis and immunoprecipitation in a gentle manner and in a physiologically relevant buffer. We do not recommend the use of sonication as a cell lysis method because this may cause protein complexes to disassemble and form aggregates.
8. Longer incubation time may increase the number of nonspecific interactions and result in the pull-down of contaminant proteins.
9. The aliquoted proteins are ready for single-molecule use. We recommend to check the enrichment of the proteins of interest in the immunoprecipitates using western blot analysis, and test the catalytic activities of the IPs using bulk assays.
10. When the cell extract is directly used without immunoprecipitation of the protein complex of interest, additional steps should be considered to avoid nonspecific absorption of the cell extract to the surface.
    (a) First, maximize the quality of the surface by treating it with 5% Tween 20, wash the surface with T50 buffer before adding the streptavidin into the chamber.
    (b) Second, the cell extract should be diluted (approximately 500 times) to limit the nonspecific adsorption of RNA-interacting proteins in the extract to the surface. We advise to reduce the incubation time of the cell extract in the imaging chamber to 30 s or less).
11. The interaction between the biotinylated proteins and streptavidin is very stable and can last for several hours without any noticeable dissociation.
12. Oxygen scavenging system was used to reduce photobleaching and 1 mM Trolox was used to reduce photoblinking of the dyes [29].
13. If a biotinylated antibody of interest is not commercially available, two antibodies immobilization scheme can be used. A biotinylated secondary antibody (e.g., anti-mouse, rabbit, or goat) is first attached to the surface through the interaction with steptavidin. Then, a primary antibody that recognises a protein tag (or epitope) can immobilized on the surface of the imaging chamber via the interaction with the secondary antibody. This two-antibody immobilization scheme is similar to the one used in ELISA or western blotting assays.

Acknowledgments

We thank C.J. lab members for technical help and discussions. We thank Luuk Loeff, Malwina Szczepaniak, Anna C. Haagsma, Kyu-Hyeon Yeom for their help and support throughout the development of our single-molecule pulldown technique. This work was supported by a European Research Council Starting Grant under the European Union's Seventh Framework Programme [FP7/2007–2013/ERC grant 309509 to C.J]; and the Fondation pour la Recherche Medicale [SPE20120523964 to M.F].

References

1. Krogan NJ, Cagney G, Yu H, Zhong G, Guo X, Ignatchenko A, Li J, Pu S, Datta N, Tikuisis AP et al (2006) Global landscape of protein complexes in the yeast Saccharomyces cerevisiae. Nature 440:637–643
2. Giot L, Bader JS, Brouwer C, Chaudhuri A, Kuang B, Li Y, Hao YL, Ooi CE, Godwin B, Vitols E et al (2003) A protein interaction map of Drosophila melanogaster. Science 302:1727–1736
3. Li S, Armstrong CM, Bertin N, Ge H, Milstein S, Boxem M, Vidalain PO, Han JD, Chesneau A, Hao T et al (2004) A map of the interactome network of the metazoan C. elegans. Science 303:540–543
4. Srihari S, Yong CH, Patil A, Wong L (2015) Methods for protein complex prediction and their contributions towards understanding the organisation, function and dynamics of complexes. FEBS Lett 589:2590–2602
5. Zhang X, Yan C, Hang J, Finci LI, Lei J, Shi Y (2017) An atomic structure of the human spliceosome. Cell 169:918–929.e914
6. Galej WP, Wilkinson ME, Fica SM, Oubridge C, Newman AJ, Nagai K (2016) Cryo-EM structure of the spliceosome immediately after branching. Nature 537:197–201
7. Kwon SC, Nguyen TA, Choi YG, Jo MH, Hohng S, Kim VN, Woo JS (2016) Structure of human DROSHA. Cell 164:81–90
8. Hoskins AA, Friedman LJ, Gallagher SS, Crawford DJ, Anderson EG, Wombacher R, Ramirez N, Cornish VW, Gelles J, Moore MJ (2011) Ordered and dynamic assembly of single spliceosomes. Science 331:1289–1295
9. Lee HW, Ryu JY, Yoo J, Choi B, Kim K, Yoon TY (2013) Real-time single-molecule coimmunoprecipitation of weak protein-protein interactions. Nat Protoc 8:2045–2060
10. Lee HW, Kyung T, Yoo J, Kim T, Chung C, Ryu JY, Lee H, Park K, Lee S, Jones WD et al (2013) Real-time single-molecule co-immunoprecipitation analyses reveal cancer-specific Ras signalling dynamics. Nat Commun 4:1505
11. Jain A, Liu R, Ramani B, Arauz E, Ishitsuka Y, Ragunathan K, Park J, Chen J, Xiang YK, Ha T (2011) Probing cellular protein complexes using single-molecule pull-down. Nature 473:484–488
12. Fareh M, Yeom KH, Haagsma AC, Chauhan S, Heo I, Joo C (2016) TRBP ensures efficient Dicer processing of precursor microRNA in RNA-crowded environments. Nat Commun 7:13694
13. Fareh M, Loeff L, Szczepaniak M, Haagsma AC, Yeom KH, Joo C (2016) Single-molecule pull-down for investigating protein-nucleic acid interactions. Methods 105:99–108
14. Yeom KH, Heo I, Lee J, Hohng S, Kim VN, Joo C (2011) Single-molecule approach to immunoprecipitated protein complexes: insights into miRNA uridylation. EMBO Rep 12:690–696
15. Green MR, Sambrook J (2012) Molecular cloning: a laboratory manual. Cold Spring Harbor Laboratory Press, Cold Spring Harbor, NY
16. Chandradoss SD, Haagsma AC, Lee YK, Hwang JH, Nam JM, Joo C (2014) Surface passivation for single-molecule protein studies. J Vis Exp 86. https://doi.org/10.3791/50549
17. Joo C, Ha T (2012) Single-molecule FRET with total internal reflection microscopy. Cold Spring Harb Protoc 2012(12). doi: https://doi.org/10.1101/pdb.top072058
18. Fareh M, Turchi L, Virolle V, Debruyne D, Almairac F, de-la- Forest Divonne S, Paquis P, Preynat-Seauve O, Krause KH, Chneiweiss H et al (2012) The miR 302-367 cluster drastically affects self-renewal and infiltration

properties of glioma-initiating cells through CXCR4 repression and consequent disruption of the SHH-GLI-NANOG network. Cell Death Differ 19:232–244

19. Ha M, Kim VN (2014) Regulation of microRNA biogenesis. Nat Rev Mol Cell Biol 15:509–524
20. Jinek M, Doudna JA (2009) A three-dimensional view of the molecular machinery of RNA interference. Nature 457:405–412
21. Yan KS, Yan S, Farooq A, Han A, Zeng L, Zhou MM (2003) Structure and conserved RNA binding of the PAZ domain. Nature 426:468–474
22. MacRae IJ, Zhou K, Li F, Repic A, Brooks AN, Cande WZ, Adams PD, Doudna JA (2006) Structural basis for double-stranded RNA processing by Dicer. Science 311:195–198
23. MacRae IJ, Zhou K, Doudna JA (2007) Structural determinants of RNA recognition and cleavage by dicer. Nat Struct Mol Biol 14:934–940
24. Tian Y, Simanshu DK, Ma JB, Park JE, Heo I, Kim VN, Patel DJ (2014) A phosphate-binding pocket within the platform-PAZ-connector helix cassette of human Dicer. Mol Cell 53:606–616
25. Chendrimada TP, Gregory RI, Kumaraswamy E, Norman J, Cooch N, Nishikura K, Shiekhattar R (2005) TRBP recruits the Dicer complex to Ago2 for microRNA processing and gene silencing. Nature 436:740–744
26. MacRae IJ, Ma E, Zhou M, Robinson CV, Doudna JA (2008) In vitro reconstitution of the human RISC-loading complex. Proc Natl Acad Sci U S A 105:512–517
27. Lee Y, Hur I, Park SY, Kim YK, Suh MR, Kim VN (2006) The role of PACT in the RNA silencing pathway. EMBO J 25:522–532
28. Ota H, Sakurai M, Gupta R, Valente L, Wulff BE, Ariyoshi K, Iizasa H, Davuluri RV, Nishikura K (2013) ADAR1 forms a complex with Dicer to promote microRNA processing and RNA-induced gene silencing. Cell 153:575–589
29. Rasnik I, McKinney SA, Ha T (2006) Non-blinking and long-lasting single-molecule fluorescence imaging. Nat Methods 3:891–893

Chapter 17

Preparing Frozen-Hydrated Protein–Nucleic Acid Assemblies for High-Resolution Cryo-EM Imaging

Panchali Goswami, Julia Locke, and Alessandro Costa

Abstract

High-resolution image acquisition and structure determination by cryo-electron microscopy is becoming increasingly streamlined. Preparing electron-microscopy grids of suitable quality remains, however, a critical bottleneck. Strategies to achieve particle monodispersity, optimal sample concentration and suitable ice thickness can vary from specimen to specimen. In this book chapter we describe our protocols for negative-stain grid and cryo-grid preparation, which we apply to studying protein–nucleic acid complexes.

Key words Cryo-electron microscopy, Negative staining, Protein–nucleic acid assemblies, Structural biology

1 Introduction

Single-particle cryo-electron microscopy (cryo-EM) is the structural technique of choice when studying large and dynamic macromolecular assemblies at high resolution. Using this approach, for instance, significant advances have been made in our mechanistic understanding of key DNA transactions. The recent structures of eukaryotic replicative helicases [1–3] and retroviral integrases [4, 5] are examples of how cryo-EM has changed the way we think about genome propagation in cells and viruses.

Reconstruction of a high-resolution three-dimensional volume from single particles involves the two-dimensional EM imaging of randomly oriented particles. To obtain a structure, cropped-out particle images are subjected to iterative cycles of alignment, angular assignment and three-dimensional averaging. Specimen preparation for cryo-EM appears comparatively less challenging than for macromolecular crystallography in that (i) sample-concentration requirements can be one to two orders-of-magnitude lower,

Panchali Goswami and Julia Locke have contributed equally to this work.

Yuri L. Lyubchenko (ed.), *Nanoscale Imaging: Methods and Protocols*, Methods in Molecular Biology, vol. 1814,
https://doi.org/10.1007/978-1-4939-8591-3_17, © Springer Science+Business Media, LLC, part of Springer Nature 2018

(ii) imaging occurs in solution with no crystallization needed and (iii) a (limited) degree of compositional and conformational heterogeneity can be tolerated, potentially informing on macromolecular dynamics [6]. During cryo-EM grid preparation, particles stored in an aqueous solution are trapped in a thin layer of amorphous (noncrystalline) ice. This "vitreous"-ice support preserves the particle native state, while also providing a robust medium that survives high vacuum in the electron microscope column [7].

Preparing grids that are suitable for high-resolution imaging presents a unique set of challenges. In the early stages of sample preparation, a minimal degree of homogeneity must be ascertained to ensure that two-dimensional images of individual particles will likely be averaged in three dimensions. A useful approach in this screening process is negative-stain electron microscopy. Here, particles adhering onto a hydrophilic carbon support are exposed to a heavy atom solution, which stains the carbon background but not the particles. Negative staining represents a quick and easy tool for early assessment of sample quality, and can provide structural information in the 2-nm resolution range [8]. Transitioning from negative stain to cryo-EM imaging of single particles can at times be complicated.

In fact, cryo-EM generally requires higher sample concentration than negative stain, especially when working in the absence of carbon support. Furthermore, different particles appear to prefer different degrees of ice thickness, adding to the parameters to be optimized. Finally, once trapped on the cryo grid, particles tend to come in contact with the air-water interface in the thin ice layer. While some particles tolerate it, the exposure to the air-water interface can induce preferential orientation of highly polarized assemblies (such as can be protein–DNA complexes), making structure determination more challenging. In the worst-case scenario, particles at the air-water interface become aggregated, making single-particle reconstruction impossible [9]. While studying a diverse range of protein–DNA complexes, we have developed a number of protocols to obtain reproducibly high-quality negative stain and cryo-EM grids. In this chapter we detail our preparation strategies and describe robust protocols to prepare negative stain EM and cryo-EM samples.

2 Materials

All solutions should be prepared with MilliQ water at room temperature. Wear personal protective equipment and strictly follow the waste disposal regulations, in particular while handling uranyl salts.

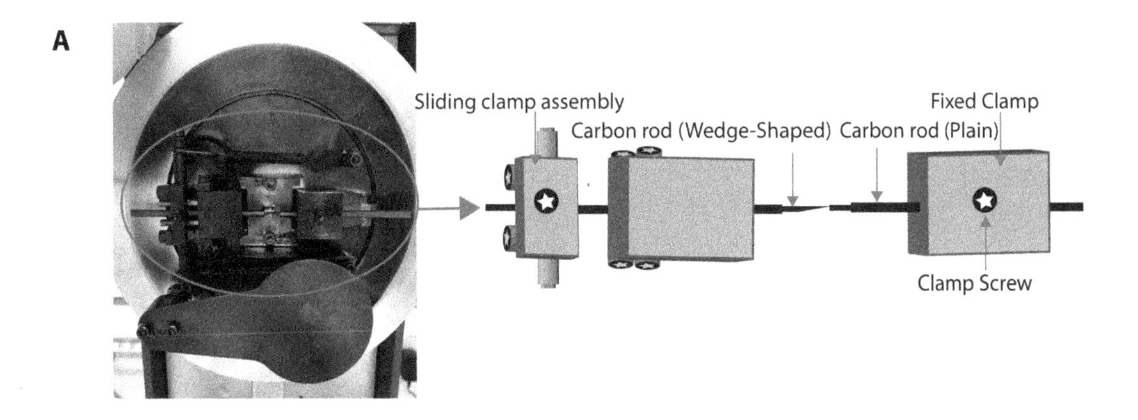

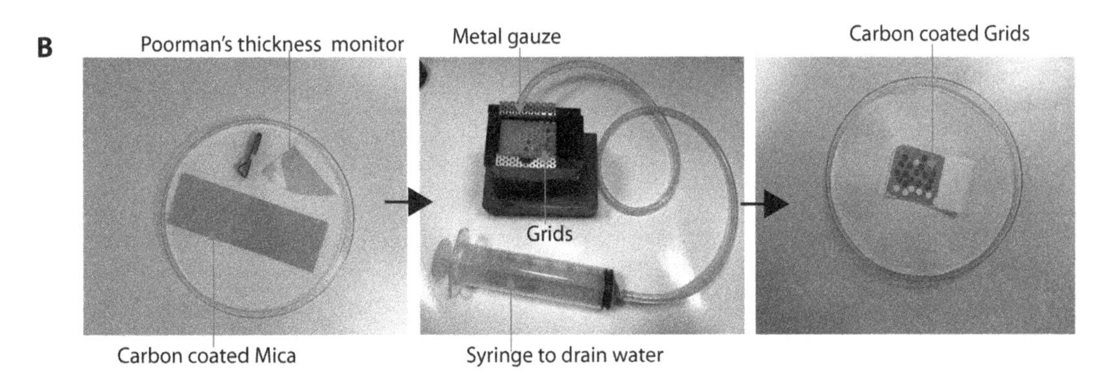

Fig. 1 Preparation of carbon coated grids for negative stain and cryo-EM. (**a**) Two carbon rods are held in contact by two clamps. A wedge-shaped carbon rod is kep under spring tension against a plain carbon rod. (**b**) *Left panel:* Carbon-coated mica next to a piece of partially shielded blotting paper used to monitor carbon thickness. *Central panel:* grid-coating trough. (**c**) *Right panel:* carbon-coated grids left to dry overnight

2.1 Carbon Coating and Glow Discharging

1. Carbon evaporation: carbon rods, carbon rod shaper, wedge tool, fine emery paper, gloves, tweezers, mica sheets, a Petri dish, Whatman blot paper for storage and a carbon coater, e.g., Q150T Turbo-Pumped Carbon Coater (Quorum Technologies, Fig. 1a).
2. Copper grids, e.g., AGG2400C Agar grids 400 mesh copper 3.05 mm (Agar Scientific) for negative stain or cryo grids, e.g., Quantifoil (R1.2/1.3 or R2/2), Protochips C-flats, lacey grids, or lacey grids with ultrathin carbon support film (we use Agar Scientific AG01824).
3. Material for floating carbon on the copper grids: film casting device (e.g., Smith Grid Coating Trough, Ladd Research) and a plastic syringe for water removal, deionized water.
4. Glow discharging grids: Glass Petri dish, glass slide, tweezers, glow discharger (e.g., K100X, Quorum Technologies)

2.2 Negative Staining

1. Uranyl acetate or uranyl formate powder.
2. Carbon-coated grids, anti-capillary tweezers, blotting paper (Whatman number 1), Parafilm, timer, grid box for storage.

2.3 Cryo-EM Grid Preparation

1. Automated plunge freezer (e.g., FEI Vitrobot) should be set up according to manufacturer's instructions.
2. Ethane preparation: Personal protective equipment, tweezers, ethane gas, liquid nitrogen.
3. Sample freezing and grid storage. Personal protective equipment, tweezers, cryo-EM grids, filter paper, cryo-EM grid boxes, large tweezers for cryo-EM grid-box transfer, screwdriver, lint-free wipes, liquid nitrogen, ethane cylinder.

3 Methods

All operations are performed at room temperature, unless otherwise specified.

3.1 Preparing Carbon Coated Grids for Negative Stain and Cryo-EM

Preparation for carbon-coated grids entails carbon evaporation on a mica sheet, followed by carbon-film floating on water and deposition onto EM grids. In the carbon coater, current is applied under vacuum through two aligned carbon rods, resulting in carbon evaporation.

1. Wearing protective gloves, cleave mica sheet (dimensions 75 × 25 × 0.15 mm, e.g., M054 from TAAB) with a razor blade (*see* **Note 1**). Place the mica sheets with the smooth surface facing upward in the carbon-coater.
2. Shape one to a spigot with approximately 5 mm length, using a carbon rod shaper. Then use the wedge tool to sharpen the end of one spigot into wedge form (Fig. 1a). Flatten the end of the second rod with emry paper (*see* **Note 2**).
3. Mount and align the two rods onto the carbon coater gun. Put the rods under spring tension by retracting the spring-loaded sliding clamp by ~0.5 cm before tightening it around the wedge-shaped rod (Fig. 1a).
4. Place mica sheets with the cleaved surface facing upward on the slide sample stage in the carbon-coater. Evaporate carbon. Let evaporated carbon rest overnight before floating (Fig. 1b).
5. To evaluate carbon thickness (in the absence of a thickness monitor), secure a piece of filter paper next to the mica, using a metal piece to hold the paper in place as vacuum is pulled. The metal piece will partially protect the paper during coating (Fig. 1b). Visual inspection of the coated/noncoated paper interface serves to estimate carbon thickness ("poor man's thickness monitor", as in Fig. 1b, *see* **Note 3**).

6. To deposit the carbon film onto the grids (copper grids for negative staining EM or Quantifoils/C-flats for cryo-EM grids) prepare a piece of filter paper and place it on a metal gauze (Fig. 1b). Submerge gauze and paper in a small water tank containing deionized water. Use tweezers to place the grids in a tightly packed pattern onto the filter paper. Slowly dip the coated mica into the water with the coated face upward, so that the carbon film can detach from the mica sheet and float on the water surface. Remove the water from the bottom of the tank by syringe suction, until the carbon film becomes deposited onto the grids (Fig. 1b). Allow the coated grids to air-dry overnight before using (*see* **Note 4**).

3.2 Preparation of Negative Stain Solutions

Weigh out uranyl powders under the fume hood and wear protective mask during the entire preparation process.

1. Preparation of a 2% uranyl acetate solution.
   (a) Dissolve 100 mg uranyl acetate (e.g., U001 form Taab) in 5 mL MilliQ water mix vigorously using a vortex for 10 min.
   (b) Spin in a microfuge at 20,000 × *g* for 5 min.
   (c) Filter the solution and wrap in aluminum foil for storage. Uranyl solutions are light sensitive.
2. Preparation of a 2% uranyl formate solution.
   (a) To prepare a 2% uranyl formate solution weigh out 100 mg of powder (e.g., U002 from Taab) and dissolve in 5 mL MilliQ water. Vortex for 5 min until dissolved.
   (b) Adjust the pH by adding slowly 20 μL of 5 N NaOH—4 μL at a time and mix in between to avoid precipitation. Vortex for 5 min and centrifuge for 5 min at 20,000 × *g* in a microfuge.
   (c) Filter the solution and wrap in aluminum foil. Filter again after 30 min, if needed, and use within 1 h.

3.3 Glow Discharging

The same glow discharge parameters apply to both negative stain and cryo-EM grids. For cryo grids coated with a thin layer of continuous carbon, shorter times might be required to maintain carbon film integrity.

1. Place carbon coated grids on a glass slide.
2. Transfer the glass slide into the glow discharge chamber and start the vacuum pump. Once a minimum vacuum of 0.07 mbar has been reached, glow-discharge the grids with a current of 45 mA for 1 min.
3. Store glow-discharged grids in a Petri dish to protect from dust.

3.4 Carbon-Coated Grid Staining

Avoid bench contamination, work on a double-layered bench coat and keep it clean.

1. Cut a piece of Parafilm and score it with four columns, and one row for each grid.
2. On each line intersection pipette a 75-μL stain droplet.
3. Pick up a glow-discharged grid with anticapillary tweezer.
4. Pipette 4 μL of sample solution onto the glow-discharged side of the carbon grid.
5. Incubate for 30 s to 2 min, then blot ~3 μL of sample solution, while leaving the grid visibly wet on the carbon side.
6. Touch a droplet of stain with the wet side of the grid and stir vigorously for 10 s. Repeat for the remaining three droplets (*see* **Note 5**).
7. Blot the grid dry using a wedge-shaped piece of Whatman paper and store inside a grid box.

3.5 Cryo-EM Grid Preparation

3.5.1 Preparing the Vitrobot for Automated Plunge Freezing

1. Replace paper on the blotting pads and turn on humidifier.
2. Switch on the Vitrobot and set parameters. We generally freeze DNA-protein complexes in 100% humidity and at 22 °C (*see* **Note 6**).
3. Assemble the coolant container and insert the grid boxes in the designated holders as shown in Fig. 2a.

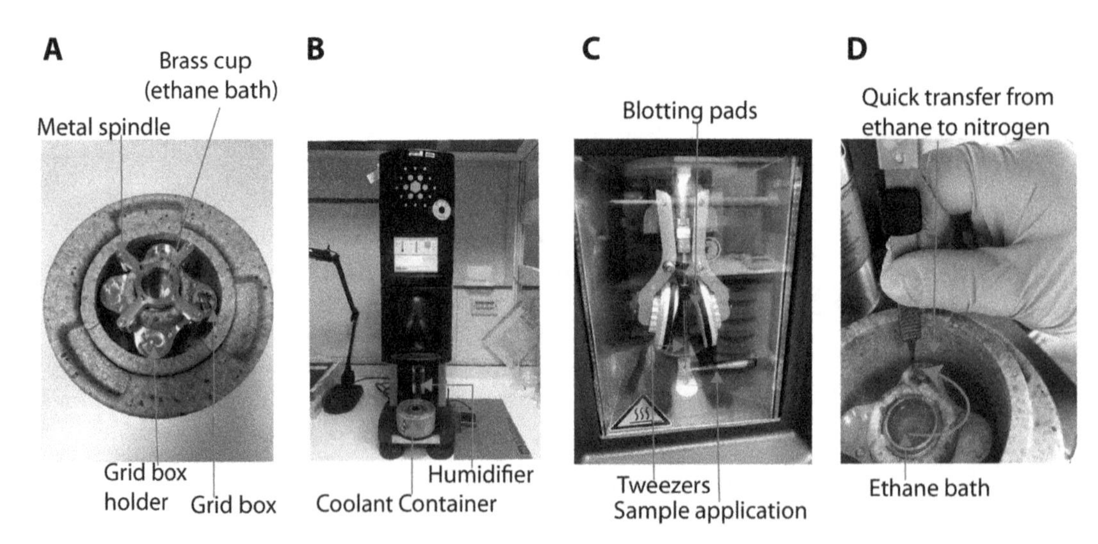

Fig. 2 Cryo-EM grid preparation. (**a**) Coolant container prepared for ethane liquefaction. (**b**) Automated plunge-freezer (Vitrobot). (**c**) Sample application in the climate chamber. (**d**) Transfer from ethane bath into liquid nitrogen

3.5.2 Preparing the Liquid Ethane

1. To speed up condensation of liquid ethane, place the metal spindle onto the ethane container (brass cup) and fill initially the entire coolant container (including the brass cup) with liquid nitrogen. Allow nitrogen to slowly evaporate from the ethane container. Once the ethane has evaporated, refill the outer ring (but not the ethane container) with liquid nitrogen (*see* **Note 7**) and maintain its level throughout the freezing process. Once temperature is reached and vigorous nitrogen bubbling has ceased, insert the tip of the ethane supply tube and direct it against the bottom of the brass cup. Open the regulator valve on the ethane tank. Ethane condensation will promptly start.
2. Fill the ethane container to the top rim.
3. The metal spindle should be removed promptly when the first signs of ethane solidification appear at the bottom of the brass cup. At this stage the liquid ethane surface will appear slightly opaque (*see* **Note 8**).

3.5.3 Freezing Cryo-EM Grids

1. Use the Vitrobot tweezers to grip a cryo-EM grid from its far edge. Secure the grid with the clamping ring on the second dent of the notched tweezer-handle. Do not to slide the clamping ring any further, as it will interfere with the blotting process (*see* **Note 9**).
2. Place the coolant container on the platform ring (Fig. 2b).
3. Rotate the lid of the cryo-grid box to ensure that one empty slit is exposed. This will facilitate grid transfer at a later stage.
4. Secure the tweezers to the central rod. The carbon side of the grid should face the side port.
5. "Start process" to raise the tweezers with the grid. Apply 4 μL of the sample through the side port (Fig. 2c).
6. Proceed with sample incubation, protting and plunge freezing. We generally incubate the sample for at least 30 s and up to 2 min, blotting time varies according to the blotting force parameter and the type of grids used. (*For blotting strategies for poorly concentrated samples see* **Note 10.** *If particles suffer from preferential orientation or aggregate once exposed to the air-water interface, you might interested in reading* **Note 11**. *For tips on how to control ice thickness, see* **Note 12**.)

3.5.4 Cryo-Grid Transfer and Storage

1. Once the grid is plunged into liquid ethane and the coolant container descends, promptly top up the liquid in the container.
2. Carefully disengage the tweezers from the central rod, while leaving the tweezers submerged in liquid ethane. In a rapid movement transfer the tweezers from the ethane into the liquid

nitrogen bath, while keeping the grid in nitrogen fumes during the transfer (Fig. 2d).

3. Disengage the clamping ring from the tweezers dent and move the grid inside the cryo-grid-box slit.
4. All tools used to handle the grids during transfer should be cooled to liquid nitrogen temperature before touching the cryo-grid box.
5. Close the grid box. Transfer cryo-grid box from the coolant container into 50 mL plastic tube filled with liquid nitrogen. The plastic tube should be perforated.
6. Store in liquid nitrogen before your next cryo-EM imaging session.

4 Notes

1. For best results use only freshly cleaved mica sheets with a smooth surface. Use of protective gloves is important to maintain smoothness of the mica sheet. Finger grease prevents even carbon coating.
2. Clean the carbon gun with a duster before inserting both rods. Clean glass chamber using an ethanol-soaked wipe.
3. Using a Q150T Turbo-Pumped Carbon Coater (Quorum Technologies) we pull a ~10^{-5} Torr vacuum and evaporate carbon by ramping the current up to 46 mA, over the course of 2 min for negative stain grids. Reduce the amperage to 37 mA for preparing a thinner carbon support for cryo grids.
4. A thin carbon film suitable for cryo-EM grids (~2 nm thickness) will hardly be visible on the water surface, making floating particularly challenging. The carbon film will appear visible on the blotting paper after drying. An excellent alternative to thin carbon is graphene oxide. Like ultrathin carbon, graphene oxide can be used as an exceptionally thin support for poorly concentrated samples [10]. For an excellent description of graphene oxide grid preparation refer to the video link https://doi.org/10.6084/m9.figshare.3178669, from Sjors Scheres's laboratory.
5. Glow-discharged grids should be used shortly after glow discharging. We normally use the grids within 1 h.
6. If using the Vitrobot at 4 °C allow ~40 min until the climate chamber reaches the temperature.
7. We keep liquid nitrogen level with the lid of the cryo-grid box. Lower liquid nitrogen levels work as well provided that the boxes are kept in nitrogen fumes. Fully submerged boxes

generally prevent contamination but can make grid handling more difficult.

8. An ethane heater can be used to maintain the ethane temperature constant.
9. Give the tweezers a gentle shake to ensure that the cryo-EM grid is securely gripped.
10. In case of exceptionally low specimen concentration, continuous carbon ultrathin continuous carbon on lacey grids can be used. This support can tolerate multiple iterations of sample application and (partial) blotting, resulting in increased particle concentrations on the grid. Multiple applications can be performed manually before mounting the tweezers onto the Vitrobot central rod, or otherwise written as a Vitrobot program. For example, two subsequent 1-min incubations followed by 0.5 s blotting can be performed, before a last 1 min incubation followed by more extensive blotting and plunge-freezing. In our experience, with a temperature of 18 °C, these extended incubations are short enough to prevent excessive liquid nitrogen evaporation, hence preserving cryo-EM grid integrity. This strategy can also be used to introduce a buffer-exchange step, as described in ref. 5.
11. To alter the surface charge particles we find that a 1-to-5-min incubation in 0.01% glutaraldehyde prior to plunge freezing can be effective. Glutaraldehyde cross-links free primary amine groups (e.g. surface exposed lysine residues), also working as a stabilization agent. When used at such low concentrations for short amounts of time, glutaraldehyde has been observed to minimally affect macromolecular structures. In our experience, particles that tend to aggregate upon exposure to the air water interface on the cryo-EM grid, can remain folded and monodisperse when treated with glutaraldehyde.
12. Protein–DNA complexes such as nucleosomes can become disassembled when exposed to thin ice. To ensure that ice is sufficiently thick to maintain the native structure, we find that coating open-hole grids with an additional layer of carbon can help. By depositing the extra carbon layer onto for example R2/2 Quantifoil grids, we find that ice coverage can be improved (carbon hydrophilicity induced by glow discharging appears to increase).

Acknowledgments

We thank the members of our group, past and present, for contributing to the development of the protocols presented here. This work was supported by the Francis Crick Institute, which receives its core funding from Cancer Research UK (FC001065), the UK Medical Research Council (FC001065), and the Wellcome Trust (FC001065).

References

1. Abid Ali F et al (2016) Cryo-EM structures of the eukaryotic replicative helicase bound to a translocation substrate. Nat Commun 7:10708
2. Georgescu R et al (2017) Structure of eukaryotic CMG helicase at a replication fork and implications to replisome architecture and origin initiation. Proc Natl Acad Sci U S A 114(5): E697–E706
3. Li N et al (2015) Structure of the eukaryotic MCM complex at 3.8 A. Nature 524 (7564):186–191
4. Maskell DP et al (2015) Structural basis for retroviral integration into nucleosomes. Nature 523(7560):366–369
5. Ballandras-Colas A et al (2017) A supramolecular assembly mediates lentiviral DNA integration. Science 355(6320):93–95
6. Nogales E, Scheres SH (2015) Cryo-EM: a unique tool for the visualization of macromolecular complexity. Mol Cell 58(4):677–689
7. Cheng Y et al (2015) A primer to single-particle cryo-electron microscopy. Cell 161 (3):438–449
8. Booth DS, Avila-Sakar A, Cheng Y (2011) Visualizing proteins and macromolecular complexes by negative stain EM: from grid preparation to image acquisition. J Vis Exp (58). https://doi.org/10.3791/3227
9. Glaeser RM et al (2016) Factors that influence the formation and stability of thin, cryo-EM specimens. Biophys J 110(4):749–755
10. Pantelic RS et al (2010) Graphene oxide: a substrate for optimizing preparations of frozen-hydrated samples. J Struct Biol 170 (1):152–156

Chapter 18

Probing Chromatin Structure with Magnetic Tweezers

Artur Kaczmarczyk, Thomas B. Brouwer, Chi Pham, Nynke H. Dekker, and John van Noort

Abstract

Magnetic tweezers form a unique tool to study the topology and mechanical properties of chromatin fibers. Chromatin is a complex of DNA and proteins that folds the DNA in such a way that meter-long stretches of DNA fit into the micron-sized cell nucleus. Moreover, it regulates accessibility of the genome to the cellular replication, transcription, and repair machinery. However, the structure and mechanisms that govern chromatin folding remain poorly understood, despite recent spectacular improvements in high-resolution imaging techniques. Single-molecule force spectroscopy techniques can directly measure both the extension of individual chromatin fragments with nanometer accuracy and the forces involved in the (un)folding of single chromatin fibers. Here, we report detailed methods that allow one to successfully prepare in vitro reconstituted chromatin fibers for use in magnetic tweezers-based force spectroscopy. The higher-order structure of different chromatin fibers can be inferred from fitting a statistical mechanics model to the force-extension data. These methods for quantifying chromatin folding can be extended to study many other processes involving chromatin, such as the epigenetic regulation of transcription.

Key words Magnetic tweezers, Force spectroscopy, Single molecule, Chromatin, Nucleosome

1 Introduction

Chromatin is a DNA-protein complex that organizes and compacts DNA in the nucleus of eukaryotic cells. The basic unit of a chromatin is a nucleosome, which consists of 147 base pairs of DNA turning 1.65 times around a histone protein octamer. The histone octamer is composed of a $(H3\text{-}H4)_2$ histone tetramer and two H2A-H2B histone dimers [1, 2]. Nucleosomes interact with each other via histone tails and form higher-order chromatin structures [3–5].

Eukaryotic DNA is subject to many processes such as replication and transcription. To regulate these processes, the structure and composition of chromatin fibers must be very dynamic [6]. Previous studies have shown how the nucleosome positioning, post-translational histone modifications, and chromatin remodelers can

Yuri L. Lyubchenko (ed.), *Nanoscale Imaging: Methods and Protocols*, Methods in Molecular Biology, vol. 1814,
https://doi.org/10.1007/978-1-4939-8591-3_18,

alter the stability of nucleosomes and may modulate the structure of chromatin [7–10]. Using X-ray crystallography [11], electron microscopy [12–15], or atomic force microscopy [16], researchers have established detailed structures of the nucleosome and small arrays of nucleosomes. However, as the structure of chromatin fibers becomes more complex and disordered as their size increases, these techniques have their limits in contributing to an examination of higher-order chromatin structure. Moreover, they do not typically address the dynamical rearrangements of chromatin fibers that are essential for the genomic transactions occurring in eukaryotic nucleus. In contrast, single-molecule force spectroscopy techniques are capable of tracking and manipulating the extension of chromatin fibers in real time and under physiological conditions, providing unique means to study their structure, assembly, and disassembly.

Magnetic tweezers represent an especially powerful tool for the study of chromatin folding, as they permit the straightforward application of both tension and torsion [17, 18] as well as multiplexing [19, 20]. Precise control of the force combined with nanometer accuracy measurements of the extension of individual fibers will help to elucidate the molecular mechanisms that control chromatin condensation and the regulation of processes involving chromatin.

In our assay, chromatin fibers are tethered between a glass slide and a paramagnetic bead. As the distance between a pair of magnets and the beads reduces, the beads experience an increasing force, which is directly applied to the tethered fibers. In this methods chapter, we describe how a DNA substrate containing an array of nucleosome positioning sequences is prepared (Subheading 3.1) and functionalized for magnetic tweezers experiments (Subheading 3.2). This involves labelling of one end of the DNA with digoxigenin and the other end with biotin. Subsequently, the DNA substrate is loaded with histone proteins resulting in well-defined reconstituted chromatin fibers (Subheading 3.3). Though protein chaperones (i.e., NAP-1) can load histones on DNA into arrays of nucleosomes [21], we employ salt-dialysis reconstitution in which positively charged histones assemble onto the negatively charged DNA as the salt concentration is slowly decreased [22]. To ensure a well-defined composition and spacing of the nucleosomes in the chromatin fiber, we use a DNA template with tandem repeats of Widom-601 positioning sequences [23]. The quality of chromatin assembly is assessed using agarose gel electrophoresis (Subheading 3.4) which shows different mobility of incomplete and overcomplete chromatin fibers compared to stoichiometric assemblies of chromatin fibers.

We use a homebuilt magnetic tweezers microscope and custom flow cells to immobilize and manipulate individual chromatin fibers. The flow cell design constitutes an aluminum holder that fixes two cover slides and PDMS (polydimethylsiloxane) layer that seals the channel between the two slides. The holder has two channels for a buffer exchange and a large aperture for illumination of the sample (Subheading 3.5). After functionalizing the chamber (Subheading 3.6), chromatin fibers are injected into a flow cell and tethered between a glass slide and a magnetic bead. The chromatin fiber is manipulated by changing the force on the beads by moving the magnets (Subheading 3.7). The resulting changes in tether end-to-end distance are measured by a bead-tracking algorithm and are subsequently analyzed with a statistical mechanics model (Subheading 3.8). Following these procedures yields a detailed characterization of the mechanical properties of single-folded chromatin fibers.

2 Materials

2.1 DNA Cloning

1. Ampicillin-resistant pUC18 plasmid with tandem repeats of the Widom-601 sequence (*see* **Note 1**).
2. XL1-blue *E. coli* competent cells stored at −80 °C (*see* **Note 2**).
3. ThermoMixer® C Eppendorf™ 5424R Microcentrifuge (Eppendorf™).
4. Shaker KS 130 (IKA KS).
5. Luria-Bertani (LB) medium and LB agar plate.
6. NucleoBond® Xtra Midi kit (Macherey-Nagel).

2.2 DNA Digestion and Labelling

1. Restriction enzymes: BseYI, BsaI (New England Biolabs).
2. NEBuffer 3.1 (with bovine serum albumin) –10× concentrated (New England Biolabs).
3. Promega Wizard SV Gel and PCR cleanup kit (Promega).
4. BioDrop μLITE spectrophotometer (Isogen Life Science).
5. 1× Tris/Borate/EDTA buffer (TBE).
6. dNTPs (100 mM) (Roche).
7. GeneRuler DNA Ladder Mix (Thermo Fisher Scientific).
8. Klenow Fragment (2 U/μL), LC (Thermo Fisher Scientific).
9. Reaction buffer for Klenow Fragment (10×) (Thermo Fisher Scientific).
10. Digoxigenin-11-dUTP (1 mM) (Roche).
11. Biotin-16-dUTP (1 mM) (Roche).
12. UV transilluminator ChemiDoc™ XRS+ (Bio-Rad).

2.3 Chromatin Reconstitution

1. Slide-A-Lyzer mini dialysis tubes, 10,000 MWCO (Thermo Scientific).
2. Low-binding pipette tips (VWR).
3. Magnetic stirrer with a heat plate (VWR).

4. Glass beaker (1 L).
5. Econo gradient pump (Bio-Rad).
6. Histone Octamer, Recombinant Human (stored at −20 °C) (EpiCypher).
7. 1.5 L of miliQ H_2O (4 °C).
8. 50× Tris/EDTA (TE), pH 7.5.
9. 5 M NaCl.
10. High salt buffer (80 mL of 5 M NaCl +4 mL of 50× Tris/ EDTA (TE) + miliQ H_2O up to 200 mL).
11. Low salt buffer (980 mL of miliQ H_2O + 20 mL of 50× Tris/ EDTA (TE)).

2.4 Electrophoretic Band Shift Assay

1. Agarose Standard (Roth).
2. PowerPac™ Basic Power Supply (Bio-Rad).
3. KuroGEL Midi 13 Electrophoresis Horizontal (VWR).
4. GeneRuler DNA Ladder Mix (Thermo Fisher Scientific).
5. 5× Tris/boric acid (TB).
6. Loading buffer (20% glycerol, 20 mM Tris, 1 mM EDTA, bromophenol blue).
7. Ethidium bromide (10,000×) (Thermo Fisher Scientific).

2.5 Assembly of the Flow Cell Chamber

1. Aluminum flow cell frame, custom-built (Fig. 1).
2. Perspex mold, custom-built (Fig. 1).
3. Cover slip 24 × 40 mm, # 1.5 thickness (Menzel Gläser).
4. Cover slip 24 × 60 mm, # 1.5 thickness (Menzel Gläser).
5. Sylgard® 184 silicone elastomer kit PDMS (Sylgard).
6. FEP tubing (1/16″ OD × 0.020″ ID) (Upchurch Scientific).
7. Electrical wire (~0.02″ OD) (Nexans).
8. Embossing tape (Dymo).
9. N_2 Spray Gun Assembly (NCI).
10. Vacuum controller CVC 3000 and Diaphragm pump MZ 2 NT (Vacuubrand).
11. M4 screws.
12. Ultrasonic cleaning bath USC-TH (VWR).
13. Microscope slide (75 × 26 × 1 mm) (Menzel Gläser).
14. Glass beakers (100 mL + 250 mL).
15. Cover slip holder (custom-made).
16. 2-Propanol.

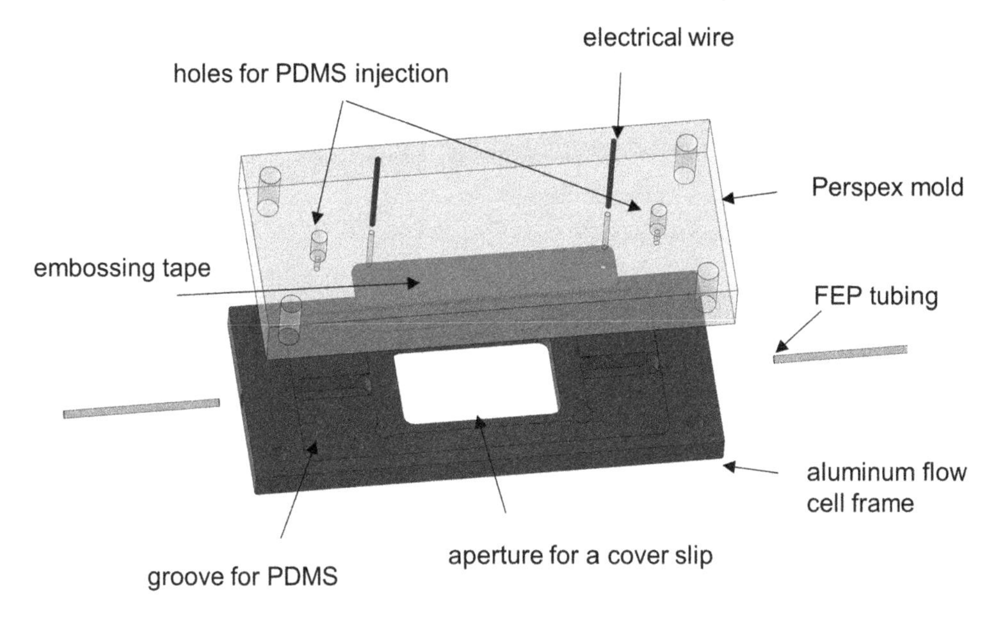

Fig. 1 Components of the flow cell chamber. A custom-built Perspex mold is mounted onto a custom-built aluminum flow cell frame that holds a 24 × 40 mm cover slip. Subsequently, PDMS is injected into the chamber between the mold and the groove of the metal frame

2.6 Flow Cell Functionalization

1. Dynabeads™ M-270 Streptavidin or MyOne Streptavidin (Invitrogen).
2. Microscope slide (75 × 26 × 1 mm).
3. Pentyl acetate, puriss. p.a., ≥98.5% (GC) (Sigma-Aldrich).
4. Nitrocellulose – 1% in pentyl acetate (Ladd Research Industries).
5. Anti-digoxigenin – from sheep (Sigma-Aldrich).
6. Bovine serum albumin (BSA) heat shock fraction, pH 7, ≥98% (Sigma-Aldrich).
7. Dynal® Invitrogen Bead Separations (Invitrogen).
8. Tween-20 (Sigma-Aldrich).
9. Sodium azide (NaN_3) (Merck).
10. 1 M HEPES, pH 7.5.
11. 2 M KCl.
12. 20 mM $MgCl_2$.
13. 1× Tris/EDTA (TE).
14. Plastipak 1 mL syringe (BD).
15. Sterican 18G single-use needle (B. Braun).
16. Tygon tubing (0.8 mm ID/0.8 mm wall).

2.7 Hardware

1. 25 Mpix Condor camera (CMOS Vision GmbH).
2. Frame grabber Camera Link PCIe-1433 (National Instruments).
3. NIKON CFI Plan Fluor objective (NA = 1.3, 40×, oil, Nikon Corporation).
4. Infinity-corrected tube lens ITL200 (Thorlabs).
5. 100 μW, 20 mA LED collimator emitting at 645 nm (IMM Photonics GmbH).
6. Computer T7610 (Dell) with a 10 core Intel Xeon 2.8 GHz processor (Intel) and 32GB DDR3 memory.
7. Multi-axis piezo scanner P-517.3CL (Physik Instrumente GmbH & Co.).
8. 5 mm cube magnets N50 (Supermagnete, Webcraft GmbH).
9. Two-phase hollow shaft stepper motor (Casun).
10. M-126.2S1 Translation Stage for a magnet (Two Physik).
11. M-126.2S1 Translation Stage for objective (Two Physik).
12. Syringe pump (Prosense B. V.).

3 Methods

3.1 DNA Cloning

1. *Cell transformation with pUC18 plasmid.* Thaw about 100 μL of XL1-blue competent cells on ice and add ~1 μL of a plasmid. Mix gently and incubate on ice for at least 30 min. Induce a 42 °C heat shock of about 90–120 s with the ThermoMixer. Place the sample back on ice for 1 min and add 900 μL of LB medium. Shake gently (240 rpm) and incubate for 30–60 min at 37 °C. Centrifuge the sample for 1 min at 14,000 rpm (18,500 × *g*), remove the supernatant, and then resuspend the sample in 100 μL LB medium. Spread the sample on an LB agar plate with ampicillin, and incubate upside-down at 37 °C overnight.
2. *Cell culture.* Select a single colony and inoculate in 500 mL flask with 250 mL LB medium with ampicillin. Incubate overnight at 37 °C while shaking (320 rpm). Next day, centrifuge the culture (14,000 rpm [18,500 × *g*] at 4 °C for 10 min). Discard the supernatant and keep the cell pellet.
3. *Plasmid DNA isolation.* Isolate and purify the plasmid DNA using NucleoBond® Xtra Midi kit using the manufacturer's protocol.

3.2 DNA Digestion and Labelling

1. *Plasmid digestion with BsaI and BseYI.* Linearize the plasmid DNA with two restriction enzymes in a buffer that contains maximally 50 ng/μL of the substrate in 1× NEBuffer 3.1 (*see*

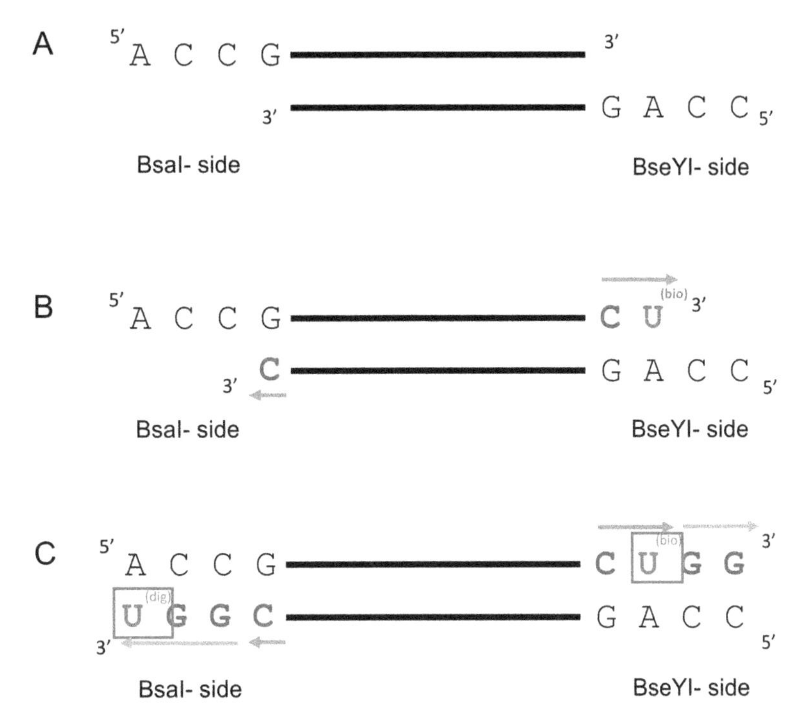

Fig. 2 Labelling of the DNA with a single biotin and digoxigenin tag. (**a**) The pUC18 plasmid digested with BsaI and BseYI has single-stranded overhangs at both sides. (**b**) A Klenow reaction with biotin-dUTP and free nucleotides other than dGTP (here dCTP) results in incorporation of a biotin only at the BseYI side. (**c**) Subsequently, the DNA is purified and a second Klenow reaction with dGTP and digoxigenin-dUTP results in binding of a digoxigenin at the BsaI side

Note 3). Digest overnight at 37 °C. Subsequently, inactivate the enzyme by incubating the reaction at 80 °C for 20 min. Purify the digested DNA with the Promega Wizard SV Gel and PCR cleanup kit (*see* **Note 4**).

2. *DNA labelling with biotin tag.* Mix the purified DNA, the Klenow fragment (5 units per 100 μL of a reaction volume), and a regular, non-tagged dNTP complementary to the first nucleotide at one side of the digested DNA, here CTP (*see* **Note 5** and Fig. 2b), together with biotin-16-dUTP (final concentration of dNTPs is 20 μM). Incubate for 2 h at 37 °C. Purify the DNA with the Promega Wizard SV Gel and PCR cleanup kit.
3. *DNA labelling with digoxigenin tag.* Mix the biotinylated DNA; the Klenow fragment (5 units per 100 μL of a reaction volume); a regular, non-tagged dNTP complementary to the first single nucleotide at the other side of the digested DNA (Fig. 2c); and digoxigenin-11-dUTP (final concentration of dNTPs is 20 μM). Incubate 2 h at 37 °C. Purify the DNA with the Promega Wizard SV Gel and PCR cleanup kit.

3.3 Chromatin Reconstitution

The DNA is mixed with histone octamers in a high salt buffer [24]. Subsequently, the salt concentration is slowly reduced by dialysis to first assemble $(H3\text{-}H4)_2$ tetrasomes on the 601 DNA. At low salt concentration, histone dimers H2A-H2B bind to the tetrasomes, and full nucleosomes are formed. The nucleosome reconstitution depends critically on the histone: DNA ratio. Excess histones yield additional tetrasomes and nucleosomes on the DNA handles, and excess DNA yields undersaturated fibers. Therefore, each reconstitution is done in a titration series.

1. *Preparation of histone : DNA titrations.* Prepare at least five batches with increasing histone : DNA ratios. Each batch contains at least 500 ng of DNA in a total volume of 50 μL. The minimal DNA concentration (m/v) is then 10 ng/μL, and the minimal molar concentration of the 601 DNA (excluding the DNA handles) is ~0.1 μM. Center the titrations at 1 : 1 molar ratio between 601 segments and histone octamer (Table 1). *See* **Note 6** for an explanation of the calculation method.
2. *Salt gradient dialysis.* Cool down all the buffers before starting the assembly (*see* **Note 7**). Insert a magnet rod into a beaker that contains 200 mL of high salt buffer. Put the dialysis tubes on a floater, and incubate for 15 min in this buffer to allow the membranes to soak. Transfer 50 μL of histone : DNA mixtures into the dialysis tubes using the low-binding pipette tips (*see* **Note 8**). Rinse the Econo gradient pump with miliQ H_2O and subsequently with the low salt buffer. Connect the pump as in Fig. 3. Set the flow rate at 0.9 mL/min to ensure a gradual

Table 1
The composition of individual titrations used for chromatin reconstitution

15 * 601 Chromatin assembly						
Component	**I**	**II**	**III**	**IV**	**V**	**VI**
180 ng/μL DNA template (μL)	7	7	7	7	7	7
200 ng/μL histone octamer (μL)	2	3	4	5	6	7
330 ng/μL competitor DNA (μL)	1	1	1	1	1	1
Final 601-DNA molar conc. (μM)	*0.179*	*0.179*	*0.179*	*0.179*	*0.179*	*0.179*
Final histone octamer molar conc. (μM)	*0.074*	*0.111*	*0.148*	*0.185*	*0.222*	*0.259*
High salt buffer (μL)	40	39	38	37	36	35
Total volume (μL)	50	50	50	50	50	50
Octamer: DNA ratio	0.4	0.6	0.8	1.0	1.2	1.4

Samples with increasing histone: DNA ratio are prepared to account for lower reproducibility while pipetting small volumes

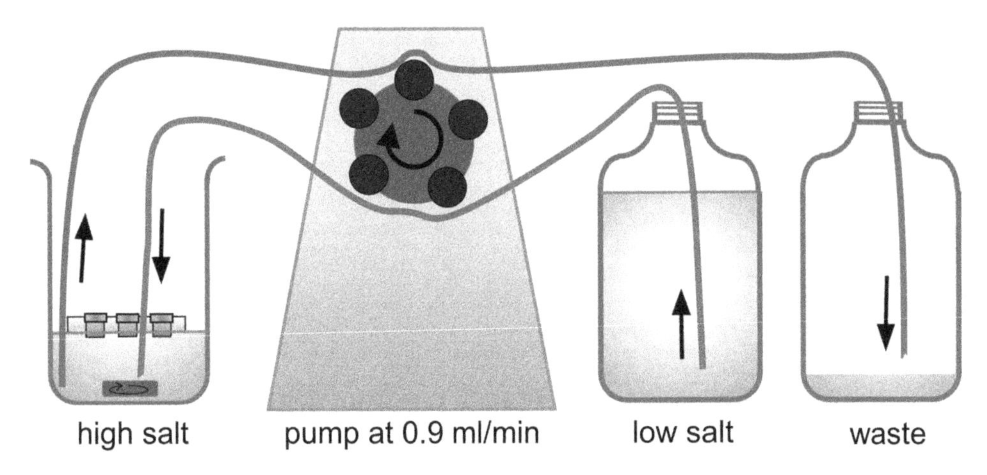

Fig. 3 Schematic representation of the dialysis system controlled by a peristaltic pump. Samples are dialyzed against a high salt buffer. Subsequently, a low salt buffer is pumped into the beaker with dialyzed samples, while an equal volume is flown out into a waste bottle to ensure a constant volume and a gradual decrease of salt concentration in the dialysis tubes

decrease of the salt concentration in the dialysis tubes. Switch on the magnet plate to properly mix the incoming low salt buffer with the high salt buffer (*see* **Note 9**).

3. *Collecting the reconstituted samples.* After approximately 18 h, 900 mL of the buffer is pumped to the waste bottle. Transfer the content of the dialysis tubes into separate low-binding tubes. Store the chromatin samples at 4 °C for maximally 5 weeks. Rinse the tubing of the pump with miliQ H_2O.

3.4 Electrophoretic Band Shift Assay

Quality assessment using agarose gel electrophoresis helps to select chromatin fibers with the optimal histone: DNA ratio prior to a single-molecule experiment. Due to their size and charge, reconstituted chromatin fibers migrate slower through a gel than a bare DNA template (Fig. 4).

1. *Gel preparation.* Dissolve 0.7 g of agarose gel in 100 mL of 0.2× TB buffer by heating the solution in a microwave (700 W, 1 min). Pour the melted agarose solution in a large gel tray. Place a 20 μL well comb in the gel. Let the agarose solidify for ~45 min.
2. *Sample loading.* Put the gel together with its tray into an electrophoresis chamber. Fill it with the 0.2× TB buffer. Gently remove the comb. Load 3 μL of GeneRuler DNA Ladder into the first and the last well. Pipette the DNA-only substrate (~200 ng) mixed with a 10× loading dye (final concentration 1×) into the second well. Load 10 μL of reconstituted chromatin samples mixed with the 10× loading dye (final concentration 1×) to subsequent wells.

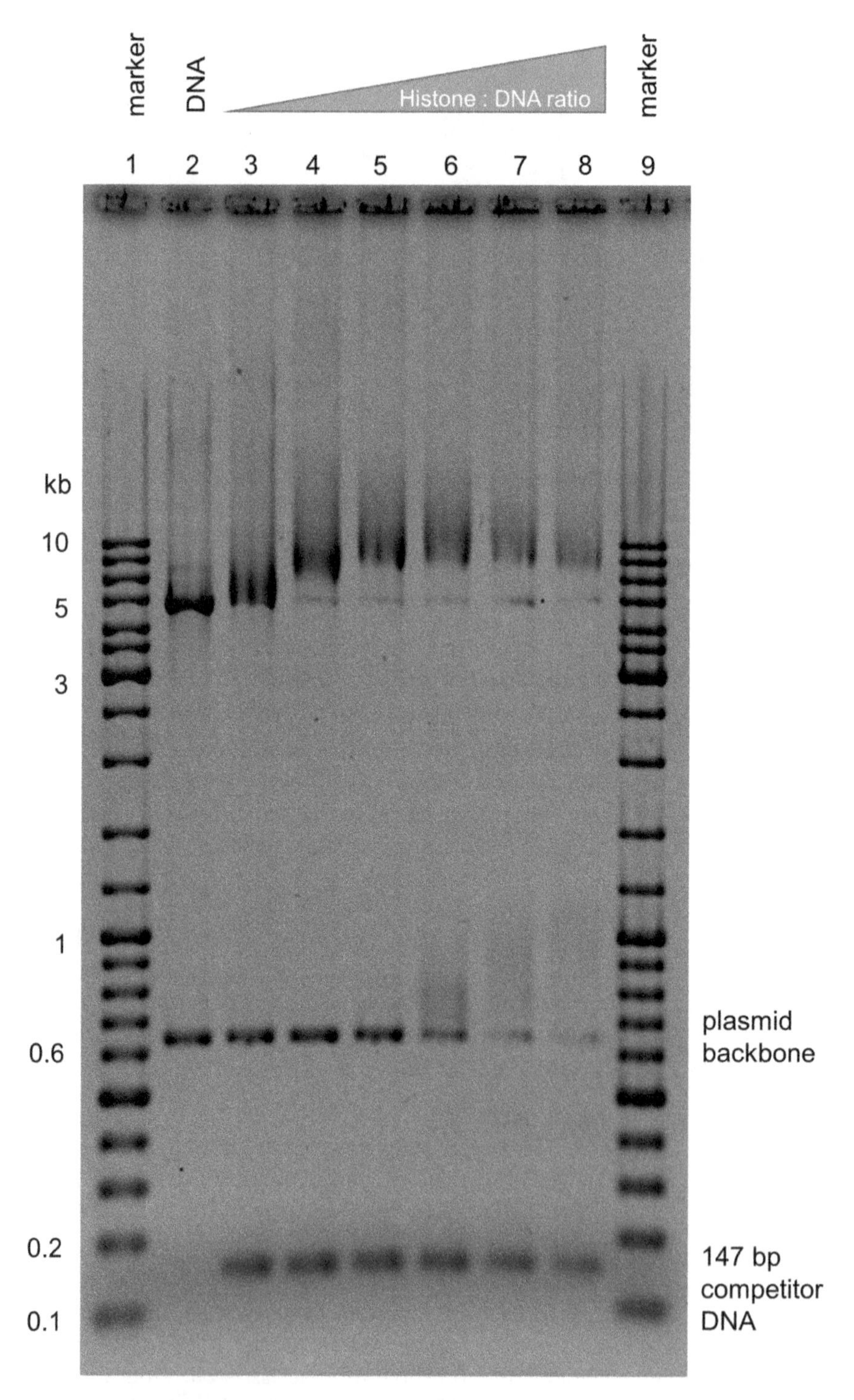

Fig. 4 Electrophoretic mobility shift assay of reconstituted chromatin fibers. With the increasing histone concentration, chromatin fibers migrate slower through the agarose gel. The consecutive bands on the gel (lanes 3–8) are gradually shifted with respect to the bare DNA band (lane 2). The plateau in the band shift reflects a full saturation of chromatin fibers and indicates that excessive histones start to assemble on the

3. *Running the electrophoresis.* Put the lid on the electrophoresis chamber, and connect to the power source. Run at 100 V for 2.5 h to ensure a good separation of bands.
4. *Staining with EtBr.* Remove the tray with the gel and place it carefully in a container with 1000 mL of 1× EtBr. Incubate while shaking for 30 min.
5. *Destaining.* Move the gel on its tray to a container with a distilled water and incubate for 15 min.
6. *Gel imaging.* Visualize the bands on the ChemiDoc UV. Select the optimal histone: DNA titration for single-molecule experiments. Choose the titration ratio with a smaller shift than the bare DNA that does not exhibit any excessive assembly on a competitor DNA (lane 5 in Fig. 4, *see* **Note 10**).

3.5 Assembly of the Flow Cell Chamber

A properly assembled flow cell chamber can be reused for numerous experiments when it is handled gently and is not contaminated during an assembly. Only the top 24 × 60 mm cover slip, with a functionalized surface, needs to be replaced.

1. *Cleaning.* Remove the top glass slide from the flow cell and any leftovers of PDMS from the Perspex molds and the aluminum holders. Wash thoroughly with miliQ H_2O and 2-propanol. Make sure that no PDMS remains inside the inlets of the molds and the holders. Dry under N_2 gas stream (Fig. 5a).
2. *PDMS preparation.* Mix PDMS (polydimethylsiloxane) base and curing agent in a 10:1 ratio (viscous and nonviscous component, respectively) (Fig. 5b). Prepare at least 2–3 mL per flow cell. Mix vigorously with a pipette tip for 2 min (*see* **Note 11**). Put under a vacuum chamber for 2 h in order to degas the viscous polymer.
3. *Assembling a flow cell chamber.* Insert a stiff electric wire into two 10-cm-long pieces of FEP tubing. Insert the tubes into a metal frame such that the electric wire protrudes out of the inner outlet of the holder, near the central aperture of the flow cell. Do the same on the other side of the flow cell (Fig. 5c, left). Put a clean glass cover slip (24 × 40 mm) onto the aperture. Mount the Perspex mold tightly onto the flow cell with M4 screws (Fig. 5c, right). Make sure that the electric wires protrude through the holes in the top of this mold and that the cover slip remains in its place. The end of the FEP tubing must remain inside the outlet near the aperture of the aluminum holder.

Fig. 4 (continued) non-601 DNA, resulting in additional band shift of the plasmid backbone and the competitor DNA (lanes 6–8). The sample in lane 5 exhibits a full saturation of the 601 arrays and no excessive assembly on the competitor DNA; therefore it was chosen for single-molecule experiments

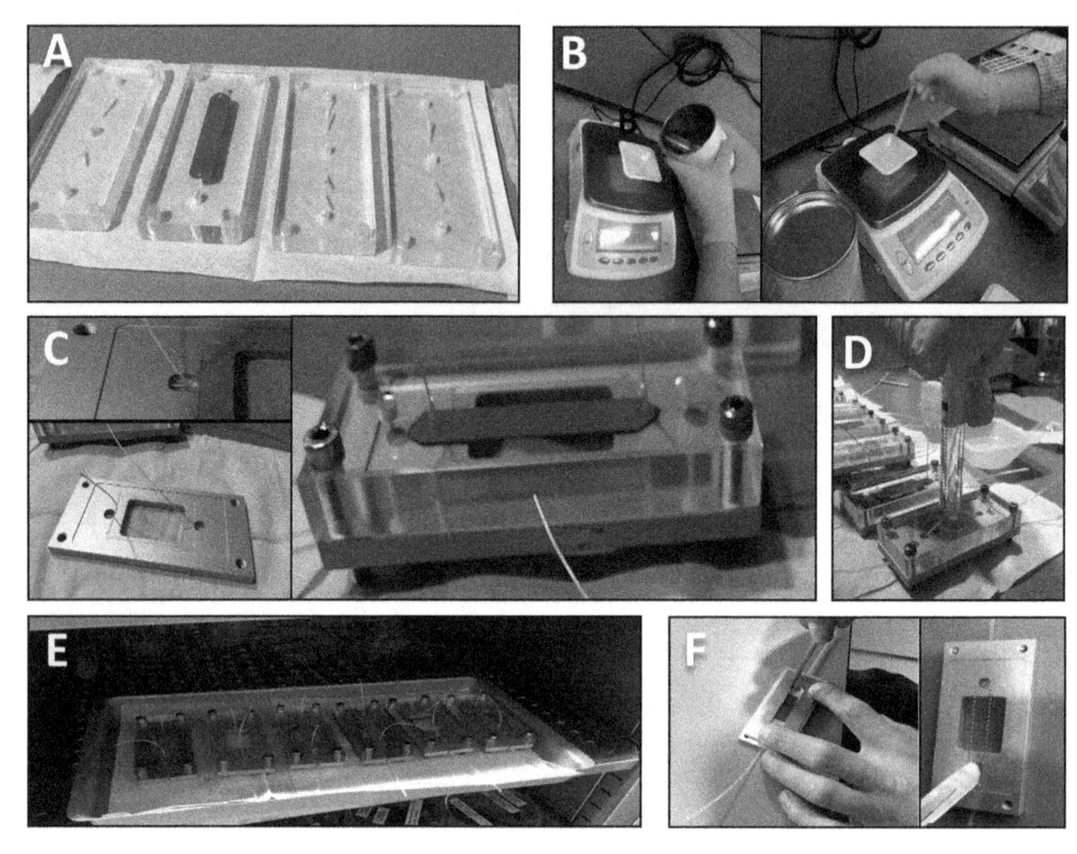

Fig. 5 Assembly of flow cell chambers. (**a**) Perspex molds are cleaned and dried. (**b**) The PDMS base and the curing agent are weighted, mixed, and degassed. (**c**) During the degassing process, aluminum flow cell holders, cover glasses, wiring, and Perspex molds are mounted. (**d**) The PDMS solution is injected into chambers. (**e**) Flow cells are cured at 65 °C. (**f**) Perspex molds are removed and a channel is cut out from a flat, cured PDMS layer (red dotted line indicates the part of PDMS that was cut out)

4. *Casting PDMS.* Gently inject the degassed PDMS with a syringe into the assembled flow chambers through one of the holes on top of the Perspex molds. Push the syringe slowly until the whole chamber is filled with the PDMS and the excess of polymer starts to leak out of the second hole in the mold (Fig. 5d). Prevent incorporating air bubbles into the chamber. Cure the assembly in an oven at 65 °C for at least 2 h to initiate the polymerization of PDMS (Fig. 5e).
5. *Cutting the channel in a cured PDMS.* Loosen the screws and gently remove the electric wires and the Perspex mold such that the FEP tubing and the cured PDMS remain intact. The large groove of the aluminum holder should be filled with a transparent, smooth, and solid layer of PDMS that covers its whole surface and, as a result, immobilizes the 24 × 40 mm cover slip firmly on the aluminum holder. Carefully cut a rectangular

channel in the PDMS layer above the cover slip (Fig. 5f) (*see* **Note 12**). The channel connects the two holes in the PDMS that are left after removing the electric wire from the FEP tubes.

3.6 Flow Cell Functionalization

A cover slip (24 × 60 mm) coated with anti-digoxigenin is put on top of the cured PDMS layer, closing the cut out channel. Buffer can be flushed in and out through the FEP tubes that connect the two sides of the flow channel. After an experiment, the cover slip can be replaced, and the flow cell may be reused for another experiment.

3.6.1 Cover Slip Functionalization

1. *Cleaning*. Put several 24 × 60 mm cover slips in a holder, and sonicate in 2-propanol for 15 min in an ultrasonic cleaner. Dry the cover slips under N_2 gas stream.
2. *Coating with nitrocellulose*. Clean a 100 mL beaker and a 1 mm microscopy slide with 2-propanol. Cover the bottom of the beaker with 5 mL of a 0.1% nitrocellulose in pentyl acetate. Put the slide vertically into the beaker, immersing its bottom edge. Gently bring a 24 × 60 mm cover slip into the liquid, parallel to the microscopy slide, and make use of capillary forces to fill the gap in between the two glasses with nitrocellulose solution. Remove the cover slip after 30 s and dry the glasses under a stream of N_2. Store the slides in a closed container with the coated side up.

3.6.2 Flow Cell Assembly Prior to an Experiment

1. *Cleaning*. When recycling a used flow cell, remove and discard the 24 × 60 mm cover slip with a scalpel. Clean the PDMS surface, the bottom 24 × 40 mm cover slip, and the FEP tubing with miliQ H_2O and subsequently with 2-propanol.
2. *Sealing the flow cell.* Put a new functionalized 24 × 60 mm cover slip in the middle of the PDMS layer, covering the channel, with the nitrocellulose-coated side facing inward. Gently press the glass to seal the flow cell channel. Connect one side of the FEP tubing to a 1 mL syringe. Insert the tubing on the opposite side of the flow cell into an Eppendorf tube containing 1 mL of miliQ H_2O, and suck the liquid into the flow cell. Prevent any air to get inside the flow channel.
3. *Incubation of anti-digoxigenin.* Insert 300 μL of 10 ng/μL anti-digoxigenin (dissolved in miliQ H_2O) into a flow cell the same way as in the previous step. Incubate for 2 h at 4 °C.
4. *Passivation of the glass slide surface.* Mix 1 mL of 4% BSA (dissolved in miliQ H_2O) with 50 μL of 2% Tween-20. Rinse the flow cell with the solution as previously. Seal the inlet and outlet tubes and incubate overnight at 4 °C. At this stage, the flow cell can be stored for maximally 1 week.

3.6.3 Tethering Chromatin Fibers

1. *Preparation of the measurement buffer.* Use a 10× concentrated solution of the stock buffer (Table 2), and prepare 10 mL of the measurement buffer (Table 3).
2. *Washing the flow cell.* Flush out the passivation buffer with 1 mL of the measurement buffer.
3. *Introducing DNA/chromatin.* Dilute 1 μL of reconstituted chromatin fibers (~20 ng/μL) in 500 μL of the measurement buffer. Gently flush in the chromatin into the flow cell with a 1 mL syringe. Incubate for 10 min at room temperature.
4. *Washing the magnetic beads.* Add 20 μL of magnetic beads from the stock to 20 μL of 1× TE in a low-retention Eppendorf tube, and mix by filling and releasing the pipette a couple of times. Pull down the beads from the solution using a magnet stand. Discard the supernatant. Resuspend the beads in 20 μL of 1× TE. Repeat the washing **step three** times.
5. *Attaching beads.* Dilute 1 μL of washed M270 or 0.5 μL of MyOne beads in 500 μL of the measurement buffer. Gently flush the beads into the flow cell with a syringe. Incubate for 10 min at room temperature.

Table 2
The composition of the 10× measurement buffer

Component		Final concentration
KCl	37.28 g	1 M
NaN_3	3.25 g	100 mM
Tween-20	5 mL	1%
1 M HEPES pH 7.5	50 mL	100 mM
mili Q	Fill up to 500 mL	

Table 3
The composition of the measurement buffer

Component		Final concentration
10× Measurement buffer	1 mL	100 mM KCl, 10 mM NaN_3, 10 mM HEPES pH 7.5, 0.1% Tween-20
20 mM $MgCl_2$	1 mL	2 mM
4% BSA	0.5 mL	0.2%
miliQ H_2O	7.5 mL	

3.7 Dynamic Force Spectroscopy on Chromatin Fibers

1. *Magnetic tweezer initialization.* Start up the magnetic tweezers microscope, including motor controllers, CMOS camera, and LED (Fig. 6). Move the magnet to its highest position relative to the objective.
2. *Mounting the functionalized flow cell on the microscope stage.* Put the flow cell on the stage such that the topside of the 24 × 60 mm cover slip is in focus (add immersion oil if an oil objective is used). Tighten the flow cell with screws. Connect one of the tubes to the pump and the other to a reservoir containing the measurement buffer. Rinse the flow chamber with the measurement buffer to remove loose beads. Use a minimal flow rate (<0.1 mL/min) to prevent sample degradation by excessive drag forces.

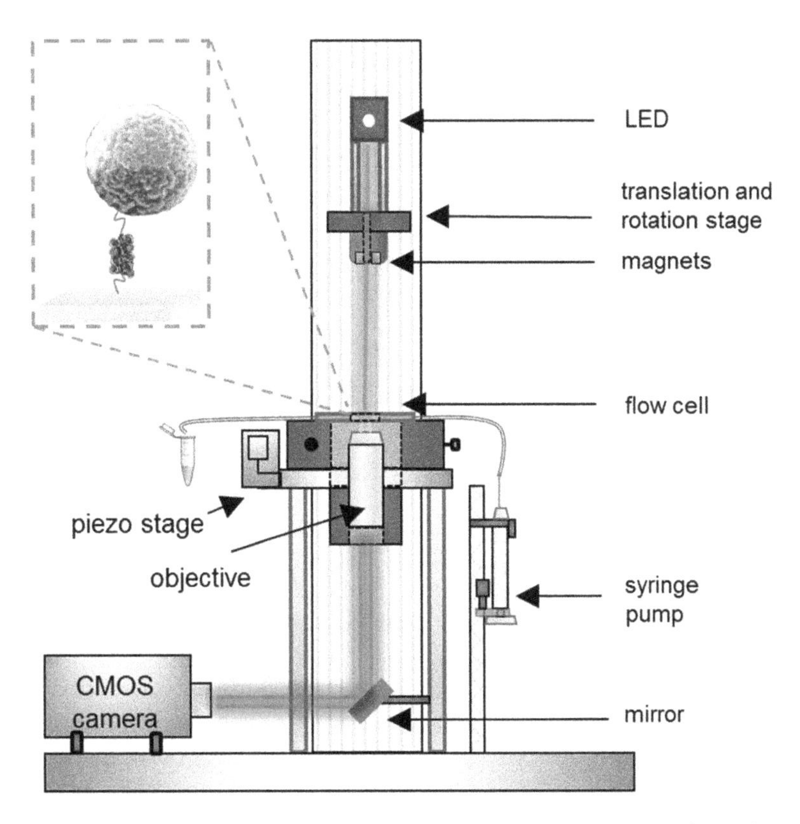

Fig. 6 Schematic representation of the magnetic tweezers setup (a front view with the cross section of the microscope stage). The sample is illuminated by a collimated LED light beam that is guided through the objective and reflected by the mirror toward the CMOS camera. Forces and torques are generated by two cubic magnets mounted on a translation and rotation stage. A piezo stage allows controlling the focus of the microscope. A syringe pump is used to exchange buffers in the flow cell during an experiment. Inset: a chromatin fiber flanked by short DNA handles, tethered in a flow cell between a cover slip and a magnetic bead (not to scale)

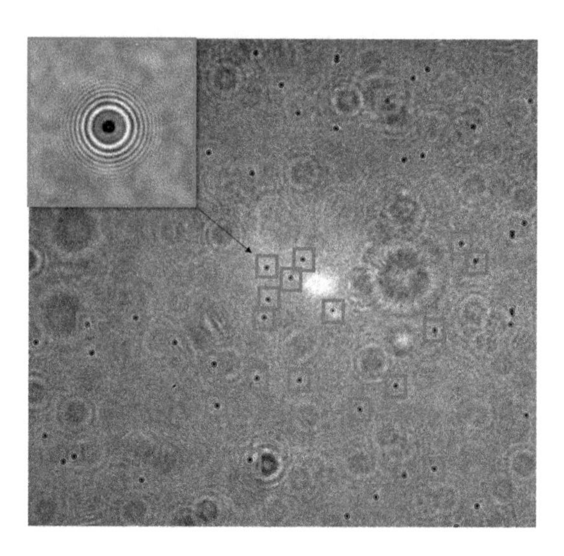

Fig. 7 Selection of beads before tracking. Distribution of tethered magnetic beads on the imaged region of a flow cell. Small regions of interest (ROIs) were selected to track the diffraction rings of individual magnetic beads. Inset: a zoom of one of the selected ROIs

3. *Adjusting the objective position.* Lower the objective until the beads appear out of focus and multiple diffraction rings are visible.
4. *Finding a field of view with tethered beads.* Move the microscope table manually to a region of the flow cell where multiple beads are distributed throughout the field of view (Fig. 7). Select regions of interest (ROIs) around the beads manually or using an automatic bead finder algorithm (Fig. 7, inset). Set a size of the ROIs such that multiple diffraction rings are included, depending on the bead size, objective magnification, and pixel size of the camera.
5. *Tracking calibration.* Using the piezo stage, move the sample through the focus, and record the diffraction ring pattern of a bead. Subsequent image analysis relates the objective position to the changes of the diffraction profile. Correct for the difference in refraction index between the aqueous flow channel and the immersion medium of the objective. For oil objectives, this is a factor 0.88 (*see* **Note 13**).
6. *Force calibration.* Calibrate the force as a function of the magnet height by measuring the lateral thermal fluctuations of a bead and the tether height at several magnet positions. Using equipartition theorem, the force can be calculated as the product of the bead height, Boltzmann's constant, and absolute temperature divided by the variance of the lateral position [25, 26] (*see* **Note 14**).

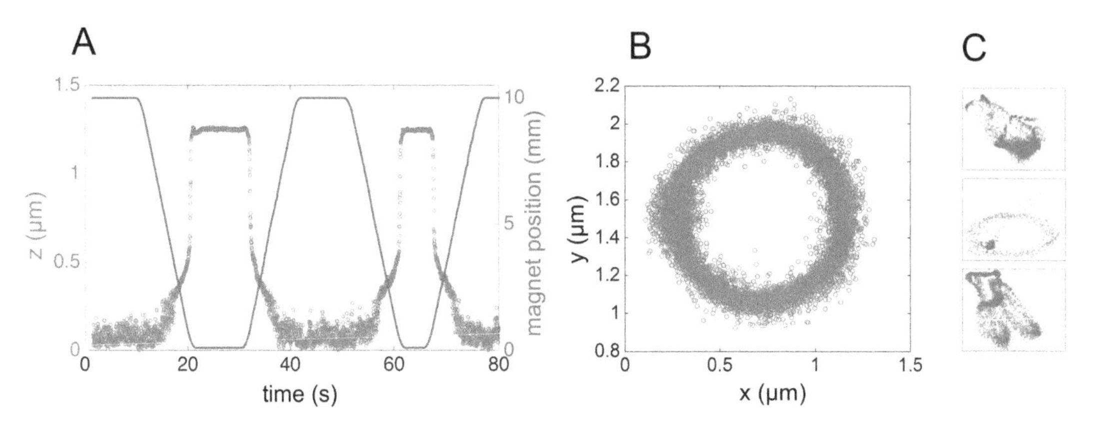

Fig. 8 Quality assessment of a selected tether. (**a**) Force-induced unfolding and refolding of a chromatin fiber tethered with a MyOne bead. Multiple stretching and refolding cycles with low forces (<10 pN) are not destructive to the sample. (**b**) A bead follows a circle in the X, Y plane upon magnet rotation unless more than one chromatin fiber is attached to a single bead. Note that the thicker spot on the left side of the circle results from an extended period before and after rotation of the magnets. (**c**) Anomalous XY tracks upon rotation of the magnets that indicate multiple tethers on a bead

7. *Setting the magnet trajectory to define a force ramp.* Define a trajectory of the magnet from a position where the force is close to 0 pN to a force up to ~7 pN during the first measurement. Prevent sample degradation due to excessive exposure to force. To maintain fiber integrity, minimize the duration of the high-force part in the force trajectory, so the fibers can refold to its initial state before histones dissociate (*see* **Note 15**). We use a magnet velocity of ~0.5 mm/s in both directions. Figure 8a shows the extension of a chromatin fiber tethered with a MyOne bead upon stretching twice with low forces.

8. *Rotation experiment for checking for individually tethered DNA molecules.* Subsequently, set the magnet position to a force of approximately 0.5 pN, and define a new trajectory that controls the magnet rotation. Turn the magnet at least 15 times in both directions. Because the tether is rarely attached to the very bottom of the bead, each bead trajectory should describe a circle in the X, Y plane, even when the DNA is not rotationally constrained (Fig. 8b). However, when more than one tether is attached to the same bead, the trajectory will deviate from a circle (Fig. 8c).

9. *Selection of good tethers.* Discard beads that exhibit too large extensions, indicating incompletely reconstituted chromatin fibers or bare DNA. Also discard too short tethers and beads that have irregular X, Y fluctuations caused by nonspecific sticking of the bead or the chromatin to the surface. The extension of the tether should be close to the contour length of the DNA substrate minus the length of the 601 array.

10. *Probing chromatin folding*. Fiber folding is reversible up to forces of ~7 pN. Multiple force trajectories can be applied to the same tethers.
11. *Complete chromatin unfolding*. Perform a high-force stretching experiment to induce the complete unpeeling of histone octamers/tetramers from the DNA template. This is informative as the unwrapping of the final wrap of DNA from the histone core results in distinctive 25 nm steps that are indicative for each nucleosome or tetrasome. Define a magnet trajectory to apply a force ramp from 0 to 50–60 pN (*see* **Note 16**) to observe these discrete steps in extension. The number of steps should correlate with the number of the 601 sequence repeats on the DNA template. Analyze the results (Subheading 3.8).
12. *Select another field of view*. Move to another field of view in the same flow cell. Make sure that the displacement is large enough to ensure that the beads in the new field of view have not been exposed to high forces during the previous experiment. Displacements should typically be larger than the magnet size, i.e., more than 5 mm.

3.8 Data Analysis

The force-extension graphs are characterized by transitions from a folded fiber through an array of partially unwrapped nucleosomes to fully unwrapped nucleosomes (Fig. 9a). We fit the force-extension curves with a statistical mechanics model developed by Meng et al. [27] (Fig. 9b). The model quantifies all conformational changes of DNA and chromatin fibers upon stretching. By fitting the data to the model, we obtain parameters like fiber's stiffness, unfolding energy, number of nucleosomes, and number of tetrasomes (Table 4).

The model contains a large number of parameters that capture the complexity of the fiber. It includes a degeneracy factor that is not easily implemented in analytical formulae, but is straightforward to implement numerically. We typically observe discrete variations of fiber compositions, indicating one or more missing or additional nucleosomes, despite best efforts to reconstitute the fibers. Moreover, the high-force transitions are not in equilibrium, whereas the model is only valid for equilibrium transitions. For these reasons, one cannot simply fit the curves using a standard Levenberg-Marquardt algorithm. Instead, model parameters are fit sequentially, using only the parameters which are relevant in a limited force regime. Subsequent fits on the same fiber in different force regimes use the thus obtained parameters as fixed constants.

1. *Correction for drift*. A difference in the extension between the beginning and the end of the time trace (Fig. 10a) may result from mechanical drift of the microscope (*see* **Note 17**). We use

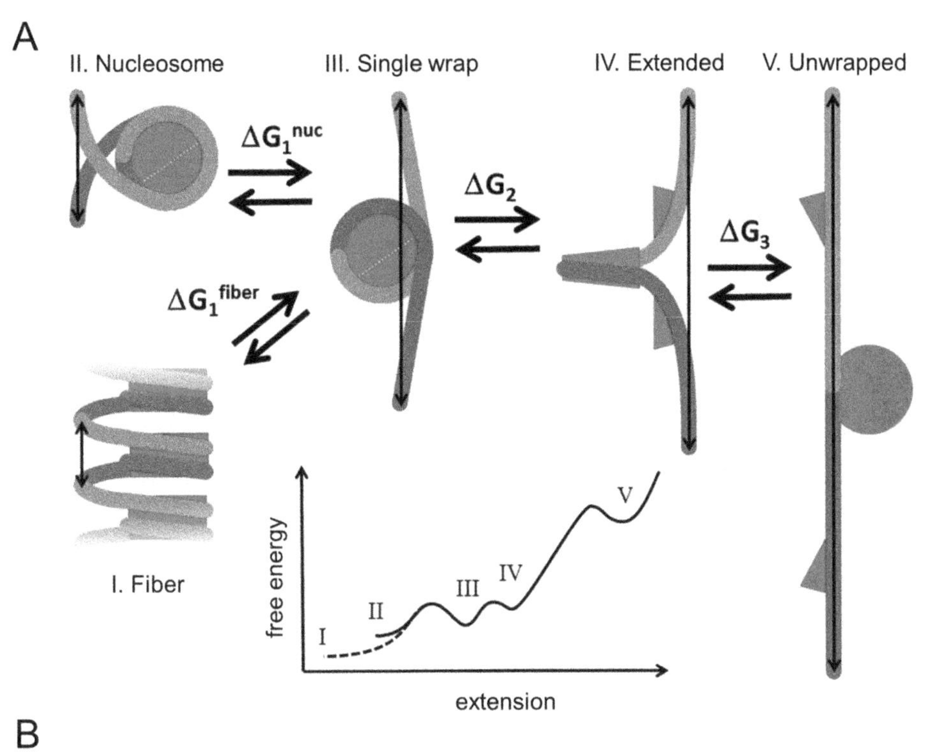

B

	extension	energy
DNA	$z_{DNA}(f,L) = L\left[1 - \frac{1}{2}\sqrt{\frac{k_BT}{fA}} + \frac{f}{S}\right]$	$G_{DNA}(f,L) = -L\left[f - \sqrt{\frac{fk_BT}{A}} + \frac{f^2}{2S}\right] + fz_{DNA}$
fiber	$z_{fiber}(f) = \frac{f}{k} + z_0$	$G_{fiber}(f) = \frac{f^2}{2k}$
total	$z_{tot}(f) = \sum_i n_i z_i(f) + z_{DNA}(f)$	$G_{tot}(f) = \sum_i n_i G_i(f) + G_{DNA}(f)$

degeneracy	$D(\text{state}) = \prod_{i<j} \binom{n_i + n_j}{n_i}$

mean equilibrium extension	$\langle z_{tot}(f)\rangle = \frac{\sum_{states} z_{tot}(f) D(state) e^{-(G_{tot} - Fz_{tot})/k_BT}}{\sum_{states} D(state) e^{-(G_{tot} - Fz_{tot})/k_BT}}$

Fig. 9 Quantification of force-induced structural transitions in chromatin fibers using a statistical mechanics model (figure adapted from Meng et al. [27]). (**a**) Fiber stretching proceeds through four stages of nucleosome unfolding. (**b**) Analytical formulas that describe the extension of each nucleosome state as a function of force. *A* persistence length (nm), *f* force (pN), *G* free energy (k_BT), *k* stiffness (pN/nm), k_B Boltzmann's constant, *z* tether extension (nm), *L* contour length (nm), *S* DNA stretching modulus (pN), *T* temperature (K), *D* degeneracy factor, n_i number of nucleosomes in a particular state *i*

Table 4
Fixed parameters of the statistical mechanics model used in data analysis

Parameter	
DNA persistence length	50 nm
DNA stretch modulus	1100 pN
Contour length	Length of the entire DNA substrate (bp)
Nucleosome Repeat Length (NRL)	Length of the 601 repeat (bp)
Folded length of a single nucleosome	1.5 nm
Unwrapped base pairs of the 1st (outer) nucleosomal turn	56 bp
Unwrapped length in the intermediate transition	5 nm

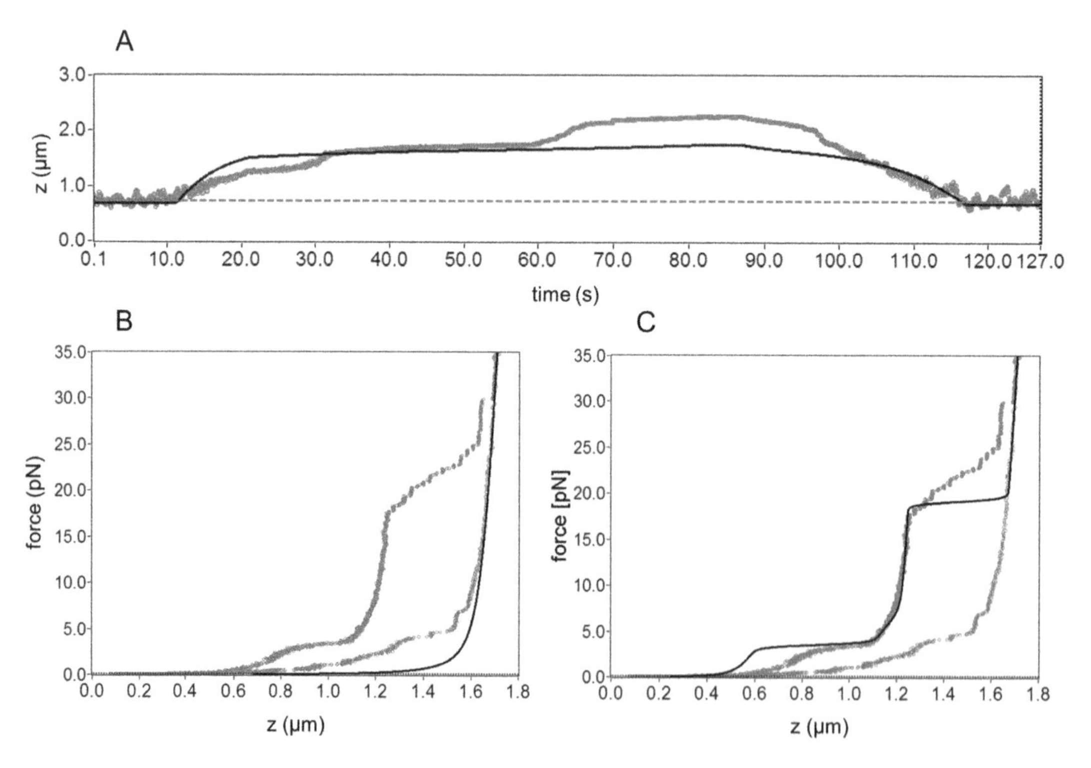

Fig. 10 Initial procedures of the data analysis. (**a**) Quantifying the mechanical drift. Data points from the beginning and the end of the time trace (red) are recorded at the same magnet position and should align with an arbitrary straight, horizontal line. The difference in height of selected regions before and after the alignment is divided by the duration of the measurement to estimate the linear drift factor (nm/s). (**b**) Finding the z-offset. Data points with the highest z-coordinate (red) are aligned with the fitted WLC model with a persistence length of 50 nm and a given contour length (Table 4). (**c**) Quantifying the number of nucleosomes. The staircase-like curve is selected (red), and the distance between the first and the last step is measured. The obtained extension is divided by 25 nm to estimate the number of nucleosomes and tetrasomes assembled on the tether

a linear drift term to enforce maximal overlap and a horizontal trend in the extension versus time plot.

2. *Determining the offset.* Determine the offset in the z-direction by fitting an extensible WLC with fixed contour length, persistence length, and stretch modulus (Table 4) to the most extended part of the curve, i.e., at forces above 30 pN (Fig. 10b).
3. *Fitting the number of nucleosomes.* Start with the number of nucleosomes equal to the number of tandem repeats of the Widom-601 sequence on the DNA used for chromatin assembly. Adjust or fit the number of nucleosomes to obtain the best fit in the range between 8 and 15 pN (Fig. 10c) (*see* **Note 18**). When fitting, the number of nucleosomes should approximate an integer number.
4. *Fitting the unfolding steps.* Check that the number of steps in the force range between 8 and 40 pN agrees with the number of nucleosomes. A Student's t-test is used to compare the 3–6 neighboring data points to identify discrete steps (Fig. 11a). The step size corresponds with the difference between the extension of the last partially wrapped nucleosome and the fully unwrapped nucleosome conformation, i.e., states IV and V in Fig. 9a (*see* **Note 19**).
5. *Fitting the low-force regime.* Fit the part of the trace between 0.5 and 8 pN to obtain the fiber stiffness, unstacking free energy and the free energy of the intermediate transition (k_BT) (Fig. 11b). The plateau width at 3–5 pN depends on the number of nucleosomes that stack and form a folded fiber. We observed that the plateau width is not always consistent with the number of nucleosomes as determined from the high-force region, described in the previous step [27]. In all experiments, the plateau size indicated that the number of stacked nucleosomes is less or equal to the number of high-force unwrapping events. The difference is attributed to tetrasomes that feature the same unwrapping steps at high force but lack the ability to fold into more condensed structures [5]. The number of tetrasomes is also fitted from the low-force region.

4 Notes

1. We use plasmid constructs with multiple repeats of the Widom-601 sequence (tggagaatcccggtgccgaggccgctcaattggtcgtagacagc tctagcaccgcttaaacgcacgtacgcgctgtcccccgcgttttaaccgccaaggggat tactccctagtctccaggcacgtgtcagatatatacatcctgt) that are spaced with 20 bp or 50 bp of non-601 linker DNA (nucleosome repeat length = 167 bp or 197 bp, respectively). In addition

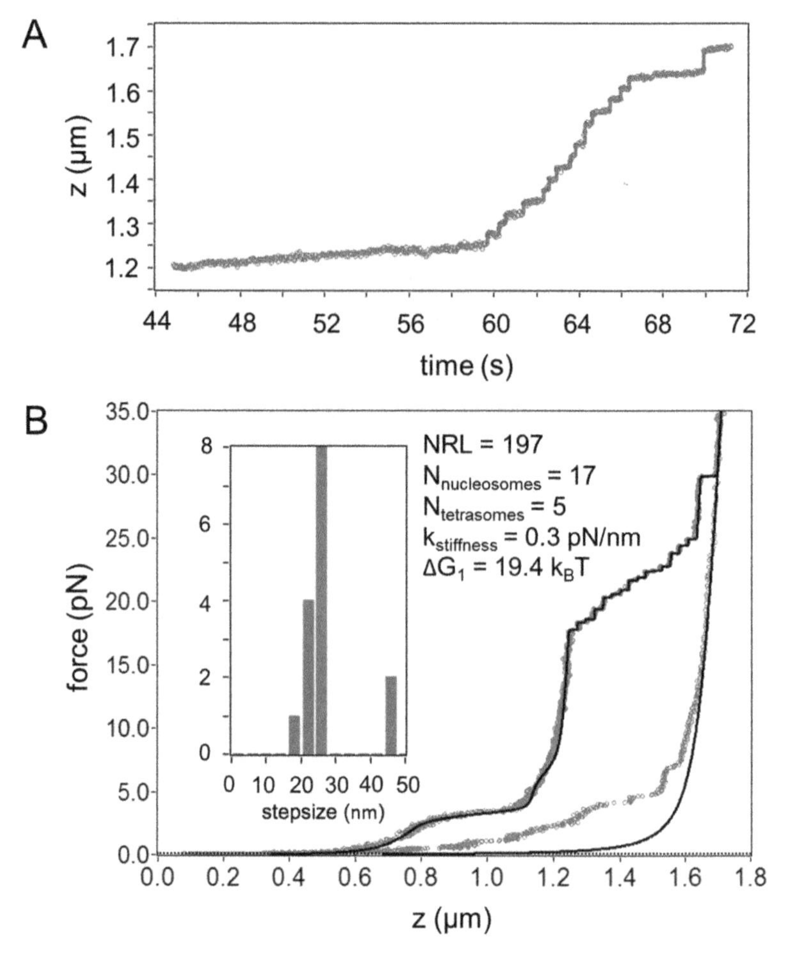

Fig. 11 Fitting the data to the statistical mechanics model. (**a**) Step size analysis. The same part of the trace remains selected as in Fig. 10c. A Student's t-test is performed to quantify the sizes of steps that occur in the 10–40 pN force regime. (**b**) Fitting the plateau of the force-extension curve at the low-force regime to characterize the chromatin fiber folding. Inset: histogram of the measured step sizes and fitted parameters – *NRL* nucleosome repeat length, $N_{nucleosomes}$ fitted number of assembled nucleosomes and tetrasomes, $N_{tetrasomes}$ fitted number of tetrasomes, $k_{stiffness}$ stiffness of a folded fiber (pN/nm), ΔG_1 free energy of the first transition ($k_B T$)

to the 601 repeats, we keep about 1 kbp of flanking DNA on both sides, yielding the following DNA substrates:

15 × 601_167–5185 bp

15 × 601_197–5635 bp

30 × 601_167–7690 bp

25 × 601_197–7605 bp

The flanking DNA helps to keep the chromatin fiber away from the surface of the flow chamber or the bead and facilitate selecting tethers that are not stuck to one of the surfaces. On the downside, additional nucleosomes can form on this DNA, putting high demands on the reconstitution stoichiometry.

2. When handling frozen competent cells, it is important to keep them cold. Frozen cells are very sensitive to temperature fluctuations, and therefore plasmid transformation may not work if the cells are not incubated on ice.
3. The reaction volume for digestion of 1 μg of isolated plasmid DNA is 20–50 μL. A minimum of 5–10 units of each enzyme is required for complete digestion of 1 μg of DNA in 1 h. For an overnight incubation, the amount of enzyme can be therefore lower. Enzyme should not exceed 10% of the total reaction volume.
4. The maximum binding capacity is approximately 40 μg per column. Multiple elutions from the spin columns and longer incubation time are recommended to improve the yield of DNA purification.
5. Selection of nucleotides is based on the sequences of the "sticky end" resulting from an enzymatic digestion. A regular, non-tagged nucleotide other than UTP should be complementary to the first nucleotide at one side of the digested DNA substrate. Absence of this nucleotide in the first Klenow reaction prevents incorporation of tagged UTP on the other side of the DNA substrate. The presence of other nucleotides could result in labelling the DNA with the same tag on both sides. This would not allow for DNA tethering in a flow cell as the bead and the surface have orthogonal binding proteins.
6. Typically, we use a 4535 bp DNA construct containing about 2200 bp of 601 sequence repeats. The rest is a random sequence of the plasmid backbone. A typical DNA concentration after labelling is ~250 ng/μL. Since the nucleosome contains roughly equal weights of protein and DNA, one can easily set up a good titration scheme:
   (a) Use 4 μL of DNA per dialysis tube (in total 1000 ng of DNA).
   (b) DNA handles are 50% of the total DNA; therefore in total there is 500 ng of 601 DNA that should fold into nucleosomes. This amount divided by the molar mass of a single 601 repeat (~10^5 g/M) and then by the total volume ($5*10^{-6}$ l) results in ~ 0.1 μM molar concentration of the DNA substrate that is supposed to form nucleosomes.

(c) We use 200 ng/μL aliquots of histone octamer, so 2.5 μL of this aliquot needs to be added to end up with a solution with 500 ng of protein.

(d) Add high salt buffer for samples to obtain a final volume of 50 μL.

(e) Prepare tubes with the same amount of DNA, and vary the volume of histones (1.5, 2, 2.5, 3, 3.5 μL) to accommodate a loss of histones by sticking to tubes and pipette tips and uncertainties in histone and DNA concentrations.

7. Keep all the buffers at 4 °C and perform the dialysis in a cold room.
8. Low-binding (siliconized) pipette tips and tubing prevent sticking of histone octamers and reconstituted chromatin fibers. This is especially important when small volumes are used.
9. To enforce that the volume in the beaker with dialysis tubes remains constant, make sure that the same amount of the low salt buffer is pumped into the beaker as the high salt buffer is pumped out. This can conveniently be achieved using a peristaltic pump with a double head.
10. Competitor DNA starts to reconstitute into nucleosomes when all 601 sequences are occupied with histones. A band shift of the competitor DNA therefore indicates saturation of the chromatin fibers. Fibers with a minimal oversaturation, i.e., having a small excess of histones assembled on the flanking DNA around 601 repeats, are preferred for magnetic tweezers experiments over subsaturated, incompletely folded fibers.
11. Mix the viscous and nonviscous components of PDMS intensively for at least 2 min. Incomplete mixing will impede homogenous polymerization.
12. Make sure that the FEP tubing does not get loose while removing the Perspex molds and the electrical wires. After its removal, the flow cell should contain two fixed pieces of the FEP tubing and a flat PDMS surface that completely covers and immobilizes the 24 × 40 mm cover slip in the central aperture of the holder. The cured PDMS is a flat, transparent layer that fills the cavity between the aluminum holder and the Perspex mold. Some Perspex molds have a blue embossing tape stripe attached to its bottom side. The void that is left by the embossing tape after curing the PDMS and the removal of the Perspex mold forms the actual flow channel. This channel can be increased in height by cutting out the remaining layer of cured PDMS that was left underneath the embossing tape, with a scalpel. When no embossing tape is used, the entire channel can be cut out at the location where the embossing

tape would be. The channel dimension should be smaller than the dimensions of the cover slip in order to keep the glass in its place. The aluminum holder remains an intrinsic part of the flow cell for convenient mounting on the microscope and connections with tubing.

13. Various methods have been published to quantify changes in the diffraction pattern of the bead [28–30]. We use the 3D FFT algorithm to correlate the measured ROIs with a computer-generated model image (Brouwer et al., in prep.). The empirical relation between the objective position and the diffraction pattern is interpolated for tracking the bead in the z-direction.

14. For a particular magnetic tweezer setup and magnets, the force calibration only needs to be performed once. As chromatin features large changes in extension as a function of force, which impedes thermal fluctuations, the calibration is done with a bare DNA molecule, prior to chromatin experiments. Typically, a double exponential decay of the force with magnet height is obtained, which is characteristic for a given batch of beads and a particular pair of magnets with a fixed configuration. In force spectroscopy experiments on chromatin, the force-magnet position relation for the appropriate bead size is used to convert magnet position to force.

    The Invitrogen beads feature a rather constant magnetic content. The variation is sufficiently small to use the same force calibration for all beads from the same batch. Other beads may have larger variations in magnetic strength.

    Our flow cells have a reproducible height. The offset of the magnet position can be obtained from measuring the height at which the magnets touch the top of the flow channel. This is easily recognized as beads will abruptly move out of focus.

15. A first experiment involves checking the quality of tethers. Some beads are directly stuck to the surface. Others feature multiple tethers or nonspecific interactions between the chromatin and surface. An initial force ramp is used to select tethers that feature force-extension relation that is indicative of a proper chromatin fiber.

    In the buffer conditions used in our method and at the pulling rate lower than 0.2 pN/s, the structural transitions in the fiber remain in equilibrium. Unstacking of the nucleosomes in the chromatin fiber can be reversed upon releasing the force. This may not be the case when a chromatin fiber is not properly folded (a hysteresis is observed between a pulling and a release curve) or in other experimental conditions. Importantly, extended exposure to forces above 3 pN should be avoided, as histone H2A/H2B dimers tend to dissociate above this threshold.

16. The pulling rate is higher in the high-force regime (above 10 pN) compared to the low-force regime. Depending on the pulling rate, unwrapping steps occur at different forces.
17. Mechanical drift can be measured more precisely by tracking the z-extension of a bead stuck to the flow cell surface. When the z-position of such a stuck bead does not remain constant, its position can be subtracted from other extension traces to correct for drift.
18. Individual chromatin fibers may have more or less nucleosomes than the number of 601 repeats on the DNA template. The measured change in extension between the first unwrapping step and the last unwrapping step, divided by 25 nm, should correspond with the fitted number of nucleosomes. A force plateau at 3–5 pN that is smaller than expected indicates the presence of tetrasomes, next to nucleosomes.
19. The average step size should be close to 25 nm. Occasionally, two unwrapping events can occur simultaneously, resulting in a double-sized step. When the number of nucleosomes obtained in the previous step does not match the number of steps observed in this force region, the amount of wrapped DNA and/or the extension of the extended conformations can be adjusted. Typically, step sizes and extensions are within 0.5 nm of the average values obtained from multiple experiments. In this force regime, the non-equilibrium steps in the force-extension curves and all points in between are assigned to discrete levels. Each level corresponds to a WLC with a contour length that is reduced by an integer number times the amount of DNA that is wrapped in a single-wrapped nucleosome or tetrasome.

Acknowledgments

We are grateful to Nicolaas Hermans, He Meng, Kurt Andresen, Orkide Ordu, and Ineke de Boer for the help in establishing these methods. This work was supported by the Netherlands Organisation for Scientific Research (NWO/OCW), as part of the Frontiers of Nanoscience program, by the NWO-VICI research program, project 680-47-616.

References

1. Kornberg RD (1977) Structure of chromatin. Ann Rev Biochem 46:931–954
2. Richmond RK, Sargent DF, Richmond TJ, Luger K, Mader AW (1997) Crystal structure of the nucleosome resolution core particle at 2.8 A. Nature 389:251–260
3. Dorigo B et al (2004) Nucleosome arrays reveal the two-start organization of the chromatin fiber. Science 306:1571–1573
4. Chen Q, Yang R, Korolev N, Fa Liu C, Nordenskiöld L (2017) Regulation of nucleosome stacking and chromatin compaction by the

histone H4 N-terminal tail -H2A acidic patch interaction. J Mol Biol 429(13):2075–2092
5. Kaczmarczyk A, Allahverdi A, Brouwer TB, Nordenskiöld L, Dekker NH, van Noort J (2017) Single-molecule force spectroscopy on histone H4 tail cross-linked chromatin reveals fiber folding. J Biol Chem 292:17506–17513
6. Gilbert N, Ramsahoye B (2005) The relationship between chromatin structure and transcriptional activity in mammalian genomes. Brief Funct Genomic Proteomic 4:129–142
7. Workman JL, Kingston RE (1998) Alteration of nucleosome structure as a mechanism of transcriptional regulation. Annu Rev Biochem 67:545–579
8. Widom J (2001) Role of DNA sequence in nucleosome stability and dynamics. Q Rev Biophys 34:269–324
9. Bowman GD, Poirier MG (2015) Post-translational modifications of histones that influence nucleosome dynamics. Chem Rev 115:2274–2295
10. Luger K, Richmond TJ (1998) The histone tails of the nucleosome. Curr Opin Genet Dev 8:140–146
11. Schalch T, Duda S, Sargent DF, Richmond TJ (2005) X-ray structure of a tetranucleosome and its implications for the chromatin fibre. Nature 436:138–141
12. Grigoryev SA (2012) Nucleosome spacing and chromatin higher-order folding. Nucleus 3:493–499
13. Robinson PJJ, Fairall L, Huynh V a T, Rhodes D (2006) EM measurements define the dimensions of the '30-nm' chromatin fiber: evidence for a compact, interdigitated structure. Proc Natl Acad Sci U S A 103:6506–6511
14. Routh A, Sandin S, Rhodes D (2008) Nucleosome repeat length and linker histone stoichiometry determine chromatin fiber structure. Proc Natl Acad Sci U S A 105:8872–8877
15. Song F et al (2014) Cryo-EM study of the chromatin fiber reveals a double helix twisted by tetranucleosomal units. Science 344:376–380
16. Krzemien KM et al (2017) Atomic force microscopy of chromatin arrays reveal non-monotonic salt dependence of array compaction in solution. PLoS One 12:e0173459
17. Kruithof M, Chien F, de Jager M, van Noort J (2008) Subpiconewton dynamic force spectroscopy using magnetic tweezers. Biophys J 94:2343–2348
18. Ordu O, Lusser A, Dekker NH (2016) Recent insights from in vitro single-molecule studies into nucleosome structure and dynamics. Biophys Rev 8:33–49
19. Ribeck N, Saleh OA (2008) Multiplexed single-molecule measurements with magnetic tweezers. Rev Sci Instrum 79:94301
20. De Vlaminck I et al (2011) Highly parallel magnetic tweezers by targeted DNA tethering. Nano Lett 11:5489–5493
21. Lusser A, Kadonaga JT (2004) Strategies for the reconstitution of chromatin. Nat Methods 1:19–26
22. Huynh VAT, Robinson PJJ, Rhodes D (2005) A method for the in vitro reconstitution of a defined '30 nm' chromatin fibre containing stoichiometric amounts of the linker histone. J Mol Biol 345:957–968
23. Lowary PT, Widom J (1998) New DNA sequence rules for high affinity binding to histone octamer and sequence-directed nucleosome positioning. J Mol Biol 276:19–42
24. Flaus A (2011) Principles and practice of nucleosome positioning in vitro. Front Life Sci 5:5–27
25. Strick TR, Allemand JF, Bensimon D, Bensimon A, Croquette V (1996) The elasticity of a single supercoiled DNA molecule. Science 271:1835–1837
26. Yu Z et al (2014) A force calibration standard for magnetic tweezers. Rev Sci Instrum 85:123114
27. Meng H, Andresen K, van Noort J (2015) Quantitative analysis of single-molecule force spectroscopy on folded chromatin fibers. Nucleic Acids Res 43:3578–3590
28. Harada Y et al (2001) Direct observation of DNA rotation during transcription by Escherichia coli RNA polymerase. Nature 409:113–115
29. Lansdorp BM, Tabrizi SJ, Dittmore A, Saleh OA (2013) A high-speed magnetic tweezer beyond 10,000 frames per second. Rev Sci Instrum 84:44301
30. Cnossen JP, Dulin D, Dekker NH (2014) An optimized software framework for real-time, high-throughput tracking of spherical beads. Cit Rev Sci Instruments 85:103712

Chapter 19

Single-Molecule and Ensemble Methods to Probe Initial Stages of RNP Granule Assembly

Jaya Sarkar and Sua Myong

Abstract

Ribonucleoprotein (RNP) granules are membraneless organelles, consisting of high local concentrations of RNA and proteins bearing intrinsically disordered regions (IDRs). They are formed by liquid-liquid phase separation (LLPS). In neurodegenerative diseases such as ALS, mutations in granule proteins such as FUS and TDP-43 accelerate abnormal liquid to solid transition of RNP granules, leading to formation of fiber-like structures. Methods to study granules must be carefully selected based on the stage of granule's life. Here we describe a strategic combination of single-molecule biophysical and ensemble biochemical techniques that may be employed to extract insightful information about early stages of RNP granule formation. Protein-RNA interaction and stoichiometry of the complex in the early soluble stage of RNP assembly can be probed by *single-molecule FRET (smFRET) assay* and *electrophoretic mobility shift assay (EMSA)*, respectively. RNP-RNP interaction that likely contributes to RNP nucleation can be reported on by a smFRET-based RNA *annealing assay*. The next stage in the assembly pathway, that is, phase separation from diffused to liquid-like droplets, may be monitored by a *phase separation assay*. Finally, RNP granules isolated from mammalian cells can be investigated using a unique *single-molecule pull-down (SiMPull) assay*.

Key words RNP granules, Single-molecule FRET (smFRET), Phase separation assay, Electrophoretic mobility shift assay (EMSA), Single-molecule pull-down (SiMPull), LAF-1

1 Introduction

Stress granules (SGs) are a subclass of RNP granules that assemble in eukaryotic cells under stress. They are composed of RNAs and RNA-binding proteins (RBPs) [1]. Many of the SG proteins such as FUS and TDP-43 contain long stretches of intrinsically disordered regions (IDRs) and RNA recognition motifs (RRMs). The low complexity and the nature of the amino acid composition of the IDRs render the SG proteins highly interactive. Multivalent RNA-protein and protein-protein interactions drive liquid-liquid phase separation (LLPS), forming liquid droplet-like SGs [2]. However, nature of SGs can change, potentially losing their fluidity and maturing to hydrogels and eventually to more solid-like fibrillar structures that are a hallmark of neurodegenerative diseases [3].

Yuri L. Lyubchenko (ed.), *Nanoscale Imaging: Methods and Protocols*, Methods in Molecular Biology, vol. 1814, https://doi.org/10.1007/978-1-4939-8591-3_19, © Springer Science+Business Media, LLC, part of Springer Nature 2018

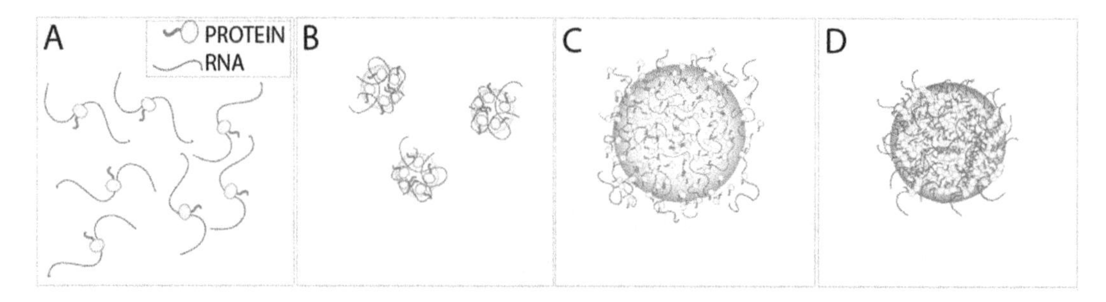

Fig. 1 Stages of granule development. (**a**) Soluble, (**b**) nucleation, (**c**) LLPS, (**d**) gel or solid-like

Indeed, FUS and TDP-43 inclusions are associated with disease pathology in ALS patients. Aberrant liquid-solid transition is accelerated by ALS-linked mutations, numerous of which have been identified in these proteins [4]. Example of RNP granules in other organisms include P granules in *C. elegans* [5]. Similar to SGs, P granules are also composed of IDR proteins such as LAF-1 (a DEAD-box helicase). Understanding the molecular mechanisms that contribute to disease onset and progression requires identifying the molecular differences between the normal wild-type SG proteins and the disease mutants, starting from the very early stages of granule assembly.

In this chapter, we describe single-molecule biophysical and ensemble biochemical techniques, an appropriate blend of which may be employed to probe early-stage molecular events of RNP granule formation. Furthermore, this combinatory approach enables connecting molecular mechanisms with biophysical properties of in vitro droplets and cellular granules. We envision the early stages to entail (a) *the soluble phase*, where proteins and RNA are soluble and remain diffused; (b) *the nucleation phase*, when RNA-protein, protein-protein, and RNP-RNP interactions tune RNP assembly; (c) *the growth phase*, in which nucleated granules continue to grow; and (d) *the maturation phase*, in which granules start losing their liquid nature and convert to solid-like aggregates, a process likely accelerated by aberrant interactions when IDR granule proteins bear disease mutations (Fig. 1). Accordingly, we categorize our methods as per their utility in probing each of these stages. In addition, we also describe a unique single-molecule method that can probe granules isolated from cells at the early stages of nucleation.

1.1 Phase Separation

This assay clearly defines the phase boundary, moving from the soluble to the phase-separated space. It allows probing biophysical properties of RNP droplets, shedding light into both the *growth and maturation phase* of the droplets. Recombinant protein expression constructs must be carefully designed for this assay. Since IDR proteins are inherently prone to aggregation, in some cases, a protein solubility-enhancing tag such as MBP or GST may need to be fused to the gene encoding the IDR protein, via a linker that encodes a protease cleavage site. This will enable purification of the soluble protein. In these cases, when conducting phase separation assays [6], the solubility tag can be cleaved off by using an appropriate protease. Once the protein is liberated from its tag, it is expected to start phase separating and forming demixed liquid droplets, under the correct buffer conditions. These droplets can then be microscopically imaged. Scoring positive or negative for droplet formation can generate a phase diagram [7]. More detailed analyses such as assessment of droplet number, size, and growth rate may also be conducted. Furthermore, FRAP (fluorescence recovery after photobleaching) assay may be employed to determine droplet fluidity, a change of which over time will provide insight into the process of droplet "aging." Hence, the phase separation assay may be used to test droplet tuning parameters including salt and protein concentration, RNA, ATP, and temperature.

1.2 smFRET

Interaction of RNA with the IDR proteins has been demonstrated to be a critical tuning parameter for both in vitro droplets and cellular granules [7–11]. In order to understand such interactions, it is important to first probe the nature of the RNA-protein interaction in the *soluble phase*. For this, we employ a smFRET binding assay based on total internal reflection fluorescence (TIRF) microscopy (Fig. 2) [12]. An RNA substrate that is dual labeled with a pair of FRET fluorophores is immobilized on a PEG-passivated quartz slide (that is used on the TIRF microscope) via biotin-NeutrAvidin linkage. Changes in FRET signal, as indicated by distance changes between the fluorophores, can report not only on protein binding to the RNA but also on the RNA conformational changes that are being induced by protein binding. IDR protein-induced

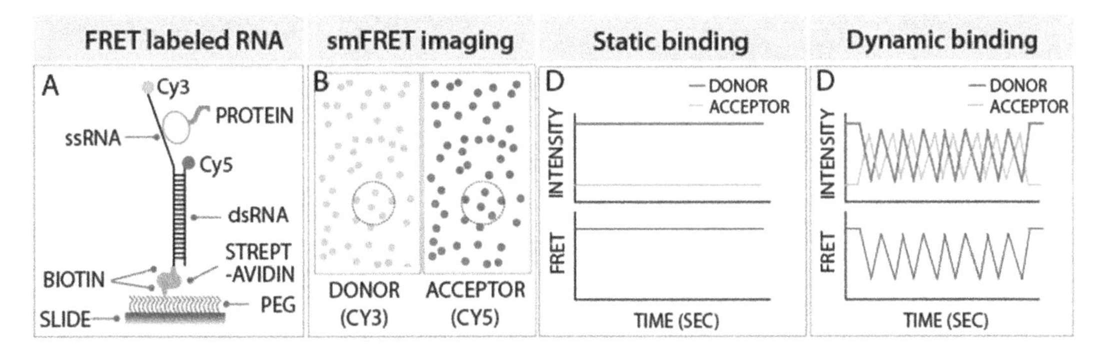

Fig. 2 smFRET detection. (**a**) FRET-RNA or DNA substrate immobilized to PEG surface. (**b**) Cy3 and Cy5 signal from same set of molecules (circle). (**c**, **d**) Individual smFRET traces reporting static (**c**) versus dynamic protein binding to single strand RNA

conformational dynamics on the RNA have been liked to granule fluidity [7, 8]. Thus, this assay can report on differences in protein-induced RNA dynamics between a wild type and a disease mutant variant of an SG protein, thereby identifying early-stage molecular differences that may impact eventual disease onset. These early differences can be potential target for drug design.

1.3 EMSA

EMSA has conventionally been used to characterize protein-nucleic acid interactions [13]. We utilize this assay to probe the stoichiometry of the protein-RNA complexes that form in the *soluble phase* preceding LLPS. By mixing a fluorophore-labeled RNA (same as the smFRET assay) and the IDR protein in an appropriate reaction buffer, we can monitor the different species of varying sizes formed, based on their differential migration on native gel electrophoresis (Fig. 3). Stoichiometric information obtained from EMSA can be co-related to RNA conformational dynamics information from smFRET assay, thereby providing insights into the stoichiometry and nature of soluble phase interactions that may ultimately govern properties of the phase-separated space.

1.4 RNA Annealing

RNP-RNP interaction is a critical factor regulating granule properties both pre and post granule *nucleation phase*. We describe here an smFRET-based annealing assay [8] that can report on

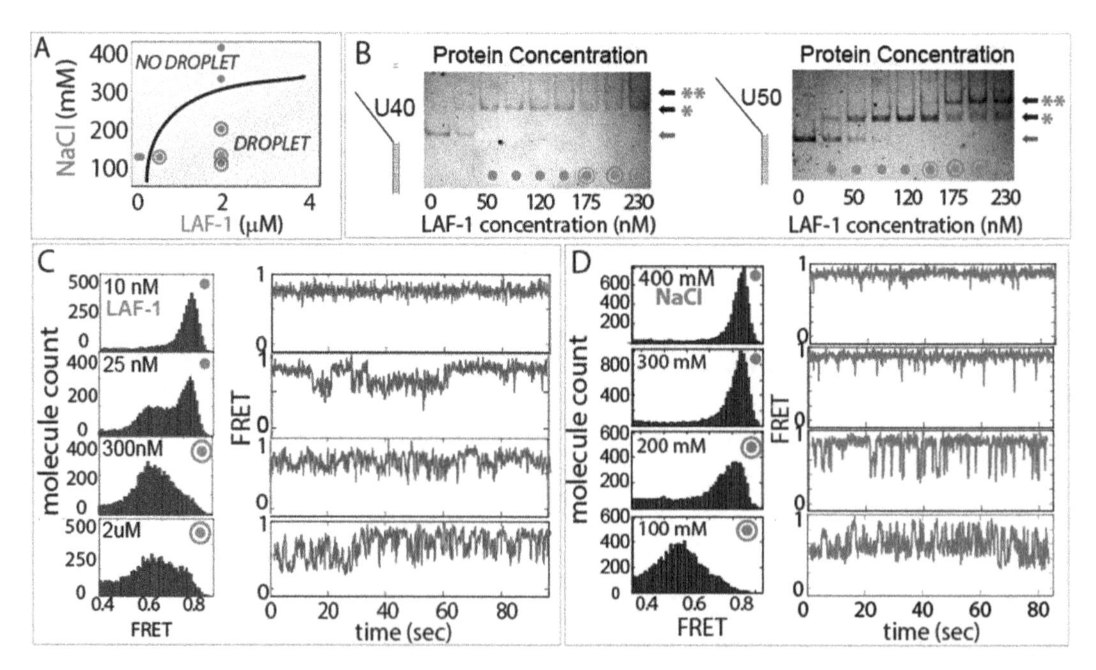

Fig. 3 LAF-1 induces dynamics on ssRNA in droplet forming condition. (**a**) Experimental conditions cutting across phase boundary in [LAF-1] and [NACl]. (**b**) Droplet forming conditions coincides with dimerization of LAF-1 denoted by double red asterix. (**c**, **d**) In droplet forming condition of high [LAF-1] LAF-1 induces dynamic mobility on ssRNA, evidenced by FRET fluctuation

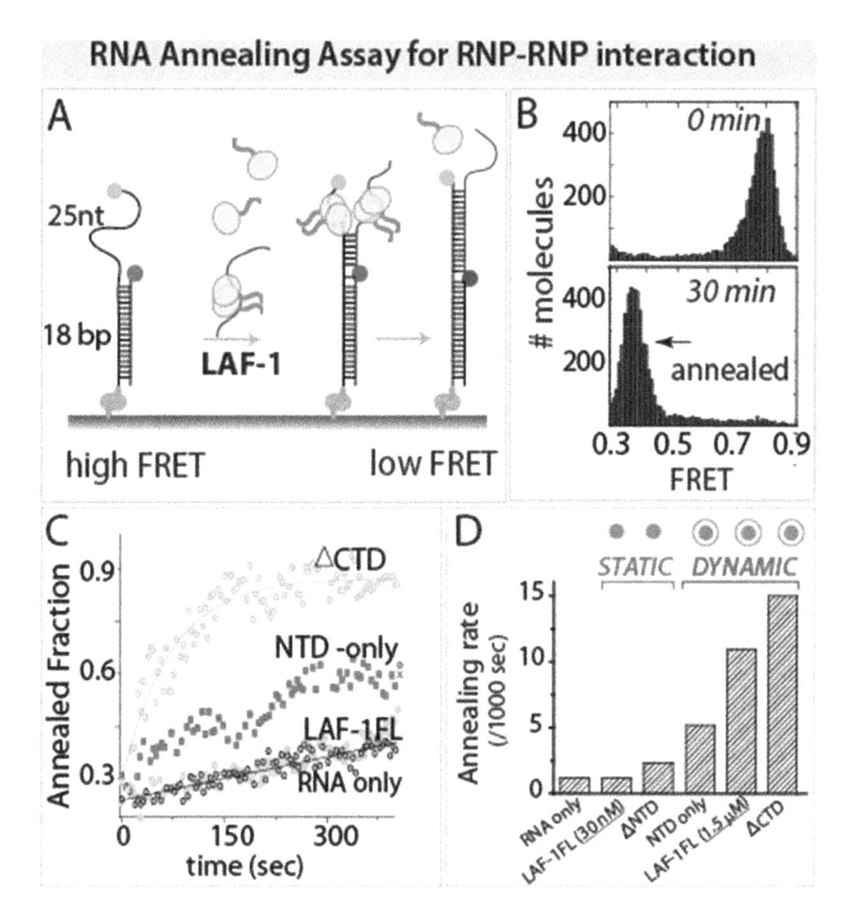

Fig. 4 LAF-1-RNA dynamics promote RNA annealing. (**a**) High FRET converts to low FRET upon RNA annealing. (**b**) FRET histogram before (top) and after (bottom) annealing. (**c**) Kinetic analysis of RNA annealing reaction. (**d**) Annealing rate for various mutants that represent static vs. dynamic LAF-1-RNA interaction

hybridization of two complementary RNA strands, an act that likely requires association between two sets of RNA-protein complexes, i.e., RNP-RNP contact. A FRET-pair fluorophore-labeled RNA substrate is immobilized on the TIRF microscope PEG-passivated slide (similar to the aforementioned smFRET assay). A preincubated mixture of a complementary RNA strand and IDR protein is applied on the slide (Fig. 4). We then monitor changes in FRET signal produced by annealing of the two complementary strands, thus reporting on RNP-RNP interaction.

1.5 SiMPull

Size and/or oligomeric state of cellular granules at the very early stage of the *nucleation phase* can be reliably measured using a uniquely customized pull-down assay, called single-molecule pull-down (SiMPull) assay [14, 15] (Fig. 5). This technique combines traditional pull-down assay principles with TIRF-based single-molecule fluorescence microscopy. It allows scope for probing cellular granules in a non-perturbing fashion. Fluorescently tagged SG protein(s) may be expressed in mammalian cells, followed by

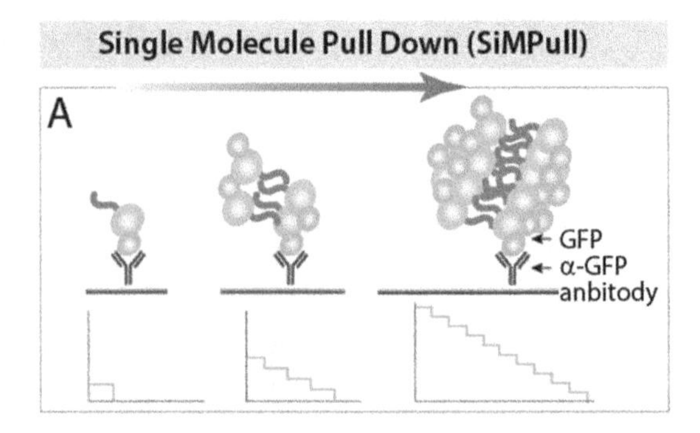

Fig. 5 Probing molecular assembly of granules. (**a**) Single-molecule pull-down assay can reveal the multimeric state of target proteins in cellular granules by quantifying photobleaching steps

application of stress, and cell lysis. The cell lysate is then applied on a PEG-passivated TIRF microscope slide that is coated with an antibody against the expressed SG protein. This allows for the fluorescently tagged SG protein molecules to be surface immobilized. Higher protein oligomers will be manifested as bright spots in fluorescence TIRF imaging, as opposed to low intensity spots exhibited to lower oligomers or monomers. Counting the number of photobleaching steps provides insight into the oligomeric state of the cellular granules.

2 Materials

2.1 Instrument and Common Reagents

1. For TIRF-based measurements, we use a home-built *TIRF-FRET microscope* [12].
2. Reagents for *quartz slides (passivated with a mixture of biotin-PEG and m-PEG)* to be used on the TIRF microscope are described elsewhere [16, 17] and will not be discussed here.
3. *Single-molecule imaging buffer*: 1 mg/mL glucose oxidase, 0.2% glucose, 2 mM 6-hydroxy-2,5,7,8-tetramethylchromane-2-carboxylic (Trolox), and 0.01 mg/mL catalase.
4. *Single-molecule wash buffer*: T50 buffer (10 mM Tris-HCl, pH 8.0, 50 mM NaCl).
5. NeutrAvidin (Thermo Fisher). Reconstitute to a stock of 5 mg/mL in T50 buffer.
6. *Purified IDR proteins* (*see* **Note 1**). We will use LAF-1 as an example here.

2.2 Phase Separation Assay

1. Inverted microscope with 60× or 100× objective (Nikon).
2. Chambered cover glass (Lab-Tek or Grace Biolabs).
3. *Phase separation buffer*: 100 mM NaCl, 50 mM Tris-HCl, pH 7.5 (*see* **Notes 2** and **3**).

2.3 Electrophoretic Mobility Shift Assay (EMSA)

1. *EMSA reaction buffer*: 50 mM Tris-HCl, pH 7.5, 125 mM NaCl, 100 mM β-mercaptoethanol, 0.1 mg/mL BSA.
2. Precast 6% DNA retardation (polyacrylamide) gels (Invitrogen).
3. *Electrophoresis buffer*: 0.5× TBE.
4. Typhoon scanner (GE) (fluorescence mode).

2.4 smFRET Binding Assay & Annealing Assay

1. *RNA substrate*: Partial duplex RNA with single strand (ss) overhang, labeled with a pair of FRET-suitable fluorophores such as Cy3 (donor) and Cy5 (acceptor) (Fig. 2) (*see* **Notes 4–6**). For the smFRET binding assay, the ss overhang of the RNA substrate is a poly(U) sequence to prevent secondary structure formation. We refer to this substrate as "binding substrate." However, for the annealing assay, the ss overhang may be a mixed sequence. We refer to this substrate as the "annealing substrate."
2. *ssRNA* that is complementary to the ss overhang of the partial duplex annealing RNA substrate. This will be used in the annealing assay.
3. *Single-molecule reaction buffer*: 50 mM Tris-HCl, pH 7.5, 125 mM NaCl.

2.5 Single-Molecule Pull-Down Assay (SiMPull)

1. HEK293 or HeLa cells (ATCC) and appropriate growth media.
2. 6-well cell culture plates (Corning).
3. Plasmids expressing fluorescently tagged SG protein of interest.
4. Transfection reagent. We use Lipofectamine 2000 (Thermo Fisher).
5. *SiMPull cell lysis buffer*: RIPA buffer (Thermo Fisher) + protease inhibitor cocktail tablet (Roche).
6. Antibody against the SG protein of interest that has been overexpressed in cells.

3 Methods

Perform all assays at room temperature, unless otherwise stated.

3.1 Phase Separation Assay

1. Prepare 50–100 μL reactions by mixing IDR protein and phase separation buffer (*see* **Notes 2** and **3**) in chambered cover glass imaging chambers.
2. Image droplets using DIC imaging with a 100× objective. Time points for imaging will depend on droplet nucleation and growth rate of the IDR protein of interest.

3. Construct the phase diagram, based on positive/negative score for droplet formation.
4. Several parameters may be systematically varied in this assay to test for its impact on LLPS of IDR protein and hence on the phase diagram. These include salt and protein concentration, presence/absence of RNA, temperature, crowding agents, etc.
5. When using fluorescently labeled IDR protein or RNA, droplets may be imaged by fluorescence imaging, apart from DIC.

3.2 Single-Molecule FRET (smFRET) Binding Assay

1. Assemble PEG-passivated slide and coverslip into sample chambers, as described in [16]. Let slide and coverslip thaw to reach room temperature before assembling.
2. Dilute stock NeutrAvidin solution 1:100, and then apply (~30 μL) on the biotin-PEG-slide chambers. Wait for ~2 min, ensuring biotin-NeutrAvidin attachment, and then wash each chamber with T50 buffer (~100 μL).
3. Flow in 30–100 pM of the binding RNA substrate (Fig. 2) in the chambers. Wait for ~2 min for the RNA molecules to be immobilized on the slide via biotin (on the RNA substrate)-NeutrAvidin linkage. Wash away excess RNA substrate with T50 buffer (*see* **Notes** 7 and **8**).
4. Flow in different concentrations of IDR protein, for example, LAF-1 (one concentration per slide chamber) (*see* **Note 9**) in single-molecule reaction buffer supplemented with single-molecule imaging buffer.
5. Record movies (~20 short movies each of 20 frames; and ~5 long movies each of 1200 frames) at different time points, using the TIRF microscope. As a control, always perform these measurements for RNA substrate alone.
6. *Data analysis*: We analyze the RNA-protein smFRET binding data in two ways—(i) For each condition, a representative smFRET histogram is generated by averaging the initial ten frames from thousands of single RNA molecules recorded over 20 movies. This shows the FRET signal distribution of all collected molecules. In Fig. 3c, the U50 RNA binding substrate alone shows a low FRET peak. Upon addition of low (~10 nM) LAF-1 concentration, the peak shifts to high FRET, whereas addition of intermediate (~25 nM) LAF-1 generates both a mid and high FRET peak. This mixed behavior shifts to a complete high FRET peak when high (>300 nM) LAF-1 concentrations are applied to the U50 RNA substrate. (ii) For more in-depth analysis, we then look at the time traces of individual RNA molecules. These show how the donor (Cy3) and acceptor (Cy5) intensities are changing and the resulting FRET signal changes. In Fig. 3c, d, individual time traces show the following behavior, under different conditions—time

traces of U50 RNA alone show static low FRET signal (corresponding histogram showed low FRET peak); low concentration LAF-1 binding to U50 shows time traces with static high FRET signal (corresponding histogram showed high FRET peak); and increasing concentrations of LAF-1 binding to U50 show emerging dynamics in FRET signal, fluctuating between high and mid FRET (corresponding histogram showed a broad mid FRET peak). *Thus, the smFRET data (derived from histograms and individual time traces), shed light into unprecendented details of RNA conformational dynamics induced by LAF-1 concentrations varying across the soluble to the phase-separated droplet phase.*

7. *Combined data interpretation from EMSA and smFRET binding assays*: However, it is the combination of these two ensemble and single-molecule methods that paints the emerging picture of molecular events occurring as an IDR protein like LAF-1 transitions across its phase boundary. In presence of low concentrations of LAF-1, when EMSA showed a single shifted band, smFRET traces exhibited static FRET signal. This indicates tight compaction of the RNA (the two dyes are now close to each other, *see* Fig. 3b–d). When LAF-1 concentrations correspond to within the phase boundary (high concentrations), EMSA primarily showed double shift and smFRET traces exhibited FRET fluctuations. This suggests that multimerization of LAF-1 induces conformational dynamics on the RNA (the distance between the two dyes changing). *Thus, overall, this combinatory approach helps us understand the molecular details of how monomer LAF-1-bound tightly wrapped RNA converts to multimer LAF-1-bound dynamic RNA, a stage likely primed for nucleation of droplets.*

3.3 Electrophoretic Mobility Shift Assay (EMSA)

1. Fluorescently labeled RNA binding substrate (the same as used for smFRET binding assay) and varying concentrations of IDR protein (e.g., LAF-1) (*see* **Note 7**) were mixed in the EMSA reaction buffer and incubated for 20 min at room temperature.
2. The reaction mixtures were mixed with loading dye, followed by electrophoresis on 6% DNA retardation polyacrylamide gels at 150 V for 50 min.
3. The gel was then scanned using fluorescence mode on a Typhoon scanner.
4. *Data analysis*: RNA binding substrate with 30 poly(U) ss overhang (U30) showed one shifted band (relative to the unbound RNA band), across low to high LAF-1 concentrations, indicating monomer LAF-1 binding to the RNA. RNA binding substrate with 40 and 50 poly(U) ss overhang (U40 and U50, respectively) showed a second super shifted band, indicative of

multimer LAF-1 binding to these longer RNA substrates. *Thus, this data informs us about the oligomeric state of LAF-1-RNA complexes across the phase boundary.*

3.4 Annealing Assay

1. Perform **steps 1** and **2** of smFRET binding assay, described above.
2. Flow in 30–100 pM of the *annealing RNA substrate* (Fig. 4) in the chambers. Wait for ~2 min for the RNA molecules to be immobilized on the slide via biotin (on the RNA substrate)-NeutrAvidin linkage. Wash away excess RNA substrate with T50 buffer (*see* **Notes 7** and **8**).
3. Flow into the chambers a preincubated (for 5 min) mixture of LAF-1 and 1 nM ssRNA (complementary to ss overhang of the annealing RNA substrate) in single-molecule reaction buffer supplemented with single-molecule imaging buffer (*see* **Note 10**).
4. Record movies (~20 short movies each of 20 frames) at different time points, using the TIRF microscope. As a control, always perform these measurements for RNA substrate alone.
5. *Data interpretation*: The smFRET histograms show that the RNA annealing substrate alone shows high FRET (since due to the mixed sequence in the ss overhang of this substrate, the dyes are close) (Fig. 4a). The preincubated mixture of LAF-1 with ss complementary RNA (that was applied in **step 3**) is expected to contain LAF-1 in complex with ssRNA as well as some free LAF-1 that can interact with the immobilized RNA substrate on surface (**step 2**). Annealing between the immobilized RNA and the ssRNA is expected to decrease FRET since the two dyes (on the immobilized RNA) will now be far apart separated by the annealed duplex RNA region (Fig. 4a). We find that RNA annealing (indicated by FRET peak shift from high to low in the smFRET histogram) is enhanced by LAF-1 concentrations within LLPS conditions, i.e., conditions that promote dynamic LAF-1-RNA interaction in the smFRET binding assay (Fig. 4c). Thus, LAF-1, in granule forming conditions, can promote RNP-RNP interaction, which likely contributes to the nucleation stage of granules.

3.5 Single-Molecule Pull-Down Assay (SiMPull)

1. Seed and growth HEK293 cells to ~70% confluency. We usually use 6-well dishes.
2. Transfect cells with fluorescently tagged granule forming protein of interest. We use Lipofectamine transfection reagent, as per manufacturer's protocol. Let protein express for 24 h.

3. Apply desired stress reagent to cells, for the recommended time. As a control, keep a sample without stress.
4. Check for protein expression (fluorescence) by cellular imaging under a fluorescent microscope.
5. Lyse cells using SiMPull cell lysis buffer on ice for 30 min (*see* **Note 11**). Keep lysates on ice till ready to be used.
6. To prepare TIRF microscope slide: Perform **steps 1–2** of smFRET binding assay.
7. Apply the biotin-conjugated antibody (*see* **Note 12**) against the granule protein of interest (that has been overexpressed in the cells). This coats the slide surface with the antibody, via biotin-NeutrAvidin linkage. Wait ~2 min. Wash away excess antibody with T50 buffer.
8. Apply diluted cell lysate on the slide (*see* **Note 13**). Wait ~2 min. Fluorescently tagged overexpressed granule protein of interest will be pulled down to the surface via its antibody (that was surface tethered in **step 6**). The goal is to achieve single-molecule density of fluorescently tagged protein on surface.
9. Using the appropriate laser excitation (depending on the fluorescent tag on the granule protein of interest), record ~5–10 movies each of ~900 frames, for each sample condition.
10. *Data analysis and interpretation*: The fluorescently tagged target granule protein of interest will be captured on the slide and manifested as a single fluorescent spot in the acquired image (Fig. 5). Analyzing the time traces of each of these spots will show the photobleaching events of the fluorescent tag. Combined information from the total intensity of each spot and counting the number of photobleaching steps individual traces provide us insight into the oligomeric state of the granule protein of interest that has been pulled down. For example, in absence of stress, these spots are expected to be low intensity, with time traces showing single-step photobleaching, indicative of monomeric proteins being pulled down. However, in presence of stress, the spots may become high intensity, and corresponding time traces will show multiple photobleaching steps (proportional to the number of fluorescent protein units present in each spot), suggesting multimeric protein complexes. Thus, analysis of SiMPull data provides unique information on oligomeric state of cellular protein clusters in the early stages of RNP formation, when they are potentially on their way to granule formation.

4 Notes

1. IDR proteins are inherently aggregation-prone and hence can be challenging to purify. Yet, high yield and purity are critical factors that must be met for conducting the assays described here. To prevent aggregation during purification and afterward during storage, these proteins are often purified in high salt (~1 M) buffer also containing low concentrations of urea (~1 M). Strategies to purify these proteins have been described by several groups [6, 7, 10, 18].
2. If IDR protein contains a solubility tag that needs to be cleaved off, add the appropriate protease to the abovementioned phase separation buffer and the requisite protease buffer (as per manufacturer's protocol).
3. Specific salt and protein concentrations at which IDR proteins will phase separate into droplets will depend on the protein. This concentration needs to be optimized from the phase diagram of the corresponding IDR protein. *For LAF-1 the phase diagram has been defined* [7].
4. We order RNA strands from IDT (Coralville, IA). For labeling the RNA strands, we use Cy3 or Cy5-NHS ester with 3′- or 5′-amine modified RNA strands, also ordered from IDT.
5. One strand of the RNA substrate must be conjugated to a biotin at one end so that the RNA substrate can be immobilized on the PEG-slide via biotin-NeutrAvidin linkage.
6. Typical position of the Cy3 and Cy5 labels is at either end of the ss overhang, which generally ranges from 40 to 70 nucleotides to allow multimerization of IDR proteins. The duplex region is typically 18-mer.
7. Meaningful concentrations to be used for IDR proteins range from low to high, corresponding to the transition from the soluble to LLPS-dependent droplet formation, deduced from the phase diagram of the protein. For example, for LAF-1, based on its phase diagram, we varied protein concentration from 10 nM to 2 μM.
8. RNA substrates are usually prepared and stored as 10 μM stocks at −80 °C. This is then serially diluted to one-time use 10 nM aliquots that are stored at −20 °C.
9. The exact concentration of RNA substrate to be applied on the slide chambers will need to be optimized. The number of RNA substrate molecules immobilized on the slide will depend on the density of biotin-PEG on the slide. *The goal is to achieve single-molecule density of RNA substrate on slide surface.*

10. Varying LAF-1 concentrations may be applied. We suggest using a range that corresponds to the static (low LAF-1) and dynamic (high LAF-1) LAF-1-RNA interaction behavior observed in the smFRET binding assay.
11. To ensure preservation of structure and oligomeric state of cellular granules, we prefer using mild lysis conditions. These include mild detergent (NP-40 in RIPA buffer), gently scraping attached cells off plate, and no centrifugation after lysis.
12. The concentration of antibody to be applied on surface will have to be empirically determined. For example, for a biotin-conjugated GFP antibody (Rockland), 10 nM is a good starting point for application on surface.
13. Dilution factor of cell lysate will have to be empirically determined. As an example, for cells harvested from one well of a 6-well plate (~0.8×10^6 cells), exhibiting ~50% transfection efficiency, we use 1:10,000 as a starting dilution.

Acknowledgments

We thank the Myong lab members for helpful discussions.

References

1. Protter DS, Parker R (2016) Principles and properties of stress granules. Trends Cell Biol 26:668–679
2. Guo L, Shorter J (2015) It's raining liquids: RNA tunes viscoelasticity and dynamics of membraneless organelles. Mol Cell 60:189–192
3. Li YR, King OD, Shorter J, Gitler AD (2013) Stress granules as crucibles of ALS pathogenesis. J Cell Biol 201:361–372
4. Aulas A, Vande Velde C (2015) Alterations in stress granule dynamics driven by TDP-43 and FUS: a link to pathological inclusions in ALS? Front Cell Neurosci 9:423
5. Updike D, Strome S (2010) P granule assembly and function in Caenorhabditis elegans germ cells. J Androl 31:53–60
6. Burke KA, Janke AM, Rhine CL, Fawzi NL (2015) Residue-by-residue view of in vitro FUS granules that bind the C-terminal domain of RNA polymerase II. Mol Cell 60:231–241
7. Elbaum-Garfinkle S, Kim Y, Szczepaniak K, Chen CC, Eckmann CR, Myong S, Brangwynne CP (2015) The disordered P granule protein LAF-1 drives phase separation into droplets with tunable viscosity and dynamics. Proc Natl Acad Sci U S A 112:7189–7194
8. Kim Y, Myong S (2016) RNA remodeling activity of DEAD box proteins tuned by protein concentration, RNA length, and ATP. Mol Cell 63:865–876
9. Zhang H, Elbaum-Garfinkle S, Langdon EM, Taylor N, Occhipinti P, Bridges AA, Brangwynne CP, Gladfelter AS (2015) RNA controls PolyQ protein phase transitions. Mol Cell 60:220–230
10. Schwartz JC, Wang X, Podell ER, Cech TR (2013) RNA seeds higher-order assembly of FUS protein. Cell Rep 5:918–925
11. Daigle JG, Lanson NA Jr, Smith RB, Casci I, Maltare A, Monaghan J, Nichols CD, Kryndushkin D, Shewmaker F, Pandey UB (2013) RNA-binding ability of FUS regulates neurodegeneration, cytoplasmic mislocalization and incorporation into stress granules associated with FUS carrying ALS-linked mutations. Hum Mol Genet 22:1193–1205
12. Roy R, Hohng S, Ha T (2008) A practical guide to single-molecule FRET. Nat Methods 5:507–516
13. Hellman LM, Fried MG (2007) Electrophoretic mobility shift assay (EMSA) for detecting protein-nucleic acid interactions. Nat Protoc 2:1849–1861

14. Jain A, Liu R, Ramani B, Arauz E, Ishitsuka Y, Ragunathan K, Park J, Chen J, Xiang YK, Ha T (2011) Probing cellular protein complexes using single-molecule pull-down. Nature 473:484–488
15. Jain A, Liu R, Xiang YK, Ha T (2012) Single-molecule pull-down for studying protein interactions. Nat Protoc 7:445–452
16. Joo C, Ha T (2012) Preparing sample chambers for single-molecule FRET. Cold Spring Harb Protoc 2012:1104–1108
17. Rothenberg E, Ha T (2010) Single-molecule FRET analysis of helicase functions. Methods Mol Biol 587:29–43
18. Patel A, Lee HO, Jawerth L, Maharana S, Jahnel M, Hein MY, Stoynov S, Mahamid J, Saha S, Franzmann TM, Pozniakovski A, Poser I, Maghelli N, Royer LA, Weigert M, Myers EW, Grill S, Drechsel D, Hyman AA, Alberti S (2015) A liquid-to-solid phase transition of the ALS protein FUS accelerated by disease mutation. Cell 162:1066–1077

Chapter 20

Correlative Atomic Force and Single-Molecule Fluorescence Microscopy of Nucleoprotein Complexes

Herlinde De Keersmaecker, Wout Frederickx, Yasuhiko Fujita, Steven De Feyter, Hiroshi Uji-i, Susana Rocha, and Willem Vanderlinden

Abstract

Correlative imaging by fluorescence and atomic force microscopy provides a versatile tool to extract orthogonal information on structurally heterogeneous biomolecular assemblies. In this chapter, we describe an integrated setup for correlative fluorescence and force microscopy. We present factors influencing data quality, as well as step-by-step protocols for sample preparation, data acquisition, and data processing that yield nanoscale topographic resolution, high image registration accuracy, and single-fluorophore sensitivity. We demonstrate the capabilities of the approach through simultaneous characterization of mesoscale geometry and composition in a multipart nucleoprotein complex.

Key words Correlative imaging, Atomic force microscopy (AFM), Single-molecule localization microscopy (SMLM), Protein-DNA interactions, Stepwise photobleaching

1 Introduction

1.1 Orthogonal Probing by Correlative Atomic Force and Single-Molecule Fluorescence Localization Microscopy

Biological processes are often carried out in the context of macromolecular assemblies. In addition, arrangements of these complexes can be dynamic, resulting in a heterogeneous ensemble [1–3]. Single-molecule techniques can resolve distinct populations in heterogeneous systems, in contrast to bulk experiments where heterogeneity is averaged out. In turn, mechanistic details of biomacromolecular interactions can be uncovered.

Atomic force microscopy (AFM) is a technique that can generate 3D reconstructions of individual biomolecules and complexes thereof in a label-free fashion and with ~ nm resolution [4–6]. To this end a very sharp tip, mounted on a flexible cantilever, scans a sample surface in a raster pattern using a piezo scanner, while keeping the interaction force between sample and tip constant.

Herlinde De Keersmaecker and Wout Frederickx contributed equally to this work.

Yuri L. Lyubchenko (ed.), *Nanoscale Imaging: Methods and Protocols*, Methods in Molecular Biology, vol. 1814, https://doi.org/10.1007/978-1-4939-8591-3_20,

Quasi-constant tip-sample interactions are typically achieved by monitoring changes in the dynamics of the cantilever and by counteracting these changes via the application of a voltage that alters the *z*-position of the piezo scanner. In every pixel (x, y) of the scanned area, the *z*-position is recorded. Consequently, a 3D representation of the surface topography can be reconstructed.

An alternative way to study single molecules is by fluorescence microscopy [7]. The molecule of interest is labeled with a fluorescent tag providing high contrast. Emission of the tag after excitation is detected through an optical system. Due to the wave character of light, the emitted light is spread out on the detector described by the point spread function (PSF) of the optical system [8]. This effect limits the resolution achieved with optical microscopy, referred to as the diffraction limit. However, when the signal of a single molecule is detected, the position of this molecule can be determined by fitting of the recorded fluorescence signal with a mathematical approximation of the PSF such as a two-dimensional Gaussian function [9–11]. This principle underlies single-molecule localization microscopy (SMLM).

AFM and SMLM are highly complementary technologies: AFM can provide insight in topographic features at a nanometer resolution, while SMLM is sensitive toward specifically labeled molecules in complex samples [12]. Integrated setups combining both technologies can therefore provide orthogonal information at the single-molecule level.

In this chapter, we introduce the factors that determine data quality (*see* Subheading 1.2), the design of the microscope used (*see* Subheadings 1.3 and 2.1), and the procedures and protocols to prepare (*see* Subheadings 2.2, 2.3, 2.4, and 3.1), acquire (*see* Subheading 3.2), and analyze (*see* Subheadings 1.4 and 3.3) correlative AFM-SMLM measurements on static (dried) samples of nucleoprotein complexes.

1.2 Factors that Define Data Quality

The quality of correlated AFM-SMLM images is a function of different parameters: resolution in the topography channel, localization precision and sensitivity in the fluorescence channel, and image registration accuracy between both channels.

1.2.1 Image Registration Accuracy

Correlative imaging requires accurate registration of fluorescence and AFM data channels. A generally applicable approach for image overlay employs fiducial markers, reference points that are stably observed in both data channels [13–15]. These are fluorescent nanoscale particles and their properties are subject to several criteria. Fiducial markers should (1) not be too bright in comparison to organic fluorophores to avoid saturation of the EM-CCD camera, (2) exhibit uninterrupted and long-lived fluorescence to ensure appropriate drift correction, (3) be inert toward biomolecules, and (4) adsorb stably on the substrate. The coordinates of these fiducials are extracted from the fluorescence channel and AFM image using an intensity threshold and further fitting the fluorescence and height signals of the beads with a 2D Gaussian function. Image registration is then performed through an affine transformation using the positions of fiducial markers.

1.2.2 Resolution in the Topography Channel

An AFM image is acquired by measuring the height in every *x,y*-coordinate. Therefore, a distinction must be made between vertical and lateral resolution. Vertical resolution is defined as the minimum step height variation measured on the surface and is thus limited by the vertical noise in the system. Lateral resolution can be defined as the minimal separation between two objects for which the depression depth between their convoluted images is larger than the vertical noise [16]. Consequently, suppression of vertical noise improves lateral resolution as well. Therefore, high-resolution topographic images of biological macromolecules benefit from a supporting surface with minimal surface roughness, such as muscovite mica.

In the context of correlative fluorescence and force microscopy, absorption throughout the UV and visible spectrum, as well as its birefringent properties, complicates the use of mica as a substrate [17]. To establish substrate transparency and flatness, while ensuring mechanical robustness and a proper optical working distance, very thin slices of mica can be glued to glass cover slides [18].

Further, it is important to mention that accurate image overlay requires recording large-scale (~10 μm) AFM topographs. Given the maximum number of pixels that can be acquired per image (in our setup 2048 × 2048), pixel size becomes a limiting factor for lateral resolution. It is however possible to overlay high-resolution topographs of smaller regions of interest, and thus smaller pixel size, with overview topographs, without the loss of image registration accuracy.

1.2.3 Sensitivity in the Fluorescence Channel

The signal to noise ratio *S/N* in the fluorescence channel directly affects the localization precision of individual emitters [10, 11]. Further, single-molecule sensitivity can be exploited in the application of correlative imaging. For instance, stepwise photobleaching allows calculating the number of molecules present in a protein complex [19, 20]. To maximize the sensitivity in the fluorescence channel, an appropriate choice of bio-orthogonal fluorophores, as well as fiducial markers, is required. We use a scheme wherein single fluorophores and fiducials are simultaneously excited at the wavelength of maximum absorption of the fluorophore, resulting in suboptimal excitation of the fiducials. This approach ensures that fiducials and single fluorophores are simultaneously detected on the same region of the CCD camera with approximately equal sensitivity. In addition, it is beneficial to increase S/N of the fluorescence signal by using a moving average over several frames, in particular with regard to the broad absorption spectrum of the mica support.

1.2.4 Localization Precision in the Fluorescence Channel

When fluorescence signals are well separated from each other, individual molecules can be localized with a precision in the order of 10 nm by fitting their fluorescence signal to a 2D Gaussian. The precision by which single molecules can be localized is influenced by the number of photons collected N, the standard deviation of the background intensity b, and the pixel size of the camera a as described by the localization precision in a single dimension [10, 11]:

$$\sigma_{\mathrm{i}} = \sqrt{2 \times \left(\frac{s_{\mathrm{i}}^2}{N}\right) \times \left(\frac{16}{9} + \frac{8\pi s_{\mathrm{i}}^4 b^2}{a^2 N}\right)}$$

In this equation σ_{i} is the standard error of the localization and $S_{\mathrm{i}}^2 = S^2 + \frac{a^2}{12}$, where S is the width of the Gaussian peak. This definition holds for the localization of a single fluorescence detection. In static samples, however, a single molecule can be detected multiple times in fluorescence time-lapse images. When a single molecule is localized multiple times, the localization precision can be experimentally determined by calculating the variance of the PSF centroid distribution [21].

$$\sigma_{x,y}{}^2 = s_x{}^2 + s_y{}^2$$

with $s_{x,\ y}$ the standard deviation of the centroid positions in the x and y direction. Accordingly, drift will affect the spread of the PSF centroids and therefore accurate drift correction is essential.

1.3 The Microscope: Commercial Solutions and Approaches to Reduce Vibrations

Correlative fluorescence and force microscopy can be performed on a commercial fluorescence microscopic setup combined with a commercial AFM head. Nowadays, several manufacturers provide AFM scan heads that are compatible with commercial inverted microscopes of major brands like Nikon (TE2000, Ti line), Zeiss (Axio Observer, Axio Vert 200, Axio Vert A1), Olympus (IX line), and Leica (DMi and DMI lines). For instance, Bruker offers the Dimension Icon Head®, JPK Instruments offers the NanoWizard® series, and Oxford Instruments offers the MFP-3D® and MFP-3D-BIO® AFM scan heads.

For demanding applications, it is crucial to reduce vibrations such that optimal AFM imaging is guaranteed. Our setup is built on an anti-vibration table (Fig. 1) to isolate low-frequency vibrations and to stabilize the optical beam by passive damping. In addition, we introduced a second, smaller active damping system (the smaller

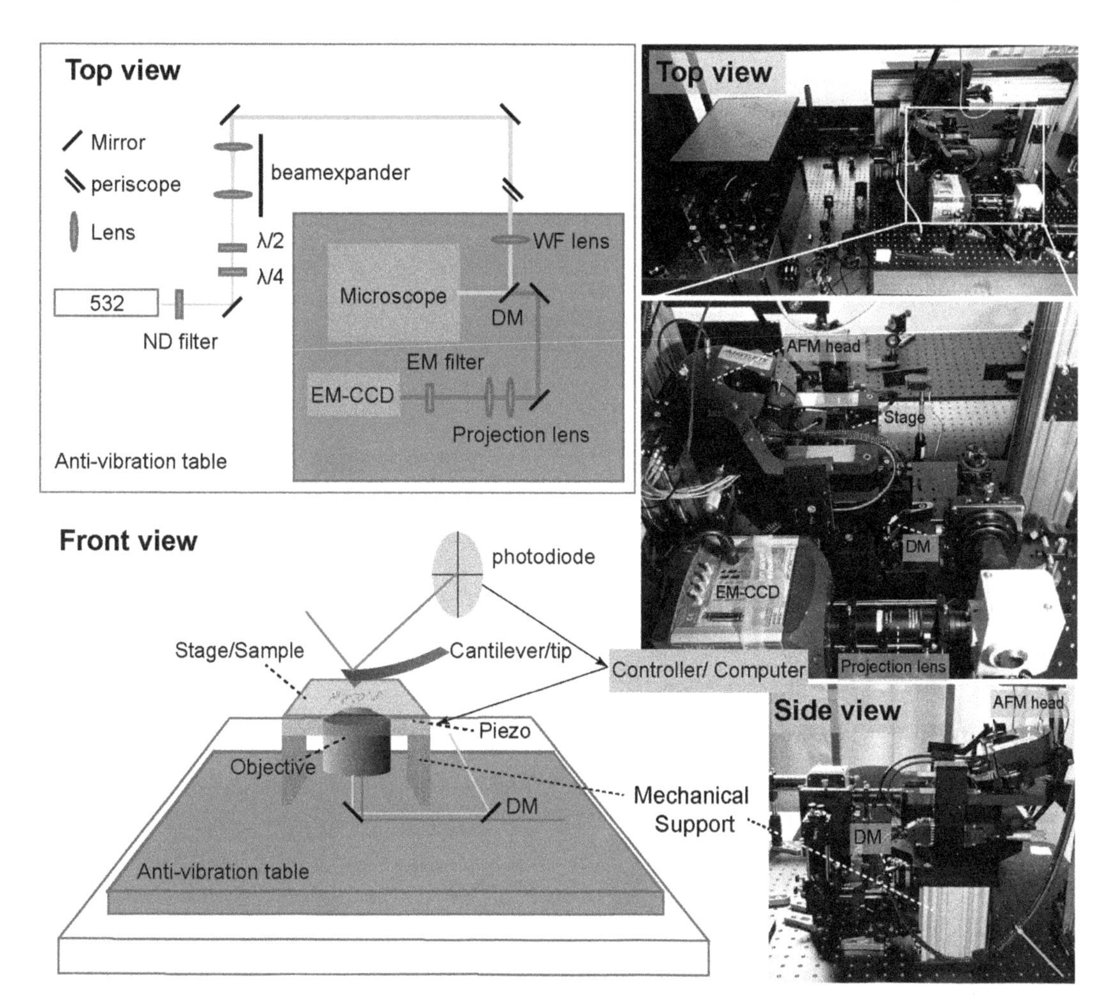

Fig. 1 Configuration of the correlative fluorescence-AFM setup. Light of a 532-nm diode laser is guided toward the sample. The laser power is controlled by a neutral density filter (ND). Wave plates (λ/4 and λ/2) ensure circular polarized light, and the beam is expanded by a beam expander. Using a periscope, the light is guided onto the second anti-vibration table and hits the wide-field (WF) lens as a collimated beam. The wide-field lens focuses the light on the back focal plane of the objective. Emitted light is separated from the excitation light by a dichroic mirror and further filtered by an emission (EM) filter. The magnification is further controlled by the projection lens whereafter the emission light is detected by an EM-CCD camera. On top of the microscope, a commercial AFM head and stage is installed. The AFM head and stage are mechanically uncoupled from the optical train by a mechanical support

anti-vibration table in Fig. 1) to damp high-frequency vibrations to ensure high-quality AFM imaging. The smaller anti-vibration table in turn required a compact design of part of the excitation and detection optics. Therefore, instead of a commercial microscope body, a homebuilt fluorescence microscope setup was assembled. It is further important to mechanically uncouple the AFM head from the optical microscope body. This is done by implementing a mechanical support for the AFM head on the anti-vibration table, under which the microscope objective is positioned. In addition, we found that EM-CCD cooling and mechanical vibrations by the microscope body are transferred through the microscope objective are factors that contribute significantly to the vertical noise level.

1.4 Data Processing

Image analysis encompasses five steps as depicted in Fig. 2. In the first step, both the positions of single-molecule emitters and reference beads are determined in the fluorescence channel.

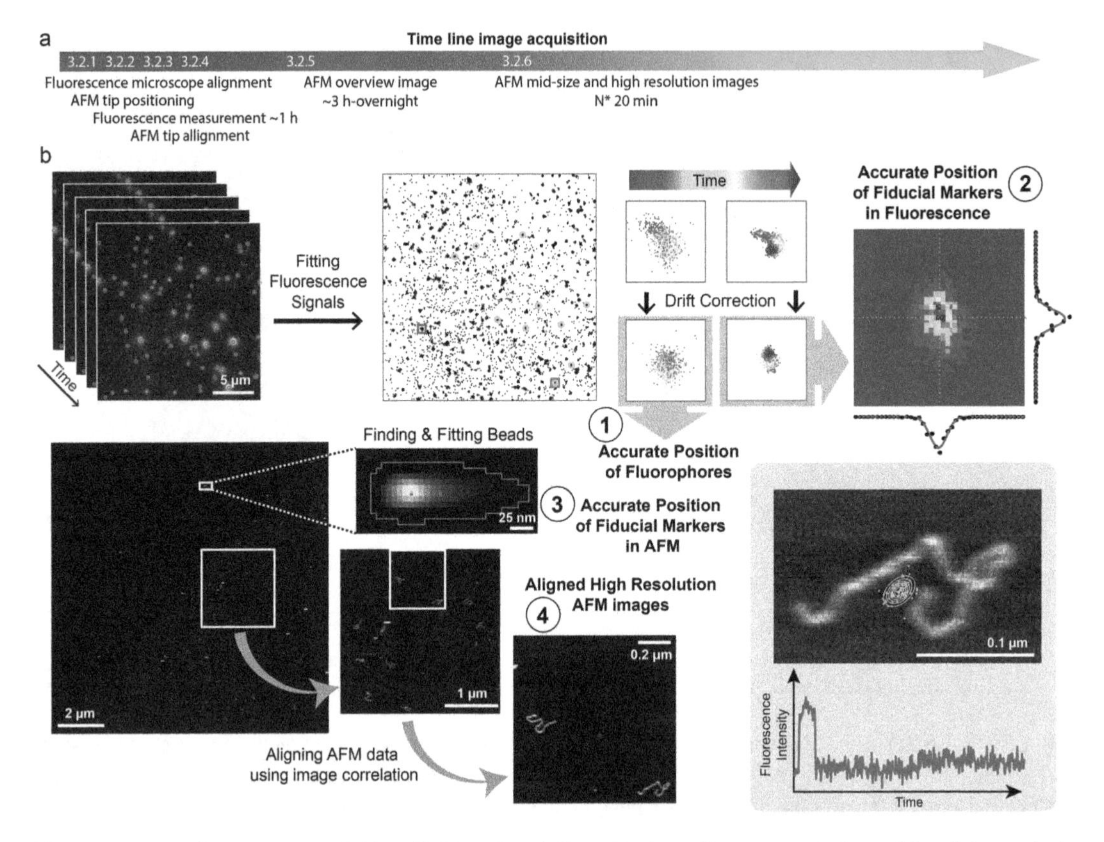

Fig. 2 Time line for image acquisition (Subheading 3.2) and schematic representation of the data analysis procedure (Subheading 3.3). (**a**) Sequence of steps during image acquisition. The indicated times are approximations. *N* indicates the number of high-resolution AFM images. (**b**) *Fluorescence channel*: Accurate positions of the fluorophores are determined (1). Therefore recorded fluorescence signals are accumulation over N frames to increase the signal to noise ratio. Then the accumulated fluorescence signals are fitted by a 2D Gaussian function. Fiducial markers are detected as the fluorescent signals in the last 15% of the recorded frames (2). Both the positions of the fiducial markers and the fluorophores are corrected for drift. *Topography channel*: After background correction, fiducial markers are recognized based on an intensity threshold (white lines delimited the region above the predefined threshold value, the red dot marks the centroid of this region) and fitted to determine the bead position (blue spot, maximum height; green spot, calculated position after fitting a 2D Gaussian) (3). *Image registration*: Both channels are registered based on the positions of the fiducial markers. *AFM image correlation:* The high-resolution AFM image is correlated with the AFM overview image based on distinguishable features, and a transformation matrix is calculated (4). To overlay the locations of the fluorescently labeled molecules with the high-resolution AFM image, the transformations for image registration and for image correlation are applied to the positions calculated from the fluorescence images. Finally, stepwise photobleaching analysis can be applied on the fluorescence signals and correlated with the topography channel

The position of the reference beads is used to correct for *x*, *y* drift during fluorescence image acquisition. Second, bead positions are determined in the overview topographic image. Based on the bead positions in both the fluorescence and the topography channel, a transformation matrix for image correlation is calculated. Then, the correlation between the high-resolution AFM images and the low-resolution overview AFM image is determined. This correlation together with the transformation matrix for image registration allows to overlay the fluorescence signals with high-resolution topography data. Next, high-resolution AFM scans are registered with the correlated fluorescence-AFM image. Finally, the fluorescence signal over time can be analyzed to detect bleaching steps reflecting the number of dyes present for each diffraction limited fluorescence signal.

2 Materials

2.1 Microscope

Measurements were performed on a commercial AFM (AIST-NT CombiScope) mounted on a homebuilt fluorescence wide-field microscope. The whole system was placed on an anti-vibration table. To minimize vibrations, the optical train starting from the wide-field lens and the AFM head were installed on a second optical table. Important to note is that the AFM head was mechanically uncoupled from the optical train by mounting the head and objective separately on the anti-vibration table. The configuration of the system is shown in Fig. 1. For the fluorescence measurement, light of a diode laser was circularly polarized, expanded, and collimated. Thereafter the light was focused on the back focal plane of the objective to achieve wide-field illumination. Emission light was separated from excitation light by an appropriate dichroic mirror and collected through an emission filter by an EM-CCD camera. A commercial AFM head and stage containing a piezo scanner were mounted above the objective. Specification of the components of this microscope was as follows:

1. Laser: 532-nm diode laser, 100 mW (Spectra-Physics).
2. $\lambda/2$ and $\lambda/4$ @532 wave plates (Thorlabs) to ensure circular polarized excitation light.
3. Neutral density (ND) filter to adjust the laser power.
4. Beam expander.
5. Dichroic mirror (Chroma Technology, Z532RDC) to separate excitation and emission light.
6. Wide-field lens ($f = 300$ mm, ARB2 VIS, Linos) to focus the excitation light on the back focal plane of the objective to achieve wide-field illumination.

7. Projection lens 2.5× (Thorlabs) to magnify the image projected on the camera (*see* **Note 1**).
8. Emission filter (HQ572LP, Chroma Technology Inc.) to selectively detect the emission light in the emissive region of the fluorophore.
9. Oil-immersion objective lens (Nikon, CFI S Fluor, 100×, N. A.1.3).
10. Piezo scanner (Physik Instrumente, PI).
11. Electron multiplying charge-coupled device (EM-CCD) camera (Andor, iXon 897) operated at −65 °C.
12. AFM head and stage: CombiScopeTM 1000 (AIST-NT).
13. AFM tip: AC240TS probes (Olympus).
14. Large anti-vibration table (Newport, RS 4000™).
15. Small anti-vibration table (Accurion, Halcyonics-micro).
16. Computer.
17. Software accompanying the CCD camera (SOLIS imaging) and AFM head (AIST).
18. Transmission lamp for positioning of the AFM tip.

2.2 (Fluorescent) DNA Fragment

1. pBR322 plasmid (NEB).
2. Primers (to amplify DNA fragment containing EcoRV recognition site) AATGCGCTCATCGTCATCC and CGACGCTCTCCCTTATGC (Integrated DNA Technologies) or (to generate a 1000 bp end-labeled fluorescent DNA fragment) Atto532-AATGCGCTCATCGTCATCC and CTGCCAAGGGTTGGTTTG (Integrated DNA Technologies).
3. Components to perform PCR and a PCR cleanup kit.

2.3 EcoRV Purification and Labeling

1. *E. coli* cells containing the expression vector encoding mutant EcoRV protein (C21S/K58C) [22, 23].
2. Lysis buffer (30 mM KH_2PO_4-KOH pH 7.2, 800 mM NaCl, 0.5 mM EDTA, 1 mM dithiothreitol (DTT)).
3. Sonicater.
4. Centrifuge.
5. Affinity chromatography using a HisTrap HP 5 mL column (GE Healthcare).
6. 250 mM imidazole (Sigma-Aldrich).
7. HiTrap heparin HP column (GE Healthcare).
8. Atto532-maleimide (ATTO-tec).
9. Purification buffer: 30 mM KH_2PO_4-KOH, 0.5 mM EDTA, 1 mM DTT, 0.01% (v/v) Lubrol, 250 mM imidazole, pH 7.2.

10. Amicon©ultra-0.5 centrifugal filters (10 KDa MWCO).
11. NanoDrop 2000 (Thermo Fisher Scientific).

2.4 Substrate and Sample Preparation

1. Mica (Agar Scientific; V1 grade).
2. Glass slide nr 1, size: 22 × 22 mm (Menzel-Gläser).
3. Tweezers.
4. Teflon holder (homemade, holder for the preparation of clean glass cover slides).
5. 1 M NaOH (Sigma-Aldrich).
6. UV-ozone photoreactor (UVP, Analytik Jena).
7. Glass vials and covers.
8. Poly(dimethylsiloxane) (PDMS) (Sigma-Aldrich).
9. Oven (should reach a temperature of 80 °C).
10. ~10 cm clean glass petri dish (Sigma-Aldrich).
11. Thick aluminum foil (Sigma-Aldrich).
12. ~5 by 5 cm piece of expanded polystyrene with round edge to distribute PDMS between cover slide and mica.
13. Scotch tape.
14. Scissors.
15. EcoRV binding buffer: 10 mM Tris buffer containing 10 mM Ca^{2+}, 100 mM NaCl, and 1 mM DTT (pH 7.6).
16. FluoSpheres® carboxylate-modified microscopheres (0.02 μm, crimson fluorescent, Thermo Fisher Scientific) as fiducial markers.
17. 0.01 w/v% poly-L-lysine (PLL) solution in Milli-Q water (poly-L-lysine hydrobromide, MW 500–2000, Sigma-Aldrich).
18. Deposition buffer: 10 mM 2-amino-2-(hydroxymethyl-propane)-1,3-diol (Tris) buffer containing 200 mM NaCl (pH 7.6).
19. Milli-Q water.

2.5 Software

1. Scanning Probe Image Processor (SPIP, Image Metrology).
2. In-house written routine in MATLAB (version 8.5.0.197613 (R2015a), MathWorks), available from the authors upon request.

3 Methods

3.1 Sample Preparation

A glass cover slide is used as support for a thin mica layer on which the sample is deposited. To avoid background fluorescence, the cover slides are first thoroughly cleaned. Thereafter, mica is glued onto this glass support and further cleaved until it is as thin as possible to permit single-molecule fluorescence imaging while still covering ~50–60% of the glass support.

3.1.1 Cleaning of Glass Cover Slides

1. Place glass cover slides in a Teflon holder. Manipulation of the individual cover slides can be done with a tweezer. Avoid overlap between the cover slides.
2. Place the Teflon holder with the cover slides in a glass container which can be covered.
3. Add acetone to the glass container.
4. Sonicate for 15 min at room temperature.
5. Discard the acetone.
6. Rinse the Teflon holder containing the cover slides with Milli-Q water.
7. Add 1 M NaOH in the glass container and immerse the Teflon holder with cover slides.
8. Sonicate for 5 min at room temperature.
9. Discard the 1 M NaOH solution.
10. Repeat **steps 6–9** with fresh 1 M NaOH solution.
11. Rinse the Teflon holder with the cover slides with Milli-Q water.
12. Add Milli-Q water in the glass container and immerse the Teflon holder with cover slides.
13. Sonicate for 15 min at room temperature.
14. Discard the Milli-Q water.
15. Repeat **steps 11–14** with fresh Milli-Q water.
16. Clean each cover slide individually with Milli-Q water while holding it using tweezers, and dry with argon.
17. Put each cover slide separately in a small glass vial.
18. Treat the cover slide with UV-ozone for 30 min.
19. Remove the glass vials from the UV-ozone generator and cover with the appropriate cap.

3.1.2 Assembly of a Glass-Mica Substrate

1. Cut a 2.5 by 2.5 cm piece of mica. (The size of the mica should be larger as the glass cover slide.)
2. Scrape along the cut edge with a scissor blade.
3. Split the mica layers in 3–4 pieces by using a sharp tweezer tip as a wedge.
4. Cleave each piece of mica on both sides with scotch tape.
5. Cover each piece of mica against dust.

6. Place a cleaned cover slide (*see* Subheading 3.1.1) onto a clean surface such as a piece of fresh aluminum foil attached to the working surface.
7. Add a small droplet of PDMS using a sterile needle on the bottom half of the glass cover slide.
8. Place a piece of mica with the freshly cleaved side on the glass cover slide.
9. Carefully spread the PDMS over the whole surface. This can be done with a rounded piece of expanded polystyrene (EPS).
10. Place the glass-mica substrate with the glass surface pointing upward in a large, clean glass petri dish covered with thick aluminum foil.
11. Cure the substrates for 1 h on 80 °C.
12. Cover your working surface with clean aluminum foil.
13. After curing, cleave the mica away from the cut edge. The layer removed last should be homogeneous, and the mica should extend over at least one side of the glass cover slide. The layer should be as thin as possible to allow single-molecule fluorescence detection, while the mica surface should still be large enough to allow sample manipulation without distortion of the sample. In general, the mica layer is thin enough for single-molecule fluorescence detection if no more than 50–60% of the glass cover slide remains covered with mica (*see* **Note 2**).

3.1.3 Construction of DNA Fragments

Primers are designed to amplify the desired DNA fragment by PCR. In this protocol a pBR322 plasmid was used as template, and two primers were chosen to amplify a DNA segment comprising the EcoRV recognition site. More specifically, we used AATGCGCTCATCGTCATCC and CGACGCTCTCCCTTATGC as forward and reverse primer, respectively. Amplification of the pBR322 vector with these primers yields a DNA fragment of 500 bp (*see* **Note 3**: alternative with labeled DNA fragments).

3.1.4 Purification and Labeling of EcoRV

In order to determine the number of EcoRV molecules in the nucleoprotein complex, the enzyme needs to be fluorescently labeled. This was done by labeling the purified protein in vitro, using thiol-maleimide chemistry. To this end, an EcoRV double mutant (C21S/K58C) containing a single, surface-accessible cysteine per protein monomer, far from the active site, is used [23]. This EcoRV mutant can be purified by affinity chromatography using a HisTrap HP 5 mL column and a HiTrap heparin HP column through a standard protein purification protocol. For a detailed protocol for purification and labeling of EcoRV, we refer to [22–24]. Fluorescence labeling of the purified proteins is as follows:

1. Incubate 50 μM of the purified protein with a tenfold excess of Atto532-maleimide in purification buffer for 3 h at room temperature.
2. Remove unbound dye with an Amicon©ultra-0.5 centrifugal filters (10 KDa MWCO) according to the manufacturer instructions. Before each centrifugation cycle, add fresh purification buffer. After each centrifugation, collect the flow-through and measure the absorbance. Repeat until no more absorbance by Atto532 in the flow-through is detectable.
3. Determine the absorption at 280 and 532 nm to calculate the degree of labeling (*DOL*) according to the following formula:

$$\mathrm{DOL} = \frac{A_{532} - \varepsilon_{\mathrm{EcoRV}}}{(A_{280} - \mathrm{CF}_{280}{}^{*}A_{532}) * \varepsilon_{\mathrm{atto532}}}$$

with $\varepsilon_{\mathrm{EcoRV}} = 51{,}800\ \mathrm{M}^{-1}\ \mathrm{cm}^{-1}$, $\varepsilon_{\mathrm{atto532}} = 11{,}500\ \mathrm{M}^{-1}\ \mathrm{cm}^{-1}$, and $\mathrm{CF}_{280} = 0.11$ which corrects for the fact that Atto532 also absorbs at 280 nm. Typically, labeling efficiencies of ~75% are obtained.

3.1.5 Preparation of a Protein-DNA Sample Containing Fluorescent Beads

1. 1 nM of Atto532-labeled EcoRV (Subheading 3.1.4) was mixed either with a 0.125 ng/μL DNA fragment comprising the EcoRV recognition site or with 0.125 ng/μL pBR322 plasmid DNA (Subheading 3.1.3) in EcoRV binding buffer and incubated for 15 min at room temperature (*see* **Note 4**).
2. Mix an appropriate amount of beads (diluted in the solution provided by the supplier) in 10 μL of the protein/DNA mixture. The amount of beads added should be optimized to give at least seven beads per field of view in the topographic overview image corresponding to ~40 beads in the field of view of the fluorescence channel (35 by 35 μm). Mix the beads with the protein-DNA mixture just prior to sample deposition in order to prevent interaction.

3.1.6 Sample Deposition

1. Deposit 20 μL of 0.01% poly-L-lysine solution on the mica-glass substrate (Subheadings 3.1.1 and 3.1.2) while holding the substrate by a part of the naked glass with a pair of clean tweezers (*see* **Note 5**).
2. Incubate for 30 s.
3. Rinse gently using 50 mL Milli-Q water. To prevent contamination from the PDMS, only wet the part covered with mica and decant excess Milli-Q water from the side of the substrate with a mica overhang.
4. Dry the substrate gently using a gentle stream of inert gas (Ar or N_2) pointed to the center of the substrate at ~1 cm distance.

5. Dropcast 10 μL of the sample (*see* Subheading 3.1.5).
6. Incubate for 30 s.
7. Rinse with 20 mL of Milli-Q water and dry the sample as described in **steps 3–4**.

3.2 Image Acquisition

First, the fluorescence microscope is aligned in wide-field mode and an AFM tip is installed. Thereafter fluorescence time-lapse images are recorded. By acquisition over time, bleaching traces can be analyzed and reference beads assigned. A region of interest is chosen to scan a low-resolution overview image by AFM. The region is chosen near the center area of the optical path to minimize chromatic aberrations. Further, the area should contain at least seven beads for image registration, and the fluorescent signals of the fluorophores should be sufficiently separated to allow single-molecule localization. The low-resolution topographic image registered with the fluorescence signals is used as a guide to acquire high-resolution topographic images of small regions of interest. An outline of the sequence of events in the image acquisition procedure is depicted in Fig. 2a.

3.2.1 Fluorescence Microscope Alignment in Wide-Field Mode

1. Align the laser line parallel to the optical table through the center of the lenses of the beam expander.
2. Remove the wide-field lens to focus the laser at the sample position.
3. Find the focal plane of the objective by focusing and position the focal spot of the laser in the center of the CCD camera.
4. Reinstall the wide-field lens.
5. Align the wide-field lens to focus the laser line onto the back focal plane of the objective to ensure wide-field illumination. Therefore, verify that the laser light follows a straight path above the objective and is collimated. The CCD camera should now be homogeneously illuminated.

3.2.2 AFM Tip Alignment

1. Install an AFM tip in the tip holder.
2. Find the tip position by scanning with the deflecting laser light.
3. Align the deflection laser (1300 nm) close to the tip apex, while maintaining sufficiently high intensity in the quadrant photodiode.
4. Adjust the quadrant photodiode so that the laser is reflected to its center.
5. Find the tip resonance frequency (between 50 and 80 kHz) by ramping the frequency and recording the RMS voltage of the driven oscillation.

3.2.3 Fluorescence Measurement

1. Place the sample on the sample holder and fix using metal springs.
2. Operate the CCD camera in electron multiplier mode (EM). Since single molecules are measured, use highest multiplication. In order to detect with high confidence single bleaching events, the frame rate should be sufficiently high. However, this will also decrease *S/N*. We find that for Atto532, an EM gain of 1000 and an exposure time of 70 ms together with a 5× preamplifier gain and a respective vertical and horizontal frame shift of 0.5 μm/s and 10 MHz resulting in a total acquisition time of ~9 Hz are sufficient to observe single bleaching steps while at the same time detecting single-molecule fluorescence with moderate sensitivity.
3. Adjust the laser power to approximately 1 mW at the sample.
4. Bring the sample approximately in the focal plane by manually adjusting the height of the objective lens.
5. Adjust the correct position on the axial direction using the *z*-piezo stage controlled by the AFM software. Focus on the single-molecule signals (dimmer/smaller diameter), this might bring the fluorescence from the internal reference slightly out of focus, which does not significantly affect the localization of the beads.
6. Close the shutter.
7. Move to a new *x*, *y*-position.
8. Start recording. To ensure detection of all bleaching steps, recording is started before illumination of the sample.
9. Open the shutter.
10. Record sufficient number of frames (≥1000). This will allow every single molecule to bleach, while internal references fluoresce throughout the movie.
11. Make sure sufficient number of beads are present (~10 on a 10 × 10 μm^2 area). If not, repeat **steps 6–10**.

3.2.4 AFM Tip Positioning

1. Install the AFM head.
2. Turn off the EM gain of the CCD camera.
3. Switch to transmission mode by turning on the lamp.
4. Move the AFM tip in *y*-direction until the tip is visible on the CCD camera. The shadow of the tip should be clearly visible in the transmission image.
5. Then move the AFM tip in the *x*-direction until the tip is positioned in the middle of the region of interest.

3.2.5 Acquisition of the AFM Overview Image

1. Switch off the lamp, turn on the EM gain, and verify that the sample was not moved due to positioning of the AFM tip. In case it has moved, reposition the sample to its original location.
2. Switch off the excitation laser and the EM-CCD. Make sure that the EM-CCD cooling is switched off. Preferably, disconnect the cooling module from the camera.
3. To prevent vibrations from the microscope piezo scanner, withdraw the objective from the sample.
4. Approach the sample with the AFM tip using the stepper motor.
5. Find again the resonance frequency of the tip close to the surface.
6. Engage the AFM tip to the surface.
7. Optimize the integral gain of the feedback loop.
8. Scan the surface. The image size, pixel density, and scanning speed should be carefully selected. Ideally, chosen parameters should ensure that specific features of the single DNA molecules are still visible in the overview image. These are necessary to obtain a good correlation with the high-resolution AFM images as discussed in the analysis section. In addition, the scanning speed should be slow enough to minimize parachuting effects to allow fitting of the beads as discussed in the analysis section (*see* Subheading 3.3, and Point 3 in Fig. 2). The larger the overview image, the more molecules can be studied, but the slower the scanning speed has to be. In our experience, an overview image of $12 \times 12\ \mu m^2$ scanned at an acquisition rate of 0.2 Hz and 2048×2048 pixels meets these criteria. Taken together, this results in a total acquisition time of ~3 h. AFM overview images can be recorded overnight.

3.2.6 Acquisition of High-Resolution AFM Images

1. Without changing the optical configuration, switch on the excitation laser.
2. Reconnect the cooling with the EM-CCD camera and switch on the camera.
3. Remove the AFM head and discard the tip.
4. Install a new AFM tip (*see* Subheading 3.2.2).
5. Reinstall the AFM head.
6. Approach the sample with the objective.
7. Bring the sample in the focal plane of the objective by adjusting the position of the objective.
8. If the sample position has changed, reposition it to its original location.

9. Position the AFM tip in the field of view of the EM-CCD camera (*see* AFM tip positioning).
10. To prevent vibrations, switch off the laser, EM-CCD camera, and cooling. Withdraw the objective lens again (*see* Subheading 3.2.5).
11. Record AFM images of molecules of interest. For efficient high-resolution AFM imaging, it can help to first record an AFM image of intermediate size, typically $3 \times 3\ \mu m^2$ (256×256 pixels, scan rate 0.5 Hz). Although no detailed structural information can be obtained, these low-resolution AFM images can guide the acquisition of high-resolution images of molecules of interest and locate their position in the overview image. Again, a trade-off should be made between acquisition time and resolution (e.g., the image size, pixel density, and scanning speed). For our system, high-resolution data was acquired at a scan area of $750 \times 750\ nm^2$ and a scan rate of 0.5 Hz and with 512×512 pixels, resulting in an acquisition time of ~20 min per image.

3.3 Image Analysis

Image analysis is performed with in-house written MATLAB routines, except if stated otherwise. These routines are available upon request.

3.3.1 Fluorescence Microscopy Analysis

1. To increase S/N in the fluorescence channel, accumulate the fluorescence frames. Typically, frames are accumulated using a sliding window average of ten frames to achieve a $S/N \sim 15$.
2. Calculate the x,y-coordinates of each individual fluorescent signal in every frame of the accumulated frames by a least-square 2D Gaussian fitting.
3. Beads are recognized as fluorescent particles that are detected in the last 15% of the recorded frames. Drift in the xy plane is corrected using the average coordinates of the detected beads in each frame with respect to the first frame (Points 1 and 2 in Fig. 2). A unique position for each bead is determined as the mean position of all drift-corrected localizations. These positions are used for image registration (*see* Subheading 3.3.3, and Point 2 in Fig. 2).

3.3.2 Topography Analysis

1. All AFM images are first processed (background correction) by the Scanning Probe Image Processor (SPIP) software. After processing, each corrected AFM is saved as an ASCII file which can be loaded in MATLAB. Depending on the resolution of the AFM image, a different approach is followed to correct the background of the AFM images:
   (a) Each line in the AFM overview image is separately corrected by fitting to a nth-order polynomial. For images of $12 \times 12\ \mu m^2$, n is typically 9.

   (b) Background correction for high-resolution AFM images is typically performed by elevating the individual *x*-profiles so that their height distribution obtains the best match (histogram alignment).

2. Beads are recognized in the background-corrected overview AFM image by defining an intensity threshold based on the triangular method. To correct for parachuting effects, the highest pixel is selected as the starting point for fitting a 2D Gaussian function, which is used to determine the bead positions in the topography image (Point 3 in Fig. 2).

3.3.3 Image Registration

1. The image size of the fluorescence channel is typically much larger than the AFM overview image. Therefore, select the region of interest in the fluorescence channel that corresponds with the AFM overview image.
2. Select a minimum of six internal references in each channel.
3. The two channels are then registered by an affine transformation using the previously calculated positions of the fiducial beads (*see* Subheadings 3.3.1 and 3.3.2). Thereafter, the coordinates in the fluorescence channel are transformed. Save the registered image and transformation matrix.
4. Verify the registration accuracy for the internal references by calculating the mean distance between the beads position in the fluorescence and AFM image.

3.3.4 AFM Image Correlation

1. Define the area of interest in the AFM overview image that corresponds to the high-resolution AFM image.
2. Select at least four clearly distinguishable features in both the overview AFM image and the high-resolution AFM image. These can be fluorescent beads (when present), the ends of DNA fragments, crossing/loops in DNA molecules or protein complexes detected on the DNA. The correlation is then calculated in two steps. First, the two images are roughly correlated by an affine transformation (Point 4 in Fig. 2). Then, fine alignment is performed with an intensity-based image registration function (build in function of MATLAB, "imregtform"). Save the correlated high-resolution AFM image and transformation between high- and low-resolution AFM images.
3. To overlay the fluorescence locations with the high-resolution AFM image, apply both the transformations for image registration (Subheading 3.3.3) and for image correlation to the fluorescence positions.
4. At this stage, a number of parameters can be calculated to quantify the registration quality (*see* **Note 6**).

3.3.5 Stepwise Photobleaching Analysis

1. Load the registered high-resolution AFM image and the raw fluorescence data. If the raw data is uploaded for the first time, it is automatically corrected for drift based on the correction matrix of the high *S/N* fluorescence data (accumulated data), and a drift-corrected file is created which can be used in subsequent analysis.
2. Select an area with the desired topographic feature with corresponding fluorescent signals based on the overlay between the locations detected in the fluorescence images and the registered high-resolution topographic image.
3. Calculate the position of the topographic feature (bead, DNA-end or protein complex). This calculation is done similar as the determination of the beads position in the overview topographic image (Subheading 3.3.2).
4. The corresponding position in the raw fluorescence data is retrieved, and the background fluorescence signal is determined as close as possible to the fluorescent signal.
5. Fluorescence intensity traces are calculated and fitted as described in [20]. Briefly, both the background intensity and fluorescence intensity time traces are calculated. The signal and background intensity can be separately averaged to increase the *S/N*. Steps are assigned based on the intensity change compared with the standard deviation of the background signal. The size of the step can be chosen as the minimal number of standard deviations that the signal has to differ from the baseline intensity. Typically, both signal and background fluorescence are averaged over three frames, and a minimal step size of two times the standard deviation of the background signal is chosen. Figure 3 depicts a typical EcoRV-DNA nucleoprotein complex recorded by correlative fluorescence-AFM

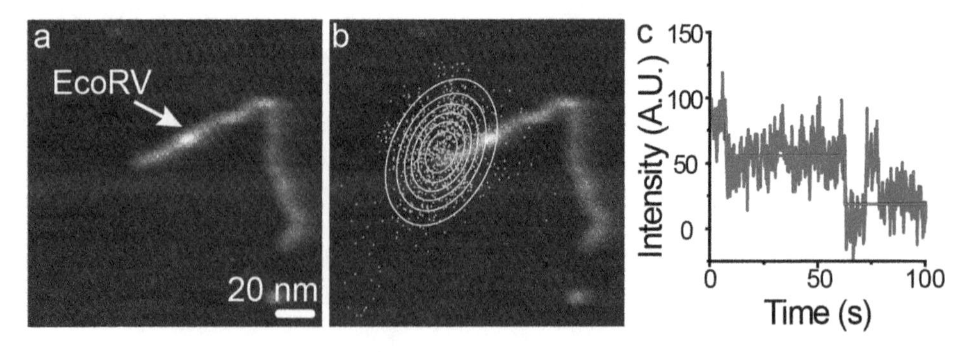

Fig. 3 Correlative imaging of EcoRV bound to DNA. (**a**) High-resolution topographic image of fluorescently labeled EcoRV bound to DNA. (**b**) The recorded fluorescence signal from fluorescently labeled EcoRV correlated with the topographic image. (**c**) The recorded fluorescence signal displays stepwise bleaching over time

microscopy, together with an intensity time trace of the Atto532-labeled EcoRV. The dimeric nature of EcoRV is reflected in the number of bleaching steps.

4 Notes

1. The image pixel size in the fluorescence image is determined by the total magnification together with the camera pixel size according to Image pixel size = Camera pixel size/Total magnification. The image pixel size is important for a proper fitting of the fluorescence signal. For the system described here, the pixel size corresponds to 64 nm (16 μm/(2.5*100) = 64 nm).
2. Mica distorts the fluorescent signal due to its birefringent and absorption properties. Therefore, the mica layer should be as thin as possible to avoid fluorescence distortion, while at the same time, it should cover a sufficiently large area of the glass support to avoid sample deposition artifacts. As a rule of thumb, a coverage of ~50–60% of the glass support meets both requirements. The use of a glass cover slide is necessary to provide sufficient mechanical support for the thin layer of mica.
3. As an alternative for a fluorescent protein bound to DNA, fluorescently end-labeled DNA can be used (e.g., to quantify the registration accuracy as described in [12, 25]). Therefore, fluorescently labeled primers can be used to amplify a DNA fragment by PCR. After PCR cleanup these fragments can be diluted in deposition buffer to a final concentration of 0.125 ng/μL. After mixing the DNA with beads, the sample can be deposited as described in Subheading 3.1.6.
4. To ensure good fitting of the fluorescence signal, molecules have to be sufficiently diluted to prevent overlap of their fluorescence signal. This dilution factor will depend on many parameters such as the affinity of the protein for DNA, protein affinity to the surface, and length of the DNA. In our experience, a concentration ranging from 1 to 5 nM can be used. We found that for EcoRV, a final protein concentration of ~1 nM and DNA concentration of 0.125 ng/μL was appropriate to obtain a sufficiently dispersed sample.
5. Alternative surface functionalization strategies have been reported, for example, deposition from a buffer containing ~mM concentrations of Mg^{2+} on freshly cleaved, untreated mica. However, we found that chemical surface functionalization influences the fluorescence behavior, which is crucial when following fluorescence intensity traces where sufficient long survival times and high *S/N* are needed. In this context, we

tested the spectroscopic properties of Atto532 covalently attached to DNA and adsorbed on mica functionalized with poly-L-lysine or Mg^{2+}. We noticed that modifying the mica surface with poly-L-lysine extended the fluorophore survival time more than twofold compared to Mg^{2+} functionalization, which in turn results in a higher quality of the correlative data.

6. Image registration can be quantified using a number of parameters. The global registration accuracy can be determined by the target registration error (*TRE*): which is the difference between corresponding points (other than the fiducial markers) after registration. In a sample of *N* reference beads, this parameter is typically determined by calculating the registration for $N-1$ beads and calculating the registration error for the bead not used for registration calculation [13–15]. However, the most important parameter is the registration error of the single organic dyes, also termed the localization registration error (*LRE*). We quantified the *LRE* by analysis of the fluorescence and topography of end-labeled DNA strands (*see* Subheadings 2.2 and 3.1.3) and found *LRE* ~25 nm. To detect systematic errors in the transformation matrices, the direction between the mean fluorescence position and the DNA end in the topographic image can be determined. The direction can be defined as the angle between the line connecting the topographic and mean fluorescent position and a horizontal line. In the absence of systematic errors, this angle should be randomly distributed. In addition, the performance of drift correction in the fluorescence channel can be examined by the ratio of the standard deviation of the fluorescence positions in x and y. This ratio should approach unity.

Acknowledgments

We thank Dr. Wolfgang Wende for kindly providing the EcoRV (C21S/K58C) expression plasmid. We acknowledge funding from KU Leuven through the IDO program for financial support; WF, SR, and WV like to thank Fonds Wetenschappelijk Onderzoek (FWO) for personal fellowships.

References

1. Engelkamp H, Hatzakis NS, Hofkens J, De Schryver FC, Nolte RJ, Rowan AE (2006) Do enzymes sleep and work? Chem Commun 9:935–940
2. Solomatin SV, Greenfeld M, Herschlag D (2011) Implications of molecular heterogeneity for the cooperativity of biological macromolecules. Nat Struct Mol Biol 18:732–734
3. van Oijen AM, Blainey PC, Crampton DJ, Richardson CC, Ellenberger T, Xie XS (2003) Single-molecule kinetics of lambda

exonuclease reveal base dependence and dynamic disorder. Science 301:1235–1238
4. Xiao J, Dufrene YF (2016) Optical and force nanoscopy in microbiology. Nat Microbiol 1:1–13
5. Binnig G, Quate CF, Gerber C (1986) Atomic Force Microscope. Phys Rev Lett 56:930–933
6. Lyubchenko YL (2011) Preparation of DNA and nucleoprotein samples for AFM imaging. Micron 42:196–206
7. Sahl SJ, Moerner WE (2013) Super-resolution fluorescence imaging with single molecules. Curr Opin Struct Biol 23:778–787
8. Abbe E (1881) VII.—On the estimation of aperture in the microscope. J R Microsc Soc 1:388–423
9. Yildiz A, Forkey JN, McKinney SA, Ha T, Goldman YE, Selvin PR (2003) Myosin V walks hand-over-hand: single fluorophore imaging with 1.5-nm localization. Science 300:2061–2065
10. Thompson RE, Larson DR, Webb WW (2002) Precise nanometer localization analysis for individual fluorescent probes. J Biophys 82:2775–2783
11. Mortensen KI, Churchman LS, Spudich JA, Flyvbjerg H (2010) Optimized localization analysis for single-molecule tracking and super-resolution microscopy. Nat Meth 7:377–381
12. Frederickx W, Rocha S, Fujita Y, Kennes K, De Keersmaecker H, De Feyter S, Uji-i H, Vanderlinden W (2018) Orthogonal probing of single molecule heterogeneity by correlative fluorescence and force microscopy. ACS Nano 12 (1):168–177
13. Churchman LS, Okten Z, Rock RS, Dawson JF, Spudich JA (2005) Single molecule high-resolution colocalization of Cy3 and Cy5 attached to macromolecules measures intramolecular distances through time. Proc Natl Acad Sci 102:1419–1423
14. Cohen EA, Ober RJ (2013) Analysis of point based image registration errors with applications in single molecule microscopy. IEEE Trans Signal Process 61:6291–6306
15. Cohen EAK, Kim D, Ober RJ (2015) Cramer-Rao lower bound for point based image registration with heteroscedastic error model for application in single molecule microscopy. IEEE Trans Med Imaging 34:2632–2644
16. Bustamante C, Rivetti C (1996) Visualizing protein-nucleic acid interactions on a large scale with the scanning force microscope. Ann Rev Biophys Biomol Struct 25:395–429
17. El-Bahrawi MS, Nagib NN, Khodier SA, Sidki HM (1998) Birefringence of muscovite mica. Opt Laser Technol 30:411–415
18. Rocha S, Hutchison JA, Peneva K, Herrmann A, Mullen K, Skjot M, Jorgensen CI, Svendsen A, De Schryver FC, Hofkens J, Uji-i H (2009) Linking phospholipase mobility to activity by single-molecule wide-field microscopy. ChemPhysChem 10:151–161
19. Das SK, Darshi M, Cheley S, Wallace MI, Bayley H (2007) Membrane protein stoichiometry determined from the step-wise photobleaching of dye-labelled subunits. Chembiochem 8:994–999
20. Kerssemakers JWJ, Laura Munteanu E, Laan L, Noetzel TL, Janson ME, Dogterom M (2006) Assembly dynamics of microtubules at molecular resolution. Nature 442:709–712
21. Deschout H, Cella Zanacchi F, Mlodzianoski M, Diaspro A, Bewersdorf J, Hess ST, Braeckmans K (2014) Precisely and accurately localizing single emitters in fluorescence microscopy. Nat Meth 11:253–266
22. Schulze C, Jeltsch A, Franke I, Urbanke C, Pingoud A (1998) Crosslinking the EcoRV restriction endonuclease across the DNA-binding site reveals transient intermediates and conformational changes of the enzyme during DNA binding and catalytic turnover. EMBO J 17:6757–6766
23. Bonnet I, Biebricher A, Porte PL, Loverdo C, Benichou O, Voituriez R, Escude C, Wende W, Pingoud A, Desbiolles P (2008) Sliding and jumping of single EcoRV restriction enzymes on non-cognate DNA. Nucleic Acids Res 36:4118–4127
24. Wenz C, Jeltsch A, Pingoud A (1996) Probing the indirect readout of the restriction enzyme EcoRV. Mutational analysis of contacts to the DNA backbone. J Biol Chem 271:5565–5573
25. Sanchez H, Kertokalio A, van Rossum-Fikkert-S, Kanaar R, Wyman C (2013) Combined optical and topographic imaging reveals different arrangements of human RAD54 with presynaptic and postsynaptic RAD51–DNA filaments. Proc Natl Acad Sci 110:11385–11390

Part IV

Imaging and Probing of Cells and Tissues and Embryos

Chapter 21

Sensing the Ultrastructure of Bacterial Surfaces and Their Molecular Binding Forces Using AFM

Yoo Jin Oh and Peter Hinterdorfer

Abstract

In this protocol, we provide a detailed step-by-step bacterial surface imaging and molecular analysis procedure. With SPM (scanning probe microscopy)-based dynamic force microscopy (DFM) imaging, we achieved a so far unprecedented resolution of ~1 nm on the outer surface layer of *Tannerella forsythia* and monitored the production of curli fibers on *Escherichia coli* in physiological conditions. Moreover, using these immobilization methods, single-molecule force spectroscopy experiments were conducted on living bacterial cells.

Key words Bacteria, AFM, Imaging, Single-molecule force spectroscopy

1 Introduction

Until recently microbiology has mainly focused on the influence of the chemical environment on bacterial behavior. Hence, for decades, growth in batch cultures and on agar plates were the methods of choice for studies of bacterial physiology. Just in the last few years, it has been recognized that mechanical interaction also plays a significant role in the microbiology on surfaces. Since the two most influential features of bacterial existence are cell-cell interaction and cell-surface interaction in natural environments, characterizing the surface structure and the binding mechanism are expected to provide crucial information for understanding fundamental processes such as bacterial adhesion, surface recognition, biofilm development, and the incipient stage of infection. Despite the high relevance in understanding the bacterial environment, many details still need to be discovered regarding organization and physiological function for identifying the main molecular players involved in the interaction of bacteria with abiotic and cellular surfaces in their adhesion.

Recently, scanning probe microscopy (SPM) techniques have emerged from biological sciences to microbiology. The latest

Yuri L. Lyubchenko (ed.), *Nanoscale Imaging: Methods and Protocols*, Methods in Molecular Biology, vol. 1814,
https://doi.org/10.1007/978-1-4939-8591-3_21,

advances in SPM techniques provide us with powerful tools [1–3] to explore the physical and mechanical properties of bacteria in unprecedented detail [4–7]. SPM has in the meantime developed to a sophisticated method with respect to biological imaging, biomolecular recognition, and localization of specific binding sites [8]. It also shows high potential and superior performance for force-probing the strength of receptor-ligand bonds [9–11] and nano-mechanical properties of biomolecules at the single molecular level [12]. Although the potential of DFM imaging at such resolution has been demonstrated before, performances on microbial cells have been limited due to the mobility and flexibility of the bacteria.

Thus, in addition to the imaging protocol, the procedure for immobilizing bacteria is another key factor to obtain high-resolution SPM images. We thus have refined three immobilization methods and tailored them for different bacteria types and properties, by keeping their natural properties intact [13–15].

Most of gram-negative bacteria own negatively charged surfaces due to the presence of lipopolysaccharides in the outer membrane. Using electrostatic interaction with positively charged surfaces (e.g., poly-L-lysine, gelatin-coated surface [16, 17]), we immobilized bacteria onto mica coated with gelatin.

For neutral bacteria and gram-positive bacteria, we developed a chemical attachment method through multiple covalent bonds using aldehyde groups on top of a poly-L-lysine coating.

When the ultrastructure of bacterial membranes is easily damaged by chemical treatment, bacteria can be mechanically trapped into micron-size holes, which finally leads to stable immobilization for high-resolution SPM imaging.

Moreover, these immobilization methods are also appropriate for performing force spectroscopy experiments on living bacterial cells [15, 18]. In this measurement mode, raw data voltage-displacement signals arising from the SPM cantilever deflection are converted to force-distance signals, so as to calculate mechanical properties and unbinding forces. The strategies for detecting single-molecule interaction forces will be described here, by using chemically modified SPM probes to quantify antibody-antigen and ligand-receptor interactions.

2 Materials

Prepare all solutions using ultrapure water (Milli-Q water prepared by purifying deionized water, to attain a sensitivity of 18 MΩ-cm at 25 °C). All preparations of solution, reactions, and measurements of pH value, weight, and forces, etc. were at a room temperature (20–25 °C, unless indicated otherwise). The cleaning and functionalization of the cantilevers was performed in a well-ventilated hood.

2.1 Bacterial Immobilization

1. Glass cover slip: Clean commercially available glass cover slips 3 times with ethanol. Subsequently, dry either using flowing nitrogen gas or air-dry under sterile conditions (in a laminar flow cabinet).
2. 0.1% (w/v) poly-L-lysine solution: Dissolve 150–300 kDa poly-L-lysine (Sigma-Aldrich, St. Louis, MO, USA) in Milli-Q water to obtain a final concentration of 0.1% (w/v). Thereafter, sterilize by passing through a 0.2 μm filter. The solution can be stored in a −20 °C freezer for ~1 year.
3. 0.1% (w/v) glutaraldehyde solution (Sigma-Aldrich, St. Louis, MO, USA) dissolved in Milli-Q water. The solution can be stored in a −20 °C freezer for ~1 year.
4. Phosphate buffer saline (PBS, pH 7.0).
5. Polycarbonate membrane: Isopore membrane filter (Sigma-Aldrich, St. Louis, MO, USA).
6. 10% PBS: Dilute PBS with Milli-Q water.
7. Gelatin (from porcine skin, G6144, Sigma-Aldrich, St. Louis, MO, USA).

2.2 Bacteria Cell Culture

1. Bacteria strain: *Escherichia coli* K-12 strain W3110 (wild type) and CsgA overexpressing mutant (CsgA(+)) obtained from the National BioResource Project (NBRP, Japan).
2. Bacteria culture medium: Luria-Bertani (LB) medium (Sigma-Aldrich, St. Louis, MO, USA) in Milli-Q water adjusted to pH 7.2.
3. 1 mM isopropyl-β-D-thiogalactoside (IPTG, Sigma-Aldrich).

2.3 Cantilever Preparation

1. Imaging: Commercially available magnetic AC (MAC) mode lever (type VII MAC lever, Keysight Technologies, Santa Rosa, CA, USA).
2. Functionalized cantilever: Clean commercially available AFM cantilever (MSCT, Bruker, CA, USA) 3 times with chloroform.
3. Covalently link the 3-Aminipropyl-triethoxysilane (APTES) coated AFM cantilever with aldehyde-PEG-NHS dissolved in chloroform, and add trimethylamine for 2–4 h. Wash with chloroform and dry with nitrogen gas. Subsequently, incubate with antibody or fibronectin solution for 2 h, and wash the tips in PBS buffer (*see* **Note 1**).

3 Methods

Carry out all procedures at room temperature unless otherwise specified.

3.1 Bacteria Culture

1. Grown in 15 mL round polypropylene tube containing LB medium for 16–18 h overnight (approximately 10^9 CFU) with aeration at 37 °C and 220 rpm. Subsequently, dilute 100-fold in fresh LB medium and continuously grow at 37 °C with aeration until optical density at 600 nm reaches 0.4.
2. CsgA(+) mutant: Grown for an additional 2 h after adding 1 mM IPTG solution.
3. Bacterial suspension: Bacterial cells from 1 mL of culture are collected by centrifugation with 1500 × *g* for 2 min and then resuspended in PBS. This washing step is repeated 3 times. The final bacterial collection is resuspended in 100 μL of PBS.

3.2 Chemical Treatment Method

1. Drop 100 μL of 0.1% concentration of poly-L-lysine solution onto cleaned glass cover slips.
2. Let the droplet dried at a room temperature (*see* **Note 2**).
3. Thereafter, immerse the glass cover slip in 0.1% glutaraldehyde solution for 15 min.
4. After incubation, wash the glass cover slip with Milli-Q water at least 10 times (*see* **Note 3**).
5. Let the glass cover slip dry by soaking liquid off the edge with a paper towel and lightly blowing the liquid off with a pure nitrogen gas.
6. Allow the glass cover slips to sit in a covered clean Petri dish for a few minutes until any remaining water had dried.
7. Drop 100 mL of bacterial suspension onto the glass cover slip and incubate at least 30 min.
8. Dip the glass cover slip in 5 mL of working buffer (PBS) and shake slightly to remove nonadherent bacteria from the glass cover slip. Repeat 2 times in fresh working buffer. Verify the proper attachment of the remaining organisms. (Fig. 1, *see* **Note 4**).

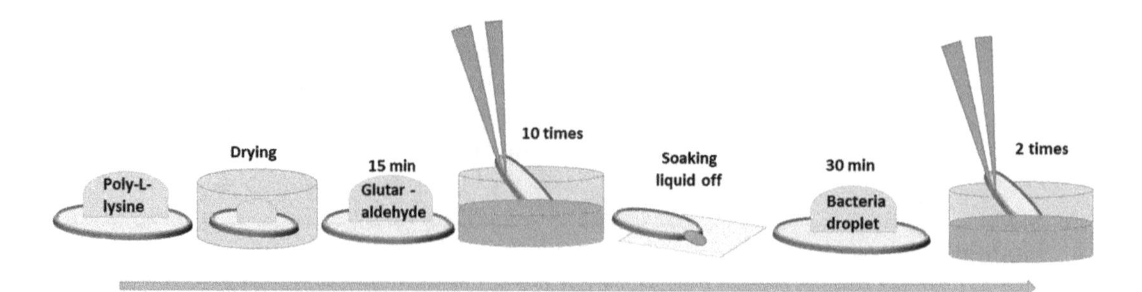

Fig. 1 Schematic design of chemical treatment method

3.3 Trapping or Filtering Method

1. Collect bacterial cells from 1 mL of culture (~OD_{600} 0.4) by centrifugation (1500 × *g*) for 3 min.
2. Resuspend in 1 mL of PBS.
3. Repeat 3 times, and finally resuspend in 1 mL of PBS or working buffer.
4. Filter this bacterial suspension through 0.8 μm polycarbonate membranes with nitrogen gas pressure (~ 0.4 atm) (*see* **Note 5**).
5. Rinse the membrane as gently shaking in 5 mL of working buffer (e.g., PBS) holding one edge with a tweezer.
6. Attach the bacteria-filtered membrane to the AFM sample holder using a double-sided adhesive tape (*see* **Note 6**).

3.4 Electrostatic Interaction Method

1. Dissolve 500 mg gelatin in 100 mL of Milli-Q water at 60 °C. After cooling to 40 °C, dip freshly cleaved mica vertically into the gelatin solution, and allow to air-dry by putting it vertically on a paper towel (*see* **Note 7**).
2. Collect bacterial cells from 1 mL of culture (~OD_{600} 0.4) by centrifugation (1500 × *g*) for 3 min.
3. Resuspend with 10% PBS.
4. Drop 100 μL of bacterial suspension on the gelatin-coated mica surface.
5. Incubate bacterial suspension on the gelatin-coated mica surface for 30 min.
6. Rinse gently and carefully in a Milli-Q water or 10% diluted PBS.

3.5 Imaging Living Bacteria

Place the AFM instrument on an electrically and mechanically isolated platform to avoid electronic noise, mechanical disturbance, and thermal drift.

To reduce aging processes of the sample, start measurements right after the sample preparation.

1. Clean tweezers and AFM sample holder with mild detergent (e.g., 1% SDS) and rinse carefully with Milli-Q water and ethanol.
2. Place the sample on the AFM holder, assemble the liquid cell, and fill measurement buffer into the liquid cell.
3. Choose cantilever according to your selected imaging mode and mount it to the AFM scanner (*see* **Note 8**).
4. Position the AFM cantilever over the selected bacteria with the assistance of a CCD camera or fluorescence microscope. Thereafter, align and focus the laser beam onto the end of cantilever and adjust the photodiode signal to zero.

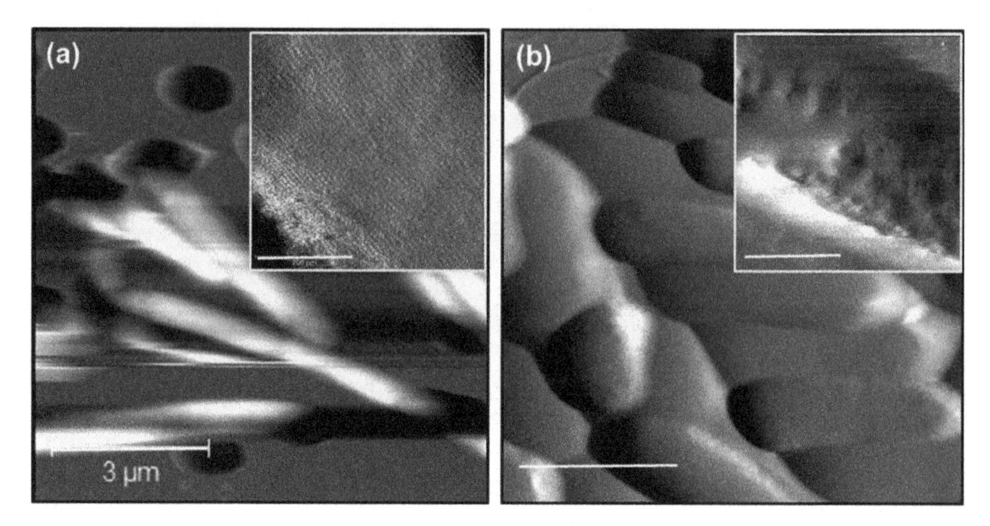

Fig. 2 AFM images of bacteria immobilized on filter- and gelatin-coated surface. (**a**) Topography of *T. forsythia* wild-type bacteria immobilized on the filter (inset; magnified amplitude image that shows a lattice structure of surface layer (S-layer) proteins with a periodicity of around 9 nm and an angle of about 90°) [15] and (**b**) amplitude image of *E. coli* immobilized on the gelatin-coated mica (inset; magnified amplitude image of CsgA overexpressed mutant bacteria that shows the protruded protein domains with diameters of 50–100 nm) [13] (scale bar = 3 μm, inset = 200 nm)

5. Approach the AFM tip with noncontact mode (e.g., tapping mode, acoustic mode, MAC mode) and scan the surface (Fig. 2).

3.6 Single Molecule Force Spectroscopy Measurement

1. Insert the functionalized cantilever into holder of AFM scanner.
2. After carefully aligning and focusing the laser beam on the end of the cantilever, adjust the photodiode signal to zero (*see* **Note 9**).
3. Approach close to the surface using the contact mode.
4. Start force-distance measurement and adjust contact forces until you achieve a good contact point. Observe the unbinding events.
5. Record between 500 and 1000 consecutive force-distance curves at 1 Hz sweep rate (*see* **Note 10**) (Fig. 3).

3.7 Data Analysis

1. After measurements, determine the cantilever sensitivity (given as photodiode output voltage per nm height change of the AFM tip) on hard surfaces (e.g., silicon surface, glass) for calibrating the spring constant of the cantilever. The latter requires also evaluating the thermally driven mean-square bending of the cantilever using the equipartition theorem in ambient environment. Most of commercial AFM setups provide this method in an automated customer application.

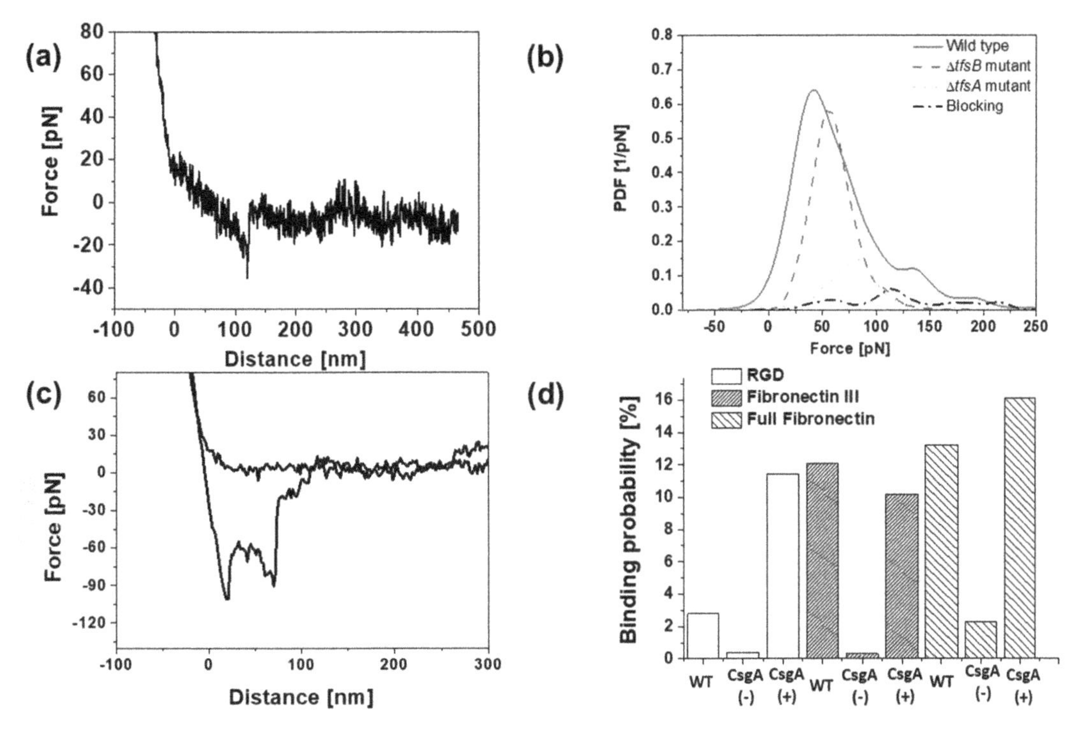

Fig. 3 Typical force-distance curve recorded on the immobilized bacteria and probability density function (PDF). (**a**) Force curve measured between anti-TfsA antibody on the AFM tip and *T. forsythia* immobilized on the filter [15], (**b**) PDF distribution of (**a**) (*see* **Note 11**), (**c**) typical force curves measured between CsgA overexpressed *E. coli* immobilized on the gelatin-coated mica and full fibronectin on the AFM tip [14], and (**d**) distribution of binding probability carried out from the measurements of different bacteria mutants on the AFM tip and different size of fibronectin constructs

2. To obtain force-distance curves, use the AFM software to convert raw data voltage-displacement curves into force-distance curves, so as to calculate the unbinding forces.
3. From the retraction curve, determine the unbinding force from the force difference between the force after the rupture and the force before rupture.
4. Probability density function (PDF): the PDF of unbinding force is constructed [19] from every unbinding event on the same cell at the same pulling speed. For each unbinding force value, a Gaussian of unitary area with its center representing the unbinding force and the width (standard deviation) reflecting its measuring uncertainty (square root of the variance of the noise in the force curve) is computed. All Gaussians from one experimental setting are accordingly summed and normalized with its binding activity to yield the experimental PDF of unbinding force (Fig. 3b) [20] (*see* **Note 11**).
5. Determine binding probability: defined as the percentage of force-distance curves displaying unbinding events (Fig. 3d).

4 Notes

1. Details of AFM tip functionalization procedures are described in instructions found at http://www.jku.at/biophysics/content/e257042. All functionalization procedures were performed according to the method "AFM tip or Support with PEG-Acetal for Coupling of Proteins and Small Amines". In this protocol, anti-TfsA antibody (0.2 mg/mL concentration in PBS) [15] and human fibronectin (YO Protein AB, Huddinge, Sweden) [14] were used for the final step of protein incubation, respectively.
2. Some of bacterial species bind to poly-L-lysine treated glass cover slip and stay sufficiently stable for further SPM imaging. In this case, the bacterial incubation step can be followed after poly-L-lysine treatment step without further glutaraldehyde treatment procedure.
3. Glutaraldehyde binds covalently to poly-L-lysine and offers an additional aldehyde group to another potential binding partner. Yet high amounts of glutaraldehyde can influence or damage to the cell surface. Thus, careful cleaning should be done until unbound glutaraldehyde is washed away.
4. The sample should be washed gently by dipping the glass cover slip into the washing solution. A harsh rinsing step can cause mechanical detachment of bacterial cells from the surface. With this procedure, the glass cover slip is sparsely covered with bacteria (within a surface concentration of ~5–10 cells/100 μm^2, as confirmed by AFM imaging in physiological conditions on planar silicon surfaces).
5. Too high nitrogen gas pressure or a higher number of bacteria can damage the membrane. Ideal pore size of polycarbonate membrane is similar with size of bacteria. Membranes with different pore sizes are commercially available.
6. During AFM measurements in a buffer condition, the double-sided adhesive tape may not be sufficiently strong to hold a polycarbonate membrane over the time. Sometimes, another fixation step can be required, depending on the AFM fluid cell structure (e.g., using larger membranes fitting to the size of the O-ring in the fluid cell).
7. Gelatin surfaces should be sufficiently flat for further AFM imaging. The gelatin solution becomes solid below 40 °C. Thus, keep the solution between 40 and 60 °C during this step.
8. Magnetically coated AFM cantilevers (type VII MAC lever, Keysight Technologies, Santa Rosa, CA, USA) with a nominal spring constant of 0.1 N/m are used in magnetic AC mode (MAC mode). The driving frequency of the cantilever is

9–11 kHz (close to resonance frequency) in liquid, and a 20% amplitude reduction is used for imaging. Amplitude-distance curves should be used to adjust the free cantilever oscillation amplitude reduction value (set point). Both the low value of the free oscillation amplitude and the low amplitude reduction used as feedback signal are of particular importance for preventing disruption of the sample and for enabling stable imaging [21]. The scan speed is a critical parameter for imaging *E. coli* in physiological condition; between 0.3 and 1 Hz is recommended. Too fast scanning can cause bacterial detachment from the surface.

9. This adjustment is crucial for a smooth approach of the tip toward the sample surface and for setting the baseline to zero.
10. For statistics, the measurements should be repeated using different cantilevers and different sample surfaces. For dynamic force spectroscopy measurement, different pulling speed are applied.
11. PDFs contain the original data and can be viewed as the equivalent of continuous force histogram, as shown in Fig. 3b. Their maxima represent the most probable measured unbinding force values, and the uncertainties (widths) reflect the stochastic nature of the unbinding process.

Acknowledgments

This work was supported by an APART (Austrian Programme for Advanced Research and Technology) fellowship of the Austrian Academy of Science.

References

1. Engel A, Muller DJ (2000) Observing single biomolecules at work with the atomic force microscope. Nat Struct Biol 7(9):715–718. https://doi.org/10.1038/78929
2. Martinez-Martin D, Herruzo ET, Dietz C, Gomez-Herrero J, Garcia R (2011) Noninvasive protein structural flexibility mapping by bimodal dynamic force microscopy. Phys Rev Lett 106(19):198101
3. Rico F, Su C, Scheuring S (2011) Mechanical mapping of single membrane proteins at submolecular resolution. Nano Lett 11 (9):3983–3986. https://doi.org/10.1021/nl202351t
4. Arnoldi M, Fritz M, Bauerlein E, Radmacher M, Sackmann E, Boulbitch A (2000) Bacterial turgor pressure can be measured by atomic force microscopy. Phys Rev E Stat Phys Plasmas Fluids Relat Interdiscip Topics 62(1 Pt B):1034–1044
5. Kuznetsova TG, Starodubtseva MN, Yegorenkov NI, Chizhik SA, Zhdanov RI (2007) Atomic force microscopy probing of cell elasticity. Micron 38(8):824–833. https://doi.org/10.1016/j.micron.2007.06.011
6. Mortensen NP, Fowlkes JD, Sullivan CJ, Allison DP, Larsen NB, Molin S, Doktycz MJ (2009) Effects of colistin on surface ultrastructure and nanomechanics of Pseudomonas aeruginosa cells. Langmuir 25(6):3728–3733. https://doi.org/10.1021/la803898g
7. Touhami A, Nysten B, Dufrêne YF (2003) Nanoscale mapping of the elasticity of microbial cells by atomic force microscopy. Langmuir 19(11):4539–4543. https://doi.org/10.1021/la034136x

8. Chtcheglova LA, Waschke J, Wildling L, Drenckhahn D, Hinterdorfer P (2007) Nanoscale dynamic recognition imaging on vascular endothelial cells. Biophys J 93(2):L11–L13. https://doi.org/10.1529/biophysj.107.109751
9. Florin EL, Moy VT, Gaub HE (1994) Adhesion forces between individual ligand-receptor pairs. Science (New York, NY) 264 (5157):415–417
10. Lee GU, Chrisey LA, Colton RJ (1994) Direct measurement of the forces between complementary strands of DNA. Science (New York, NY) 266(5186):771–773
11. Hinterdorfer P, Baumgartner W, Gruber HJ, Schilcher K, Schindler H (1996) Detection and localization of individual antibody-antigen recognition events by atomic force microscopy. Proc Natl Acad Sci 93(8):3477–3481
12. Benoit M, Gabriel D, Gerisch G, Gaub HE (2000) Discrete interactions in cell adhesion measured by single-molecule force spectroscopy. Nat Cell Biol 2(6):313–317. https://doi.org/10.1038/35014000
13. Oh Yoo J, Cui Y, Kim H, Li Y, Hinterdorfer P, Park S (2012) Characterization of Curli a production on living bacterial surfaces by scanning probe microscopy. Biophys J 103 (8):1666–1671. https://doi.org/10.1016/j.bpj.2012.09.004
14. Oh YJ, Hubauer-Brenner M, Gruber HJ, Cui Y, Traxler L, Siligan C, Park S, Hinterdorfer P (2016) Curli mediate bacterial adhesion to fibronectin via tensile multiple bonds. Sci Rep 6:33909. https://doi.org/10.1038/srep33909
15. Oh YJ, Sekot G, Duman M, Chtcheglova L, Messner P, Peterlik H, Schäffer C, Hinterdorfer P (2013) Characterizing the S-layer structure and anti-S-layer antibody recognition on intact Tannerella forsythia cells by scanning probe microscopy and small angle X-ray scattering. J Mol Recognit 26(11):542–549. https://doi.org/10.1002/jmr.2298
16. Allison DP, Mortensen NP, Sullivan CJ, Doktycz MJ (2010) Atomic force microscopy of biological samples. Wiley Interdiscip Rev Nanomed Nanobiotechnol 2(6):618–634. https://doi.org/10.1002/wnan.104
17. Doktycz MJ, Sullivan CJ, Hoyt PR, Pelletier DA, Wu S, Allison DP (2003) AFM imaging of bacteria in liquid media immobilized on gelatin coated mica surfaces. Ultramicroscopy 97 (1–4):209–216
18. Dufrêne YF (2002) Atomic force microscopy, a powerful tool in microbiology. J Bacteriol 184 (19):5205–5213. https://doi.org/10.1128/jb.184.19.5205-5213.2002
19. Baumgartner W, Hinterdorfer P, Schindler H (2000) Data analysis of interaction forces measured with the atomic force microscope. Ultramicroscopy 82(1):85–95
20. Zhu R, Gruber HJ, Hinterdorfer P (2018) Two ligand binding sites in Serotonin transporter revealed by nanopharmacological force sensing. Nanoscale imaging: Methods Mol Biol Vol.1814
21. Puntheeranurak T, Neundlinger I, Kinne RKH, Hinterdorfer P (2011) Single-molecule recognition force spectroscopy of transmembrane transporters on living cells. Nat Protocols 6(9):1443–1452

Chapter 22

Nanoscale Visualization of Bacterial Microcompartments Using Atomic Force Microscopy

Jorge Rodriguez-Ramos, Matthew Faulkner, and Lu-Ning Liu

Abstract

Bacterial microcompartments (BMCs) are polyhedral protein organelles in many prokaryotes, playing significant roles in metabolic enhancement. Due to their self-assembly and modularity nature, BMCs have gained increased interest in recent years, with the intent of constructing new nanobioreactors and scaffolding to promote cellular metabolisms and molecule delivery. In this chapter, we describe the technique of atomic force microscopy (AFM) as a method to study the self-assembly dynamics and physical properties of BMCs. We focus on the sample preparation, the measurement procedure, and the data analysis for high-speed AFM imaging and nanoindentation-based spectroscopy, which were used to determine the assembly dynamics of BMC shell proteins and the nanomechanics of intact BMC structures, respectively. The described methods could be applied to the study of other types of self-assembling biological organelles.

Key words Atomic force microscopy, High-speed AFM, Force spectroscopy, Nanoindentation, Bacterial microcompartment, Carboxysome, Self-assembly, Nanomechanics

1 Introduction

Bacterial microcompartments (BMCs) are assembled entirely by proteins, involving several thousand polypeptides of 10–20 different types. In their native context, BMCs form very large and sophisticated complexes, approximately 100–200 nm in size. The function of BMCs is to optimize metabolic processes that produce intermediates that are toxic or poorly retained by the cell envelope. Three types of BMCs have been extensively characterized: the carboxysomes for CO_2 fixation, the PDU for 1,2-*p*ropane*d*iol *u*tilization, and the EUT for *e*thanolamine *ut*ilization. One of the remarkable properties of BMCs is the self-assembly of building blocks to create a well-defined structure. All BMCs consist of interior enzymes that catalyze sequential metabolic reactions, surrounded by a single-layer proteinaceous shell, 3–4 nm thick [1]. The protein shell, structurally resembling virus capsids, is made of hexameric and trimeric proteins and vertices capped by

Yuri L. Lyubchenko (ed.), *Nanoscale Imaging: Methods and Protocols*, Methods in Molecular Biology, vol. 1814, https://doi.org/10.1007/978-1-4939-8591-3_22, © Springer Science+Business Media, LLC, part of Springer Nature 2018

pentameric proteins [2–4]. These shell building blocks are from the same family of proteins. In addition to confining the interior organization and packing, the intact shell also acts as a physical barrier that has the specific permeability to control the passage of substrates and products of enzymatic reactions. Metabolic enzymes and intermediates are physically concentrated within the self-assembling compartments, which accelerates catalysis, diminishes wasteful O_2-fixing photorespiration by enhancing CO_2 fixation, and effectively confines toxic/volatile metabolites to mitigate cellular damage and carbon loss [5]. As such, there is tremendous interest in engineering BMCs to produce nanobioreactors and factories for enhancing cellular metabolism and creating new molecular scaffolding systems.

Carboxysomes are found in most cyanobacteria and some chemoautotrophs and are currently the most intensively studied group of BMCs. They contain an outer polyhedral shell composed primarily of a few protein paralogs encapsulating the primary carboxylation enzyme, ribulose-1,5-bisphosphate carboxylase oxygenase (Rubisco), and carbonic anhydrase (CA) [6]. The interior CA dehydrates HCO_3^- to CO_2 and supplies a high concentration of CO_2 in the vicinity of Rubisco; the protein shell acts as a selectively permeable boundary to CO_2 and HCO_3^- efflux and O_2 influx, promoting significant CO_2 accumulation in the compartment. Thus, the combination of specialized enzymes and exquisite nanoscale organization of carboxysomes enhances the efficiency of CO_2 fixation and reduces the wasteful side reaction of photorespiration. The evolution of carboxysomes and powerful CO_2-concentrating mechanisms enables cyanobacteria to contribute up to 25% of global carbon fixation [7].

There are open questions in the study of BMCs. For example, what is the self-assembly mechanism, and what are the structural and mechanical properties of functional BMCs. Recently, atomic force microscopy (AFM) has been increasingly used to study the dynamics of protein assemblies [8–11] and to extract the mechanical characteristics of biological specimens [12, 13], such as viruses [14, 15]. AFM imaging can be performed in a buffer solution, at ambient temperature and pressure, allowing the study of biological specimens under near physiological conditions [16, 17]. Compared to conventional AFM imaging, high-speed AFM (HS-AFM) allows for the capture of molecular-resolution images at high frame rates. It has been used to dissect the lateral organization and self-assembly dynamics of shell proteins, HOCH_5815, in the synthetic BMC shell facets [1]. HOCH_5815 possesses a hexameric structure and serves as the major shell protein of BMCs in *Haliangium ochraceum* [1, 18, 19]. When expressed heterologously, HOCH_5815 self-assemble to form sheets in *E. coli* cells. Recording subsequent AFM images revealed the dynamic diffusion of individual HOCH_5815 proteins and the formation and disassembly processes of BMC shell facets. The observed structural dynamics could have implications to the in vivo biosynthesis flexibility of BMC structures.

AFM-based nanoindentation has been exploited in studying the viral capsid mechanics [14, 15, 20]. A recent study has reported the first determination of the topography and intrinsic mechanics of functional β-carboxysomes using AFM and AFM-based nanoindentation [21]. It illustrated the flexible organization and soft mechanical properties of β-carboxysomes compared to rigid viruses and represents a method to inspect the structural and mechanical features of synthetic metabolic organelles and protein scaffolds in bioengineering.

This chapter describes the procedure to prepare samples for AFM and outlines the high-resolution AFM imaging and nanoindentation experiments on BMC shell proteins and intact BMCs, as well as data analysis. The method can be modified to study other self-assembling biological systems as well.

2 Materials

2.1 Sample Preparation

1. For protein dynamics studies, a shell facet protein (HOCH_5815) of the BMC from a myxobacterium *Haliangium ochraceum* was expressed [1, 19].
2. For nanomechanical analysis, functional β-carboxysomes were isolated from a freshwater cyanobacterium *Synechococcus elongatus* PCC 7942 [21].

2.2 Atomic Force Microscopy

1. Atomic force microscopy: For HS-AFM, use an AFM system with the capacity of scanning at 30 Hz or above, e.g., JPK NanoWizard® NanoScience AFM with the Ultra-S head (JPK Instruments, Germany). For AFM nanoindentation, any commercial AFM system should be valid, as far as the applied force during scanning can be kept around 100 pN (*see* **Notes 1** and **2**).
2. AFM cantilever: For HS-AFM, use a sharp cantilever ($R < 10$ nm) which is capable of being operated at a high scan rate, e.g., USC-F0.3-k0.3 (NanoWorld, Switzerland). For AFM nanoindentation, use a soft cantilever ($k < 0.4$ N/m) with the tip radius R between 15 and 20 nm, e.g., DNP-D (Bruker, USA) (*see* **Note 3**).
3. Mica disc (diameter: ~8 mm) mounted on the appropriate substrate with a water-resistant glue (Crystal Clear Araldite, Switzerland). Depending on the AFM system, the substrate could typically be a glass slide or a metal disc.
4. Translucent scotch tape, e.g., Scotch Magic tape (3M, USA).

5. Buffer solutions: adsorption buffer (10 mM Tris, 25 mM $MgCl_2$, 150 mM KCl, pH 7.5) and imaging buffer (10 mM Tris, 150 mM KCl, pH 7.5) [17, 22, 23] (*see* **Note 4**).

3 Methods

3.1 AFM Sample Preparation

The following steps are common for both HS-AFM imaging and AFM nanoindentation experiments:

1. Cleave the mica with the translucent tape (*see* **Note 5**).
2. Deposit an aliquot of adsorption buffer on mica (volume suggested, 40 μL).
3. Add 2 μL of HOCH_5815 or β-carboxysomes into the adsorption buffer drop. HOCH_5815 proteins were incubated for 5 min, and β-carboxysomes were incubated for at least 30 min at room temperature (*see* **Note 6**).
4. Wash three times with the imaging buffer to remove weakly bound shell proteins/BMCs from the mica surface.
5. Mount the samples onto the AFM sample stage.
6. Mount the cantilever in the tip holder, and add 20 μL of the imaging buffer to diminish the probability of having air bubbles (*see* **Note 7**).
7. Wait until the system stabilizes and engage (typically $t > 30$ min) (*see* **Note 8**).

3.2 High-Speed AFM Imaging to Study the Self-Assembly Dynamics of Shell Proteins

1. Amplitude modulation AFM (AM-AFM) is generally used to scan the samples at high frame rates. Tune the cantilever to its resonant frequency in liquid, and select a free amplitude near the surface $A_0 \leq 2$ nm. During the AFM scan, use a setpoint value as high as possible (i.e., as close to the free amplitude A_0 as possible) to minimize the applied force (*see* **Note 9**).
2. Scan a large region at a low resolution to identify the area of interest (e.g., $5 \times 5\ \mu m^2$, 512 pixels/line, 5 Hz).
3. Scan the area while optimizing parameters to apply minimal forces and gradually zoom in (*see* **Note 10**).
4. Once the desired area is framed (e.g., $100 \times 100\ nm^2$), gradually increase the scan rate (e.g., up to ≥30 Hz), while optimizing the scan parameters to obtain the molecular-level resolution (*see* **Note 11**).
5. Capture a series of images at the desired speed and image quality (Fig. 1).
6. Repeat **steps 2–5** to scan different regions of your sample (e.g., $n = 5$).
7. Process the image series to produce a video of the image sequence, and identify the dynamic processes or events.

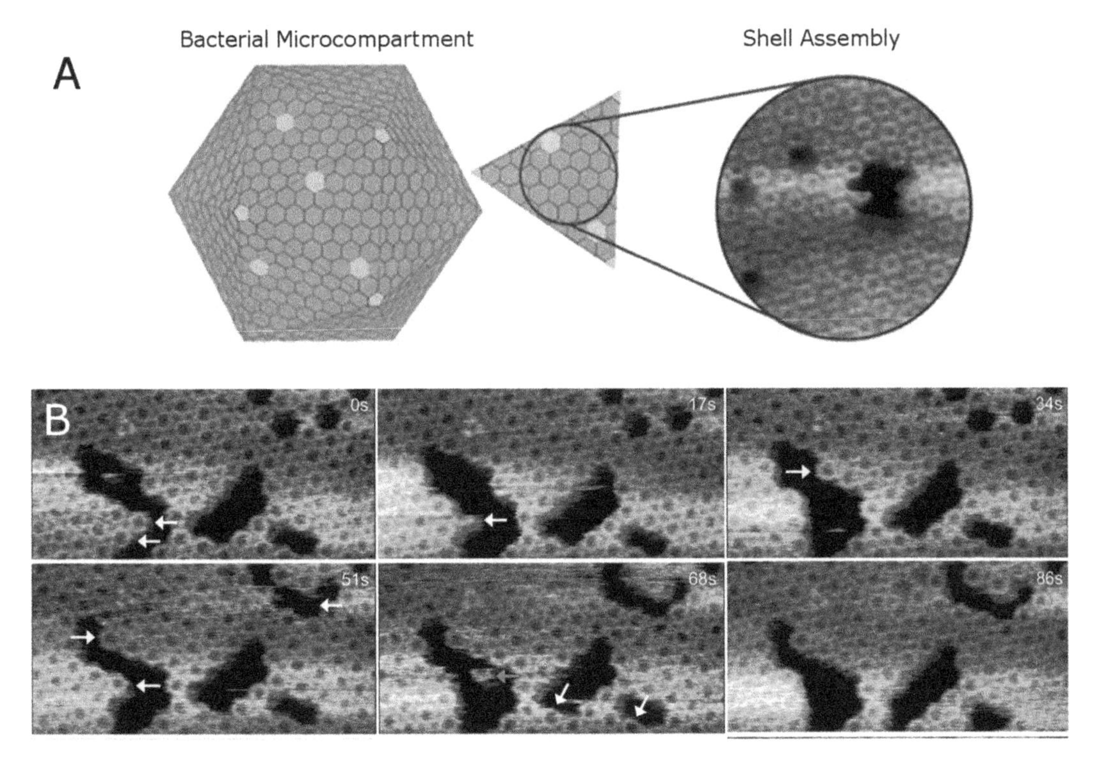

Fig. 1 Visualization of the self-assembly dynamics of BMC shell proteins using the JPK NanoWizard® 3 NanoScience AFM with the Ultra-S head. (**a**) Representation of a full BMC and one of its facets; highlighted, AFM image of the hexagonal assembly of HOCH_5815 from *Haliangium ochraceum*. (**b**) Series of frames captured at HS-AFM, 512 pixels/line, captured at 30 Hz. The arrows show individual proteins in motion (blue), joining (yellow) or leaving the patch between scans (white). Reproduced from [1] (http:/pubs.acs.org/doi/full/10.1021/acs.nanolett.5b04259, further permissions related to the material excerpted should be directed to the ACS)

3.3 AFM Nanoindentation and Spectroscopic Analysis to Study the Mechanics of Carboxysomes

1. Calibrate the cantilever stiffness before starting the nanoindentation measurements (*see* **Note 12**).
2. Locate isolated β-carboxysomes by scanning at gentle forces ($F \leq 100$ pN) in a noninvasive mode like the Quantitative Imaging (QI, JPK) or the PeakForce Tapping (Bruker) (*see* **Note 13**).
3. Perform 5–10 static force curves (FC) on the central region of the β-carboxysome (Fig. 2). The FCs should be in the elastic regime of the particle (for β-carboxysomes, $F_{max} \approx 300$ pN) (*see* **Note 14**). These curves are processed to obtain the equivalent stiffness and Young's modulus of β-carboxysomes.
4. Repeat **steps 2** and **3** until you have enough particles and force curves for statistical analysis (e.g., $n = 25$ particles) (*see* **Note 15**).

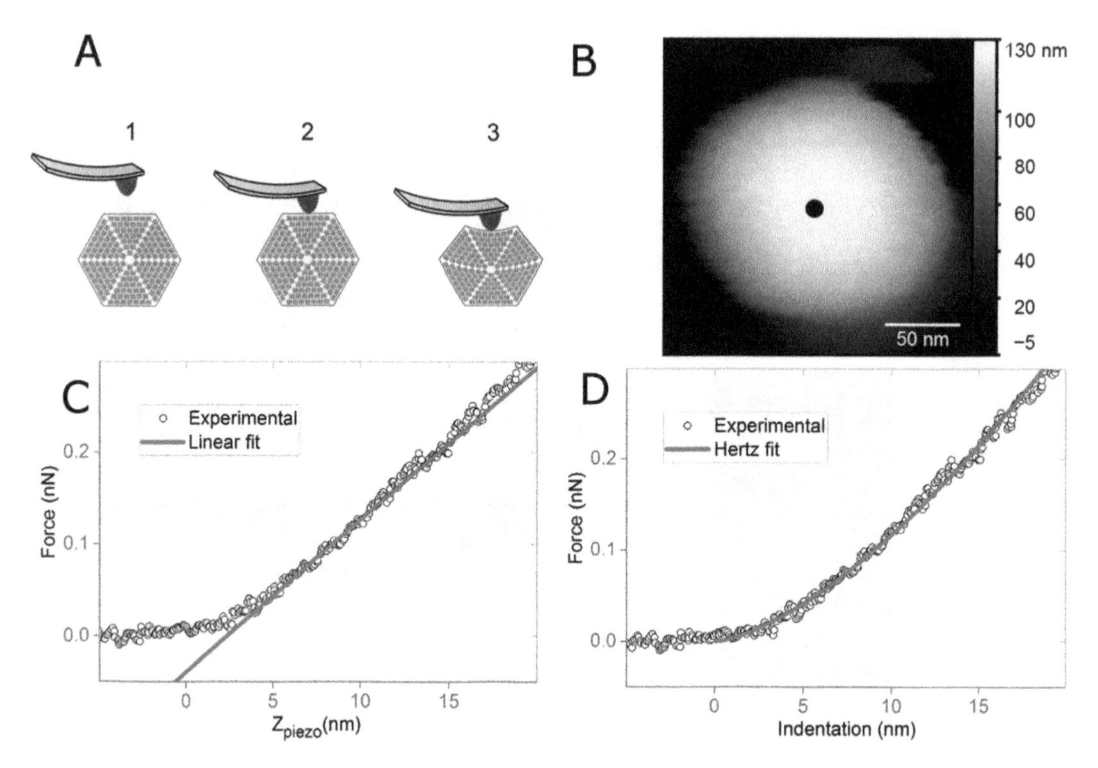

Fig. 2 Mechanical characterization of β-carboxysomes using AFM nanoindentation. (**a**) Schematic of an AFM spectroscopy experiment, including AFM tip engagement (1), tip-carboxysome contact (2) and indentation (3) with increasing force. (**b**) AFM image of a single β-carboxysome isolated from *Synechococcus elongatus*. The black dot represents the indentation position on the carboxysome. (**c**) Force-displacement curve. The equivalent stiffness is recovered from the linear fitting on the 0.05–0.15 nN region of the force curve. (**d**) Force-indentation curve. The Young's modulus is recovered from the fitting using the Hertzian model based on the 0–10 nm indentation region. Reproduced from [21] with permission from the Royal Society of Chemistry

5. Estimate the equivalent stiffness of β-carboxysomes, by fitting the linear range of the force F versus piezo displacement Z_p curve (Fig. 2c):

$$k_s = \frac{k_{cs} k_c}{k_c - k_{cs}} \quad (1)$$

where:

k_s is the estimated spring constant of β-carboxysomes.

k_{cs} is the measured slope in the fitting region.

k_c is the pre-calibrated stiffness of the cantilever (**step 1**) (*see* **Note 16**).

6. Transform the piezo displacement Z_p into indentation depth d by using the following equation:

$$d = z_p - z_0 - \frac{F}{k_c} \quad (2)$$

where:

Z_0 is the initial contact distance.

k_c is the cantilever's spring constant.

The expression F/k_c represents the expected deflection of the cantilever interacting with an infinitely stiff surface (*see* **Note 17**).

7. Estimate Young's modulus E of β-carboxysomes by fitting the force F versus indentation d curve to the Hertz contact model (Fig. 2d):

$$F = \frac{4}{3}\frac{E}{1 - \nu_{CB}^2}R_t^{1/2}d^{3/2} \quad (3)$$

where:

R_t is the tip radius.

ν_{CB} is the Poisson coefficient of β-carboxysomes. For β-carboxysomes, we consider $\nu_{CB} = 0.5$.

Make the fitting in a region that represents up to 10% of the particle's height (for β-carboxysomes, $d \approx 10$ nm) [21] (*see* **Note 18**).

4 Notes

1. Not all AFM systems are capable to measure at a high scan rate. Please check if your equipment meets the requirements.
2. β-carboxysomes could be damaged when scanned in contact mode. This could be a general feature for other types of BMCs or protein organelles. For this reason, it is recommended to use a gentle imaging mode, such as the QI mode (JPK Instruments), the PeakForce Tapping (Bruker, USA), or the Jumping mode (Nanotec Electronica, Spain). All of them allow for scanning samples under the forces ≤100 pN with the appropriate cantilevers.
3. HS-AFM imaging requires small cantilevers with high resonance frequencies. In the case of JPK with Ultra-S, the resonance frequency in liquid should be above 30 kHz. For example, the small USC-F0.3-k0.3 cantilever (NanoWorld, Switzerland) is a sharp tip with a low spring constant $k = 0.3$ N/m, resonance in liquid around 150 kHz, and wearing resistance. However, its small dimension (length = 20 μm, width = 10 μm) makes it unsuitable for many commercial AFMs.

4. All solutions must be freshly prepared and filtered (0.22 μm) to avoid contamination.
5. It is important to cleave a full mica sheet. If not cleaved areas are visible, the process must be repeated, since the tape's glue can act as the contaminant for the sample.
6. The specific concentration of samples for AFM studies varies and should be adjusted depending on the properties of samples and incubation conditions.
7. If air bubbles appear on the cantilever, one can pipet an aliquot of imaging buffer between the cantilever and the holder several times to get rid of the air bubbles.
8. It is important to keep the AFM system at a stable temperature. It is a good practice to turn on the system a few hours before AFM measurements.
9. AM-AFM has different denomination for different systems, e.g., AC mode in JPK, tapping mode in Bruker, and intermittent contact in NT-MDT. It is recommended to read the work of Miller et al. [24] for high-resolution imaging using AM-AFM.
10. Optimize the AFM imaging parameters while framing the area of interest to improve the image quality. Features in the patch should start to be visible while decreasing the scan size (~500 nm^2 for HOCH_5815).
11. Individual proteins are visible at relatively low scan rates (i.e., 4–8 Hz) [25, 26]. However, the time required to record an image is too long in comparison with the dynamic processes that take place. On the other hand, a frame recorded at 30 Hz, 512 pixels/line requires ~17 s. If we use 256 pixels/line, it requires ~7 s, but if we increase the scan rate to 60 Hz, a 256×256 pixels frame is recorded in ~2 s. The higher scan rates decrease the quality of the image, which makes harder to optimize the scan parameters. The main limitation comes from your specific instrument. In the case of HOCH_5815 proteins, with a JPK AFM with the Ultra-S head (JPK Instruments, Germany), protein dynamic events were observed at the relatively low frame rate of 30 Hz [1].
12. Most commercial AFM systems have implemented calibration methods to estimate the cantilever's stiffness. If this option is not available for your AFM, you may use the Sader's method [27] or the thermal tune method [28] to calibrate the cantilever's stiffness.
13. Contact mode and AM-AFM are not recommended for AFM imaging of soft protein complexes, since they can be aggressive

to the flexible structures. Successive scans (over 20 times) on carboxysomes could produce permanent deformations in the structure.

14. Perform the FC analysis based only on the elastic regime of the particles (i.e., where the approach and retract lines coincides). The threshold for the maximum applied force should be established for each type of BMCs and self-assembling protein organelles. When the applied force exceeds the elastic regime, plastic deformations may occur. This could produce irreversible damage to the structure and may cause an inaccurate estimation of the material properties.
15. The cantilever might be contaminated during the experiment session. Although there are methods to clean the cantilever tips [24], other problems, like tip wear, can affect the measurement [29]. If a cantilever shows symptoms of contamination, it is recommended to replace it with a new cantilever prior to the spectroscopy measurement, to ensure a similar starting point for every set of experiments.
16. The linear model was adopted in the past for the study of viruses because the FCs deviate from the classical Hertz contact model [30]. In the case of β-carboxysomes, the response is not as linear as in the case of viruses. Due to the fact that the outer shell of the β-carboxysome structurally resembles virus capsids, it is interesting to estimate the equivalent stiffness using the linear region after the initial nonlinear zone of the curve [21].
17. The analysis software of most commercial AFMs implements the conversion to indentation described in Eq. (2). In some cases, it is called vertical tip position (JPK), or separation (Bruker), but it refers to the same term.
18. The tip radius R_t is one of the main sources of uncertainty in Young's modulus estimation. It can be checked by scanning electron microscopy. In general, it is recommended to use the value reported by the manufacturer and use brand-new cantilevers during the experiments. The Hertz model is implemented in many commercial AFMs. Otherwise, the curves can be processed externally with the software of your preference.

Acknowledgments

This work was supported by a Royal Society University Research Fellowship (UF120411), a Royal Society Research grant for University Research Fellowship (RG130442), a Royal Society Challenge grant (CH160004), a Biotechnology and Biological Sciences Research Council grant (BB/R003890/1), and a Biotechnology and Biological Sciences Research Council grant (BB/M024202/

1). We acknowledge the Liverpool Centre for Cell Imaging for technical assistance and access to confocal/TIRF microscopes (Biotechnology and Biological Sciences Research Council, BB/M012441/1).

References

1. Sutter M, Faulkner M, Aussignargues C, Paasch BC, Barrett S, Kerfeld CA, Liu L-N (2016) Visualization of bacterial microcompartment facet assembly using high-speed atomic force microscopy. Nano Lett 16 (3):1590–1595. https://doi.org/10.1021/acs.nanolett.5b04259
2. Klein MG, Zwart P, Bagby SC, Cai F, Chisholm SW, Heinhorst S, Cannon GC, Kerfeld CA (2009) Identification and structural analysis of a novel carboxysome shell protein with implications for metabolite transport. J Mol Biol 392(2):319–333. https://doi.org/10.1016/j.jmb.2009.03.056
3. Tanaka S, Kerfeld CA, Sawaya MR, Cai F, Heinhorst S, Cannon GC, Yeates TO (2008) Atomic-level models of the bacterial carboxysome shell. Science 319(5866):1083–1086. https://doi.org/10.1126/science.1151458
4. Kerfeld CA, Sawaya MR, Tanaka S, Nguyen CV, Phillips M, Beeby M, Yeates TO (2005) Protein structures forming the shell of primitive bacterial organelles. Science 309 (5736):936–938. https://doi.org/10.1126/science.1113397
5. Yeates TO, Crowley CS, Tanaka S (2010) Bacterial microcompartment organelles: protein shell structure and evolution. Annu Rev Biophys 39:185–205. https://doi.org/10.1146/annurev.biophys.093008.131418
6. Rae BD, Long BM, Badger MR, Price GD (2013) Functions, compositions, and evolution of the two types of carboxysomes: polyhedral microcompartments that facilitate CO_2 fixation in cyanobacteria and some proteobacteria. Microbiol Mol Biol Rev 77(3):357–379
7. Behrenfeld MJ, Randerson JT, McClain CR, Feldman GC, Los SO, Tucker CJ, Falkowski PG, Field CB, Frouin R, Esaias WE, Kolber DD, Pollack NH (2001) Biospheric primary production during an ENSO transition. Science 291(5513):2594–2597. https://doi.org/10.1126/science.1055071
8. Ando T (2017) Directly watching biomolecules in action by high-speed atomic force microscopy. Biophys Rev 9:421. https://doi.org/10.1007/s12551-017-0281-7
9. Uchihashi T, Scheuring S (2017) Applications of high-speed atomic force microscopy to real-time visualization of dynamic biomolecular processes. Biochim Biophys Acta 1862:229. https://doi.org/10.1016/j.bbagen.2017.07.010
10. Preiner J, Horner A, Karner A, Ollinger N, Siligan C, Pohl P, Hinterdorfer P (2015) High-speed AFM images of thermal motion provide stiffness map of interfacial membrane protein moieties. Nano Lett 15(1):759–763. https://doi.org/10.1021/nl504478f
11. Uchihashi T, Kodera N, Ando T (2012) Guide to video recording of structure dynamics and dynamic processes of proteins by high-speed atomic force microscopy. Nat Protoc 7 (6):1193–1206. https://doi.org/10.1038/nprot.2012.047
12. Sicard D, Fredenburgh LE, Tschumperlin DJ (2017) Measured pulmonary arterial tissue stiffness is highly sensitive to AFM indenter dimensions. J Mech Behav Biomed Mater 74:118–127. https://doi.org/10.1016/j.jmbbm.2017.05.039
13. Ramos JR, Pabijan J, Garcia R, Lekka M (2014) The softening of human bladder cancer cells happens at an early stage of the malignancy process. Beilstein J Nanotechnol 5 (1):447–457. https://doi.org/10.3762/bjnano.5.52
14. Marchetti M, Wuite G, Roos WH (2016) Atomic force microscopy observation and characterization of single virions and virus-like particles by nano-indentation. Curr Opin Virol 18:82–88
15. Mateu MG (2012) Mechanical properties of viruses analyzed by atomic force microscopy: a virological perspective. Virus Res 168 (1–2):1–22. https://doi.org/10.1016/j.virusres.2012.06.008
16. Liu LN, Scheuring S (2013) Investigation of photosynthetic membrane structure using atomic force microscopy. Trends Plant Sci 18 (5):277–286. https://doi.org/10.1016/j.tplants.2013.03.001
17. Liu LN, Duquesne K, Oesterhelt F, Sturgis JN, Scheuring S (2011) Forces guiding assembly of light-harvesting complex 2 in native membranes. Proc Natl Acad Sci U S A 108 (23):9455–9459

18. Sutter M, Greber B, Aussignargues C, Kerfeld CA (2017) Assembly principles and structure of a 6.5-MDa bacterial microcompartment shell. Science 356(6344):1293–1297. https://doi.org/10.1126/science.aan3289
19. Lassila JK, Bernstein SL, Kinney JN, Axen SD, Kerfeld CA (2014) Assembly of robust bacterial microcompartment shells using building blocks from an organelle of unknown function. J Mol Biol 426(11):2217–2228. https://doi.org/10.1016/j.jmb.2014.02.025
20. Roos WH (2011) How to perform a nanoindentation experiment on a virus. In: Peterman EJG, Wuite GJL (eds) Single molecule analysis: methods and protocols. Humana Press, Totowa, NJ, pp 251–264. https://doi.org/10.1007/978-1-61779-282-3_14
21. Faulkner M, Rodriguez-Ramos J, Dykes GF, Owen SV, Casella S, Simpson DM, Beynon RJ, Liu L-N (2017) Direct characterization of the native structure and mechanics of cyanobacterial carboxysomes. Nanoscale 9 (30):10662–10673. https://doi.org/10.1039/C7NR02524F
22. Liu LN, Sturgis JN, Scheuring S (2011) Native architecture of the photosynthetic membrane from *Rhodobacter veldkampii*. J Struct Biol 173 (1):138–145. https://doi.org/10.1016/j.jsb.2010.08.010
23. Liu LN, Duquesne K, Sturgis JN, Scheuring S (2009) Quinone pathways in entire photosynthetic chromatophores of *Rhodospirillum photometricum*. J Mol Biol 393(1):27–35. https://doi.org/10.1016/j.jmb.2009.07.044
24. Miller EJ, Trewby W, Farokh Payam A, Piantanida L, Cafolla C, Voitchovsky K (2016) Sub-nanometer resolution imaging with amplitude-modulation atomic force microscopy in liquid. J Vis Exp 118:54924
25. Kumar S, Cartron ML, Mullin N, Qian P, Leggett GJ, Hunter CN, Hobbs JK (2017) Direct Imaging of Protein Organization in an intact bacterial organelle using high-resolution atomic force microscopy. ACS Nano 11 (1):126–133. https://doi.org/10.1021/acsnano.6b05647
26. Scheuring S, Nevo R, Liu LN, Mangenot S, Charuvi D, Boudier T, Prima V, Hubert P, Sturgis JN, Reich Z (2014) The architecture of Rhodobacter sphaeroides chromatophores. Biochim Biophys Acta 1837(8):1263–1270. https://doi.org/10.1016/j.bbabio.2014.03.011
27. Sader JE, Chon JWM, Mulvaney P (1999) Calibration of rectangular atomic force microscope cantilevers. Rev Sci Instrum 70 (10):3967–3969. https://doi.org/10.1063/1.1150021
28. Cook S, Schaffer TE, Chynoweth KM, Wigton M, Simmonds RW, Lang KM (2006) Practical implementation of dynamic methods for measuring atomic force microscope cantilever spring constants. Nanotechnology 17 (9):2135–2145. https://doi.org/10.1088/0957-4484/17/9/010
29. Ramos JR (2014) Tip radius preservation for high resolution imaging in amplitude modulation atomic force microscopy. Appl Phys Lett 105(4):043111. https://doi.org/10.1063/1.4892277
30. Carrasco C, Carreira A, Schaap IAT, Serena PA, Gómez-Herrero J, Mateu MG, de Pablo PJ (2006) DNA-mediated anisotropic mechanical reinforcement of a virus. Proc Natl Acad Sci 103(37):13706–13711. https://doi.org/10.1073/pnas.0601881103

Chapter 23

Time-Resolved Imaging of Bacterial Surfaces Using Atomic Force Microscopy

Haig Alexander Eskandarian, Adrian Pascal Nievergelt, and Georg Ernest Fantner

Abstract

Time-resolved atomic force microscopy (AFM) offers countless new modes by which to study bacterial cell physiology on relevant time scales, from mere milliseconds to hours and days on end. In addition, time-lapse AFM acts as a complementary tool to optical fluorescence microscopy (OFM), for which the combination offers a correlative link between the physical manifestation of bacterial phenotypes and molecular mechanisms obeying those principles. Herein we describe the essential materials and methods necessary for conducting time-resolved AFM and dual AFM/OFM experiments on bacteria.

Key words Atomic force microscopy, Dual AFM/Optical fluorescence microscopy, Time-lapse imaging, Bacterial morphogenesis, Bacterial surface dynamics, Sample immobilization, Off-resonance tapping (ORT)

1 Introduction

Imaging dynamic processes of individual bacteria has long been pursued by optical fluorescence and phase microscopy. Although the resolution limit of conventional optical microscopy is on the order of the size of many bacteria, countless discoveries into how molecular mechanisms function have been made by exploring time-lapse imaging. Atomic force microscopy has proven to be a very powerful method for studying microbes, especially since it enables imaging the surface of living bacteria with exquisite resolution [1–3]. The multi-parametric measurements of AFM can also be greatly enhanced by adding a temporal aspect through continuous time-lapse imaging, thereby enabling the dynamic study of bacterial physiology by imaging the cell surface at high resolution. Therefore, time-lapse AFM imaging offers a unique opportunity to understand how the bacterial cell surface evolves in time. However, challenges remain to technically enable AFM imaging at relevant

Yuri L. Lyubchenko (ed.), *Nanoscale Imaging: Methods and Protocols*, Methods in Molecular Biology, vol. 1814, https://doi.org/10.1007/978-1-4939-8591-3_23,

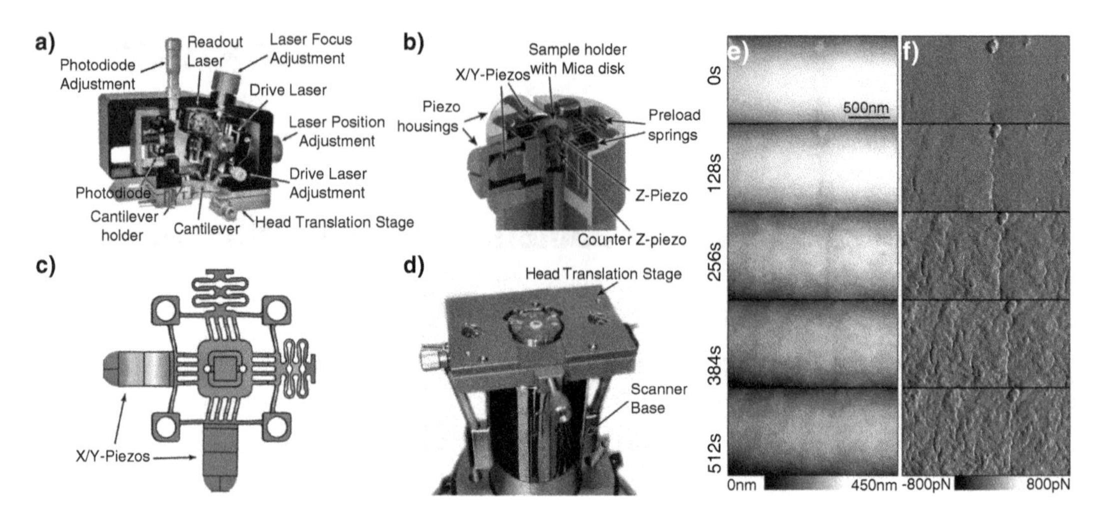

Fig. 1 High-speed AFM imaging of bacterial processes. (**a**) Section view through a custom small cantilever AFM head with photothermal excitation. Both the drive and readout laser paths are depicted in red. The laser spot is displaced manually in X, Y, and Z. (**b**)–(**d**) Flexure-based custom high-speed AFM scanner compatible with Bruker Multimode and the HS-AFM head of (**a**). The AFM head (**a**) is magnetically set atop the AFM scanner (**b** and **d**). (**c**) Finite element analysis, color showing relative uniaxial displacement of the central flexure and piezos depicted in (**b**). (**d**) Photography of the built scanner, mounted on a vertical engage base with sample translation stage. (**e**) and (**f**) AFM image sequence (**e**, topography, and **f**, error signal) of the action of the antimicrobial peptide CM-15 on the surface of *E. coli*. Images were taken at 12.6 s/frame

scan rates for suitably long periods of time, so as to measure accurately fast or slow kinetic processes.

Short short-term time-lapse imaging for up to a couple of hours offers the ability to track cell events occurring on the time scale shorter than one cell division. Recent advances in high-speed AFM offer the ability to image cells with second and sub-second frame rates and maintain a sharp image, with high spatial resolution [3–5], (Fig. 1). On the other end of the spectrum, long-term time-lapse imaging offers the ability to trace numerous cell phenotypes manifesting over multiple generations, with continuous experiments lasting for several days (Figs. 2, 3, and 4) [2]. While these two examples are seemingly very different, many of the same principles and methods apply to both high-speed AFM and long-term time-lapse AFM imaging of bacteria. We therefore discuss both temporal scales in this chapter. The two imaging modes we highlight are amplitude modulation mode (tapping mode) and off-resonance tapping (ORT).

Time-resolved AFM imaging of bacteria depends heavily on bacterial immobilization on the sample surface to withstand lateral forces, but without impacting cell viability and behavior. Several immobilization strategies have been developed. The type of immobilization strategy to be used depends on the type of bacterium (shape, surface chemistry), the solution in which the experiment has to be performed (ionic strength), as well as the type of experiment (long-term vs. high-speed).

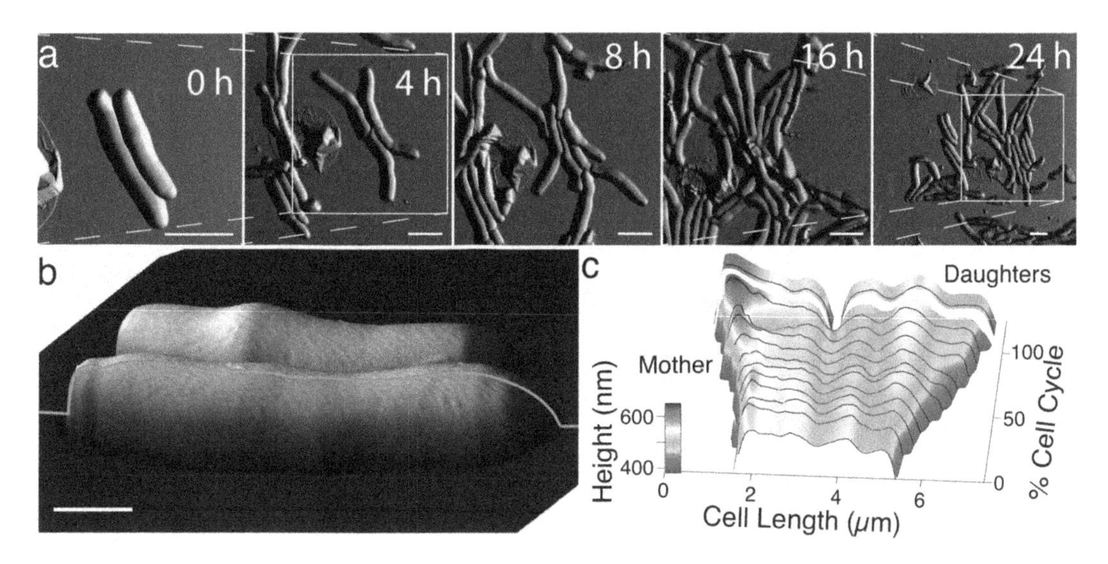

Fig. 2 Long-term time-lapse AFM imaging of *Mycobacterium smegmatis*. (**a**) Time series of 3D AFM height images overlaid with AFM peak force error images for wild-type *M. smegmatis*. Scale bar, 3 μm. (**b**) Mycobacterial surface topology. Yellow trace of the cell profile highlights the undulating mycobacterial surface morphology. Scale bar, 1 μm. (**c**) Kymograph of the cell surface height of a representative cell from birth to division, showing that division occurs within the center-most wave trough. (**b** and **c**) Images are representative of $n = 270$. Reprinted with permission from [2]

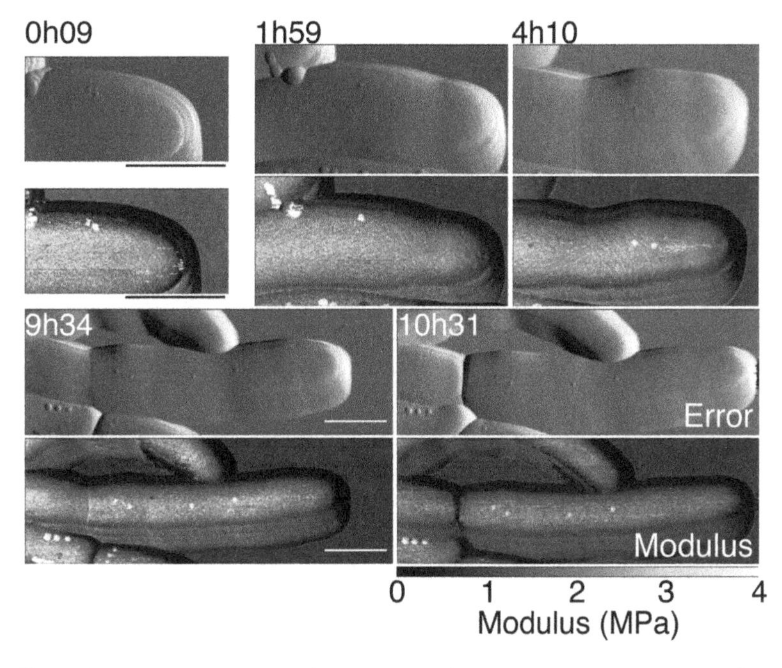

Fig. 3 Growth dynamics revealed by measuring the mycobacterial cell surface Young's modulus, imaged at high resolution, by off-resonance tapping, during long-term time-lapse AFM imaging. Time is represented in hours and minutes. Color bar represents the DMT modulus, used to calculate the rigidity modulus of an ideal isotropic elastic surface. Scale bar, 1 μm. Adapted and reprinted with permission from [2]

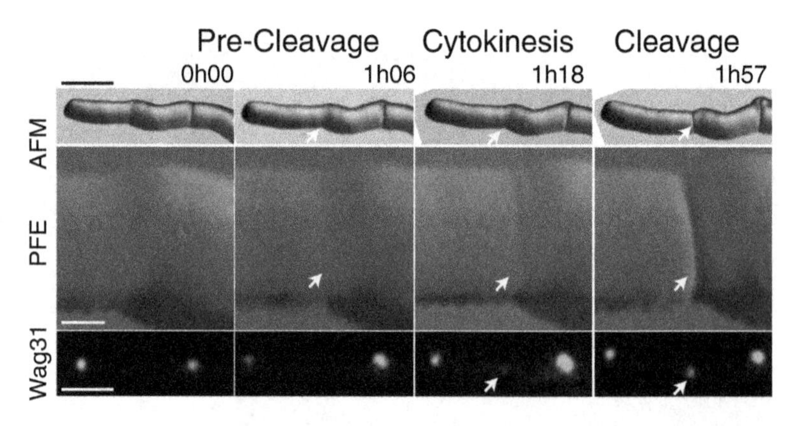

Fig. 4 Sequence of events in the mycobacterial cell cycle. Cells imaged by correlated dual AFM (upper and middle panels) and OFM (bottom panels). Upper panels depict the three-dimensional representations of topography images. Middle panels depict peak force error images, which highlights the pre-cleavage furrow (white arrowheads). Numbers indicate the time elapsed since birth (hours, minutes). Scale bars: upper and bottom, 3 μm; middle, 500 nm. Cells expressing the cytokinesis marker Wag31-GFP show that cytokinesis completes before cell cleavage. Reprinted with permission from [2]

1.1 Bacterial Immobilization

Bacterial adherence is dependent on the interaction between molecules on the bacterial cell surface and the sample surface. Some immobilization strategies consist of depositing ligands on the sample surface to promote bacterial surface receptor adherence [6]. Alternatively, specific ligand-directed covalent immobilization can be utilized [7]. Adherent polyphenolic proteins offer a non-specific and powerful adhesion in aqueous conditions between biochemically diverse bacteria.

Non-specific immobilizing interactions such as electrostatic interactions between sample surfaces and ligands can be used [8]. The ionic strength of the medium the cells are imaged in strongly affects bacterial-surface interactions [9–12]. Recent evidence suggests that the ubiquitous and continuous interaction of poly-L-lysine (PLL) with the bacterial surface negatively impacts cell growth and viability [13]. For time-resolved AFM experiments, the electrostatic interactions between the sample surface and bacteria are often modified by the growth medium used to culture bacteria in time-lapse experiments [7, 14]. Coated surfaces utilizing electrostatics to promote charge-based immobilizing interactions between a bacterium and sample surface are suitable for simplified time-lapse experiments, while problematic for more complex time-lapse experiments in which growth conditions are varied. Non-specific immobilizing interactions maintain a homogenous interaction between bacteria and the sample surface in a non-biased manner. A low-melting temperature agarose gel provides a sufficient local aqueous environment to maintain the structural integrity of the outer cell membrane while exchanging nutrient conditions [15]. PDMS immobilizes rod-shaped mycobacteria exhibiting a hydrophobic waxy outer surface composed of mycolic acids [7, 16], in a non-specific manner and permissive to growth [2, 17]. The transparency of PDMS enables for dual AFM/OFM imaging. A nanometers-thin PDMS coat allows for the minimization of the working distance between the sample and an inverted optical fluorescence microscope objective.

An alternative method for immobilization, physical entrapment of bacteria has been used [18]. These methods physically overcome the lateral forces generated by AFM scanning of the cantilever [19] by holding the cells in pores, pits, or traps.

1.2 Suitability of Imaging Modes for Time-Lapse Imaging of Bacteria

The way in which the cantilever contacts the cell surface is controlled by the movement of the cantilever and has a major impact on the information that can be gathered from the bacterial sample. While some modes are amenable to high-speed AFM imaging, others are more permissive for long-term time-lapse AFM imaging experiments.

Contact mode, which comprises of continuously maintaining a stable tip-sample force, creates large lateral forces that are often too large for imaging live bacteria that are weakly or even strongly adherent to a surface [7]. Image bacteria by AFM in contact mode using a scan rate between 0.2 and 2 Hz.

Amplitude modulation mode (also called tapping mode) is a more “gentle” mode of imaging that comprises of exciting a cantilever close to its resonance frequency and measuring the change in the amplitude as the cantilever is brought closer to the sample surface [20]. Intermittent contact between the cantilever tip and the bacterial cell surface in tapping mode minimizes lateral forces generated during scanning across the sample. High-speed AFM studies report frame rates of seconds or tens of seconds and require resonating cantilevers at frequencies 1–2 orders of magnitude higher than normal levers (greater than 100 kHz–1 MHz in fluid) [1, 3–5, 21, 22].

Tapping mode also requires intensive manual manipulation of the imaging parameters in order to maintain a stable scan over prolonged periods of time or in varying physiological conditions [10]. In long-term imaging, freely floating bacteria or other particles that adhere to the cantilever cause changes to the cantilever resonance frequency or damping, such that the scan gradually fades away from the bacterial surface. Continuously monitor to ensure proper surface tracking [23].

Force volume mode, where the AFM performs force ramps at a defined number of positions, provides sample topography and information about the sample mechanical properties. Force volume is often too slow to image growing bacteria in time-lapse, since the rate at which the force ramps can be acquired is limited by the fluid drag, and the AFM/cantilever dynamics.

Fast force-distance curve-based imaging techniques have emerged that periodically tap on the surface but with frequencies much lower than the resonance frequency of the cantilever. Commercial implementations of this technique include Pulsed Force Mode™, PeakForce Tapping™, or HybriD™ mode. To avoid favoring any trade names, we will refer to these modes in this text as off-resonance tapping (ORT) modes. These modes can measure the mechanical and related properties (such as adhesion, stiffness, or dissipation) of the bacteria (Fig. 3) at normal AFM imaging speeds. An additional advantage of these modes is that they are more stable due to the continuous background subtraction which makes them very well suited for long-term, time-lapse imaging of bacterial process.

Because of inherent physiological differences between bacterial species and the different requirements for high-speed vs. long-term AFM experiments, it is not possible to make a clear recommendation which imaging mode is the best for imaging time-resolved processes of bacteria. In general, for high-speed short-term AFM experiments, tapping mode has emerged as the preferred technique, since it is, at the moment, faster than ORT methods. For slower scan rates but long-term experiments, off-resonance tapping modes are preferable due to their higher stability and the additional mechanical properties information that can be gained.

1.3 Instrument Considerations

AFMs come in a large variety of shapes and sizes. Generally, there are three groups. For AFMs with sample scanners, the AFM cantilever stays stationary, and the sample is moved in the XYZ directions to scan and track the topography. For AFMs with tip scanners, the sample remains stationary and the tip is scanned in XYZ directions. Other forms of AFMs have a combination of sample and tip scanner, where, for example, the tip is moved in the Z direction and the sample in the XY direction. Which type of instrument is best suited depends very much on the particular experiment. For high-resolution imaging on small samples, sample scanners have proven to be superior, whereas for larger samples it is imperative that the sample remains stationary. For integration of the AFM with other techniques such as optical microscopy, the tip scanning AFM or a combination of tip-scanning and sample-scanning AFM has proven advantageous.

High-speed AFM imaging generally requires specialized custom AFM equipment [1, 21, 24–26] (Fig. 1a–d). Recently, commercial instruments have become available that can be used for these tasks. Long-term time-lapse imaging can, in principle, be performed with most commercial AFMs. With long experiment durations, there emerge additional issues that require special experimental conditions or improved instrumentation. Extra functionality is required to maintain the concentration of nutrients and temperature as per the experimental conditions as well as to eliminate freely floating bacteria in the sample medium where they can disrupt the AFM cantilever.

AFM setups are available also in combination with inverted optical microscopes, especially with optical fluorescence microscopes (OFM). The combination of AFM and optical fluorescence microscopy offers complementary sets of measurements [15, 27]. Time-resolved imaging by dual AFM/OFM offers insight into the dynamics of physical properties and molecular mechanisms of bacteria (Fig. 4). From a practical standpoint, an optical microscope can provide molecular justification to target a bacterial sample for imaging at high resolution by time-resolved AFM. Unexpectedly, with time-resolved dual AFM/OFM arise new challenges.

Commercial systems function to stabilize the AFM on an inverted optical fluorescence microscope. Maintaining optimal image acquisition in both AFM and optical fluorescence microscopy is essential to time-resolved dual imaging. To that end, the most important drawback is maintaining a low level of noise in the AFM as a result of imaging bacteria by optical microscopy. This can be overcome by mounting the AFM on a mechanical support structure that stabilizes the AFM above the sample and mechanically isolates it from the vibrating optical microscope base, localized below the sample (Fig. 5). For large samples that do not require high magnification, air objectives minimize the transfer of mechanical noise through the sample to the AFM. However, with bacteria, submicron resolution imaging requires the use of a 100× oil objective. In noise-sensitive measurements, the oil objectives should be withdrawn until the oil meniscus is broken.

For dual imaging, it is important to take into account the impact of the wavelength of the AFM laser, which can interfere with experimental dyes or fluorescent biological markers by bleaching excited fluorophores. To avoid this, an infrared laser can be used to read out the cantilever deflection, or fluorophores can be used that are not excited by the wavelength at which the AFM laser emits. Another significant challenge to account for when imaging by OFM is the overexposure to light, which can result in phototoxicity to bacteria. Phototoxicity can limit the OFM imaging frame rate. For optimal optical fluorescence images in dual AFM/OFM experiments, the position of the cantilever must be displaced out of the field of view of the objective in order to reduce optical distortions due to light reflecting off of the cantilever. Finally, maintaining the image registration is an issue that must be properly accounted for by using fiducial markers, like fluorescent beads, which can be imaged both by AFM and OFM. Commercial AFM/OFM setups often provide calibration procedures and software routines for the registration.

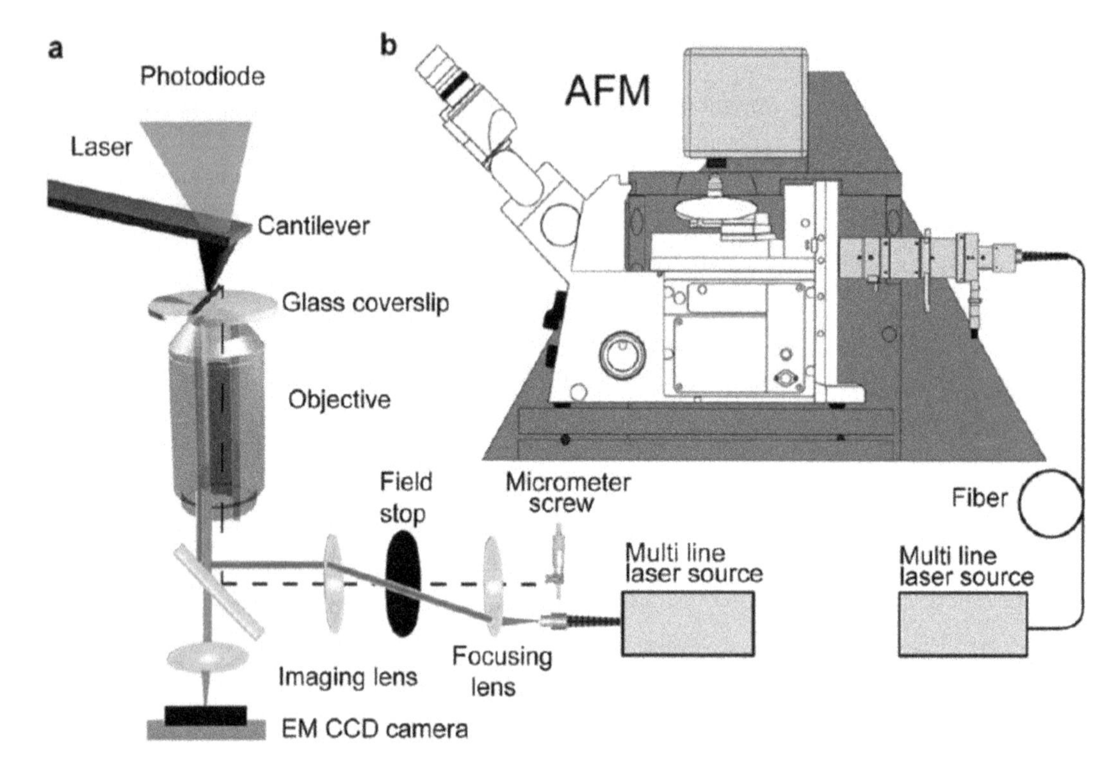

Fig. 5 Schematic representation of a dual AFM/OFM setup. (**a**) Schematic of the aligned optical path with the AFM cantilever. The AFM cantilever is centered in the field of view by adjusting the position of the inverted optical fluorescence microscope, mounted on a x/y-translation stage. (**b**) The AFM is mechanically stabilized using a rigid and massive aluminum apparatus, which is mechanically isolated from the optical microscope body. The dual instrument setup is supported on a vibration isolation table and insulated within an acoustically isolated box that equally function as a dark space void of parasitic light sources. Reprinted with permission from [42]

We describe below the materials and methods necessary for conducting time-lapse AFM and dual AFM/OFM experiments on live bacteria in order to resolve both high-speed and long-term cell processes.

2 Materials

2.1 Instrumentation

1. For HS-AFM imaging we use a homebuilt HS-AFM system based on the Bruker Multimode platform [26, 28, 29]; *see* Fig. 1. The setup can be used with a small HS-AFM scanner as depicted in Fig. 1c, d or, when larger ranges are required, with a Bruker EV scanner. In the latter case, feed-forward controllers should be used to suppress the scanner resonances [30–32]. The method described in this chapter has also been tested with a Bruker Dimension FastScan and should be easily transferred to other HS-AFM systems.

2. For long-term time-lapse AFM imaging as well as combined AFM/optical microscopy imaging, we used a modified Bruker Dimension Icon or a Bruker Dimension FastScan, each mounted on an inverted optical microscope (*see* Fig. 5). The methods described here should be easily transferred to other combined AFM/optical microscopy systems.

2.2 Bacterial Immobilization

1. Round glass coverslip. Use 6–10 mm for high-speed AFM and 12–25 mm for long-term time-lapse AFM.
2. 2.5 M hydrochloric acid (HCl).
3. Autoclaved, distilled water.
4. Poly-L-lysine hydrobromide (PLL, mol wt $\geq$ 300,000 Da) [3, 15, 33]: dissolve 1 mg/mL PLL hydrobromide in autoclaved, distilled water and bring pH to 8.0. Sterilize PLL solution by passing through 0.2 μm filter. Store solution at −20 °C.
5. Tris buffer: 1 M, pH 8.0.
6. Polydimethylsiloxane elastomer (PDMS, Sylgard 184) and curing agent.
7. Vacuum desiccator.
8. Spin coater (MicroTec LabSpin6) with vacuum chuck matching the coverslip size.
9. Polyvinylidene difluoride syringe filters (0.2 and 5 μm).

2.3 Bacterial Growth Medium

1. Luria Broth for *Escherichia coli* (*E. coli*). 950 mL H_2O, 10 g tryptone, 10 g NaCl, 5 g yeast extract, pH adjusted to 7.0 with 5 N NaOH (~0.2 mL), final volume adjusted to 1 L with H_2O.
2. 7H9 medium for *Mycobacterium smegmatis* (*M. smegmatis*) supplemented with 0.5% albumin, 0.2% glucose, 0.085% NaCl, 0.5% glycerol, and 0.05% Tween-80.
3. Imaging medium:
   - Use distilled water or phosphate buffer solution (PBS) for *E. coli* adhered to PLL.
   - Use 7H9 growth medium for *M. smegmatis* adhered to PDMS.
4. Antibiotics, antimicrobial peptides, as well as small molecule inhibitors or inducers, if needed to trigger the bacterial process to be studied.

2.4 AFM and OFM Imaging

1. For HS-AFM imaging in tapping mode: small cantilevers with spring constant of ca 0.1 N/m and resonance frequency in water of 500 kHz (e.g., Olympus AC-10).

2. For time-lapse AFM imaging in off-resonance tapping: cantilevers with spring constant of 0.5–1 N/m, resonance frequency ca 40 kHz (e.g., Bruker ScanAsyst-Fluid).
3. Temperature controller (230 W programmable system) (*see* **Note 1**).
4. Round, fluorescent microparticle beads (1 μm diameter).

3 Methods

The following methods section describes different protocols depending on the bacterial type and the speed of the process that is to be imaged. For high-speed AFM imaging of *E. coli*, follow Subheadings 3.1–3.5 and 3.8. For time-lapse AFM imaging of *M. smegmatis*, follow Subheadings 3.1, 3.3–3.4, 3.6, and 3.8. For time-lapse imaging of *M. smegmatis* with correlated AFM/fluorescence microscopy, follow Subheadings 3.1, 3.3–3.4, and 3.6–3.8.

3.1 Coverslip Cleaning

Boil round coverslips in 100 mL 2.5 M HCl solution for 10 min.

Remove coverslips and rinse in distilled water (6 times).

Place coverslips vertical to air-dry, and cover with aluminum foil to prevent the deposition of dust.

3.2 Bacterial Immobilization with Poly-L-Lysine

1. Immerse coverslips vertically into a solution of poly-L-lysine hydrobromide (PLL, 0.05 mg/mL, pH 8.0) and 10 mM Tris and incubate at room temperature for 10 min.
2. Vertically remove the newly coated coverslip from solution and rinse with distilled water.
3. Place coverslips vertically and let dry, overnight.
4. Coverslips can be stored at room temperature for up to 1 week.

3.3 Bacterial Immobilization Using Polydimethylsiloxane (PDMS)

1. Mix PDMS elastomer with curing agent at a weight ratio of 15:1.
2. Place in a desiccator and apply vacuum until no air bubbles are visible in the PDMS mixture (~10 min).
3. Dilute the 15:1 PDMS mixture with hexane at a ratio of 1:10 (PDMS mixture: hexane).
4. Place a droplet of the PDMS mixture onto a round coverslip.
5. Spin-coat the PDMS droplet on a coverslip at a spin velocity of 8000 rpm for 30 s.
6. Bake PDMS-coated coverslips at 80 °C for 10 min, and place a cover over the coverslips to prevent dust from settling on the PDMS.
7. Store PDMS-coated coverslips at room temperature for up to 1 month.

3.4 Sample Preparation

1. Prepare 5–10 mL of growth medium (e.g., Luria Broth for *E. coli* or 7H9 for *M. smegmatis*) in a round bottom test tube, if necessary in the presence of antibiotic.
2. Inoculate growth medium with an isogenic population of bacteria.
3. Grow cells to the desired growth phase (e.g., OD_{600nm} of 0.6–1), incubated in a heated shaker-incubator (*see* **Notes 2–4**).
4. If cells are prone to clump, filter cells through a 5 μm pore size PVDF filter (Millipore) or, respectively, appropriate filter pore size (*see* **Note 5**).
5. Spin a 2 mL aliquot of bacterial solution at 6000 rpm (~ 3500 × *g*) for 10 min in a centrifuge.
6. Remove the supernatant above the pelleted cells and resuspend in 100 μL of imaging medium.
7. If a heated sample holder is used, place coverslip in sample holder (*see* **Note 1**).
8. Drop 50 μL of bacterial sample on the coated coverslip.
9. Allow for bacteria in solution to sediment on the sample surface (e.g., 15–30 min).
10. Gently wash the bacterial sample with imaging medium to remove non-adherent cells that are freely floating.
11. Add sufficient medium (circa 2 mL, depending on AFM system used) to maintain both adherent cells and the AFM cantilever and the underside of the cantilever holder immersed in liquid medium.
12. Connect the heating controller to the sample holder and set to the desired temperature (Fig. 6).

3.5 High-Speed Time-Lapse AFM Imaging of Bacteria

1. Load appropriate cantilever in cantilever holder of high-speed AFM (*see* **Notes 6, 7**, and **10**).
2. Add a droplet of imaging solution on the cantilever.
3. Align laser over the cantilever and zero the photodetector (according to high-speed AFM user manual).
4. Approach the sample and set imaging gains according to AFM user manual.
5. Find an appropriate area by first performing a low-speed (0.5–1 Hz line rate), overview scan of the sample to choose an appropriate experimental field to image (*see* **Note 11**).
6. Scan along the long axis of rod-shaped bacteria to help prevent the possibility that the cell will detach from the surface.

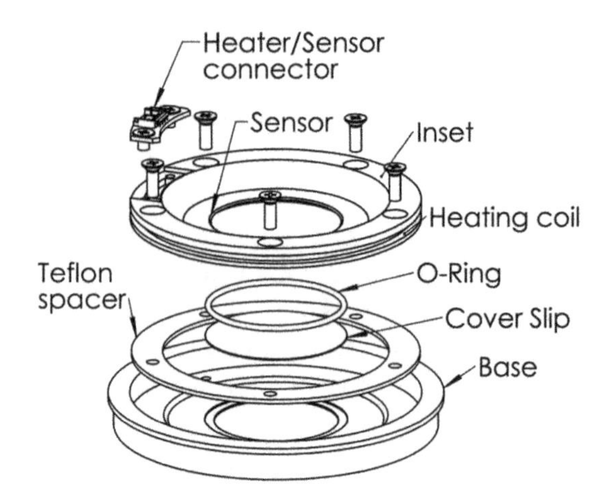

Fig. 6 Schematic representation of a heated coverslip sample holder for dual AFM/OFM imaging. A glass 22 mm coverslip sits directly upon a metal base (anodized aluminum). A Teflon spacer isolates the heating from the upper metal piece (also made of anodized aluminum) with an integrated resistive heating coil, sensor, and connector in order to reduce the heat sink represented by the base. A rubber O-ring mechanically stabilizes the glass coverslip in place, sandwiched between the base and upper heated metal inset

7. When imaging small features at high speed, choose the scan area such that only the upper part of the cell is imaged at high-speed to limit lateral forces.
8. Tune the imaging settings properly before increasing the scan speed such that when scanning at higher speeds, only minor changes need to be made to optimize the image quality.
9. If required, trigger the process of interest (e.g., by adding reagents).

3.6 Long-Term Time-Lapse AFM Imaging in ORT

1. Load appropriate cantilever for ORT in the AFM cantilever holder (e.g., ScanAsyst-Fluid cantilever (Bruker) exhibiting a spring constant 0.7 N m^{-1} and a resonance frequency in fluid at ~40 kHz) (*see* **Notes 6, 8–10**).
2. Add a droplet of imaging solution on the cantilever.
3. Align laser over the cantilever and zero the photodetector (according to AFM user manual).
4. Approach the sample and set imaging gains according to AFM user manual.
5. If available, use an optical image to choose an area with adherent bacteria (*see* **Note 11**).
6. Engage with a zero scan size.

7. Choose a scan size that is appropriate for the expected final colony size.
8. Set scan rate appropriate such that bacteria are well tracked at the final scan size. Images acquired with 512 samples per line and 256 lines at 0.3–0.5 Hz produce an image acquisition rate of 5–10 min for image aspect ratios of 2:1 and 1:1, respectively.
9. If required, trigger the process of interest (e.g., by adding reagents) (*see* **Note 3**).
10. Exchange the imaging fluid every ~2 h (depending on imaging fluid volume) to maintain a constant physiological condition.
11. If the ORT force curve no longer exhibits the typical shape, lift the cantilever from the surface and recalibrate the background.
12. In prolonged experiments lasting longer than several bacterial cell generations, detached, free-floating cells in the sample fluid can interfere with the cantilever and impede the stability of the AFM scan. In that case, thoroughly wash the sample environment by exchanging the image fluid 5 times. If washing is no longer effective to remedy the problem, change the cantilever, exchange the imaging fluid, and find the original imaging area.

3.7 Time-Resolved Dual Atomic Force and Optical Fluorescence Microscopy

1. Prior to beginning bacterial imaging, align the AFM cantilever with the optical microscope with a 100× oil objective. Intermediate power objectives (e.g., 20× or 40×) are helpful to find the region of interest.
2. Add fluorescent beads on the sample surface as fiducial markers for off-line registration of the AFM and optical microscopy images.
3. Use the bright field optical microscope to find non-clumped, adherent bacteria so as to reduce the time spent finding an experimental sample to image by AFM.
4. To acquire an optical fluorescence image, move the AFM cantilever out of the field of view in order to minimize optical distortions due to light reflecting off of the cantilever.
5. Turn the AFM laser off during optical imaging to reduce parasitic light affecting the OFM imaging.
6. Reengage the cantilever to the original imaging position.
7. Continue imaging for the duration of the experiment, taking fluorescent images at appropriate intervals (*see* **Notes 12** and **13**).

3.8 AFM Image Analysis

Multiple programs exist for AFM image analysis, several of which are freely available for download, including Gwyddion (Department of Nanometrology, Czech Metrology Institute), Image

SXM, WSxM, or GXSM. Below are described the most basic steps to be performed for image processing. For more information regarding more advanced and uniform automatic image processing of AFM image sequences, *see* **Note 14**.

1. Mask the area in the image that is to be considered the flat image background (excluding all the bacteria).
2. Subtract the plane of the background across the image using a plane fit subtraction.
3. If necessary, flatten the scan lines to remove temporal drifts and laser mode hops.
4. Reset the background level to near 0 nm.
5. If there are regular oscillations in the image that are an artifact (e.g., fan noise from the optical microscope), a filter based on fast Fourier transform (2D FFT) can be used to remove these artifacts.
6. For image sequences, make sure that all images have the same Z data scale as well as the same background value.
7. If necessary, remove image drift using a drift correction plugin Fiji (e.g., BisQuit) (*see* **Note 15**) or a drift correction feature in the AFM image processing software.

4 Notes

1. Temperature control is a critical factor in the investigation of many biological phenomena, like virulence [34]. Therefore, maintaining an experimental sample at a biologically relevant temperature is a necessary addition to investigating dynamic processes by time-lapse AFM imaging. General solutions exist by which to control the temperature of the experimental sample: (1) Heat the fluid bath in which the sample is directly immersed. (2) Flow through pre-heated medium using a mechanically controlled fluidic system. (3) Control the environmental temperature around the AFM and sample. An example of a custom implementation containing a heating element (a heating wire for resistive heating), a temperature-sensing element (a thermistor), and a way to seal the area around the sample is shown in Fig. 6. The system is connected to a temperature controller.
2. Culture bacteria originating from a clonal population of cells isolated from one colony grown on an agar plate or an aliquot of a cryogenic frozen subculture. A small number of bacteria are inoculated into a larger volume of medium, generally representing an optical density at 600 nm (OD_{600}) of 0.05, such that the bacteria may multiply to an optical density of around 0.8

OD_{600} indicative of vegetative growing cells multiplying in a logarithmic manner per increment of time. Bacteria in log phase are believed to exhibit the most homogenous behavior. Bacteria grown to stationary phase exhibit variable growth rates and doubling times, such that phenotypic variability increases.

3. Bacterial mutant strains may require the addition of antibiotics in order to select for the desired genotype.
4. Bacteria that agglomerate into a mass can eventually develop into a biofilm. Cell clumps confound AFM imaging, as cells do not remain uniformly adherent to the sample surface. To reduce clumping, supplement bacterial growth medium with detergent, like Tween-80.
5. In case of continued clumping, pass bacteria through a syringe filter to recuperate individual non-clumped bacterial cells. A syringe filter pore size should be selected for its capacity to allow single cells to pass through while blocking clumps of preferably as small as two cells attached along their long axis.
6. Choosing a cantilever depends strongly on the AFM operating mode, as well as the mechanical properties of the sample. Bacteria with a cell wall exhibit an equivalent Young's modulus on the order of hundreds of kPa to tens of MPa [35–37]. For imaging in contact mode, spring constants ranging from 0.05 to 0.3 N/m are well suited. For imaging in amplitude modulation mode, spring constants of usable cantilevers range from 0.1 to 0.5 N m^{-1}, depending on the size and resonance frequency of the cantilever in fluid.
7. For high-speed AFM, small cantilevers are preferable with k = 0.1–0.2 N m^{-1} [5, 22]. Commercially available small cantilevers include Olympus BL-AC7DS-KU5, Olympus BL-AC10DS-A2, and NanoWorld UCS-1.2, Neuchâtel, Switzerland.
8. For off-resonance tapping, cantilevers with a slightly higher spring constant of 0.2–0.7 N m^{-1} (like the Bruker ScanAsyst-Fluid) have been successfully used for long-term time-lapse AFM imaging of *Mycobacterium smegmatis* [2].
9. For nano-manipulation or dissection, cantilevers with a larger spring constant can be used, such as the Bruker RESP-40 and RTESPA-150 cantilevers with $k = 5$ N m^{-1} spring constant or OTESPA-R3, which exhibits $k = 26$ N m^{-1}; however imaging bacteria with these cantilevers is difficult and can lead to damage to the bacteria.
10. The cantilever tip shape can impact imaging of the bacterial sample. Many bacteria are either round, rod-shaped, or ovococcoid. When imaging such shapes with an AFM tip, by definition, at maximum, only the upper half of the bacterium

can be reached. Most cantilever tips are either pyramidal or conical in shape. At sharp edges or overhangs (such as the case for the lower part of the bacterium), these shapes result in well-known imaging artifacts manifested as regions with parallel stripes in the image [38]. For live bacteria, the area with image artifacts can represent a substantial part of the apparent width of the cell. If accurate size measurement of bacterial cells is important, a cantilever should be used with a high aspect ratio tip. Such cantilevers are commercially available, or a high-aspect ratio tip can be deposited using focused electron beam-induced deposition [39].

11. Avoid very crowded sample fields, as cells tend to settle atop one another.
12. Dual time-lapse AFM and optical fluorescence microscopy requires that images be acquired in close succession to one another, in order to minimize dynamic changes in bacterial physiology that could occur within the successive acquisition of AFM and optical fluorescence images. Therefore, the speed with which the images are successively taken must be appropriate to the experiment being conducted.
13. The frequency with which one can acquire fluorescence images is limited by both phototoxicity and photobleaching. For example, for long-term time-lapse experiments of *Mycobacterium smegmatis* strains expressing fluorescent fusion proteins, the frequency of optical fluorescence imaging is limited to around one exposure (200 ms of a 130 W mercury lamp, excitation filter 460–480 nm) every 10 min [2].
14. Several routines have been developed for more advanced and uniform automatic image processing of AFM image sequences [40]. For a more detailed description, see Eaton and West [41].
15. Fiji is a freely available image analysis program (also known as NIH Image, ImageJ) that can be used to harvest data in time-lapse image sequences and for overlaying AFM and optical fluorescence images.

Acknowledgments

This work was supported by the Swiss National Science Foundation under grant agreement numbers 205321_134786 and 205320_152675. H.A. Eskandarian acknowledges the support of an EMBO advanced long-term fellowship (LTF 191-2014 & ALTF 750-2016).

References

1. Ando T, Uchihashi T, Scheuring S (2014) Filming biomolecular processes by high-speed atomic force microscopy. Chem Rev 114 (6):3120–3188. https://doi.org/10.1021/cr4003837
2. Eskandarian HA et al (2017) Division site selection linked to inherited cell surface wave troughs in mycobacteria. Nat Microbiol 2 (9):17094. https://doi.org/10.1038/Nmicrobiol.2017.94
3. Fantner GE et al (2010) Kinetics of antimicrobial peptide activity measured on individual bacterial cells using high-speed atomic force microscopy. Nat Nanotechnol 5(4):280–285. https://doi.org/10.1038/nnano.2010.29
4. Watanabe H et al (2013) Wide-area scanner for high-speed atomic force microscopy. Rev Sci Instrum 84(5):053702. https://doi.org/10.1063/1.4803449
5. Yamashita H et al (2012) Single-molecule imaging on living bacterial cell surface by high-speed AFM. J Mol Biol 422 (2):300–309. https://doi.org/10.1016/j.jmb.2012.05.018
6. Suo Z et al (2009) Antibody selection for immobilizing living bacteria. Anal Chem 81 (18):7571–7578. https://doi.org/10.1021/ac9014484
7. Meyer RL et al (2010) Immobilisation of living bacteria for AFM imaging under physiological conditions. Ultramicroscopy 110 (11):1349–1357. https://doi.org/10.1016/j.ultramic.2010.06.010
8. Dufrene YF (2008) Atomic force microscopy and chemical force microscopy of microbial cells. Nat Protoc 3(7):1132–1138. https://doi.org/10.1038/nprot.2008.101
9. Butt HJ, Downing KH, Hansma PK (1990) Imaging the membrane protein bacteriorhodopsin with the atomic force microscope. Biophys J 58(6):1473–1480. https://doi.org/10.1016/S0006-3495(90)82492-9
10. Camesano TA, Natan MJ, Logan BE (2000) Observation of changes in bacterial cell morphology using tapping mode atomic force microscopy. Langmuir 16(10):4563–4572. https://doi.org/10.1021/La990805o
11. Hoh JH et al (1993) Structure of the extracellular surface of the gap junction by atomic force microscopy. Biophys J 65(1):149–163. https://doi.org/10.1016/S0006-3495(93)81074-9
12. Velegol SB, Logan BE (2002) Contributions of bacterial surface polymers, electrostatics, and cell elasticity to the shape of AFM force curves. Langmuir 18(13):5256–5262. https://doi.org/10.1021/La011818g
13. Colville K et al (2010) Effects of poly(L-lysine) substrates on attached Escherichia coli bacteria. Langmuir 26(4):2639–2644. https://doi.org/10.1021/la902826n
14. Liu Y, Strauss J, Camesano TA (2008) Adhesion forces between Staphylococcus epidermidis and surfaces bearing self-assembled monolayers in the presence of model proteins. Biomaterials 29(33):4374–4382. https://doi.org/10.1016/j.biomaterials.2008.07.044
15. Micic M et al (2004) Correlated atomic force microscopy and fluorescence lifetime imaging of live bacterial cells. Colloids Surf B Biointerfaces 34(4):205–212. https://doi.org/10.1016/j.colsurfb.2003.10.020
16. Hett EC, Rubin EJ (2008) Bacterial growth and cell division: a mycobacterial perspective. Microbiol Mol Biol Rev 72(1):126–156, table of contents. https://doi.org/10.1128/MMBR.00028-07
17. Wakamoto Y et al (2013) Dynamic persistence of antibiotic-stressed mycobacteria. Science 339(6115):91–95. https://doi.org/10.1126/science.1229858
18. Mendez-Vilas A, Gallardo-Moreno AM, Gonzalez-Martin ML (2007) Atomic force microscopy of mechanically trapped bacterial cells. Microsc Microanal 13(1):55–64. https://doi.org/10.1017/S1431927607070043
19. Mendez-Vilas A et al (2008) AFM probing in aqueous environment of Staphylococcus epidermidis cells naturally immobilised on glass: physico-chemistry behind the successful immobilisation. Colloids Surf B Biointerfaces 63 (1):101–109. https://doi.org/10.1016/j.colsurfb.2007.11.011
20. Fritz M et al (1994) Visualization and identification of intracellular structures by force modulation microscopy and drug-induced degradation. J Vac Sci Technol B 12 (3):1526–1529. https://doi.org/10.1116/1.587278
21. Ando T et al (2001) A high-speed atomic force microscope for studying biological macromolecules. Proc Natl Acad Sci U S A 98 (22):12468–12472. https://doi.org/10.1073/pnas.211400898
22. Casuso I et al (2012) Characterization of the motion of membrane proteins using high-speed atomic force microscopy. Nat Nanotechnol 7(8):525–529. https://doi.org/10.1038/nnano.2012.109

23. Kindt JH et al (2002) Atomic force microscope detector drift compensation by correlation of similar traces acquired at different setpoints. Rev Sci Instrum 73(6):2305–2307. https://doi.org/10.1063/1.1475352
24. Fantner GE et al (2005) Data acquisition system for high speed atomic force microscopy. Rev Sci Instrum 76(2):026118. https://doi.org/10.1063/1.1850651
25. Fantner GE et al (2006) Components for high speed atomic force microscopy. Ultramicroscopy 106(8–9):881–887. https://doi.org/10.1016/j.ultramic.2006.01.015
26. Nievergelt AP, et al. (2017) Components for high-speed atomic force microscopy optimized for low phase-lag. 2017 I.E. International Conference on Advanced Intelligent Mechatronics (AIM). p 731–736. doi:https://doi.org/10.1109/AIM.2017.8014104
27. Hu DH et al (2003) Correlated topographic and spectroscopic imaging beyond diffraction limit by atomic force microscopy metallic tip-enhanced near-field fluorescence lifetime microscopy. Rev Sci Instrum 74 (7):3347–3355. https://doi.org/10.1063/1.1581359
28. Adams JD et al (2014) High-speed imaging upgrade for a standard sample scanning atomic force microscope using small cantilevers. Rev Sci Instrum 85(9):093702. https://doi.org/10.1063/1.4895460
29. Nievergelt AP et al (2014) High-frequency multimodal atomic force microscopy. Beilstein J Nanotechnol 5:2459–2467. https://doi.org/10.3762/bjnano.5.255
30. Burns DJ, Youcef-Toumi K, Fantner GE (2011) Indirect identification and compensation of lateral scanner resonances in atomic force microscopes. Nanotechnology 22 (31):315701. https://doi.org/10.1088/0957-4484/22/31/315701
31. Kammer CM, et al. Data-driven controller design for atomic-force microscopy In: 20th World Congress of IFAC2017: Toulouse, France
32. Nievergelt AP et al (2015) Studying biological membranes with extended range high-speed atomic force microscopy. Sci Rep 5:11987. https://doi.org/10.1038/srep11987
33. Lonergan NE, Britt LD, Sullivan CJ (2014) Immobilizing live Escherichia coli for AFM studies of surface dynamics. Ultramicroscopy 137:30–39. https://doi.org/10.1016/j.ultramic.2013.10.017
34. Johansson J et al (2002) An RNA thermosensor controls expression of virulence genes in Listeria monocytogenes. Cell 110(5):551–561
35. Deng Y, Sun M, Shaevitz JW (2011) Direct measurement of cell wall stress stiffening and turgor pressure in live bacterial cells. Phys Rev Lett 107(15):158101. https://doi.org/10.1103/PhysRevLett.107.158101
36. Polyakov P et al (2011) Automated force volume image processing for biological samples. PLoS One 6(4):e18887. https://doi.org/10.1371/journal.pone.0018887
37. Raman A et al (2011) Mapping nanomechanical properties of live cells using multi-harmonic atomic force microscopy. Nat Nanotechnol 6 (12):809–814. https://doi.org/10.1038/nnano.2011.186
38. Velegol SB et al (2003) AFM imaging artifacts due to bacterial cell height and AFM tip geometry. Langmuir 19(3):851–857. https://doi.org/10.1021/la026440g
39. Kindt JH et al (2004) Automated wafer-scale fabrication of electron beam deposited tips for atomic force microscopes using pattern recognition. Nanotechnology 15(9):1131–1134. https://doi.org/10.1088/0957-4484/15/9/005
40. Erickson BW et al (2012) Large-scale analysis of high-speed atomic force microscopy data sets using adaptive image processing. Beilstein J Nanotechnol 3:747–758. https://doi.org/10.3762/bjnano.3.84
41. Eaton PJ, West P (2010) Atomic force microscopy, vol viii. Oxford University Press, Oxford; New York, p 248
42. Odermatt PD et al (2015) High-resolution correlative microscopy: bridging the gap between single molecule localization microscopy and atomic force microscopy. Nano Lett 15(8):4896–4904. https://doi.org/10.1021/acs.nanolett.5b00572

Chapter 24

Probing Bacterial Adhesion at the Single-Molecule and Single-Cell Levels by AFM-Based Force Spectroscopy

Sofiane El-Kirat-Chatel and Audrey Beaussart

Abstract

Functionalization of AFM probes with biomolecules or microorganisms allows for a better understanding of the interaction mechanisms driving microbial adhesion. Here we describe the most commonly used protocols to graft molecules and bacteria to AFM cantilevers. The bioprobes obtained that way enable to measure forces down to the single-cell and single-molecule levels.

Key words Atomic force microscopy, Functionalization, Single-molecule, Single-cell, Interactions

1 Introduction

In the last two decades, atomic force microscopy (AFM) modes have evolved and now offer a wealth of new possibilities to address pertinent biological questions. As AFM is particularly suitable to determine sample surface topography at high resolution, the first adaptation to biological systems was to image cells under liquid, i.e., in physiological conditions. This approach allowed imaging the surface of live mammalian or microbial cells at the nanoscale. As non-exhaustive examples, the organization of the bacterial peptidoglycan [1–5], the surface of macrophages [6, 7], the S-layer of *Corynebacterium glutamicum* [8], or the rodlets of *Aspergillus fumigatus* [9, 10] have been revealed with unprecedented resolution. Although powerful for topographical analysis, AFM is not limited to imaging but can also serve to probe interaction forces. Taking advantage of the sharpness of the AFM tip, several groups have developed approaches to detect single molecular interaction by the so-called single-molecule force spectroscopy (SMFS). In SMFS, AFM tips are functionalized with biomolecules and used to probe purified targets immobilized on abiotic surfaces or receptors expressed at the surface of living cells (Fig. 1a, b) [11–15]. By means of SMFS, biomolecules could be mapped on microbial

Yuri L. Lyubchenko (ed.), *Nanoscale Imaging: Methods and Protocols*, Methods in Molecular Biology, vol. 1814,
https://doi.org/10.1007/978-1-4939-8591-3_24,

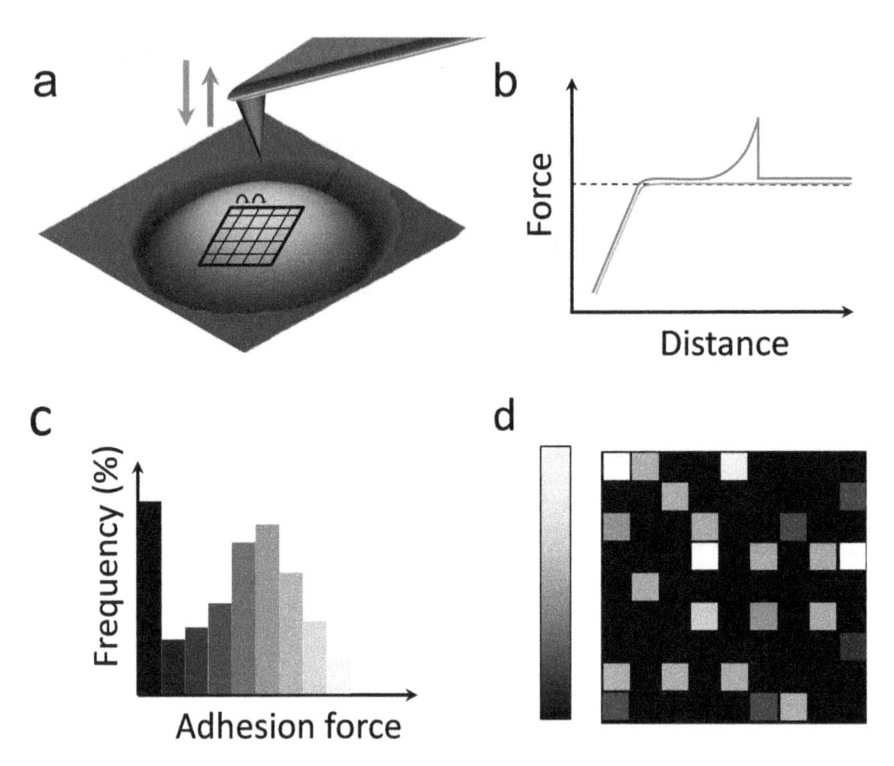

Fig. 1 Principle of spatially resolved force spectroscopy to map the distribution of individual molecules. (**a**) An area (black square) is defined at the surface of a single microbial cell mechanically trapped in a pore of a polycarbonate filter membrane. For force-volume measurement, the area is virtually split in a mesh of 32 × 32 pixels on which individual approach/retract force-distance curves will be recorded. (**b**) Typical approach (blue) and retract (red) force-distance curve showing an adhesive event reflected by the hysteresis on the retract curve. (**c**) Histogram showing the distribution of the adhesion forces extracted from the force-distance curves. (**d**) Adhesion map reconstructed from a force volume. Black pixels correspond to nonadhesive curves, while adhesive events are represented with white/gray pixels reflecting the adhesion strength (force in pN)

surfaces to determine whether their distribution is random and spatially organized or can change upon external stimuli (Fig. 2a, b) [16–18]. Additionally, SMFS allows to decipher the intrinsic mechanical properties of biomolecules as they are manipulated and unfolded after detection (Figs. 1b and 2d). In that sense, fitting the unfolding curves with commonly used models such as the worm-like chain (WLC) or free jointed chain (FJC) permits to quantitatively describe the elasticity of flexible molecules [19–23]. Microbial adhesion is driven by a complex interplay of physicochemical forces. Receptors-ligands interactions are often coupled to non-specific hydrophobic and electrostatic forces. Deciphering such component of cellular adhesion has been made possible with the development of chemical force microscopy (CFM), where AFM tips are decorated with specific functional groups, such as hydrophobic thiols [24–28].

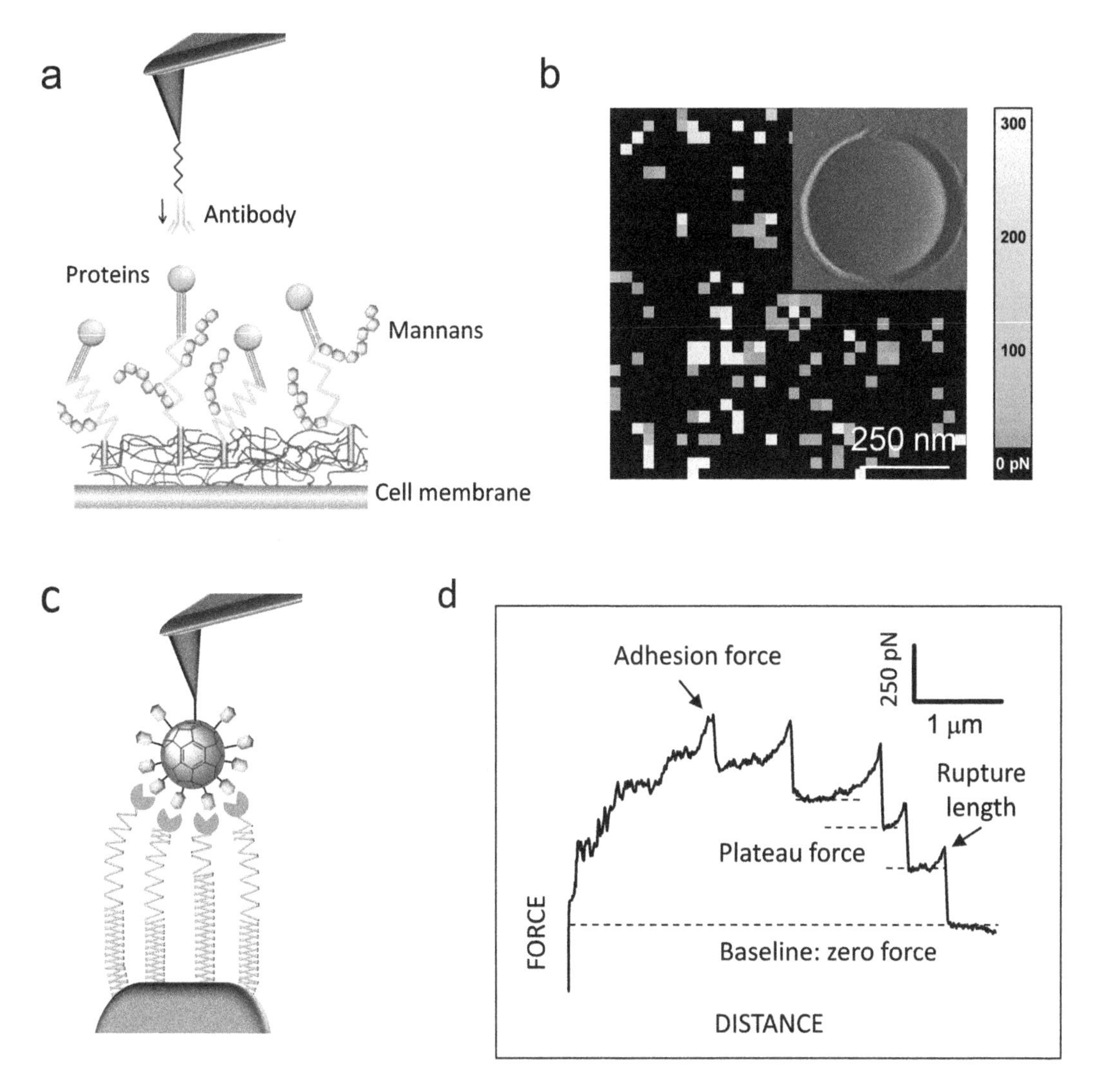

Fig. 2 AFM tip functionalization and single-molecule force spectroscopy. (**a**) Silicon nitride AFM tip functionalized via PEG linkers to covalently attach an antibody for the specific detection of mannoproteins at the surface of yeast cells. (**b**) Adhesion map recorded at the surface of a living yeast cell and showing the distribution of cell wall mannoproteins detected with the antibody tip in (**a**) (insert correspond to the deflection image of the probed yeast). (**c**) Gold-coated AFM tip functionalized with nanoparticles binding to pili appendages at the surface of living bacteria (green). (**d**) Representative retraction force curve recorded at the surface of a bacteria and documenting the unfolding of pili. The force signature reveals the interaction of multiple pili with the nanoparticle terminated tip. Reprinted with permission from "Single-Molecule Imaging and Functional Analysis of Als Adhesins and Mannans during *Candida albicans* Morphogenesis, Beaussart *et al.*, ACS Nano, 2012, 6, 10950" and "Force Nanoscopy as a Versatile Platform for Quantifying the Activity of Antiadhesion Compounds Targeting Bacterial Pathogens, Beaussart *et al.*, Nano Letters, 2016, 16, 1299." Copyright 2017 American Chemical Society

Further understanding of how these different forces act together for microbial cells to attach to each other, to solid substrates, or to host cell was not possible until the recent development of single-cell force spectroscopy (SCFS) adapted to microorganisms

[29–31]. Indeed, if the development of SCFS for large cells (mammalian cells, fungi) was quite straight forward with the use of tipless cantilever [30, 32, 33], probing interactions of smaller cells such as bacteria had needed a dedicated design that allows proper positioning of the cell. Here, we detailed how to proceed to perform force spectroscopy at the single-molecule and the single-cell levels, and we give insights on the choice of functionalization procedures the best suitable for specific desired applications. The methods described hereafter are commonly used in bio-AFM studies, but other ways to functionalize AFM probes with biomolecules or cells also exist, e.g., with dendritips for SMFS [34] or with microchanneled cantilevers for SCFS in the FluidFM mode [35].

2 Materials

Prepare all solutions using ultrapure water and analytical grade reagents.

2.1 Microorganisms Immobilization for Force Spectroscopy with Functionalized Tips

1. Polycarbonate porous membranes (it4ip or Millipore) of pore diameter similar to the bacterial size (if the goal is to image/probe the side of rod-shaped bacteria, Millipore is more suitable as these membranes present jointed pores making holes bigger and with adapted shapes).
2. Glass filtration unit (Sartorius).
3. AFM magnetic steel puck (Bruker).
4. Double-sided tape (3 M).
5. Nanoscope VIII multimode atomic force microscope (Bruker).
6. Silicon nitride triangular-sharpened cantilevers with a nominal spring constant of 0.01 N m^{-1} (MSCT, Bruker) or gold-coated cantilevers with a nominal spring constant of 0.02 N m^{-1} (OMLC-TR400PB-1, AtomicForce, Olympus).
7. Piranha solution: 3:1 solution of concentrated H_2SO_4 and 30% H_2O_2.

2.2 Single-Cell Force Spectroscopy

1. Bioscope catalyst atomic force microscope (Bruker) mounted on an inverted optical microscope (Zeiss Axio Observer ZI) equipped with a Hamamatsu C1000 camera with a 40× objective and a 100× oil-immersion objective. Other instruments, such as JPK NanoWizard can also be considered.
2. UV-curable glue NOA 63 (Norland, Edmund Optics).
3. LIVE/DEAD *Bac*light viability kit (Invitrogen).
4. Silica microspheres, 6.1 μm diameter (Bangs laboratories).
5. Triangular-shaped tipless cantilever with a nominal spring constant of 0.06 N m^{-1} (NP-010, Microlevers, Bruker).

6. Dopamine hydrochloride, 99% (Sigma).
7. Tips and gold surface cleaning: UV-ozone treatment (Jetlight)
8. Glass-bottomed Petri dishes (WillCo-dish).
9. UV lamp, 365 nm, 6 W (handheld UV lamp UVL-56; UVP).

3 Methods

Carry out all procedures at room temperature unless otherwise stated.

3.1 Force Measurements at the Surface of Microorganisms with (Bio)-molecules Grafted Tips

1. Grow bacterial cell cultures in the recommended conditions (stationary or exponential phase; with or without shaking for aerobic and anaerobic conditions, respectively; at appropriate temperature; with the proper antibiotics).
2. Transfer 20 mL of the bacterial suspension in a 50 mL falcon, and rinse twice or 3 times by centrifugation (2 min at 5000 × *g* or other recommended parameters for harvesting) in the experimental buffer.
3. To properly immobilize mechanically the microorganisms, filter 1 mL of a 10^6 cells mL^{-1} suspension of the rinsed suspension through a polycarbonate membrane whose pore size approximate the bacterial size, using a filtration unit.
4. Cut a 1 cm × 1 cm square in the porous membrane, gently rinse to remove untrapped cells, and stick it to an AFM magnetic steel puck using double-sided tape (*see* **Note 1**).
5. Immediately cover the membrane with a 100 μL drop of buffer, and transfer to the AFM liquid setup without dewetting.
6. Image the membrane under low applied force (~100 pN) using a bare soft cantilever (MSCT of nominal spring constant 0.01 N m^{-1}, Bruker Corporation, are recommended).
7. Once bacteria are identified and located on the image, remove the bare AFM tip, and place the functionalized tip (*see* details below) exactly at the same position above the membrane (*see* **Note 2**).
8. Several methodologies can be chosen to decorate the tip with molecules. The most common ones are described here (**steps 9–13**):
9. **Covalent attachment of single protein** using PEG-benzaldehyde linkers as described by Ebner et al. [36–38] (*see* **Note 3**).
   (a) Briefly, cleaned MSCT (silicon nitride) cantilevers are decorated by amine groups via overnight immersion in an ethanolamine solution.

(b) The ethanolamine-coated cantilevers are then dipped in an acetal-PEG-NHS solution, and the acetal groups are further transformed into aldehyde functions by immersion in a citric acid solution.

(c) The tips are covered with a droplet of buffer containing 0.2 mg mL^{-1} of the proteins to be attached (*see* **Note 4**).

(d) Coated cantilevers are then immersed in ethanolamine solution in order to passivate unreacted aldehyde groups, washed with and stored in buffer.

10. **Covalent attachment of single protein using NHS-EDC** (*see* **Note 5**).

(a) Gold-coated cantilevers are rinsed with ethanol, cleaned for 15 min with UV-ozone, rinsed again with ethanol, dried with a gentle flow of N_2, immersed overnight in a 1 mM solution of 10% $HS(CH_2)_{16}COOH$ (Sigma-Aldrich) and 90% $HS(CH_2)_{11}OH$ (Sigma-Aldrich), and rinsed with ethanol.

(b) They are then immersed for 30 min in a solution containing 10 mg mL^{-1} N-hydroxysuccinimide (NHS) (Sigma-Aldrich) and 25 mg mL^{-1} 1-ethyl-3-(3-dimethylamino-propyl)-carbodiimide (EDC) (Sigma-Aldrich) and rinsed 3 times with ultrapure water.

(c) The activated cantilevers are incubated with 0.2 mg mL^{-1} of protein solution, followed by rinsing and storage in buffer.

11. **Oriented attachment of His-tagged-protein** (*see* **Note 6**).

(a) Gold-coated cantilevers are rinsed with ethanol, cleaned for 15 min with UV-ozone, rinsed again with ethanol, dried with a gentle flow of N_2, and immersed overnight in a 0.1 mM solution of 90% HS-C11-$(EG)_3$-OH thiols (ProChimia) and 10% HS-C11-$(EG)_3$-NTA thiols (ProChimia).

(b) The tips are then rinsed with ethanol, dried with N_2, and immersed in a 40 mM aqueous solution of $NiSO_4$ (pH 7.2) for 30 min.

(c) They are then incubated in a 200 μL droplet of a 0.2 mg mL^{-1} His6-tagged proteins solution for 1 h, rinsed, and stored in buffer until use [39] (*see* **Note 7**).

12. **Tip coating with thiol molecules** (*see* **Note 8**). Gold-coated cantilevers are rinsed with ethanol, cleaned for 15 min with UV-ozone, rinsed again with ethanol, dried with a gentle flow of N_2, and immersed overnight in a 0.1 mM solution of HS-$(CH_2)_{11}$-CH_3 or $HS(CH_2)_{16}COOH$ thiols (Sigma-Aldrich), for hydrophobic or negatively charged probes, respectively [24, 26, 27].

13. **Grafting of nanoparticles.**

    Given the wild range of NPs chemical compositions that exist, no systematic protocol can be given. Here, one example is detailed for the attachment of mannofullerene NPs containing one thiol group.

    (a) Briefly, thiol-mannofullerenes (disulfide-containing bis-mannofullerene displaying 20 peripheral mannose residues) were synthetized from a [5:1] heterovalent fullerene scaffold [40].

    (b) Prior to tip immersion, 1 mM TCEP (tris(2-carboxyethyl) phosphine hydrochloride)) solution is added to the thiol solution (50% vol) for reduction of the disulfide bond of the mannofullerenes.

    (c) Gold-coated cantilevers are rinsed with ethanol, cleaned for 15 min with UV-ozone, rinsed again with ethanol, dried with a gentle flow of N_2, and immersed overnight in a 1 mM solution of thiol-mannofullerenes, rinsed with ethanol and dried with N_2.

14. Choose an appropriate size of the surface area to be mapped, depending on the size of the cell (for instance, 500 nm × 500 nm for bacteria; 1 μm × 1 μm for yeast cells) (Fig. 1a).

15. Record force-distance curves on the selected area using the force-volume mode (Fig. 1a, b), with a maximum applied force of 250 pN and approach and retract velocities of around 1 μm/s (*see* **Note 9**).

16. To generate adhesion force and rupture length histograms (Fig. 1c), use the AFM software to calculate the adhesion force and rupture distance for each curve. Color-scaled map of the adhesion values can be recreated using the Matlab software (Fig. 1d) (*see* **Note 10**).

3.2 Single-Cell Force Spectroscopy

1. Deposit a small amount of UV-curable glue on a glass slide stuck to a magnetic steel puck using double-sided tape (*see* **Note 11**).
2. Transfer the puck to an AFM setup equipped with an optical microscope. Dip an AFM soft tipless cantilever into the glue by manual engagement (Fig. 3a) (*see* **Note 12**).
3. Deposit a small amount of silica microsphere on another glass slide stuck to a magnetic puck.
4. Using the AFM, bring the glue-covered cantilever down to a single microsphere (Fig. 3b).
5. After a short contact of few seconds, disengage the cantilever, and cure the glue by putting the colloidal probe under the UV lamp for 10 min.

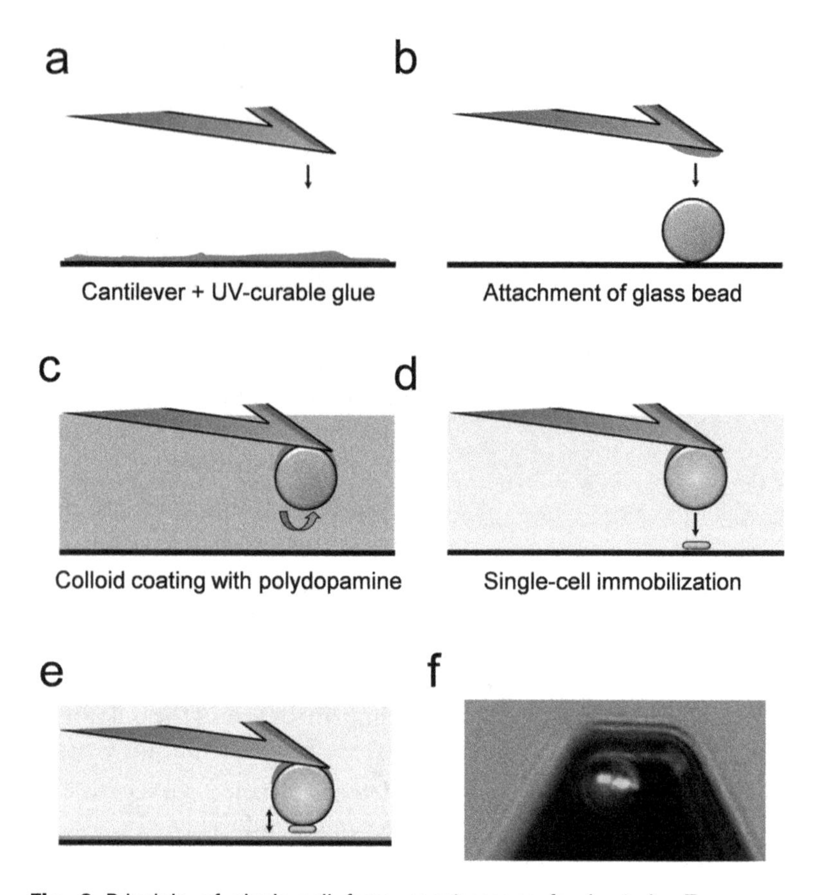

Fig. 3 Principle of single-cell force spectroscopy for bacteria. To prepare colloidal probes, a tipless cantilever is immersed in a drop of UV-curable glue (**a**) and then brought into contact with an isolated glass bead before UV exposure (**b**). For bacterial attachment, the colloidal probe cantilever is first immersed in a solution of polydopamine bioadhesive (**c**), and then a single cell is attached to the bead (**d**) before force measurement (**e**). The proper positioning of the bacteria and its viability are assessed by fluorescence imaging with the *Bac*light LIVE/DEAD stain (**f**). Reproduced with permission from "Quantifying the forces guiding microbial cell adhesion using single-cell force spectroscopy, Beaussart *et al.*, Nature Protocols, 2014, 9, 1049"

6. Prepare a 10 mM Tris-HCl buffer solution at pH 8.5 containing 4 mg mL^{-1} of polydopamine hydrochloride, and immediately immerse the colloid probe for 1 h in the solution (Fig. 3c).
7. Rinse the probe by immersion in ultrapure water and dry under gentle flow of nitrogen.
8. For attaching a single bacterium, first deposit 50 μL of diluted bacterial suspension (cultured as in **step 1** of Subheading 3.1) on a glass-bottom Petri dish, and let the bacteria settle.
9. Mark the bacteria viability using a LIVE/DEAD kit, and complete the volume of the Petri dish on settled-stained bacteria.

10. Prepare the sample whose interactions with the bacteria will be measured, and insert it in the AFM liquid setup.
11. Using an AFM mounted on an inverted fluorescence microscope, bring the dopamine-coated colloidal probe into contact with an isolated bacteria marked in green for 1–3 min (Fig. 3d, e).
12. Withdraw the colloidal probe from the glass Petri dish, and ensure that the bacteria are properly attached by taking an optical/fluorescent image (Fig. 3f).
13. Engage the cell probe on the substrate of interest, and record consecutive force-distance curves on different spot of the surface of interest.
14. To generate adhesion force and rupture length histograms, use the AFM software to calculate the maximum adhesion force and rupture distance for each curve.

4 Notes

1. In order to facilitate the sticking of the wet membrane to the double-sided tape, the membrane can be very gently deposited briefly on a wiper (for instance Kimwipes, KimTeck Science) before adhesion to the tape. In this case ensure that the top surface layer of the membrane remains moist.
2. To properly place the functionalized tip, locate a peculiar pattern on the porous membrane, and record an optical image of the position of your bare tip relatively to this pattern. Once the functionalized tip is placed in the setup, move it to the same location where the bare tip was imaged, and engage and record a rapid scan of the area until finding the bacteria you had previously selected.
3. This chemistry allows grafting covalently a single molecule to the apex of the AFM tip. The use of a PEG linker lowers unspecific interactions and ensures the flexibility of the attached protein for a better accessibility to the target.
4. Addition of sodium cyanoborohydride ($NaCNBH_3$) is required for this step. Care should be taken given the toxicity of the cyanide compound, in particular as dangerous vapor may emanate in case of contact with acid.
5. This protocol allows to covalently attach the protein to the tip and to modulate the grafting density by varying the ratio of COOH/OH terminated thiols. The NHS and EDC have to be mixed just before use.

6. The main advantage here is that grafting the protein via a tag permits to control its orientation. In case of enzyme, for instance, the tag would be located outside the catalytic domain for it to remain active.
7. Given the weakness of the His-tag/NTA bond, we recommend to use the tip directly after functionalization and to avoid storage.
8. Decorating the tip with thiol molecules presents the advantage to obtain a probe whose chemistry can be fully controlled, to decipher the surface physical-chemistry properties (for instance, hydrophobicity [26, 27] or charges distribution [24]).
9. In case of ligand/receptor interactions, the contact time between the tip and the cell surface may be varied from 0 to 1 s. When analyzing the adhesion properties of different strains, for instance, the contact time should be equivalent for comparison to be possible.
10. Typical retract force-distance curve obtained by SMFS between a NP-functionalized tip and the surface of a bacteria decorated with pili is depicted on Fig. 2d. The unfolding of each pili under constant force results in a plateau. The combination of consecutive adhesion of several pili at the microbial surface generates the peculiar signature observed on the force-distance curve.
11. To obtain a small quantity of glue on the surface, spread the glue droplet using another glass slide or a pointed tweezer.
12. To avoid tip damage, bring the cantilever down to the glue step by step very gently. Once the contact is reached, the cantilever reflective color (observed via the optical camera) will slightly change. Lift up the cantilever as soon as the color starts to change.

References

1. Beaussart A, Pechoux C, Trieu-Cuot P et al (2014) Molecular mapping of the cell wall polysaccharides of the human pathogen *Streptococcus agalactiae*. Nanoscale 6:14820–14827
2. Dover RS, Bitler A, Shimoni E et al (2015) Multiparametric AFM reveals turgor-responsive net-like peptidoglycan architecture in live streptococci. Nat Commun 6:7193
3. Hayhurst EJ, Kailas L, Hobbs JK et al (2008) Cell wall peptidoglycan architecture in *Bacillus subtilis*. Proc Natl Acad Sci U S A 105:14603–14608
4. Turner RD, Ratcliffe EC, Wheeler R et al (2010) Peptidoglycan architecture can specify division planes in *Staphylococcus aureus*. Nat Commun 1:1
5. Andre G, Kulakauskas S, Chapot-Chartier MP et al (2010) Imaging the nanoscale organization of peptidoglycan in living *Lactococcus lactis* cells. Nat Commun 1:1
6. El-Kirat-Chatel S, Dufrene YF (2012) Nanoscale imaging of the *Candida* - macrophage interaction using correlated fluorescence-atomic force microscopy. ACS Nano 6:10792–10799
7. Labernadie A, Thibault C, Vieu C et al (2010) Dynamics of podosome stiffness revealed by

atomic force microscopy. Proc Natl Acad Sci U S A 107:21016–21021

8. Dupres V, Alsteens D, Pauwels K et al (2009) In vivo imaging of S-layer Nanoarrays on *Corynebacterium glutamicum*. Langmuir 25:9653–9655
9. Alsteens D, Aimanianda V, Hegde P et al (2013) Unraveling the nanoscale surface properties of chitin synthase mutants of *Aspergillus fumigatus* and their biological implications. Biophys J 105:320–327
10. Bayry J, Beaussart A, Dufrene YF et al (2014) Surface structure characterization of *Aspergillus fumigatus* conidia mutated in the melanin synthesis pathway and their human cellular immune response. Infect Immun 82:3141–3153
11. Chtcheglova LA, Hinterdorfer P (2011) Simultaneous topography and recognition imaging on endothelial cells. J Mol Recognit 24:788–794
12. El-Kirat-Chatel S, Boyd CD, O'Toole GA et al (2014) Single-molecule analysis of *Pseudomonas fluorescens* footprints. ACS Nano 8:1690–1698
13. Heinisch JJ, Lipke PN, Beaussart A et al (2012) Atomic force microscopy – looking at mechanosensors on the cell surface. J Cell Sci 125:4189–4195
14. Hinterdorfer P, Dufrene YF (2006) Detection and localization of single molecular recognition events using atomic force microscopy. Nat Methods 3:347–355
15. Muller DJ, Dufrene YF (2011) Atomic force microscopy: a nanoscopic window on the cell surface. Trends Cell Biol 21:461–469
16. Alsteens D, Garcia MC, Lipke PN et al (2010) Force-induced formation and propagation of adhesion nanodomains in living fungal cells. Proc Natl Acad Sci U S A 107:20744–20749
17. Beaussart A, Alsteens D, El-Kirat-Chatel S et al (2012) Single-molecule imaging and functional analysis of Als Adhesins and Mannans during *Candida albicans* morphogenesis. ACS Nano 6:10950–10964
18. Heinisch JJ, Dupres V, Wilk S et al (2010) Single-molecule atomic force microscopy reveals clustering of the yeast plasma-membrane sensor Wsc1. PLoS One 5:e11104
19. Alsteens D, Dupres V, Klotz SA et al (2009) Unfolding individual Als5p adhesion proteins on live cells. ACS Nano 3:1677–1682
20. El-Kirat-Chatel S, Beaussart A, Boyd CD et al (2014) Single-cell and single-molecule analysis deciphers the localization, adhesion, and mechanics of the biofilm Adhesin LapA. ACS Chem Biol 9:485–494
21. Francius G, Alsteens D, Dupres V et al (2009) Stretching polysaccharides on live cells using single molecule force spectroscopy. Nat Protoc 4:939–946
22. Marszalek PE, Oberhauser AF, Pang YP et al (1998) Polysaccharide elasticity governed by chair-boat transitions of the glucopyranose ring. Nature 396:661–664
23. Rief M, Oesterhelt F, Heymann B et al (1997) Single molecule force spectroscopy on polysaccharides by atomic force microscopy. Science 275:1295–1297
24. Beaussart A, Ngo TC, Derclaye S et al (2014) Chemical force microscopy of stimuli-responsive adhesive copolymers. Nanoscale 6:565–571
25. Alsteens D, Dague E, Rouxhet PG et al (2007) Direct measurement of hydrophobic forces on cell surfaces using AFM. Langmuir 23:11977–11979
26. Alsteens D, Dupres V, Yunus S et al (2012) High-resolution imaging of chemical and biological sites on living cells using peak force tapping atomic force microscopy. Langmuir 28:16738–16744
27. Dague E, Alsteens D, Latge JP et al (2007) Chemical force microscopy of single live cells. Nano Lett 7:3026–3030
28. Dufrene YF (2008) Atomic force microscopy and chemical force microscopy of microbial cells. Nat Protoc 3:1132–1138
29. Beaussart A, El-Kirat-Chatel S, Sullan RMA et al (2014) Quantifying the forces guiding microbial cell adhesion using single-cell force spectroscopy. Nat Protoc 9:1049–1055
30. Alsteens D, Beaussart A, Derclaye S et al (2013) Single-cell force spectroscopy of Als-mediated fungal adhesion. Anal Methods 5:3657–3662
31. Beaussart A, El-Kirat-Chatel S, Herman P et al (2013) Single-cell force spectroscopy of probiotic Bacteria. Biophys J 104:1886–1892
32. El-Kirat-Chatel S, Dufrene YF (2016) Nanoscale adhesion forces between the fungal pathogen *Candida albicans* and macrophages. Nanoscale Horiz 1:69–74
33. Helenius J, Heisenberg CP, Gaub HE et al (2008) Single-cell force spectroscopy. J Cell Sci 121:1785–1791
34. Jauvert E, Dague E, Severac M et al (2012) Probing single molecule interactions by AFM using bio-functionalized dendritips. Sens Actuators B Chem 168:436–441
35. Meister A, Gabi M, Behr P et al (2009) FluidFM: combining atomic force microscopy and nanofluidics in a universal liquid delivery

system for single cell applications and beyond. Nano Lett 9:2501–2507
36. Ebner A, Wildling L, Kamruzzahan ASM et al (2007) A new, simple method for linking of antibodies to atomic force microscopy tips. Bioconjug Chem 18:1176–1184
37. Wildling L, Unterauer B, Zhu R et al (2011) Linking of sensor molecules with amino groups to amino-functionalized AFM tips. Bioconjug Chem 22:1239–1248
38. http://www.jku.at/biophysics/content
39. Dupres V, Alsteens D, Wilk S et al (2009) The yeast Wsc1 cell surface sensor behaves like a nanospring in vivo. Nat Chem Biol 5:857–862
40. Beaussart A, Abellan-Flos M, El-Kirat-Chatel S et al (2016) Force nanoscopy as a versatile platform for quantifying the activity of antiadhesion compounds targeting bacterial pathogens. Nano Lett 16:1299–1307

Chapter 25

Fluorescence Correlation Spectroscopy on Genomic DNA in Living Cells

Cameron Hodges and Jens-Christian Meiners

Abstract

Fluorescence correlation spectroscopy (FCS) is a powerful technique used to measure diffusion, fluctuations, and other transport processes in biomolecular systems. It is, however, prone to artifacts and subject to considerable experimental difficulties when applied to living cells. In this chapter, we provide protocols to conduct quantitative FCS measurements on DNA inside living eukaryotic and prokaryotic cells. We discuss sample preparation, dye selection and characterization, FCS data acquisition, and data analysis, including a method to com pensate for photobleaching to obtain quantitatively meaningful spectra.

Key words Fluorescence correlation spectroscopy, Live cell microscopy, Chromosomal DNA, Photobleaching

1 Introduction

Fluorescence correlation spectroscopy (FCS) is a powerful tool used to quantitatively study the diffusion and fluctuations of fluorescent or fluorescently labeled molecules. Conventionally, this method has mostly been used to measure diffusion coefficients and observe molecular binding events in free solution. There is considerable interest in expanding this method to measurements inside living cells, for instance, to study membrane structure and dynamics [1], transcription factor binding to nuclear DNA [2], protein diffusion [3], or the dynamics of chromosomal DNA [4]. In all these instances, there are substantial complications. Most stem from photobleaching of the dyes [5], an effect which can substantially distort the FCS spectra to the point that quantitative analysis is no longer meaningful.

We have developed a protocol to observe the fluctuations of DNA in living eukaryotic and prokaryotic cells using an adaptation of FCS. It starts with sample preparation that leaves the cells physiologically intact, but well immobilized for imaging, and properly stained with dyes that minimize artifacts. We compensate for

Yuri L. Lyubchenko (ed.), *Nanoscale Imaging: Methods and Protocols*, Methods in Molecular Biology, vol. 1814, https://doi.org/10.1007/978-1-4939-8591-3_25, © Springer Science+Business Media, LLC, part of Springer Nature 2018

the deleterious effects of photobleaching by implementing a data analysis procedure that computes the FCS spectrum from the arrival times of the photons. In this scheme we assign later photons a greater weight than earlier ones—thus accounting for already lost fluorophores. This allows further quantitative analysis to determine, for instance, microrheological parameters of the chromosomal DNA inside its cellular environment.

2 Materials

2.1 MOPS Media

1. 100 mL of 10× MOPS buffer for EZ Rich defined medium kit sterile solution from Teknova.
2. 10 mL 0.132 M K_2PO_4, sterile.
3. 10 mL of 20% glucose, sterile.
4. 880 mL sterilized water.

2.2 Hela Cell Media (1 L, Adjust as Needed)

1. 940 mL Dulbecco's Modified Eagle Medium, Thermo Fisher, sterile.
2. 50 mL fetal bovine serum, heat inactivated, Thermo Fisher, sterile.
3. 10 mL Pen-Strep, 10,000 U/mL, Thermo Fisher, sterile.

2.3 Dyes

1. Enzo Nuclear-ID Red dye (HeLa cells), stock concentration 2.5–5 μM.
2. Thermo Fisher Syto 62 DNA dye (E. coli), stock concentration 5 mM diluted down to 5 μM, or other preferred dye (*see* **Note 1**).

2.4 λ-DNA for Dye Characterization

500 μg/mL, New England Biolabs #N3011S.

2.5 TBE Buffer (pH ~8.0 5x Stock)

54 g Tris Base, 27.5 g boric acid, 20 mL 0.5 M EDTA.

2.6 Microscope Slides and Coverslips

75 × 25 mm Slides (0.96–1.06 mm thickness) and 22 × 22 mm No. 1.5 Coverslips (0.17 mm thickness).

2.7 Antibody-Covered Coverslips

Instead of agarose pads, antibody-coated coverslips may be used for the immobilization of the *E.* coli cells as well. To coat the coverslips, clean coverslips with 15% KOH solution and air dry. Completely cover coverslip with *E. coli* antibody at stock concentration, typically 4–5 mg/mL, and let sit for 1 h. Wash coated coverslip three times with deionized water and air dry.

2.8 Poly-D-Lysine-Coated Coverslip Petri Dish

MatTek 35 mm dish, No. 1.5 coverslip, 14 mm glass diameter, poly-D-lysine coated.

2.9 Trypan Blue Stain

0.4%, from Thermo Fisher.

3 Methods

3.1 HeLa Cell Culture

HeLa cells are cultured using standard procedures. Always prewarm all media/solutions used with cell cultures to 37 °C in water bath.

1. Incubate at 37 °C with 5% CO_2. Determine total cell count by taking 10 μL of cell culture mixed with 10 μL trypan blue, and count on a hemocytometer.
2. Spin down cell culture at 500× g at room temperature for 5 min.
3. Pour off supernatant, and resuspend in media to make the final concentration 10^6 cells/mL, allowing for easier plating.
4. For next day imaging, plate out 50 K cells, or for 2 days out, plate 25 K cells onto poly-D-lysine-coated coverslip petri dishes, and gently add 1 mL of media while being careful to try and keep the sample on the 14 mm glass coverslip in the center of the petri dish.
5. Immediately prior to staining, pour off old HeLa media and wash 2–3 times with fresh media.

3.2 *E. coli* Cell Culture

1. Grow *E. coli* culture overnight in MOPS media at 37 °C and shaken at a frequency of 285 rpm to a typical OD of 0.700.
2. Dilute overnight culture to optical density of roughly 0.003. This will typically be 15 μL of cell culture into 6 mL of media.
3. Allow culture to grow back to OD of 0.25–0.30, typically around 7 h after diluting.

3.3 Staining HeLa Cells

1. Add enough dye at a concentration of roughly 250 nM to cover all of the poly-D-lysine-coated coverslip petri dish, and incubate for 20–30 min at 37 °C, 5% CO_2. Avoid as much light exposure as possible (*see* **Note 2**). Adjust dye concentration as needed to obtain optimal FCS spectra (*see* **Note 3**).
2. Wash off excess dye three times with 1 mL Hela media. Pour out all media and image.

3.4 Dye Characterization

If a new dye with unknown DNA-binding properties is used, it is important to ascertain that the off-rate of the dye is slower than a typical measurement time window. We use a gel imaging assay based on [6] for this purpose, as an agarose gel mimics the cellular environment better than free solution.

1. Prepare a λ-DNA-dye solution by mixing 50 μL stock λ-DNA with concentrated dye (for instance, 10 μL 5 μM Syto62) to achieve a final ratio of 1 to 0.1 dye molecules per 100 base pairs (*see* **Note 4**). Then add TBE buffer to bring the final sample amount to 200 μL.
2. Prepare a 0.8% agarose gel with TBE buffer. Add 20 μL LDNA sample to the first well and run for 5 min at roughly 100 V. Add LDNA sample to subsequent wells at 5-min intervals, making sure to turn off the power supply when adding sample to the subsequent wells. After the final well has been loaded, run for an additional 5 min. The resulting gel will now have eight lanes of LDNA sample that have been started at staggered 5-min intervals, with the first lane being run for 60 min total and the final lane being run for 25 min. The intervals between the addition of subsequent wells and the number of wells can be adjusted as needed. The image of such a gel is shown as the inset in Fig. 1.
3. Image the gel on a GE Amersham Molecular Dynamics Typhoon 9410 Molecular Imager, or similar instrument, using excitation and emission wavelengths suitable for the selected dye. We analyze the images using GE ImageQuant TL software. After automatic or manual identification of the bands and background subtraction, the software calculates the total number of photon counts in each band by integrating

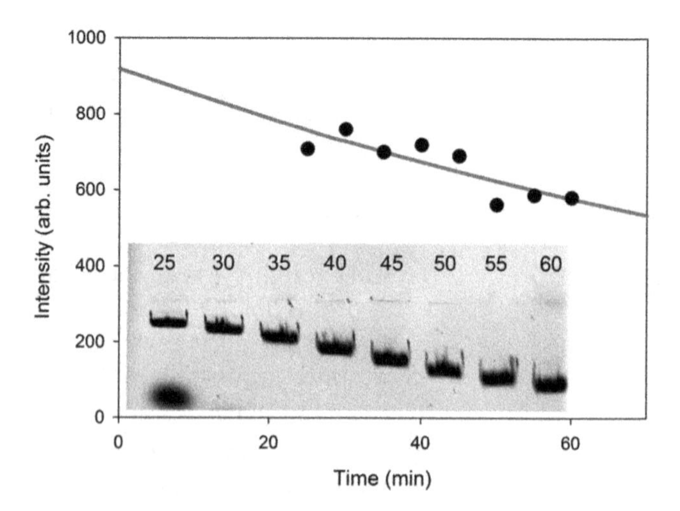

Fig. 1 Dissociation of Nuclear Red dye from DNA in a gel. The fluorescence intensity as a measure for the amount of bound dye decreases exponentially over an hour. The inset shows the gel with the DNA-dye complex loaded at different times

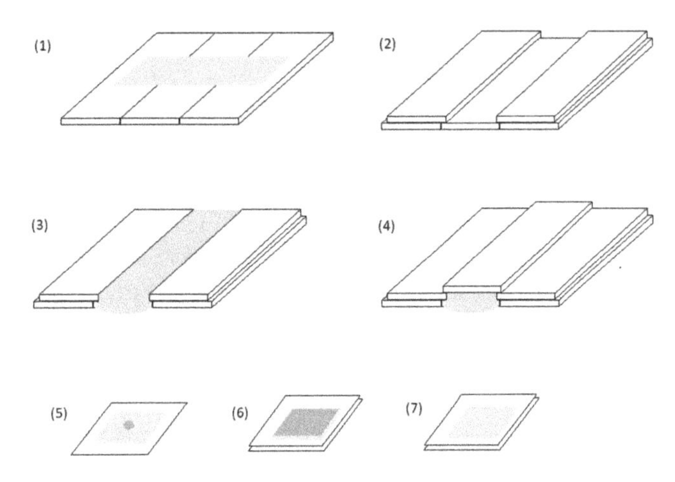

Fig. 2 Preparation of the agarose pads for the immobilization of *E. coli* during imaging and FCS data acquisition

over the area of each individual band. An exponential decay can be fitted to the intensities to quantify the dissociation time constant.

3.5 *Staining* E. coli *Cells*

1. Mix 95 μL of culture with 5 μL of 5 μM Syto 62 dye (or other preferred dye). Final dye concentration should be roughly 250 nM.
2. Incubate for 20–30 min at 37 °C and shaken at a frequency of 285 rpm.

3.6 *Immobilizing* E. coli *Cells*

1. To immobilize *E. coli* cells during imaging, we prepare agarose pads according to a modified protocol by Skinner et al. [7] (Fig. 2)
   (a) Tape three cover slides together in a row and then flip the slides so the tape is on the bottom.
   (b) Place two additional slides on top, positioned such that there is a gap between them that is slightly less than the width of an individual slide.
   (c) Pour 1% molten agarose made with MOPS media into the trench that is formed. Agarose that has over flowed is okay.
   (d) Carefully place another cover slide on top, while avoiding any air bubbles. Place a weight on top to hold everything in place and let sit for 45 min.
   (e) Remove excess agarose from the ends with a razor blade, then cut a square of agarose pad off that is roughly 1 × 1 cm in size, and place it onto a coverslip. Pipette 5 μL of sample onto the agarose pad.

(f) Carefully place another coverslip on top of the agarose pad with sample, again being careful to avoid any air bubbles, and let sit for 3–5 min to allow the cells to settle into the agarose.

(g) Carefully flip so that the sample is now underneath the agarose pad, and image.

2. To immobilize the stained *E. coli* cells on the agarose pads, cut off a section of the gel that is no larger than the coverslip being used. Slide the gel onto the coverslip and place 5 μL of bacteria sample onto the gel. Place another coverslip on top, and wait 2–3 min. This will allow the bacteria to settle into the agarose. Flip the sample and image.
3. For immobilization using the antibody-coated coverslips instead, load 25–30 μL sample and image after cells have settled for about 5 min.

3.7 Confocal Imaging

1. Confocal images and FCS data were collected using an ALBA time-resolved confocal microscope, with an IX-81 Olympus microscope body and a U-Plan S-APO 60× water immersion objective (1.2 NA, 0.28 mm working distance). A Fianium supercontinuum laser with an acousto-optical filter was used to generate picosecond-excitation pulses at a wavelength of 630 nm for Syto 62, and 561 nm for Nuclear-ID Red, at a repetition rate of 20 MHz. Typical laser power ranged from roughly 100 to 10 μW going into the back aperture of the objective. Fluorescence was collected through a 100 μm pinhole and a 700 nm bandpass filter onto a low-noise avalanche photodiode. Arrival times were recorded in time-tagged mode for each photon.
2. Each cell is imaged before and after FCS spectra are collected, with an image size that is somewhat larger than the cell itself. The first image is used to determine the location of interest inside the cell where the FCS spectra will be taken. The second image is compared to the first in order to determine if the cell has moved during data acquisition. If the cell has moved noticeably, typically more than 100 nm, the FCS data is discarded.

3.8 Fluorescence Correlation Spectroscopy

1. Prior to imaging and FCS data collection, the excitation volume of the FCS beam is calibrated by using a dye with known photo-physical properties at similar excitation/emission wavelengths. Cy5 was used when calibrating for Syto 62 samples, and Cy3 was used when calibrating for Nuclear-ID Red samples (*see* **Note 5**). Beam parameters for the excitation volume were determined by taking FCS data for the known dye in TBE buffer at 50 nM, 5 nM, 0.5 nM, and 50 pM concentrations. A

program within the Vista Vision software package assigns a global simultaneous least-square fit to these curves, which in turn yields the beam parameters.

2. The alignment of the instrument is verified by using the fluorescence lifetime correlation spectrometry to compare the signal-to-noise ratio. Ideally signal size should be three orders of magnitude higher than noise levels.
3. Several FCS spectra are acquired at the previously selected spot for 60s each. The intensity of the laser beam is set such that typical count rates of 30,000 to 100,000 are obtained. The photon arrival times are saved in time-tagged mode for further analysis. Typically, we can acquire one to two FCS spectra at one spot before photobleaching reduces the signal intensity below our desired threshold and a new spot must be selected. New spots can be selected within the same cell if another image is taken to confirm there is still a significant amount of dye fluorescing within the cell; otherwise a new cell should be located and selected.

3.9 Data Analysis

The inputs for the data analysis algorithm are the arrival times t_i of every photon, as recorded by the FCS detector in time-tagged mode as collected in Subheading 3.8.

1. The starting point for the photobleaching correction is to determine the overall extent of the photobleaching in the data. For this aim, the photon arrival data is binned into a histogram that provides the intensity as a function of time $I(t)$. This function is then fitted to the double-exponential decay

$$I(t) = a_{\text{fast}} e^{-k_{\text{fast}} t} + a_{\text{slow}} e^{-k_{\text{slow}} t} + y_0$$

which in turn yields the amplitudes and time constants of the observed photobleaching.

2. The intensity autocorrelation function is calculated directly from the photon arrival times using a method based on an algorithm by Wahl et al. [8] using custom Matlab code. In this scheme the autocorrelation function is calculated as the number of all pairs of photons falling into a particular lag time interval τ_k. As the algorithm progresses to larger lag time intervals, time-scale coarsening is implemented by combining multiple photons into one entry and giving them a weight w_i that corresponds to the number of combined photons. This algorithm forms the basis for our correction method [9]: To account for photobleaching, we modify this scheme by assigning each photon a starting weight $w_i = I^{-1}(t_i)$, instead of one, in order to compensate for missing photons from fluorophores that have already been lost to photobleaching. Time-scale

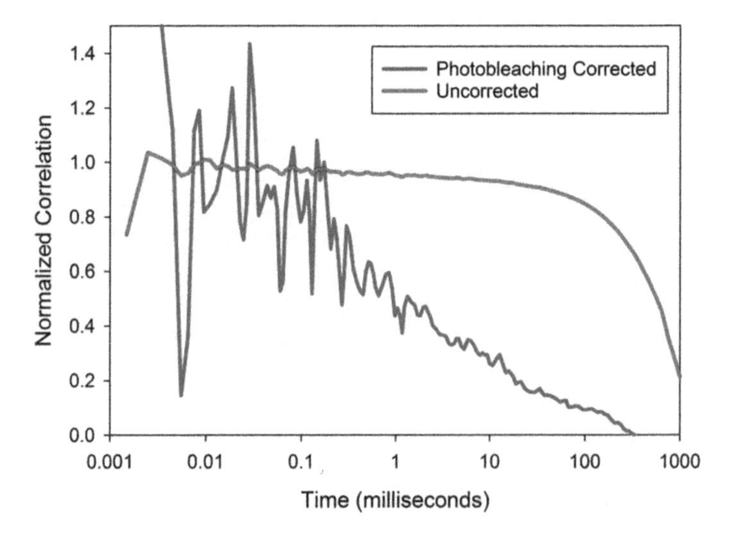

Fig. 3 Normalized FCS autocorrelation function of Nuclear Red-stained genomic DNA in live HeLa cells, with and without correction for photobleaching. The uncorrected curve is dominated by an artificial slow correlation from photobleaching. After correction for this effect, the curve reflects the underlying fluctuations of the DNA in the nucleus

coarsening proceeds as usual by adding the weights of all combined photons.

3. The photobleaching-corrected time correlation functions are compiled and normalized. The $t = 0$ data point is discarded, as all instrumental noise accumulates here. The curves are further smoothed as necessary and normalized.
4. Normalized correlation functions from the same spot can be averaged together to further reduce the noise. The resultant curves are now available for further quantitative analysis, such as the extraction of mean-square displacements of the fluorophores and the determination of diffusion coefficients and anomalous diffusion exponents. Examples of FCS correlation functions for the genomic DNA of HeLa cells and *E. coli* are shown in Fig. 3.

4 Notes

1. *Dye Selection*

   Dye selection is critical for the success of these measurements. When considering the suitability of a dye, we look for the following characteristics:

(a) Permeability of a dye into a specific cell type. Nuclear Red, for instance, penetrates eukaryotic cells well, but does not stain live *E. coli*.

(b) High fluorescence enhancement upon binding to the DNA.

(c) Low off-rate to ensure each fluorophore stays bound to the DNA longer than a typical measurement time. We check this by running stained DNA on an agarose electrophoresis gel and measure dye dissociation by observing the reduction of fluorescence as a function of time as described above in Subheading 3.4 and [4, 6].

2. *Dye Handling*

The fluorescent dyes used here are quite sensitive to degradation by ambient light. Therefore, we store the dye solution in aluminum-foil wrapped tubes and handle them with the lights turned down. Cells are incubated in the dark during staining.

3. *Dye Concentration*

The dye concentration directly affects the quality of the FCS spectra. The amplitude of the fluorescence fluctuations scales as the square root of the number of fluorophores in the excitation volume. The background fluorescence signal, however, scales linearly. Therefore, a lower number of fluorophores will result in a better signal-to-noise ratio in the FCS spectrum. If the concentration is too low, however, there is no meaningful signal to analyze left. Ideally there should be a small number of fluorophores, less than 10 and more than 1, in the excitation volume. To achieve this under typical staining conditions, some optimization through trial and error may be required.

4. *Dye to DNA Ratio*

The ideal ratio of dye to DNA for the gel dissociation experiments is somewhat subject to trial and error. It should give a strong fluorescent signal, while avoiding saturation of the detector or resulting in a potentially significant alteration of the properties of the DNA. For the latter, a ratio of no more than one dye molecule per 100 base pairs is generally considered safe.

5. *FCS Calibration*

When we calibrate the excitation volume of the FCS instrument, we watch for signs of dye sticking to the surface of the coverslip. Any such stuck molecules would show up in the correlation functions as a second, immotile species. To reduce the likelihood for this to happen, a surfactant like Tween 20 can be added to the solution.

Acknowledgments

This work was funded through internal resources of the University of Michigan. The FCS measurements were carried out at the Single-Molecule Analysis in Real Time (SMART) Center with the help of J. D. Hoff on equipment acquired through NSF MRI-ID grant DBI-0959823 to Nils G. Walter. Michael Jones contributed to the dye characterization measurements. The authors would like to thank Jörg Enderlein from the University of Göttingen for generously sharing computer code for the calculation of correlation functions from photon arrival times with us.

References

1. Chiantia S, Ries J, Schwille P (2009) Fluorescence correlation spectroscopy in membrane structure elucidation. Biochim Biophys Acta 1788:225–233
2. Michelman-Ribeiro A, Mazza D, Rosales T, Stasevich TJ, Boukari H, Rishi V, Vinson C, Knutson JR, McNally JG (2009) Direct measurement of association and disassociation rates of DNA binding in live cells by fluorescence correlation spectroscopy. Biophys J 97:337–346
3. Gröner N, Capoulade J, Cremer C, Wachsmuth M (2010) Measuring and imaging diffusion with multiple scan speed image correlation spectroscopy. Opt Express 18:21225–21236
4. Kafle RP, Liebeskind MR, Bourg JT, White E, Meiners JC (2015) Artifacts from photobleaching and dye dissociation in fluorescence correlation spectroscopy. Proc SPIE 9548:94581M
5. Windengreen J, Rigler R (1996) Mechanisms of photobleaching investigated by fluorescence correlation spectroscopy. Bioimaging 4:149–157
6. Eriksson M, Karlsson HJ, Westman G, Akerman B (2003) Groove-binding unsymmetrical cyanine dyes for staining of DNA: dissociation rates in free solution and electrophoresis gels. Nucleic Acids Res 31:6235–6242
7. Skinner S, Sepúlveda L, Xu H, Golding I (2013) Measuring mRNA copy number in individual Escherichia coli cells using single-molecule fluorescent in situ hybridization. Nat Protoc 8:1100–1113
8. Wahl M, Gregor I, Patting M, Enderlein J (2003) Fast calculation of fluorescence correlation data with asynchronous time-correlated single photon counting. Opt Express 11:3583–3591
9. Hodges C, Kafle RP, Hoff JD, Meiners JC (2018) Fluorescence correlation spectroscopy with photobleaching correction in slowly diffusing systems. J Fluoresc:1–7 https://doi.org/10.1007/s10895-018-2210-y

Chapter 26

Three-Dimensional Tracking of Quantum Dot-Conjugated Molecules in Living Cells

Lucia Gardini, Martino Calamai, Hiroyasu Hatakeyama, Makoto Kanzaki, Marco Capitanio, and Francesco Saverio Pavone

Abstract

Here, we describe protocols for three-dimensional tracking of single quantum dot-conjugated molecules with nanometer accuracy in living cells using conventional fluorescence microscopy. The technique exploits out-of-focus images of single emitters combined with an automated pattern-recognition open-source software that fits the images with proper model functions to extract the emitter coordinates. We describe protocols for targeting quantum dots to both membrane components and cytosolic proteins.

Key words Single-molecule localization, Three-dimensional tracking, Living cells, Out-of-focus imaging, Quantum dots, Plasma membrane proteins, Lipids, Cytosolic proteins

1 Introduction

Since their introduction a couple of decades ago, optical techniques that allow the investigation of single biological molecules have represented a revolution for biological research. Thanks to the possibility to study the behavior of molecules *in singulo*, current knowledge of many biological processes has grown considerably, complementing information derived from bulk experiments and providing new insight about the effect of the natural variability between molecules [1, 2].

Recent applications of single-molecule tracking revealed the behavior of fundamental proteins driving the machinery of cell life, such as molecular motors [3–8], DNA-binding proteins [9–11], and membrane proteins [12, 13]. On the other hand, super-resolution microscopy based on single-molecule localization uncovered the structure of intracellular components that were previously hindered by resolution limits owing to light diffraction [14, 15]. Nevertheless, most of these studies were based on two-dimensional imaging, whereas intracellular processes are

Yuri L. Lyubchenko (ed.), *Nanoscale Imaging: Methods and Protocols*, Methods in Molecular Biology, vol. 1814, https://doi.org/10.1007/978-1-4939-8591-3_26,

strongly influenced by the three-dimensional structure of the cellular environment [16]. In fact, analyzing subcellular structures and 3D motion of single molecules from their 2D projection on the focal plane can lead to dramatic misinterpretation of phenomena. For this reason, there is an increasing interest in finding sensitive techniques to reach nanometer localization in all three dimensions.

In recent years, different techniques have been proposed for accurate measurement of the axial position of single molecules, either based on off-focus imaging [17–22], point-spread-function (PSF) engineering [23, 24], or astigmatism [25–30]. Each technique shows its own strengths and limitations. Bifocal imaging [17] has been demonstrated to reach few nanometers axial localization accuracy on fluorescent beads but limited to about half micron depth in the sample. On the other hand, multiplane imaging [18–22] can span thicker volumes but at the expense of the axial localization precision. Tracking of photoactivatable dyes and quantum dots (QDs) with 10 nm localization accuracy in 3D was obtained by PSF engineering as a double helix [23, 24]. Astigmatic detection [25, 26], which has been fruitfully applied in super-resolution techniques [27–30], has allowed 3D localization within few tens of nanometers over 1 μm thickness. Finally Parallax, a technique based on two split images corresponding to the view of the object from different sides, could reach 3D nanometer localization within 1 μm axial range on fluorescent beads [31].

Regardless of the technique, single-molecule studies of biomolecules inside living cells are limited by the efficiency of the chromophore of choice in terms of brightness and photostability. Although fluorescent proteins are extensively used due to the possibility to be genetically encoded inside cells, their concentration cannot be easily controlled and are not efficient as single emitters. Organic dyes are brighter and more stable emitters that are widely used in single-molecule localization-based super-resolution microscopy but still relatively short-lived to be tracked for long period of times as single molecules. To this end, fluorescent beads and QDs are currently the brightest and most photostable fluorescent probes available. QDs are usually preferred for live cell studies due to their small size and long photostability [32, 33]. Although introducing QDs inside living cells can be difficult, several protocols have been developed to deliver them into the cytoplasm and the nucleus- and target-specific proteins. For example, QDs with exposed surface groups have been delivered inside the cell through electroporation [34] or targeted selectively to multiple genetically encoded tags by microinjection of QD-streptavidin conjugates precomplexed with cognate biotinylated hapten molecules [35]. Furthermore, modular peptides displaying both QD bioconjugation motifs and specific subcellular targeting domains were constructed to facilitate their specific delivery to either the plasma membrane, the endosomes, the cytosol, or the mitochondria of target cells [36].

These developments have opened the possibility to investigate intracellular events, such as the dynamics of molecular motors [37–41], virus infection [42–44], endocytic receptors trafficking [45], and intracellular import by nuclear pore complexes [46]. Moreover, in recent years new researches have been carried out to further reduce QD size to target them toward confined intracellular and extracellular regions [47–49]. Despite this progress, the application of 3D tracking techniques to anisotropic emitters such as single chromophores or elongated QDs has been challenged so far by the observation that their PSF is rotationally dependent [50]. More generally, the requirement of custom setups has represented an obstacle to the wide spreading of three-dimensional tracking techniques.

We recently developed a technique for three-dimensional tracking of QDs with few nanometers accuracy in living cells using commercial wide-field microscope [51].

In a standard epifluorescence microscope, point-like isotropic emitters produce an intensity profile in the image plane that is roughly Gaussian, with a radius of ~$\lambda/2NA$ (~250 nm for visible light). According to the distance of the source from the objective focal plane, the image shape changes into a diffraction pattern consisting of a central Gaussian profile surrounded by Gaussian rings. The number of rings and the radius of the diffraction pattern increase depending on the increasing distance of the object from the focal plane. Based on this observation, it is possible to infer the axial position of the probe by measuring the radius of the circular pattern, after an appropriate calibration. In the past, this technique was applied to fluorescent nanobeads demonstrating nanometer axial localization precision [52], while its application to QDs was not possible due to the anisotropy of off-focus PSF [50]. Moreover, the lack of an automated procedure to measure the radius of the diffraction pattern made the application of this method difficult and time-consuming. We extended the applicability of this technique to QDs, provided that an appropriate ligation strategy that leaves rotational freedom to QDs is adopted. Additionally, we developed a software, called PROOF (Pattern-Recognition Out-Of-Focus), that automatically recognizes the number of rings of an out-of-focus PSF, generates an appropriate function, and uses it to fit the PSF image. In this way, we can readily determine the axial (z) position of the probe from the radius of the outermost ring as well as the x, y center of the probe over a range of several microns, depending on the signal-to-noise ratio of the image.

In this chapter, we describe protocols for single-molecule tracking of protein-conjugated QDs on the membrane and in the cytosol of living cells using our technique. We first give protocols for calibration and three-dimensional localization of single QDs from out-of-focus images (*see* Subheading 3.1). Then, we describe protocols for 3D tracking of plasma membrane proteins (*see* Subheading 3.2), lipids (*see* Subheading 3.3), and cytosolic proteins in living cells (*see* Subheading 3.4).

2 Materials

2.1 Calibration Sample

1. Cultured SH-SY5Y human neuroblastoma cells in 1:1 DMEM/F-12 supplemented with 10% FBS and 1% penicillin streptomycin antibiotics.
2. 1× PBS.
3. 4% paraformaldehyde (PFA) (Sigma, dilution in 1× PBS).
4. Phenol red-free Leibovitz's L-15 medium (Thermo Fisher).
5. Quantum dot wheat germ agglutinin conjugate, 655 nm emission wavelength (QDots655-WGMA) (Life Technologies).
6. DL-Dithiothreitol (Sigma).
7. Imaging buffer: 1.2 μM glucose oxidase (Sigma), 0.2 μM catalase (Sigma), 4.2 μM alpha-casein (Sigma), 17 mM glucose (Sigma), and 20 mM DL-dithiothreitol (Sigma) in phenol red-free Leibovitz's L-15 medium (Thermo Fisher).

2.2 Live Cell Labeling for Tracking of Plasma Membrane Proteins

1. Heating block.
2. Vacuum pump for washes (optional).
3. Tweezers.
4. 12 wells plate (Sigma).
5. 18 mm diameter round glass coverslips (VWR).
6. Imaging chamber (*see* **Note 1**).
7. Phenol red-free Leibovitz's L-15 medium (Thermo Fisher).
8. 4× solution of quantum dot-binding buffer (QDBB): add 2 g sodium tetraborate decahydrate (EMD MILLIPORE) and 4.9 g boric acid (Sigma) to 400 mL of ddH_2O. Adjust the pH to 8.0 and bring the volume to 500 mL (*see* **Note 2**).
9. Bovine serum albumin (BSA) (Sigma).
10. Sodium azide (Sigma).
11. Sucrose (Sigma).
12. Quantum dot streptavidin conjugate (Thermo Fisher Qdot™ 655).
13. Primary antibodies against the extracellular epitopes of molecule of interest.
14. Fetal bovine serum (FBS) (Thermo Fisher).
15. F(ab')2 IgG (H + L) secondary antibody directly coupled to QDs (Thermo Fisher F(ab')2-goat anti-rabbit IgG (H + L) secondary antibody, Qdot 655) or a biotin-conjugated F(ab')2 IgG (H + L) (Abcam rabbit F(ab')2 anti-mouse IgG H&L biotin).

2.3 Live Cell Labeling for Tracking of Membrane GM1 Gangliosides

1. Cultured SH-SY5Y human neuroblastoma cells in 1:1 DMEM/F-12 supplemented with 10% FBS and 1% penicillin streptomycin antibiotics.
2. Heating block.
3. Vacuum pump for washes (optional).
4. Tweezers.
5. 12 wells plate (Sigma).
6. 18 mm diameter round glass coverslips (VWR).
7. Imaging chamber (*see* **Note 1**).
8. Phenol red-free Leibovitz's L-15 medium (Thermo Fisher).
9. Fetal bovine serum (FBS) (Thermo Fisher).
10. 10 μg/mL biotinylated B subunit of cholera toxin-ctxb (Sigma).
11. Quantum dot streptavidin conjugate (Thermo Fisher Qdot™ 655).

2.4 Live Cell Labeling for Tracking of Cytosolic Proteins

1. Electroporator and adherent cell electrode (BEX CUY21 EDI-TII electroporator and electrode; Fig. 4a).
2. Expression vector for protein of interest fused to HaloTag protein; various constructs can be obtained from Promega.
3. HaloTag succinimidyl ester (O_2) ligand (Promega; formula $C_{18}H_{29}ClN_2O_7$, molecular weight 420.89).
4. Qdot™ 655 ITK™ amino (PEG) quantum dots (Thermo Fisher; containing 250 μL of 8 μM QDs in borate buffer, i.e., it contains 2 nmol of QDs).
5. Anhydrous DMSO (Thermo Fisher).
6. Lipofection reagent (Lipofectamine 3000, Thermo Fisher).
7. Cell-permeable HaloTag fluorescent ligands (HaloTag TMR ligand or HaloTag Oregon Green ligand).
8. Ultrafiltration device (Amicon Ultra-0.5, 100 kDa, Merck)
9. Spectrophotometer (NanoDrop 2000, Thermo Fisher).
10. Fluorescent gel imager (Molecular Imager PharosFX systems, Bio-Rad).
11. Electroporation buffer: 150 mM trehalose, 5 mM K-phosphate buffer, 5 mM $MgCl_2$, 2 mM EGTA, 2 mM ATP, 25 mM HEPES-KOH (pH 7.3), and 1% DMSO.
12. PBS (pH 7.0–7.6).
13. Borate buffer: 50 mM borate, pH 8.5.
14. 0.5× TAE buffer: 20 mM Tris, 10 mM acetic acid, 0.5 mM EDTA, pH 7.6.

15. Imaging buffer of choice (150 mM NaCl, 5 mM KCl, 2 mM $CaCl_2$, 1 mM $MgCl_2$, 10 mM HEPES-NaOH, and 5.5 mM D-glucose, pH 7.4).

2.5 Setup for Single-Molecule Tracking

The microscope for three-dimensional tracking of single QDs can be any commercial inverted fluorescence microscope, but components should be carefully selected to get best signal-to-noise ratio (SNR) images, minimize optical aberrations, and accurately calibrate axial localization. The optical scheme of the setup we use for 3D single-molecule localization is shown in Fig. 1 [51, 53].

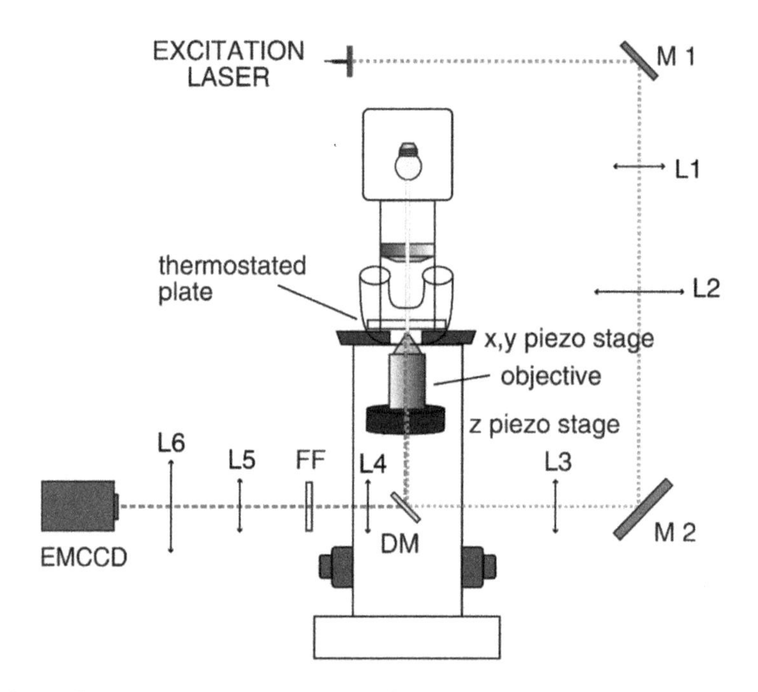

Fig. 1 Scheme of the optical setup. Our optical setup is composed by a commercial inverted fluorescence microscope equipped with a 488 nm laser source for QDs-655 fluorescence excitation and EMCCD camera for detection (current sCMOS camera usually provide better SNR). The excitation laser is focused on the back focal plane of the microscope objective after magnification by 3× achromatic doublet telescope (L1, L2). The objective is moved along its optical axis with nanometer accuracy through a piezo translator, while the sample can be displaced along x, y axis through a piezo stage. Fluorescence from the sample is collected by the objective, separated from excitation through a dichroic mirror (DM), and filtered through a band-pass filter (FF). The image is then projected onto the EMCCD camera, after an additional 3× magnification through an achromatic doublet telescope (L5, L6) to create a field of view of $40 \times 40\ \mu m^2$ area and a pixel size of about 90 nm, which is a good compromise between sampling accuracy and pixel signal intensity. The microscope stage is equipped with a thermostated plate to keep the sample at 37 °C during experiments. Reproduced from Gardini 2015 with permission from Nature Research

1. Laser sources for QDs excitation (Laser Physics argon laser, 488 nm in this case).
2. Inverted fluorescence microscope (e.g., Nikon Eclipse TE300).
3. High NA oil immersion objective (e.g., Nikon 60× Plan Apo TIRF, 1.45 oil immersion).
4. Piezo stage to move the objective along the optical axis, positioning error $\leq$ 2 nm is required (e.g., P-721.C PIFOC Physik Instrumente).
5. Piezo stage for x, y displacement (optional) (e.g., P-527.2CL Physik Instrumente).
6. Dichroic mirror with appropriate cutoff wavelength to separate fluorescence excitation from emission (Semrock FF500/646 in case of 655-QDs).
7. Fluorescence emission filter (e.g., Semrock FF01–655/40).
8. Achromatic doublet telescope (L5, L6 in Fig. 1) in the detection path (pixel size should be in the ~80–100 nm range).
9. EMCCD camera for single-molecule fluorescence detection (e.g., Andor iXon X3).
10. Thermostated plate mounted on the microscope stage to keep the sample at 37 °C during experiments.

3 Methods

3.1 Calibration

The aim of the calibration procedure is to empirically measure the relationship between the radius of the outermost diffraction ring of the QD PSF and the distance of the QD from the focal plane of the objective. Once this relationship is established and fitted with an appropriate calibration function, the axial position of QDs can be inferred from the measured radius of their intensity profile. In this section we show how to prepare a calibration sample (1–10) and how to use it for calibration (11–15).

1. Plate cells on glass coverslips in 12 wells plates at 37 °C, 5% CO_2 in 1:1 DMEM/F-12 supplemented with 10% FBS and 1% penicillin streptomycin at least 24 h before imaging. Specific treatments or transfections can be carried out 1 day after plating.
2. Warm up the heating block at 37 °C. Place a drop of water on the surface of the heating block and place a strip of parafilm on it (Fig. 3a). The parafilm will be useful later to prevent leakage of the medium and for easy removal of the glass coverslips (*see* **Note 3**).

3. Transfer the coverslip with attached cells on the parafilm, remove the medium with a pipette or tip attached to a vacuum pump, and replace it with 200 μL phenol red-free 10% FBS Leibovitz's L-15 medium (*see* **Note 4**).
4. Wash again the coverslip with 200 μL 10% FBS Leibovitz's L-15 medium.
5. Discard it and replace with 200 μL 4% PFA, and incubate for 10 min at room temperature.
6. Wash cells three times with 10% FBS Leibovitz's L-15 medium, 5 s each.
7. Incubate QDots655-WGMA for 1 min (previously diluted 105 times in Leibovitz's medium +10 mM DTT).
8. Wash cells five times with 10% FBS Leibovitz's L-15 medium, 5 s each.
9. Incubate in 10% FBS Leibovitz's L-15 medium for 10 min (*see* **Note 5**).
10. Wash with imaging buffer, and mount on the imaging chamber (Fig. 3d) for positioning on the microscope stage.
11. Choose a single isolated QD (*see* **Note 6**).
12. Starting from the in-focus position, displace the objective (or the sample) along the axial direction by 3 μm in 50 nm steps using the piezo stage (*see* Subheading 2.5). At each step, acquire a series of ten images, and record the actual axial position of the stage (*see* **Notes** 7 and **8**).
13. Once the images of the QD at different objective positions are acquired, analyze them using our custom software PROOF [51] to extract the radius of the outermost ring (R) as well as the x, y coordinates of the PSF center (*see* **Note 9**).
14. Given the axial position of the focal plane for each image of the stack and the correspondent value of the radius of the outermost ring R for the QD of interest, a stepwise relationship such as that shown in Fig. 2b is found (*see* **Note 10**).
15. Fit data with a polynomial function with order high enough to reduce the systematic calibration error below other error sources. In fact, the main error in the axial localization Δz originates from the error in the determination of R (σz_R), which is obtained from the standard deviation of R between multiple images sharing the same axial position, after conversion in nm using the calibration curve. We find an average $\sigma z_R = 6.9$ nm in the range between 750 and 1900 nm depth, while it increases rapidly for larger distances from the focal plane (Fig. 2c). Conversely, the systematic calibration error is calculated as the mean squared error between the measured z

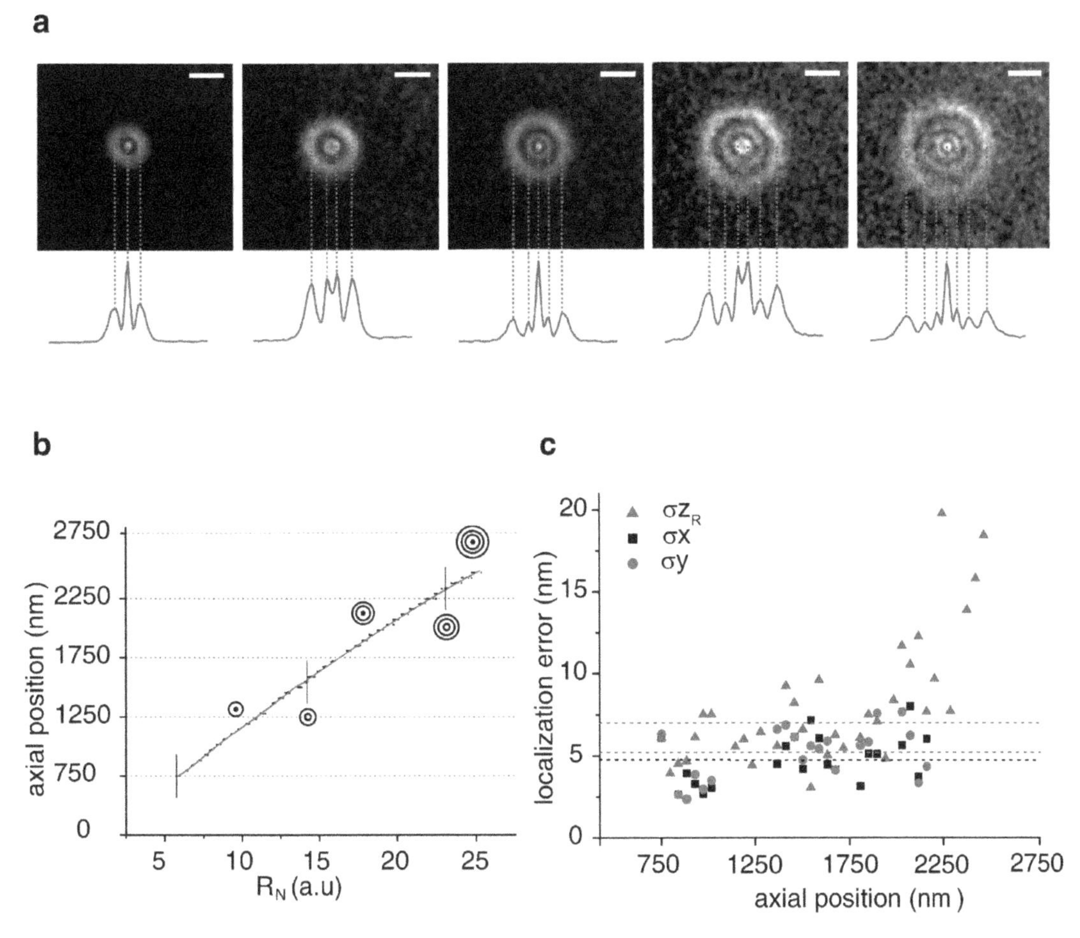

Fig. 2 Calibration of axial localization from out-of-focus images. (**a**) Diffraction patterns of an out-of-focus QD (QD655-WGMA, wheat germ agglutinin-QD conjugated) targeted to the membrane of a fixed human neuroblastoma cell, at different depths. Blue lines represent average radial profiles. Profiles are radially isotropic similarly to those obtained with fluorescent nanobeads but present a central "hole" at specific axial positions. Our algorithm recognizes these cases and chooses a proper fitting function. Scale bar = 2 μm. (**b**) Calibration curve obtained for QDs. The cartoons represent the number of rings in different axial regions as indicated by vertical lines. Colored curves are nonlinear fits to the data in the different regions. (**c**) Localization errors in all three dimensions are plotted as a function of the axial position of the probe. Blue, black, and red dotted lines represent the average errors, which are, respectively, $\sigma z_R = 6.9$ nm (for axial positions in the range 750–1900 nm), $\sigma x = 4.7$, and $\sigma y = 5.2$ nm (over the whole calibration range). Reproduced from Gardini 2015 with permission from Nature Research

position and the z position predicted by the calibration curve at a given R. This systematic error should be reduced below σz_R by a proper choice of the calibration function (*see* **Note 11**). The x, y localization error is calculated as the standard deviations of x_0, y_0 at each z, after subtraction of local drifts. In this case, we obtain σ_x, $\sigma_y = 5.2$ nm over the whole calibration range.

3.2 Live Cell Labeling for Tracking of Plasma Membrane Proteins

In this section we report protocols for the labeling of transmembrane proteins with specific antibodies directed against the extracellular epitopes in order to run 3D single-molecule tracking experiments. By labeling membrane recycling proteins (e.g., the insulin-responsive glucose transporter GLUT4 and transferrin receptor) with specific antibodies against the extracellular epitopes or specific ligands conjugated to QDs followed by tracking after allowing internalization, we can also successfully investigate "intracellular" behavior of these molecules [54–58].

1. Plate cells on glass coverslips in 12 wells plates at 37 °C, 5% CO_2 at least 24 h before imaging. Specific treatments or transfections can be carried out 1 day after plating. Use medium specific for the cell type of your choice (*see* **Note 12**).
2. Warm up the heating block at 37 °C. Place a drop of water on the surface of the heating block and place a strip of parafilm on it (Fig. 3a). The parafilm will be useful later for preventing leakage of the medium and for removing the glass coverslips more easily (*see* **Note 3**).
3. Prepare 50 mL of sterile phenol red-free Leibovitz's L-15 medium 10% FBS. Leibovitz's L-15 medium is a CO_2-independent medium that allows keeping the cells without additional CO_2 in the atmosphere.
4. Prepare 10 mL 1× solution of QDBB with ddH_2O supplemented with 2% (w/v) bovine serum albumin and 0.05% (w/v) sodium azide. This solution prevents nonspecific binding (*see* **Note 13**).
5. Prepare 10 mL 1 M stock solution of sucrose. At the time of the experiment, mix 200 μL of sucrose stock solution with 800 μL of 1× QDBB. 200 mM sucrose is used to make the solution isosmolar with the cells.
6. To obtain a 0.1 nM QDs solution, pre-dilute 1 μL of 1 μM streptavidin QDs in 100 μL sucrose/QDBB solution, and then take out 9 μL and mix them with the remaining 900 μL of sucrose/QDBB solution (*see* **Note 14**).
7. Dilute the primary antibody (better if monoclonal) directed against the extracellular domain of the membrane protein of interest in 10% FBS Leibovitz's L-15 medium (*see* **Note 15**). To start, use the concentration suggested by the manufacturer for immunofluorescence experiments; the concentration of the primary antibody can then be adjusted to reach a satisfactory low number of labeled molecules (*see* **Note 16**).
8. Dilute 1:500 the secondary antibody in 10% FBS Leibovitz's L-15 medium. You can either use a F(ab')2 IgG (H + L) secondary antibody directly coupled to QDs or a biotin-conjugated F(ab')2 IgG (H + L).

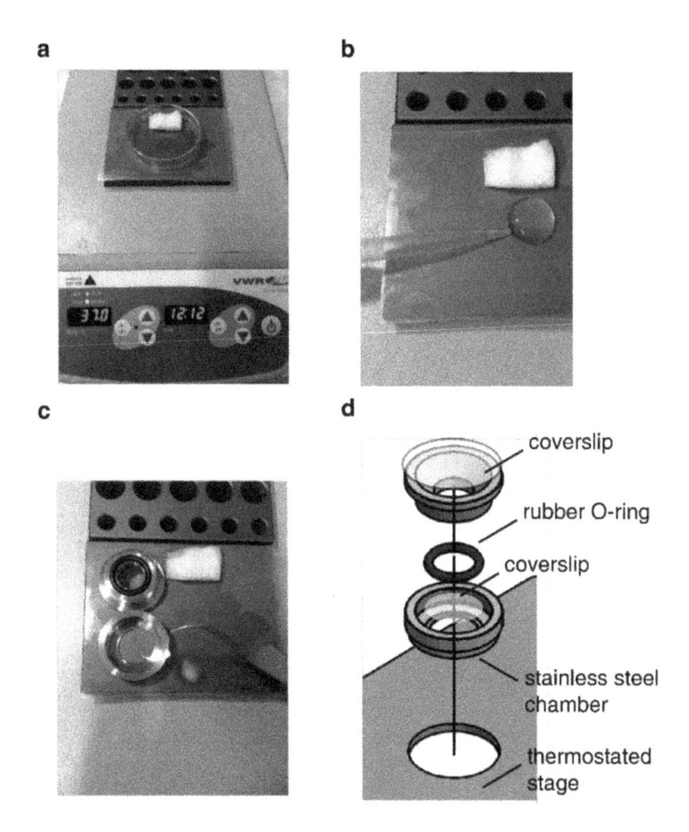

Fig. 3 Labeling of molecules on the plasma membrane of living cells. (**a**) Turn on the heat block at 37 °C, transfer cells previously plated on a glass coverslip on a layer of parafilm, and set a moist chamber with the bottom of a petri dish and tissue paper to avoid evaporation of the liquid in the sample. (**b**) Washing and incubation steps: careful aspiration of medium from one side with a tip connected to a vacuum pump, followed by the addition of solution from one side. (**c**) Use tweezers to gently transfer the coverslip on a live imaging chamber such as the one depicted schematically in (**d**) to perform measurements

9. Set up a moist chamber by using the top or the bottom of a petri dish and wet tissue paper to avoid sample drying (Fig. 3a).
10. Transfer the coverslip with attached cells on the parafilm, remove the medium with a pipette or tip attached to a vacuum pump, and replace it with 200 μL phenol red-free 10% FBS Leibovitz's L-15 medium (*see* **Note 4**).
11. Wash again the coverslip with 200 μL 10% FBS Leibovitz's L-15 medium.
12. Remove and replace with 200 μL of the primary antibody solution. Incubate for 5–10 min.
13. Wash cells three times with 10% FBS Leibovitz's L-15 medium, 5 s each.

14. Incubate with 200 μL of the secondary antibody solution for 5 min.
15. Wash cells three times with 10% FBS Leibovitz's L-15 medium, 5 s each, or six if you have used the QDs conjugated secondary antibody.
16. If you have used the biotin-conjugated secondary antibody, incubate the cells for 1 min with 100 μL of the 0.1 nM QDs/sucrose/QDBB solution (*see* **Note 17**).
17. Wash cells six times with 10% FBS Leibovitz's L-15 medium, 5 s each.
18. Mount the coverslip on the imaging chamber (Fig. 3c, d).
19. Gently add 1 mL of 10% FBS Leibovitz's L-15 medium on the side of the chamber. Cover with a larger glass coverslip to reduce evaporation of the medium during measurements.
20. Mount the chamber on the thermostated plate of the microscope (*see* Subheading 2.5) preheated at 37 °C (*see* **Note 18**).

3.3 Live Cell Labeling for Tracking of Membrane GM1 Gangliosides

In this section we give instructions for the labeling of the GM1 ganglioside through the coupling of the biotinylated B subunit of cholera toxin to streptavidin quantum dots.

1. Culture SH-SY5Y human neuroblastoma cells on a glass coverslip in 1:1 DMEM/F-12 supplemented with 10% FBS and 1% penicillin streptomycin antibiotics.
2. Transfer the coverslip with attached cells on the parafilm, remove the medium with a pipette or tip attached to a vacuum pump, and replace it with 200 μL phenol red-free 10% FBS Leibovitz's L-15 medium (*see* **Note 4**).
3. Wash again the coverslip with 200 μL 10% FBS Leibovitz's L-15 medium.
4. Incubate cells for 10 min with 100 μL of 10 μg/mL biotinylated ctxb.
5. Wash cells three times with 10% FBS Leibovitz's L-15 medium, 5 s each.
6. Incubate cells with 100 μL streptavidin-QD655 at a 1:1000 dilution for 1 min.
7. Wash cells six times with 10% FBS Leibovitz's L-15 medium, 5 s each.
8. Mount the coverslip on the imaging chamber (Fig. 3c, d).
9. Gently add 1 mL of 10% FBS Leibovitz's L-15 medium on the side of the chamber. Cover with a larger glass coverslip to reduce evaporation of the medium during measurements.
10. Mount the chamber on the microscope on the thermostated plate (*see* Subheading 2.5) preheated at 37 °C (*see* **Note 18**).

3.4 Live Cell Labeling for Tracking of Cytosolic Proteins

In this section we describe the protocols for labeling molecules inside living cells [41]. This approach allows us to track cytosolic protein of interest fused to HaloTag protein by using HaloTag ligand-QDs and electroporation (*see* **Note 19**). First, we describe the protocol for preparing HaloTag ligand-QDs, then we give a method for validating the prepared HaloTag ligand-QDs to bind with HaloTag-fusion proteins, and finally we describe the protocol for introducing HaloTag ligand-QDs inside living cells.

Steps 1–7 illustrate the protocol for the preparation of QD-conjugated HaloTag ligand.

1. Take 250 μL of the Qdot™ 655 ITK™ amino (PEG) QDs stock and replace the QDs storage buffer with PBS by ultrafiltration, and adjust the final volume to 250 μL.
2. Dissolve 5 mg HaloTag succinimidyl ester (O_2) ligand in 990 μL anhydrous DMSO (12 mM stock solution).
3. Mix QDs and HaloTag succinimidyl ester (O_2) ligand: for 1:12 conjugation, mix 250 μL of 8 μM QDs (2 nmol) with 2 μL of 12 mM HaloTag succinimidyl ester (O_2) ligand (24 nmol) (*see* **Note 20**).
4. Incubate for 30 min at room temperature with continuous gentle agitation.
5. Remove the unreacted ligand by ultrafiltration with borate buffer.
6. Determine the concentration of HaloTag ligand-QD with the formula $A = \varepsilon cL$, where A, ε, c, and L are the absorbance at the specified wavelength, molar extinction coefficient, molar concentration, and path length, respectively. Molar extinction coefficient (ε) at the specified wavelength can be obtained from the manufacturer's information.
7. Aliquot the HaloTag ligand-QD and store the solution at 4 °C (*see* **Note 21**).
8. Prepare cell lysates containing HaloTag-fusion proteins with the standard protein expression systems.
9. Divide the lysates into two parts, and boil one part to denature the proteins (used as negative control of binding with HaloTag ligands).
10. Mix the lysates with 50 nM HaloTag ligand-QD for 1 h at 4 °C.
11. Perform electrophoresis of the mixed samples with 0.5% agarose gel in 0.5× TAE buffer.
12. Detect the QD fluorescence with a fluorescent gel imager (Fig. 4b).

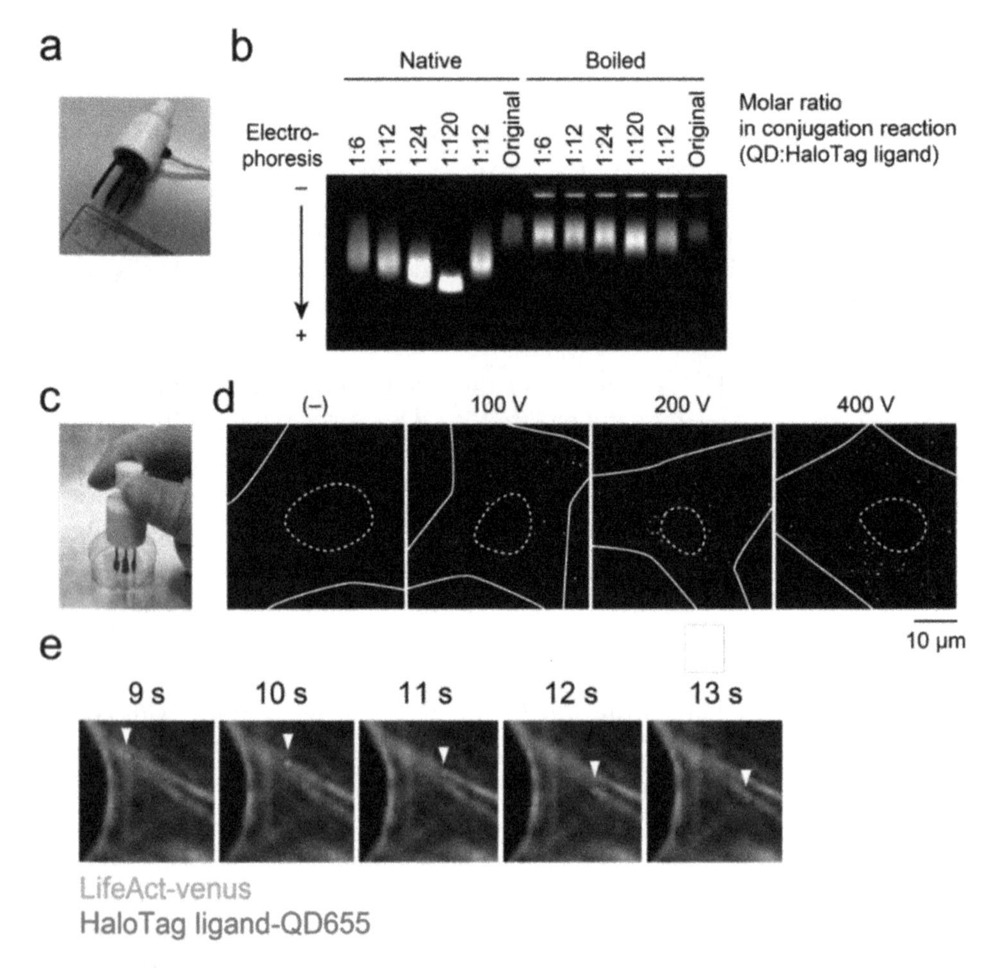

Fig. 4 Electroporation of HaloTag ligand-QDs into living cells. (**a**) Adherent cell electrode. (**b**) Binding of HaloTag ligand-QD with HaloTag proteins in vitro. HaloTag ligand-QDs were incubated with the lysate of KRX cells expressing HaloTag-14-3-3β/α proteins and separated on an agarose gel [41]. Samples on the right half of the gel were denatured by boiling the lysates before mixing with HaloTag ligand-QDs. "Original" refers to QDs with no conjugation reaction. QD fluorescence was detected with a fluorescent gel imager at an excitation wavelength of 488 nm. (**c**) Placing the electrode onto the cells attached to the glass surface. (**d**) Images of QD fluorescence in 3T3-L1 fibroblasts. Cells were immersed in QD-containing solution, and electroporation was performed with the indicated poration pulse voltages. Solid and dashed lines represent plasma and nuclear membranes, respectively. The Laplacian of Gaussian-filtered images were shown. (**e**) Linear movement of HaloTag-myosin Vb labeled with HaloTag ligand-QDs (magenta) along F-actin, which is labeled by using Lifeact-Venus (green). Panels (**b**), (**d**), and (**e**) are reproduced from Hatakeyama (2017) with permission from the American Society for Cell Biology

Steps 13–23 describe a protocol for simultaneous electroporation of expression vectors and HaloTag ligand-QDs (*see* **Note 22**).

13. Plate cells onto coverslips or in glass-bottom dishes.
14. Culture cells overnight in appropriate conditions (e.g., in a humidified atmosphere of 5% CO_2/air at 37 °C).

15. Prepare 100 μL of electroporation buffer containing 10–25 μg HaloTag expression vector and 1–5 nM HaloTag ligand-QD (*see* **Note 23**).
16. Rinse the cells with ice-cold PBS once.
17. Put cells in the electroporation buffer prepared at **step 15** and incubate on ice for 3 min.
18. Place the electrode onto the glass surface (Fig. 4c) and perform electroporation; typical setting with CUY21 EDITII electroporator in 3T3-L1 fibroblasts was as follows [41]: poration pulse at 100–400 V for 10 ms, followed by five pulses at −30 V for 10 ms at 50 ms intervals (*see* **Note 24**) (Fig. 4d).
19. Rinse cells with culture media twice.
20. Culture cells overnight in appropriate conditions.
21. Stain the cells with HaloTag TMR or Oregon Green ligand according to the manufacturer's instructions in order to visualize the cells appropriately expressing HaloTag-fused protein of interest (*see* **Note 25**).
22. Wash cells with the imaging buffer.
23. Perform single-molecule imaging, and track individual QD fluorescence with the method described in Subheading 3.5.

    Steps 24–34 illustrate a protocol for sequential introduction of expression vectors and HaloTag ligands-QDs (*see* **Note 22**).

24. Plate cells onto coverslips or in glass-bottom dishes.
25. Culture cells overnight in appropriate conditions (e.g., in a humidified atmosphere of 5% CO_2/air at 37 °C).
26. Transfect HaloTag expression vectors into the cells with a transfection reagent such as Lipofectamine 3000 according to the manufacturer's instructions.
27. Incubate cells overnight in appropriate conditions.
28. Prepare 100 μL of electroporation buffer containing HaloTag ligand-QD (1–5 nM).
29. Rinse cells with ice-cold PBS once.
30. Put cells in the electroporation buffer prepared at **step 15** and incubate on ice for 3 min.
31. Perform electroporation as in **step 18**.
32. Rinse cells with culture media twice.
33. Incubate in appropriate conditions for at least 3 h.
34. Stain the cells with HaloTag TMR or Oregon Green ligand according to the manufacturer's instructions in order to visualize the cells appropriately expressing HaloTag-fused protein of interest (*see* **Note 25**) and immerse the cells with the imaging buffer.

3.5 Image Acquisition and Data Analysis

Proper setup of acquisition parameters is pivotal to reach the SNR needed for fitting out-of-focus PSFs. Quantum Dots 655 can be efficiently excited with any laser source below 550 nm. We suggest using a 488 nm laser source, with about 3 mW power on the sample illuminating a $40 \times 40\ \mu m^2$ field of view in wide-field configuration. Semrock FF500/646-Di01–25 × 36 dichroic mirror and Semrock FF01–655/40 emission filter or any equivalent mirror/filter set can be used for imaging. Consider that accurate three-dimensional trajectories of QDs are achievable once an optimal distance between single molecules is guaranteed over the entire acquisition time. Overlapping of intensity patterns belonging to different molecules is going to perturb PSF fitting and localization of the emitters.

1. Adjust the integration time of your camera to about 100 ms and the EM gain in EMCCD cameras to 300×.
2. Choose a single isolated QD moving on the cell membrane (samples prepared as in Subheadings 3.2 and 3.3) or inside the cell (samples prepared as in Subheading 3.4).
3. Starting from the upper in-focus position, displace the objective (or the sample) along the axial direction by about 500–750 nm using of the piezo stage, until a single ring with a central Gaussian peak is clearly observed. This will assure to fully exploit the tracking range and avoid that the QD moves into the "blind zone" (Fig. 5a).

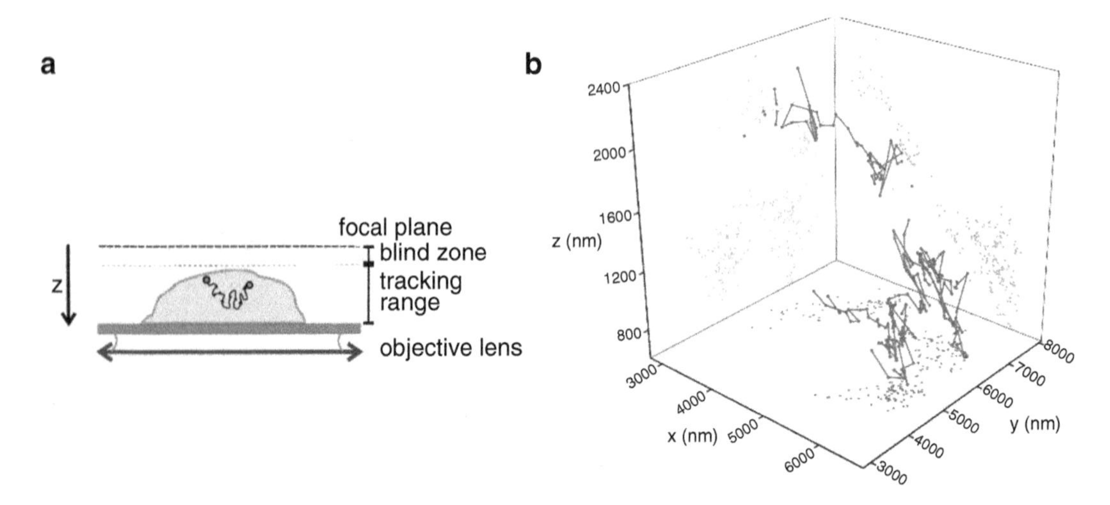

Fig. 5 Acquisition and analysis of three-dimensional trajectories of single QD-labeled molecules. (**a**) Sketch of the accessible tracking range. Proper probe defocusing can be obtained by adjusting the position of the focal plane above the sample. In this way, the entire volume of the cell is potentially accessible for tracking. (**b**) Three-dimensional trajectory of GM1 in a living neuroblastoma cell (blue line). GM1 on the membrane of living neuroblastoma was targeted with biotinylated cholera toxin conjugated with streptavidin-QD655. QD655-conjugated GM1 was tracked in a neuroblastoma cell in three dimensions at 100 ms integration time over 23 s within a volume of about (3 × 4 × 1.4) μm (x, y, z). Gaps in the QD trajectory correspond to blinking events. Reproduced from Gardini 2015 with permission from Nature Research

4. Start acquisition.
5. Analyze image sequences of moving QDs through the PROOF software [51] to extract x, y coordinates (*see* **Note 26**) and the radius of the outermost diffraction ring R (*see* Subheading 3.1).
6. Using the calibration curve obtained in Subheading 3.1, convert R to z.
7. 3D trajectories are then obtained by importing the coordinates into OriginPro (or analogous software) and plotting them using the "trajectory tool" (Fig. 5b) (*see* **Note 27**).

4 Notes

1. We have built a custom-made imaging chamber (Fig. 3d) to fit our microscope stage; alternatively commercial chambers are available such as Thermo Fisher Attofluor™ Cell Chamber A7816.
2. QDBB 4× can be kept at 4 °C for several years.
3. If you use a permanent ink pencil to label the parafilm, avoid the contact with the medium of the cells, as a tiny amount of it could be diluted and internalized by the cells, increasing the background fluorescence.
4. Carefully aspirate the medium from one side of the coverslip, and then add the new medium also from the side (Fig. 3b). Dropping the medium directly on the coverslip might damage and detach the cells.
5. Final wash is very important to remove QDots655-WGMA that did not insert on the membrane.
6. The labeling procedure described in Subheading 3.1 guarantees well-separated, immobile single QDs on the membrane of fixed cells.
7. The effective axial position of the focal plane in the sample is determined by correcting for the refractive index mismatch between water and the coverslip glass:

$$z_{\text{real}} = z_{\text{piezo}} n_{\text{water}} / n_{\text{glass}};\ n_{\text{water}} = 1.333;\ n_{\text{glass}} = 1.515$$

8. Note that the emission pattern of slightly out-of-focus QDs has been reported to be anisotropic [50], which would challenge the use of rotationally immobile QDs with this technique. However, if QDs are sufficiently rotationally mobile such that they explore much of the orientation space within a single acquisition, this effect is averaged away, and accuracy can be recovered [59]. We therefore use commercially available QDs (QDots655-WGMA), where a PEG-linker strategy is utilized

to link the QD to the attached protein (WGMA in this case). PEG linkers have high rotational mobility and the additional advantage of high-quality staining with low nonspecific binding. We observe radially isotropic diffraction patterns reproducible between different QDs (Fig. 2a). Information on the length of the PEG linker connecting QD to wheat germ agglutinin is not publicly available. The readers can reproduce the experiments as long as they use the exact same linker.

9. Details on the algorithm, the open-source software, and a comprehensive user guide can be found elsewhere [51]. Briefly, the algorithm first calculates an average radial profile (blue curve, Fig. 6a) from radial profiles along different directions in $\Delta\theta$ steps. The profile peaks are then automatically detected through local maxima recognition (five peaks for the diffraction profile shown in Fig. 6a). Finally, a proper function (red surface, Fig. 6b) is chosen to fit experimental data (green surface), based on the number, position, and amplitude of the peaks detected previously. The function is usually composed by a central Gaussian peak surrounded by n Gaussian rings. However, the QD emission pattern shows a central Gaussian ring (a central "hole") at axial locations where the number of rings changes (Fig. 2a) [50]. The algorithm recognizes these cases and accounts for the absence of the central peak. Weighted least squares fitting is performed assuming Poisson noise in the data

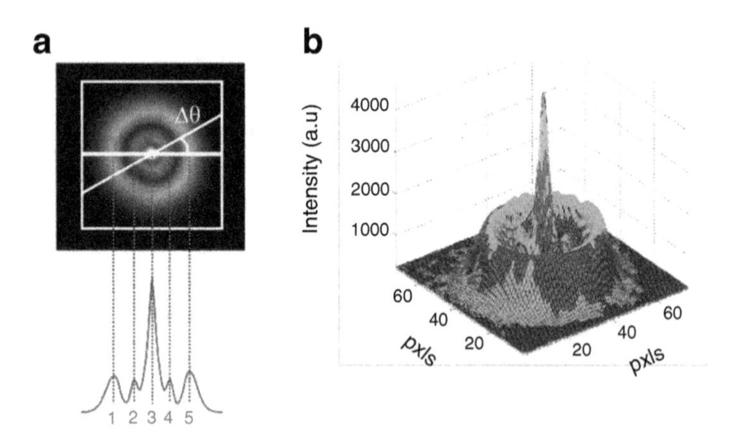

Fig. 6 Automatic out-of-focus pattern recognition by PROOF. (**a**) PROOF algorithm automatically recognizes and analyses the diffraction pattern emitted by defocused chromophores. An average radial profile (blue curve) is calculated over a region of interest (white square) by averaging radial profiles along different directions in $\Delta\theta$ steps. The profile peaks are automatically detected through local maxima recognition (five peaks for the diffraction profile shown here). (**b**) From the analysis described in panel (**a**), a proper function (red surface plot) is chosen to fit experimental data (green surface plot). Reproduced from Gardini 2015 with permission from Nature Research

giving the radius of the diffraction rings and x, y coordinates of the defocused emitter.

10. As emerges from the plot, there is a region of about half micron from the focal plane, in which the axial localization is prevented, due to the absence of defined rings. Actually, this "blind zone" does not prevent tracking in specific axial ranges, but only constrains the position of the focal plane (Fig. 5a). On the other hand, at axial positions deeper than about 2.5 microns, accurate measurement of the axial position is hindered by the low S/N, resulting from the spreading of the emitted photons over many rings.
11. In particular, the stepwise relationship obtained by calibration in Fig. 2b was fitted with two different polynomial functions, depending on the defocusing range. Data in the axial range from 0.75 to 1.5 μm were fitted by a fifth-degree polynomial ($z = z_0 + ax + bx^2 + cx^3 + dx^4 + ex^5$, red curve), while data in the axial range from 1.5 to 2.5 μm were fitted by third-degree polynomial ($z = z_0 + ax + bx^2 + cx^3$, blue curve).
12. For rat hippocampal neurons, coverslips need to be pre-coated with polyornithine or polylysine.
13. QDBB 1× can be kept at 4 °C for several months.
14. We use QDs emitting at 655 nm as they appear to be brightest than other QDs emitting at lower wavelengths [60]
15. If a suitable antibody against your protein is not available, you can genetically add a small peptide tag (e.g., HA or AP) and use a primary antibody against the tag. The AP tag is especially convenient as it can be specifically biotinylated [61], allowing to directly bind streptavidin QDs without the use of antibodies.
16. It is advisable to have 20–50 labeled molecules per cell for single-molecule tracking experiments. 200 μL of solution are used for one 18 mm coverslip. Do not store this solution for more than 1 day.
17. It is advisable to prepare duplicates of each experiment, as it is quite common to break the coverslip while tightening the chamber.
18. Recording sessions should not last more than 30 min. If Leibovitz's L-15 medium is used, it is possible to avoid using a CO_2 chamber. When the session is finished, wash the chamber with demineralized water and ethanol to remove traces of mineral oil.
19. Other tag proteins such as SNAP-tag may be applicable instead of HaloTag.

20. We prepared HaloTag ligand-QDs at various QD:HaloTag ligand ratios and confirmed that QDs with a higher abundance of HaloTag ligand labeled to HaloTag-14-3-3β/α proteins showed greater mobility in agarose gel electrophoresis (Fig. 4b). According to the Promega Corporation, a 12-fold molar excess works well to ensure at least one label per antibody in case of antibody labeling with small molecules, and therefore, we mainly used the HaloTag ligand-QD prepared in 1:12 conjugation reactions in our previous experiments [41]. We found no obvious differences in intracellular movement of HaloTag-myosin-5B labeled with HaloTag ligand-QD among these HaloTag ligand-QDs which are thought to have different numbers of HaloTag ligands on single QD.
21. QDs seem to easily form aggregates over time, therefore, it is better to make small aliquots of the prepared solution, stock them without excess dilution, and dilute one of the aliquots with electroporation buffer just before electroporation. The diluted solution can be stored for a short time, but discard the working solution soon after a visible pellet appears.
22. Choose either 13–23 or 24–34. Both protocols worked well in 3T3-L1 fibroblasts [41], and we successfully detected linear HaloTag ligand-QDs movement along F-actin (Fig. 4e). However, with protocol 13–23, we found that only ~5% of HaloTag ligand-QDs showed linear movement and most of the QDs were almost immobile. Slight increases in the numbers of mobile HaloTag-QDs can be observed with protocol 24–34, but they still comprised no more than 10% of the total QD signals. This observation is consistent with previous reports [39, 40], and this may be attributed to several possibilities including nonspecific binding of HaloTag ligand-QDs and binding with HaloTag ligand-QDs to inactive myosin motors.
23. This solution can be used for at least three rounds of electroporation.
24. Increasing the poration pulse voltage can increase the numbers of QDs inside the cells (Fig. 4d). Optimization of the electroporation parameters will be necessary for optimal labeling of intracellular molecules for precise single-molecule tracking.
25. This procedure is optional but strongly recommended to confirm whether the HaloTag-fused proteins of interest have been appropriately expressed.
26. The pixel size of your image should be calibrated using standard methods to convert the x, y coordinates from pixel to nm.
27. Blinking events are recognized by the PROOF software that gives a *Null* position that results in a break in the final reconstructed trajectory.

Acknowledgments

This work was supported by the European Union's Horizon 2020 research and innovation program under grant agreement no 654148 Laserlab-Europe, by EMPIR under the European Union's Horizon 2020 research and innovation program grant agreement 15HLT01 MetVBadBugs, by the Italian Ministry of University and Research (FIRB "Futuro in Ricerca" 2013 grant n. RBFR13V4M2 and Flagship Project NANOMAX), by Ente Cassa di Risparmio di Firenze and by the Program for Fostering Researchers for the Next Generation in a Project for Establishing a Consortium for the Development of Human Resources in Science and Technology, Japan.

References

1. Capitanio M, Pavone FS (2013) Interrogating biology with force: single molecule high-resolution measurements with optical tweezers. Biophys J 105(6):1293–1303. https://doi.org/10.1016/j.bpj.2013.08.007
2. Lord SJ, Lee HLD, Moerner WE (2010) Single-molecule spectroscopy and imaging of biomolecules in living cells. Anal Chem 82 (6):2192–2203. https://doi.org/10.1021/Ac9024889
3. Yildiz A, Forkey J, McKinney S, Ha T, Goldman Y, Selvin P (2003) Myosin V walks hand-over-hand: single fluorophore imaging with 1.5-nm localization. Science (New York, NY) 300(5628):2061–2065. https://doi.org/10.1126/science.1084398
4. Churchman LS, Okten Z, Rock RS, Dawson JF, Spudich JA (2005) Single molecule high-resolution colocalization of Cy3 and Cy5 attached to macromolecules measures intramolecular distances through time. Proc Natl Acad Sci U S A 102(5):1419–1423. https://doi.org/10.1073/pnas.0409487102
5. Sakamoto T, Webb M, Forgacs E, White H, Sellers J (2008) Direct observation of the mechanochemical coupling in myosin Va during processive movement. Nature 455 (7209):128–132. https://doi.org/10.1038/nature07188
6. Capitanio M, Canepari M, Maffei M, Beneventi D, Monico C, Vanzi F, Bottinelli R, Pavone FS (2012) Ultrafast force-clamp spectroscopy of single molecules reveals load dependence of myosin working stroke. Nat Methods 9(10):1013–1019. https://doi.org/10.1038/nmeth.2152
7. Capitanio M, Canepari M, Cacciafesta P, Lombardi V, Cicchi R, Maffei M, Pavone FS, Bottinelli R (2006) Two independent mechanical events in the interaction cycle of skeletal muscle myosin with actin. Proc Natl Acad Sci U S A 103(1):87–92
8. Gardini L, Tempestini A, Pavone FS, Capitanio M. (in press) High-speed optical tweezers for the study of single molecular motors. In: Molecular motors, methods in molecular biology
9. Mónico C, Belcastro G, Capitanio M, Vanzi F, Pavone FS (2011) Combined optical trapping and nanometer-precision localization for the single-molecule study of DNA-binding proteins. In: BioPhotonics, 2011 International Workshop on, Parma, 8–10 June 2011. Institute of Electrical and Electronics Engineers, Parma, pp 1–3. https://doi.org/10.1109/IWBP.2011.5954832
10. Monico C, Belcastro G, Vanzi F, Pavone FS, Capitanio M (2014) Combining single-molecule manipulation and imaging for the study of protein-DNA interactions. J Vis Exp 90:51446. https://doi.org/10.3791/51446
11. Monico C, Capitanio M, Belcastro G, Vanzi F, Pavone FS (2013) Optical methods to study protein-DNA interactions in vitro and in living cells at the single-molecule level. Int J Mol Sci 14(2):3961–3992. https://doi.org/10.3390/ijms14023961
12. Calamai M, Pavone FS (2011) Single molecule tracking analysis reveals that the surface mobility of amyloid oligomers is driven by their conformational structure. J Am Chem Soc 133 (31):12001–12008. https://doi.org/10.1021/ja200951f
13. Calamai M, Pavone FS (2013) Partitioning and confinement of GM1 ganglioside induced by amyloid aggregates. FEBS Lett 587

(9):1385–1391. https://doi.org/10.1016/j.febslet.2013.03.014

14. Betzig E, Patterson GH, Sougrat R, Lindwasser OW, Olenych S, Bonifacino JS, Davidson MW, Lippincott-Schwartz J, Hess HF (2006) Imaging intracellular fluorescent proteins at nanometer resolution. Science 313 (5793):1642–1645
15. Rust MJ, Bates M, Zhuang X (2006) Sub-diffraction-limit imaging by stochastic optical reconstruction microscopy (STORM). Nat Methods 3(10):793–795
16. Oswald F, Bank ELM, Bollen YJM, Peterman EJG (2014) Imaging and quantification of trans-membrane protein diffusion in living bacteria. Phys Chem Chem Phys 16 (25):12625–12634. https://doi.org/10.1039/c4cp00299g
17. Toprak E, Balci H, Blehm B, Selvin P (2007) Three-dimensional particle tracking via bifocal imaging. Nano Lett 7(7):2043–2045. https://doi.org/10.1021/nl0709120
18. Ram S, Prabhat P, Chao J, Ward ES, Ober RJ (2008) High accuracy 3D quantum dot tracking with multifocal plane microscopy for the study of fast intracellular dynamics in live cells. Biophys J 95(12):6025–6043. https://doi.org/10.1529/biophysj.108.140392
19. Dalgarno PA, Dalgarno HIC, Putoud A, Lambert R, Paterson L, Logan DC, Towers DP, Warburton RJ, Greenaway AH (2010) Multiplane imaging and three dimensional nanoscale particle tracking in biological microscopy. Opt Express 18(2):877–884. https://doi.org/10.1364/Oe.18.000877
20. Juette MF, Gould TJ, Lessard MD, Mlodzianoski MJ, Nagpure BS, Bennett BT, Hess ST, Bewersdorf J (2008) Three-dimensional sub-100 nm resolution fluorescence microscopy of thick samples. Nat Methods 5 (6):527–529. https://doi.org/10.1038/Nmeth.1211
21. Ram S, Kim D, Ober RJ, Ward ES (2012) 3D Single molecule tracking with multifocal plane microscopy reveals rapid intercellular transferrin transport at epithelial cell barriers. Biophys J 103(7):1594–1603. https://doi.org/10.1016/j.bpj.2012.08.054
22. Hajj B, Wisniewski J, El Beheiry M, Chen JJ, Revyakin A, Wu C, Dahan M (2014) Whole-cell, multicolor superresolution imaging using volumetric multifocus microscopy. Proc Natl Acad Sci U S A 111(49):17480–17485. https://doi.org/10.1073/pnas.1412396111
23. Pavani S, Thompson M, Biteen J, Lord S, Liu N, Twieg R, Piestun R, Moerner W (2009) Three-dimensional, single-molecule fluorescence imaging beyond the diffraction limit by using a double-helix point spread function. Proc Natl Acad Sci U S A 106 (9):2995–2999. https://doi.org/10.1073/pnas.0900245106
24. Thompson MA, Lew MD, Badieirostami M, Moerner WE (2010) Localizing and tracking single nanoscale emitters in three dimensions with high spatiotemporal resolution using a double-helix point spread function. Nano Lett 10(1):211–218. https://doi.org/10.1021/Nl903295p
25. Kao H, Verkman A (1994) Tracking of single fluorescent particles in three dimensions: use of cylindrical optics to encode particle position. Biophys J 67(3):1291–1300. https://doi.org/10.1016/S0006-3495(94)80601-0
26. Laurent H, Tobias M, Thomas S (2007) Nanometric three-dimensional tracking of individual quantum dots in cells. Appl Phys Lett 90:053902. https://doi.org/10.1063/1.2437066
27. Huang B, Wang W, Bates M, Zhuang X (2008) Three-dimensional super-resolution imaging by stochastic optical reconstruction microscopy. Science (New York, NY) 319 (5864):810–813. https://doi.org/10.1126/science.1153529
28. Xu K, Babcock HP, Zhuang XW (2012) Dual-objective STORM reveals three-dimensional filament organization in the actin cytoskeleton. Nat Methods 9(2):185–188. https://doi.org/10.1038/Nmeth.1841
29. Xu K, Zhong GS, Zhuang XW (2013) Actin, spectrin, and associated proteins form a periodic cytoskeletal structure in axons. Science 339(6118):452–456. https://doi.org/10.1126/science.1232251
30. Zanacchi FC, Lavagnino Z, Donnorso MP, Del Bue A, Furia L, Faretta M, Diaspro A (2011) Live-cell 3D super-resolution imaging in thick biological samples. Nat Methods 8(12):1047. https://doi.org/10.1038/Nmeth.1744
31. Sun Y, McKenna JD, Murray JM, Ostap EM, Goldman YE (2009) Parallax: high accuracy three-dimensional single molecule tracking using split images. Nano Lett 9 (7):2676–2682. https://doi.org/10.1021/Nl901129j
32. Pinaud F, Clarke S, Sittner A, Dahan M (2010) Probing cellular events, one quantum dot at a time. Nat Methods 7(4):275–285. https://doi.org/10.1038/Nmeth.1444
33. Pierobon P, Cappello G (2012) Quantum dots to tail single bio-molecules inside living cells. Adv Drug Deliv Rev 64(2):167–178. https://doi.org/10.1016/j.addr.2011.06.004

34. Sun C, Cao ZN, Wu M, Lu C (2014) Intracellular tracking of single native molecules with electroporation-delivered quantum dots. Anal Chem 86(22):11403–11409. https://doi.org/10.1021/ac503363m
35. Saurabh S, Beck LE, Maji S, Baty CJ, Wang Y, Yan Q, Watkins SC, Bruchez MP (2014) Multiplexed modular genetic targeting of quantum dots. ACS Nano 8(11):11138–11146. https://doi.org/10.1021/nn5044367
36. Delehanty JB, Blanco-Canosa JB, Bradburne CE, Susumu K, Stewart MH, Prasuhn DE, Dawson PE, Medintz IL (2013) Site-specific cellular delivery of quantum dots with chemoselectively-assembled modular peptides. Chem Commun 49(72):7878–7880
37. Nan X, Sims P, Chen P, Xie X (2005) Observation of individual microtubule motor steps in living cells with endocytosed quantum dots. J Phys Chem B 109(51):24220–24224. https://doi.org/10.1021/jp056360w
38. Courty S, Luccardini C, Bellaiche Y, Cappello G, Dahan M (2006) Tracking individual kinesin motors in living cells using single quantum-dot imaging. Nano Lett 6 (7):1491–1495. https://doi.org/10.1021/nl060921t
39. Pierobon P, Achouri S, Courty S, Dunn AR, Spudich JA, Dahan M, Cappello G (2009) Velocity, processivity, and individual steps of single myosin V molecules in live cells. Biophys J 96(10):4268–4275. https://doi.org/10.1016/j.bpj.2009.02.045
40. Nelson S, Ali M, Trybus K, Warshaw D (2009) Random walk of processive, quantum dot-labeled myosin Va molecules within the actin cortex of COS-7 cells. Biophys J 97 (2):509–518. https://doi.org/10.1016/j.bpj.2009.04.052
41. Hatakeyama H, Nakahata Y, Yarimizu H, Kanzaki M (2017) Live-cell single-molecule labeling and analysis of myosin motors with quantum dots. Mol Biol Cell 28(1):173–181. https://doi.org/10.1091/mbc.E16-06-0413
42. Liu SL, Zhang LJ, Wang ZG, Zhang ZL, Wu QM, Sun EZ, Shi YB, Pang DW (2014) Globally visualizing the microtubule-dependent transport behaviors of influenza virus in live cells. Anal Chem 86(8):3902–3908
43. Zhang Y, Ke XL, Zheng ZH, Zhang CL, Zhang ZF, Zhang FX, Hu QX, He ZK, Wang HZ (2013) Encapsulating quantum dots into enveloped virus in living cells for tracking virus infection. ACS Nano 7(5):3896–3904
44. Herod MR, Pineda RG, Mautner V, Onion D (2015) Quantum dot labelling of adenovirus allows highly sensitive single cell flow and imaging cytometry. Small 11(7):797–803
45. Rajan SS, Liu HY, Vu TQ (2008) Ligand-bound quantum dot probes for studying the molecular scale dynamics of receptor endocytic trafficking in live cells. ACS Nano 2 (6):1153–1166. https://doi.org/10.1021/nn700399e
46. Lowe AR, Siegel JJ, Kalab P, Siu M, Weis K, Liphardt JT (2010) Selectivity mechanism of the nuclear pore complex characterized by single cargo tracking. Nature 467 (7315):600–603. https://doi.org/10.1038/nature09285
47. Zhan NQ, Palui G, Mattoussi H (2015) Preparation of compact biocompatible quantum dots using multicoordinating molecular-scale ligands based on a zwitterionic hydrophilic motif and lipoic acid anchors. Nat Protoc 10 (6):859–874
48. Cai E, Ge PH, Lee SH, Wang Y, Jeyifous O, Lim SJ, Smith AM, Green WN, Selvin PR (2014) Development of stable small quantum dots for AMPA receptor tracking at neuronal synapses. Biophys J 106 ((2):605a–606a
49. Xu JM, Ruchala P, Ebenstain Y, Li JJ, Weiss S (2012) Stable, compact, bright biofunctional quantum dots with improved peptide coating. J Phys Chem B 116(36):11370–11378
50. Toprak E, Enderlein J, Syed S, McKinney SA, Petschek RG, Ha T, Goldman YE, Selvin PR (2006) Defocused orientation and position imaging (DOPI) of myosin V. Proc Natl Acad Sci U S A 103(17):6495–6499. https://doi.org/10.1073/pnas.0507134103
51. Gardini L, Capitanio M, Pavone FS (2015) 3D tracking of single nanoparticles and quantum dots in living cells by out-of-focus imaging with diffraction pattern recognition. Sci Rep 5:16088. https://doi.org/10.1038/srep16088
52. Speidel M, Jonás A, Florin E-L (2003) Three-dimensional tracking of fluorescent nanoparticles with subnanometer precision by use of off-focus imaging. Opt Lett 28(2):69–71
53. Gardini L, Capitanio M, Pavone FS (2016) Single molecule study of Processive myosin motors. IET Digital Library 4:1. https://doi.org/10.1049/cp.2016.0940
54. Fujita H, Hatakeyama H, Watanabe TM, Sato M, Higuchi H, Kanzaki M (2010) Identification of three distinct functional sites of insulin-mediated GLUT4 trafficking in adipocytes using quantitative single molecule imaging. Mol Biol Cell 21(15):2721–2731. https://doi.org/10.1091/mbc.E10-01-0029

55. Hatakeyama H, Kanzaki M (2011) Molecular basis of insulin-responsive GLUT4 trafficking systems revealed by single molecule imaging. Traffic 12(12):1805–1820. https://doi.org/10.1111/j.1600-0854.2011.01279.x
56. Hatakeyama H, Kanzaki M (2013) Regulatory mode shift of Tbc1d1 is required for acquisition of insulin-responsive GLUT4-trafficking activity. Mol Biol Cell 24(6):809–817. https://doi.org/10.1091/mbc.E12-10-0725
57. Hatakeyama H, Kanzaki M (2017) Heterotypic endosomal fusion as an initial trigger for insulin-induced glucose transporter 4 (GLUT4) translocation in skeletal muscle. J Physiol 595(16):5603–5621. https://doi.org/10.1113/JP273985
58. Hatakeyama H, Kanzaki M (2013) Development of dual-color simultaneous single molecule imaging system for analyzing multiple intracellular trafficking activities. Conf Proc IEEE Eng Med Biol Soc 2013:1418–1421. https://doi.org/10.1109/EMBC.2013.6609776
59. Backlund MP, Lew MD, Backer AS, Sahl SJ, Grover G, Agrawal A, Piestun R, Moerner WE (2012) Simultaneous, accurate measurement of the 3D position and orientation of single molecules. Proc Natl Acad Sci U S A 109 (47):19087–19092. https://doi.org/10.1073/pnas.1216687109
60. Calamai M, Evangelisti E, Cascella R, Parenti N, Cecchi C, Stefani M, Pavone F (2016) Single molecule experiments emphasize GM1 as a key player of the different cytotoxicity of structurally distinct Abeta1-42 oligomers. Biochim Biophys Acta 1858 (2):386–392. https://doi.org/10.1016/j.bbamem.2015.12.009
61. Howarth M, Takao K, Hayashi Y, Ting AY (2005) Targeting quantum dots to surface proteins in living cells with biotin ligase. Proc Natl Acad Sci U S A 102(21):7583–7588. https://doi.org/10.1073/pnas.0503125102

Chapter 27

AFM Indentation Analysis of Cells to Study Cell Mechanics and Pericellular Coat

Igor Sokolov and Maxim E. Dokukin

Abstract

Atomic force microscopy (AFM) indentation analysis of cells is a unique method of measuring stiffness of the cell body and physical properties of its pericellular coat. These cell parameters correlate with cells of abnormality and diseases. Viable biological cells can be studied with this method directly in a culture dish with no special preparation. Here we describe a step-by-step method to analyze the AFM force-indentation curves to derive cell mechanics (the modulus of elasticity of the cell body) and the parameters of the pericellular coat (density and the thickness of the coat layer). Technical details, potential difficulties, and points of special attention are described.

Key words Cell mechanics, Physical properties of pericellular coat, Atomic force microscopy, Indentation analysis, Physics of cells

1 Introduction

Mechanical properties of cells define different physical responses of the cells to external stresses. Besides just pure fundamental interest, correlation between elasticity of cells and different human diseases or abnormalities has been recently found. It has been implicated in the pathogenesis of many progressive diseases, including vascular and kidney diseases, cancer, malaria, cataracts, Alzheimer, complications of diabetes, cardiomyopathies [1–3], and even aging [4–7]. It was also discovered that the motility and spreading of cancer cells may be controlled by the application of external forces, which may alter the mechanics of a tumor [8, 9]. Recently the modulus of elasticity of cancer cells was suggested to be used for cancer diagnosis [10, 11].

Atomic force microscopy (AFM) is a unique technique allowing to study mechanics of cells with a high spatial resolution and large range of deformation stresses. Analysis of force-indentation curves collected over the cell surface is typically done with the help of the standard Hertz model. However, it was recently

Yuri L. Lyubchenko (ed.), *Nanoscale Imaging: Methods and Protocols*, Methods in Molecular Biology, vol. 1814, https://doi.org/10.1007/978-1-4939-8591-3_27, © Springer Science+Business Media, LLC, part of Springer Nature 2018

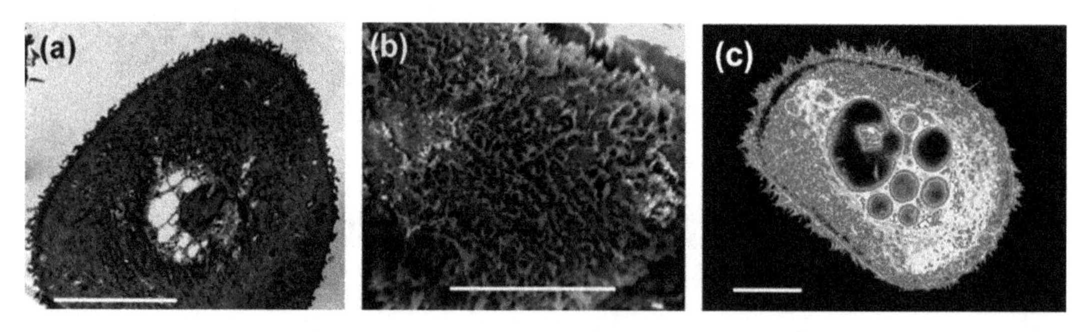

Fig. 1 An example of non-flat pericellular boundary of a cell surface. The surface of human cervical epithelial cells is shown in (**a**) and (**b**) are electron microscopy images, (**c**) optical confocal image. Scale bars are 10 μm. Reproduced from Ref. [33] with permission from Elsevier (C) 2013

demonstrated that the Hertz model cannot be used to describe cell mechanics because the cell is too far from homogeneous and isotropic material, which is a requirement in the Hertz model [12]. If the Hertz model is used, the value of the elastic modulus (the Young's modulus) depends on the indentation depth, which should not. The same work demonstrated that a more sophisticated "brush model" results in the elastic modulus which is practically independent on the indentation depth. The brush model is treating cell as isotropic homogeneous material (cell body) which is coated with a pericellular layer that is mechanically very different from the elastic cell body [13]. The presence of non-flat pericellular boundary is clearly seen in optical and electron microscopy (Fig. 1).

Besides being self-consistent, the brush model allows extracting information not only about the mechanics of the cell body but also the parameters of the pericellular coat surrounding cells. The latter is important, for example, to distinguish between cancer and normal cells [14–19]. Molecular entropic brushes are known to surround neurofilaments to maintain interfilament spacing [20, 21]. The pericellular brush layer is responsible for the cell-cell interaction [22], cell migration [23], differentiation, and proliferation [24, 25]. The brush is important in embryonic development [26], wound healing [27], inflammation [28, 29], and mammalian fertilization [30]. It is involved in epithelial-mesenchymal transition [26], resistance to apoptosis (cell death), and multidrug resistance [31].

It is important to note that smaller stiffness of cancer cells reported in the literature [10, 11] might be confused with the different properties of pericellular coat of cancer cells [19] when the latter is not taken into account. Another interesting application of the brush model is the study of viscoelastic response of cells. As was demonstrated in [32], viscoelastic response of cells, which had been reported previously in the literature, was entirely attributed to the viscoelastic behavior of the pericellular coat. Cell body was reacting as a pure elastic material within the load rate 1–10 μm/s.

The previous development of the brush model was described in [33]. Since then there have been several important developments of the method, which expand the brush model to calculation of the elastic modulus for lower indentation depth [12], analyzing cells which are loosely attached to the dish surface [34], the analysis of double brush surrounding cells (large layer of glycosaccharide and smaller layer of membrane protrusions), and the analysis of the case when a large pericellular layer can touch the AFM cantilever (not only the AFM probe) [35]. Here we present a step-by-step description of the brush model including all recent developments. Potential difficulties and points of special attention are also discussed.

2 Materials

2.1 Cell Preparation

1. Clean physiological buffer (e.g., PBS or HBSS at 1× standard (physiological) molarity and desired pH) or growth medium (preferably without serum) filtered using 200 nm filter (the filter pore size).
2. Cell plated in 60 or 30 mm culture dishes (*see* **Note 1**).

2.2 Atomic Force Microscopy and Image Processing

1. 3100 or Dimension Icon or Bioscope Catalyst (Bruker/Veeco, Inc., Santa Barbara, CA) atomic force microscopes equipped with at least NanoScope (version IIIa or above) controller. Other microscopes capable of working in the force-volume mode (*see* the next item) will also work.
2. If not included in the basic AFM configuration, a closed-loop scanner with the range of at least 15–20 microns for the vertical displacement (more is desirable for the analysis of fibroblasts generating collagen layer) is a must. nPoint scanners (Middleton, WI, USA) can be used as add-ons for the microscopes mentioned above.
3. Force-volume imaging modes.
4. Standard AFM cantilever holders for operation in fluids (Bruker/Veeco, Inc., Santa Barbara, CA).
5. Colloidal AFM probes (*see* **Note 2**).
6. Standard tip-check sample TIP001 (Aurora NanoDevices, Canada).
7. NanoScope Analysis (Bruker/Veeco, Inc., Santa Barbara, CA) or similar software (*see* **Note 3**).
8. Matlab (MathWorks, Natick, MA, USA) or OriginPro software (OriginLab Corp., Northampton, MA, USA) or similar software capable of nonlinear curve fitting.

3 Methods

To study the physical properties of cells and its pericellular coat, individual cells should be grown in a culture dish or, alternatively, be attached to any stable substrate (*see* **Note 4**).

3.1 Cell Preparation for AFM Study

1. Gently wash cells cultured in a culture dish two times with PBS to remove not attached and dead cells (*see* **Note 5**).
2. Attach double sticky tape to the bottom of the culture dish (for dimension and icon type of microscopes) and place the culture dish on the sample holder. Use a magnetic holder for microscopes, like bioscope, to firmly fix the culture dish in the sample holder.
3. If cells are planned to be studied a long time (more than 3–4 h in room temperature in physiological buffer), plan to use the growth medium and mini incubators to keep cells viable for longer time.
4. Specific cells used in these examples are MCF7 cell line provided by ATCC; primary skin fibroblasts, cells were isolated from the underarm skin of outbred multicolored guinea pigs (tissues were obtained from a frozen collection) as described in [36]. Primary cultures of human epithelial normal and cancer cells were collected from tissue of the cervix of healthy and cancer patients, respectively. All human tissue was obtained from the Cooperative Human Tissue Network (CHTN, National Cancer Institute, National Institutes of Health, Bethesda, MD). Informed consent was obtained from patients according to their published guidelines (http://chtn.nci.nih.gov/phspolicies.html).

3.2 Collecting the Force Curves over Cell Surface

1. Calibrate the deflection sensitivity and the spring constant of the AFM cantilever (*see* **Note 6**).
2. Measure the radius of the AFM probe: Collect the image of the standard TIP001 tip-check sample. Use NanoScope Analysis to calculate the radius of the AFM probe (*see* **Note 7**).
3. Locate the AFM probe over the surface of a cell using an optical microscope which is typically built in or attached to virtually all AFMs.
4. Choose the contact mode of operation for the force microscope (if it does not have a separate force-volume workspace or experiment to load).
5. Choose an appropriate load force to ensure a reliable contact (typically in the order of a few nano Newtons).
6. Put the scan size to zero (to avoid lateral motion of the probe after engaging the cell).

7. Engage the surface.
8. Immediately switch to the force calibration mode. Choose the appropriate maximum load force (relative trigger). It will be the maximum force in the force-volume imaging mode. The key criteria in the choice of the right force are (a) sufficiently large to squeeze the pericellular coat and (b) do not produce the highest deformation which would result in strains of the cell body more than 10–15% (can be estimated to retrospectively, only after some initial measurements; 10nN is a good starting point).
9. Choose the appropriate (vertical) ramp size to record the force curves. It should be larger than the size of the pericellular coat. If it is not known in advance, it is advisable to use the maximum possible ramp size. Ten or more microns are recommended (*see* **Note 8**). If closed-loop operation is an option, be sure to have it switched on.
10. Choose an appropriate vertical ramping speed. Typically, 1–10 μm/s is a frequently used range of speeds in this type of study (*see* **Note 9**). If the control software allows doing that, it is advisable to increase the retraction speed to accelerate overall scanning. For example, the retraction speed can be increased to 100 μm/s.
11. Switch back to the force-volume mode (*see* **Note 10**).
12. Change the scan size to obtain the image of the entire cell. The number of pixels (force curves) should be sufficient to get enough statistics—remember not all of the force curves will be used for the analysis, see below. As an example, $50 \times 50\ \mu m^2$ at the resolution of 32×32 pixels or 16×16 pixels (and 1024 points per force curve) could be used.
13. Press start. After the start of the scanning, press "start from the top" or from the bottom—it saves you half of the imaged time (only the full square image can be saved in the force-volume mode).
14. Press capture to save the file for the next offline processing. Disengage after completion of the imaging.
15. Repeat the collecting of force-volume images for the necessary number of cells required to obtain sufficient statistics.

3.3 Analysis of Images Collected in Force-Volume Mode

1. Open a recorded force-volume file in one of the software described above.
2. Identify points/pixels on the cell surface that are appropriate for the analysis. It can be either a flat part of the cell or the cell top (a spherical top generally above the cell nucleus). Following the previous works [13, 19, 37], we take the force curves in the surface points around the top when the incline of the surface is

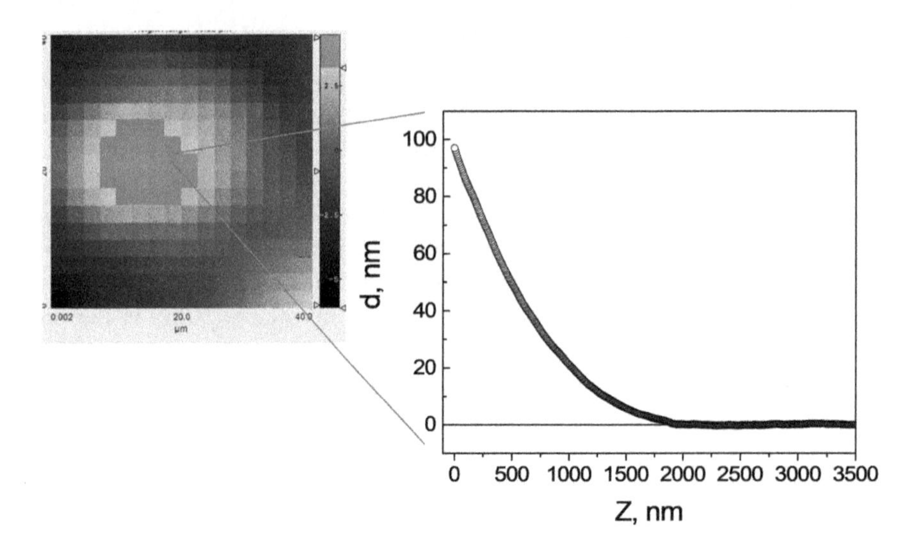

Fig. 2 An example of the force-volume image. Each pixel of the recorded image is a force curve associated with it

<10–15 degrees. These areas can be identified by the analysis of the force-volume image (height part of the force-volume image). An example of the force-volume image is shown in Fig. 2. Each pixel of the image is a force curve associated with it.

3. Calculate/extract the cell radius using the heightened part of the force-volume image, *see* Fig. 3. Take two perpendicular cross sections in the cell top. Use NanoScope Analysis (Bruker/Veeco, Inc., Santa Barbara, CA) or similar software to extract the curvature radii of both cross sections. Cell radius can be calculated as a geometrical average of both curvature radii (*see* **Notes 11** and **12**).
4. Highlight the pixels appropriate for the analysis and export the approaching part of each force curve (*see* **Note 13**).
5. Open each force curve using software capable of nonlinear curve fitting. An example of such a curve is shown in Fig. 4. Different parts of this curve correspond to the different indentation stages. For small indentations, the AFM probe deforms mostly the pericellular coat (Fig. 4a). For intermediate indentations, both the pericellular coat and cell body are indented (Fig. 4b). And for large indentations, when the pericellular coat is almost entirely squeezed, the deformation consists only of that of the cell body. This is shown in Fig. 4c.

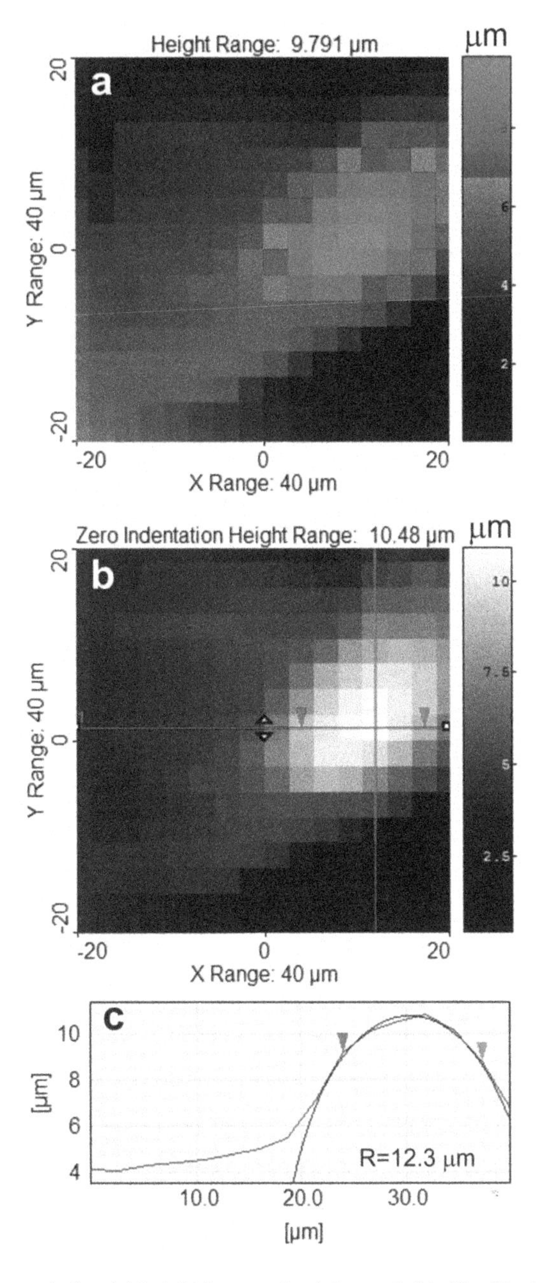

Fig. 3 A representative (**a**) height image of a deformed skin fibroblast cell used to find the cell radius. Area in the center (on top) of the cell where force curves were extracted is highlighted by blue color, (**b**) restored undeformed cell topography, (**c**) radius of an undeformed cell calculated from the horizontal cross section. The radius used for the final calculation was derived as a geometrical average from radii taken from the vertical and horizontal cross sections. Reproduced from Ref. [35] with permission from Elsevier (C) 2016

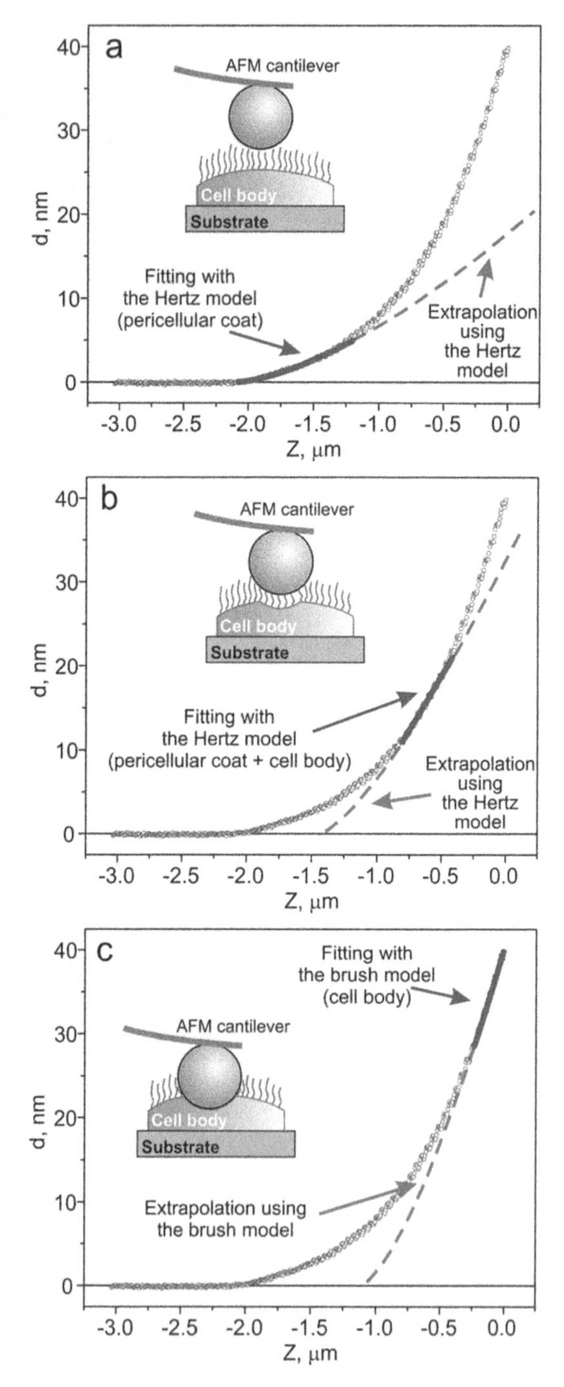

Fig. 4 Different parts of this curve corresponding to the different indentation stages. For small indentations, the AFM probe deforms mostly the pericellular coat (**a**). For intermediate indentations, both the pericellular coat and the cell body are indented (**b**). And for large indentations, when the pericellular coat is almost entirely squeezed, the deformation consists only of that of the cell body (**c**)

6. *Finding the elastic modulus of the cell body.* Assuming that the maximum load force is sufficient to almost entirely compress the pericellular coat layer (typically the force of 4–8 nN when using a colloidal probe of 5 μm in diameter), we can apply the standard Hertz model equations. It is critical, however, to use the appropriate equation, which takes into account the position of non-deformed surface. This equation reads

$$Z_0 - Z = \left[\frac{9}{16}\frac{k}{E}\sqrt{\frac{R_{\text{probe}} + R_{\text{cell}}}{R_{\text{probe}} R_{\text{cell}}}}\right]^{2/3} d^{2/3} + d, \qquad (1)$$

   where the parameters are defined in Fig. 5a.

   If one deals with a cell that is not well attached to the dish surface, another equation has to be used:

$$Z = Z_0 - d^{2/3}\left(\left[\frac{9}{16}\frac{k}{E}\sqrt{\frac{R_{\text{probe}} + R_{\text{cell}}}{R_{\text{probe}} R_{\text{cell}}}}\right]^{2/3} + \left[\frac{9}{16}\frac{k}{E\sqrt{R_{\text{cell}}}}\right]^{2/3}\right) - d. \qquad (2)$$

   Here the parameters are defined in Fig. 5b. Figure 5c shows an optical example of MFC-7 cells which have both firmly and loosely attached cells within one culture dish [34].

7. Using the appropriate formula above, one can find the unknown elastic modulus of cell E by nonlinear curve fitting of the raw force curve data collected in the force-volume mode (Z vs. d dependence) while treating E and Z_0 as unknown parameters.
8. To verify self-consistency of the model, one needs to check if elastic modulus is reasonably independent of the indentation depth. To do that, one has to do nonlinear curve fitting for small intervals of the load force (or deflection of the cantilever) within the vicinity of the maximum load. If the fitting reaches the region in which the brush is not fully squeezed (still unknown), the modulus of elasticity should decline. Ideally, one should observe a plateau of the modulus within the vicinity of the maximum load. An example of such behavior is shown in Fig. 6.
9. If the plateau doesn't exist and the modulus keeps increasing with the indentation depth, the experiment has to be repeated with a higher maximum indentation/load force. If this is the case, repeat the previous steps to verify the presence of plateau.
10. The value of the elastic modulus on the plateau is the true value of the elastic modulus of the cell body.
11. *Extracting the parameters of the pericellular coat layer.* When the elastic modulus of the cell body is chosen (at the plateau), it automatically fixes nondeformed position of the cell body Z_0.

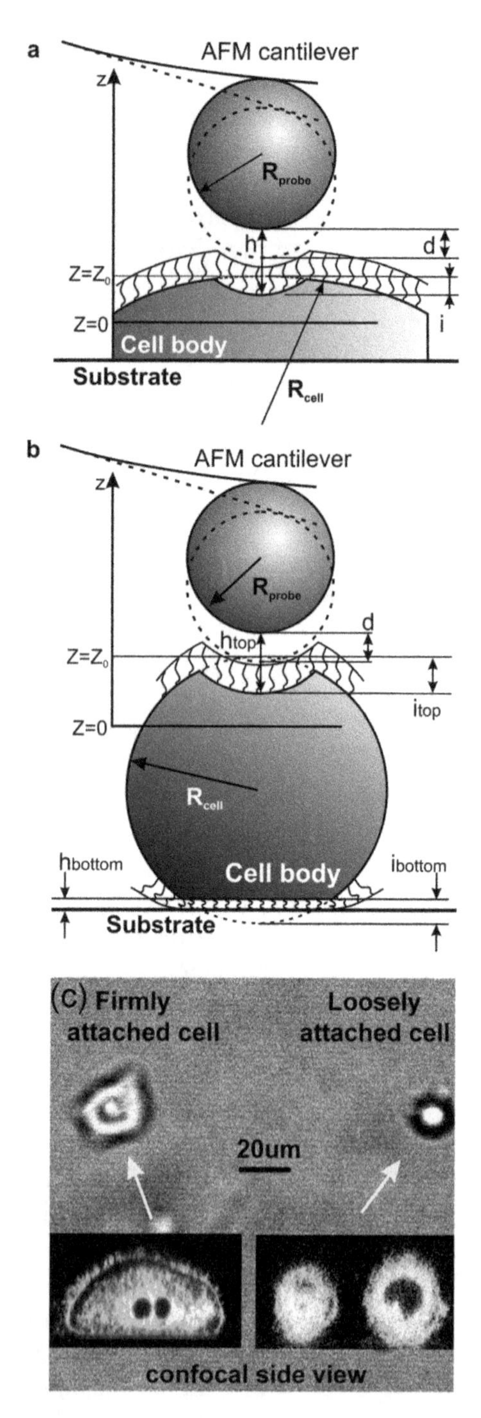

Fig. 5 The schematics of the interaction between an AFM spherical indenter (probe) and (**a**) firmly and (**b**) loosely attached cells. Deformations of both the cell body and surrounding cellular brush are shown. Z is the relative position of the cantilever; d is the cantilever deflection; Z_0 is nondeformed position of the cell body; i, i_{top}, and i_{bottom} are the deformations of the cell body; h and h_{top} are the separation distances between the cell body and the probe; and h_{bottom} is the

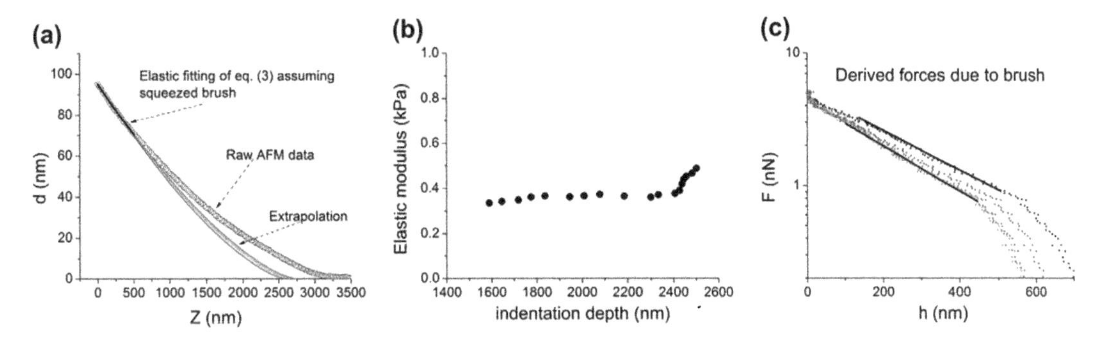

Fig. 6 An example of processing raw indentation data $d(Z)$ for a MFC-7 epithelial cell. The model allows separating deformations of the brush layer and the elastic response of the cell body, (**a**) fitting the rigid contact to determine Z_0 and the elastic modulus of the cell body (several lines are shown to demonstrate variability of the fitting taking different numbers of points near the maximum deflection, the region where Z ranges from 0–200 to 0–500 nm). (**b**) The derived elastic modulus as a function of the indentation depth; one can see a clear plateau. (**c**) The forces F due to brush as a function of h obtained from Eqs. (3a) and (3b). Reproduced from Ref. [34] with permission from Elsevier (C) 2013

In further calculations Z_0 has to be fixed to that value (*see* **Note 14**).

12. Finding of the force dependence due to brush (F vs. h) from the recorded force curves (raw data, d vs. Z). The sought force dependence $d(h)$ can be found using the calculated above elastic modulus E as follows:

$$h(d) = Z - Z_0 + \left[\frac{9}{16}\frac{k}{E}\sqrt{\frac{R_{\text{probe}} + R_{\text{cell}}}{R_{\text{probe}}R_{\text{cell}}}}\right]^{2/3} d^{2/3} + d \qquad (3a)$$

or

$$h^*(d) = Z - Z_0 + d^{2/3}\left(\left[\frac{9}{16}\frac{k}{E}\sqrt{\frac{R_{\text{probe}} + R_{\text{cell}}}{R_{\text{probe}}R_{\text{cell}}}}\right]^{2/3} + \left[\frac{9}{16}\frac{k}{E\sqrt{R_{\text{cell}}}}\right]^{2/3}\right) + d \qquad (3b)$$

where $h^* = h_{top} + h_{bottom}$ (size of the squeezed brush layer at the top and bottom together).

Fig. 5 (continued) distance between the cell body and substrate. (**c**) Representative optical images of firmly (look like well-spread) and loosely (look like spherical) cells. An insert in the figure demonstrates a confocal "side view" of both well-spread and spherically looking cells. Confocal images are not to scale. Reproduced from Ref. [34] with permission from Elsevier (C) 2013

13. The force can be found from the cantilever deflection by multiplying it by the spring constant of the cantilever $F(h) = kd$. An example of such force recorded on MFC-7 epithelial cell is shown in Fig. 6c. Note some dependence of the derived forces on different parts of the force curve used to extract the elastic modulus.

14. It is instructional to look at the logarithm of the derived force dependence. Typically, one will see a part of the force curve as a straight line in logarithmic scale, which implies the exponential force dependence (Fig. 6c). This is a typical behavior of the polymer brush model or irregular asperities. Following the literature, one can characterize the obtained force dependence by the brush parameters, effective grafting density of the brush constituents, and the equilibrium size/length of the brush, by using the following equation:

$$F(h) \approx 100 k_B \mathrm{TR}^* N^{3/2} \exp\left(-2\pi \frac{h}{L}\right) L, \tag{4a}$$

$$F(h_{top}) = F(h_{bottom}) = kd \approx 100 k_B \mathrm{TR}_1 N_1^{3/2} L_1 \exp\left(-2\pi \frac{h_{top}}{L_1}\right) = \\ = 100 k_B \mathrm{TR}_2 N_2^{3/2} L_2 \exp\left(-2\pi \frac{h_{bottom}}{L_2}\right), \tag{4b}$$

where k_B is the Boltzmann constant, T is temperature, $R_1 = R_{probe} \cdot R_{cell}/R_{probe} + R_{cell}$ is the effective cell-probe radius, $R_2 = R_{cell}$ is a cell radius, N_1 and N_2 are the effective surface densities of the brush constituents (grafting density or effective molecular surface density of the brush), and L_1 and L_2 are the sizes of the top and bottom brushes. Equation (4a) should be used for cells firmly attached to the culture dish, whereas Eq. (4b) (*see* **Note 15**) is for cells loosely attached to the dish.

15. Using an appropriate choice of Eqs. (4a) and (4b), one needs to do nonlinear curve fitting with respect to the unknown pericellular brush layer parameters L and N. It should also be noted that one has to verify that derived brush length is within the validity of this formula ($0.1 < h/L < 0.8$) [38]. This can easily be cross verified using the logarithmic plot of the obtained force dependence (exponential dependence in logarithmic plot is just a straight line).

16. In some cases, the pericellular coat can be comprised of two distinctive brushes schematically presented in Fig. 7a. Figure 7b shows force dependence due to brush, which is

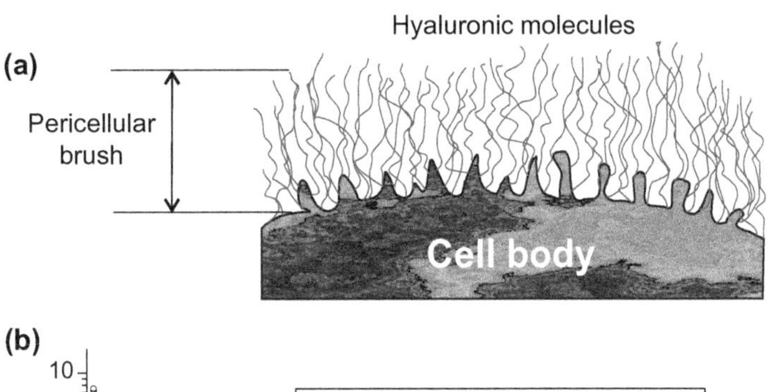

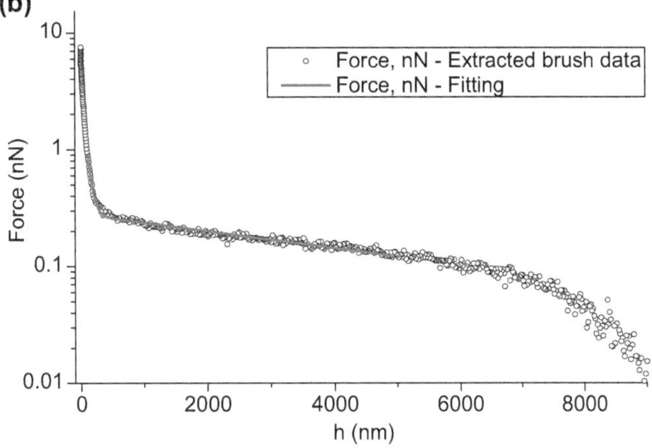

Fig. 7 (**a**) Two distinctive brushes schematically shown. (**b**) Shows a force dependence due to the double brush; a fibroblast cell surrounded by a large collagen layer is shown (the derived values of L_1 and L_2 are 420 and 38,000 nm; grafting densities N_1 and N_2 are 380 and 3.8 μm^{-2}, respectively)

typical for such a case. Then, the parameters of both brushes can be extracted using the following equation:

$$F_{\text{double brush}} \approx 100k_B\text{TR}^* \left[N_1{}^{3/2}\exp(-2\pi h/L_1)\, L_1 + N_2{}^{3/2}\exp(-2\pi h/L_2)\, L_2 \right], \quad (5)$$

(*see* **Note 16**).

17. In some cases, for example, fibroblasts generating collagen layer, the brush can be very large. As a result, the brush molecules will touch directly the AFM cantilever, not only its probe. This is schematically shown in Fig. 8. In this case, Eqs. (4a) and (4b) should be changed for the following:

$$F_{\text{Large single brush}} = 100k_b\text{TR}^* N^{3/2}e^{-\frac{2\pi h}{L}}L + 100k_b\text{TR}_{\text{cell}} N^{3/2}e^{-\frac{2\pi\left(h+2R_{\text{probe}}+\delta\right)}{L}}L. \quad (6)$$

If we are dealing with a double brush and the larger brush $L_1 > 2R_{\text{probe}}$, the AFM probe cell force can now be written as:

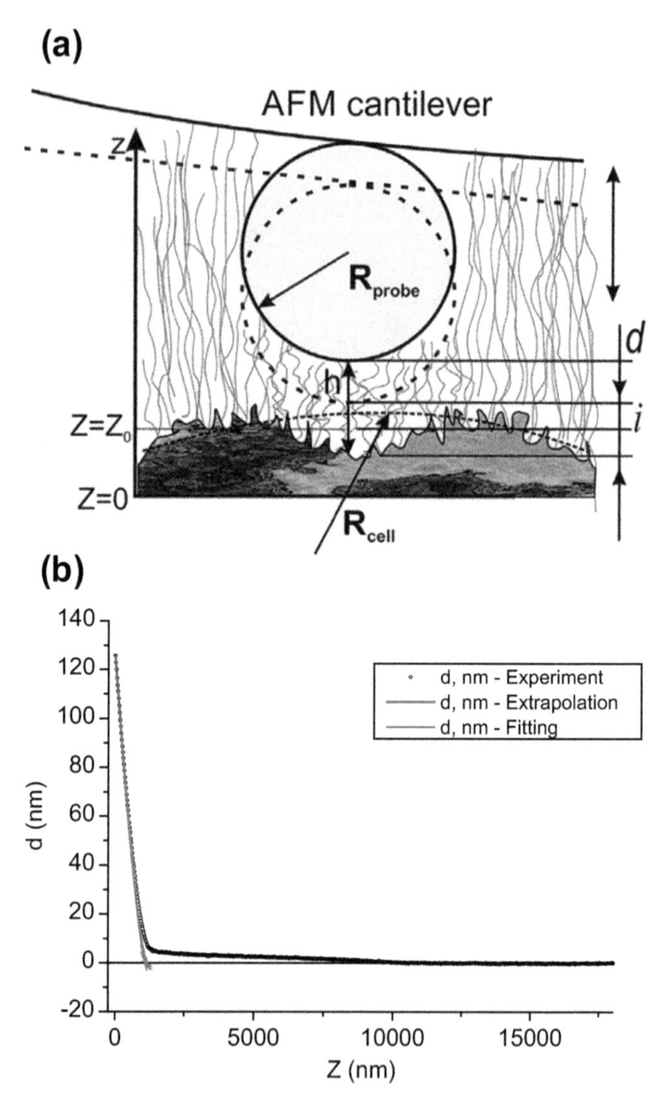

Fig. 8 (**a**) Schematics of the AFM probe deforming a large brush layer which touches the AFM cantilever. (**b**) A typical force dependence curve recorded in such a case

$$F_{\text{Large double brush}} = 100 k_{\text{b}} \text{TR}^* \left[{N_1}^{3/2} \text{e}^{-\frac{2\pi h}{L_1}} L_1 + {N_2}^{3/2} \text{e}^{-\frac{2\pi h}{L_2}} L_2 \right] + 100 k_{\text{b}} \text{TR}_{\text{cell}} {N_1}^{3/2} \text{e}^{-\frac{2\pi \left(h + 2R_{\text{probe}} + \delta \right)}{L_1}} L_1, \tag{7}$$

where index 1 is assigned to the large brush. In both equations, $\delta = R_{\text{cell}} - \sqrt{R_{\text{cell}}^2 - R_{\text{probe}}^2}$. Note that $R_{\text{cell}} > R_{\text{probe}}$ for all cases observed in this work.

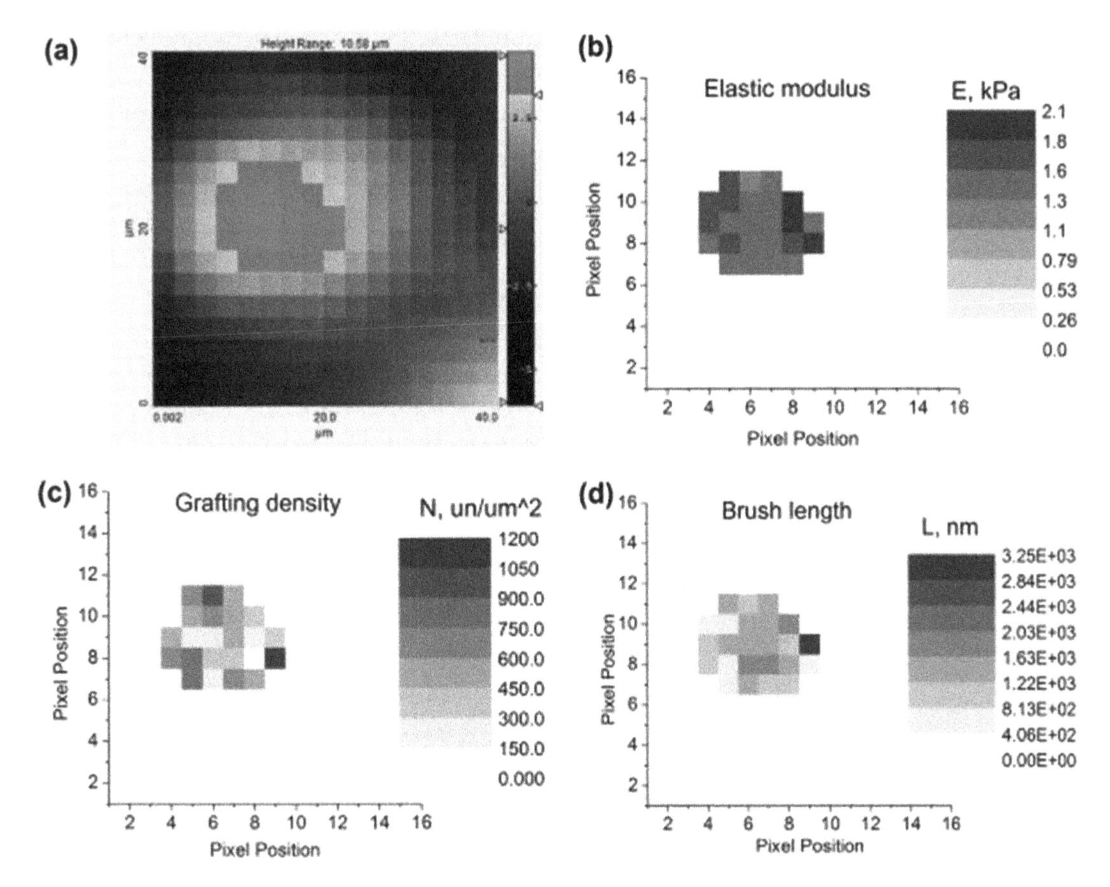

Fig. 9 A representative example of the use of the force-volume mode on a human cervical epithelial cell. (**a**) A topographic image collected in this mode. Results of processing of the raw data recorded at each pixel: (**b**) the elastic modulus, (**c**) grafting density, and (**d**) length of the cellular brush. The lateral size of each pixel in all images is 2.5 μm. Reproduced from Ref. [33] with permission from Elsevier (C) 2013

18. The procedure described above has to be repeated for all extracted force curves. This is important because of intrinsic heterogeneity of the cell surface. Figure 9 shows an example of distribution of the elastic modulus grafting density and the brush length obtained for human cervical cells.

4 Notes

1. A glass-bottom culture dish is recommended if high-resolution optical measurements are planned in parallel with the AFM study.
2. A 5 μm diameter silica spheres (Bangs Labs, Inc.) were glued to the cantilevers using either the AFM built-in micromanipulator (Dimension 3100 microscope) or a micromanipulator station

6000 (by Micromanipulator, Inc.). This method is described, for example, in [4, 39] in great detail.

3. Any software capable of extracting force curves from the force-volume files in any editable format will work.
4. Mechanical properties of cells will depend on multiple factors [40], for example, the cell passage [41], aging [4, 37], mitosis stage [42], etc.
5. In principle, the described method can be used for fixed cells as well. Then the viability preservation stages are not applicable.
6. Calibration of the AFM cantilever sensitivity. The deflection of the AFM cantilever on a rigid surface, for example, a clean silicon wafer or glass slide, should be recorded (*see* the manual of the microscope for more detail; for Bruker microscopes it can be recorded in the force calibration mode). The spring constant of the AFM cantilever can be found using the thermal tuning method, a popular approach used in virtually all new AFMs.
7. Alternatively, electron microscopy imaging is also possible.
8. The ramp size should be substantially larger if using fibroblasts which started to generate collagen around them.
9. Strictly speaking, to derive the elastic modulus of the cell body (also known as the Young's modulus), one needs to use an infinitely slow speed. However, besides being unrealistic with the finite time of experiment, slow indentation can alter cell. Additionally, cells and the AFM probe can develop time-dependent adhesion [43], which would alter the measurements substantially. At the same time, the increase in the imaging speed will bring more contribution from the viscoelastic effects. 1–10 μm/s ramping speed seems to be a reasonable compromise to minimize the influence of these factors.
10. For Bruker AFMs, switch back to the imaging mode and then quickly to the force-volume mode. This is just a historical (though logically unnecessary) step required by the software implementation of the force-volume mode in Bruker microscopes.
11. When calculating the radius of the cell, it is important to restore each pixel to its undeformed height. This is done by adding the value of deformation of the cell i at that particular pixel using the following formula for a cell firmly attached to the surface:

$$i = Z_0 - d - Z = d^{2/3} \cdot \left[\frac{9}{16} \frac{k}{E} \sqrt{\frac{R_{\text{probe}} + R_{\text{cell}}}{R_{\text{probe}} R_{\text{cell}}}} \right]^{2/3}, \qquad \text{(8a)}$$

and for a loosely attached cell:

$$i^* = d^{2/3}\left(\left[\frac{9}{16}\frac{k}{E}\sqrt{\frac{R_{\text{probe}} + R_{\text{cell}}}{R_{\text{probe}}R_{\text{cell}}}}\right]^{2/3} + \left[\frac{9}{16}\frac{k}{E\sqrt{R_{\text{cell}}}}\right]^{2/3}\right), \quad (8b)$$

where $i^* = i_{\text{top}} + i_{\text{bottom}}$ is the total deformation of the cell body in the case of a loosely attached cell (which is deformed from the top and the bottom), k is the spring constant of the AFM cantilever, and R_{probe} and R_{cell} are the radii of the AFM probe and cell body, respectively. In the case of a flat cell, R_{cell} tends to infinity. The Poisson ratio ν of a cell is typically chosen to be 0.5 (because of a small range of possible variations of ν, the error due to the uncertainty its definition is small and can be ignored).

12. Obviously, the value of the deformation can be found only after calculation of the elastic modulus of the cell. Thus, the radius of the cell has to be found retrospectively, after using the deformed value as a first approximation. Fortunately, the error due to the deformation is not large, and it is sufficient to do only a single iteration.
13. It is the approaching part of the force curve that is considered for the analysis. This is justified by the concern that the AFM probe may alter the cell sufficiently and the retraction curve will contain artifacts because of incorrect assumption of fast recovery of the area of contact between the AFM probe and cell surface as well as the fast expansion of the pericellular brush. The latter is clearly a wrong assumption (it is known from polymer physics that a large polymer brush takes at least several seconds to restore its expansion).
14. Note that Z_0 may be different for each force interval taking for the nonlinear curve fitting. However, Z_0 is the position of nondeformed cell, which is obviously just one value. Therefore, when the appropriate fitting interval (plateau position of the modulus) is found, Z_0 has to be fixed to that value.
15. For simplicity and less ambiguity, Eq. (4b) can be rewritten as [34]:

$$F(h^*) \approx 100k_B\text{TR}_1 N_1^{3/2}\exp\left(-\frac{2\pi L_2 M}{L_1 + L_2}\right)\exp\left(-\frac{2\pi h^*}{L_1 + L_2}\right)L_1, \quad (9)$$

where

$$M = -\frac{1}{2\pi}\ln\left(\frac{50k_B\text{TR}_1 N_1^{3/2} L_1}{50k_B\text{TR}_2 N_2^{3/2} L_2}\right).$$

From Eq. (9), one can see that it is impossible to extract separate parameters of the top and bottom brushes.

Furthermore, Eq. (9) can be reduced to a simple single steric brush equation, provided we assume that the brush lengths of L_1 and L_2 are equal. Introducing an effective brush length $L^*(L_1 = L_2 = L^*/2)$ and an effective grafting density $N^* = \sqrt{N_1 N_2}$, Eq. (9) can be rewritten as follows:

$$F(h^*) \approx 100 k_B \mathrm{TR}^* N^{*3/2} \exp\left(-2\pi \frac{h^*}{L^*}\right) L^* \tag{10}$$

where the effective radius is $R^* = 1/2\sqrt{R_1 R_2}$.

This reduced form of the brush description allows us to extract the effective parameters of both top and bottom brushes unambiguously.

16. Validity of this approach was experimentally demonstrated in [35].

Acknowledgments

I.S. gratefully acknowledges funding for this work by NSF CMMI-1435655.

References

1. Ulrich P, Zhang X (1997) Pharmacological reversal of advanced glycation end-product-mediated protein crosslinking. Diabetologia 40:S157–S159
2. Perry G, Smith MA (2001) Active glycation in neurofibrillary pathology of Alzheimer's disease: N-(Carboxymethyl) lysine and hexitol-lysine. Free Radic Biol Med 31:175–180
3. Bucala R, Cerami A (1992) Advanced glycosylation: chemistry, biology, and implications for diabetes and aging. Adv Pharmacol 23:1–34
4. Berdyyeva TK, Woodworth CD, Sokolov I (2005) Human epithelial cells increase their rigidity with ageing in vitro: direct measurements. Phys Med Biol 50(1):81–92. https://doi.org/10.1088/0031-9155/50/1/007
5. Suresh S (2007) Biomechanics and biophysics of cancer cells. Acta Biomater 3(4):413–438
6. Cross SE, Jin YS, Rao J, Gimzewski JK (2007) Nanomechanical analysis of cells from cancer patients. Nat Nanotechnol 2(12):780–783. https://doi.org/10.1038/nnano.2007.388
7. Zahn JT, Louban I, Jungbauer S, Bissinger M, Kaufmann D, Kemkemer R, Spatz JP (2011) Age-dependent changes in microscale stiffness and mechanoresponses of cells. Small 7 (10):1480–1487. https://doi.org/10.1002/smll.201100146
8. Lu S, Long M (2005) Forced dissociation of selectin-ligand complexes using steered molecular dynamics simulation. Mol Cell Biomech 2 (4):161–177
9. Maidment SL (1997) The cytoskeleton and brain tumour cell migration. Anticancer Res 17(6B):4145–4149
10. Paszek MJ, Zahir N, Johnson KR, Lakins JN, Rozenberg GI, Gefen A, Reinhart-King CA, Margulies SS, Dembo M, Boettiger D, Hammer DA, Weaver VM (2005) Tensional homeostasis and the malignant phenotype. Cancer Cell 8(3):241–254
11. Huang S, Ingber DE (2005) Cell tension, matrix mechanics, and cancer development. Cancer Cell 8(3):175–176
12. Guz N, Dokukin M, Kalaparthi V, Sokolov I (2014) If cell mechanics can be described by elastic modulus: study of different models and probes used in indentation experiments. Biophys J 107(3):564–575. https://doi.org/10.1016/j.bpj.2014.06.033
13. Sokolov I, Iyer S, Subba-Rao V, Gaikwad RM, Woodworth CD (2007) Detection of surface brush on biological cells in vitro with atomic force microscopy. Appl Phys Lett 91:023902–023901–023903

14. Guz NV, Dokukin ME, Woodworth CD, Cardin A, Sokolov I (2015) Towards early detection of cervical cancer: fractal dimension of AFM images of human cervical epithelial cells at different stages of progression to cancer. Nanomedicine 11(7):1667–1675. https://doi.org/10.1016/j.nano.2015.04.012
15. Gaikwad RM, Dokukin ME, Iyer KS, Woodworth CD, Volkov DO, Sokolov I (2011) Detection of cancerous cervical cells using physical adhesion of fluorescent silica particles and centripetal force. Analyst 136 (7):1502–1506. https://doi.org/10.1039/c0an00366b
16. Dokukin ME, Guz NV, Gaikwad RM, Woodworth CD, Sokolov I (2011) Cell surface as a fractal: normal and cancerous cervical cells demonstrate different fractal behavior of surface adhesion maps at the nanoscale. Phys Rev Lett 107(2):028101
17. Gaikwad RM, Iyer S, Guz N, Volkov D, Dokukin M, Woodworth CD, Sokolov I (2010) Atomic force microscopy helps to develop methods for physical detection of cancerous cells. In: IEEE 2010 Fourth International Conference on Quantum, Nano and Micro Technologies. pp 18–22. https://doi.org/10.1109/ICQNM.2010
18. Sokolov I (2009) Interaction between silica particles and human epithelial cells: atomic force microscopy and fluorescence study. In: Jelinek R (ed) Cellular and biomolecular recognition. Wiley-VCH Verlag GmbH & Co, NY, pp 69–96
19. Iyer S, Gaikwad RM, Subba-Rao V, Woodworth CD, Sokolov I (2009) AFM detects differences in the surface brush on normal and cancerous cervical cells. Nat Nanotechnol 4:389–393
20. Kumar S, Hoh JH (2004) Modulation of repulsive forces between neurofilaments by sidearm phosphorylation. Biochem Biophys Res Commun 324(2):489–496
21. Brown HG, Hoh JH (1997) Entropic exclusion by neurofilament side arms: a mechanism for maintaining interfilament spacing. Biochemistry 36(49):15035–15040
22. Kosaki R, Watanabe K, Yamaguchi Y (1999) Overproduction of hyaluronan by expression of the hyaluronan synthase Has2 enhances anchorage-independent growth and tumorigenicity. Cancer Res 59(5):1141–1145
23. Itano N, Atsumi F, Sawai T, Yamada Y, Miyaishi O, Senga T, Hamaguchi M, Kimata K (2002) Abnormal accumulation of hyaluronan matrix diminishes contact inhibition of cell growth and promotes cell migration. Proc Natl Acad Sci U S A 99(6):3609–3614. https://doi.org/10.1073/pnas.052026799
24. Toole B (1982) Glycosaminoglycans in morphogenesis. In: Hay E (ed) Cell biology of the extracellular matrix. Plenum Press, New York, pp 259–294
25. Zimmerman E, Geiger B, Addadi L (2002) Initial stages of cell-matrix adhesion can be mediated and modulated by cell-surface hyaluronan. Biophys J 82(4):1848–1857
26. Camenisch TD, Schroeder JA, Bradley J, Klewer SE, McDonald JA (2002) Heart-valve mesenchyme formation is dependent on hyaluronan-augmented activation of ErbB2-ErbB3 receptors. Nat Med 8(8):850–855. https://doi.org/10.1038/nm742
27. Chen WY, Abatangelo G (1999) Functions of hyaluronan in wound repair. Wound Repair Regen 7(2):79–89
28. Jiang D, Liang J, Fan J, Yu S, Chen S, Luo Y, Prestwich GD, Mascarenhas MM, Garg HG, Quinn DA, Homer RJ, Goldstein DR, Bucala R, Lee PJ, Medzhitov R, Noble PW (2005) Regulation of lung injury and repair by toll-like receptors and hyaluronan. Nat Med 11(11):1173–1179. https://doi.org/10.1038/nm1315
29. de la Motte CA, Hascall VC, Drazba J, Bandyopadhyay SK, Strong SA (2003) Mononuclear leukocytes bind to specific hyaluronan structures on colon mucosal smooth muscle cells treated with polyinosinic acid: polycytidylic acid: inter-alpha-trypsin inhibitor is crucial to structure and function. Am J Pathol 163 (1):121–133
30. Richards JS (2005) Ovulation: new factors that prepare the oocyte for fertilization. Mol Cell Endocrinol 234(1–2):75–79. https://doi.org/10.1016/j.mce.2005.01.004
31. Toole BP (2004) Hyaluronan: from extracellular glue to pericellular cue. Nat Rev Cancer 4 (7):528–539. https://doi.org/10.1038/nrc1391
32. Simon M, Dokukin M, Kalaparthi V, Spedden E, Sokolov I, Staii C (2016) Load rate and temperature dependent mechanical properties of the cortical neuron and its Pericellular layer measured by atomic force microscopy. Langmuir 32(4):1111–1119. https://doi.org/10.1021/acs.langmuir.5b04317
33. Sokolov I, Dokukin ME, Guz NV (2013) Method for quantitative measurements of the elastic modulus of biological cells in AFM indentation experiments. Methods 60 (2):202–213. https://doi.org/10.1016/j.ymeth.2013.03.037

34. Dokukin ME, Guz NV, Sokolov I (2013) Quantitative study of the elastic Modulus of loosely attached cells in AFM indentation experiments. Biophys J 104(10):2123–2131. https://doi.org/10.1016/j.bpj.2013.04.019
35. Dokukin M, Ablaeva Y, Kalaparthi V, Seluanov A, Gorbunova V, Sokolov I (2016) Pericellular brush and mechanics of Guinea pig fibroblast cells studied with AFM. Biophys J 111(1):236–246. https://doi.org/10.1016/j.bpj.2016.06.005
36. Seluanov A, Vaidya A, Gorbunova V (2010) Establishing primary adult fibroblast cultures from rodents. J Vis Exp 44:pii: 2033. https://doi.org/10.3791/2033
37. Sokolov I, Iyer S, Woodworth CD (2006) Recovery of elasticity of aged human epithelial cells in-vitro. Nanomedicine 2:31–36
38. Butt HJ, Kappl M, Mueller H, Raiteri R, Meyer W, Ruhe J (1999) Steric forces measured with the atomic force microscope at various temperatures. Langmuir 15 (7):2559–2565
39. Volkov DO, Dandu PRV, Goodman H, Santora B, Sokolov I (2011) Influence of adhesion of silica and ceria abrasive nanoparticles on chemical-mechanical planarization of silica surfaces. Appl Surf Sci 257(20):8518–8524
40. Sokolov I (2007) Atomic force microscopy in Cancer cell research. In: Webster HSNaT (ed) Cancer nanotechnology – nanomaterials for Cancer diagnosis and therapy. APS, Los Angeles, pp 43–59
41. Dokukin ME, Guz NV, Sokolov I (2017) Mechanical properties of cancer cells depend on number of passages: atomic force microscopy indentation study. Jpn J Appl Phys 56:08LB01
42. Matzke R, Jacobson K, Radmacher M (2001) Direct, high-resolution measurement of furrow stiffening during division of adherent cells. Nat Cell Biol 3(6):607–610. https://doi.org/10.1038/35078583
43. Iyer S, Woodworth CD, Gaikwad RM, Kievsky YY, Sokolov I (2009) Towards nonspecific detection of malignant cervical cells with fluorescent silica beads. Small 5(20):2277–2284

Chapter 28

Imaging of Soft and Biological Samples Using AFM Ringing Mode

Igor Sokolov and Maxim E. Dokukin

Abstract

Ringing mode of atomic force microscopy (AFM) enables imaging the surfaces of biological samples, cells, tissue, biopolymers, etc. to obtain unique information, such as the size of molecules pulled by the AFM probe from the sample surface, heights of the sample at different load forces, etc. (up to eight different imaging channels can be recorded simultaneously, which is in addition to five channels already available in other rival modes). The imaging can be done in both air (gases) and liquid (buffers). In addition, the images obtained in ringing mode do not have several common artifacts and can be collected up to 20× faster compared to the rival imaging modes. Here we describe a step-by-step approach to collect images in ringing mode applied to biological and soft materials in general. Technical details, potential difficulties, and points of special attention are described.

Key words AFM imaging, Ringing mode, Cell imaging, Physical properties of cells, Biophysics of cells, Tissues, Biopolymers

1 Introduction

Atomic force microscopy (AFM) is a technique allowing the study of samples, including biological ones, cells, molecules with a small probe, and AFM tip [1]. The lateral resolution on soft samples can be as high as single molecules but practically is of the order of 1–10 nm [2–4]. The vertical resolution is 0.01–0.1 nm. AFM allows not only recording of the cell topography but also extraction of a lot of physical information about the surfaces of interest [5–9] using various modes of operation. Recently novel sub-resonant tapping modes were introduced by different manufacturers (under different trade names): Digital Pulse Mode™ (WITec), PeakForce Tapping ™ and ScanAsyst™ (Bruker Nano), HybriD Mode™ (NT-MDT), QI™ mode (JPK), etc. The sub-resonant modes allow very stable and gentle basic imaging as well as delivery of quantitative maps of various sample surface physical properties, such as adhesion (between the AFM probe and sample surface),

Yuri L. Lyubchenko (ed.), *Nanoscale Imaging: Methods and Protocols*, Methods in Molecular Biology, vol. 1814,
https://doi.org/10.1007/978-1-4939-8591-3_28, © Springer Science+Business Media, LLC, part of Springer Nature 2018

stiffness of the sample, and viscoelastic energy losses of the sample (during indentation/deformation cycle).

Ringing mode™ (NanoScience Solutions, Inc.) is a further development of sub-resonant tapping modes, which allows simultaneous recording of eight additional physical parameters of samples: restored (averaged) adhesion, adhesion height, disconnection height, pull-off neck height, disconnection distance, disconnection energy loss, dynamic creep phase shift, and zero-force height (this channel, though, is available in some commercial AFMs). It should be noted that the ringing mode channels are recorded independently of the channels recorded in standard sub-resonant AFM modes. The meanings of all these parameters are described in the present work. Knowledge of high-resolution maps of the physical parameters mentioned above may be useful for multiple applications, which is hard to outline at this moment because such images have never been recorded before. We only give a few examples in which there are definite benefits of having these novel images.

Ringing mode can visualize surfaces of various biological samples, including cells. It is known that the cell surface changes physically in various cell abnormalities and diseases. It has been shown that the pericellular coat surrounding cells is substantially different for human epithelial cells when they become cancerous [10–17]. The difference of the pericellular layer was observed for leukemia cells, including quiescent and proliferating stem cancer cells [18, 19]. It is also known that molecular layers surround neurofilaments to maintain interfilament spacing [20, 21]. The pericellular layer is responsible for the cell-cell interaction [22], cell migration [23], cell differentiation, and cell proliferation [24, 25]. It is important in embryonic development [26], wound healing [27], inflammation [28, 29], mammalian fertilization [30], epithelial-mesenchymal transition [26], resistance to apoptosis (cell death), and multidrug resistance [31]. Ringing mode allows visualizing, for example, distribution of molecules on the cell surface, which will be useful in the study of all above phenomena.

Besides having the ability to record novel information of the sample surface, the images are collected substantially faster (up to 20 times) in ringing mode compared to the rival sub-resonant tapping modes. Furthermore, it is possible to avoid some artifacts when imaging in ringing mode [32].

Here we present a step-by-step description on how to work with ringing mode and explain the meaning of the recorded information. Potential difficulties and points of special attention are also noted.

2 Materials

2.1 Sample Preparation

Any substrate capable of having the sample attached to it as exampled in **items 1–3** below:

1. A glass slide. Standard glass slides (#1 by Thermo Fisher Scientific, Waltham, MA, USA).
2. Culture dish (60 mm polystyrene culture dishes by Thermo Fisher Scientific, Waltham, MA, USA; *see* **Note 1**).
3. Cuderm Dark (carbon-filled) adhesive tape strips for collecting of skin flakes (D-squame skin indicator D200, CuDerm, Corp.).
4. If the sample is to be fixed, any suitable fixative (*see* **Note 2**). Karnovsky fixative was used here (by SPI Supplies, West Chester, PA, USA).

2.2 Hardware and Software

1. Dimension Icon or BioScope Catalyst or MultiMode 8 (Bruker/Veeco, Inc., Santa Barbara, CA, USA) atomic force microscopes equipped with NanoScope V controller (*see* **Note 3**).
2. Ringing mode attachment A (by NanoScience Solutions, Inc., VA, USA) if the AFM has a sub-resonant mode included (either ScanAsyst or PeakForce QNM for Bruker AFMs).
3. Ringing mode attachment B (by NanoScience Solutions, Inc., VA, USA) if the AFM does not have a sub-resonant mode included.
4. Basic AFM control software included with the AFM (for Bruker AFMs, NanoScope basic software of any version).
5. Ringing mode software supplied with the ringing mode attachment (currently version 1.12).
6. Standard AFM cantilever holders (Bruker/Veeco, Inc., Santa Barbara, CA, USA).
7. AFM probes for work in sub-resonance tapping mode (*see* **Note 4**).
8. Standard tip-check sample TIP001 for quantitative measurements (Aurora NanoDevices, Canada).
9. NanoScope Analysis (Bruker/Veeco, Inc., Santa Barbara, CA, USA) or similar software.
10. Labconco Lyph-Lock 12 (Labconco, Kansas City, MO, USA) is used here for freeze-drying.

3 Methods

Ringing mode allows obtaining up to eight different channels of information simultaneously in real time. The number of the ringing mode channels that can be stored at the same time is limited at present by the AFM setup, which allows visualization and storage of a limited number of auxiliary input channels. For example, the number of auxiliary channels in Bruker NanoScope AFM is three. Thus, one can record three ringing mode channels chosen out of eight available. These channels are additional to the standard AFM channels.

3.1 Sample Preparation for AFM Study

1. The sample should be immobilized to be steady during imaging. No other special treatment of samples is required.
2. If sample fixation is needed, either freeze-drying or subcritical drying method is typically required to avoid drying artifacts. A standard procedure of sample preparation for either freeze-drying or subcritical dryer should be used.

3.2 Imaging with Ringing Mode

1. Connect the ringing mode add-on as described in the instrument manual.
2. Start both AFM and the ringing mode add-on. Start AFM software on the AFM computer and the ringing mode software on the ringing mode computer.
3. Install an AFM cantilever; align the laser on the cantilever as described in standard AFM protocols.
4. For quantitative measurements, calibrate the deflection sensitivity, the spring constant, and the resonance quality factor of the AFM cantilever (*see* **Note 5**). Measure the radius of the AFM probe. Image a standard TIP001 tip-check sample using any imaging mode (tapping and sub-resonance are recommended). Use NanoScope Analysis software to calculate the radius of the AFM probe (*see* **Note 6**).
5. Input the parameters needed for ringing mode software: the deflection sensitivity, spring constant, and resonance quality factor of the cantilever, which were found in **step 4**. For several AFM models, with sub-resonant mode included, the sensitivity of the AFM Z piezo sensor is required to be inserted in the ringing mode software (consult the ringing mode manual).
6. Locate the AFM probe over a desirable surface area using an optical microscope which is typically built in or attached to AFMs.
7. When using microscopes:

   With a sub-resonant mode installed: Choose the sub-resonance mode of operation for the force microscope.

Without a sub-resonant mode installed: Choose the mode as instructed in the manual supplied with the ringing mode attachment.

8. Choose an appropriate load force to ensure a reliable contact (a standard one, the same as without ringing mode).
9. Using the ringing mode software, choose desirable three (out of eight available) channels to be visualized and recorded in the AFM software as auxiliary channels 1, 2, and 3.
10. Press engage. After the engagement with the surface, perform imaging using controls of regular AFM software. For advanced users: Depending of the sample, additional ringing mode parameters may be changed to fine-tune/improve the image quality or to remove some possible scanning artifacts (*see* the manual supplied with the ringing mode attachment for detail).
11. To verify the absence of artifacts, use a standard procedure: try several scan speeds, parameters of the feedback, peak (imaging) forces, oscillation amplitude, and zooms. The artifact-free image should be reasonably independent on these scanning parameters.

3.3 Interpretation of Images Collected in Ringing Mode

The images are recorded in standard AFM format and can be open and studied with any appropriate software (e.g., NanoScope Analysis (Bruker/Veeco, Inc., Santa Barbara, CA, USA, or SPIP™, Image Metrology A/S, Denmark). The following images can be recorded in ringing mode (*see* **Note 7**):

1. Restored adhesion. The adhesion force which is restored from the AFM ringing signal. It is typically less than the regular adhesion. The difference comes from the viscous losses and molecular rupturing during disconnection of the AFM probe from the sample surface. A comparative example is shown in Fig. 2 (*see* **Note 8**).
2. Zero-force height. It is the height of the sample at zero deflection of the cantilever. An example is shown in Fig. 3b (*see* **Note 9**).
3. Adhesion height. It is the height of the sample at the maximum negative deflection of the cantilever (when the deflection force is equal to the force of adhesion). An example is shown in Fig. 3c (*see* **Note 10**).
4. Disconnection height. It is the height of the sample when the force acting on the AFM cantilever is equal to the force of restored adhesion. An example is shown in Fig. 3d (*see* **Note 11**).
5. Pull-off neck height. It is the maximum height of the neck pulled from the sample surface by the AFM probe. An example is shown in Fig. 4a (*see* **Note 12**).

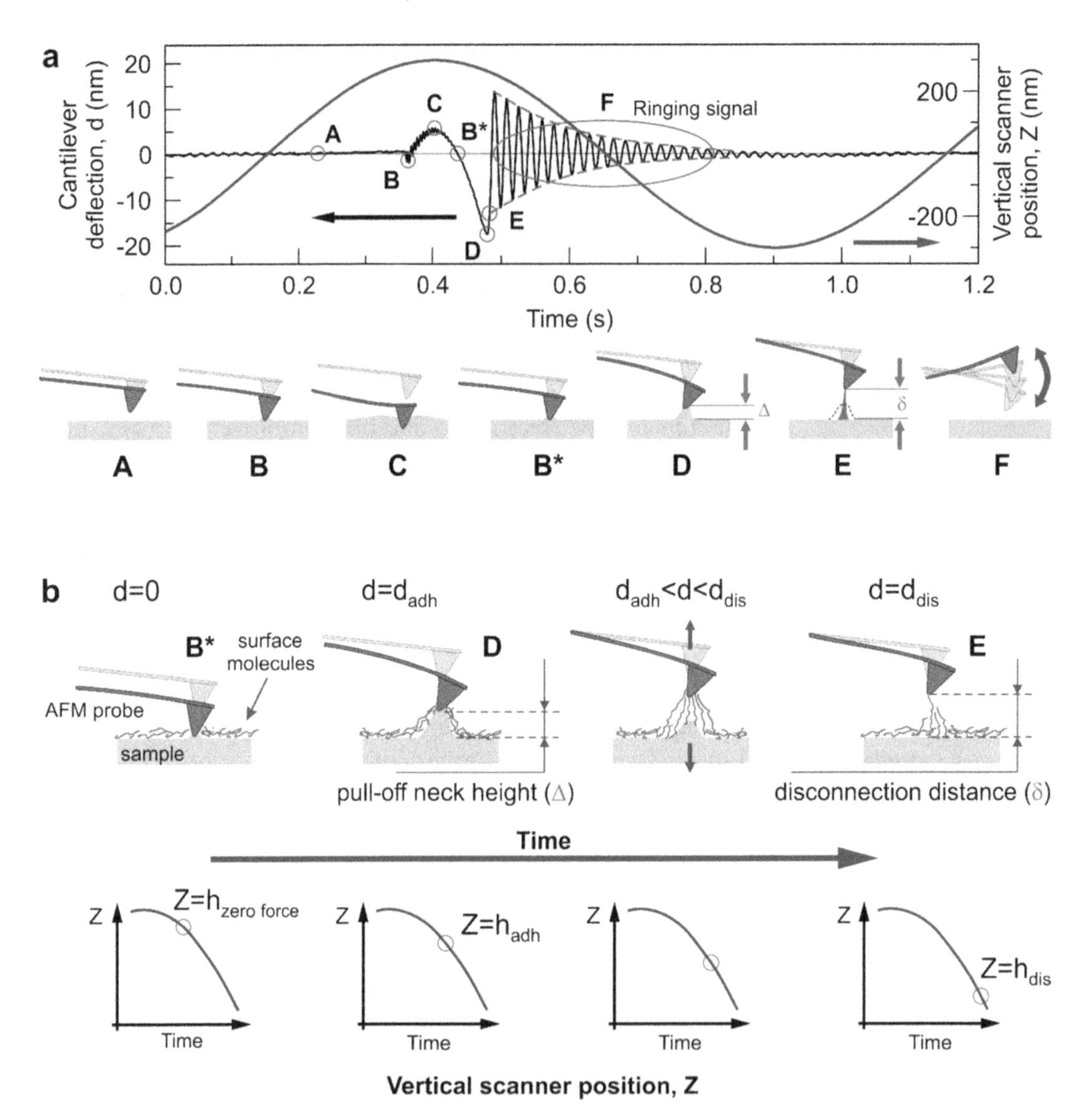

Fig. 1 A schematic of the signal exploited in ringing mode. The cantilever deflection d and the vertical position of the AFM scanner Z versus time are shown. (**a**) One full cycle of 1 kHz vertical oscillation of the Z scanner is presented. A typical unfiltered signal of the cantilever deflection d (or force $= kd$, where k is the spring constant of the cantilever) as a function of time is shown. The dash-envelope lines show the decrease of the oscillating amplitude of the AFM cantilever due to dissipation. Positive values of the cantilever deflection d are due to forces of repulsion, and vice versa, and negative values stand for the attraction between the AFM probe and sample surface. Specific points during the cantilever motion indicate the following probe positions: (A) far from the sample surface, (B) touches the sample, (C) deforms the sample surface, (B*) touches the surface with zero force when retracting, (D) starts a fast detachment (pull-off) from the sample (the point defining adhesion force), (E) completely detaches from the sample (the point defining the restored adhesion force), and (F) starts free oscillations above the sample surface. (**b**) Definition of the neck height Δ, the size of the neck of the pull-off deformation caused by the AFM probe right before disconnection from the surface; definition of disconnection distance δ, the size of the molecular tails pulled from the surface by the AFM probe during the disconnection from the sample surface. The figure was reproduced from Ref. [32]. An open-access publication: permission has been distributed under a Creative Commons CC-BY license (Adapted from [32])

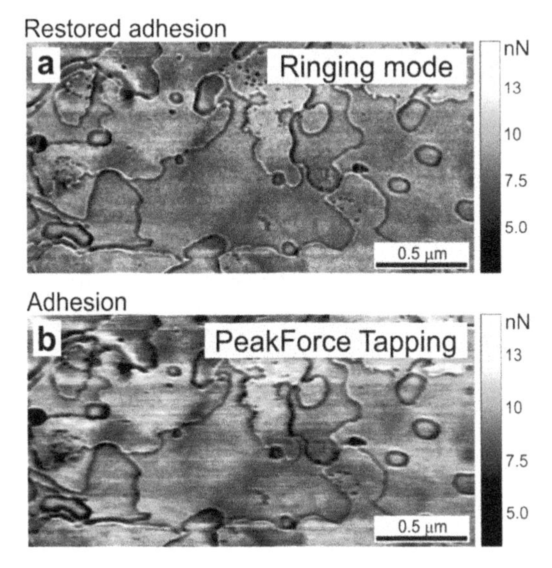

Fig. 2 An example of the averaged restored adhesion channel (**a**) compared to the simultaneously imaged regular adhesion channel (**b**) of the Bruker PeakForce Tapping mode. AFM map of human skin flake surface is shown. The scan rate is 0.5 Hz (2 s per line). The figure was reproduced from Ref. [32]. An open-access publication: permission has been distributed under a Creative Commons CC-BY license

6. Disconnection distance. It is the length of molecules on the sample surface, which are disconnecting from the AFM probe during the pull-off of the AFM probe from the sample surface. An example is shown in Fig. 4c (*see* **Note 13**).
7. Disconnection energy loss. It is the energy losses due to dissipative disconnection of the AFM probe from the sample surface. A comparative example is shown in Fig. 5 (*see* **Note 14**).
8. Dynamic creep phase shift. It is the phase difference between the oscillation of the sample or cantilever holder (depending on the AFM model) and AFM cantilever. An example is shown in Fig. 6 (*see* **Note 15**).

4 Notes

1. A glass-bottom culture dish is recommended if high-resolution optical measurements are planned in parallel with the AFM study.
2. It is recommended to use fixative utilized in electron microscopy to preserve nanometer structure of samples.
3. Other microscopes may also be used soon.

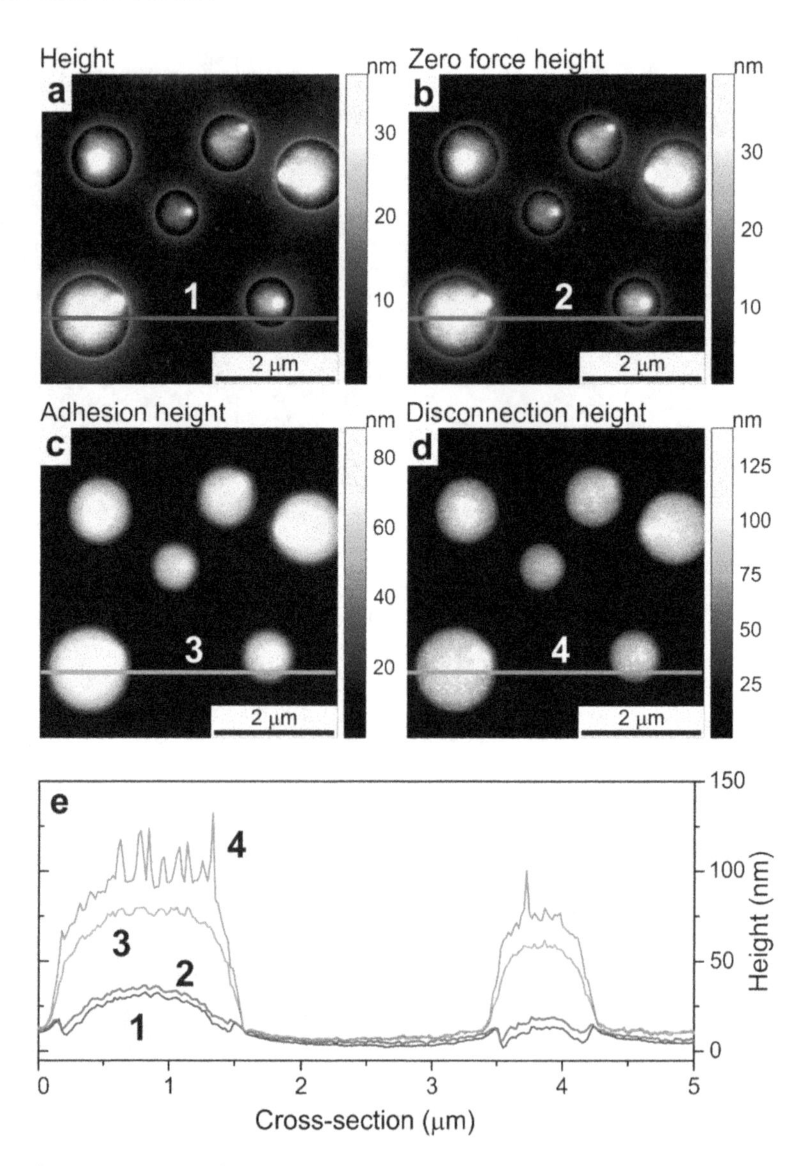

Fig. 3 An example of simultaneous imaging of four different heights: (**a**) regular, (**b**) zero force, (**c**) adhesion, and (**d**) disconnection heights. The cross-sectional lines shown in the images are presented in panel (**e**). All images are collected on a Bruker calibration sample, a blend of two polymers, and low-density polyethylene and polystyrene. The scan rate of 1 Hz (1 s per line) is used. The figure was reproduced from Ref. [32]. An open-access publication: permission has been distributed under a Creative Commons CC-BY license

4. Bruker ScanAsyst probes for working in air and water can be used. To produce better adhesion, probes with larger radius (e.g., by NanoScience Solutions, Inc.) should be used.
5. To calibrate the sensitivity of the AFM cantilever, one needs to measure the cantilever deflection on a rigid surface, for

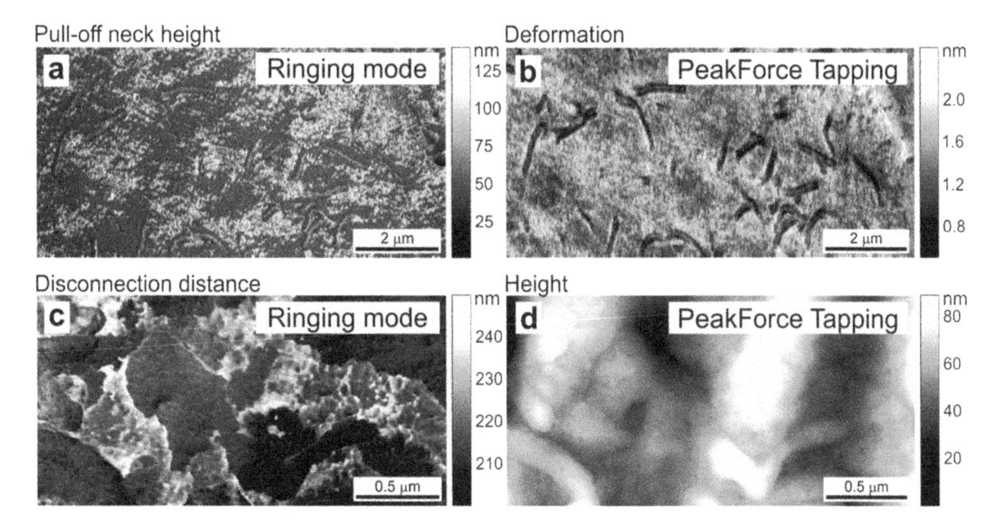

Fig. 4 Pull-off neck height (**a**) and disconnection distance (**c**) channels of ringing mode are shown. An example of human melanoma cell is presented. Images of surface deformation (**b**) and regular cell height (**d**) are shown as reference (simultaneously recorded in the Bruker PeakForce Tapping mode). The scan rate is 0.5 Hz (2 s per line). The figure was reproduced from Ref. [32]. An open-access publication: permission has been distributed under a Creative Commons CC-BY license

example, a clean silicon wafer or glass slide (*see* the manual of the microscope for more detail; for Bruker microscopes, it can be recorded in the force calibration mode). The spring constant of the AFM cantilever can be found using the thermal tuning method, a popular approach used in virtually all new AFMs. The quality factor is also measured using the thermal tuning method. It has to be done in the vicinity of the sample surface (smaller than 1 micron; the quality factor is practically constant below this distance even in air; *see* [32], Figure S3 for detail).

6. Alternatively, electron microscopy imaging is also possible. Note that the exact probe radius is required if one needs to compare results with another AFM probe or other experiments, or to measure the elastic modulus of the sample.
7. Definition of parameters imaged in ringing mode is shown in Fig. 1.
8. Physical meaning of the restored adhesion. After the pull-off event (point D, Fig. 1), the AFM cantilever jumps from the surface, and the ringing starts. Each oscillation has smaller amplitude because of dissipation. Using the known quality factor of the cantilever (should be calibrated before measurements separately; *see* [32] for details), it is possible to extrapolate the value of the adhesion force to the moment of pull-off (when the extrapolation curve crosses the deflection curve,

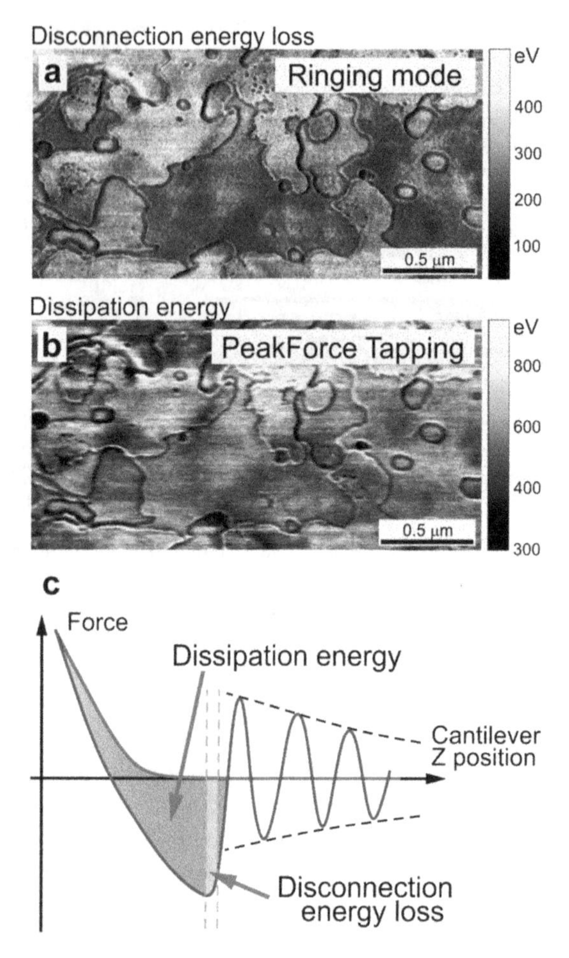

Fig. 5 An example of the disconnection energy loss channel (**a**) compared to the simultaneously imaged regular Bruker PeakForce Tapping dissipation energy channel (**b**). A schematic of the disconnection energy loss is shown in panel (**c**). The scan rate is 0.5 Hz (2 s per line). The figure was reproduced from Ref. [32]. An open-access publication: permission has been distributed under a Creative Commons CC-BY license

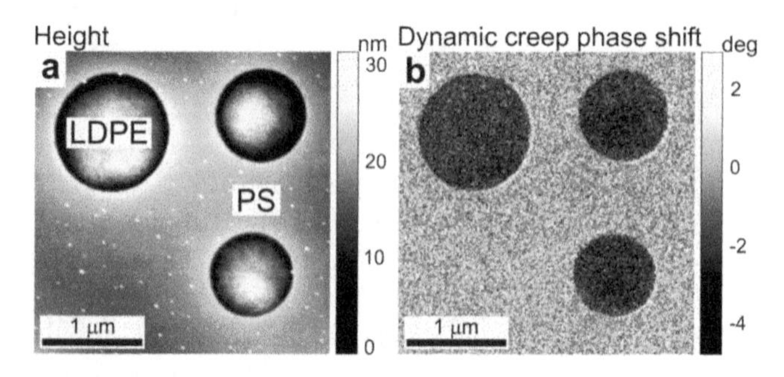

Fig. 6 Maps of the regular height channel (**a**) and simultaneously collected dynamic creep phase shift channel (**b**). Images are collected on a standard Bruker calibration sample, a blend of two polymers, and low-density polyethylene (LDPE) and polystyrene (PS). The scan rate of 0.5 Hz (2 s per line) is used

point E, Fig. 2). This is the value of the restored adhesion. Furthermore, one can calculate the restored adhesion using multiple points of the ringing signal. Doing averaging of the obtained values, one obtains lower noise in the measured value of the restored adhesion and gets rid of several artifacts (*see* [32] for details). This value is recorded by AFM in ringing mode as the restored adhesion (could also be called averaged restored adhesion).

9. Physical meaning of the zero-force height (this channel, though, is available in some commercial AFMs). It is the position of the sample at the moment when the deflection of the AFM cantilever is zero, and the AFM probe is retracting from the sample. It can be interpreted as the height of almost non-deformed sample if the adhesion force is low.
10. Physical meaning of the adhesion height. During disconnection of the AFM probe from the sample surface, the surface is pulled off by the adhesive action of the probe. So the sample height is higher at the moment when the probe disconnects from the sample. This is the adhesion height.
11. Physical meaning of the disconnection height. It is the position of the sample at the moment when the last molecule of the sample surface is disconnected from the AFM probe.
12. Physical meaning of the pull-off neck height. As was defined, this is the maximum height of the neck pulled from the sample surface by the AFM probe due to its adhesive action. This is rather different from disconnection distance (*see* the next Note) and can be unambiguously identified due to its different mechanical behaviors; *see* Fig. 7.
13. Physical meaning of the disconnection distance. As was defined, it is the length of molecules on the sample surface, which are pulled off the surface during retraction of the AFM probe from the sample. Note that this is rather different from the pull-off neck height (*see* the previous Note) and can be unambiguously identified due to its different mechanical behaviors; *see* Fig. 7.
14. Physical meaning of the disconnection energy loss. This energy loss comes from the energy of the viscoelastic pull of the sample material under the probe and from rupturing the molecular contacts between the probe and sample. The latter can include the breakage of the water meniscus if the scanning is done in air (and nonzero humidity).
15. Physical meaning of the dynamic creep phase shift. It occurs due to dissipative interaction between the AFM probe and sample during both the sample deformation and probe-sample disconnection.

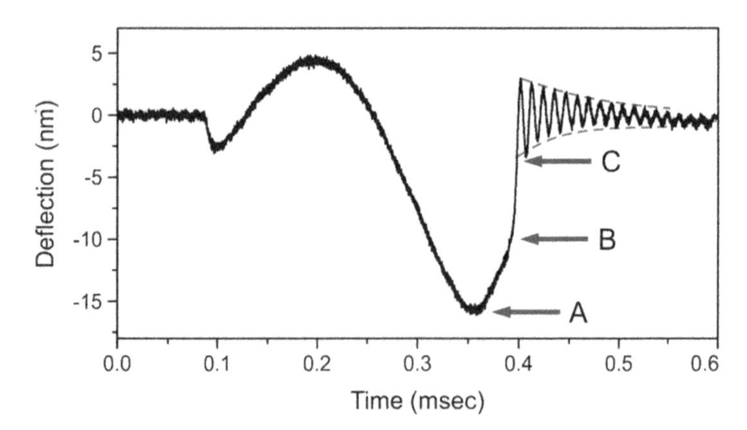

Fig. 7 An example of the unfiltered signal of the cantilever deflection as a function of time measured on a low-density polyethylene (LDPE) sample. The dash-envelope lines show the decrease of the oscillating amplitude of the AFM cantilever due to dissipation. Points of pull-off (adhesion) force (**A**), beginning of rapturing of molecular bonds (**B**), and the complete disconnection (**C**) are shown. Average vertical speed is 300 μm/s. The figure was reproduced from Ref. [32]. An open-access publication: permission has been distributed under a Creative Commons CC-BY license

Acknowledgments

I.S. gratefully acknowledges funding for this work by NSF CMMI-1435655.

References

1. Binnig G, Quate CF, Gerber C (1986) Atomic force microscope. Phys Rev Lett 56 (9):930–933
2. Trohalaki S (2012) Multifrequency force microscopy improves sensitivity and resolution over conventional AFM. MRS Bull 37 (6):545–546
3. Yang J (2004) AFM as a high-resolution imaging tool and a molecular bond force probe. Cell Biochem Biophys 41(3):435–450
4. Sokolov I, Firtel M, Henderson GS (1996) In situ high-resolution AFM imaging of biological surfaces. J Vac Sci Technol B 14:674–678
5. Haase K, Pelling AE (2015) Investigating cell mechanics with atomic force microscopy. J R Soc Interface 12(104):artn 20140970. https://doi.org/10.1098/rsif.2014.0970
6. Gaikwad RM, Dokukin ME, Iyer KS, Woodworth CD, Volkov DO, Sokolov I (2011) Detection of cancerous cervical cells using physical adhesion of fluorescent silica particles and centripetal force. Analyst 136 (7):1502–1506. https://doi.org/10.1039/c0an00366b
7. Sokolov I, Iyer S, Subba-Rao V, Gaikwad RM, Woodworth CD (2007) Detection of surface brush on biological cells in vitro with atomic force microscopy. Appl Phys Lett 91:023902–023901. 023903
8. Stan G, Solares SD (2014) Frequency, amplitude, and phase measurements in contact resonance atomic force microscopies. Beilstein J Nanotech 5:278–288. https://doi.org/10.3762/bjnano.5.30
9. Herruzo ET, Perrino AP, Garcia R (2014) Fast nanomechanical spectroscopy of soft matter. Nat Commun 5:artn 3126. https://doi.org/10.1038/Ncomms4126
10. Sokolov I (2015) Fractals: a possible new path to diagnose and cure cancer? Future Oncol 11

(22):3049–3051. https://doi.org/10.2217/fon.15.211

11. Guz NV, Dokukin ME, Woodworth CD, Cardin A, Sokolov I (2015) Towards early detection of cervical cancer: fractal dimension of AFM images of human cervical epithelial cells at different stages of progression to cancer. Nanomedicine 11(7):1667–1675. https://doi.org/10.1016/j.nano.2015.04.012
12. Dokukin ME, Guz NV, Woodworth CD, Sokolov I (2015) Emerging of fractal geometry on surface of human cervical epithelial cells during progression towards cancer. New J Phys 17:033019
13. Iyer KS, Gaikwad RM, Woodworth CD, Volkov DO, Sokolov I (2012) Physical labeling of papillomavirus-infected, immortal, and cancerous cervical epithelial cells reveal surface changes at immortal stage. Cell Biochem Biophys 63(2):109–116. https://doi.org/10.1007/s12013-012-9345-2
14. Dokukin ME, Guz NV, Gaikwad RM, Woodworth CD, Sokolov I (2011) Cell surface as a fractal: normal and cancerous cervical cells demonstrate different fractal behavior of surface adhesion maps at the nanoscale. Phys Rev Lett 107(2):028101
15. Sokolov I (2009) Interaction between silica particles and human epithelial cells: atomic force microscopy and fluorescence study. In: Jelinek R (ed) Cellular and biomolecular recognition. Wiley-VCH Verlag GmbH & Co, NY, pp 69–96
16. Iyer S, Woodworth CD, Gaikwad RM, Kievsky YY, Sokolov I (2009) Towards nonspecific detection of malignant cervical cells with fluorescent silica beads. Small 5(20):2277–2284
17. Iyer S, Gaikwad RM, Subba-Rao V, Woodworth CD, Sokolov I (2009) AFM detects differences in the surface brush on normal and cancerous cervical cells. Nat Nanotechnol 4:389–393
18. Guz NV, Patel SJ, Dokukin ME, Clarkson B, Sokolov I (2016) Biophysical differences between chronic myelogenous leukemic quiescent and proliferating stem/progenitor cells. Nanomedicine 12(8):2429–2437. https://doi.org/10.1016/j.nano.2016.06.016
19. Guz NV, Patel SJ, Dokukin ME, Clarkson B, Sokolov I (2016) AFM study shows prominent physical changes in elasticity and pericellular layer in human acute leukemic cells due to inadequate cell-cell communication. Nanotechnology 27(49):494005. https://doi.org/10.1088/0957-4484/27/49/494005
20. Kumar S, Hoh JH (2004) Modulation of repulsive forces between neurofilaments by sidearm phosphorylation. Biochem Biophys Res Commun 324(2):489–496
21. Brown HG, Hoh JH (1997) Entropic exclusion by neurofilament sidearms: a mechanism for maintaining interfilament spacing. Biochemistry 36(49):15035–15040
22. Kosaki R, Watanabe K, Yamaguchi Y (1999) Overproduction of hyaluronan by expression of the hyaluronan synthase Has2 enhances anchorage-independent growth and tumorigenicity. Cancer Res 59(5):1141–1145
23. Itano N, Atsumi F, Sawai T, Yamada Y, Miyaishi O, Senga T, Hamaguchi M, Kimata K (2002) Abnormal accumulation of hyaluronan matrix diminishes contact inhibition of cell growth and promotes cell migration. Proc Natl Acad Sci U S A 99(6):3609–3614. https://doi.org/10.1073/pnas.052026799
24. Toole B (1982) Glycosaminoglycans in morphogenesis. In: Hay E (ed) Cell biology of the extracellular matrix. Plenum Press, New York, pp 259–294
25. Zimmerman E, Geiger B, Addadi L (2002) Initial stages of cell-matrix adhesion can be mediated and modulated by cell-surface hyaluronan. Biophys J 82(4):1848–1857
26. Camenisch TD, Schroeder JA, Bradley J, Klewer SE, McDonald JA (2002) Heart-valve mesenchyme formation is dependent on hyaluronan-augmented activation of ErbB2-ErbB3 receptors. Nat Med 8(8):850–855. https://doi.org/10.1038/nm742
27. Chen WY, Abatangelo G (1999) Functions of hyaluronan in wound repair. Wound Repair Regen 7(2):79–89
28. Jiang D, Liang J, Fan J, Yu S, Chen S, Luo Y, Prestwich GD, Mascarenhas MM, Garg HG, Quinn DA, Homer RJ, Goldstein DR, Bucala R, Lee PJ, Medzhitov R, Noble PW (2005) Regulation of lung injury and repair by toll-like receptors and hyaluronan. Nat Med 11(11):1173–1179. https://doi.org/10.1038/nm1315
29. de la Motte CA, Hascall VC, Drazba J, Bandyopadhyay SK, Strong SA (2003) Mononuclear leukocytes bind to specific hyaluronan structures on colon mucosal smooth muscle cells treated with polyinosinic acid:polycytidylic acid: inter-alpha-trypsin inhibitor is crucial to structure and function. Am J Pathol 163(1):121–133
30. Richards JS (2005) Ovulation: new factors that prepare the oocyte for fertilization. Mol Cell

Endocrinol 234(1–2):75–79. https://doi.org/10.1016/j.mce.2005.01.004

31. Toole BP (2004) Hyaluronan: from extracellular glue to pericellular cue. Nat Rev Cancer 4 (7):528–539. https://doi.org/10.1038/nrc1391

32. Dokukin ME, Sokolov I (2017) Nanoscale compositional mapping of cells, tissues, and polymers with ringing mode of atomic force microscopy. Sci Rep 7(1):11828. https://doi.org/10.1038/s41598-017-12032-z

Chapter 29

Probing Single Virus Binding Sites on Living Mammalian Cells Using AFM

Martin Delguste, Melanie Koehler, and David Alsteens

Abstract

In the last years, atomic force microscopy (AFM)-based approaches have evolved into a powerful multi-parametric tool that allows biological samples ranging from single receptors to membranes and tissues to be probed. Force-distance curve-based AFM (FD-based AFM) nowadays enables to image living cells at high resolution and simultaneously localize and characterize specific ligand-receptor binding events. In this chapter, we present how FD-based AFM permits to investigate virus binding to living mammalian cells and quantify the kinetic and thermodynamic parameters that describe the free-energy landscape of the single virus-receptor-mediated binding. Using a model virus, we probed the specific interaction with cells expressing its cognate receptor and measured the affinity of the interaction. Furthermore, we observed that the virus rapidly established specific multivalent interactions and found that each bond formed in sequence strengthens the attachment of the virus to the cell.

Key words Atomic force microscopy, Single-molecule force spectroscopy, Fluorescence microscopy, Virus, Virus-host interactions, Force-distance curve, Receptor-ligand bonds, Free-energy landscape, Tip functionalization, Dynamic force spectroscopy

1 Introduction

1.1 Virus-Cell Interactions

Viruses are small and simple parasitic agents that cannot reproduce themselves. Because of their simplicity, they strictly depend on a host organism in nearly all steps of the infection cycle. Through the evolution viruses acquired the relevant molecular "passwords" or "entrance tickets" enabling to control and hijack cellular functions [1]. Consequently, nearly all viruses are species-specific and only infect a narrow range of organisms.

The infection pathway of a virus particle from its binding to the cell surface, its entry into the cytosol, and the delivery of their genetic cargo within the nucleus consists of a series of consecutive steps tightly regulated [1]. The first step starts with the virus

Martin Delguste and Melanie Koehler are contributed equally to this work.

Yuri L. Lyubchenko (ed.), *Nanoscale Imaging: Methods and Protocols*, Methods in Molecular Biology, vol. 1814,
https://doi.org/10.1007/978-1-4939-8591-3_29, © Springer Science+Business Media, LLC, part of Springer Nature 2018

landing or "touchdown" on the cell surface via interactions, whether specific or not, between virion-exposed glycoproteins and cell surface proteins. These preliminary interactions are followed by the engagement of specific receptors. These first interactions already define the consecutive complex series of processes to which viruses have to face to gain access to the intracellular compartment [2]. Such processes include virus uptake, intracellular trafficking, and, finally, penetration to the cytosol. Tremendous effort has been made to characterize the cellular receptors and entry pathways [3], but the molecular details by which these interactions determine cellular binding and uptake are poorly understood because of the lack of suitable techniques that allow to gain information on the molecular interactions that occur at the single virus-receptor level. The understanding and exploration of the first steps of receptor-mediated endocytosis of viruses, from receptor binding to the physical internalization of the viral particle into host cells and their dynamics, is an important challenge in virology. A full picture of these interactions would provide insights valuable to medicine, cell biology, molecular biology, neurobiology, structural biology, biochemistry, and biophysics and offers novel potential therapeutic strategies [4–6].

1.2 Current Methods to Study Virus Binding to Cell Surface Receptors

In the context of virus host interactions, the cell imposes multiple barriers to the virus entry. However, viruses exploit fundamental cellular processes to gain entry to the cell and deliver their genetic cargo. Virus entry is largely defined by the first interactions that take place at the cell surface. These first interactions determine the mechanism of virus attachment to the cell surface, the penetration of the virus, and ultimately the transfer to the cytosol. Methods to study viral infection, especially virus binding, have undergone rapid development over the last decade, in particular since viruses can be propagated and purified. It now becomes easier to obtain purified viruses for studies using biochemical and biophysical techniques [7–9]. Among the well-established methods, most of the techniques rely mainly on ensemble studies that give an average response of a population of virions, failing to account for biological variability or on methods that do not preserve the physiological state of the cells or the virus (performed either on fixed cells or with isolated receptors). Moreover, most of the methods developed so far are based on binding assays with long incubation periods, thus lacking time resolution to decipher the dynamic character of the first binding steps. As an example, solid-phase binding assays (Fig. 1a) are used to measure or screen virus binding to a variety of receptor molecules such as glycan moieties [10], in which the investigated receptor is coupled to a flat surface and is allowed to interact with intact viruses [11]. Thermodynamic properties of virus-receptor binding can be obtained using surface plasmon resonance (SPR) (Fig. 1b) [12, 13]. SPR consists in flushing receptor

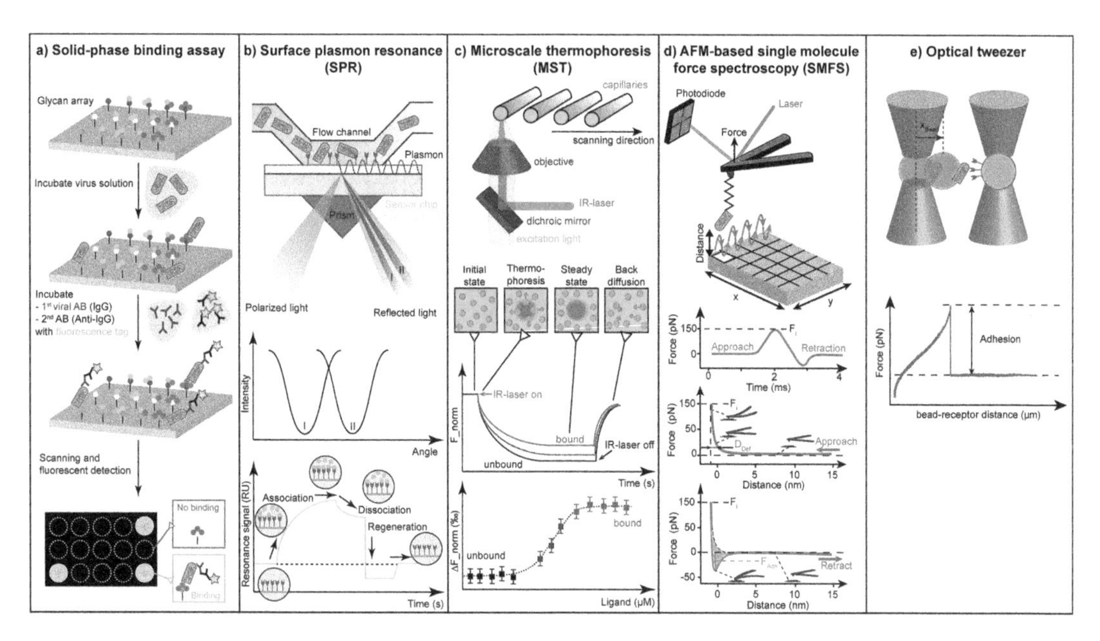

Fig. 1 Overview of current methods available to study virus-receptor interaction. (**a–c**) Ensemble binding assays: (**a**) solid-phase binding assay, where the investigated receptor is coupled to a flat surface and allowed to interact with intact viruses. By utilizing a "sandwich-building" approach, virus binding is detected via fluorescently labeled secondary antibodies. (**b**) Surface plasmon resonance (SPR) is used to obtain thermodynamic properties of virus-receptor binding. The receptor molecules are flushed into a chamber, where they interact with a gold-coated sensor chip. Subsequent flushing of the virus into this chamber allows determining association and dissociation kinetics by following the time vs. resonance signal curve. (**c**) Microscale thermophoresis makes it possible to measure binding and unbinding kinetics in solution under defined, controlled conditions. Microscale thermophoresis is the directed movement of particles in a microscopic temperature gradient. Any change in the hydration shell of biomolecules due to changes in their structure/conformation (e.g., after virus binding) results in a negative change of the movement along the temperature gradient and thus can be used to determine binding affinities. (**d**, **e**) Single-molecule techniques: (**d**) AFM-based single-molecule force spectroscopy can be utilized to determine intermolecular forces by repeatedly approaching and withdrawing a virus-functionalized AFM tip to their respective (cell surface) receptors in z-direction. (**e**) Optical tweezers are instruments that use a highly focused laser beam to provide an attractive or repulsive force, depending on the refractive index mismatch to physically hold and/or move microscopic dielectric objects similar to tweezers

molecules into a chamber, where they interact with a gold-coated sensor chip. Subsequent flushing of the virus into this chamber allows determining association and dissociation kinetics. The main limitation about these methods is the poor control of ligand density and orientation, which presumably affects binding [14]. More importantly, an error in SPR can arise from the multivalency of the interaction leading to an underestimation of the dissociation rate due to the local high concentrated ligand on the viral surface. Recently, microscale thermophoresis (MST) (Fig. 1c) was applied to the study of receptor-virus interactions [15]. Besides the advantage that binding and unbinding kinetics can be measured in solution under defined and controlled conditions, this method requires a complex environment of a 3D host cell plasma membrane, with the receptors of interest incorporated. This could give rise to difficulties in isolating the effect of specific molecules. Also radioactive labeling of structural viral components and electron microscopy (EM) of infected cells have been used to investigate virus binding [16, 17]. Even though EM techniques are able to give visual insights into virus entry and even spectacular three-dimensional images of the samples, the identification of cellular factors and pathways involved in the uptake process is difficult. Moreover, to characterize virus binding to cells by EM usually requires high virus concentrations and can only be operated under vacuum, which does not reflect physiological conditions and lacks dynamics.

For these reasons, compared to conventional ensemble methods, single-molecule experiments offer distinct advantages. First, conducting many sequential measurements enables to determine the distribution of molecular properties of inhomogeneous systems. Second, being direct records of the fluctuations of the system, single-molecule trajectories provide dynamic and statistical information, which are often hidden in ensemble-averaged results. Finally, they permit real-time observation of rarely populated transients, which are difficultly captured using conventional methods [18, 19]. Atomic force microscopy (AFM)-based single-molecule force spectroscopy (SMFS) (Fig. 1d) and optical tweezers (Fig. 1e) or magnetic tweezers provide powerful tools to measure forces with single-molecule resolution and high temporal resolution [20, 21]. Well-developed, specific grafting protocols allow the attachment of single viral particles on AFM cantilevers or beads [22–26]. Such measurements allow characterizing the binding of intact viruses on living cells that are kept close to physiological conditions and have been used on a variety of viruses in the past [9, 22, 23, 27, 28]. However, assignment of forces to their corresponding molecular interactions remains difficult. To circumvent this limitation, the use of mutant virions (i.e., lacking individual glycoproteins) or the comparison of unbinding parameters with molecular dynamic (MD) simulation [22] is a valuable control. Yet, regular MD simulation, such as steered molecular dynamic (SMD) simulation, is capable of modeling force spectroscopy experiments, but the modeling is limited to very fast pulling rates not amenable to AFM, including a unique high-speed AFM (10^7 times higher). To overcome this limitation, Zhang et al. described an alternative computational all-atom Monte Carlo pulling (MCP) approach that enables to model results at pulling rates comparable to those used in AFM pulling experiments [29]. Just very recently, Booth et al. introduced boxed molecular dynamics (BXD) simulation as another tool to reach the appropriate time scales required for AFM force spectroscopy experiments [30].

1.3 Atomic Force Microscopy

Since its invention by Binnig et al. in 1986 [31], the AFM has become a powerful tool in biology, physics, chemistry, and medicine. Being a multi-versatile imaging platform, it enables the visualization and manipulation of biological samples, from single molecules to living cells with sub-nanometer lateral resolution and under quasi-physiological conditions [32–36]. In addition to high-resolution imaging, the high sensitivity of force measurements allows the determination of inter- and intramolecular forces (piconewton (pN) range) at the single-molecule level [37]. Moreover, data obtained from force spectroscopy include physical parameters (e.g., stiffness, friction, elasticity) not measurable by other methods and open new perspectives in exploring the regulation of the dynamics of biological processes [38]. It can also capture dynamic features of individual molecules in the millisecond time scale [39]. The proof-of-principle stage of the pioneering experiments has already evolved into established methods for exploring kinetic and structural details of interactions and molecular recognition processes.

Compared to conventional ensemble methods, single-molecule experiments offer several advantages as already mentioned in the section before [18, 19]. AFM has successfully complemented electron microscopy and X-ray diffraction studies of viruses [40]. Moreover, force spectroscopy measurements have been used to study the mechanics between viral envelope proteins and host cell receptors at the single-molecule level in living cells [9, 22, 27, 41]. Taken together, the unique flexibility of AFM to image, probe, and manipulate materials and biological systems (under quasi-physiological conditions) [36] made it a highly versatile instrument in nanoscience and nanotechnology as well as biology and stimulated numerous discoveries and technologies [42]. Thus, it makes it an optimal tool to explore the mechanisms by which virus-cell surface receptor bonds are formed as a starting point of cell entry and which properties they possess in vivo. This being so, the specific binding of a particular virus to cell surface receptors should be best characterized by the lifetime, affinity, and free energy of the virus-receptor bonds. While AFM alone is an appropriate quantitative method to characterize binding properties, it lacks the capacity to identify host cell receptors. In this context, light microscopy has been a standard tool in cell biology for decades, but bringing both techniques together in a sealed physical environment remained difficult for a long time. To address this challenge, we recently introduced the combination of FD-based AFM [33] and confocal microscopy under cell culture conditions to simultaneously image animal cells and topographically map the specific binding events of single viruses [41, 43].

1.3.1 AFM Imaging of Viral Particles and Living Cells

In conventional AFM imaging mode, a sharp tip placed at the free end of a cantilever contours the sample surface and generates a 3D image. Different operating modes allow to image biological samples. In contact mode, topographic images of biological specimens are obtained by maintaining the tip in contact with the sample [44–49]. Changes in the cantilever deflection are monitored and kept constant using an electronic feedback loop [50, 51]. The image consists of the calibrated height information about the sample relief. However, contact mode imaging turned out to be less suitable for weakly attached and soft samples, as biomolecules are often pushed away or get damaged by the AFM stylus during imaging [52]. To overcome this disadvantage, dynamic force microscopy (DFM, originally termed tapping or oscillation mode) was invented to minimize the friction and the force applied between tip and sample [53]. In its simplest application, the cantilever is oscillated close to its resonance frequency as it raster-scans over the surface and touches the sample only at the end of its downward movement, resulting in amplitude reductions at positions of elevated objects. The reduction in oscillation amplitude is used as the feedback-control signal to measure the surface topography. As the lateral forces are greatly reduced during imaging, the study of biological specimens has therefore exceptionally benefited from the development of DFM and has been applied to a variety of biological objects that are only weakly adsorbed to supports [53–56] or are highly corrugated, such as living cells [57].

With AFM, virus particles can be visualized in appropriate buffers and at room temperature [58–61]. Most importantly, AFM yields three-dimensional images and does not rely on symmetry averaging. In contrast to EM, the resolution of AFM is very good in the vertical direction (less than a nanometer). AFM has been utilized to study various viruses and their substructures by topographical imaging [62], their mechanical properties [63–65] and human immunodeficiency virions were imaged on lymphocytes at high resolution, and considerable details of the process of virus-cell attachment were obtained [66].

Living mammalian cells are very fragile and complex systems protected from the external environment by a highly dynamic and flexible barrier, called the plasma membrane. This very sophisticated structure contains a wide variety of biomolecules, such as lipids (phospholipids, cholesterol), (glycol-) proteins, and small quantities of carbohydrates (sugar, polysaccharides), and plays key roles in fundamental cellular processes, such as signaling, communication, adhesion, and sensing. Depending on the physiological cell state, the structural and functional assembly of cellular surfaces can be adapted, changing dynamically its chemical and biophysical properties [67]. Therefore, cellular structures should be investigated close to their native state. High-resolution imaging of cellular surfaces usually requires fixed or frozen cells, failing in probing the dynamic character of the molecular events occurring at the plasma membrane (e.g., cryo-electron microscopy [68]) or to specifically label the studied molecules with fluorophores (e.g., far-field optical nanoscopy methods [69]). Nowadays, AFM provides a powerful tool to image the surface topography of living cells with a nanometer precision, in real time and under physiologically relevant conditions [70]. AFM has been successfully applied to gain insights into the surface morphology of microbes such as bacteria [71], fungi [72], or yeasts [73] with a resolution up to 10 nm [74]. Dynamic functional cellular processes were also observed using AFM imaging, such as bacterial pore germination [75] and cell division [76].

1.3.2 Single-Molecule Force Spectroscopy for Virus-Cell Interactions

The AFM can be used not only to image but also to manipulate biological samples. In the so-called SMFS (single-molecule force spectroscopy) mode, the AFM tip is approached and retracted from the sample while recording a force-distance (FD) curve (Fig. 2a). From the approach curve, structural height, surface forces, and mechanical deformation of the sample can be quantified, whereas from the retraction curve, the elastic modulus, dissipation, and adhesion can be extracted [77]. A tip functionalized with a certain molecule (e.g., lectin, antibody, or virus) is upgraded to a biosensor, able to measure specific interaction forces between the tip-linked molecule and cell surface receptors [78–80]. The interaction (unbinding) force is measured by following the deflection of the

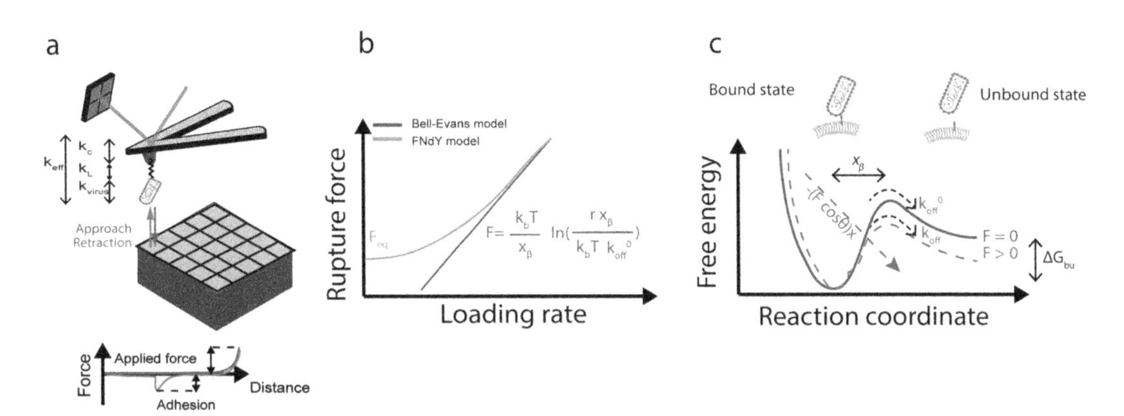

Fig. 2 Force-distance curve-based AFM and extraction of kinetic and thermodynamic properties of receptor-ligand bonds. (**a**) During each approach-retraction cycle, the AFM tip grafted with a single viral particle records a force versus distance curve, on which adhesion events between the probe and the sample can be observed. (**b**) Reporting the measured rupture force as a function of the loading rate that allows quantitative kinetic and thermodynamic parameters of the virus-receptor interaction to be extracted. To do so, the effective spring constant of the force transducer system can be evaluated by considering its different components as springs in series (a) (see text for details). (**c**) Free-energy potential separating bound and unbound states of a virus-cell receptor system in the absence (purple) and in the presence of an externally applied pulling force (dashed, pink). The external force tilts the energy landscape toward the dissociation of the interacting complex, reducing the energy barrier

cantilever, which behaves like a Hookean spring, where the force F is exerted on the sample by the AFM tip scales linearly with the cantilever deflection Δz, according to Hooke's law:

$$F = -k_c \cdot \Delta z$$

In this equation, k_c refers to the cantilever's spring constant and Δz to the deflection. By pulling the AFM tip away from the surface, an increasing force is applied onto the receptor-ligand complex until the pair dissociates. For a detailed protocol describing the application of single-molecule force spectroscopy on living cells, please refer to Puntheeranurak et al. [81].

However, because rupture forces observed between a tip-linked ligand and cell surface receptors depend on the rate at which force load on the bond is applied, the quantification of forces is relative [82]. In combination with theoretical models, probing the force dependency on the loading rate enables the receptor-ligand's free-energy landscape to be determined and the quantification of unique structural, kinetic, and energetic parameters of viruses interacting with cell surface receptors (Fig. 2b, c) [83]. In the past, SMFS measurements have been used to study the mechanics between viral envelope proteins and host cell receptors at the single-molecule level in living cells [9, 22, 27, 41], which are presented and explained in great detail in Subheading 1.3.3.

Another mode, which combines AFM imaging and SMFS, is called FD-based AFM, where each FD curve records a local stochastic unbinding event from which these properties can be inferred and directly mapped to the sample topography as FD curves are obtained at each pixel on the sample surface (Fig. 2a). Just very recently, two important limitations related to the lateral and temporal resolution were circumvented, allowing the imaging and force probing of biological samples with high lateral (~50–100 nm) and temporal resolution (~ms per FD curve) [33, 84, 85]. The latter will be described in detail in Subheading 1.3.4, as this mode is the most reliable and best way to study virus binding sites quantitatively on living cells.

Reconstruction of the binding free-energy landscape can be obtained using appropriate biophysical models giving access to the kinetic and thermodynamic parameters of biomolecular interactions. To this end, we need to relate how measurements performed out-of-equilibrium can give access to the equilibrium free-energy and kinetic parameters. The first phenomenological description of how an external force pulling on a bond reduces the activation energy barrier toward dissociation was described by Bell in a seminal article on cell adhesion (Fig. 2c) [86]. This description was later formulated by Evans and Ritchie (Bell-Evans model) [83]. Focused on the bond dissociation kinetics, the Bell-Evans model predicts that the force of a single energy barrier in the thermally activated regime linearly increases with the logarithm of the force loading rate (L_R): $F \sim \ln (L_R)$ (Fig. 2b) [83]. A rigorous theoretical framework for the estimation of thermodynamic parameters is provided by fluctuation theorems such as the Jarzynski equality (JE) and its generalization, the Crooks fluctuation theorem [87]. Previous studies have shown that the free-energy difference between initial and final equilibrium states could be calculated via a non-equilibrium, irreversible process that connects them, thus bridging the gap between equilibrium and non-equilibrium statistical mechanics [87]. A more recent theoretical approach, the Friddle-Noy-de Yoreo (FNdY) model, describes the force spectrum of a ligand-receptor bond attached to a force transducer, consisting in two primary regimes [88]. First, at low loading rate, a close-to-equilibrium regime exists, where rebinding events can occur, which is characterized by an equilibrium force, F_{eq}. At higher loading rate, a kinetic regime characterized by a fast non-equilibrium bond rupture is described. The F_{eq} defines the equilibrium force for the bond-transducer system:

$$F_{eq} = \sqrt{2k_{eff}\Delta G_{bu}}.$$

The mean rupture force is defined as

$$F = F_{eq} + F_{\beta} \ln \left(1 + e^{-\gamma} R(F_{eq})\right),$$

with

$$F_{\beta} = \frac{k_B T}{x_{\beta}} \text{ and } R(F_{eq}) = \frac{L_R}{k_{off}(F_{eq}) F_{\beta}}.$$

F_{β} is the thermal force, γ the Euler constant, $k_B T$ the thermal energy, k_{eff} the effective spring constant of the transducer, ΔG_{bu} the equilibrium free-energy between the bound and unbound state, and $k_{off}(F_{eq})$ the unbinding rate scaled by the Boltzmann-weighted energy of a spring extended between the barrier location and the relative displacement of the spring minimum under F_{eq} (for further details, *see* [88]).

1.3.3 Virus Interactions with Purified Receptors and Cells Using AFM

Virus-host cell surface interactions mark the first critical step of infection. Hence, forces involved in this process are essential. AFM-based single virus force spectroscopy (SVFS), whether combined with imaging or not, has become a robust, accurate, and reliable technique within the past decade. It has been applied to study virus interactions with purified receptors and on cells. A few selected examples of SMFS studies are described here in more detail, and a summary can be found in Table 1 although the authors raise no claim to completeness.

Rankl et al. [23] showed that human rhinovirus (HRV) forms multiple parallel interactions with living host cells utilizing AFM-

Table 1
Overview of SVFS studies on purified receptors and/or living cells in the years 2005–2017 (no claim to completeness)

Virus	Purified receptors, (living) cells, membranes, etc.	Study	Author, year, references
Vesicular stomatitis virus (VSV)	Membranes of different phospholipid compositions	Carneiro et al. studied the interactions between VSV and phosphatidylserine and found out that binding forces dramatically depend on the membrane phospholipid composition	Carneiro et al., 2005 [110]
HIV-1	Receptor expressing cells	Monitoring of early fusion dynamics of HIV-1 at single virus level	Dobrowsky et al., 2008 [28]
Human rhinovirus	– Very low-density lipoprotein receptor (VLDLR) – Cells expressing LDLR	Discovery of multiple receptors involved in human rhinovirus attachment to living cells and comparison to experiments on purified receptor surfaces	Rankl et al., 2008 [23]
Tobacco mosaic virus		They show the possibility to study nucleic acid-protein interactions in more complicated systems using SVFS on tobacco mosaic virus, where they investigate RNA-coat protein interactions by pulling genetic RNA step by step out of the virus	Liu et al., 2010 [111]
Influenza virus	Different types of living cells	Sieben et al. showed that influenza virus binds its host cell using multiple dynamic interactions	Sieben et al., 2012 [22]
– Vaccinia virus – Influenza virus – Bacteriophage AP22	– Erythrocyte monolayer – Bacterial film of phage-sensitive *Acinetobacter baumannii*	They show that it is of great importance in terms of reliability of the results to verify the presence of the virus particle attached to the tip by complementary methods such as electron microscopy and/or dielectrophoresis	Korneev et al., 2016 [112]
Influenza A virus (H1N1)	– CHO cells – Human alveolar basal epithelial cells	This group used SVFS as a supporting technique to genetically characterize an adapted pandemic 2009 H1N1 influenza virus that reveals improved replication rates in human lung epithelial cells	Wörmann et al., 2016 [113]
Pseudotyped rabies virus	Living MDCK cells expressing receptor for virus binding	Alsteens et al. introduced an advanced AFM probing technique in combination with optical microscopy to nanomechanically map the first binding steps of a virus to animal cells	Alsteens et al., 2017 [41]

based SVFS. Moreover, the binding forces were confirmed by in vitro experiments on artificial receptor surfaces, and an estimation of the number of receptors involved in binding was extracted. Furthermore, estimation of k_{on}/k_{off} describing the kinetics of the interaction between HRV2 and plasma membrane-anchored receptors was obtained. In another application of SVFS, the variability of single-molecule interactions for influenza virus was studied using AFM measurements on different living host cell types [22]. Using various cell types that differ with respect to their sialic acid surface composition, the study revealed that hemagglutinin (HA, viral envelope spike protein) receptor specificity might not be a direct indicator for binding to living cells. Moreover, sequential unbinding events were observed with each individual event following a unique unbinding trajectory with different kinetic and thermodynamic parameters. Also for HIV-1 (human immunodeficiency virus type 1), the interaction of the spike protein with co-receptors was studied using SVFS. It has been shown that engagement with the primary receptor CD4 is very stable but only for a short lifetime until the viral glycoprotein gp120, organized with gp41 in a homotrimeric complex, finds its co-receptor molecule [28]. More recently, Alsteens et al. [41] introduced an AFM-confocal microscopy setup that allows imaging cell surfaces and simultaneously probing virus binding events within the first millisecond of contact. Moreover, they present theoretical approaches to contour the free-energy landscape of early binding events between an engineered virus (rabies virus, RABV) and cell surface receptors.

SVFS enables to decipher the role of individual viral constituents during their very first interactions with target cells. Furthermore, the time scale of force spectroscopy experiments allows gaining insights into the dynamics of the molecular processes involved in virus-receptor interactions and virus internalization. Overall, this innovative method addresses the molecular mechanism of virus binding with high spatial and high temporal resolution and, moreover, provides quantitative insights into the kinetics and thermodynamics of individual binding steps.

1.3.4 Multiparametric Imaging and Quantitative Mapping of Virus Binding Sites

For a long time, AFM investigations of cellular processes have been limited by their poor spatial and temporal resolution. In addition, AFM imaging of mammalian cells is still a challenging task, as cell surface components can be easily deformed and damaged by the vertical and lateral forces applied by the scanning probe. Therefore, FD-based AFM methods that vertically oscillate the cantilever in the kilohertz range on top of the sample have been developed. Recording FD curves at frequencies much lower than the resonant frequency of the cantilever allows precise control of the applied force in the piconewton range [89] with a high positional accuracy. Thereby, cellular membranes can be imaged at high spatial resolution while probing dynamic molecular events occurring in the millisecond range. Furthermore, the reduced contact time between the AFM tip and the sample (~ms) limits damaging lateral forces on the examined structures, yielding topographs of cell surfaces closer to their native state.

Using FD-based AFM, the interactions occurring between viruses and their receptors on the plasma membrane can be studied directly on living cells. Using AFM tips functionalized with a single viral particle, the surface of living cells can be imaged while recording virus adhesion events simultaneously. Both topography and adhesion parameters extracted from each point of the probed surface can be displayed in correlated maps with high resolution. This allows locating and evaluating the number and density of cell surface receptors that interact with the single viral particle. Furthermore, a FD curve can be displayed from each pixel of the recorded maps, so that FD curves corresponding to virus binding events can be extracted for further DFS analysis. Thereby, kinetic and thermodynamic parameters governing the complex virus-receptor interactions in close-to-physiological conditions can be extracted. As the attachment of viral particles to cellular surfaces is usually a multistep process involving multiple glycoprotein-receptor binding events, this method allows to gain insights into the mechanistic processes involved in the initial events of viral infection. By combining this approach with the tools of genetic engineering, the individual role of viral glycoproteins can be deciphered using mutant viruses deficient in various glycoprotein expressions. This provides indications on how viruses modulate and optimize their attachment to cellular surfaces to efficiently gain access to the cytoplasm.

In this chapter, we provide a step-by-step description of experiments to investigate interactions between viral particles and cell surface receptors. The preparation of required biological materials is first described, going from virus production to the preparation of cell monolayers. Then, the grafting of viral particles to AFM tips is explained, following a protocol developed by Prof. H. Gruber (JKU Linz, Austria). This method consists in covalently attaching virions to the tip apex via a three-step cross-linking procedure using a polymeric linker. Then, we describe how to perform FD-based AFM to probe virus interactions with cells maintained alive under the microscope. This goes together with the use of fluorescence confocal microscopy, to provide an internal control pointing out virus interactions with a specific receptor of interest. Finally, the processing of recorded FD curves is explained in order to extract quantitative maps of adhesion events, together with the use of biophysical models to determine kinetic and thermodynamic parameters of the probed interactions. The protocol is based on the use of a Bruker AFM system which enables to record FD curves with a sinusoidal movement. This unique movement constantly varies the speed at which interactions are probed, facilitating the acquisition of dynamic force spectroscopy data.

2 Materials

2.1 Viruses, Cell Lines, and Reagents

1. Phosphate-buffered saline (PBS) buffer.
2. Virus solution: suspension of virions in buffer (e.g., PBS) (~10^8 PFU mL^{-1}).
3. Animal cell lines for cell culture (e.g., MDCK or CHO cells) expressing and non-expressing the receptor of interest.
4. Cell culture medium: buffer (e.g., Dulbecco's modified Eagle medium with 4500 mg/L glucose, L-glutamine, sodium pyruvate, and sodium bicarbonate), 10% serum (e.g., fetal bovine serum, sterile filtered), and antibiotics (e.g., 100 units mL^{-1} penicillin and 100 μg mL^{-1} streptomycin).
5. Trypsin-EDTA 1x.
6. Dulbecco's phosphate-buffered saline (DPBS) buffer.

2.2 Equipment

1. Atomic force microscope with required capabilities (e.g., BioScope Catalyst or BioScope Resolve (Bruker Nano, Santa Barbara, CA) with PeakForce QNM mode).
2. Inverted optical microscope (Observer Z1, Zeiss, Germany) equipped with epifluorescence or confocal microscopy (LSM 800, Zeiss, Germany).
3. Upright bench top microscope for examining probes (e.g., Stemi DV4, Zeiss, or equivalent).
4. Si_3N_4 cantilevers with spring constants of ~0.08 N/m (e.g., PeakForce QNM- Live Cell (PFQNM-LC), Bruker Nano, Santa Barbara, CA).
5. Active vibration isolation table (TS 150, HWL Scientific Instruments GmbH, Germany).
6. Acoustic enclosure for the AFM and inverted optical microscope with thermoregulation at 35 ± 1 °C.
7. Silicon hollow fiber membrane module (PDMSXA-2500 cm^2, PermSelect, MedArray, Ann Arbor, MI).
8. Synthetic air supplemented with 5% CO_2 gas bottle.
9. Pressure-reducing regulator with flow rate control (0.1–1 l/min) (Swiss Opto varius, Gloor, Switzerland).
10. Resin tubing (Cole-Parmer).
11. Glass-bottom Petri dishes.
12. Sofware for FD curve analysis (e.g., NanoScope software (NanoScope 9.3R1; Bruker)).
13. Ultraviolet radiation and ozone (UV-O) cleaner (Jelight).

2.3 Chemicals for Tip Functionalization

1. Ethanolamine hydrochloride ($H_2NC_2H_4OH$).
2. Dimethyl sulfoxide anhydrous (>99.9% $(CH_3)_2SO$).
3. Molecular sieves, 3 Å (beads, 8–12 mesh).
4. Heterobifunctional cross-linker (e.g., NHS-PEG_{27}-acetal provided by H. Gruber, JKU, Linz, Austria).
5. Triethylamine (>99.5% $(C_2H_5)_3N$).
6. Sodium cyanoborohydride ($NaCNBH_3$).
7. Citric acid ($HOC(COOH)(CH_2COOH)_2$).
8. Chloroform (>99.9% $CHCl_3$).
9. Ethanol absolute, G Chromasolv (>99.9% C_2H_5OH).
10. Milli-Q water (Millipore).
11. Acidic piranha solution (70% sulfuric acid, 30% oxygen peroxide).

3 Methods

3.1 Virus Production

To ensure successful functionalization of AFM tips, a highly pure virus solution is required. This ensures a solution of viral particles that is free from cellular aggregates or macromolecules that would adsorb to the AFM probe. This tip contamination can modify the shape of FD curves and alter the topography images by lowering the resolution and introducing imaging artefacts [33]:

1. To reach a high purity of viral solutions, perform ultracentrifugation through sucrose, cesium chloride, iodixanol, and/or potassium tartrate density gradients [90, 91].
2. Suspend the amplified virions in buffer solution (e.g., PBS), and use as such for tip functionalization (*see* Subheading 3.2). Depending on the size of virions and the tip radius of curvature, use concentrations around $\sim 10^7$–10^{10} plaque-forming units (PFU)/mL.

3.2 Cell Culture and Sample Preparation

1. Grow the animal cell lines in Dulbecco's modified Eagle medium supplemented with serum and antibiotics within cell culture flasks at 37 °C in a humidified, 5% CO_2-supplemented atmosphere.
2. Perform regular cell passages by detaching confluent monolayers of cells from the bottom of the flask using trypsin and seeding them in a less concentrated fashion.
3. Prior to AFM measurements, perform the last cell passage between 1 and 3 days before the experiment, depending on the rate at which the cells grow. Seed the cells in a glass bottom

Petri dish in an adequate concentration to reach confluence on the day when the experiment is planned. Cells should be well adhered on the surface and form a continuous monolayer.

3.3 Tip Selection

1. Cantilevers with a small spring constant ($k_c \sim 0.01$–$0.1\ \text{N m}^{-1}$) are required for AFM imaging or force spectroscopy of biological samples in order to achieve high lateral resolution and to measure adequately forces arising from interactions between single biomolecules ($F \sim 5$–250 pN) [32].
2. To allow the tip to precisely follow the vertical movement applied by the piezoelectric scanner, the resonance frequency of the cantilever has to be at least five times higher than the frequency at which the probe is oscillated to record FD curves [85].
3. For the abovementioned reasons, and since the detection of fast and dynamic biomolecular interactions requires to oscillate the probe in the kilohertz range, cantilevers with high resonance frequencies (>100 kHz) are utilized [92].
4. Due to the complexity of the structure of cellular samples, specially designed probes (e.g., PeakForce QNM-Live Cell (PF-LC) probes, Bruker) are used to image living cells [89]. Usually, height differences between adjacent cells cause imaging problems because AFM probes have a tip height below 5 μm. This leads to shadowing and blind spots in the images where the tip was not able to reach the surface and the cantilever comes into contact with the cell body. Specially designed tips with height ~17 μm enable to image cell surfaces with large height differences. The high resolution is maintained, thanks to a protruding area at the tip apex, displaying a radius of curvature of ~65 nm.

3.4 Tip Functionalization

In the past decade, a lot of progress has been made in developing and optimizing coupling strategies for single-molecule force spectroscopy [93]. The most common functionalization method is tethering the virus on an AFM tip in a multiple-step cross-linking procedure, including the creation of reactive amino groups on the chemically inert tip surface [94], followed by the covalent binding of the cross-linker (with different lengths) and finally coupling of the sensor molecule. The inert water-soluble PEG cross-linker is designed heterobifunctional [95] with one end being an N-hydroxysuccinimide (NHS) group to bind free amino groups on the AFM tip and the other end specifically designed to react with the desired virus for coupling (Fig. 3). Since lysine residues (amino acid with a pending NH_2 group) are abundant in most of the glycoproteins of virus surfaces, the acetal-PEG_{27}-NHS linker for virus coupling is utilized [93]:

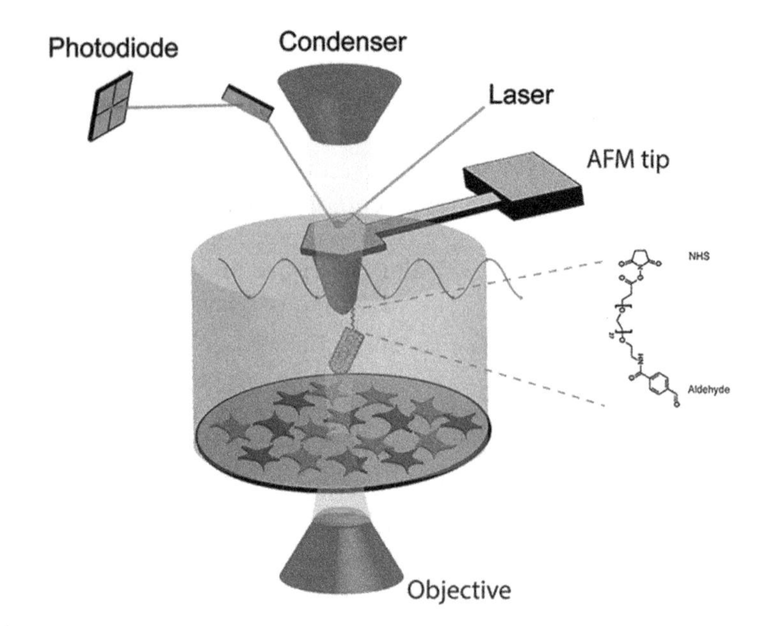

Fig. 3 Combined atomic force and fluorescence microscopy setup for the investigation of virus-cell receptor interactions. A Petri dish containing mixed fluorescent cells expressing a particular receptor and non-fluorescent cells that do not express the receptor is placed on an inverted optical fluorescence microscope. A virus particle is grafted to the AFM tip by means of a hetero-bifunctional PEG linker, which can then be oscillated over the confluent mono-layer, to record virus-cell interactions

1. *Aminofunctionalization of the AFM tip* (**steps 1**–7): Wash cantilevers in chloroform for 10 min, rinse with ethanol, and dry them in a gentle stream of argon or nitrogen gas. After that, further clean the cantilevers for 10 min in an ultraviolet radiation and ozone (UV-O) cleaner (Jelight).
2. Dissolve 3.3 g ethanolamine hydrochloride in 6.6 mL DMSO in a crystallization dish, cover with lid, heat to ~70 °C for complete dissolution, and let cool to room temperature after everything is dissolved. In addition, a magnetic stirrer can be added to the solution to speed up the dissolving process.
3. Immerse a Teflon block for the cantilevers in the center of the dish, and add 3 Å molecular sieve beads to cover the surrounding area (~ 25% of the total volume of the liquid).
4. Put the dish into a desiccator (or vacuum chamber), and degas the solution and the molecular sieve beads by applying an aspirator vacuum for ~30 min.
5. Place the cantilevers on the Teflon block, cover with lid, and incubate at room temperature overnight.
6. Wash cantilevers in DMSO (3×1 min) and ethanol (3×1 min).

7. Dry tips with a gentle stream of nitrogen or argon gas, and store under argon in a dust box for up to 3 weeks (preferably <1 week), if tips are not used immediately.
8. *NHS-PEG$_{27}$-acetal coupling* (**steps 8–10**): Dissolve one portion of NHS-PEG$_{27}$-acetal (1 mg) in chloroform (0.5 mL), transfer the solution into a Teflon reaction chamber, add triethylamine (30 μL), and mix by pipetting.
9. Immediately place ethanolamine-coated cantilevers in the reaction chamber, cover the chamber, and incubate for 2 h.
10. Wash with chloroform (3 × 10 min) and dry with nitrogen gas. Store cantilevers for up to several months under argon, or continue with next step.
11. *Virus coupling to the free acetal end* (**steps 11–21**): Immerse the cantilevers functionalized with the NHS-PEG$_{27}$-acetal cross-linker for 10 min in 1% (wt/vol) citric acid (in water).
12. Wash the cantilevers in Milli-Q water (3 × 5 min), and dry with nitrogen or argon gas.
13. Freshly prepare a 1 M solution of sodium cyanoborohydride containing 20 mM NaOH from the following components: 13 mg of $NaCNBH_3$, 20 μL of 100 mM NaOH, and 180 μL of Milli-Q water.
14. Place the AFM chips in a radial arrangement (with the tips in the center and facing upward) on Parafilm in a polystyrene Petri dish.
15. Thaw an 80 μL aliquot of virus solution, and centrifuge at 1677 × *g* for 5 min at room temperature to remove aggregates.
16. Fill a Styrofoam box with ice, and place this under a laminar flow. **Steps 17–20** should all be performed on ice under a laminar flow.
17. Place the polystyrene Petri dish with the cantilevers from **step 14** on ice.
18. Pipet an 80 μL droplet of the virus solution supernatant carefully onto the tip of each cantilever, so that they are extended into the virus drop.
19. Add 2 μL of the freshly prepared 1 M $NaCNBH_3$ stock solution to the virus solution droplet, mix carefully by pipetting, cover with a lid, close the ice box, and incubate for 1 h at +4 °C.
20. Add 5 μL of ethanolamine solution (1 M, pH 8.0) to the drop on each cantilever, mix cautiously by pipetting, cover with a lid, close the ice box, and incubate for 10 min at +4 °C to quench the reaction.
21. Remove the cantilevers from the dish, wash them once in ice-cold PBS or any other buffer, and store them in individual wells

of a clear polystyrene 12-well plate containing 2 mL of ice-cold buffer per well until used in AFM experiments.

22. During these functionalization steps, the virus-functionalized cantilevers are never allowed to dry. Transfer the functionalized AFM cantilevers to PBS buffer and then to the AFM very rapidly so that they do not dry (within 20 s). The cantilevers for AFM experiments should be used the same day they are functionalized with the virus.
23. *Control of the successful virus grafting:* Record confocal microscopy images of the tip functionalized with a fluorescently tagged virus according to Alsteens et al. [41]. Alternatively, take scanning electron microscopy (SEM) images of the probe to visualize the virions at the tip apexes. The latter is more challenging, as it requires to dehydrate the sample without destroying the delicate tips to allow SEM imaging in vacuum. Perform this through immersing the tips in graded ethanol baths (e.g., 30, 50, 75, 90, 100% for 10 min each) [41].

3.5 AFM Imaging and Probing of Biophysical Properties on Live Cells

Measuring interactions between functionalized tips and cellular membranes can lead to the detection of nonspecific adhesion events with other cell surface molecules. Therefore, as an internal control, cells containing the endogenous receptor should be mixed with cells lacking the receptor of interest enabling direct comparison. To this end, different cell types are fluorescently labeled (Fig. 3):

1. *Sample, tip, and setup preparation* (**steps 1**–7): Split both cell lines (cell line expressing the fluorescent receptor and the parental-control cell line), and add 1/8 (if splitting from near-confluent T-25 flasks) of the cells of each type to a single 15 mL conical tube in a total of 2 mL of fresh cell culture medium. Mix well with a serological pipette.
2. Seed both cell lines together in a glass-bottomed dish, and culture them in cell culture conditions (37 °C, 5% CO_2, 95% RH) in order to reach confluency on the day of the AFM experiment.
3. Transfer the dish to the AFM heated at 37 °C.
4. Set up a 95% relative humidity atmosphere of 5% CO_2 in air by blowing the gas mixture through the silicon hollow fiber membrane filled with Milli-Q water at a flow rate of 0.1 L/min.
5. Mount the functionalized AFM tip into the cantilever holder, and place it in the optical head.
6. Align the laser spot at the free end of the cantilever (in some setups, this can be done prior to mount the AFM head on the microscope), and maximize the sum into the photodetector (*see* **Note 1**).

7. Leave the system for ~15 min to equilibrate, until the signal into the photodiode is relatively stable, indicating a low thermal drift [85].
8. *Calibration of the force transducer and optical detection pathways* (**steps 8** and **9**): The output data from the AFM device has to be treated adequately, in order to accurately extract the force exerted between the tip and the sample. To do so, determine the sensitivity factor (nm V^{-1}) of the optical detection system (relation of voltage output from photodiode and deflection of the cantilever) from the tangent of the approach curve of the tip on a rigid substrate on which the sample vertical displacement is equal to the cantilever vertical deflection. The software can thereby determine the cantilever deflection required to generate a certain voltage difference into the position-sensitive photodiode.
9. Determine the spring constant of the cantilever, to allow identifying the force acting on the probe from the cantilever deflection. This is necessary, because the nominal spring constant of cantilevers estimated from the dimensions and material mechanical properties, which is furnished by manufacturers for a batch of probes, may differ significantly from the actual individual k_c value. To accurately relate the cantilever deflection to the force acting on the tip, the spring constant of each cantilever has to be determined experimentally [20]. The most commonly used method is based on relating the average energy of thermal vibrations of the cantilever with the absolute temperature of the system using the equipartition theorem [96–98]. This so-called thermal noise (or thermal tune) method is implemented in many commercial atomic force microscopes and is applicable for the calibration of cantilevers in liquid. It was however shown that the error in determining the spring constant of cantilevers can be as high as 20% [99], with large user-dependent variations in the calibration of the same cantilevers [100]. Therefore, some manufacturers provide AFM probes that are individually pre-calibrated using laser Doppler vibrometry, providing more accurate values for cantilever spring constants [101]. This allows performing a "no touch" calibration of the deflection sensitivity of the cantilever. This method is preferable for virus-derivatized probes, to avoid pressing the cantilever on a hard surface, which could damage the functionalized tip. To do so, introduce the pre-calculated value of the cantilever spring constant into the software, and perform a thermal tune (at a position where the tip is removed at least 100 μm from the sample surface) to calculate the deflection sensitivity.

10. *Combined FD-based AFM and confocal microscopy* (**steps 10–12**): Take a fluorescence image to distinguish between the different cell types on the monolayer.
11. Approach the AFM tip on top of the area of interest, and bring the tip gently in contact with the cell surface, using parameters minimizing the force (<500 pN) applied by the probe during the first contact (*see* **Note 2**) (Fig. 4a–c). Well-defined images are best taken on cells that form a relatively flat monolayer. The recorded image should contain at least two cells (one expressing the receptor of interest and one lacking the receptor) (*see* **Note 3**) to facilitate comparison of adhesion events and other biophysical properties (Fig. 4d, e).
12. Optimize the FD-based AFM imaging parameters in order to extract high-resolution topographs and adequate FD curves for subsequent analysis. This is largely a trial-and-error process and will differ between samples. The choice of imaging parameter values can be guided by the quality of the image, as well as the shape of recorded FD curves. The latter should exhibit a low noise level, a flat baseline (from the contact point), and a low hysteresis between the approach and retract curves. Extracted topography data should be similar from both trace and retrace scanning lines. Here are a few indications on fundamental parameters that can be tuned, together with typical value ranges used for animal cell imaging (*see* **Note 4**).
13. The *maximum force* applied is the imaging *set point* that is used as a feedback for the movement of the piezoelectric scanner. Set this force low enough to limit damages on cell surfaces but sufficiently high to allow tip-linked virions to reach the plasma membrane and interact with cellular receptors. Typical imaging forces are in the region of 300–500 pN, depending on the cell type, properties, and shape.
14. The *oscillation frequency* determines the number of approach-retraction cycles exerted by the AFM probe in a period of time and thus defines the contact time of tip-bound particles with the cellular surface. Lower frequency increases the contact time, which provides a longer time frame for virus-receptor interactions to take place. It is advised to use a 0.125 or 0.250 kHz oscillation frequency in order to allow sufficient time (~1 ms) for virions to bind receptors adequately as the tip-sample contact approximately occurs during one fourth of a scanning cycle. The speed of the scanning movement of the AFM probe should be adapted to the oscillation frequency used and be approximately 2000 times less. For example, for a 256 × 256 pixel image recorded using an oscillation frequency of 0.25 kHz, the scanning speed should be set to 0.125 Hz.

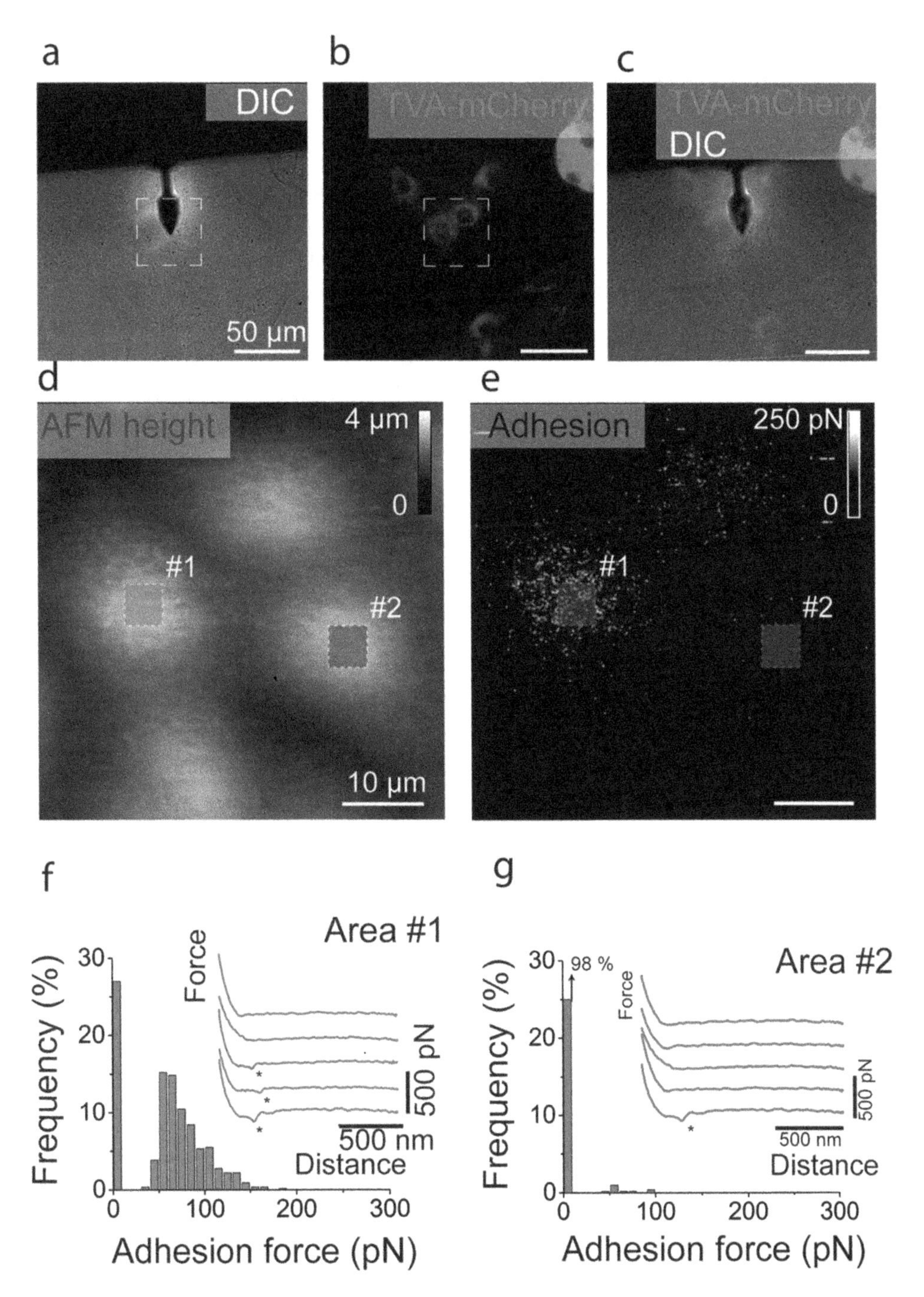

Fig. 4 FD-based AFM investigation of virus binding sites to living MDCK cells (**a–c**), DIC image, fluorescence channel, and superimposition of both images showing the confluent layer of cells expressing TVA-mCherry receptor surrounded with control cells. The AFM tip is placed above the region of interest. (**d**, **e**) FD-based AFM height image and adhesion map of cells recorded in the dashed square shown in (**a**, **b**). (**f**, **g**) Distribution of adhesion forces measured between the virus-derivatized AFM tip and two areas of cells highlighted in (**d**, **e**). The adhesion force of the last peak (asterisk) can be extracted for further analysis. The red data (area #1) are extracted from cells expressing the TVA receptor (as shown on the fluorescence channel), and the blue data (area #2) are from control cells. Insets show representative force-distance curves with asterisks indicating maximum adhesion peaks

15. The *oscillation amplitude* defines the height at which the tip is retracted from the imaged surface during the oscillation movement. When imaging soft and sticky samples such as cellular membranes, we recommend that amplitudes of at least 500 nm should be used to definitely pull the probe out of contact from the cell surface. Increased amplitudes induce a greater tip velocity and result in higher hydrodynamic drag. This problem can be partially circumvented by using specially designed probes, with a tip height > 15 μm. This reduces the hydrodynamic drag force variation, since the cantilever is moved far from the sample surface.
16. *Feedback gains* determine how quick the piezoelectric scanner will react to maintain the maximum measured force equal to the force set point. Increased gains allow imaging with a high resolution and accurately tracking the sample surface. Increase the gains until the point where the system oscillates and then reduced to a value slightly below that point to ensure maximal contrast imaging.

3.6 Data Processing

The data extracted from the raster scanning of cell monolayers comprise a topography image of the sample surface together with high amount of FD curves (e.g., 65,536 curves for a 256 × 256 pixels image) that locally quantify biophysical properties and interactions between the tip-linked viral particle and the plasma membrane of investigated cells (Fig. 4d).

1. *Reconstruction of FD-based height images and adhesion maps* (**steps 1–6**): To adequately reconstruct multiparametric maps and match the measured intrinsic physical properties to the topography of the sample, some off-line analysis is required to provide the software with corrected FD curves that eliminate unwanted effects due to the recording conditions. To do so, first upload the FD-based AFM height and adhesion maps into an adequate analysis software.
2. Improve the height images (Fig. 4d) by using filters such as flatten or plane fit. Users should keep in mind that application of any filter will modify the raw value of the file to some extent and that quantitative parameters should be extracted before their use.
3. For the adhesion map (Fig. 4e), perform a correction of the drag force acting on the cantilever when the approach and retract curves do not coincide at high separation distance [102, 103]. For very heterogeneous samples such as living mammalian cells, viscous drag is not homogeneous over the sample (top of a cell versus within cell boundaries). This error can be corrected using a subtraction of a linear fit to the last 30% of the baseline segment (*see* **Note 5**).

4. Evaluate the specificity of adhesion events by taking into account that rupture events occur at a certain distance from the cell surface. Depending on the extended length of the PEG linker when bond breakage occurs, the size and localization on the tip of the attached viral particle, and the mechanical properties of the cell membrane, specific binding events should appear at distances between 50 and 300 nm from the sample surface (Fig. 4f, g).
5. When investigating biological samples, defining the exact contact point between the tip and the probed surface can be particularly complex, due to the high deformability of soft cells, the structural heterogeneities of the surface, and the long-range surface forces involved. Therefore, software analyses only provide an estimate (e.g., by linearly extrapolating the contact region to zero force) of the first contact point on FD curves that usually lead to negligible approximation errors.
6. From the adhesion map, calculate binding probabilities and distribution of forces from specific retraction curves. To do so, select an area of interest based on the height image and/or adhesion map, and extract the individual force curves for further analysis. The FD curves may also be extracted and saved as text or ASCII files for further analysis by user-adaptable software (e.g., Origin). Individual retraction curves might show the rupture of single or multiple virus-receptor bonds, no or nonspecific adhesion events, or a number of abnormal phenomena. To extract curves of interest from the pool of FD curves recorded, a second witness for the specificity of interactions lies in the elongation pattern of the PEG linker, i.e., the shape of the retraction curve from the contact point to the bond rupture point. Fitting this part of the curve with the worm-like chain (WLC) model [104] for polymer extension ensures that the bond rupture observed corresponds to the breaking of an interaction occurring between the sample surface and a species attached at the free end of the PEG linker.
7. *Analysis of force curves* (**steps** 7 and **8**): In addition, further analysis of specific FD curves allows to extract quantitative information on kinetic and thermodynamic parameters of the probed interactions. The sinusoidal movement of the tip results in a continuum of velocities. To evaluate the dependency of the bond rupture force on the loading rate, extract the adhesion force together with the loading rate from each individual binding event. To rigorously extract the loading rate of each rupture event, measure the slope of the force versus time curve just before the rupture. For extraction of the rupture force, take the value on the *y* axis (force axis) right on the (negative) maximum peak of the rupture event before it returns to baseline level.

In the case of curves containing multiple rupture events, each rupture should be considered separately.

8. If desired, extract the effective spring constant from the force-distance curve by means of the slope, dividing the force segment ΔF through the distance segment Δx of the very last part of the rupture event.

9. *Extraction of kinetic and thermodynamic parameters* (**steps 9–11**): Display both values in a DFS plot (force vs. loading rate) (Fig. 5a). Using typical oscillation frequencies of 0.125 and 0.250 kHz, the loading rate range applied to the virus-receptor bond usually varies between 10^4 and 10^7 pN s^{-1}, depending on when rupture occurs on the tip trajectory and the elongation of the PEG linker at that time.

10. Fit the loading rate-dependent interaction forces with the fitting model of choice to extract the kinetic parameters of the interaction (*see* Subheading 1.3.2). Different strategies can be used to probe the LR dependency of the rupture force over the whole LR spectrum [105]. For example, all the data points in the DFS plot can be fitted with the best fit straight line. However, this could fail to capture all the information contained in the DFS plot, such as the presence of multiple interactions. An elegant way to proceed with the analysis is to separate the DFS plot in ~4–8 narrow LR ranges and plot the distribution of rupture forces as histograms [41] (Fig. 5a, b). Fitting these histograms with Gaussian distributions allows to determine the presence of single or multiple peaks, corresponding to one or several parallel virus-receptor interactions (*see* **Note 6**). For each LR range, the mean rupture force of each peak can be reported together with the corresponding mean LR value in a new DFS plot. These data points can then be fitted with either a linear iterative algorithm (Levenberg-Marquardt) along with the Bell-Evans model (Fig. 5c, d) [83] or a nonlinear iterative algorithm (Levenberg-Marquardt) along with the FNdY model (Fig. 5d) [88] that has to be used only if the forces measured do not scale linearly with the logarithm of the loading rate. An extensive overview on analysis and fitting dynamic force spectroscopy (DFS) data can be found in ref. [106] as well in ref. [88], while a comparison of the different models was performed by Hane et al. [107].

11. Fitting the data with the FNdY model requires calculating the effective spring constant of the probed setup (Fig. 1a), since the finite near-equilibrium unbinding force F_{eq} depends on the stiffness of the force transducer. This is equivalent to the stiffness of the whole system, i.e., the cantilever, the linker, the virion, and the cell surface, acting as springs in series. This value can either be extracted by fitting individual force versus piezo

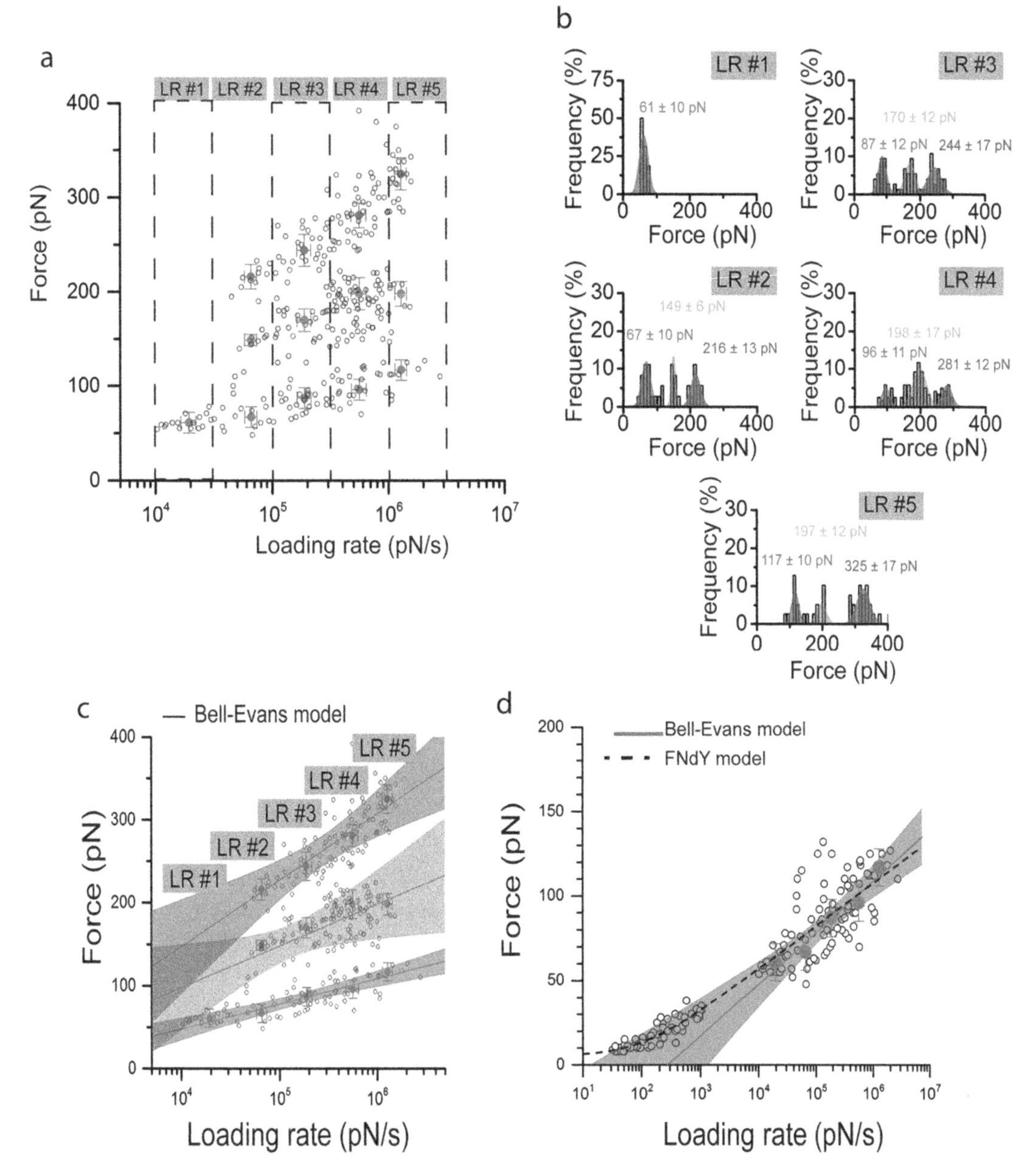

Fig. 5 Extraction of kinetic and thermodynamic parameters of virus-receptor interactions. (**a**) Dynamic force spectroscopy (DFS) plot showing the force required to separate the virus from the cell surface. The forces of single virus-receptor rupture events (circles) are plotted against the *LR*. (**b**) Small *LR* ranges #1–#5 are binned, and the distributions of the rupture forces are plotted as histograms. This classification reveals multiple force peaks with average values corresponding to single (red), double (green), or triple (blue) simultaneously established virus-receptor interactions. The average values of the force distributions are extracted and plotted on the DFS plot (**a**), enabling their analysis using the Bell-Evans model (**c**). (**d**) Probing the rupture force of the system at reduced loading rate allows to gain insights into thermodynamic properties of the virus-receptor bonds, by fitting the data with the Friddle-Noy-de Yoreo model (see text for details)

movement curves or theoretically modeled [43, 108, 109], since each individual spring behavior can be estimated using the appropriate biophysical model (e.g., worm-like chain model for protein extension [104] or PEG elasticity models for the linker extension [109]).

4 Notes

1. For accurate force measurements, the laser should be aligned at the free end of the cantilever and correctly sent in the middle of the position-sensitive photodetector. If the laser spot cannot be located close to the cantilever, there might be an air bubble trapped between the cantilever and the tip holder. In this case, the AFM head should be removed from the liquid environment and then re-submerged. Alternatively, the cantilever can be removed from the holder and replaced. If the sum is zero when the laser spot is correctly located on the cantilever, align the mirrors manually, so that a signal from the laser is detected in the photodiode.
2. Engaging the probe on the cell surface is a critical step, as the tip has to come in close contact with the cell membrane, without altering its shape or destroy its constituents. Therefore, the engaging force set point should be low enough to avoid deteriorating the probed biological materials. However, a too low engage set point can lead to a "false engage" when the tip is too far from the surface. In this case, the engaging force should be increased to ensure proper engagement.
3. When choosing an appropriate area for imaging, it might be difficult to find a zone were two different cells (i.e., expressing and not expressing the receptor of interest) are close to each other. This might be due to an imbalance between the number of cells from the two types in the confluent cell layer. To overcome this issue, different ratios should be used to mix the cells before seeding. It may be that one cell line grows faster than the other (generally, mutant cells divide slower) yielding a confluent layer containing more cells of one type. An appropriate ratio should then be found to generate a monolayer containing ~50/50% of each cell type.
4. Flat and "featureless" cells are intrinsically easier to image than round and/or rough cell surfaces and should thus be chosen for imaging, based on their appearance on the optical image of the sample. Cell division may however occur while imaging, during which the cells round up and detach from the surface. In this case, imaging should be stopped and started again on other cells.

5. Adhesion maps allow to localize interactions with cellular receptors and provide an estimate of the abundance of these interactions. In some cases, however, adhesion maps can look different than expected. If it shows no interactions at all, it is probably because there is no viral particle grafted at the apex of the tip (e.g., the virus is too high on the tip, preventing interactions to occur). Experiments should then be performed with another tip. If this happens with most of the probes, the concentration of the virus solution used to functionalize the tip should be increased, to increase the probability of grafting a virion at the tip apex. If adhesion events are observed all over the cell surface, the tip is probably contaminated with cellular debris from the viral solution. To overcome this, insist on the centrifugation step prior to the virus coupling to the PEG spacer, or use a virus solution with a higher purity.
6. Depending on the viral specie, one or multiple glycoproteins from the virus coat can be expected to interact with specific receptors on the cell surface, resulting in different patterns of the Gaussian rupture force distributions. When the virus-cell system probe contains only one type of glycoprotein-receptor pair, the establishment of multiple bonds should appear as peaks in the force distribution located at values that are multiple from the first (lowest force) peak, with the latter that should correspond to single interactions. When multiple viral glycoproteins are involved in cell binding, the attribution of different peaks to a specific glycoprotein is a challenging task, since multiple interactions can occur with one or a combination of different glycoproteins.

Acknowledgments

The research of the authors is financially supported by the Fonds National de la Recherche Scientifique (F.R.S.-FNRS grant number: PDR T.0070.16 to D.A.), the Université catholique de Louvain (Fonds Spéciaux de Recherche), and the "Erwin Schroedinger Fellowship Abroad" (Austrian Science Fund (FWF), grant number J-4052 to M.K.). M.D. and D.A. are Research Fellow and Research Associate at the FRS-FNRS, respectively.

References

1. Pelkmans L, Helenius A (2003) Insider information: what viruses tell us about endocytosis. Curr Opin Cell Biol 15(4):414–422
2. Boulant S, Stanifer M, Lozach PY (2015) Dynamics of virus-receptor interactions in virus binding, signaling, and endocytosis. Virus 7(6):2794–2815
3. Carette JE, Guimaraes CP, Varadarajan M, Park AS, Wuethrich I, Godarova A, Kotecki M, Cochran BH, Spooner E, Ploegh HL (2009) Haploid genetic screens in human

cells identify host factors used by pathogens. Science 326(5957):1231–1235

4. Dimitrov DS (2004) Virus entry: molecular mechanisms and biomedical applications. Nat Rev Microbiol 2(2):109–122
5. Brandenburg B, Zhuang X (2007) Virus trafficking–learning from single-virus tracking. Nat Rev Microbiol 5(3):197
6. Smith AE, Helenius A (2004) How viruses enter animal cells. Science 304 (5668):237–242
7. Schnell MJ, Mebatsion T, Conzelmann KK (1994) Infectious rabies viruses from cloned cDNA. EMBO J 13(18):4195–4203
8. Ghanem A, Kern A, Conzelmann KK (2012) Significantly improved rescue of rabies virus from cDNA plasmids. Eur J Cell Biol 91 (1):10–16
9. Herrmann A, Sieben C (2015) Single-virus force spectroscopy unravels molecular details of virus infection. Integr Biol (Camb) 7 (6):620–632
10. Matrosovich MN, Gambaryan AS (2012) Solid-phase assays of receptor-binding specificity. Methods Mol Biol 865:71–94
11. Watanabe T, Kiso M, Fukuyama S, Nakajima N, Imai M, Yamada S, Murakami S, Yamayoshi S, Iwatsuki-Horimoto K, Sakoda Y et al (2013) Characterization of H7N9 influenza a viruses isolated from humans. Nature 501(7468):551–555
12. Shi Y, Zhang W, Wang F, Qi J, Wu Y, Song H, Gao F, Bi Y, Zhang Y, Fan Z et al (2013) Structures and receptor binding of hemagglutinins from human-infecting H7N9 influenza viruses. Science 342(6155):243–247
13. Suenaga E, Mizuno H, Penmetcha KK (2012) Monitoring influenza hemagglutinin and glycan interactions using surface plasmon resonance. Biosens Bioelectron 32(1):195–201
14. Papp I, Sieben C, Ludwig K, Roskamp M, Bottcher C, Schlecht S, Herrmann A, Haag R (2010) Inhibition of influenza virus infection by multivalent sialic-acid-functionalized gold nanoparticles. Small 6(24):2900–2906
15. Xiong X, Coombs PJ, Martin SR, Liu J, Xiao H, McCauley JW, Locher K, Walker PA, Collins PJ, Kawaoka Y et al (2013) Receptor binding by a ferret-transmissible H5 avian influenza virus. Nature 497(7449):392–396
16. Roingeard P (2008) Viral detection by electron microscopy: past, present and future. Biol Cell 100(8):491–501
17. Mercer J, Helenius A (2009) Virus entry by macropinocytosis. Nat Cell Biol 11 (5):510–520
18. Ando T, Uchihashi T, Kodera N (2013) High-speed AFM and applications to biomolecular systems. Annu Rev Biophys 42:393–414
19. Kienberger F, Mueller H, Pastushenko V, Hinterdorfer P (2004) Following single antibody binding to purple membranes in real time. EMBO Rep 5(6):579–583
20. Hinterdorfer P, Dufrêne YF (2006) Detection and localization of single molecular recognition events using atomic force microscopy. Nat Methods 3(5):347–355
21. Neuman KC, Nagy A (2008) Single-molecule force spectroscopy: optical tweezers, magnetic tweezers and atomic force microscopy. Nat Methods 5(6):491–505
22. Sieben C, Kappel C, Zhu R, Wozniak A, Rankl C, Hinterdorfer P, Grubmüller H, Herrmann A (2012) Influenza virus binds its host cell using multiple dynamic interactions. Proc Natl Acad Sci U S A 109(34):13626–13631
23. Rankl C, Kienberger F, Wildling L, Wruss J, Gruber HJ, Blaas D, Hinterdorfer P (2008) Multiple receptors involved in human rhinovirus attachment to live cells. Proc Natl Acad Sci U S A 105(46):17778–17783
24. Joo K-I, Lei Y, Lee C-L, Lo J, Xie J, Hamm-Alvarez SF, Wang P (2008) Site-specific labeling of enveloped viruses with quantum dots for single virus tracking. ACS Nano 2 (8):1553–1562
25. Pandori MW, Hobson DA, Sano T (2002) Adenovirus–microbead conjugates possess enhanced infectivity: a new strategy for localized gene delivery. Virology 299(2):204–212
26. Tian P, Yang D, Jiang X, Zhong W, Cannon J, Burkhardt Iii W, Woods J, Hartman G, Lindesmith L, Baric R (2010) Specificity and kinetics of norovirus binding to magnetic bead-conjugated histo-blood group antigens. J Appl Microbiol 109(5):1753–1762
27. Chang MI, Panorchan P, Dobrowsky TM, Tseng Y, Wirtz D (2005) Single-molecule analysis of human immunodeficiency virus type 1 gp120-receptor interactions in living cells. J Virol 79(23):14748–14755
28. Dobrowsky TM, Zhou Y, Sun SX, Siliciano RF, Wirtz D (2008) Monitoring early fusion dynamics of human immunodeficiency virus type 1 at single-molecule resolution. J Virol 82(14):7022–7033
29. Zhang Y, Lyubchenko YL (2014) The structure of misfolded amyloidogenic dimers: computational analysis of force spectroscopy data. Biophys J 107(12):2903–2910
30. Booth JJ, Shalashilin DV (2016) Fully atomistic simulations of protein unfolding in low

speed atomic force microscope and force clamp experiments with the help of boxed molecular dynamics. J Phys Chem B 120 (4):700–708

31. Binnig G, Quate CF, Gerber C (1986) Atomic force microscope. Phys Rev Lett 56 (9):930–933
32. Muller DJ, Helenius J, Alsteens D, Dufrene YF (2009) Force probing surfaces of living cells to molecular resolution. Nat Chem Biol 5(6):383–390
33. Dufrene YF, Martinez-Martin D, Medalsy I, Alsteens D, Muller DJ (2013) Multiparametric imaging of biological systems by force-distance curve-based AFM. Nat Methods 10(9):847–854
34. Hörber J, Miles M (2003) Scanning probe evolution in biology. Science 302 (5647):1002–1005
35. Engel A, Müller DJ (2000) Observing single biomolecules at work with the atomic force microscope. Nat Struct Mol Biol 7 (9):715–718
36. Dufrene YF, Ando T, Garcia R, Alsteens D, Martinez-Martin D, Engel A, Gerber C, Muller DJ (2017) Imaging modes of atomic force microscopy for application in molecular and cell biology. Nat Nanotechnol 12 (4):295–307
37. Kienberger F, Kada G, Mueller H, Hinterdorfer P (2005) Single molecule studies of antibody–antigen interaction strength versus intra-molecular antigen stability. J Mol Biol 347(3):597–606
38. Radmacher M (2002) Measuring the elastic properties of living cells by the atomic force microscope. Methods Cell Biol 68(1):67–90
39. Viani MB, Pietrasanta LI, Thompson JB, Chand A, Gebeshuber IC, Kindt JH, Richter M, Hansma HG, Hansma PK (2000) Probing protein-protein interactions in real time. Nat Struct Biol 7(8):644–647
40. Kuznetsov YG, Malkin A, Lucas R, Plomp M, McPherson A (2001) Imaging of viruses by atomic force microscopy. J Gen Virol 82 (9):2025–2034
41. Alsteens D, Newton R, Schubert R, Martinez-Martin D, Delguste M, Roska B, Muller DJ (2017) Nanomechanical mapping of first binding steps of a virus to animal cells. Nat Nanotechnol 12(2):177–183
42. Gerber C, Lang HP (2006) How the doors to the nanoworld were opened. Nat Nanotechnol 1(1):3–5
43. Sieben C, Herrmann A (2017) Single virus force spectroscopy: the ties that bind. Nat Nanotechnol 12(2):102–103
44. Henderson E, Haydon PG, Sakaguchi DS (1992) Actin filament dynamics in living glial cells imaged by atomic force microscopy. Science 257(5078):1944–1946
45. Hoh JH, Schoenenberger CA (1994) Surface morphology and mechanical properties of MDCK monolayers by atomic force microscopy. J Cell Sci 107(Pt 5(5)):1105–1114
46. Hoh JH, Lal R, John SA, Revel JP, Arnsdorf MF (1991) Atomic force microscopy and dissection of gap junctions. Science 253 (5026):1405–1408
47. Mou J, Yang J, Shao Z (1995) Atomic force microscopy of cholera toxin B-oligomers bound to bilayers of biologically relevant lipids. J Mol Biol 248(3):507–512
48. Schabert FA, Henn C, Engel A (1995) Native Escherichia coli OmpF porin surfaces probed by atomic force microscopy. Science 268 (5207):92–94
49. Hansma HG, Vesenka J, Siegerist C, Kelderman G, Morrett H, Sinsheimer RL, Elings V, Bustamante C, Hansma PK (1992) Reproducible imaging and dissection of plasmid DNA under liquid with the atomic force microscope. Science 256(5060):1180–1184
50. Hoh JH, Sosinsky GE, Revel JP, Hansma PK (1993) Structure of the extracellular surface of the gap junction by atomic force microscopy. Biophys J 65(1):149–163
51. Müller D, Schabert FA, Büldt G, Engel A (1995) Imaging purple membranes in aqueous solutions at sub-nanometer resolution by atomic force microscopy. Biophys J 68 (5):1681–1686
52. Karrasch S, Dolder M, Schabert F, Ramsden J, Engel A (1993) Covalent binding of biological samples to solid supports for scanning probe microscopy in buffer solution. Biophys J 65(6):2437–2446
53. Putman CA, Van der Werf KO, De Grooth BG, Van Hulst NF, Greve J (1994) Tapping mode atomic force microscopy in liquid. Appl Phys Lett 64(18):2454–2456
54. Wegmann S, Jung YJ, Chinnathambi S, Mandelkow E-M, Mandelkow E, Muller DJ (2010) Human tau isoforms assemble into ribbon-like fibrils that display polymorphic structure and stability. J Biol Chem 285 (35):27302–27313
55. Ido S, Kimura K, Oyabu N, Kobayashi K, Tsukada M, Matsushige K, Yamada H (2013) Beyond the helix pitch: direct visualization of native DNA in aqueous solution. ACS Nano 7(2):1817–1822
56. Ido S, Kimiya H, Kobayashi K, Kominami H, Matsushige K, Yamada H (2014)

Immunoactive two-dimensional self-assembly of monoclonal antibodies in aqueous solution revealed by atomic force microscopy. Nat Mater 13(3):264–270

57. Hansma HG, Hoh JH (1994) Biomolecular imaging with the atomic force microscope. Annu Rev Biophys Biomol Struct 23 (1):115–139
58. Ohnesorge F, Hörber J, Häberle W, Czerny C, Smith D, Binnig G (1997) AFM review study on pox viruses and living cells. Biophys J 73(4):2183–2194
59. YuG K, Malkin A, Land T, DeYoreo J, Barba A, Konnert J, McPherson A (1997) Molecular resolution imaging of macromolecular crystals by atomic force microscopy. Biophys J 72(5):2357–2364
60. Drygin YF, Bordunova OA, Gallyamov MO, Yaminsky IV (1998) Atomic force microscopy examination of tobacco mosaic virus and virion RNA. FEBS Lett 425(2):217–221
61. Kienberger F, Zhu R, Moser R, Rankl C, Blaas D, Hinterdorfer P (2004) Dynamic force microscopy for imaging of viruses under physiological conditions. Biol Proced Online 6 (1):120
62. Malkin AJ, Plomp M, McPherson A (2005) Unraveling the architecture of viruses by high-resolution atomic force microscopy. Methods Mol Biol 292:85–108
63. Kasas S, Longo G, Dietler G (2013) Mechanical properties of biological specimens explored by atomic force microscopy. J Phys D Appl Phys 46(13):133001
64. Mateu MG (2012) Mechanical properties of viruses analyzed by atomic force microscopy: a virological perspective. Virus Res 168 (1):1–22
65. Martinez-Martin D, Carrasco C, Hernando-Perez M, De Pablo PJ, Gomez-Herrero J, Perez R, Mateu MG, Carrascosa JL, Kiracofe D, Melcher J (2012) Resolving structure and mechanical properties at the nanoscale of viruses with frequency modulation atomic force microscopy. PLoS One 7(1):e30204
66. Kuznetsov YG, Victoria J, Robinson W, McPherson A (2003) Atomic force microscopy investigation of human immunodeficiency virus (HIV) and HIV-infected lymphocytes. J Virol 77(22):11896–11909
67. Parachoniak CA, Park M (2012) Dynamics of receptor trafficking in tumorigenicity. Trends Cell Biol 22(5):231–240
68. Matias V, Beveridge T (2005) Cryo-electron microscopy reveals native polymeric cell wall structure in Bacillus subtilis 168 and the existence of a periplasmic space. Mol Microbiol 56(1):240–251
69. Hell SW (2007) Far-field optical nanoscopy. Science 316(5828):1153–1158
70. Fornasiero EF, Rizzoli SO (2014) Super-resolution microscopy techniques in the neurosciences. Springer, Berlin
71. Boonaert CJ, Rouxhet PG (2000) Surface of lactic acid bacteria: relationships between chemical composition and physicochemical properties. Appl Environ Microbiol 66 (6):2548–2554
72. Dufrêne YF, Boonaert CJ, Gerin PA, Asther M, Rouxhet PG (1999) Direct probing of the surface ultrastructure and molecular interactions of dormant and germinating spores of Phanerochaete chrysosporium. J Bacteriol 181(17):5350–5354
73. Kasas S, Ikai A (1995) A method for anchoring round shaped cells for atomic force microscope imaging. Biophys J 68(5):1678–1680
74. Dague E, Alsteens D, Latge JP, Verbelen C, Raze D, Baulard AR, Dufrene YF (2007) Chemical force microscopy of single live cells. Nano Lett 7(10):3026–3030
75. Dague E, Alsteens D, Latge JP, Dufrene YF (2008) High-resolution cell surface dynamics of germinating Aspergillus fumigatus conidia. Biophys J 94(2):656–660
76. Touhami A, Jericho MH, Beveridge TJ (2004) Atomic force microscopy of cell growth and division in Staphylococcus aureus. J Bacteriol 186(11):3286–3295
77. Butt H-J, Cappella B, Kappl M (2005) Force measurements with the atomic force microscope: technique, interpretation and applications. Surf Sci Rep 59(1):1–152
78. Hinterdorfer P, Baumgartner W, Gruber HJ, Schilcher K, Schindler H (1996) Detection and localization of individual antibody-antigen recognition events by atomic force microscopy. Proc Natl Acad Sci U S A 93 (8):3477–3481
79. Ludwig M, Dettmann W, Gaub HE (1997) Atomic force microscope imaging contrast based on molecular recognition. Biophys J 72(1):445–448
80. Wildling L, Unterauer B, Zhu R, Rupprecht A, Haselgrubler T, Rankl C, Ebner A, Vater D, Pollheimer P, Pohl EE et al (2011) Linking of sensor molecules with amino groups to amino-functionalized AFM tips. Bioconjug Chem 22(6):1239–1248

81. Puntheeranurak T, Neundlinger I, Kinne RKH, Hinterdorfer P (2011) Single-molecule recognition force spectroscopy of transmembrane transporters on living cells. Nat Protoc 6(9):1443–1452
82. Evans EA, Calderwood DA (2007) Forces and bond dynamics in cell adhesion. Science 316(5828):1148–1153
83. Evans E, Ritchie K (1997) Dynamic strength of molecular adhesion bonds. Biophys J 72 (4):1541–1555
84. Alsteens D, Dupres V, Yunus S, Latgé J-P, Heinisch JJ, Dufrêne YF (2012) High-resolution imaging of chemical and biological sites on living cells using peak force tapping atomic force microscopy. Langmuir 28 (49):16738–16744
85. Pfreundschuh M, Martinez-Martin D, Mulvihill E, Wegmann S, Muller DJ (2014) Multiparametric high-resolution imaging of native proteins by force-distance curve-based AFM. Nat Protoc 9(5):1113–1130
86. Bell GI (1978) Models for the specific adhesion of cells to cells. Science 200 (4342):618–627
87. Collin D, Ritort F, Jarzynski C, Smith SB, Tinoco I Jr, Bustamante C (2005) Verification of the crooks fluctuation theorem and recovery of RNA folding free energies. Nature 437 (7056):231–234
88. Friddle RW, Noy A, De Yoreo JJ (2012) Interpreting the widespread nonlinear force spectra of intermolecular bonds. Proc Natl Acad Sci U S A 109(34):13573–13578
89. Schillers H, Medalsy I, Hu S, Slade AL, Shaw JE (2016) PeakForce tapping resolves individual microvilli on living cells. J Mol Recognit 29(2):95–101
90. Huhti L, Blazevic V, Nurminen K, Koho T, Hytonen VP, Vesikari T (2010) A comparison of methods for purification and concentration of norovirus GII-4 capsid virus-like particles. Arch Virol 155(11):1855–1858
91. Hutornojs V, Niedre-Otomere B, Kozlovska T, Zajakina A (2012) Comparison of ultracentrifugation methods for concentration of recombinant alphaviruses: sucrose and iodixanol cushions. Environ Exp Biol 10:117–123
92. Viani MB, Schäffer TE, Chand A, Rief M, Gaub HE, Hansma PK (1999) Small cantilevers for force spectroscopy of single molecules. J Appl Phys 86(4):2258–2262
93. Gruber H (2016) Crosslinkers and protocols for AFM tip functionalization. http://www.jku.at/biophysics/content/e257042. Accessed 22 Aug 2016
94. Ebner A, Hinterdorfer P, Gruber HJ (2007) Comparison of different aminofunctionalization strategies for attachment of single antibodies to AFM cantilevers. Ultramicroscopy 107(10–11):922–927
95. Kienberger F, Pastushenko VP, Kada G, Gruber HJ, Riener C, Schindler H, Hinterdorfer P (2000) Static and dynamical properties of single poly(ethylene glycol) molecules investigated by force spectroscopy. Single Mol 1 (2):123–128
96. Butt HJ, Jaschke M (1995) Calculation of thermal noise in atomic force microscopy. Nanotechnology 6(1):1–7
97. Florin E-L, Rief M, Lehmann H, Ludwig M, Dornmair C, Moy VT, Gaub HE (1995) Sensing specific molecular interactions with the atomic force microscope. Biosens Bioelectron 10(9):895–901
98. Hutter JL, Bechhoefer J (1993) Calibration of atomic-force microscope tips. Rev Sci Instrum 64(7):1868–1873
99. te Riet J, Katan AJ, Rankl C, Stahl SW, van Buul AM, Phang IY, Gomez-Casado A, Schon P, Gerritsen JW, Cambi A et al (2011) Interlaboratory round robin on cantilever calibration for AFM force spectroscopy. Ultramicroscopy 111(12):1659–1669
100. Sader JE, Borgani R, Gibson CT, Haviland DB, Higgins MJ, Kilpatrick JI, Lu J, Mulvaney P, Shearer CJ, Slattery AD et al (2016) A virtual instrument to standardise the calibration of atomic force microscope cantilevers. Rev Sci Instrum 87(9):093711
101. Gates RS, Osborn WA, Shaw GA (2015) Accurate flexural spring constant calibration of colloid probe cantilevers using scanning laser Doppler vibrometry. Nanotechnology 26(23):235704
102. Janovjak H, Struckmeier J, Müller DJ (2005) Hydrodynamic effects in fast AFM single-molecule force measurements. Eur Biophys J 34(1):91–96
103. Alcaraz J, Buscemi L, Puig-de-Morales M, Colchero J, Baro A, Navajas D (2002) Correction of microrheological measurements of soft samples with atomic force microscopy for the hydrodynamic drag on the cantilever. Langmuir 18(3):716–721
104. Bustamante C, Marko JF, Siggia ED, Smith S (1994) Entropic elasticity of lambda-phage DNA. Science 265(5178):1599–1600
105. Friedsam C, Wehle AK, Kühner F, Gaub HE (2003) Dynamic single-molecule force spectroscopy: bond rupture analysis with variable spacer length. J Phys Condens Matter 15(18): S1709

106. Bizzarri AR, Cannistraro S (2010) The application of atomic force spectroscopy to the study of biological complexes undergoing a biorecognition process. Chem Soc Rev 39 (2):734–749
107. Hane FT, Attwood SJ, Leonenko Z (2014) Comparison of three competing dynamic force spectroscopy models to study binding forces of amyloid-beta (1-42). Soft Matter 10 (12):1924–1930
108. Alsteens D, Pfreundschuh M, Zhang C, Spoerri PM, Coughlin SR, Kobilka BK, Müller DJ (2015) Imaging G protein-coupled receptors while quantifying their ligand-binding free-energy landscape. Nat Methods 12(9):845–851
109. Sulchek T, Friddle RW, Noy A (2006) Strength of multiple parallel biological bonds. Biophys J 90(12):4686–4691
110. Carneiro FA, Lapido-Loureiro PA, Cordo SM, Stauffer F, Weissmüller G, Bianconi ML, Juliano MA, Juliano L, Bisch PM, Poian ATD (2006) Probing the interaction between vesicular stomatitis virus and phosphatidylserine. Eur Biophys J 35(2):145–154
111. Liu N, Peng B, Lin Y, Su Z, Niu Z, Wang Q, Zhang W, Li H, Shen J (2010) Pulling genetic RNA out of tobacco mosaic virus using single-molecule force spectroscopy. J Am Chem Soc 132(32):11036–11038
112. Korneev D, Popova A, Generalov V, Zaitsev B (2016) Atomic force microscopy-based single virus particle spectroscopy. Biophysics 61 (3):413–419
113. Wörmann X, Lesch M, Welke R-W, Okonechnikov K, Abdurishid M, Sieben C, Geissner A, Brinkmann V, Kastner M, Karner A (2016) Genetic characterization of an adapted pandemic 2009 H1N1 influenza virus that reveals improved replication rates in human lung epithelial cells. Virology 492:118–129

Chapter 30

Applications of Atomic Force Microscopy for Adhesion Force Measurements in Mechanotransduction

Andreea Trache, Leike Xie, Huang Huang, Vladislav V. Glinsky, and Gerald A. Meininger

Abstract

Adhesive interactions between living cells or ligand-receptor interactions can be studied at the molecular level using atomic force microscopy (AFM). Adhesion force measurements are performed with functionalized AFM probes. In order to measure single ligand-receptor interactions, a cantilever with a pyramidal tip is functionalized with a bio-recognized ligand (e.g., extracellular matrix protein). The ligand-functionalized probe is then brought into contact with a cell in culture to investigate adhesion between the respective probe-bound ligand and endogenously expressed cell surface receptors (e.g., integrins or other adhesion receptor). For experiments designed to examine cell-cell adhesions, a single cell is attached to a tipless cantilever which is then brought into contact with other cultured cells. Force curves are recorded to determine the forces necessary to rupture discrete adhesions between the probe-bound ligand and receptor, or to determine total adhesion force at cell-cell contacts. Here, we describe the procedures for measuring adhesions between (a) fibronectin and α5β1 integrin, and (b) breast cancer cells and bone marrow endothelial cells.

Key words Ligand-receptor adhesion force, Cell-cell adhesion force, Atomic force microscopy

1 Introduction

The mechanisms surrounding the process of cell adhesion have not been fully understood due to their complexity and continue to stimulate investigation at the cellular and molecular levels. Atomic force microscopy (AFM) is a powerful technique that has been proven to be useful for investigating adhesive interactions of living cells under physiological conditions [1–3]. AFM has the ability to detect a large range of forces, from 10 pN to 100 nN, making it possible to measure forces ranging from single-molecule adhesion to total cell-cell adhesion. Therefore, AFM is often the method of choice due to its wide applicability and precision compared to other methods used to study cell-cell adhesions such as plate-and-wash assay, flow chamber assay [4], or magnetic twisting cytometry [5].

Yuri L. Lyubchenko (ed.), *Nanoscale Imaging: Methods and Protocols*, Methods in Molecular Biology, vol. 1814,
https://doi.org/10.1007/978-1-4939-8591-3_30,

Adhesion force spectroscopy has been typically used to measure single-molecule adhesion forces between AFM probes functionalized with ligand proteins and receptor proteins expressed on the cell surface [6, 7]. By approaching and retracting the functionalized probe from the cell surface, changes of cantilever deflection are recorded as force curves (Fig. 1) [8, 9]. Discrete ligand-receptor adhesion forces (i.e., force required to separate or rupture an adhesion) are calculated from the force curves (Fig. 2). The force required to break a single ligand-receptor bond was defined here as the *adhesion force*.

AFM has also been used to study leukocyte-endothelial and tumor-endothelial cell adhesions, which are associated with extravasation of leukocytes or tumor cells during inflammation or tumor metastasis, respectively [10–12]. Interactions between multiple ligands and receptors on the cellular membrane are involved in adhesions of the leukocyte or cancer cell to vascular endothelium [13, 14]. Acquisition and quantification of cell-to-cell adhesion forces can be performed using the adhesion force spectroscopy method described above. However, the AFM probe is functionalized by attaching a single cell to the tip of a flexible cantilever. This cell-tipped probe can then be brought in contact with other cells in culture (Fig. 3). The force required to completely separate the cell-tipped probe from other cells was defined as the *total adhesion force* necessary to rupture all adhesions that form during contact between the two cell types. The total cell-cell adhesion forces are calculated similarly from the force curves (Fig. 4).

Here, as examples, we describe the experimental procedures to measure (a) single-molecule adhesions between fibronectin and endogenous $\alpha 5\beta 1$ integrin expressed on vascular smooth muscle (VSM) cells, and (b) total cell-cell adhesions between a breast cancer cell MDA-MB-435 (MB435) and bone marrow endothelial cells (HBMEC-60).

2 Materials

2.1 Equipment

By combining AFM with optical imaging modalities, we are able to precisely position the AFM probe in contact with a specific area of a cell in culture. As presented here, measuring adhesion forces between a cell-functionalized probe and cell-cell junctions, or measuring single ligand-receptor interactions in a specific cell area, is greatly facilitated by the optical system that allows continuous visualization of the AFM probe and also the cells in the culture dish [15–17].

1. AFM systems: Bioscope with XZ Hybrid head, Bruker Nano Surfaces (formerly Veeco Instruments, Santa Barbara, CA); MFP-3D-BIO, Asylum Research (Santa Barbara, CA).

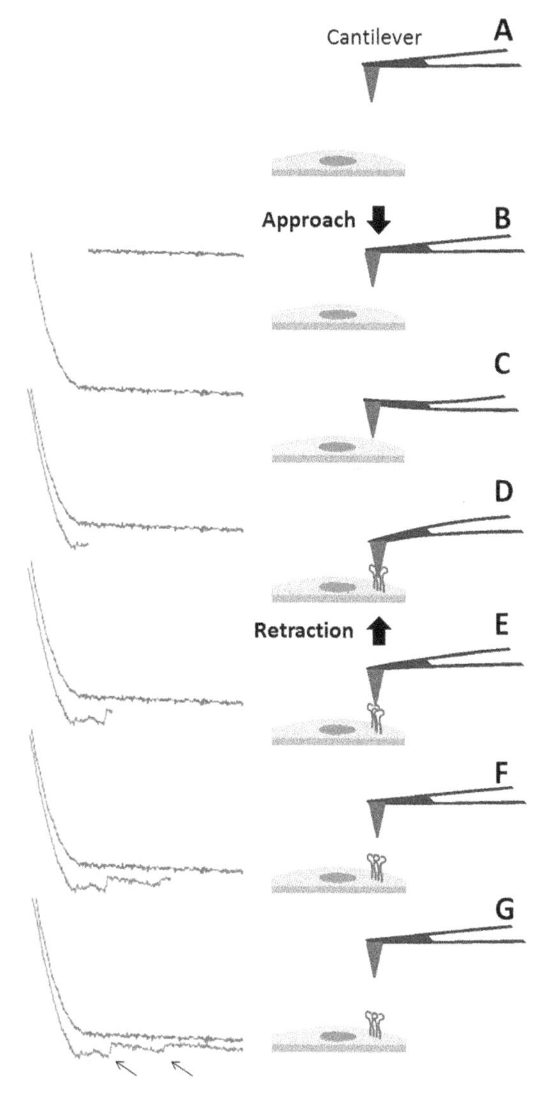

Fig. 1 Schematic diagram of ligand-receptor adhesion force measurements. For measuring ligand-receptor adhesion force, the cantilever functionalized with fibronectin (**A**) is lowered toward the cell (**B**) until contact is made (**C**). If adhesions occurred between the fibronectin and cell surface receptors (i.e., integrins) (**D**), then integrin-fibronectin bonds will be broken when the cantilever is retracted (**E-F**) returning to its original position (**G**) [8, 9]. Representative force curves, including approach (red) and retraction (blue) curves, that show cantilever deflection are also presented (left). Arrows indicate discrete unbinding events between the matrix functionalized probe and endogenous integrin receptors at the apical cell surface

2. Software: Igor Pro 6.37 (WaveMetrics, Inc.), MatLab R2014a (The MathWorks, Inc.), and NForceR [18] and PeakFit (v4.11, Systat Software Inc., Chicago, IL)).

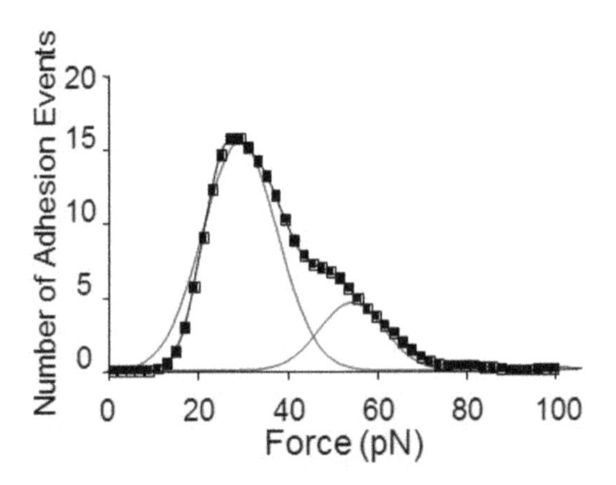

Fig. 2 Ligand-receptor adhesion force measurements. Adhesion force between fibronectin and the $\alpha 5\beta 1$ integrin on VSM cells (filled squares on black line) was plotted as a function of the frequency of occurrence (i.e., number of adhesion events). Data were analyzed by fitting with multiple Gaussian distributions to resolve single integrin-fibronectin adhesion force (indicated by the peak of the red distribution curve). The presence of a second (blue) distribution peak at two times the magnitude of the single-bond force suggests the presence and rupture of two bonds with a much lower incidence. Reproduced with permission from reference [26]

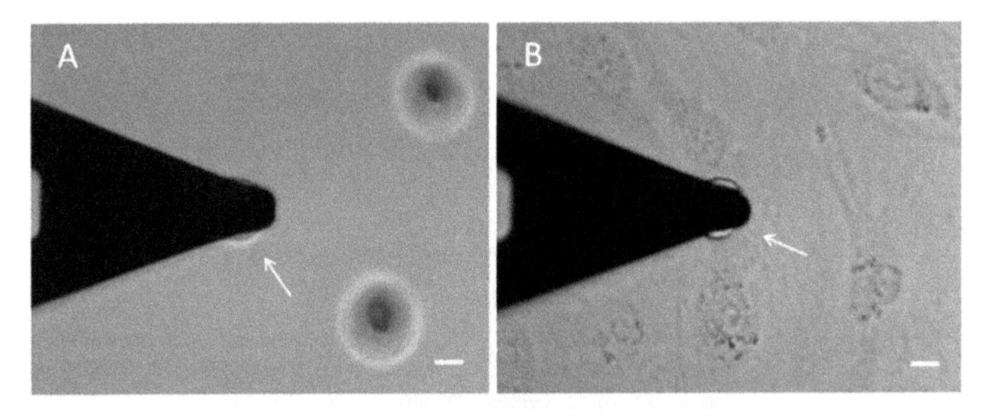

Fig. 3 MB435 cell attachment to the AFM cantilever and contact with endothelial cells. (A) A single suspended MB435 cell (indicated by arrow) is picked up by the tipless cantilever functionalized with Con A. (B) The MB435 cell that is firmly attached at the end of the cantilever is brought into contact with an endothelial cell-cell junction (see arrow) on the HBMEC-60 monolayer. Scale bar is 10 μm

2.2 Buffers, Solutions, and Cell Culture Media

Buffers and solutions are prepared using analytical grade reagents and ultrapure deionized (DDI) water with a sensitivity of 18 MΩ/cm at 25 °C. Solutions are prepared at room temperature and stored at the indicated temperature. Reagents are sodium azide free.

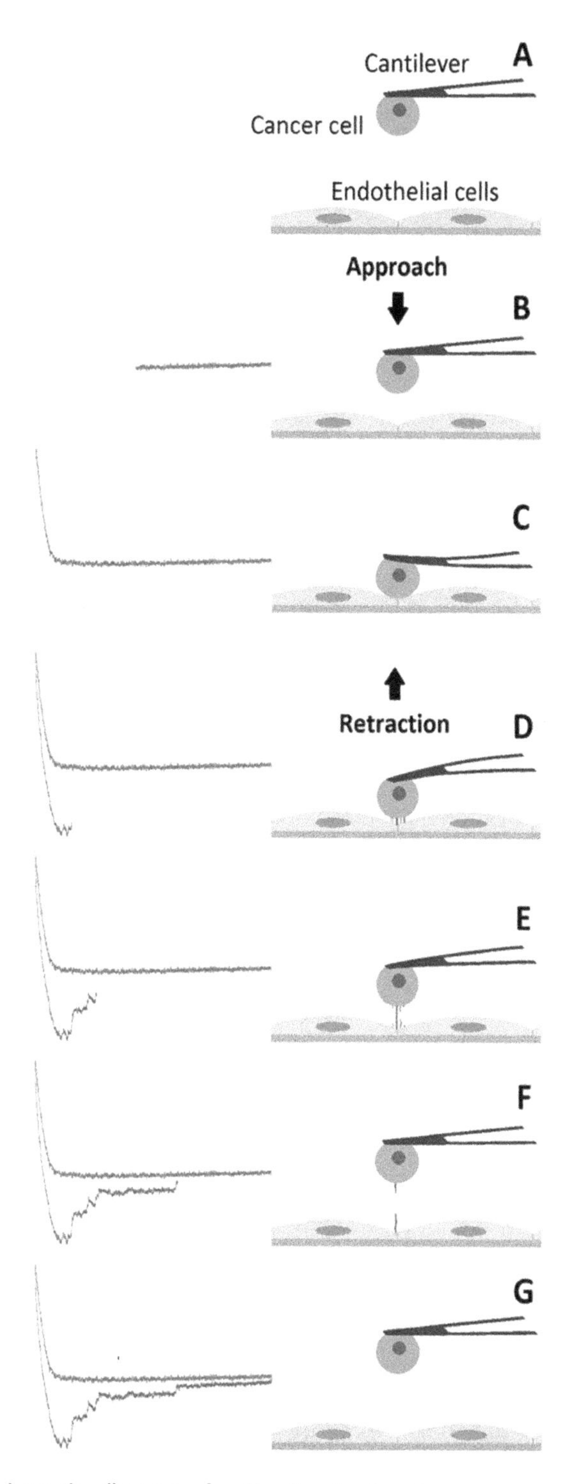

Fig. 4 Schematic diagram of cell-cell adhesion force measurements. For measuring cell-cell adhesion force, the MB435 cell attached on the cantilever (**A**) is lowered toward an endothelial cell-cell junction (**B**) until contact is made (**C**). After a defined contact time, the cantilever is retracted (**D-F**) until the MB435 cell is completely separated from the endothelial cells (**G**). Representative force curves, including approach (red) and retraction (blue) curves, that show cantilever deflection are presented (left)

2.2.1 Ligand-Receptor Adhesion Measurements

1. Polyethylene glycol (PEG) solution: Weigh 0.1 g PEG (Sigma-Aldridge), and mix with 10 mL DDI water. Mix well, and store at 4 °C (*see* **Note 1**).
2. Fibronectin solution: Reconstitute 1 vial of 5 mg lyophilized fibronectin powder in 5 mL of cold Dulbecco's phosphate-buffered saline (DPBS) for a final concentration of 1 mg/mL. Then, dispense the fibronectin solution into aliquots, and store at −20 °C (*see* **Note 2**).
3. VSM cell culture medium: Add a pouch of Dulbecco's Modified Eagle Medium (DMEM) F-12 to a beaker with 1 L of DDI water. Weigh 2.438 g NaHCO3 and 2.6 g HEPES, and transfer to the beaker. Mix and adjust to pH = 7.2 with HCL to make stock medium. Then, make the cell culture medium by adding 20 mL FBS, 2 mL L-glutamine, 2 mL sodium pyruvate, and 2 mL PSA (penicillin-streptomycin-amphotericin) into 174 mL stock medium. Mix well, and store at 4 °C (*see* **Note 3**).

2.2.2 Cell-Cell Adhesion Measurements

1. $NaHCO_3$ buffer (0.1 M, pH 8.6): Weigh 4.3 g of $NaHCO_3$, and transfer to a 500 mL graduated cylinder. Add water to a volume of 450 mL. Mix and adjust pH to 8.6 with 1 M NaOH. Add water to a total volume of 500 mL. Store at room temperature.
2. Biotin-labeled bovine serum albumin (BSA) solution (0.5 mg/mL): Weigh 5 mg of biotin-BSA, and transfer to a 10 mL graduated cylinder. Prepare a 10 mL solution by adding 0.1 M $NaHCO_3$ buffer (pH 8.6). Dispense into aliquots, and store at −20 °C [19, 20].
3. Streptavidin solution (0.5 mg/mL): Dissolve 1 vial of 5 mg streptavidin in 5 mL of Dulbecco's phosphate-buffered saline (DPBS) without Ca^{2+} and Mg^2, and transfer to a 10 mL graduated cylinder. Make up to 10 mL solution with DPBS. Dispense into aliquots, and store at −20 °C [20].
4. Concanavalin A (Con A) solution (0.5 mg/mL): Dissolve 1 vial of 10 mg biotin conjugated Con A with 5 mL of DPBS, and transfer to a 20 mL graduated cylinder. Make up to 20 mL solution with DPBS. Dispense into aliquots, and store at −20 °C.
5. MB435 cell culture medium: Add 50 mL of fetal bovine serum (FBS), 5 mL of 200 mM L-glutamine, and 1 mL of 50 mg/mL gentamicin into 444 mL of RPMI 1640. Mix well, and store at 4 °C (*see* **Note 3**).
6. HBMEC-60 cell culture medium: Add 90 mL of FBS, 10 mL of low serum growth supplement (LSGS) and 1 mL of 50 mg/mL gentamicin into 399 mL of Medium 200. Mix well, and store at 4 °C (*see* **Note 3**).

3 Methods

3.1 Cell Cultures

All cell cultures are maintained in a humidified incubator in 5% CO_2 at 37 °C.

3.1.1 MB435 Cell Culture

1. Culture MB435 cells in MB435 culture medium in 60 mm cell culture dishes. Cells reaching around 40% confluence are ready for experiment (*see* **Note 4**).
2. Wash cells three times in 2 mL of DPBS for 2 min each time.
3. Incubate cells in 1 mL cell dissociation buffer for about 5 min or until all cells are detached (*see* **Note 5**).
4. Pellet cells in a 15 mL conical tube by centrifugation at 300–400 × *g*.
5. Remove the supernatant and wash once with DPBS.
6. Gently resuspend the cells in CO_2 independent medium (CO_2IM) at a concentration of at least 1×10^5 cells/mL (*see* **Note 6**).

3.1.2 HBMEC-60 Cell Culture

1. HBMEC-60 cells are cultured in HBMEC-60 culture medium onto collagen-I coated 60 mm cell culture dishes until confluent.
2. Replace the medium with 4 mL of CO_2IM, and equilibrate cells in CO_2IM at room temperature for 15 min before experiments (*see* **Note 7**).

3.1.3 VSM Cell Culture

1. Culture VSM cells in DMEM-F12 medium in 60 mm cell culture dishes. Cell cultures reaching 50% confluence are ready for experiment.
2. Wash cells two times in 2 mL of DPBS to remove any debris in the dish.
3. Add 1.5 mL of DMEM F-12 cell culture medium, and equilibrate cells at room temperature for 15 min before experiments.

3.2 Cantilever Calibration

Cantilever calibration is performed before functionalization (*see* **Note 8**).

1. Clean the cantilever by immersing in a small glass beaker with acetone for 1 min, and then dry it out.
2. Use a blunt-end micro-tweezer to gently transfer the cantilever onto the cantilever holder.
3. Mount the holder with the cantilever onto the AFM head.
4. Place a 60 mm cell culture dish containing 1.5 mL cell culture medium on the AFM sample stage.

5. Position the AFM head over the 60 mm cell culture dish, and lower the head so that the cantilever is immersed in the cell culture medium.
6. Align laser beam toward the end of the cantilever, and maximize the intensity value on the detector by turning the X–Y knobs that adjust the laser beam position.
7. Adjust the beam position to zero deflection on the detector.
8. Run a cycle of force measurements by lowering the cantilever to make contact with the bottom of the dish at a defined speed and loading force for a zero setpoint setting. Record the approach and retraction curves. The sensitivity of the cantilever is measured as the slope of the approach curve using the built-in program in the AFM software.
9. Determine the cantilever spring constant by using the thermal noise method built-in the AFM software [21, 22].
10. Dismount the cantilever, and wash with DDI water. The cantilever is ready for functionalization.

3.3 Cantilever Functionalization

3.3.1 Cantilever Functionalization for Ligand-Receptor Measurements

1. Unsharpened silicon nitride cantilevers (MLCT probes, cantilever C) from Bruker Nano Surfaces (Santa Barbara, CA) with a spring constant of ~12 pN/nm are mounted on the cantilever holder.
2. Incubate the end of the cantilever in a droplet (10 μL) of 10 mg/mL PEG solution for 5 min.
3. Gently wash the cantilever five times with DPBS by incubating the end of the cantilever for 1 min in 10 μL water droplets.
4. Incubate the end of the cantilever in a droplet of fibronectin for 3 min.
5. Gently wash the cantilever again five times in DPBS, as mentioned in **step 3** and mount immediately on the AFM head.

3.3.2 Cantilever Functionalization for Cell-Cell Measurements

1. Tipless nitride cantilevers (MLCT-O10 probes, cantilever D) from Bruker Nano Surfaces (Santa Barbara, CA) with a spring constant of ~30 pN/nm are used.
2. Incubate the cantilever in a droplet (50 μL) of biotin-BSA solution in a humidified chamber at 37 °C overnight (*see* **Note 9**).
3. Gently wash the cantilever three times by dipping into a small beaker with DBPS.
4. Incubate the cantilever in a droplet of streptavidin solution in a humidified chamber at room temperature for 15 min.
5. Gently wash the cantilever three times in DPBS, as mentioned in **step 3**.

6. Incubate the cantilever in a droplet of Con A solution in a humidified chamber at room temperature for 15 min.
7. Wash the cantilever three times in DPBS, as mentioned in **step 3**, and store in DPBS (*see* **Note 10**).

3.4 Adhesion Force Measurements Using the AFM

3.4.1 Single Molecule Ligand-Receptor Adhesion Force Measurements

1. Place the 60 mm cell culture dish containing VSM cells already equilibrated at room temperature on the AFM sample stage (*see* Subheading 3.1.3).
2. Mount the functionalized cantilever on the cantilever holder and then mount the holder on the AFM head (*see* Subheading 3.3.1).
3. Position the head so that the cantilever is immersed in the cell culture medium.
4. Align laser beam at the end of the cantilever, and adjust the beam position to zero deflection on the detector.
5. While monitoring under the optical microscope, align the end of the cantilever with a VSM cell in an area half way between the edge of the cell and the nucleus.
6. Lower the cantilever toward the VSM cell, and start acquiring force curves for about 2 min with a preset speed at setpoint zero. Measurements performed on one cell are defined as one experimental sample.
7. Withdraw the cantilever from the cell, and randomly approach another cell and repeat the force measurements as in **step 6**. Collect force curves for at least ten cells. Repeat the experiment at least four times (*see* **Note 11**).
8. Force curves are exported first using the AFM software. Rupture forces for discrete ligand-receptor adhesion events are extracted from the force curve data using NForceR software [18]. Adhesion force between the matrix functionalized probe and endogenous integrin receptors is calculated by multiplying the deflection height associated with the unbinding event (*see* arrows in Fig. 1) and the spring constant of the AFM probe. Experimental data are further processed with PeakFit software (Fig. 2).

3.4.2 MB435 Cell-Cantilever Attachment and Cell-Cell Adhesion Force Measurements

1. Place the 60 mm cell culture dish containing HBMEC-60 monolayer already equilibrated in CO_2IM on the AFM sample stage (*see* Subheading 3.1.2).
2. Mount the functionalized cantilever on the cantilever holder, and then mount the holder on the AFM head.
3. Position the head so that the cantilever is immersed in the cell culture medium.

4. Align laser beam at the end of the cantilever, and adjust the beam position to zero deflection on the detector.
5. Retract the cantilever, and lift the AFM head. Then, transfer about 100 μL of the suspended MB435 cells (*see* Subheading 3.1.1) into the HBMEC-60 dish.
6. While monitoring under the optical microscope, align the end of the cantilever with the center of a single MB435 cell by moving the sample stage in xy directions.
7. Gently lower the cantilever toward the MB435 cell until the tip of the cantilever centrally touches the cell (*see* **Note 12**).
8. Maintain the contact for 1–2 s, and gently retract the cantilever. Keep watching if the cell remains attached on the cantilever during retraction.
9. Leave the cell-tipped cantilever undisturbed for 5–10 min to allow the MB435 cell to adhere firmly to the cantilever (*see* **Note 13**).
10. Align the attached MB435 cell with a cell-cell junction of HBMEC-60 cells (*see* **Note 14**).
11. Run a cycle of force measurement by lowering the cantilever to bring the MB435 cell into contact with the HBMEC-60 cell junction at a defined speed and force for a defined contact time, and then retracting at the same speed (*see* **Note 15**).
12. Randomly change the endothelial cell-cell junction site of measurement, and collect force curves (Fig. 4). Measurements performed with one MB435 cell-tipped probe is defined as one experimental sample (*see* **Note 16**).
13. Force curves are exported first using the AFM software. Total adhesion force required to separate the MB435 cell from the endothelial cells is extracted from the exported force curve data using Matlab software (Fig. 5a). Total adhesion force measured from each retraction curve was determined as the sum of individual unbinding events for each plateau (i.e., snap-offs) in the respective force curve. The mean of total adhesion forces for all force curves obtained with one MB435 cell-tipped probe was calculated and then averaged together for all cells at one contact time point (Fig. 5b).

4 Notes

1. Very gently shake the tube containing the PEG solution to avoid the formation of air bubbles.
2. Gently invert the bottle to mix well. Make sure that the lyophilized fibronectin is completely dissolved before making the aliquots. Keep the bottle on ice while making the aliquots.

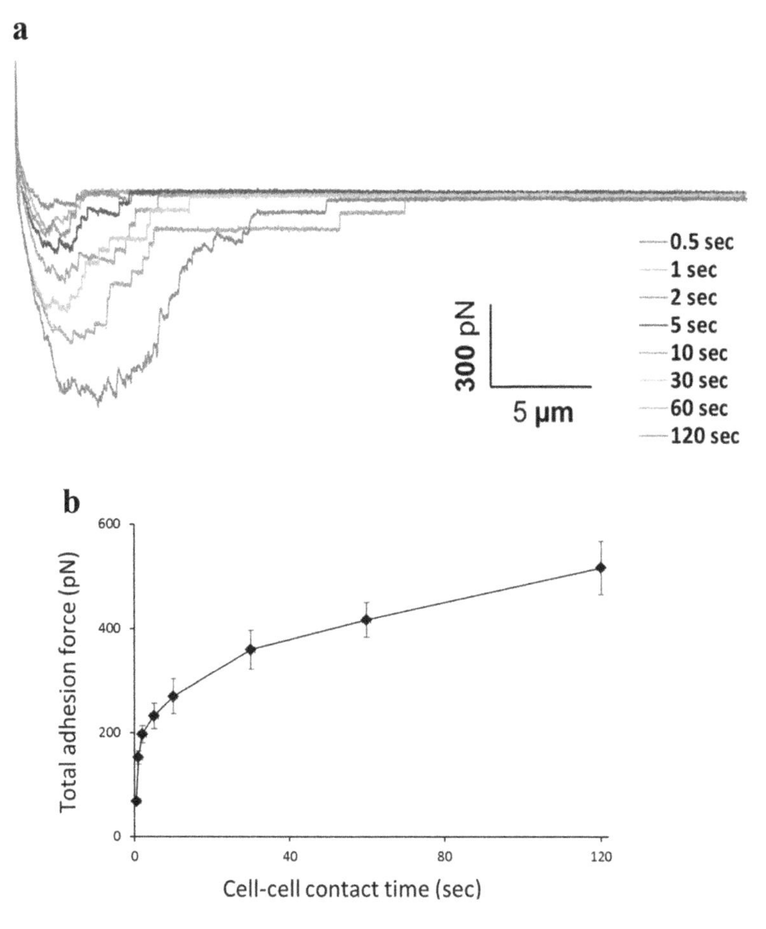

Fig. 5 (**a**) Representative retraction force curves from adhesions of MB435 to HBMEC-60 cells as a function of increasing contact time. The force curves were color coded by time as shown in the legend. The total adhesion force required to separate the MB435 cell from the junction of HBMEC-60 cells increased as a function of increasing contact time. The vertical and horizontal scales represent adhesion force and cantilever retraction distance, respectively. (**b**) Total adhesion forces of MB435 to the junction of HBMEC-60 cells as a function of increasing contact time. The contact time between the cell-tipped probe and the endothelium was set at 0.5, 1, 2, 5, 10, 30, 60, and 120 s. Total adhesion force increased progressively over 7.5-fold when the contact time was varied from 0.5 to 120 s

3. Prepare cell culture media in a laminar flow hood using sterile conditions to avoid contamination.
4. In the present cell-cell adhesion assay, a single MB435 cell was attached to the AFM cantilever. The cell-tipped probe was brought in contact with HBMEC-60 endothelial cells. This approach mimics a single MB435 cell extravasation during metastasis. It is recommended to always culture the MB435 cells at a non-confluent (individual) state before experiments so

that most cells keep their single-cell properties. Cell growth should be monitored every day. Split the MB435 cells as necessary to avoid confluence.

5. Some types of cells may be easily detached from the cell culture dish by incubation with DPBS. For those that are difficult to detach, a cell dissociation buffer with less than 0.25% trypsin-EDTA may be used instead. If using trypsin, it may be necessary to verify that the adhesion proteins to be detected on the cell membrane were not affected by trypsin [23].
6. MB435 cells in suspension in CO_2IM medium may be stored on ice or at 4 °C for 2 h.
7. Do not keep cells in CO_2IM inside a 5% CO_2 incubator, as the 5% CO_2 slightly changes the pH of the medium.
8. The nominal spring constant of a cantilever is usually specified by the manufacturer. However, the spring constant of each cantilever should be empirically determined. Be very careful handling AFM cantilevers in any processes, including cantilever cleaning, treatments, transfer, and mounting, to avoid accidental breakage.
9. Avoid drying out the cantilevers between functionalization processing steps.
10. It is recommended to use freshly functionalized cantilevers for every experiment. However, if the cantilevers are not used immediately, they can be stored in DPBS at 4 °C for several days. Tipless cantilevers can be recycled several times. To clean used cantilevers, treat them with droplets of H_2SO_4–H_2O_2 solution (three parts H_2SO_4 and one part H_2O_2) at room temperature for 15 min, wash with DDI water and then with 100% ethanol, and let them dry. The cantilevers are ready for the next functionalization cycle.
11. Ligand-receptor interaction measurements exemplified in Fig. 2 were obtained by setting the AFM probe to touch and retract from the cell surface at a speed of 800 nm/s with a frequency of 0.5 Hz. Measurements were acquired for 2 min/cell, and they were repeated for 20 cells.
12. The force applied by the cantilever on the MB435 cell is determined by the cantilever sensitivity, spring constant, and setpoint setting. After the setpoint is preset, the force is monitored and kept constantly in the AFM software. A force of ~500 pN is used for MB435 cell-cantilever attachment.
13. If the MB435 cell is loosely attached to the cantilever, keep them in the medium for a longer period of time to let the cell attach well to the cantilever, or lower the cantilever once again over the MB435 cell and maintain the force applied by the cantilever on the cell for 2–3 s. If the cell is completely

detached from the cantilever, repeat from **step 7**. Do not lift the AFM head out of cell culture media after the cell has been attached on the cantilever. If the cantilever is removed from the medium by lifting the head, the cell will be detached from the cantilever due to the surface tension at the air-water interface [1].

14. A cell-cell junction in the HBMEC-60 monolayer is chosen as a contact site for the MB435 cell because these regions are the putative sites of cancer cell extravasation [12].
15. The range of cantilever force distance in this experiment is adjusted up to 40 μm to completely separate the MB435 cell from the endothelial cells when the cantilever retracts. Measurements are performed using closed loop feedback mode in order to minimize the vertical drift of the cell-tipped probe [1]. During force measurement, the cantilever undergoes hydrodynamic drag force due to viscous friction with the medium. To decrease the hydrodynamic drag acting on the cantilever, relatively low cantilever velocity should be used [24, 25]. The cantilever velocity used for this experiment was 1.6 μm/s.
16. Usually, after collecting 3–5 force curves for a short contact time (< 60 s) or 1–2 force curves for a longer contact time, keep the cell-tipped probe retracted for 1 min to allow cell recovery. Collect at least 10 force curves for each cell-cell junction site.

Acknowledgments

This work was supported by NIH P01HL095486 (to G.A. Meininger), NIH R01CA160461 (to V.V. Glinsky), and VA BLR&D Service Award 1I01BX000609 (to V.V. Glinsky).

References

1. Friedrichs J, Legate KR, Schubert R, Bharadwaj M, Werner C, Müller DJ, Benoit M (2013) A practical guide to quantify cell adhesion using single-cell force spectroscopy. Methods 60:169–178
2. Puech PH, Poole K, Knebel D, Muller DJ (2006) A new technical approach to quantify cell-cell adhesion forces by AFM. Ultramicroscopy 106:637–644
3. Helenius J, Heisenberg CP, Gaub HE, Muller DJ (2008) Single-cell force spectroscopy. J Cell Sci 121:1785–1791
4. Khalili AA, Ahmad MR (2015) A review of cell adhesion studies for biomedical and biological applications. Int J Mol Sci 16:18149–18184
5. Lele TP, Sero JE, Matthews BD, Kumar S, Xia S, Montoya-Zavala M, Polte T, Overby D, Wang N, Ingber DE (2007) Tools to study cell mechanics and mechanotransduction. Methods Cell Biol 83:443–472
6. Willemsen OH, Snel MM, Cambi A, Greve J, De Grooth BG, Figdor CG (2000) Biomolecular interactions measured by atomic force microscopy. Biophys J 79:3267–3281

7. Wojcikiewicz EP, Zhang MVT (2004) Force and compliance measurements on living cells using atomic force microscopy (AFM). Biol Proced Online 6:1–9
8. Trache A, Meininger GA (2008) Atomic force microscopy (AFM). Curr Protoc Microbiol Chapter 2:Unit 2C 2
9. Trache A, Trzeciakowski JP, Gardiner L, Sun Z, Muthuchamy M, Guo M, Yuan SY, Meininger GA (2005) Histamine effects on endothelial cell fibronectin interaction studied by atomic force microscopy. Biophys J 89:2888–2898
10. Zhang X, Wojcikiewicz EP, Moy VT (2006) Dynamic adhesion of T lymphocytes to endothelial cells revealed by atomic force microscopy. Exp Biol Med (Maywood) 231:1306–1312
11. Reeves KJ, Hou J, Higham SE, Sun Z, Trzeciakowski JP, Meininger GA, Brown NJ (2013) Selective measurement and manipulation of adhesion forces between cancer cells and bone marrow endothelial cells using atomic force microscopy. Nanomedicine 8:921–934
12. Laurent VM, Duperray A, Rajan VS, Verdier C (2014) Atomic force microscopy reveals a role for endothelial cell ICAM-1 expression in bladder cancer cell adherence. PLoS One 9 (e98034):1–11
13. Strell C, Entschladen F (2008) Extravasation of leukocytes in comparison to tumor cells. Cell Commun Signal 6:10
14. Glinskii OV, Li F, Wilson LS, Barnes S, Rittenhouse-Olson K, Barchi JJ Jr, Pienta KJ, Glinsky VV (2014) Endothelial integrin α3β1 stabilizes carbohydrate-mediated tumor/endothelial cell adhesion and induces macromolecular signaling complex formation at the endothelial cell membrane. Oncotarget 5:1382–1389
15. Trache A, Meininger GA (2005) An atomic force – multi optical imaging integrated microscope for monitoring molecular dynamics in live cells. J Biomed Opt 10(064023):1–17
16. Trache A, SM Lim SM (2009) Integrated microscopy for real-time imaging of mechanotransduction studies in live cells. J Biomed Opt 14:034024
17. Van Vliet KJ, Bao G, Suresh S (2003) The biomechanics toolbox: experimental approaches for living cells and biomolecules. Acta Mater 51:5881–5905
18. Trzeciakowski JP, Meininger GA (2004) NForceR: Nanoscale Force Reader and AFM data analysis package (copyrighted)
19. Zhang X, Wojcikiewicz E, Moy VT (2002) Force spectroscopy of the leukocyte function-associated antigen-1/intercellular adhesion molecule-1 interaction. Biophys J 83:2270–2279
20. Franz CM, Taubenberger A, Puech PH, Muller DJ (2007) Studying integrin-mediated cell adhesion at the single-molecule level using AFM force spectroscopy. Sci STKE 406:pl5. 1–16
21. Butt HJ, Jaschke M (1995) Calculation of thermal noise in atomic force microscopy. Nanotechnology 6(1):1
22. Hutter JL, Bechhoefer J (1993) Calibration of atomic-force microscope tips. Rev Sci Instrum 64:1868–1873
23. Huang HL, Hsing HW, Lai TC, Chen YW, Lee TR, Chan HT, Lyu PC, Wu CL, Lu YC, Lin ST, Lin CW, Lai CH, Chang HT, Chou HC, Chan HL (2010) Trypsin-induced proteome alteration during cell subculture in mammalian cells. J Biomed Sci 17:36–46
24. Janovjak H, Struckmeier J, Muller DJ (2005) Hydrodynamic effects in fast AFM single-molecule force measurements. Eur Biophys J 34:91–96
25. A-Hassan E, Heinz WF, Antonik MD, D'Costa NP, Nageswaran S, Schoenenberger CA, Hoh JH (1998) Relative microelastic mapping of living cells by atomic force microscopy. Biophys J 74:1564–1578
26. Martinez-Lemus LA, Sun Z, Trache A, Trzeciakowski JP, Meininger GA (2005) Integrins and regulation of the microcirculation: from arterioles to molecular studies using atomic force microscopy. Microcirculation 12:99–112

Chapter 31

Methods for Atomic Force Microscopy of Biological and Living Specimens

Simone Dinarelli, Marco Girasole, and Giovanni Longo

Abstract

Two main precautions must be taken into account to obtain high-resolution morphological and nanomechanical characterization of biological specimens with an atomic force microscope: the tip-sample interaction and the sample-substrate adhesion. In this chapter we discuss the necessary steps for a correct preparation of three types of biological samples: erythrocytes, bacteria, and osteoblasts. The main goal is to deliver reproducible protocols to produce good cellular adhesion and minimizing the morphological alterations of the specimens.

Key words Atomic force microscopy, Living cells, Bacteria, Topography, Young's modulus, Nanomechanical information

1 Introduction

The atomic force microscope (AFM) is a characterization technique [1] which **employs a sharp tip (with radius of curvature** in the nanometer range) attached to the end of a flexible cantilever with a known elastic constant and a reflective coverage on the back. In the most common setup, a laser is focused on the back on the cantilever, and the reflected spot is monitored by a four-quadrant detector that allows the evaluation of the tip position and cantilever deflection. An xyz stage, usually made of piezoelectric materials, is guided by a feedback loop in order to scan the surface on the xy plane and maintain a constant distance/force between tip and sample in the z direction. By collecting the point-by-point deflections of the cantilever, the AFM can reconstruct the morphology of any kind of sample. The resulting image has a lateral resolution that can reach few nanometers and is intrinsically tridimensional and quantitative.

Due to the AFM's ability to perform analyses in various environmental conditions, including vacuum, air, or liquid, this technique has become extremely important to determine the high-resolution features of biological specimens. Furthermore, due to

Yuri L. Lyubchenko (ed.), *Nanoscale Imaging: Methods and Protocols*, Methods in Molecular Biology, vol. 1814,
https://doi.org/10.1007/978-1-4939-8591-3_31,

the extremely high sensitivity of the cantilever, the tip-sample interaction can be controlled down to less than a nN, allowing a very soft probing of any kind of surface. With such capabilities, the AFM can be used to reconstruct the surface topography of viable cells in biocompatible fluid and monitor their real-time evolution.

In addition to the super-resolution imaging, by exploiting the sensitivity of the cantilever, the AFM can be used to investigate other characteristics of specimens. For instance, it can determine the tribological properties of a surface by collecting the lateral torsions of the cantilever during the scan. On the other hand, since the cantilever is a very sensitive force sensor, it can provide information on the elastic properties of the samples. Virtually every category of living organisms had its stiffness and Young's moduli explored by AFM, starting from single proteins up to whole tissues or organs [2–6].

While the sample preparation for AFM imaging and characterization has very few limitations, great care must be taken in the preparation protocols. Indeed, unlike conventional optimal microscopy, in the AFM the sample plays an active role in the measurement procedure. The information on the sample arises from a direct interaction between the tip and the specimen; thus the success of a high-resolution AFM investigation often lays in the development, or optimization, of the sample's preparation and in the control of the tip-sample interaction. In practical terms, the strongest requirement is a continuous control of the tip pressure and of the sample adhesion. The tip pressure needs to be finely tuned with respect to the sample under investigation, in order to avoid sample damages or morphological artifacts in the acquired images. At the same time, the specimens must be well-attached to a substrate: every movement of the sample, especially those induced by the action of the tip, will result in a loss of lateral resolution and consequent loss in image quality. To ensure this, there are many protocols present in literature, which depend on the cells under investigation and on the experimental environment.

Here we describe some of these methods, presenting those that we use in our everyday research practice to prepare both living and dehydrated biological samples for proper morphological and ultrastructural characterizations. The paramount goal is to achieve a controlled preparation of the surface and sample immobilization to ensure that the specimens are immobilized to the substrate but do not undergo modifications caused by the procedure.

2 Materials

All the media and reagents must be sterile. To ensure buffers' effectiveness, it is advisable to employ new solutions for each experiment.

2.1 Buffers

1. Physiological solution: 0.9% NaCl solution in ultrapure water.
2. Phosphate buffer: 0.5 M: $Na_2HPO_42H_2O$ (0.5 M) and H_2NaPO_4 (0.5 M) to obtain pH 7.4.
3. RBC buffer solution: final concentrations of 0.8% NaCl, 1 mM ethylenediaminetetraacetic acid (EDTA) as anticoagulant, 10 mM of phosphate buffer (*see* **Note 1**).
4. Antibiotic mixture for RBC sterility: gentamicin 5 mL/L, penicillin 1:100 and streptomycin 1:100 (*see* **Note 2**).
5. Lysogeny broth is the most widely used nutritionally rich medium, for the culture and growth of bacteria. There are several commercial formulations, but the composition generally comprises peptides and casein peptones, trace elements, minerals, and vitamins.
6. McCoy's medium is a general purpose and commercially available medium that supports the propagation of many types of primary cells.
7. Phosphate buffer saline (PBS) is one of the most common salt solutions, containing a variety of salts. The most common commercial recipe includes NaCl, KCl, Na_2HPO_4, and KH_2PO_4.
8. Trypsin-EDTA is a commercial buffer used to detach the cells from a Petri dish or other substrates. In our case we used a 1:250 solution.

2.2 Titanium Substrates

The metabolic activity of osteoblasts, including the triggering of their adhesion and bone-forming pathways, is strongly dependent on the environmental conditions, and these cells exhibit many different surface receptors that probe the substrates' physical and chemical properties. It has been demonstrated that, in addition to the surface chemistry, ideal cellular growth requires a high control over the micro- and nano-roughness of the substrate (*see* **Note 3**):

1. Cut Titanium grade 2 rods (*see* **Note 4**) into disks 13 mm in diameter, 2 mm thick.
2. Polish with 1 μm diamond paste.
3. Blast with 120 μm microspheres for 2–10 min (*see* **Note 5**).
4. Clean thoroughly by immersion for 30 min in a 0.1% solution of Triton X-100.
5. Sonicate 10 min in ultrapure water and rinse with ultrapure water.
6. Wash three times with hexane and air dry.
7. Verify the achieved roughness using AFM and profilometer characterization [7].

2.3 Substrates for Bacteria

For bacterial immobilization, we have often employed polydimethylsiloxane (PDMS) substrates (*see* **Notes 6** and 7) or conventional cover slips:

1. Combine the monomer and curing agent in 10:1 proportion in a separate cuvette.
2. Mix the components continuously for more than 5 min to ensure the complete mixing of the components.
3. Expose the mix to vacuum for at least 1 h or until all air bubbles have been expelled (*see* **Note 8**).
4. Pour gently the mixture in the chosen container, and, if needed, press a metallic mold over the PDMS to bestow the chosen form [8].
5. Expose the mix to vacuum for a minimum of 10 min or until all air bubbles have been expelled.
6. Incubate overnight in an oven maintained at 70 °C, and verify the solidity of the resulting layer.
7. To make the substrate temporarily hydrophilic, place in plasma cleaner for 5 min at maximum power in low vacuum air atmosphere.
8. Expose the clean and hydrophilic substrates for 1.5 min to 0.2% (3-aminopropyl)triethoxysilane (APTES) (*see* **Note 9**), and rinse thoroughly in ultrapure water.

2.4 AFM Imaging

Commercial AFM instruments are extremely varied and are now routinely used in many laboratories. In addition, several groups have developed custom devices, often optimized for specific applications, whose performances can be comparable to the commercial counterparts [9, 10]. In our experience we have employed both systems to obtain excellent characterizations, and most of the measurement protocols are instrument-independent. The only main requirement for a bio-oriented AFM is the ability to work in liquid environment and the possibility to bear liquid-filled analysis chambers. A big advantage is the presence of a good optical imaging equipment coupled to the AFM, in order to allow cell preliminary investigation and tip pointing.

3 Methods

3.1 Protocols for RBC Preparation

As first example we present protocol for the cellular adhesion, which we have employed to analyze fine morphological properties of cells without affecting their features. We chose red blood cells (RBCs), which are among the simplest and most studied biological specimens (*see* **Note 10**) [11, 12]. RBCs lack all adhesion proteins (such as integrins and annexins) and react strongly to any attempt of firm chemical immobilization. This greatly limits their use in their native viable state in AFM, thus the need to perform cell attachment through dehydration [13].

3.1.1 RBC Purification

The first step involves the purification of the cells from whole blood samples, separating the RBC from white blood cells, plasma, and other blood components:

1. Withdraw the whole blood from a donor.
2. Dilute immediately 1:3 in the RBC buffer solution.
3. Centrifuge: 850 × *g* for 10 min.
4. The cells will precipitate on the bottom of the vial, while the plasma-enriched supernatant must be separated and stored at 4 °C with 1 mM of PSMF (phenyl-methyl-sulfonyl-fluoride, a protease inhibitor) (*see* **Note 11**).
5. Carefully remove the whole supernatant and re-suspend the cells in RBCs buffer solution.
6. Repeat **steps 3–5** two more times to obtain purified red blood cells in proportion 1:4 (v/v) with the buffer solution (*see* **Notes 12** and **13**).

3.1.2 Smear Procedure

1. Prepare a smear solution composed by a 1:1 (v/v) mixture of the RBC solution and the plasma fraction (*see* **Note 14**).
2. Place a small aliquot (around 5 μL) of the smearing onto a commercial poly-l-lysine-coated glass slide.
3. Spread the aliquot using a second glass slide, moving slowly to obtain a uniform covering of the surface (detailed in Fig. 1; *see* **Note 15**).
4. Dry the smeared glass slides in air and store in low-humidity atmosphere (*see* **Note 16**).

3.2 Preparation of Live Osteoblasts for AFM Characterization

Osteoblasts are the main actors in the bone formation. To achieve an optimal osseointegration of implants and prostheses, cell adhesion and proliferations are the main parameters to be monitored. To achieve this goal, high-resolution techniques are used to study the formation of filopodia and pseudopodia, which indicate a good adhesion with the substrate. In some cases, the AFM can be used directly to measure the cell-surface interaction [14]:

1. Isolate human primary osteoblasts (hOB) by collecting bone fragments (*see* **Note 17**).
2. Wash the bone fragments in sterile phosphate buffer saline (PBS).
3. Mince the fragments and treat them with 1 mg/mL collagenase type IV and 0.25% trypsin for 1 h at 37 °C with gentle agitation.

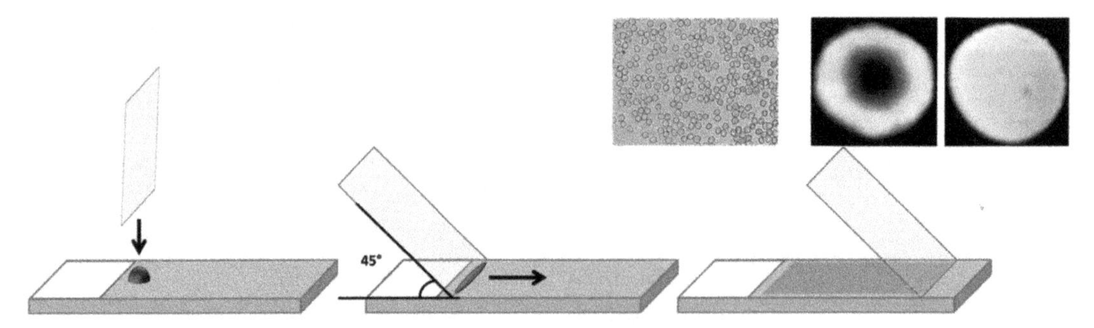

Fig. 1 Left: Place a 5 μL droplet of the smear solution on the beginning of the glass slide, and gently put a cleaned rounded edge glass slide in contact with the droplet; wait for few seconds until the drop wet the entire edge of the glass slide. Center: Without losing contact with the surface, maintain an angle of around 45° prior to the starting of the smear. Right: Complete the smear by moving horizontally with a constant speed and without losing the contact. Inset: left, resulting typical smear for optimal AFM imaging; right, AFM images of a biconcave (left) and flat RBC (right)

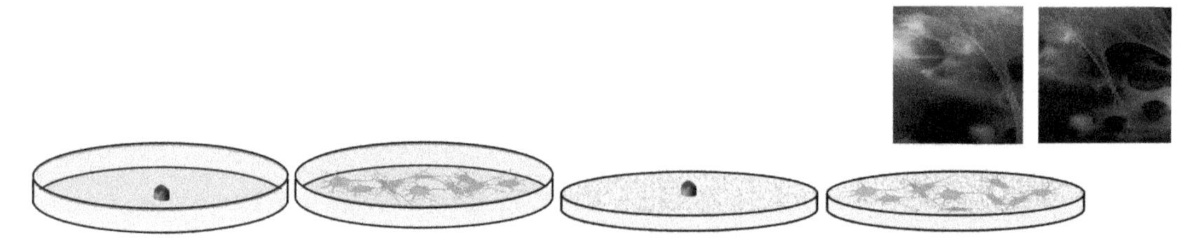

Fig. 2 Schematization of the preparation protocol for AFM imaging of live osteoblasts. Grow cells in a petri dish up to confluence, extract cells, and seed them on a titanium substrate with controlled micro- and nano-roughness. Inset: resulting typical AFM images of a living osteoblast. The surface roughness and chemical properties of the substrate induce the formation of long filopodia which ensure a good cellular attachment. The interaction between the tip and the cell causes, over time, the filopodia to retract. The images evidence the actin stress fibers on the extremities of the cell

4. Repeat the procedure and collect the cells by centrifugation at $200 \times g$.
5. Seed the cells at minimal concentration of 2.5×10^4 cells, and culture at 37 °C in 5% CO_2 in McCoy's medium supplemented with L-glutamine, penicillin/streptomycin, and 15% FBS.
6. Once the cells have reached 80% confluence, detach using trypsin-EDTA for 5 min at 37 °C.
7. Seed the cells, at minimal concentration of 2.5×10^4 on a titanium disc.
8. Culture for 6 h in McCoy medium at 37 °C, 5% CO_2 (a schematic of passages 5–8 is shown in Fig. 2).

3.3 Preparation of Live Bacteria for AFM Characterization

In a second example, we characterize the nanomechanical properties of living bacteria (*see* **Note 18**). The method described here was optimized for *Escherichia coli* (*see* **Note 19**), but it can be easily extended to other bacterial species [15–17].

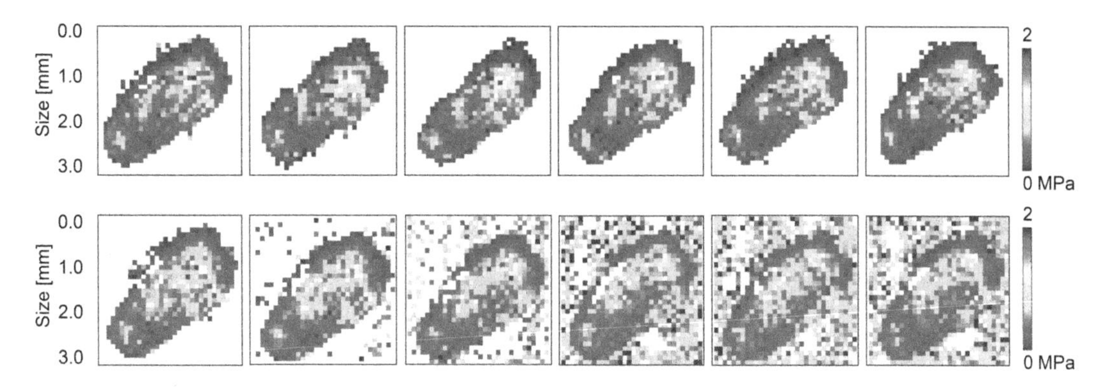

Fig. 3 Time evolution of the stiffness map of a single bacterium before (top images) and after (bottom images) the exposure to a bactericidal dose of antibiotic (ampicillin). The substrate functionalization allows a very good immobilization of the cell, enabling the determination of the effect of the drug at the nanoscale

1. Keep frozen stocks of *E. coli* at −80 °C in LB media supplemented with glycerol.
2. Streak the bacteria from stock on LB agar media (*see* **Note 20**).
3. When the first colonies appear, collect few of them from the agar plate, and incubate overnight at 37 °C in LB.
4. After incubation, wash the bacteria by centrifugation for 10 min at 2400 × *g*.
5. Resuspend the pellet in phosphate buffer saline (PBS).
6. Repeat washing three times (*see* **Note 21**).
7. Place a droplet of the bacteria-rich PBS on the APTES-functionalized substrates, and incubate for 15 min at room temperature.
8. Rinse very gently with PBS to remove non-attached cells, and cover with fresh PBS for AFM imaging.
9. During measurements, very gently introduce drugs or chemicals to monitor, at the single-cell level, their effect on the bacteria (Fig. 3).

3.4 AFM Imaging

1. The AFM tip-cantilever combination for the measurement must be chosen bearing in mind the properties of the sample under investigation (*see* **Notes 22** and **23**).
2. Typical cantilever stiffness values for contact and force investigations of a living biological system vary from 0.01 to 0.12 N/m. For fixed and dehydrated cells, cantilevers from 0.03 to 0.3 N/m can be used. For tapping or non-contact mode, the stiffness can be as high as 1 N/m.
3. Ensure that the cantilever mechanical properties and AFM detector are correctly calibrated (*see* **Note 24**).

4. Use the fine AFM motors to place the tip in the close vicinity of the chosen cell.
5. Approach on the substrate near to the cell to avoid tip damaging during the approach. Ensure a tip-sample interaction of 1–5 nN is chosen. For fixed cells, higher interactions can be used (typically not larger than 10–15 nN).
6. Ensure that the scan rate for the imaging is slow, to ensure a proper feedback of the counterreaction system to maintain a constant tip-sample interaction.
7. For force investigations, especially in the case of living specimens, keep the maximum applied force to low values (1–5 nN), but make sure that the retraction part of the force curves allows the complete detachment of the tip (*see* **Note 25**).
8. Increasing the maximum tip-sample interaction can lead to more subtle and complex investigations of the sample [18, 19] but can also lead to sample damaging.
9. Viscoelasticity must be taken into consideration, especially when performing fast force-curve measurements.

4 Notes

1. The osmolarity of this solution is around 300 mOsm in order to maintain the osmotic equilibrium of the RBC environment and avoid the action of osmotic forces. The buffer was sterilized by autoclavation of each component.
2. To be used in case of prolonged storage of the cells.
3. A smooth surface does not allow the osseointegration of a surface, while, in order for osteoblasts to carry out the bone formation, a microscopic (1–10 μm) or nanoscopic (1–10 nm) roughness must be induced, mechanically or chemically.
4. Titanium is among the most used materials for implants and prostheses, due to its small weight and high resistance. We have employed titanium rods composed of Ti = 99.85%; C = 0.0006%; Ni = 0.01%; O = 0.1%; H = 0.003%; and Se = 0.03%.
5. Sand blasting with alumina (Al_2O_3) or silica (SiO_2) particles is widely used. However, it has been found that these materials often release into the implant cytotoxic silica or aluminum ions. This drawback is eliminated by using calcium phosphate particles or zirconia (ZrO_2) particles.
6. PDMS is a silicon-based organic polymer, in two components, which solidify after mixing and heating. The material is optically transparent, highly hydrophobic, and nontoxic.

7. This material has greater versatility than conventional cover slips and is naturally hydrophobic, enabling controlled patterning of the liquid.
8. The mixture will swell during the degassing phase, and care must be taken to avoid spilling.
9. This highly effective silane coupling agent is used on a wide series of substrates to enhance cell adhesion [20].
10. In their mature form, RBCs adopt a biconcave shape with a diameter of around 8 μm and a typical height of 800 nm. Their physiological role of oxygen transporter in the bloodstream makes them a crucial component for the health of a living organism.
11. This fraction of plasma is necessary for the RBC smears and needs to be stored at 4 °C, or else it will denature in few hours.
12. Final hematocrit 20%.
13. In case of prolonged storage, add 5 μL/mL of antibiotic, to avoid contamination during the experiments.
14. The plasma must be allowed to reach room temperature prior to the mixing.
15. The dilution ratio 1:1 and the 20% hematocrit are chosen in order to obtain homogeneous smears with an optimal density of cells over the whole glass slide (critical for the AFM measurements).
16. Stored smears are stable over years if properly conserved: no direct light exposure and constant room temperature and humidity.
17. The protocols to grow and prepare commercially available cell lines are similar to the hOB. The primary cultures are more sensitive to the environment and must be grown and handled more carefully.
18. It has been demonstrated that the mechanical properties of the membrane can reflect the physiological state of the entire biosystem and that these properties can be altered by the occurrence of pathological conditions [21–24].
19. Specifically the DH5α strain, Gram-negative rodlike-shaped bacteria (typically 500 nm wide and several microns long).
20. In most cases a small scratch from the top of the frozen aliquot is sufficient to ensure a good streak without the requirement for complete thawing of the stock.
21. This protocol is required since amino acids and other components of the LB can passivate the substrate, impeding the immobilization of the bacteria.

22. A tip too stiff can damage the sample and will not be able to investigate the mechanical properties of a soft cell; on contrary, a too soft tip could be inadequate for mechanical characterization and could be damaged by the tip-sample interaction.
23. The users must control the consistency between the tip length (typically 1–20 μm) and the expected height of the specimen. Attempts to perform imaging of a cell nucleus (whose height can be 1–5 μm) with a short tip will produce artifacts due to nonoptimal tip-sample interaction.
24. Each microscope has its own protocol for this calibration; thus each user must refer to the instruction manual for the calibration.
25. Start with low maximum tip-sample interaction and long force curves (1–2 μm for each retraction curve). Once the first force curves are collected over the specimen, the parameters can be adjusted to reduce the curve length and increase the applied force.

Acknowledgments

This work was supported by the Italian Department of Health through the grant n.GR-2009-1305007.

References

1. Binnig G, Quate CF, Gerber C (1986) Atomic Force Microscope. Phys Rev Lett 56 (9):930–933. https://doi.org/10.1103/PhysRevLett.56.930
2. Cappella B, Dietler G (1999) Force-distance curves by atomic force microscopy. Surf Sci Rep 34:1–104. https://doi.org/10.1016/S0167-5729(99)00003-5
3. Kasas S, Longo G, Dietler G (2013) Mechanical properties of biological specimens explored by atomic force microscopy. J Phys D Appl Phys 46(13):133001. https://doi.org/10.1088/0022-3727/46/13/133001
4. Longo G, Kasas S (2014) Effects of antibacterial agents and drugs monitored by atomic force microscopy. Wiley Interdiscip Rev Nanomed Nanobiotechnol 6(3):230–244. https://doi.org/10.1002/wnan.1258
5. Japaridze A, Muskhelishvili G, Benedetti F, Gavriilidou AF, Zenobi R, De Los Rios P, Longo G, Dietler G (2017) Hyperplectonemes: a higher order compact and dynamic DNA self-organization. Nano Lett 17 (3):1938–1948. https://doi.org/10.1021/acs.nanolett.6b05294
6. Ruggeri FS, Mahul-Mellier A-L, Kasas S, Lashuel HA, Longo G, Dietler G (2017) Amyloid single-cell cytotoxicity assays by nanomotion detection. Cell Death Discov 3:17053. https://doi.org/10.1038/cddiscovery.2017.53. https://www.nature.com/articles/cddiscovery201753#supplementary-information
7. Zanoni R, Ioannidu CA, Mazzola L, Politi L, Misiano C, Longo G, Falconieri M, Scandurra R (2015) Graphitic carbon in a nanostructured titanium oxycarbide thin film to improve implant osseointegration. Mater Sci Eng C Mater Biol Appl 46:409–416. https://doi.org/10.1016/j.msec.2014.10.073
8. Kasas S, Radotic K, Longo G, Saha B, Alonso-Sarduy L, Dietler G, Roduit C (2013) A universal fluid cell for the imaging of biological specimens in the atomic force microscope. Microsc Res Tech 76(4):357–363. https://doi.org/10.1002/jemt.22174
9. Girasole M, Longo G, Cricenti A (2006) An alternative tapping scanning near-field optical microscope setup enabling the study of biological Systems in Liquid Environment.

Jpn J Appl Phys 45(3B):2333–2336. https://doi.org/10.1143/JJAP.45.2333

10. Longo G, Girasole M, Cricenti A (2008) Implementation of a bimorph-based aperture tapping-SNOM with an incubator to study the evolution of cultured living cells. J Microsc 229 (Pt 3):433–439. https://doi.org/10.1111/j.1365-2818.2008.01924.x
11. Girasole M, Dinarelli S, Boumis G (2012) Structure and function in native and pathological erythrocytes: a quantitative view from the nanoscale. Micron 43(12):1273–1286. https://doi.org/10.1016/j.micron.2012.03.019
12. Girasole M, Dinarelli S, Boumis G (2012) Structural, morphological and nanomechanical characterisation of intermediate states in the ageing of erythrocytes. J Mol Recognit 25 (5):285–291. https://doi.org/10.1002/jmr.2170
13. Girasole M, Pompeo G, Cricenti A, Longo G, Boumis G, Bellelli A, Amiconi S (2010) The how, when, and why of the aging signals appearing on the human erythrocyte membrane: an atomic force microscopy study of surface roughness. Nanomed-Nanotechnol Biol Med 6(6):760–768. https://doi.org/10.1016/j.nano.2010.06.004
14. Longo G, Ioannidu CA, Scotto d'Abusco A, Superti F, Misiano C, Zanoni R, Politi L, Mazzola L, Iosi F, Mura F, Scandurra R (2016) Improving osteoblast response in vitro by a nanostructured thin film with titanium carbide and titanium oxides clustered around graphitic carbon. PLoS One 11(3):e0152566. https://doi.org/10.1371/journal.pone.0152566
15. Longo G, Rio LM, Roduit C, Trampuz A, Bizzini A, Dietler G, Kasas S (2012) Force volume and stiffness tomography investigation on the dynamics of stiff material under bacterial membranes. J Mol Recognit 25(5):278–284. https://doi.org/10.1002/jmr.2171
16. Longo G, Rio LM, Trampuz A, Dietler G, Bizzini A, Kasas S (2013) Antibiotic-induced modifications of the stiffness of bacterial membranes. J Microbiol Methods 93 (2):80–84. https://doi.org/10.1016/j.mimet.2013.01.022
17. Dinarelli S, Girasole M, Kasas S, Longo G (2017) Nanotools and molecular techniques to rapidly identify and fight bacterial infections. J Microbiol Methods 138:72–81. https://doi.org/10.1016/j.mimet.2016.01.005
18. Kasas S, Dietler G (2008) Probing nanomechanical properties from biomolecules to living cells. Pflugers Archiv 456(1):13–27. https://doi.org/10.1007/s00424-008-0448-y
19. Roduit C, Sekatski S, Dietler G, Catsicas S, Lafont F, Kasas S (2009) Stiffness tomography by atomic force microscopy. Biophys J 97 (2):674–677. https://doi.org/10.1016/j.bpj.2009.05.010
20. Sunkara V, Park DK, Hwang H, Chantiwas R, Soper SA, Cho YK (2011) Simple room temperature bonding of thermoplastics and poly (dimethylsiloxane). Lab Chip 11(5):962–965. https://doi.org/10.1039/c0lc00272k
21. Cross SE, Jin Y-S, Rao J, Gimzewski JK (2007) Nanomechanical analysis of cells from cancer patients. Nat Nanotechnol 2(12):780–783. https://doi.org/10.1038/nnano.2007.388
22. Lekka M, Fornal M, Pyka-Fościak G, Lebed K, Wizner B, Grodzicki T, Styczeń J (2005) Erythrocyte stiffness probed using atomic force microscope. Biorheology 42(4):307–317
23. Gaboriaud F, Parcha BS, Gee ML, Holden JA, Strugnell RA (2008) Spatially resolved force spectroscopy of bacterial surfaces using force-volume imaging. Colloids Surf B: Biointerfaces 62:206–213. https://doi.org/10.1016/j.colsurfb.2007.10.004
24. Rossetto G, Bergese P, Colombi P, Depero LE, Giuliani A, Nicoletto SF, Pirri G (2007) Atomic force microscopy evaluation of the effects of a novel antimicrobial multimeric peptide on Pseudomonas aeruginosa. Nanomed Nanotechnol Biol Med 3(3):198–207. https://doi.org/10.1016/j.nano.2007.06.002

Chapter 32

Single Molecule Imaging in Live Embryos Using Lattice Light-Sheet Microscopy

Mustafa Mir, Armando Reimer, Michael Stadler, Astou Tangara, Anders S. Hansen, Dirk Hockemeyer, Michael B. Eisen, Hernan Garcia, and Xavier Darzacq

Abstract

In the past decade, live-cell single molecule imaging studies have provided unique insights on how DNA-binding molecules such as transcription factors explore the nuclear environment to search for and bind to their targets. However, due to technological limitations, single molecule experiments in living specimens have largely been limited to monolayer cell cultures. Lattice light-sheet microscopy overcomes these limitations and has now enabled single molecule imaging within thicker specimens such as embryos. Here we describe a general procedure to perform single molecule imaging in living *Drosophila melanogaster* embryos using lattice light-sheet microscopy. This protocol allows direct observation of both transcription factor diffusion and binding dynamics. Finally, we illustrate how this *Drosophila* protocol can be extended to other thick samples using single molecule imaging in live mouse embryos as an example.

Key words Single molecule imaging, Single molecule kinetics, Lattice light-sheet microscopy, *Drosophila melanogaster*, Live embryo imaging, Single molecule fluorescence, Transcription factor dynamics, Single particle tracking, Selective plane illumination microscopy

1 Introduction

Single molecule measurements of protein dynamics in living cells have provided key insights into how proteins explore their environment [1], how they interact with DNA [2–5], and have revealed novel regulatory mechanisms of spatiotemporal organization of proteins [6, 7]. However, due to technological limitations, such studies have been limited to monolayer cell cultures.

The critical barrier to overcome to achieve single molecule sensitivity in imaging is the signal-to-noise ratio. Improved signal comes from the ability to collect as many photons as possible from a fluorophore, while the noise contribution is mainly from out of focus fluorophores and sensor noise. In conventional wide-field microscopes, the same objective lens is used for excitation and

Yuri L. Lyubchenko (ed.), *Nanoscale Imaging: Methods and Protocols*, Methods in Molecular Biology, vol. 1814, https://doi.org/10.1007/978-1-4939-8591-3_32,

detection, and as a result the noise increases when moving deeper into a sample. For these reasons primarily, single molecule imaging studies have been restricted to regions of the sample within just a few microns of the coverslip. Lattice light-sheet microscopy (LLSM) [8] addresses this technological limitation and enables single molecule imaging in thicker specimens such as living embryos [9].

In short, LLSM works by creating a nearly diffraction-limited light sheet which only excites the slice of the sample from which useful, in-focus signal can be gathered. The emitted light is collected by a separate detection objective oriented orthogonally to the excitation objective. The thin excitation sheet and separate detection objective results in signal-to-noise ratios (SNRs) comparable to those achieved in monolayer cultures while minimizing photo-bleaching and phototoxicity.

We have recently demonstrated the first application of LLSM to perform single molecule imaging in embryos by tracking single molecules of the transcription factor Bicoid in early *Drosophila melanogaster* embryos [9]. The ability to quantify single molecule kinetics during development in terms of off-rates of DNA binding, fraction of molecules in a bound or mobile state, diffusion coefficients, and spatial distributions has a huge discovery potential as highlighted by our findings on Bicoid [9] . We expect that extending these live-embryo single molecule imaging approaches to other factors including proteins [5], RNA [10], and DNA [11] will contribute greatly to our understanding of *Drosophila* development. Accordingly, we expect that interest in performing such experiments will grow as LLSM and other complementary technologies mature and become widely available.

The procedures described here are thus meant to serve as a starting point for those who have access to a LLSM setup and require single molecule dynamics data in live *Drosophila* or other embryos suitable for imaging to answer their research question. Access to a functional LLSM and general proficiency in aligning and optimizing the microscope is assumed. Furthermore, the exact data analysis and microscope acquisition parameters are highly dependent on the particular factor you are studying, the fluorescent label that it is tagged with, and the specific questions you would like to answer. As such, this chapter emphasizes the lessons we have learned along the way in terms of sample preparation and optimizing acquisition parameters and only provides general guidelines for data analyses. We hope that these descriptions will serve as guiding principles to those performing single molecule measurements in embryos for the first time and thus minimize the amount of troubleshooting necessary to optimize their particular experimental conditions. Furthermore, to emphasize the point that the guidelines detailed here can serve as starting point to acquire single molecule data in any living embryo suitable for imaging, we also close this chapter by providing an example of single molecule imaging in mouse embryos, but as this is a proof-of-principle, we do not provide details of sample preparation and acquisition for this case as these have not yet been fully optimized.

2 Materials

2.1 Fly Husbandry in Preparation for Embryo Collection

1. Fly cages and appropriate agar lids: can be homemade or bought. For example, plastic Drosophila stock bottle, flystuff.com, or embryo collection cages. Petri dishes of an appropriate size can be used as food-containing lids. For example, *see* Fig. 1a.
2. Food preparation for cage lid: 2.4% g/w Bacto agar, 25% apple juice, 75% distilled water, and 0.001% of mold inhibitor from solution of 0.1 g/mL (Carolina 87–6165 or your favorite food mixture (e.g., grape agar plates)), smear prepared lids with yeast paste in the center (mix 1 g of dry yeast in 1 mL of water to make yeast paste).
3. Fly strain with protein of interest tagged with a suitable fluorescent label and a histone (e.g., H2B) labeled in another color for precise focusing and staging. In addition, a line with histones labeled using the same fluorescent label as is used for single molecule imaging of the protein of interest is essential to perform the necessary controls for quantitative analysis of residence times. For fast kinetics, a line containing your fluorescent label fused to a nuclear localization signal (NLS) should be used as a control.

2.2 Lattice Light-Sheet Microscopy

1. Lattice light-sheet microscope equipped with appropriate excitation laser and emission filters.
2. Phosphate-buffered saline (PBS).
3. Squirt bottle containing deionized water (from Milli-Q or equivalent system).
4. Squirt bottle containing 10% ethanol in deionized water.
5. Dye alignment solution: Phosphate-buffered saline, dye corresponding to fluorophore (e.g., Alexa Fluor 488 for GFP). Multiple dyes can be combined if using multiple laser lines for an experiment. The dye concentration should be kept at the minimum required for the alignment on your system.
6. Bead alignment slide: TetraSpeck Microspheres, 0.1 μm (Thermo Fisher T-7279), poly-D-lysine (1 mg/mL), glass coverslip 5 mm diameter (Warner Instruments). Dilute beads to 1:200 in deionized water, and sonicate for 5 min prior to use to break up any aggregates (*see* **Note 1**). Add 10 μL poly-D-lysine

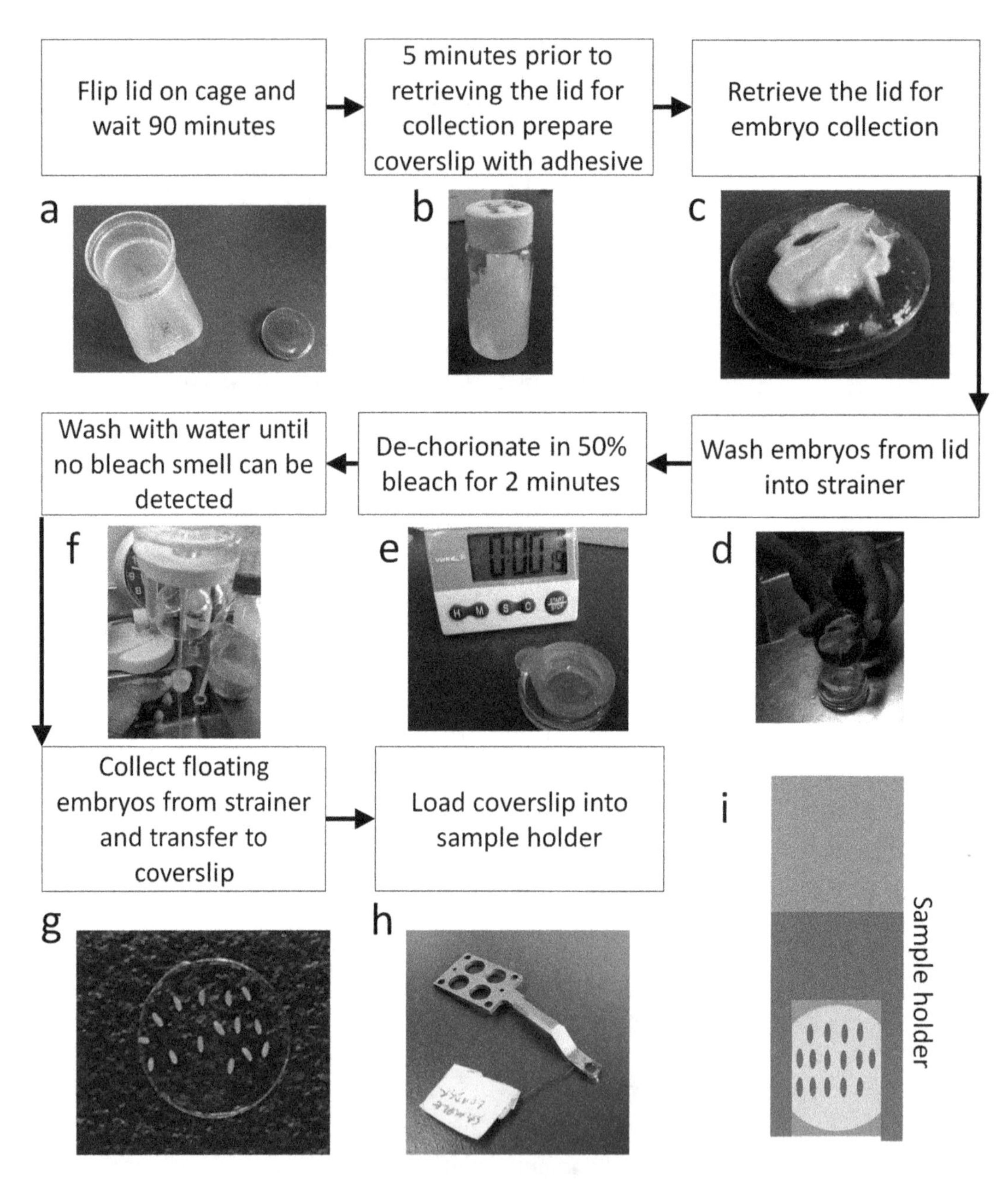

Fig. 1 Summary of embryo collection and mounting protocol. The flowchart shows the sequence of the key steps. (**a**) Empty fly cage and lid containing agar food mixture. (**b**) Double-sided scotch tape in heptane to make adhesive solution. (**c**) Lid with embryos ready for collection. (**d**) Washing embryos from lid into a strainer basket using a squirt bottle filled with water. (**e**) De-chorionation in strainer basket. (**f**) Washing bleach off after de-chorionation. (**g**) Coverslip with embryos mounted in a grid pattern. (**h**) Sample holder with wire used to hold up clip ready for loading. (**i**) Illustration showing how the coverslip should be oriented in the sample holder

to a clean coverslip for 10 min, rinse with water, and then add 10 μL of the microsphere solution, and allow to dry. Rinse again with water to wash off any beads that have not adhered.

7. Fine non-marring tweezers.

2.3 Embryo Collection and Mounting

1. Adhesive solution: Double-sided scotch tape, heptane. Prepare glue by dissolving adhesive from ~1/4 of a roll of double-sided scotch tape in heptane overnight (*see* **Note 2** and Fig. 1b). Scotch tape can be removed from vial after dissolving the adhesive.
2. Glass coverslip 5 mm diameter (Warner Instruments).
3. Kimwipes.
4. Squirt bottle containing isopropyl alcohol, 99%.
5. Squirt bottle containing deionized water (from Milli-Q or equivalent system).
6. Cell strainer 40 μm nylon (Falcon).
7. De-chorionation solution: 50% household bleach in water. Prepared fresh daily.
8. Phosphate-buffered saline 1× (PBS).
9. Dissection microscope (e.g., flystuff.com).
10. Fine haired paintbrush.
11. 60 mm petri dishes (e.g., Fisher).
12. Fine non-marring tweezers.

2.4 Lattice Light-Sheet Microscopy and Analysis

1. Computer with MATLAB® (MathWorks) and ImageJ installed.

3 Methods

Please note that preparation of the adhesive solution and bead alignment slide should be performed the day before the actual imaging experiment. The description of the LLSM alignment assumes understanding, experience, and competency in operating the system and associated control software. It is also assumed that the LLSM you are using is built to the specifications as described in the original report [8] and that you are familiar with methods common to *Drosophila* research such as preparing fly cages for embryo collection. All procedures should be performed at room temperature.

3.1 Fly Husbandry in Preparation for Embryo Collection

1. Prepare a fly cage by combining males and females of the desired strain. Approximately 100–400 flies for a 6 ounce collection bottle.

2. Cap cage with lid containing preferred food mixture and smear of yeast paste.
3. Leave cage at room temperature for at least 3 days prior to imaging, exchanging the lid for a fresh one containing the agar food mixture and yeast paste once a day (*see* **Note 3**).
4. Prior to proceeding to **step 5** (flip the lid on the fly cage for embryo collection), make sure that the microscope is aligned and configured correctly as described in Subheading 3.2.
5. Exchange the lid on the fly cage 90 min prior to collection (after completing the dye alignment steps in Subheading 3.2); do not disturb the cage during this laying period (*see* **Note 4**).

3.2 Lattice Light-Sheet Microscope Alignment

1. Turn on all hardware components of the LLSM setup (*see* **Note 5**), the desired laser lines (*see* **Note 6**), and boot up the control software. Confirm proper communication with all components before proceeding (*see* **Note 7**).
2. Fill the microscope sample chamber with the dye solution which should match the fluorophore you will be using for imaging, for example, Alexa Fluor 488 for GFP.
3. Configure the spatial light modulator (SLM) to display a single Bessel beam for your primary imaging wavelength with the minimum and maximum numerical aperture set to the same values you will be using for the lattice patterning (*see* **Note 8**). Follow the routine alignment procedures for your system with the end goal to ensure that the galvo-scanning planes are flat with your detection plane and that the focal planes for the excitation and detection objectives are the same.
4. Once the initial dye alignment is complete, switch to the lattice pattern you will be using for imaging, and ensure uniform illumination over your desired field of view. Make adjustments to the laser collimation optics as necessary (*see* **Note 9**), and return to **step 3** above if necessary.
5. At the end of the dye alignment, ensure that you have at least 10 mW of available power at the back entrance aperture of the excitation objective for the excitation laser you will be using. For photo-switchable or photo-convertible fluorophores, ensure that at least 50 μW of power is available for your switching wavelength (typically 405 nm) at the back aperture of the excitation objective (*see* **Note 9**). These power values are the maximum you should have available; the actual values to use during the experiment should be determined as explained in Subheading 3.4.
6. Remove the dye alignment solution from the chamber, and save for future use. Continuously rinse the sample chamber with deionized water for 30 s with the main drain of the chamber connected to a vacuum trap (*see* **Note 10**).

7. Fill the chamber with deionized water, lower the objectives, and let soak for 5 min. Drain, rinse with water, and repeat the soak step with the 10% ethanol solution. Drain, and wash again continuously for 30 s with deionized water to remove all residual dye.
8. Fill the chamber with 1× PBS solution or your desired imaging media (*see* **Note 11**).
9. Load the bead alignment slide into a sample holder, and mount it on the microscope (*see* **Note 12**).
10. Reduce the exposure time, EM Gain (if applicable), and laser powers to the minimum required for visualizing the beads. Find a single isolated bead, and bring it into focus in the center of the field of view.
11. Set the camera ROI to a 64 × 64 pixel window centered in the field of view. Using a single bead in the field of view, acquire a 10 micron z-stack with 100 nm slice spacing to ensure the system and detection point spread functions are as expected. Make adjustments to the correction collar on the detection objective to reduce any spherical aberrations if necessary.
12. Run the autofocus bead alignment routine and note the new offset. Run the autofocus bead routine at least every 15 min until the offset does not change by more than 50 nm between runs. For our system this stabilization typically takes around 90 min from the wash steps after dye alignment.
13. You may flip the lid for embryo collection as soon as you run the autofocus routine for the first time.

3.3 Embryo Collection and Mounting

1. As described in Subheading 3.2, exchange the lid on the fly cage 90 min prior to collection, and do not disturb the cage during the laying period.
2. After 90 min exchange the lid again for a fresh one, and keep the freshly removed one for embryo collection (Fig. 1c).
3. Rinse a clean (*see* **Note 13**) 5 mm coverslip with water, and wick dry using a Kimwipe (*see* **Note 14**).
4. Place a small drop of the glue (~20 μL) in the center of the clean coverslip, and leave it to dry in air until ready to use (*see* **Note 15**).
5. Using the squirt bottle containing water, carefully wash the embryos off the lid into the cell strainer cup over a sink or waste beaker (Fig. 1d). Be sure to also wash off the yeast paste as many embryos could be laid into the paste. Keep squirting water over the cup to break up and wash out any yeast clumps leaving behind only water and embryos (*see* **Note 16**).
6. Once clean, wick off excess water from around the embryos by patting the bottom of the basket with a Kimwipes.

7. De-chorionate embryos by placing the strainer cup in the bottom of a 60 mm petri dish and filling with the 50% bleach solution (prepared fresh daily) while gently shaking the agitating the dish for 2 min (*see* **Note 17** and Fig. 1e).
8. After 2 min remove the strainer cup from the petri dish containing bleach, and rinse with copious amounts of water (~50 mL) until no bleach smell can be detected from the strainer cup (Fig. 1f). Place the cup in a clean 60 mm petri dish, and fill with enough water so that embryos float to the top.
9. Using the dissection microscope, collect embryos that are floating at the top (*see* **Note 18**) using a paintbrush one at a time, gently touch the embryo to a Kimwipes to wick off water, and then gently place the embryo down in the desired position on the coverslip with glue. Repeat to make an array of embryos (Fig. 1g); it is useful to have as many embryos as possible on a coverslip prior to loading it onto the microscope (*see* **Note 19**). Once all embryos are mounted, put a drop of PBS on top of the slide to prevent desiccation.
10. Load the coverslip with embryos into the LLSM sample holder by first lifting up the clip using a thin wire inserted at the base of holder (Fig. 1h), placing a drop of PBS in the empty holder, and then using non-marring tweezers to place the coverslip in the holder, gently rotate the coverslip so the embryos are oriented correctly (Fig. 1i) prior to removing the thin wire from under the sample holder clip to secure the coverslip.

3.4 Lattice Light-Sheet Microscopy

1. Once the sample slide is ready and the microscope is stable as determined by the autofocus bead routine offset correct described in Subheading 3.2, load the sample holder containing the embryos onto the microscope.
2. Set the x–y stage scan speed to 100 μm/s and the z speed to 20 μm/s to quickly navigate across the slide.
3. Using the camera corresponding to the epi-objective (on the bottom port of the LLSM) along with transillumination provided by a lamp (*see* **Note 5**) or using a dedicated transillumination LED (*see* **Note 20**), locate and record the x, y coordinates of all embryos on the slide (*see* **Note 21**).
4. Change the x–y and z scan speeds to 10 μm/s for more precise positioning.
5. Once all embryo locations are marked, turn off the transillumination and switch to the continuous scan mode with lattice light-sheet illumination.
6. Use the full camera field of view, and set acquisition parameters to minimize the laser powers necessary to visualize nuclei in

your histone marker channel (*see* **Note 22**) while setting the exposure time to 50 ms.

7. Go through as many embryos as necessary to find one at the developmental stage you want as determined by the histone channel images. Correctly being able to identify the age of the embryo simply comes with practice at gauging the stage from the size and number of nuclei in your field of view.
8. Using the histone channel image, position the embryo so you are imaging as close to the surface as possible. This is to ensure that the excitation light and emission light is travelling through as little of the actual embryo as possible (Fig. 2a, b). Once the appropriate focal plane is found, and if the embryo is oriented as shown in Fig. 1i, you may navigate using the x-stage along the length of the embryo with only minimal corrections to focus required (Fig. 2b).
9. Once you find a region, you would like to perform single molecule imaging on, return to the continuous scan mode settings, and switch the laser line to the one to be used for single molecule imaging.
10. Set the exposure time according to the temporal resolution you require. For example, 100 ms or longer, where freely diffusing molecules largely "motion blur" into the background, but bound molecules are readily detected, for residence/binding time [2, 5, 9] measurements. For capturing the dynamics of both the diffusing and bound molecules, typically an exposure time of 10 ms or shorter is appropriate (*see* **Note 23**).
11. If using an EMCCD camera, set the EM Gain to 300, and reduce the number of vertical pixels in the camera ROI such that the frame transfer time is not limiting your frame rate.
12. If this is the first time you are performing single molecule imaging with this particular fluorophore-protein combination, determine the minimum laser power you need to achieve a SNR at least 4 (*see* **Note 24**) at the exposure time you determined previously. To do this acquire ~1000 frames (or the minimum necessary to have sufficient detections) in the z-scan mode with the number of z-slices set to 0 at a given laser power (*see* **Note 25**). Load the time series acquisition into ImageJ or your favorite analysis software, and draw a line profile through a few suspected single molecule detections to get the intensity profiles and draw boxes in the background region to calculate the standard deviation of the background. You can then estimate the SNR as MaxI/σ_{bg}, where the MaxI is the maximum intensity of a single molecule detection, and σ_{bg} is the standard deviation of the background pixels (Fig. 3a). Determine the minimum laser power necessary to achieve your required SNR at your desired frame rate and exposure time.

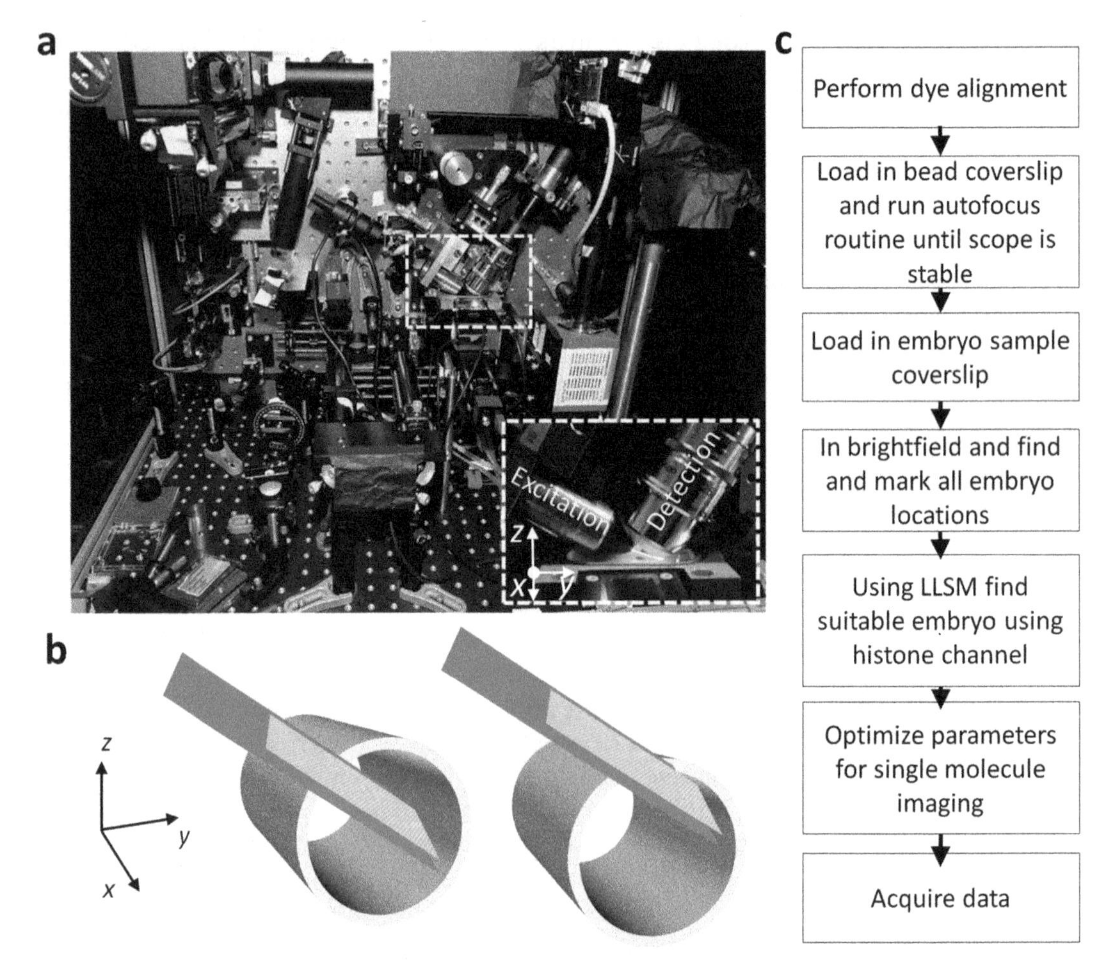

Fig. 2 Coordinate system of the sample navigation stages with respect to the detection and excitation objectives. (**a**) Picture of the lattice light-sheet microscope with inset showing a zoomed in image of the sample chamber and excitation and detection objectives. The coordinate system illustrates the motion of the stages with respect to the objectives. (**b**) Illustration of how to position the embryo within the excitation sheet. The hollow cylinder represents the embryo, the solid plane the excitation illumination, and the patterned plane shows the corresponding image plan in view of the detection objective. The first panel shows incorrect positioning where part of the image is deep inside the embryo and the second panel a better position where the excitation and detection planes skip the surface of the embryo. The coordinate system corresponds to the stage motion. Note that moving in *y* and *z* both result in a change in the axial position of the image whereas motions in *x* will provide a translation at a fixed depth. This configuration is important since both the excitation and detection objectives are on top of the sample. (**c**) Summary of workflow for LLSM alignment and data acquisition

13. Once you have estimated your acquisition parameters, navigate to a different area in the embryo or to a new embryo, and acquire data (*see* Fig. 2c for a summary of key steps). For a given set of exposure times and fluorescent labels, once you have optimized your parameters you should keep them constant for all data acquisition.

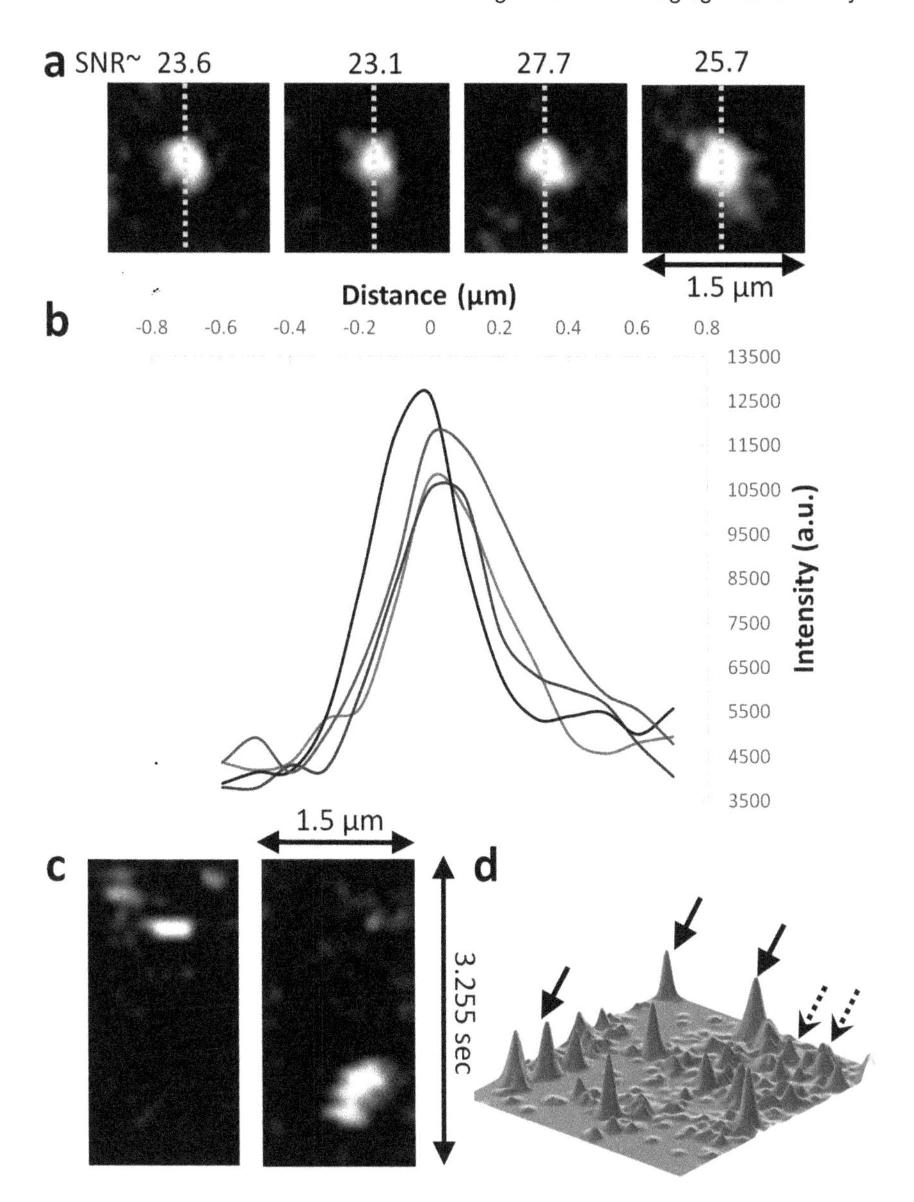

Fig. 3 Examples of verification of ability to detect single molecules on data acquired at 100 ms exposure times on Bicoid-GFP. (**a**) Examples of single molecule detections and corresponding SNR, dashed lines indicate the lines used to plot the profiles shown in (**d**). (**b**) Line profiles corresponding to the example detections shown in (**a**). (**c**) x–t slice of a single molecule detections exhibiting single step loss of signal. The first panel shows a short interaction, while the second shows a detection with some lateral motion. (**d**) Surface plot of an image with several single molecule detections with some examples indicated by the solid arrows; dashed arrows are likely signal from slightly out-of-focus or rapidly moving molecules (that is faster than the exposure time)

14. Keep track of the nuclear divisions, spatial positions, and other relevant parameters for your experiment. Time the nuclear division cycles to ensure they are proceeding as is normal for your fly line when not doing single molecule imaging to assess if you are causing any photodamage. It is good practice to

observe an embryo until it gastrulates to ensure that the data you acquired was on a healthy specimen. For each experimental condition, if performing residence time measurements, also acquire data on a sample with histones tagged with the same fluorescent protein so that you can determine the accessible dynamic range for your acquisition settings; for fast tracking utilize the labeled-NLS control (*see* **Note 26**).

3.5 Data Analysis

1. The initial step involves verifying that the detections you observe are indeed single molecules. First, the single molecule detections should exhibit single step disappearance (Fig. 3c), either through photo-bleaching or unbinding. This can be performed by re-slicing your time series to give an x–t view of the data. Second, the intensity distribution of detections of single molecules should exhibit a consistent peak height (Fig. 3d).
2. A number of tools [12] exist to analyze single molecule data (*see* **Note 27**), and we list a few options in Fig. 4. The general workflow is to first perform frame-by-frame localization of single molecule detections and then connect the detections into trajectories.
3. For any tracking software, ensure the following parameters are used: for longest "gap" frames (the number of frames during which a particle is not visible and still included in a trajectory), it should be set less than 2 and set the maximum displacements allowed (or the maximum diffusion coefficient) according to your frame rate and particle density (these parameters should be determined empirically for each set of experiments).
4. Check your dataset statistics to make sure that your trajectories are suitable for analysis. The main parameters to check are that you have sufficient number of trajectories (histograms should not be noisy), they are of sufficient lengths (median trajectory length should be greater than 3), number of particles per frame (median number should be low that is ~1 to avoid errors), and for fast tracking the mean translocation or jump length (should be significantly bigger than the localization error) (*see* **Note 28**).
5. Once you have trajectory data, there are also a number of options to extract information about binding dynamics, sub-populations, etc. as shown in Fig. 4. For example, for faster acquisition rates, you may use a model-based fitting to the distributions of displacements to estimate the fraction of the population that is bound and mobile and the diffusion coefficient of these populations. Recently our group has set up a web-based tool called spot-on where you can upload trajectory data to perform such analysis [13]. For residence time analysis

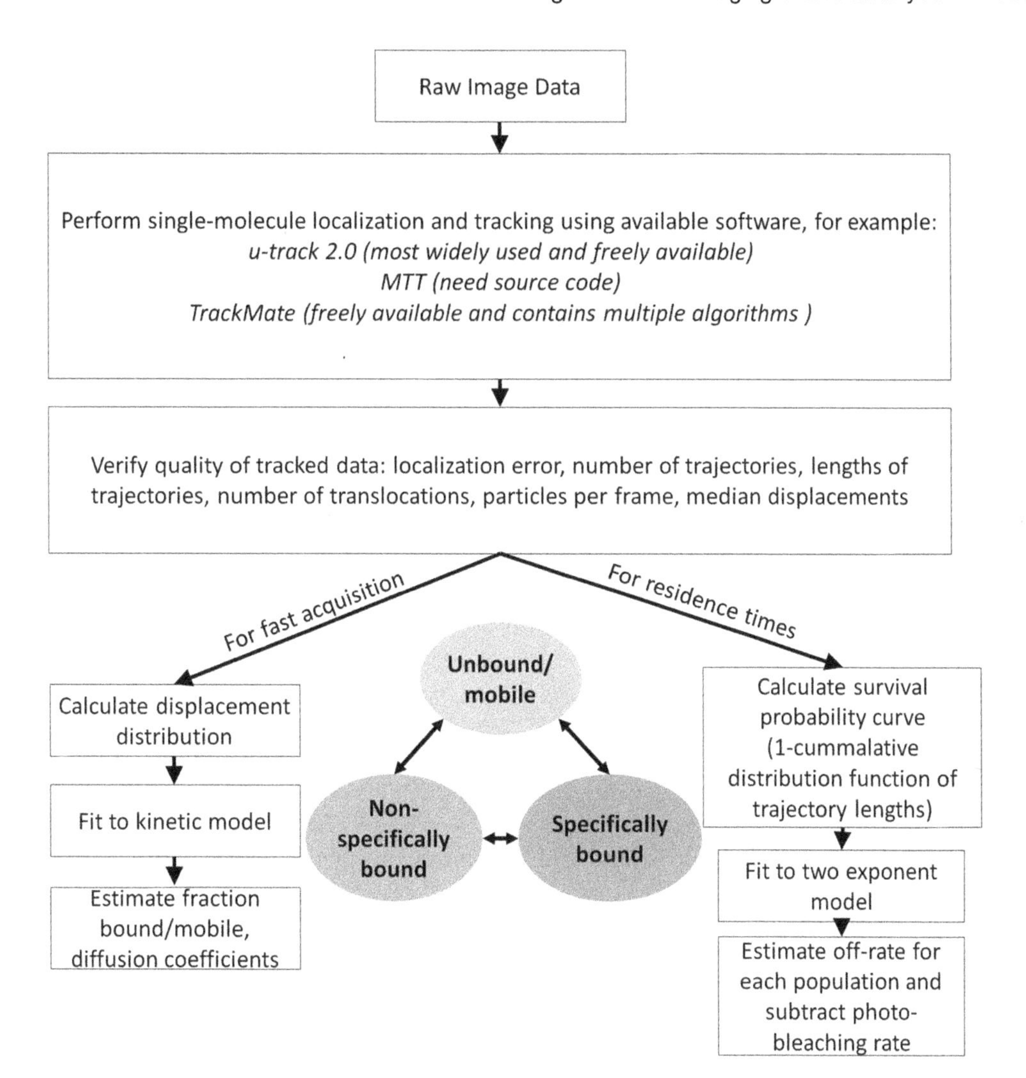

Fig. 4 Summary of key steps for data analysis. Inset shows an example of a three-state model; the population in each state and the transition rates between states can be estimated from a combination of the fast and slow acquisition and data analysis

on longer exposure time data, you may calculate the distribution of trajectory lengths of binding events, known as the survival probability distribution. This distribution may then be fit to a two-population model to determine the average lifetime of the short- and long-lived binding events. These are simply examples of the type of analyses you may perform; the data can be treated in a variety of ways, for example, calculating a distribution of angles within trajectories [1] or using hidden-Markov model-based classification [14].

6. For any particular protein-fluorophore and acquisition parameter combination you acquire data with, you should also perform appropriate controls. For example, for residence time

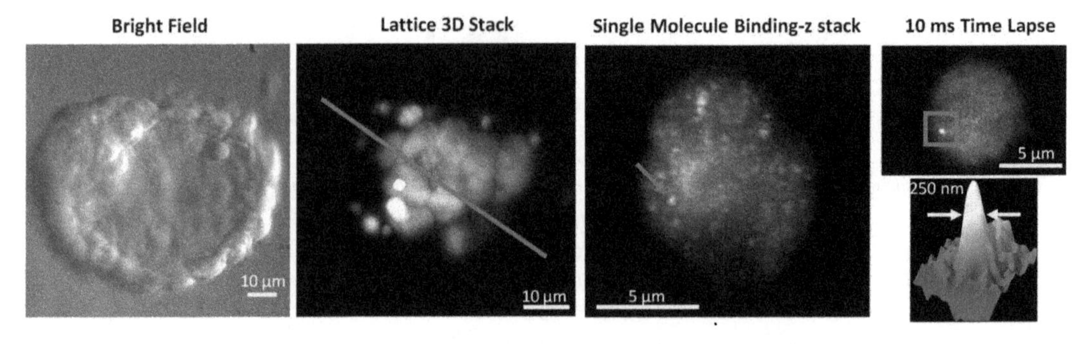

Fig. 5 Single molecule imaging in a mouse embryo. From left to right: (1) Bright-field image of an E4.5 late blastocyst mouse embryo chimera of wild type and JM8.N4 mES cells expressing endogenously tagged Halo-CTCF labeled with JF_{549}. (2) Lattice light-sheet 3D image of the same embryo; red line indicates the lattice slice used for time lapse imaging of a single nucleus. (3) Single molecules of CTCF visible in a z-stack of a single nucleus in the same embryo. (4) Frame from 10 ms time series acquisition shown a single CTCF-binding event and corresponding PSF

analysis, determine the photo-bleaching rates and whether your detections are limited by drift of the specimen or the sample by calculating the survival probability distribution of your single molecule histone line; the off-rate for your factor should then be corrected by subtracting the photo-bleaching rate constant. For displacement distributions, you can perform measuring using a fluorescent label-NLS line to determine what the distribution for a completely non-specific factor is.

3.6 Single Molecule Imaging in Mice Embryos

1. The general guidelines for imaging in Drosophila embryos can also be used for imaging in blastocyst stage mice embryos (Fig. 5).
2. Mice embryos from a line with either a fluorescent protein or a HaloTag on the protein of interest should be collected. If using Halo-tagged protein, label the embryo with your organic dye of choice, and perform washing steps before mounting the sample on the LLSM.
3. Embryos can be mounted on a coverslip and immobilized using matrigel at high concentration.
4. Imaging condition optimization and data analysis should be performed as described above. *See* Fig. 5 for an example of single molecule imaging of JF_{549}-labeled Halo-CTCF in an E4.5 late blastocyst mouse embryo.

4 Notes

1. The density of beads on the slide should be such that an isolated bead can be found for alignment easily.

2. The strength of the adhesive can be varied to your preference by increasing or decreasing the amount of scotch tape used.
3. If working with a fly line that is not laying well, the lids can be exchanged several times a day along with careful observation to determine the best time of the day for embryo collection.
4. If this is the first exchange after an overnight lay, it is good practice to perform a 90 min "clearing" lay before exchanging the lid again for your experiment. This helps reduce the occurrence of older than expected embryos in your experimental collection.
5. It is not necessary to have the objective water-heater blocks mounted for *Drosophila* experiments since they can be conducted at room temperature. Removing the blocks allows for using an appropriately positioned small LED lamp (e.g., Ikea Jansjo desk lamp) to provide bright-field-like images for quickly navigating the sample coverslip.
6. If using more than one fluorophore at a time, it is highly recommended to use a two camera setup with a separate set of filters and appropriate dichroic in between instead of recording the two colors sequentially on the same camera. Although the multicolor scanning in LLSM is sequential, commonly used fluorophores often have overlapping excitation or emission spectra or long tails, which can result in bleed through when using less than ideal filters such as multiband-pass emission filters.
7. We highly recommended using EMCCD cameras for single molecule imaging experiments at this point due to their higher quantum efficiency than most sCMOS cameras. Although the newest generation of sCMOS cameras has reported efficiencies that could be suitable for single molecule detection, we have not personally tested them.
8. Designing the correct lattice for your experiment should be done empirically. We have found that using a 30 beam pattern with a maximum numerical aperture 0.6 and minimum numerical aperture of 0.505 is suitable for single molecule imaging in fly embryos. Making the annulus thinner will result in more Bessel-like behavior (more side lobes but longer field of view) and conversely making it thicker results in more Gaussian-like illumination.
9. If using the LLSM design as originally described, it is recommended to use two Watt lasers as it will allow for uniform illumination while having sufficient power at the sample for single molecule imaging. If sufficient power cannot be achieved or if you only have lower power lasers, it is possible to adjust the laser collimation lenses to focus more of the energy in the center of the illumination pattern at the expense of the field of view over which the illumination is uniform.

10. It is very useful to have a vacuum trap connected to the main drain of the sample chamber to facilitate washing and rinse steps. If your system is not configured in this manner, it is worth to modify it to have this capability as it requires negligible effort, cost, and space.
11. Make sure to use the same medium for imaging and for alignment (e.g., for making up the dye solution for alignment). Small changes in salt concentrations result in large enough changes to the refractive index to cause misalignment.
12. It is useful and efficient to keep a reserved sample holder with a bead alignment slide loaded in for quick exchange during experiments for checking on suspected microscope drift.
13. 5 mm coverslips can be cleaned in a petri dish by rinsing with clean deionized water, followed by sonication while immersed in isopropyl alcohol. Cleaned slides can be stored in isopropyl alcohol or dried and kept in a clean padded container.
14. Kimwipes can be cut into small 1 in. squares and stored in a tub. This smaller size is useful for wicking up and drying small areas and also reduces waste.
15. It takes 5–10 min for the glue to dry. It is prudent to check this by using your manipulation paintbrush prior to mounting embryos.
16. Many users may prefer to collect the embryos by hand using an inoculation loop or paintbrush instead of washing them out.
17. Timing the de-chorionation down to second precisely is not necessary. Embryos can also be de-chorionated manually by laying down a stripe of double-sided scotch top on a flat surface and gently rolling the embryos across it. Of course, this is much more labor and time intensive than using bleach.
18. Embryos that are not damaged or "punctured" will float to the top increasing the probability that only viable developing embryos are mounted on the coverslip.
19. As it can be difficult to manipulate the embryos once they are on the adhesive surface of the coverslip without damaging, an alternative method is to make an array of embryos on a pad of agar and then transfer them via gentle contact to the adhesive surface of the coverslip.
20. To set up a dedicated transillumination source for bright-field-like imaging, install a LED source (e.g., Thorlabs MWWHL4) at the same angle as the detection objective just prior to the final flip mirror before the excitation objective. Using a lens the light from the LED should be brought to a focus in the back focal plane of the objective to have collimated light exiting the detection objective. With such a setup, it is possible to flip between the light sheet and transillumination by just flipping the final mirror.

21. We find this method of finding and marking all the embryos first more efficient as you don't have to switch between bright-field and light-sheet illumination, but if you prefer you can check each embryo with LLSM imaging as you find them.
22. For example, for H2B-RFP with a 30 beam lattice, we typically need 0.5 mW of 561 nm illumination at the back aperture of the excitation objective at 50 ms exposure times.
23. To understand how to set the exposure time to access different timescales, you can think of the population of single molecules you are imaging to exist in three states: diffusing and thus moving quickly (typical $D \sim 2\text{–}10\ \mu m^2/s$), non-specific binding (short interactions; $D < 0.1\ \mu m^2/s$), and specific binding (long interactions; $D < 0.1\ \mu m^2/s$). Thus when measuring binding interactions, you can increase the exposure time (decrease the frame rate) such that diffusive mobile population motion blurs into the background, and you only get good signal from molecules that are bound or immobile for at least the length of the exposure. To access the diffusive population as well, you need to minimize the exposure time. Decreasing the exposure time naturally requires higher laser powers to achieve a suitable signal-to-noise ratio. Increasing the laser power results in faster photo-bleaching and thus the long binding times are not quantifiable, since binding times can generally only be accurately measured if the unbinding rate is higher than the photobleaching rate.
24. The SNR you require should be determined by the localization accuracy you require for your particular application in addition to being sufficiently high to allow high-confidence localization of single molecules.
25. If you are using a photo-switchable or photo-convertible fluorophore, a brief flash (typically the length of 1 exposure time) with the activation laser (typically 405 nm) is sufficient to convert enough molecules for imaging. If acquiring lengthier movies, you may need to use scripting to flash the activation laser periodically. We are currently exploring options to have a constant low-level of activation which would be more desirable than intermittent activation. If using a regular fluorescent protein at high densities, you may need to wait to photo-bleach enough molecules before single molecule detections are possible.
26. The fast nature of the early nuclear division cycles makes these experiments challenging. With practice you will determine the best window between divisions where the nuclei are not moving where reliable single molecule data can be acquired. The histone single molecule controls you should perform will also provide you with the dynamic range of your measurements

and inform you on when you are limited by photo-bleaching and nuclear motion or drift for slow tracking or residence-time measurements. For fast tracking your fluorescent label fused with a nuclear localization signal (NLS) should be used as a control. Of course the exact sets of controls that are necessary will depend on your particular experiments.

27. In our recent work on Bicoid dynamics [9], we used an implementation of the dynamic multiple-target tracing (MTT) [15] algorithm to perform localization and tracking. A number of software options exist for both localization and tracking, and as this list is constantly evolving, it is left to the reader to decide on the appropriate one for their application. The MTT algorithm implementation that we have used will be provided upon request, and a MATLAB-based GUI of this implementation is available for download here: https://doi.org/10.7554/eLife.22280.022.
28. The localization accuracy can be calculated in a variety of ways. One way is to convert the signal to photon counts and calculate the uncertainty in the determined position using standard formulas for PALM/STORM microscopy. Second you may perform tracking on a fixed sample, calculate a mean-squared displacement curve, and determine the localization error from the intercept of the curve. If you are using spot-on [13] for analysis, it will estimate the localization error based on the data you provide.

Acknowledgments

The authors thank the Betzig lab at HHMI Janelia Research Campus for designs and advice on setting up the LLSM. We thank all members of the Darzacq, Tjian, Garcia, and Eisen labs for reagents, suggestions, and useful discussions. This work was supported by the California Institute of Regenerative Medicine (CIRM) LA1-08013 and the National Institutes of Health (NIH) UO1-EB021236 & U54-DK107980 to X.D., by the Burroughs Wellcome Fund Career Award at the Scientific Interface, the Sloan Research Foundation, the Human Frontiers Science Program, the Searle Scholars Program, and the Shurl and Kay Curci Foundation to H.G., a Howard Hughes Medical Institute investigator award to M.E., NSF Graduate Research Fellowships A.R. D.H. is a Pew-Stewart Scholar for Cancer Research supported by the Pew Charitable Trusts, the Alexander and Margaret Stewart Trust, the Siebel Stem Cell Institute, and NIH R01-CA196884.

References

1. Izeddin I, Recamier V, Bosanac L, Cisse II, Boudarene L, Dugast-Darzacq C, Proux F, Benichou O, Voituriez R, Bensaude O, Dahan M, Darzacq X (2014) Single-molecule tracking in live cells reveals distinct target-search strategies of transcription factors in the nucleus. eLife 3:e02230. https://doi.org/10.7554/eLife.02230
2. Chen JJ, Zhang ZJ, Li L, Chen BC, Revyakin A, Hajj B, Legant W, Dahan M, Lionnet T, Betzig E, Tjian R, Liu Z (2014) Single-molecule dynamics of enhanceosome assembly in embryonic stem cells. Cell 156 (6):1274–1285. https://doi.org/10.1016/j.cell.2014.01.062
3. Morisaki T, Muller WG, Golob N, Mazza D, McNally JG (2014) Single-molecule analysis of transcription factor binding at transcription sites in live cells. Nat Commun 5:4456. https://doi.org/10.1038/Ncomms5456
4. Mazza D, Abernathy A, Golob N, Morisaki T, McNally JG (2012) A benchmark for chromatin binding measurements in live cells. Nucleic Acids Res 40(15):e119. https://doi.org/10.1093/nar/gks701
5. Hansen AS, Pustova I, Cattoglio C, Tjian R, Darzacq X (2017) CTCF and cohesin regulate chromatin loop stability with distinct dynamics. eLife 6:e25776. https://doi.org/10.7554/eLife.25776
6. Cho WK, Jayanth N, English BP, Inoue T, Andrews JO, Conway W, Grimm JB, Spille JH, Lavis LD, Lionnet T, Cisse II (2016) RNA polymerase II cluster dynamics predict mRNA output in living cells. eLife 5:e13617. https://doi.org/10.7554/eLife.13617
7. Cisse II, Izeddin I, Causse SZ, Boudarene L, Senecal A, Muresan L, Dugast-Darzacq C, Hajj B, Dahan M, Darzacq X (2013) Real-time dynamics of RNA polymerase II clustering in live human cells. Science 341 (6146):664–667. https://doi.org/10.1126/science.1239053
8. Chen BC, Legant WR, Wang K, Shao L, Milkie DE, Davidson MW, Janetopoulos C, Wu XFS, Hammer JA, Liu Z, English BP, Mimori-Kiyosue Y, Romero DP, Ritter AT, Lippincott-Schwartz J, Fritz-Laylin L, Mullins RD, Mitchell DM, Bembenek JN, Reymann AC, Bohme R, Grill SW, Wang JT, Seydoux G, Tulu US, Kiehart DP, Betzig E (2014) Lattice light-sheet microscopy: Imaging molecules to embryos at high spatiotemporal resolution. Science 346(6208):439–439. https://doi.org/10.1126/science.1257998
9. Mir M, Reimer A, Haines JE, Li X-Y, Stadler M, Garcia H, Eisen MB, Darzacq X (2017) Dense Bicoid hubs accentuate binding along the morphogen gradient. Genes Dev 31 (17):1784–1794
10. Katz ZB, Wells AL, Park HY, Wu B, Shenoy SM, Singer RH (2012) Beta-actin mRNA compartmentalization enhances focal adhesion stability and directs cell migration. Genes Dev 26 (17):1885–1890. https://doi.org/10.1101/gad.190413.112
11. Lucas JS, Zhang YJ, Dudko OK, Murre C (2014) 3D trajectories adopted by coding and regulatory DNA elements: first-passage times for genomic interactions. Cell 158 (2):339–352. https://doi.org/10.1016/j.cell.2014.05.036
12. Chenouard N, Smal I, de Chaumont F, Maska M, Sbalzarini IF, Gong YH, Cardinale J, Carthel C, Coraluppi S, Winter M, Cohen AR, Godinez WJ, Rohr K, Kalaidzidis Y, Liang L, Duncan J, Shen HY, Xu YK, Magnusson KEG, Jalden J, Blau HM, Paul-Gilloteaux P, Roudot P, Kervrann C, Waharte F, Tinevez JY, Shorte SL, Willemse J, Celler K, van Wezel GP, Dan HW, Tsai YS, de Solorzano CO, Olivo-Marin JC, Meijering E (2014) Objective comparison of particle tracking methods. Nat Methods 11(3):281–U247. https://doi.org/10.1038/nmeth.2808
13. Hansen AS, Woringer M, Grimm JB, Lavis LD, Tjian R, Darzacq X (2017) Spot-On: robust model-based analysis of single-particle tracking experiments. bioRxiv. https://doi.org/10.1101/171983
14. Persson F, Linden M, Unoson C, Elf J (2013) Extracting intracellular diffusive states and transition rates from single-molecule tracking data. Nat Methods 10(3):265–269. https://doi.org/10.1038/Nmeth.2367
15. Serge A, Bertaux N, Rigneault H, Marguet D (2008) Dynamic multiple-target tracing to probe spatiotemporal cartography of cell membranes. Nat Methods 5(8):687–694. https://doi.org/10.1038/nmeth.1233

Chapter 33

Silver Filler Pre-embedding to Enhance Resolution and Contrast in Multidimensional SEM: A Nanoscale Imaging Study on Liver Tissue

Gerald J. Shami, Delfine Cheng, and Filip Braet

Abstract

Contemporarily, serial block-face scanning electron microscopy (SBF-SEM) has emerged as an immensely powerful nanoscopic imaging technique, capable of generating large-volume three-dimensional information on a variety of biological specimens in a semiautomated manner. Despite the plethora of insights and advantages provided by SBF-SEM, a major challenge inherent to the technique is that of electron charging, which ultimately reduces attainable resolution and detracts from overall image quality. In this chapter, we describe a pre-embedding approach that involves infiltration of tissue with a highly conductive silver filler suspension following primary fixation. Such an approach is demonstrated to improve overall sample conductivity, resulting in the minimization of charging under high-vacuum conditions and an improvement in lateral resolution and image contrast. The strength of this sample preparation approach for SBF-SEM is illustrated on liver tissue.

Key words 3-D reconstruction, 3-D scanning electron microscopy, 3View, Charging, Conductivity, Contrast enhancement, Hepatic, Resolution, Serial block-face scanning electron microscopy, Silver filler, Sinusoidal vasculature, Slice and view

1 Introduction

A fundamental tenet of structural biology is the requirement to generate three-dimensional (3-D) information in order to holistically reconstruct and visualize biological structures. Contemporarily, serial block-face scanning electron microscopy (SBF-SEM) has emerged as a profoundly powerful imaging modality, capable of generating large-volume (>600 μm^3) 3-D information from the tissue to subcellular level, in a semiautomated manner [1, 2]. The technique involves acquiring inverted backscattered electron images from the surface of a block of resin-embedded biological material. Once an image has been acquired, the sample is then raised between 15 and 200 nm within the chamber of a scanning electron microscope (SEM), and an ultrathin section is removed

Yuri L. Lyubchenko (ed.), *Nanoscale Imaging: Methods and Protocols*, Methods in Molecular Biology, vol. 1814,
https://doi.org/10.1007/978-1-4939-8591-3_33, © Springer Science+Business Media, LLC, part of Springer Nature 2018

from the block face, by means of an automated ultramicrotome equipped with a diamond knife. In this manner, the repeated cutting, raising, and imaging of the block results in a pre-aligned, large-volume 3-D dataset, with a lateral resolution matching the capabilities of a high-resolution field emission SEM.

Despite the plethora of advantages provided by SBF-SEM, a major challenge inherent to the technique is that of electron charging, whereby electrons accumulate within nonconductive regions of the sample that are uncontrollably and randomly discharged. This phenomenon results in a range of undesirable artifacts, such as inhomogeneous illumination of the sample, artificial shifting of gray values along the image histogram resulting in poor contrast, image deformation, beam damage, and a reduction in resolution [3].

The primary method currently used to mitigate charging and its adverse effects is the employment of sample preparation protocols that involve the repeated application of heavy metal fixatives (e.g., osmium tetroxide), stains (e.g., lead aspartate and uranyl acetate), and mordanting agents (e.g., tannic acid and thiocarbohydrazide) in an attempt to improve sample conductivity and contrasting [4, 5]. This approach effectively works for lipid-rich samples (e.g., neural tissue) and cellular regions (e.g., cellular membranes) which readily react with osmium tetroxide, a major component of such staining protocols. A shortcoming of this approach, however, is the occurrence of charging (1) in the absence of conductive material, such as in "free resin spaces" (e.g., blood and lymph vessels); (2) within electron-lucent organelles, such as cell nuclei; and (3) at high magnifications, where the electron dose per unit area is increased.

Fittingly, a feature common to many SBF-SEM systems is the ability to image samples under charge controlled, variable pressure (VP) conditions—via the introduction of various gases (e.g., nitrogen or argon) within the specimen chamber (10–2500 Pa)—in conjunction with the use of specialized high-sensitivity backscattered electron detectors [6]. The introduction of gas just above the sample results in the generation of positive ions that effectively neutralize the accumulation of negative charge within the sample, thus maintaining charge neutrality [7]. Imaging under variable pressure conditions is not without compromise, however, resulting in inferior resolution, contrast, and signal-to-noise ratio, relative to imaging under high-vacuum (HV) conditions. Furthermore, increased technical difficulties are associated with imaging under VP conditions, such as optimizing specimen focus and astigmatism correction [8].

In this chapter, we outline a novel "silver filler pre-embedding approach" that minimizes charging within extracellular spaces and increases overall specimen conductivity, by reducing the presence of "free resin spaces" (e.g., blood vessels) via the introduction of highly conductive silver nanoparticles suspended in gelatin. For this investigation, we utilized liver tissue as our standard—given its highly vascularized nature—in order to reap the improved image contrast and resolution benefits obtained, by imaging under high-vacuum conditions.

2 Materials

2.1 Animals

1. Female Wistar rats (10–12 weeks of age) were housed in plastic cages at 21 °C with a 12-h light-dark cycle and fed and watered ad libitum.

2.2 Buffer Reagents

1. 0.2 M sodium cacodylate buffer, pH 7.4: Prepare 100 mL of buffer by combining 3.27 g sodium cacodylate trihydrate, 96.4 mL MQ-(purified by ion-exchange), and 3.6 mL 0.2 M HCl. Filter through 0.2 μm Millipore filter paper.
2. 0.2 M phosphate buffer solution (PBS), pH 7.4: Prepare 100 mL of buffer by combining 162 mL of 0.2 M Na_2HPO_4 (17.8 g in 500 mL MQ-water) and 38 mL 0.2 M NaH_2PO_4 (13.8 g in 500 mL MQ-water). Filter through 0.2 μm Millipore filter paper.
3. 0.1 M PBS, pH 7.4: Prepare 100 mL by adding 50 mL 0.2 M PBS and 50 mL MQ-water.

2.3 Primary and Secondary Fixatives

1. Primary fixative: 1.5% glutaraldehyde in 0.067 M sodium cacodylate buffer, 1% sucrose and 0.05% calcium chloride, pH 7.4. Prepare 20 mL of primary fixative by adding 1.2 mL 25% glutaraldehyde solution, 6.7 mL 0.2 M cacodylate buffer, 0.2 g sucrose, 0.01 g calcium chloride powder, and 12.1 mL MQ-water.
2. Secondary fixative: 4% formaldehyde in 0.1 M PBS, pH 7.4. Prepare 10 mL of secondary fixative by combining 2.5 mL 16% aqueous formaldehyde solution, 5 mL 0.2 M PBS, and 2.5 mL MQ-water.
3. 1 mL plastic syringe.
4. 25 G needle.
5. 60 mm × 15 mm non-treated plastic petri dish.
6. 5 mL solvent-resistant plastic specimen vials.
7. Razor blade, single edge, and stainless steel.
8. Scissors.
9. Forceps.
10. Cyanoacrylate glue.
11. Vibratome® (Leica VT1200 S, Heerbrugg, Switzerland).

2.4 Silver Filler Components

1. 7.5% w/v silver nanopowder in 10% w/v gelatin: Weigh 0.075 g of silver nanopowder (<150 nm particle size, Sigma-Aldrich, NSW, Australia) in a 1.5 mL Eppendorf tube. Add 10% w/v gelatin (1 g porcine gelatin in 10 mL 0.1 M PBS) at 40 °C to the 1 mL graduation line of the Eppendorf tube. Vortex the solution until a homogenous dark-gray suspension forms.
2. Mini-centrifuge (Traditional LabMini8, Southwest Science, NJ, USA).
3. Bench vortex mixer.
4. 1.5 mL Eppendorf® tubes.
5. 75 mm × 25 mm × 1 mm glass microscope slides.
6. Razor blade, single edge, and stainless steel.
7. Fine-tipped forceps.
8. Ice/ice pack.
9. Polystyrene foam box.
10. Toothpicks.

2.5 Post-fixation and En Bloc Staining Components

1. 2% osmium tetroxide and 1.5% potassium ferrocyanide in 0.1 M PBS: Prepare 6 mL of solution by adding 3 mL 4% aqueous osmium tetroxide solution and 3 mL 3% potassium ferrocyanide (0.3 g $K_4[Fe(CN)_6]$ $3H_2O$ in 10 mL 0.2 M PBS).
2. 1% thiocarbohydrazide solution: Add 0.1 g thiocarbohydrazide in 10 mL MQ-water and place in 60 °C oven. Agitate gently every 10 min. Filter through 0.22 μm pore syringe filter equipped with a 10 mL syringe directly before use.
3. 2% aqueous osmium tetroxide: Prepare by diluting 3 mL of 4% aqueous osmium tetroxide solution with 3 mL MQ-water.
4. 1% aqueous uranyl acetate. Filter through a 0.22 μm pore syringe filter equipped with a 10 mL syringe directly before use.
5. Walton's lead aspartate solution: Prepare an aspartic acid stock solution by dissolving 0.998 g L-aspartic acid in 250 mL MQ-water. The pH should be adjusted to 3.8 with 1 M HCl to facilitate dissolution. To prepare the en bloc lead aspartate stain, add 0.066 g lead nitrate in 10 mL stock aspartic acid, and adjust the pH to 5.5 with 1 N KOH. Place the solution in a 60 °C oven for 30 min prior to use (no precipitates should form) [4].

2.6 Dehydration and Plastic Embedding Components

1. Absolute ethanol: Dilute into 5 mL aliquots (30%, 50%, 70%, 90% w/v) with MQ-water.
2. Absolute acetone.

3. Procure 812 (Epon substitute) resin kit. Prepare 42 mL of hard-grade resin by combining 20 mL procure 812, 9 mL dodecenylsuccinic anhydride (DDSA), 12 mL methylnadic anhydride (NMA), and 1.2 mL dimethylbenzylamine (BDMA). It is imperative to mix components thoroughly to ensure blocks are of uniform hardness.
4. Aclar® plastic film (cut into 48 mm × 24 mm × 0.198 mm).
5. 75 mm × 25 mm × 1 mm glass microscope slides.
6. Polystyrene petri dish (100 mm × 15 mm).
7. Toothpicks.

2.7 Sample Mounting for SBF-SEM

1. Aluminum specimen pins (Gatan, Pleasanton, CA).
2. Injector razor blades.
3. Fine-tipped forceps.
4. Silver conductive epoxy.
5. Leica EM UC7 ultramicrotome (Leica, Heerbrugg, Switzerland).
6. Glass ultramicrotomy knives.
7. Conductive silver paint.
8. Toluidine blue: 0.5% in 1% aqueous sodium tetraborate.
9. Bright-field light microscope.
10. Metal sputter coater equipped with gold target.

2.8 Serial Block-Face Scanning Electron Microscope

1. Zeiss Sigma Variable Pressure SEM (Carl Zeiss Microscopy, GmbH), equipped with Gatan 3View 2XP (Gatan, Pleasanton, CA).

3 Methods

3.1 Primary Fixation of Liver Tissue

1. Fix a resected lobe of liver tissue (≤1 cm × 1 cm × 0.5 cm) by means of injection-mediated perfusion fixation [9] (*see* **Note 1**). Directly after opening the abdominal cavity, remove a segment of liver tissue using a sharp pair of scissors and transfer to a petri dish containing 0.1 M PBS at 37 °C. Gently holding the tissue at a corner using a pair of forceps, slowly inject the primary fixative warmed to 37 °C using a 1 mL syringe equipped with a 25 G needle. The fixative should be injected slowly until discoloration and hardening of the tissue occur. Injections should be performed multiple times until blood is flushed from the tissue, rendering it a homogenous color and consistency.
2. Slice the tissue into 3 mm^3 blocks under the primary fixative solution, using a clean razor blade. Allow the blocks to react in the fixative for 1 h at room temperature (RT) (*see* **Note 2**).

3.2 Pre-embedding with Silver Filler

1. Rinse tissue blocks with 0.1 M PBS (3 × 5 min).
2. Generate 300 μm thick vibratome sections, adhering the tissue block to the vibratome holder using cyanoacrylate glue. Once dried, immerse the tissue under 0.1 M PBS prior to sectioning. Vibratome settings: knife amplitude 2 mm/s, knife travel speed 1 mm/s, knife angle 9°.
3. Transfer the vibratome sections to fresh 0.1 M PBS at RT in preparation for infiltration with silver filler.
4. Carefully place ≤5 vibratome sections (measuring ~3 mm × 3 mm × 0.3 mm) in an Eppendorf tube containing the silver filler, using fine-tipped forceps, and gently invert the tube ensuring the sections are suspended solution (*see* **Note 3**).
5. Centrifuge the tube for 30 s using a mini-centrifuge (2000 × *g* (RCF) @ 6000 rpm). Centrifugation is required for rapid and homogenous infiltration of the sample.
6. Pipette 0.75 mL of the silver filler solution from the Eppendorf tube onto a clean glass slide. Carefully retrieve the tissue sections from the bottom of the Eppendorf tube using a toothpick, and arrange on the glass slide covered in silver filler solution. Place the slide on an ice pack in a polystyrene foam box for 15 min, until the gelatin sets.
7. Using a clean, sharp razor blade, cut the tissue sections from the glass slide, leaving ~1 mm of gelated silver filler around the sections (*see* **Note 4**).
8. Place the sections in a 5 mL solvent-resistant specimen vial, and secondarily fix in 4% formaldehyde in 0.1 M PBS for 20 min at RT, in order to fix the silver filler (*see* **Note 5**).
9. Rinse tissue with 0.1 M PBS (3 × 5 min).

The resolution and contrast observed from samples prepared under the silver filler pre-embedding method are far superior relative to control samples imaged under both variable pressure and high-vacuum conditions (Figs. 1 and 2). Charging is completely eliminated within extracellular spaces, such as the hepatic sinusoidal capillaries and significantly reduced within prone-charging subcellular structures, namely, parenchymal cell nuclei (Fig. 1f).

The low viscosity of the silver filler medium, additionally, allows complete penetration of the tissue (Fig. 1I), including within such confined spaces as bile canaliculi (Figs. 1f and 2b) and the space of Disse (Fig. 3). A video revealing the distribution of silver filler pre-embedding medium and the remarkable high-contrast images of liver tissue (measuring 80 μm × 80 μm × 64 μm) can be viewed at: https://vimeo.com/235291098 [10].

3.3 Post-Fixation and En Bloc Staining for SBF-SEM

Samples were prepared under a protocol developed by Deerinck et al. [4] and validated for the use on liver tissue [8].

1. Post-fix tissues in 2% aqueous osmium tetroxide and 1.5% potassium ferrocyanide in 0.1 M PBS in darkness for 1 h at RT.
2. Rinse tissues with MQ-water (3 × 5 min).
3. Incubate tissues in filtered 1% aqueous thiocarbohydrazide solution for 20 min at RT.
4. Rinse tissues with MQ-water (3 × 5 min).
5. Secondarily post-fix tissues in 2% osmium tetroxide in darkness for 30 min at RT.
6. Rinse tissues with MQ-water (3 × 5 min).
7. Incubate tissues in filtered 1% aqueous uranyl acetate in darkness, overnight at 4 °C.
8. Rinse tissues with MQ-water (3 × 5 min).
9. Incubate tissues in Walton's lead aspartate solution in a 60 °C oven for 30 min.
10. Rinse tissues with MQ-water (3 × 5 min).

3.4 Dehydration, Infiltration, and Plastic Embedding

1. Dehydrate tissues in an ascending series of ethanol concentrations (30%, 50%, 70%, 90%, and 100%, 100%) for 5 min each at RT.
2. Dehydrate tissues in an ascending series of ethanol concentrations (30%, 50%, 70%, 90%, 100% and 100%) for 5 minutes each at RT.
3. Infiltrate tissues with hard-grade Procure 812 (Epon-substitute) resin (25%, 50%, and 75% in absolute acetone) with gentle agitation on a specimen rotor for 3 h each.
4. Place tissues in 100% resin overnight and then into fresh resin for 3 h.
5. Flat embed tissues by placing a glass microscope slide (75 mm × 25 mm × 1 mm) in the bottom of a petri dish (100 mm × 15 mm). Center a piece of Aclar film cut into 48 mm × 24 mm × 0.198 mm atop the glass slide. Syringe 0.5 mL of pure Epon on the Aclar sheet, and carefully arrange the tissue sections using a toothpick (Fig. 4a). Gently place a second Aclar sheet and glass slide on top of the sections, ensuring the resin forms an even, thin layer (Fig. 4b). Polymerize at 60 °C for 48 h (*see* **Note 6**).

3.5 Sample Mounting for SBF-SEM

1. Remove the glass slides from the Aclar sheets by pinching and sliding between thumb and index finger (Fig. 4c).
2. Carefully peel both Aclar sheets from the polymerized resin sheet using fine-tipped forceps (Fig. 4c).

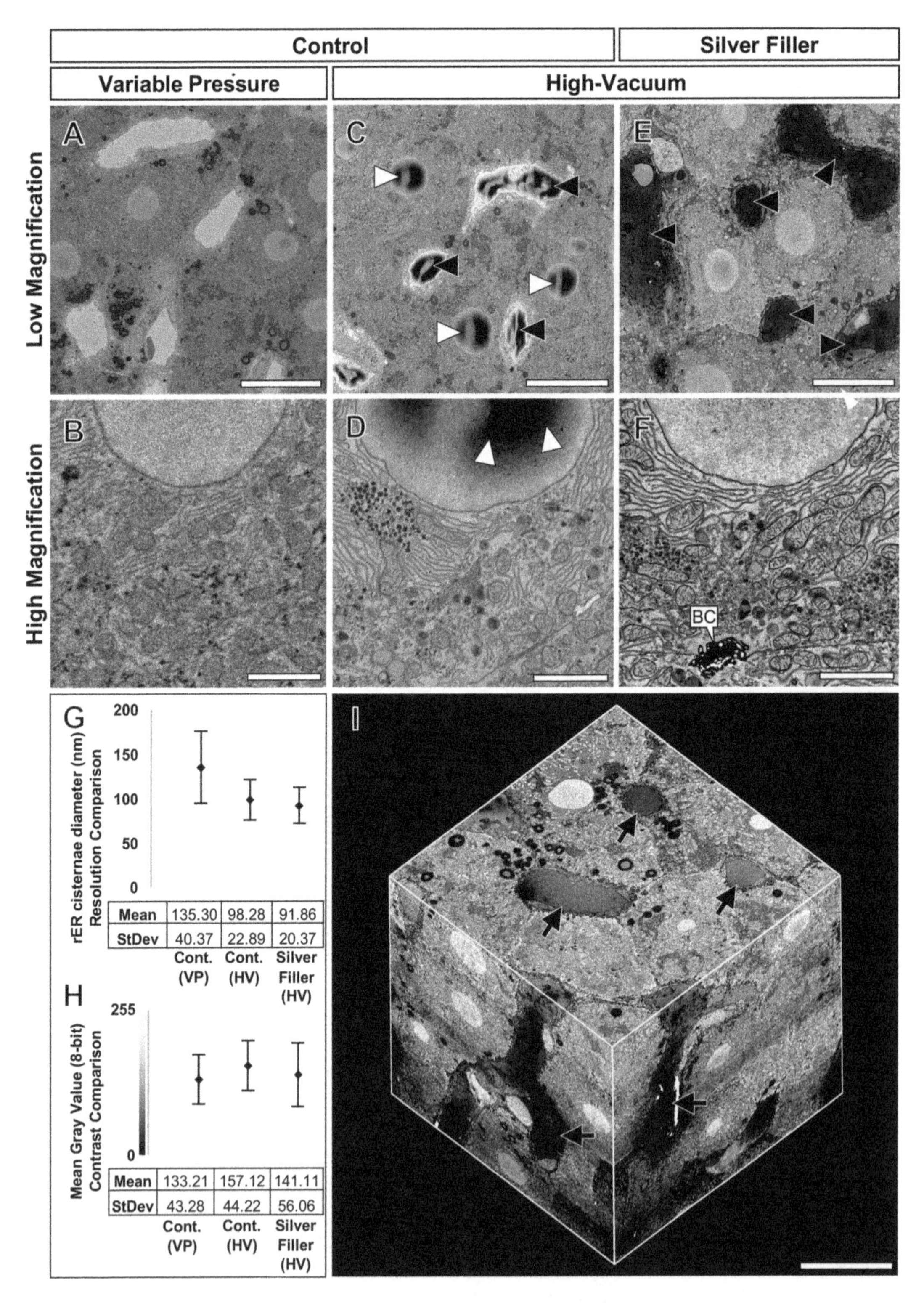

Fig. 1 Comparison of control versus rat liver tissue prepared under the silver filler method. (**a**, **b**) Control tissue imaged under VP conditions. No charging is visible; however contrast and resolution relative to (**c**, **e**) is inferior. (**c**, **d**) Control tissue imaged under HV conditions. Contrast and resolution is improved relative to (**a**, **b**);

3. Using a clean, sharp dissector razor blade, cut a tissue block measuring ~600 μm × 600 μm × 300 μm (Fig. 4d).
4. Adhere the tissue block to an aluminum specimen pin using conductive silver epoxy, under the aid of a dissection microscope or ultramicrotome binoculars.
5. Sparingly paint the exposed edges of the tissue block with conductive silver paint, in order to improve sample conductivity.
6. Face up the block face using an ultramicrotome, equipped with a glass knife.
7. Generate semithin sections (~0.5 μm thick), stain with toluidine blue, and observe with a light microscope to verify complete exposure of the tissue along the block face, and verify adequate fixation quality.
8. Sputter coat the tissue block with a 30 nm thick layer of gold to assist with alignment of the sample within the microscope (Fig. 4e).

3.6 Serial Block-Face Scanning Electron Microscopy (SBF-SEM)

For comparative purposes, inverted backscattered electron images (8192 × 8192 pixels, 16-bit, *XY* pixel size 15 nm, *Z* pixel size (slice thickness) 100 nm, pixel dwell time 3 μs) were acquired at a fixed working distance of 4.3 mm. Datasets were captured under both variable pressure (35 Pa) and high-vacuum conditions for control samples, and high-vacuum conditions for samples prepared under the silver filler pre-embedding method (Fig. 1).

3.6.1 Selection of Imaging Parameters

The selection of imaging parameters is of course dependent on the biological question at hand, and there are several considerations which should be taken into account.

1. *Higher accelerating voltages* (e.g., ≥3 kV) result in superior signal-to-noise ratio, contrast, and lateral (*XY*) resolution,

Fig. 1 (continued) however charging is present within parenchymal cell nuclei (white arrowheads) and hepatic sinusoidal capillaries (black arrowheads). (**e**, **f**) Liver tissue prepared under the silver filler method, imaged under HV conditions. The images display high contrast and resolution relative to (**a**, **c**), Charging is absent within parenchymal cell nuclei and the hepatic sinusoids, due the presence and increased conductivity provided by the silver filler (black arrowheads). The silver filler is also visible within such confined intercellular spaces as a bile canaliculus (BC) (**f**). Note: For comparative purposes, no contrast enhancement was applied to images **a–f**, with the exception of automated histogram normalization, which was applied equally to all images. (**g**) Quantitative comparison of line-profile measurements, revealing the improved resolution gained, under HV conditions. (**h**) Quantitative comparison of image contrast. Samples prepared under the silver filler method, reveal greater distribution of gray values along the image histogram. Note: The gradient bar on the *y*-axis of the graph was inserted using Adobe Photoshop as a visual aid. (**i**) 3-D volume revealing the complete penetration of silver filler throughout inter- and extracellular spaces (black arrows). Scale bar: **a**, **c**, **e**, and **i** = 20 μm and **b**, **d**, and **f** = 3 μm

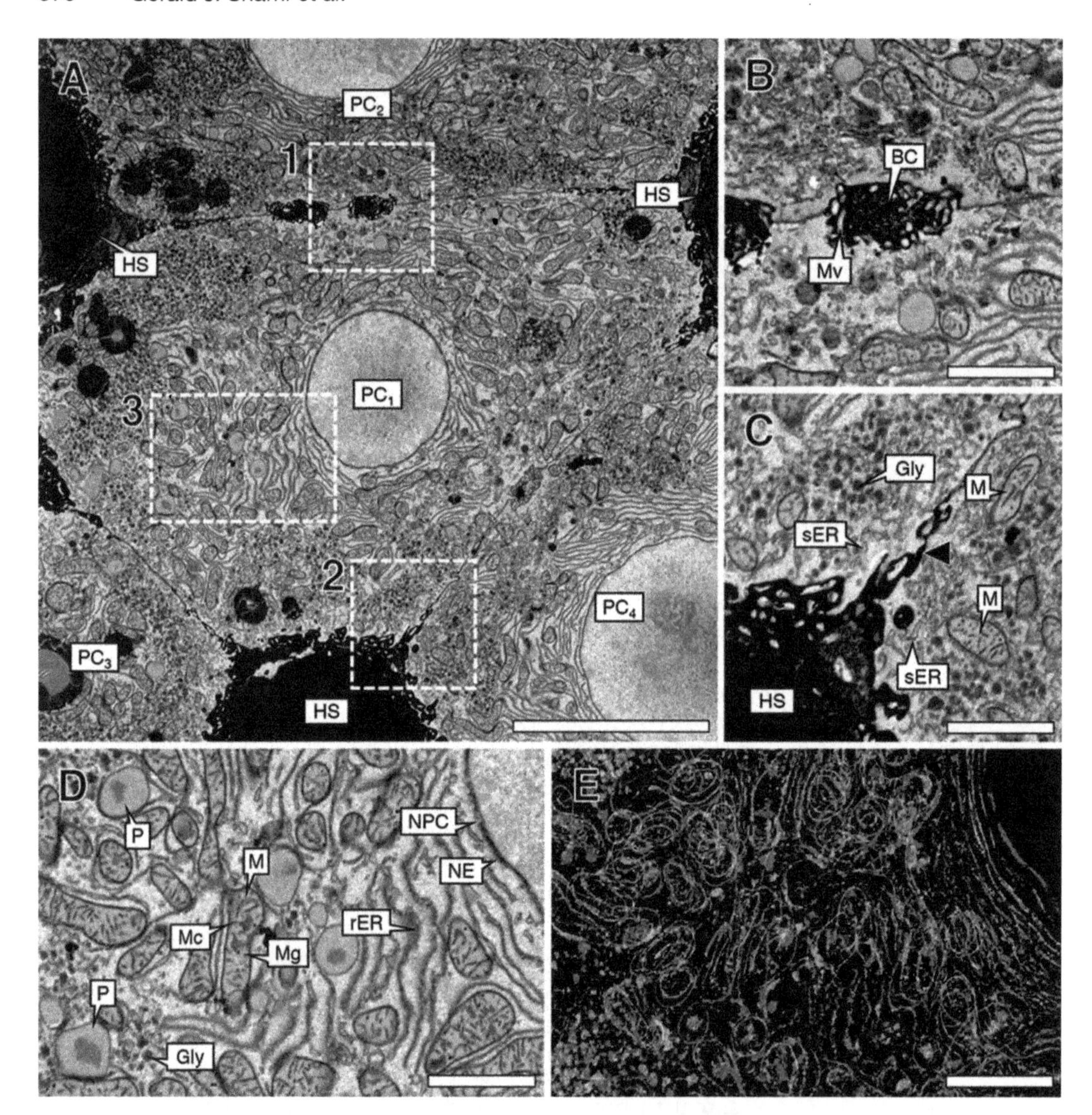

Fig. 2 (**a**) Low-magnification overview of a hepatic parenchymal cell (PC_1), bounded by three adjacent parenchymal cells (PC_2, PC_3, and PC_4) and three sinusoidal capillaries (HS), which contain silver filler. Higher-magnification insets (1–3) are shown in (**b**–**d**). (**b**) A bile canaliculus (BC) is shown, that is formed by the opposing plasma membranes of PC_1 and PC_2 in (**a**). Microvilli (Mv) are visible, projecting within the canalicular lumen. (**c**) Inset taken from a region between PC_1 and PC_4, including a part of a sinusoidal capillary (HS) in (**a**). Silver filler is visible within the sinusoidal lumen (HS), in addition to the confined intercellular spaces between the two opposing parenchymal cells (black arrowhead). Subcellular structures including mitochondria (M), glycogen (Gly), and the convoluted profiles of smooth endoplasmic reticulum (sER) are clearly visible. (**d**) Inset taken from the perinuclear region, extending toward the plasma membrane of PC_1 in (**a**). High-resolution nanoscopic information, generated by imaging under high-vacuum conditions, reveals highly contrasted subcellular structures. These include the nuclear envelope (NE), perforated by nuclear pore complexes (NPC), rough endoplasmic reticulum (rER), arranged as parallel cisternae, glycogen rosettes (Gly), peroxisomes (P) and numerous mitochondria (M), which display infoldings of the inner mitochondrial membrane to form cristae (Mc), and such diminutive inclusions as mitochondrial granules (Mg). (**e**) 3-D model of a volume corresponding to the region indicated in (D) (*Z*-axis depth $= 1\ \mu m$) achieved by means of histogram thresholding in IMOD, revealing electron-dense structures—primarily cellular membranes and glycogen rosettes—that can be selectively visualized, due to the increased resolution and contrast benefits provided by pre-embedding with silver filler. Scale bars: **a** $= 10\ \mu m$, **b**–**e** $= 2\ \mu m$

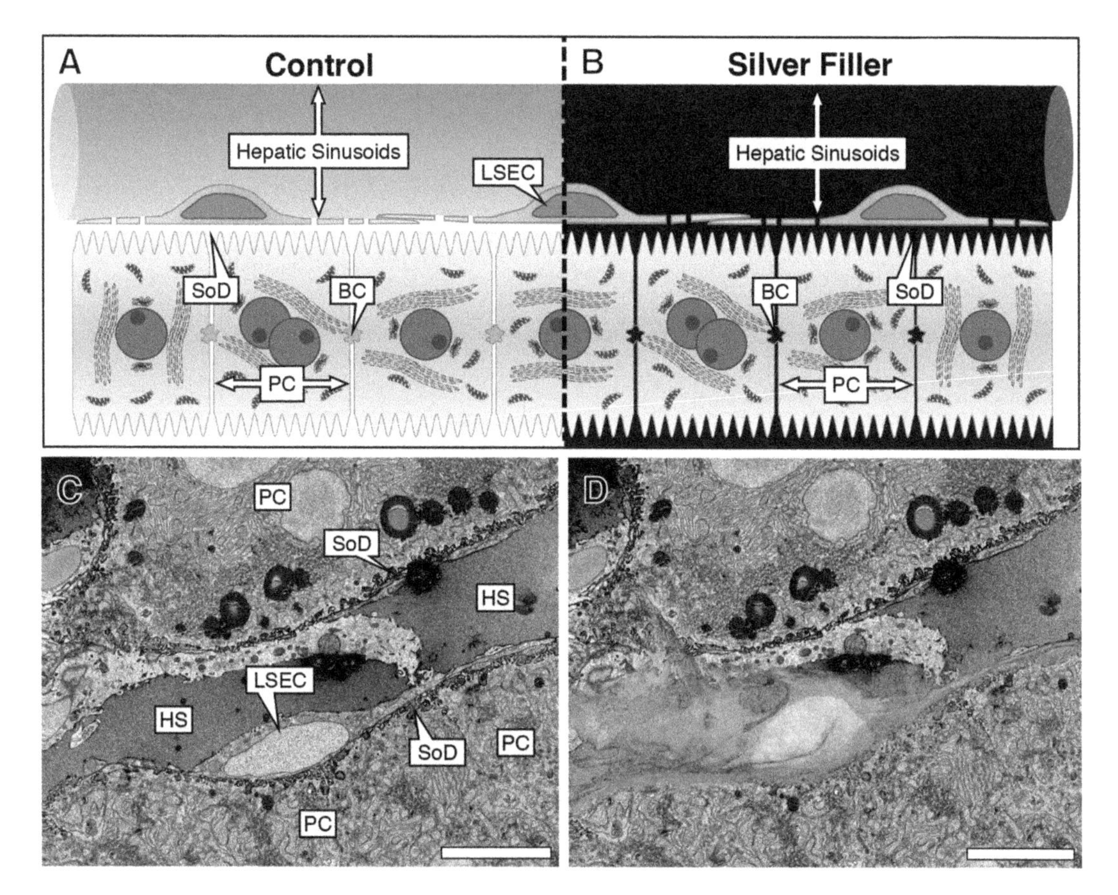

Fig. 3 (**a**, **b**) Schematic diagram depicting (**a**) control tissue versus (**b**) tissue prepared under the silver filler pre-embedding method, revealing the distribution of silver filler within intercellular spaces (e.g., the space between adjacent hepatic parenchymal cells (PC), including the bile canaliculi (BC)) and extracellular spaces, such as the lumen of the hepatic sinusoids and the space of Disse (SoD) (the interface between parenchymal cell microvilli and the basal plasma membrane of liver sinusoidal endothelial cells (LSECs)). (**c**) Inverted back-scattered electron micrograph revealing a liver sinusoidal endothelial cell (LSEC) forming the lumen of a hepatic sinusoidal capillary (HS), captured under HV conditions. Silver filler is visible within the lumen of the hepatic sinusoidal capillary (HS), in addition to such constricted extracellular spaces as the space of Disse (SoD)—between LSECs and parenchymal cells (PC). (**d**) 3-D model of a liver sinusoidal endothelial cell (green), its flattened nucleus (white), and the space of Disse (blue) superimposed with (**c**). Automated segmentation of the space of Disse (blue) was facilitated by means of histogram thresholding, due to the homogenous gray scale intensity of the silver filler. Scale bars: **c** and **d** = 10 μm

however inferior axial (Z) resolution, due to increased beam energy, and, subsequently, increased interaction volume within the sample. Samples are also increasingly prone to charging and beam damage due to the increased electron dose.

2. *Larger pixel arrays* (e.g., 8192 × 8192 pixels vs. 4096 × 4096) result in higher resolution datasets, providing low-magnification overview information (e.g., at the tissue or cellular level), whilst providing the ability to zoom into regions of interest at the subcellular level. This however comes at the expense of prolonged acquisition time.

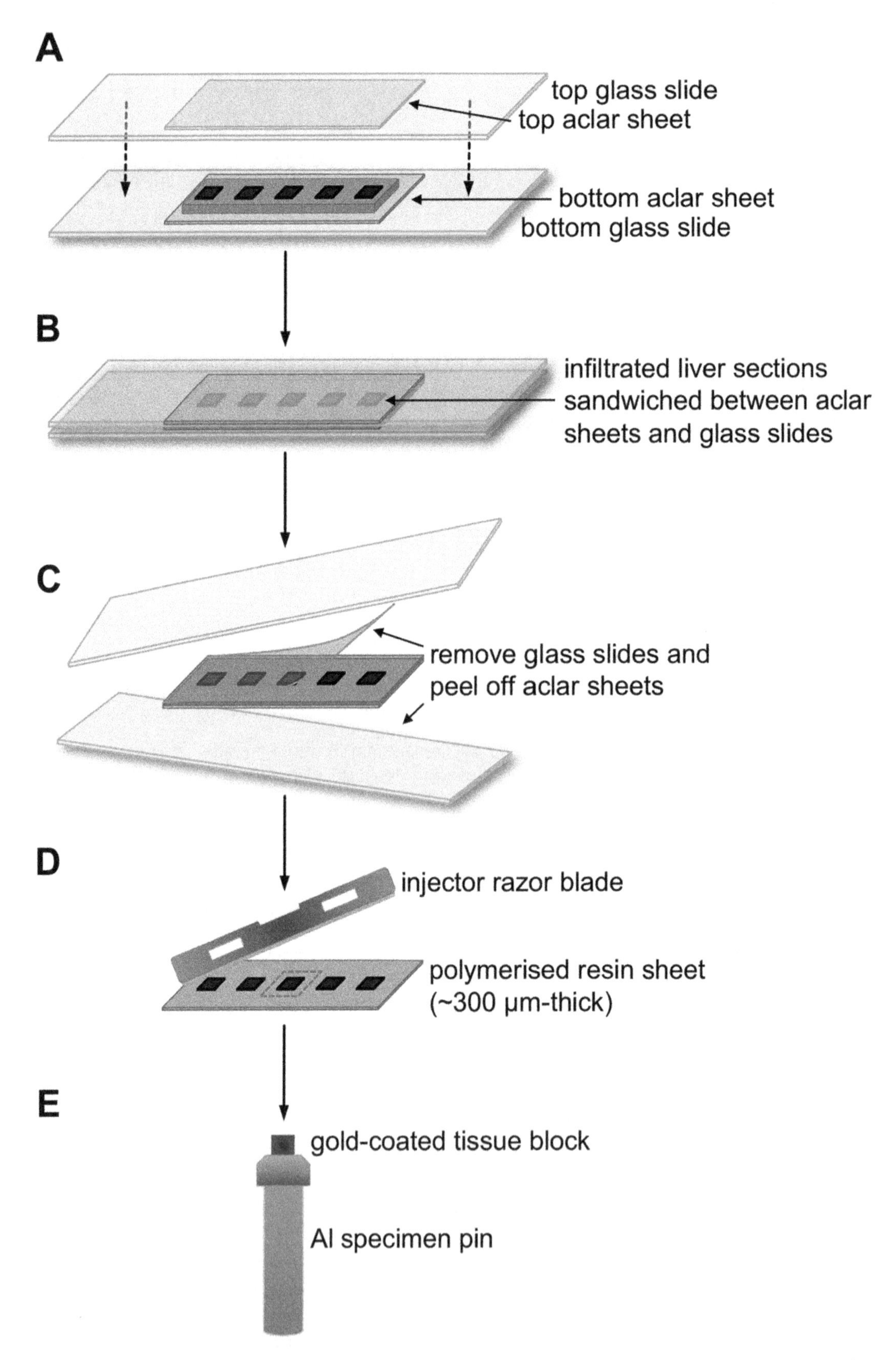

Fig. 4 Flat embedding and sample mounting for SBF-SEM. (**a**) Place a clean glass slide in the bottom of a 100 mm × 15 mm non-treated plastic petri dish. Place a sheet of Aclar atop the glass slide and syringe

3. *Section thickness* (between 15 and 200 nm) is the primary determinant of axial resolution, and its value is used for the calculation of voxel dimensions (*see* **Note** 7). Thinner sections provide superior axial resolution and are appropriate when structures change significantly between slices (e.g., membranes of the endoplasmic reticulum, vesicles, etc.). For a given volume, however, thinner sections increase both imaging time and dataset size (*see* **Note 8**).
4. *Variable pressure chamber conditions* are advantageous in combatting electron charging, however result in inferior image quality, namely resolution, contrast and signal-to-noise ratio, relative to imaging under HV conditions.
5. A large o*bjective aperture* (e.g., >60 μm) results in greater signal, however, at the expense of poorer resolution and depth of focus and increased risk of charging.

3.7 Post-Processing and Image Analysis

There are a multitude of applications—both free (e.g., ImageJ/Fiji) and those requiring an annual license subscription (e.g., Aviso)—which are suitable for image processing, analysis, segmentation, and visualization. We use the freeware software packages "Fiji" [11] for image processing and "IMOD" [12] for image segmentation, 3-D visualization, and morphometric analysis.

1. Given the sheer size of SBF-SEM datasets (*see* **Note 9**), we first perform pixel binning in order to resize datasets to a more manageable size (*see* **Note 10**). If high-resolution information is required from specific regions of interest, an unbinned sub-volume is cropped from the raw dataset.
2. The dataset is inspected for substandard images, such as those where occasional sections have fallen from the knife, back onto the block face. These sections are deleted and replaced by the "previous" or "next" image in the dataset. This step is especially important when "thresholding segmentation" is performed.
3. Image stack histogram normalization is performed, in order to make full use of gray values along image histogram, thereby improving image contrast.

Fig. 4 (continued) 0.5 mL of pure Epon onto the Aclar sheet. Carefully transfer and position tissue sections using a toothpick. Gently position a second Aclar sheet and glass slide on top of the resin ensuring an even layer, devoid of air bubbles, is formed. (**b**) Polymerize the sections in an oven at 60 °C for 48 h. (**c**) Remove the two glass slides by pinching between thumb and index finger. Carefully remove Aclar sheets using fine-tipped forceps, producing a thin resin sheet ~0.3 μm thick. (**d**) Using an injector razor blade, trim samples to desired size. (**e**) Mount trimmed specimen atop an aluminum specimen pin using conductive epoxy; paint lateral edges with silver paint and gold coat, prior to loading in the SBF-SEM

4. Files are exported as a suitable format (e.g., MRC or Tiff stack) for 3-D segmentation.
5. For 3-D modelling, segmentation, and visualization of SBF-SEM datasets, we use the freeware software package IMOD [12]. Cellular features can be segmented manually—by tracing high-contrast lines—or by means of "isosurface thresholding" of specified gray scale values.
6. Volumetric image analysis is performed using the "imodinfo" command, providing such morphometric parameters as volume, surface area and length, etc.

4 Notes

1. Successful perfusion of the tissue and removal of blood from the hepatic sinusoids are essential in order for the silver filler to penetrate these charging-prone spaces.
2. If tissue blocks are to be stored, they should remain in cacodylate buffer at 4 °C until required. If samples are fixed with a concentration of glutaraldehyde below 0.5%, then they should remain in the primary fixative.
3. It is essential to work quickly in order to prevent sedimentation of the silver nanopowder (due to its high relative density to the gelatin solution). This also prevents gelation of the solution prior to infiltration of the tissue, allowing complete penetration of the tissue block within such confined intercellular spaces as the space of Disse and bile canaliculi, as well as larger free resin spaces, such as the lumen of sinusoidal capillaries.
4. Once the tissue sections have been infiltrated with the silver filler and the gelatin has set on the slide, the sections can be difficult to see. It is advisable to shine a light (e.g., desk lamp) on the undersurface of the slide so that the tissue sections can be easily visualized, and excess silver filler is not collected with the tissue section.
5. Silver filler pre-embedding has also been successfully applied to other tissues including small intestine (jejunum), where it occupies the intervillous space and extracellular spaces of the vasculature, demonstrating the suitability of the method for 3-D investigation of a diverse range of samples. A prerequisite for successful pre-embedding with silver filler is the absence of physical cellular barriers. These include such structures as the basal lamina/basement membrane, which the silver filler cannot penetrate, due to insufficient discontinuities, in addition to the plasma membrane of cells, thereby preventing the passage of silver filler within the intracellular environment (except

where the cell has been perforated by means of dissection or vibrating blade microtomy).

6. Flat-embedding tissues reduces charging and improves conductivity due to the minimization of excess resin above and below the sample.
7. Axial resolution is also determined by accelerating voltage, however is highly variable given the heterogeneous composition of biological materials, and is also difficult to accurately calculate. As such, section thickness should be used to calculate voxel dimensions.
8. From our experience, 30 nm is the minimum section thickness that the automated microtome can reliably cut.
9. Typical datasets consisting of 500 images, acquired at 8192 × 8192 pixels, have a file size of ~64 gigabits (Fig. 11).
10. Pixel binning reduces image resolution, however results in improved signal-to-noise ratio.

Acknowledgments

The authors acknowledge the facilities as well as the technical assistance from staff of the AMMRF node at the University of Sydney. We are also grateful to Dr. Laura A. Lindsay and Professor Christopher R. Murphy from the Cell and Reproductive Biology Laboratory, the University of Sydney, for their kind provision of tissue samples.

References

1. Denk W, Horstmann H (2004) Serial block-face scanning electron microscopy to reconstruct three-dimensional tissue nanostructure. PLoS Biol 2(11):e329. https://doi.org/10.1371/journal.pbio.0020329
2. Leighton SB (1981) SEM images of block faces, cut by a miniature microtome within the SEM – a technical note. Scan Electron Microsc Pt 2:73–76
3. Titze B, Genoud C (2016) Volume scanning electron microscopy for imaging biological ultrastructure. Biol Cell 108:307. https://doi.org/10.1111/boc.201600024
4. Deerinck TJ, Bushong EA, Thor A, Ellisman MH (2010) NCMIR methods for 3D EM: a new protocol for preparation of biological specimens for serial block face scanning electron microscopy. National Center for Microscopy and Imaging Research. https://www.ncmir.ucsd.edu/sbem-protocol/
5. Hua Y, Laserstein P, Helmstaedter M (2015) Large-volume en-bloc staining for electron microscopy-based connectomics. Nat Commun 6:7923. https://doi.org/10.1038/ncomms8923
6. Griffin BJ (2007) Variable pressure and environmental scanning electron microscopy: imaging of biological samples. Methods Mol Biol 369:467–495. https://doi.org/10.1007/978-1-59745-294-6_23
7. Stokes D, Royal Microscopical S (2008) Principles and practice of variable pressure/environmental scanning electron microscopy (VP-ESEM). Book, Whole. Wiley, Chichester, UK
8. Shami GJ, Cheng D, Huynh M et al (2016) 3-D EM exploration of the hepatic microarchitecture – lessons learned from large-volume in situ serial sectioning. Sci Rep 6:36744

9. Wisse E, Braet F, Duimel H et al (2010) Fixation methods for electron microscopy of human and other liver. World J Gastroenterol 16(23):2851–2866. https://doi.org/10.3748/wjg.v16.i23.2851
10. Shami GJ, Cheng D, Braet F (2017) Silver filler pre-embedding for multidimensional nanoscopic imaging. https://vimeo.com/235291098
11. Schindelin J, Arganda-Carreras I, Frise E et al (2012) Fiji: an open-source platform for biological-image analysis. Nat Methods 9 (7):676–682. https://doi.org/10.1038/nmeth.2019
12. Kremer JR, Mastronarde DN, McIntosh JR (1996) Computer visualization of three-dimensional image data using IMOD. J Struct Biol 116(1):71–76. https://doi.org/10.1006/jsbi.1996.0013

Part V

Computational Modeling of Biomolecules

Chapter 34

Nanoscale Dynamics and Energetics of Proteins and Protein-Nucleic Acid Complexes in Classical Molecular Dynamics Simulations

Suresh Gorle and Lela Vuković

Abstract

The present article describes techniques for classical simulations of proteins and protein-nucleic acid complexes, revealing their dynamics and protein-substrate binding energies. The approach is based on classical atomistic molecular dynamics (MD) simulations of the experimentally determined structures of the complexes. MD simulations can provide dynamics of complexes in realistic solvents on microsecond time-scales, and the free energy methods are able to provide Gibbs free energies of binding of substrates, such as nucleic acids, to proteins. The chapter describes methodologies for the preparation of computer models of biomolecular complexes and free energy perturbation methodology for evaluating Gibbs free energies of binding. The applications are illustrated with examples of snapshots of proteins and their complexes with nucleic acids, as well as the precise Gibbs free energies of binding.

Key words Molecular dynamics simulations, Atomistic simulations, Protein-nucleic acid complexes, Free energy calculations, Binding free energies

1 Introduction

1.1 Overview

Molecular dynamics (MD) simulation is a powerful technique to study nanoscale dynamics of solvated biomolecules [1]. Since first simulations of biomolecules have been performed in 1970s [1], MD simulations have evolved into a high-performance mature methodology that can reveal functional behavior and events of biomolecules at the nanoscale in atomistic detail. MD simulation and visualization tools have been called a computational microscope and can today describe the systems of biomolecules in realistic environments containing from several atoms to hundreds of million atoms or more [2]. For example, MD simulations are a valuable tool for understanding the physical basis of function of biomolecular machines, due to the precision and high resolution dynamics that can be explored [3]. Simulations have proven useful in providing unprecedented level of atomistic detail for functional

Yuri L. Lyubchenko (ed.), *Nanoscale Imaging: Methods and Protocols*, Methods in Molecular Biology, vol. 1814,
https://doi.org/10.1007/978-1-4939-8591-3_34,

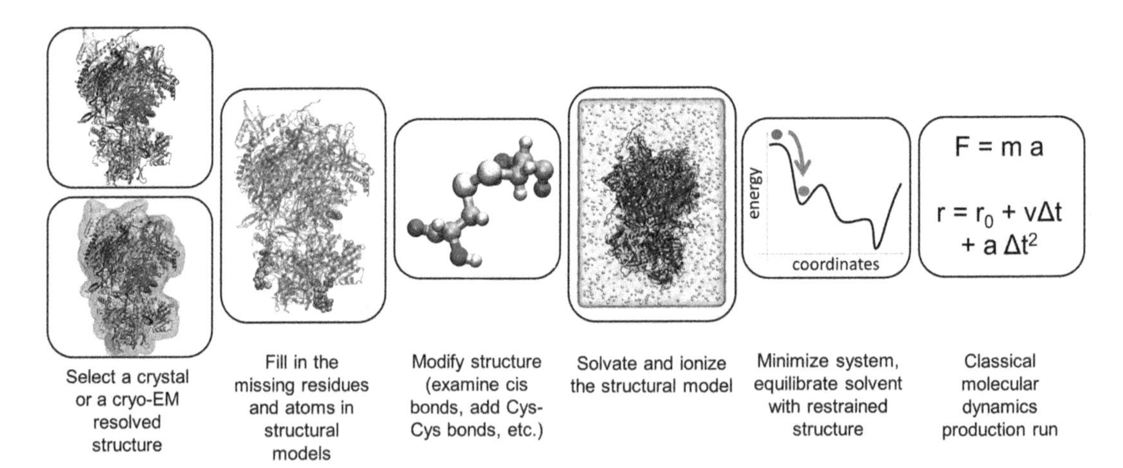

Fig. 1 Example protocol for preparing a biomolecular system for classical atomistic MD simulation. The RNA exosome complex with a bound single strand of RNA (ssRNA) is shown as an example; the preparation of this system for atomistic MD simulations is described in detail in Ref. [22]. (Reproduced from Ref. [22] with permission from ACS)

dynamics of biomolecules, such as protein folding [4, 5], complex conformational transitions [6–8], biomolecular recognition processes [9–11], conversion of chemical reaction energy to conformational changes [12, 13], transport of small molecules [14–16] and peptides [17, 18] through membrane proteins, chemical reactions involving enzymes and nucleic acids [15, 19–21], and others.

MD simulations rely on the underlying physics of atomic and molecular interactions. The key information required is the accurate atomic structure of the complexes that are to be studied. Once the structural information is available, through crystallography, cryo-electron microscopy, or NMR techniques, complete structures of interest can be completed through modeling and adjusted to have correct chemical bonding, as shown in Fig. 1. Then, the biomolecular complexes can be solvated in aqueous solvent with appropriate salt concentrations. Finally, the solvated system is minimized, so that it reaches the nearest energy minimum, and equilibrated. With computational resources available today, equilibration MD simulations can commonly reach several microseconds.

The equilibration process in MD simulations is performed by numerically integrating Newton or Langevin equations of motion for all the atoms (Fig. 2c). These equations contain the potential energy function, which represents the interactions between atoms. A set of predefined interaction potentials for the atoms present in the system is named force field. The interactions between atoms include bonded interactions and nonbonded interactions, as summarized in Fig. 2. The bonded interactions are described by bond potentials (stretching of two covalently bound atoms), angle potentials (bending of three covalently bound atoms), and dihedral

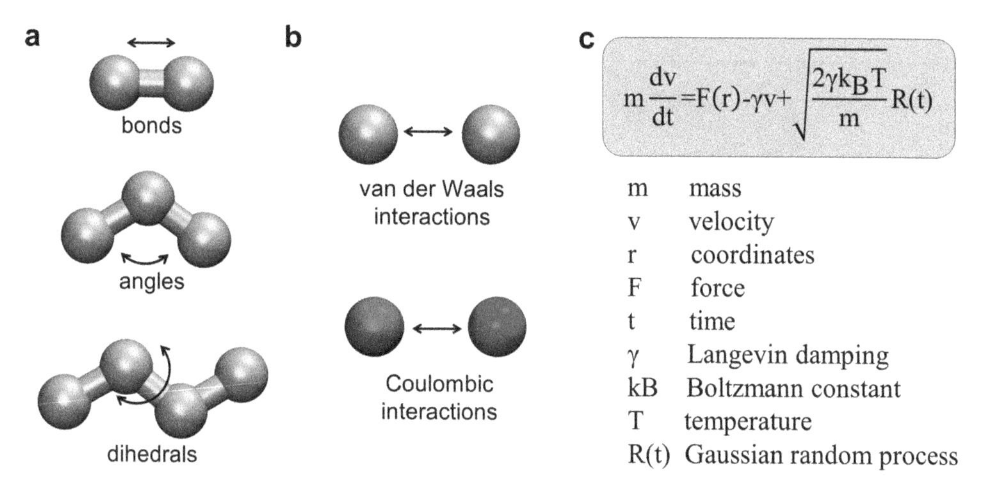

Fig. 2 A scheme of interactions presents in MD simulations. (**a**) Bonded interactions. (**b**) Nonbonded interactions. (**c**) The Langevin equation of motion commonly used in MD simulations. Forces between atoms, obtained from interaction potential energies, are used in Langevin equations of motion, which are numerically integrated to obtain positions and velocities of all atoms in time

potentials (torsional motion, which also usually includes improper torsions, of four covalently bound atoms). Nonbonded interactions are described by the Lennard-Jones potential (van der Waals interactions between two nonbonded atoms) and the Coulomb potential (charge-charge interactions between two nonbonded atoms).

1.2 Dynamics of Biomolecules in MD Simulations

Figure 3 shows an example of protein dynamics on the microsecond timescale, as obtained from MD simulations. The protein shown is a two domain protein APOBEC3G (A3G). MD simulations and high-speed AFM experiments have shown that A3G can dynamically readjust between globular and dumbbell shapes, through readjustment of its two domains with respect to each other. The MD simulations capture the transition between the two shapes, as described in detail in Ref. [23]. As MD simulations result in trajectories that capture molecular transitions in atomistic detail, protein dynamics can be analyzed and quantified. For example, Fig. 3a, b shows RMSD fluctuations and interdomain distance for structures of A3G in globular and dumbbell shapes. Furthermore, the simulations reveal which parts of the protein act as flexible linkers during shape transitions, as highlighted in Fig. 3c, d.

1.3 Alchemical Free Energy Methods for Evaluating Free Energies of Binding of Nucleic Acid Substrates to Their Protein Environment

Equilibrium MD simulations can in principle capture dynamics of both proteins and protein-nucleic acid complexes on microsecond timescales. However, at present, the drawback of MD simulations is that they are too short to capture functionally important processes. Furthermore, they are usually not able to provide sampling of all the relevant conformations of the system, from which Gibbs free energies can be obtained. For protein-nucleic acid complexes, we

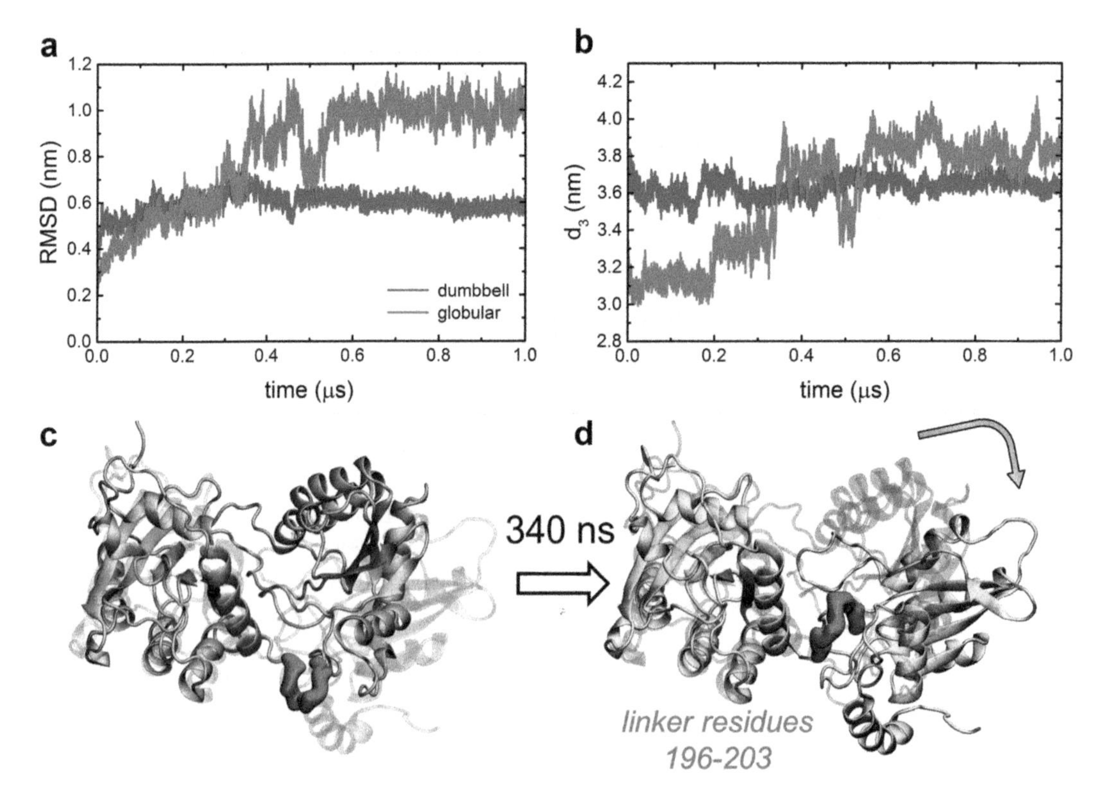

Fig. 3 Structural dynamics of A3G protein in MD simulations. (**a**) RMSDs of two A3G structural models, visually representative of a globular conformation (red) and a dumbbell conformation (blue). (**b**) Distances between two A3G domains of the two A3G structural models. Red and blue plots represent the results for globular and dumbbell conformations, respectively. (**c**, **d**) Conformational transition of the globular form of a full A3G during a 1 μs MD trajectory. The initial structure of A3G is shown in light blue, the final structure of A3G is shown in green, and the flexible linker, which repositions during the simulation course and correlates with the reorientation of two A3G domains, is shown as a red tube. For comparison, panels **c**, **d** show both opaque and transparent structures of initial and final A3G states, aligned with respect to the left domain of A3G. (Reproduced from Ref. [23] with permission from ACS)

are often interested in the binding affinity of nucleic acids to proteins. To obtain precise free energies of binding of nucleic acids (substrates) to proteins, free energy calculations should be performed.

Figure 4 shows an example system, for which the Gibbs free energies of binding were obtained for the single stranded RNA (ssRNA) substrate bound to the RNA exosome complex. The relative binding affinities of individual RNA nucleotides to the active site tunnel of Rrp44 were determined by means of free energy perturbation (FEP) calculations [22, 24]. FEP calculations were performed to obtain the transfer free energies of single RNA nucleotides from aqueous solution to different binding sites inside the exosome complex, such as those shown in Fig. 4. Structures of the exosome bound to ssRNA pieces placed in different positions are based on relaxed initial systems and prepared by removing RNA

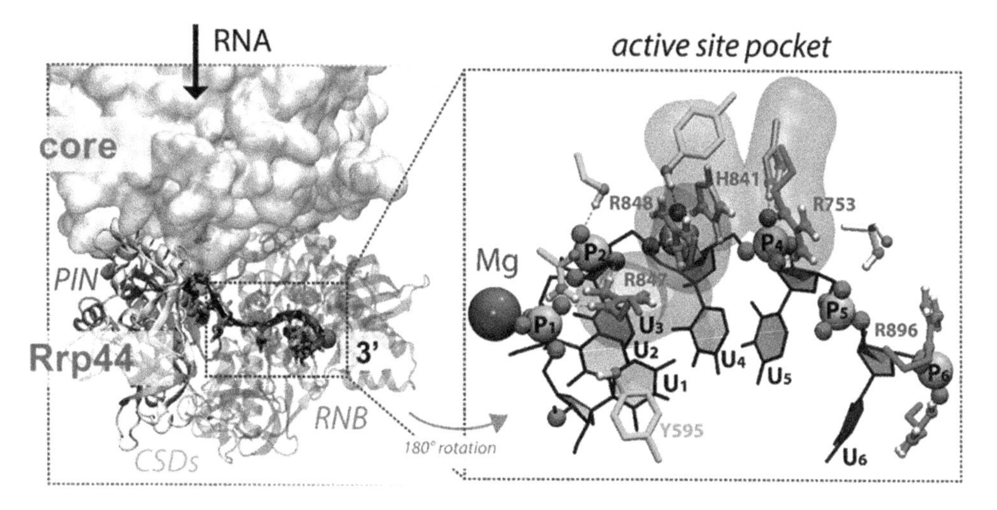

Fig. 4 Structure of the RNA exosome complex bound to an RNA substrate (inset). The RNA bound to the active site tunnel in the RNB domain of the Rrp44 enzyme. (Reproduced from Ref. [22] with permission from ACS)

nucleotides from unwanted positions gradually, along with a corresponding number of counterions. The gradual removal of single RNA nucleotides is performed through a parameter λ, which defines the extent of nucleotide's coupling to its environment. The parameter λ smoothly couples the initial state of the system (where the nucleotide exists fully) to the final state of the system (where the nucleotide is fully removed).

Free energies of transfer from water to protein environment were obtained for 3′-end or 5′-end single RNA nucleotides (covalently bound to longer pieces of ssRNA) in different positions within the active site tunnel, according to the thermodynamic cycle shown in Fig. 5a. The procedure employed ensured that the binding free energies of single nucleotides take into account the presence of longer covalently bound ssRNA fragments in the active site tunnel of Rrp44, i.e., a realistic environment of single nucleotides is studied. The FEP transformations for all single nucleotides were carried out bidirectionally (by changing the nucleotides to and from nothing, via the parameter λ; Fig. 5b, c).

Binding energies of RNA nucleotides to the active site tunnel of the exosome complex are reported in Fig. 6. Transfer of nucleotides from water to the protein environment is favorable at all examined positions of the extended RNA-binding pathway; in fact, the transfer free energies range from −5 to −19 kcal/mol. The results in Fig. 6 demonstrated that the exonuclease active site tunnel contains several sites binding RNA very strongly, including positions P1, P4, and P5, where the transfer free energies are as large as −14 kcal/mol, akin to the energy gained through ATP hydrolysis. The weaker binding sites for RNA are at positions P2, P7, P8, and P9. All details of the calculations and the results, as well as the implications

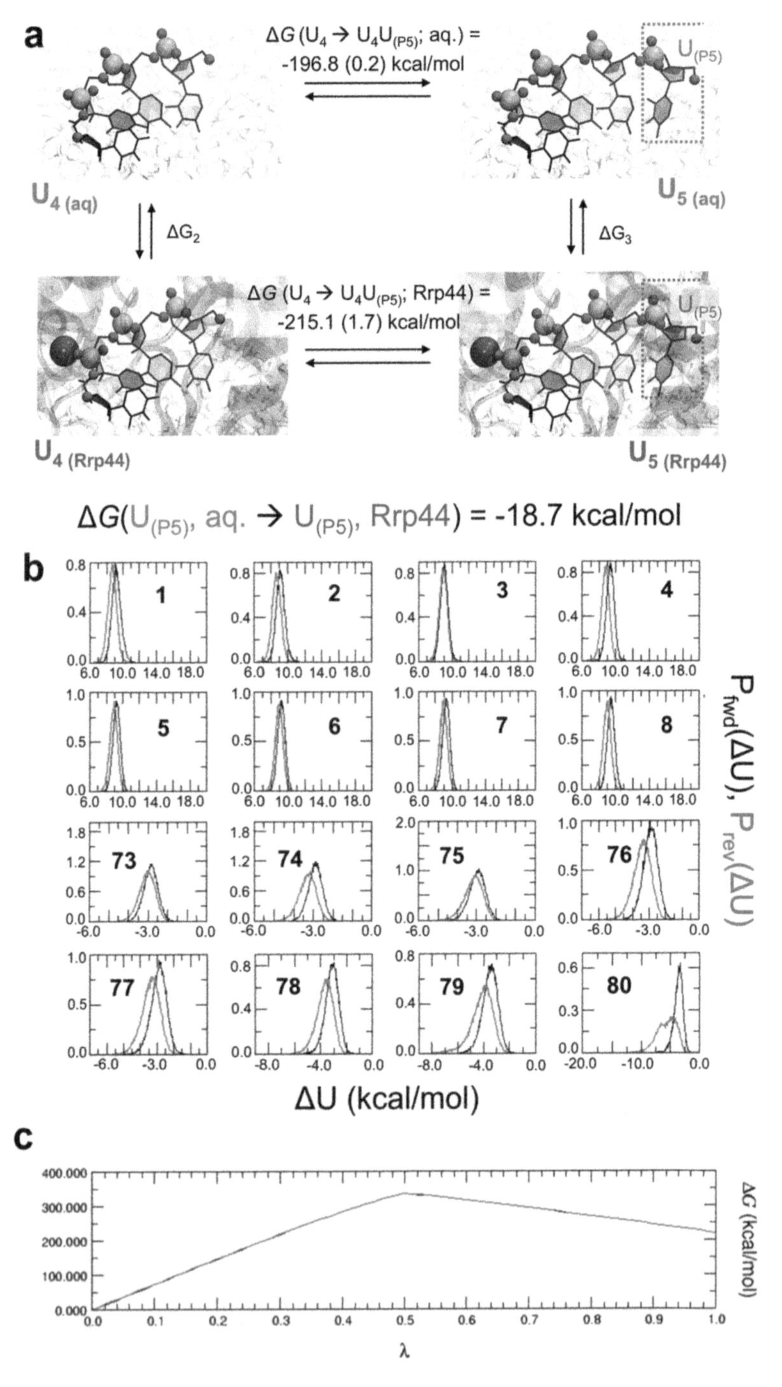

Fig. 5 (**a**) Example thermodynamic cycle in FEP calculations, shown for the addition of the fifth nucleotide, labeled U(P_5), to a 4-nt RNA segment, U_4. RNA atoms are shown in black, and P, O, and Mg atoms are shown as orange, red, and purple spheres. (**b**) Probability distribution functions for selected λ-windows (1–8 and 73–80) in FEP transformations, shown in bottom of panel (**a**). (**c**) Free energy change as a function of λ confirms calculation convergence. In (**b**, **c**), data from forward (red) and backward (black) transformations are shown. (Reproduced from Ref. [22] with permission from ACS)

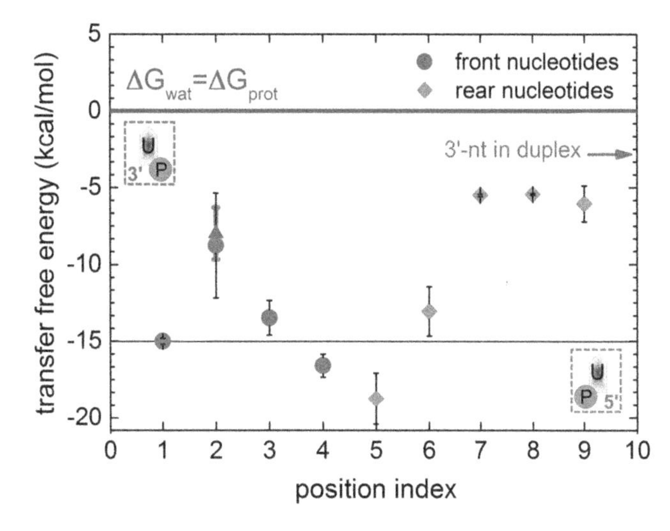

Fig. 6 Interactions of RNA with the active site tunnel of Rrp44. Shown are transfer free energies of single RNA nucleotides, covalently bound to longer ssRNAs, from water into Rrp44. (Reproduced from Ref. [22] with permission from ACS)

for the free energy profile for ssRNA translocation inside the exosome active site tunnel, are reported in Ref. [22].

The quality of FEP calculations can be evaluated by examining the errors. Most of the errors in Fig. 6 are small or moderate (<1.7 kcal/mol). The largest error of 3.4 kcal/mol is obtained for the binding free energy of a 3′-end nucleotide at position P2. This error demonstrates the sensitivity of results on the presence of divalent ions close to nucleic acids. This large error is most likely related to variations in coordination of Mg^{2+} in different λ-windows of FEP calculations, which should have a significant effect on the 3′-end nucleotide in position P2: the 3′-end hydroxyl group of this nucleotide is very close to Mg^{2+}. The binding free energies at all other positions are accompanied by small errors, since Mg^{2+} coordination is either well determined with respect to RNA (when the 3′-nucleotide is at position P1, as it occurs in the initial crystal structure) or not as influential (in all other positions, 3′- end nucleotides do not directly interact with Mg^{2+} and its coordination shell). To ascertain that the binding free energy at position P2 is reproducible despite the magnitude of the error, an independent FEP calculation at this position was performed starting from a different initial state. The second simulation confirmed reproducibility of the binding free energy at position P2, as shown in Fig. 6.

2 Materials

1. Molecular dynamics software. Commonly used software codes are NAMD [25], AMBER [26], CHARMM [27], GROMACS [28], among others.
2. Visualization software. Commonly used software codes are VMD [29], PYMOL [30], among others.
3. Computer. MD simulations can be run on a wide range of computers, including basic workstations, computer clusters, and supercomputers. The codes can use both processors and graphical processing units (GPU).
4. Trajectory analysis. Wide range of software codes can be used to analyze the trajectories. Commonly used codes are VMD [29], CHARMM [27], AMBER tools [26], GROMACS tools [28], among others.

3 Methods

The method below is presented for preparation of systems simulated with the NAMD2 code [25]. Simulations with other codes require similar preparation steps but may differ in details. All the simulations were performed at room temperature unless otherwise specified. The basic knowledge about using VMD and performing simulations using NAMD2 code is very helpful in modeling and performing MD simulations described here.

3.1 System Preparation for MD Simulations

1. Model structure. MD simulation procedure starts with preparation of the structural model. The structure is usually obtained from crystallography, cryo-electron microscopy, or NMR experiments and is usually found and downloaded from the PDB database [31] (*see* **Notes 1** and **2**).
2. Check for any missing atoms or amino acids in the structures. Add the missing residues and atoms in the structure by loading into the visualization software such as VMD [29] or with modeling software, such as MODELLER [32, 33] by using another available protein structure as a reference (*see* **Notes 1** and **2**).
3. Choose a suitable force field for the system like the Amber force field with SB36 and BSC037 corrections [26] or the CHARMM36 [34] force field (*see* **Note 3**) to calculate energy during MD simulations. Other force fields can be used, depending on the nature of biomolecules that the system contains.

4. Charge neutralization. The completed systems are then load into VMD, open *Addions* tab via *Extension* and *Modeling* tabs and calculate the number of ions required to make the system's net charge to zero and add them to the system.
5. Solvation. The resulting structures are solvated in water by going to *Solvate* tab via *Extension* and *Modeling* tabs in VMD [29]. Then add salt NaCl to this system by opening *Addions* tab, as mentioned above. This step will result in a system with water and ions/salt in pdb and psf format files (*see* **Notes 4** and **5**).

3.2 Molecular Dynamics Simulations

1. The prepared system needs to be energy minimized. The system, as described in pdb and psf files, is minimized for 2000 steps using NAMD2 software [25] (*see* **Note 2**). All the heavy atoms of the protein or protein-nucleic acid complexes are constrained by setting their beta column values to 1 within the pdb structure file. The constrained pdb file is a copy of the initial pdb file, and it needs to be specified separately in the NAMD configuration file using constraints option.
2. Simulation parameters need to be specified in NAMD configuration file. The particle mesh Ewald (PME) method [PME] is commonly used for the evaluation of long-range Coulomb interactions [35]. The time step is set to 2.0 fs; all bonds involving hydrogen are constrained with SHAKE algorithm [36]. All simulations are performed in the isobaric-isothermal ensemble, at a constant temperature of 310 K, with a friction constant γ of 1.0 ps^{-1} and at a constant pressure of 1 bar. The equations of motion are integrated with the r-RESPA multiple time-step propagator with an effective time step of 2 and 4 fs for short- and long-range interactions.
3. Equilibration. Take the minimized system with water and ions and perform equilibration for short time scales like 2–5 ns by restraining the protein or protein-nucleic acids system using harmonic forces with a spring constant of 1 kcal/(mol$Å^2$).
4. Production simulations. The complete models are then simulated in equilibrium MD simulations; the timescales that can be reached depend on the size of the system and typically range from several nanoseconds to several microseconds. MD simulations record time evolution of all the atomic positions within the structure and write these atomic positions into trajectory files. These files are key to obtain the information required to understand the systems behavior. The macroscopic properties of the system can be calculated from trajectory files by using statistical mechanics equations (*see* **Notes 6** and 7).

3.3 Free Energy Perturbation Calculations

FEP calculations can be performed to obtain the binding free energies of nucleic acid substrates (partitioned as single nucleotides) to their protein environments. The binding free energies in the example above were obtained by identifying first transfer free energies of single nucleotides from aqueous solution to their binding sites of the exosome complex.

1. Generate dual-topology file representing the disappearing nucleotide and appearing 3′-end of RNA in exosome-RNA complexes by combining the standard nucleotide and 3′-end topologies (*see* **Note 8**). This dual-topology file is called together with the standard topology files, while generating pdb and psf files with VMD.
2. The structures of the exosome bound to ssRNA pieces in different positions were based on relaxed initial systems and prepared by removing RNA nucleotides from unwanted positions, along with a corresponding number of counterions (*see* **Note 9**) similar to the protocol described before.
3. NAMD2 uses a coupling parameter λ, from $\lambda = 1$ (complete presence) to $\lambda = 0$ (complete removal) to represent the disappearance of the nucleotide from exosome-ssRNA complex. The systems at λ-values are called windows and exosome-ssRNA complexes must be generated separately for all these λ-values. 80 λ-windows were considered in the present study (*see* **Note 10**).
4. The systems corresponding to each window are minimized and equilibrated using the similar protocol described before.
5. Generate fep file, a copy of the initial pdb file, where the beta column marks the atoms that are disappearing and appearing in the FEP method. The disappearing atoms have beta column values set to -1.0, appearing atoms have beta column values set to 1.0, and the remaining atoms to 0.0.
6. Turn on the *alch* option in NAMD configuration file along with other options that are described before. The output energies will be written to files with fepout extensions.
7. The single nucleotide was unrestrained, whereas the position of the P atoms of the RNA phosphate groups of the remaining nucleotides was restrained with the force constant $k = 1$ kcal/(mol Å^2) during the MD simulations.
8. The alchemical transformations for all single nucleotides were carried out bidirectionally (by changing to and from nothing) and used 80 λ-windows of even width, with 250,000 steps of equilibration followed by 500,000 steps of data collection. In total, each single nucleotide simulation (backward and forward) lasted 240 ns.

9. The ParseFEP plugin in VMD [37] are used to calculate the free energy changes from the FEP simulations by means of the Bennett acceptance ratio (BAR) algorithm. The fepout files corresponding each window are combined separately for the forward and backward transformations and uploaded into ParseFEP plugin in VMD via *Extensions*, *Analysis*, and *Analyze FEP simulation* tabs. As all the FEP simulations involved a creation or annihilation of charged nucleotides, the computed free energies were corrected, as detailed in Ref. [22].
10. Check the probability distribution functions for the forward and backward transformations exhibit a good overlap or not, as shown in a representative example in Fig. 5b, c. This is a good method to check convergence of the free energy calculations (*see* **Note 11**).
11. For comparison and in order to obtain transfer free energies of nucleotides from water to protein environments, absolute solvation free energies of single 3′-end and 5′-end RNA nucleotides in water were obtained in separate simulations.
12. The transfer free energies are obtained for 3′-end or 5′-end single RNA nucleotides (covalently bound to longer pieces of ssRNA; *see* **Note 12**) in eight different positions, according to the thermodynamic cycle shown in Fig. 5a.

4 Notes

1. Correct protein structures should be selected for modeling the proteins via homology modeling or for adding the missing parts in experimental structures. Much attention should be paid on pH conditions, ionic states of amino acids, etc. when modeling the proteins.
2. Energy minimization should be performed to remove steric interactions. Restraints must be placed on the protein core atoms while performing energy minimization in order to avoid the chances of losing the major protein structure. The minimization will be considered successful if the maximum force on any atom is less than 0.0001 kcal/molÅ^2.
3. Appropriate force field must be chosen to study protein dynamics and protein-nucleic acid interactions. Do not mix different force fields to represent a single system. All the tasks including system preparation, energy minimization, MD simulations, and trajectory analysis must be performed using the same force field.
4. The dimensions of water box must be minimum 1 nm larger than protein dimensions in order to avoid the protein to

interact its periodic image across the boundary of the box during the simulations.

5. The environmental conditions around the protein structures must be modeled close to the native experimental conditions.
6. The translation and rotational motions of the protein structures must be removed by simply performing a best-fit translation and rotation calculation on each frame of the trajectory using the starting coördinates as the reference.
7. Converged portion of the protein trajectory should be used for calculating macroscopic properties. Root mean square deviations, total energy of the system, change in temperature, and system density are some of the good measurements to check convergence.
8. When preparing dual-topology file, care must be taken for total charge of the residue in which the modification has occurring. Ignoring this charge imbalance the results in wrong free energy values.
9. When charged species, such as nucleotides, are disappearing through the FEP procedure, the analytical correction needs to be applied to the free energy result in order to account for the net charge of initial or final states of the system in the simulation (details are provided in Ref. [22]).
10. The FEP calculation is performed by gradually changing the single nucleotide (or the molecule of interest) via a series of λ windows, where $\lambda = 1$ marks the system where the nucleotide fully exists, and $\lambda = 0$ marks the system where the nucleotide has vanished. The selected series of λ-windows needs to be examined for convergence by performing separately forward (λ changing from 1 to 0) and backward (λ changing from 0 to 1) FEP transformations. If forward and backward transformations give different free energy results by more than several kcal/mol, the density of λ-windows needs to be increased. ParseFEP plugin in VMD [37] can provide detailed insight in the quality of the FEP calculation performed, as shown in Fig. 5b, c.
11. ParseFEP plugin in VMD [37] can provide detailed insight in the quality of the FEP calculation performed.
12. The quality of FEP calculation depends on conformations of molecules in the system. The behavior of molecules during FEP transformations should be carefully examined, and unusual conformational changes during these transformations should be eliminated by cautious use of restraints.

References

1. McCammon JA, Gelin BR, Karplus M (1977) Dynamics of folded proteins. Nature 267:585–590
2. Perilla JR, Schulten K (2017) Physical properties of the HIV-1 capsid from all-atom molecular dynamics simulations. Nat Commun 8:15959
3. Karplus M, McCammon JA (2002) Molecular dynamics simulations of biomolecules. Nat Struct Mol Biol 9:646–652
4. Freddolino PL, Liu F, Gruebele M, Schulten K (2008) Ten-microsecond molecular dynamics simulation of a fast-folding WW domain. Biophys J 94:L75–L77
5. Gumbart J, Chipot C, Schulten K (2011) Free energy of nascent-chain folding in the translocon. J Am Chem Soc 133:7602–7607
6. Fajer M, Meng Y, Roux B (2016) The activation of c-Src tyrosine kinase: conformational transition pathway and free energy landscape. J Phys Chem B 121:3352–3363
7. Bernardi RC, Melo MCR, Schulten K (2015) Enhanced sampling techniques in molecular dynamics simulations of biological systems. BBA-Gen Subjects 1850:872–877
8. Shukla D, Meng Y, Roux B, Pande VS (2014) Activation pathway of Src kinase reveals intermediate states as targets for drug design. Nat Commun 5:3397
9. Levy Y, Onuchic JN, Wolynes PG (2007) Fly-casting in protein-DNA binding: frustration between protein folding and electrostatics facilitates target recognition. J Am Chem Soc 129:738–739
10. Zhang Y, Vukovic L, Rudack T, Han W, Schulten K (2016) Recognition of poly-ubiquitins by the proteasome through protein refolding guided by electrostatic and hydrophobic interactions. J Phys Chem B 120:8137–8146
11. Vukovic L, Koh HR, Myong S, Schulten K (2014) Substrate recognition and specificity of double-stranded RNA binding proteins. Biochemistry 53:3457–3466
12. Ma W, Schulten K (2015) Mechanism of substrate translocation by a ring-shaped ATPase motor at millisecond resolution. J Am Chem Soc 137:3031–3040
13. McCullagh M, Saunders MG, Voth GA (2014) Unraveling the mystery of ATP hydrolysis in actin filaments. J Am Chem Soc 136:13053–13058
14. Dhakshnamoorthy B, Rohaim A, Rui H, Blachowicz L, Roux B (2016) Structural and functional characterization of a calcium-activated cation channel from Tsukamurella paurometabola. Nat Commun 7:12753
15. Moradi M, Enkavi G, Tajkhorshid E (2015) Atomic-level characterization of transport cycle thermodynamics in the glycerol-3-phosphate:phosphate antiporter. Nat Commun 6:8393
16. Moradi M, Tajkhorshid E (2013) Mechanistic picture for conformational transition of a membrane transporter at atomic resolution. Proc Natl Acad Sci U S A 110:18916–18921
17. Gumbart JC, Chipot C, Schulten K (2011) Free-energy cost for translocon-assisted insertion of membrane proteins. Proc Natl Acad Sci U S A 108:3596–3601
18. Gumbart JC, Chipot C (2016) Decrypting protein insertion through the translocon with free-energy calculations. Biochim Biophys Acta Biomembr 1858:1663–1671
19. Rosta E, Nowotny M, Yang W, Hummer G (2011) Catalytic mechanism of RNA backbone cleavage by ribonuclease H from quantum mechanics/molecular mechanics simulations. J Am Chem Soc 133:8934–8941
20. Ganguly A, Thaplyal P, Rosta E, Bevilacqua PC, Hammes-Schiffer S (2014) Quantum mechanical/molecular mechanical free energy simulations of the self-cleavage reaction in the Hepatitis Delta virus ribozyme. J Am Chem Soc 136:1483–1496
21. Dittrich M, Hayashi S, Schulten K (2004) ATP hydrolysis in the ATP and ADP catalytic sites of F1-ATPase. Biophys J 87:2954–2967
22. Vuković L, Chipot C, Makino DL, Conti E, Schulten K (2016) Molecular mechanism of processive 3′ to 5′ RNA translocation in the active subunit of the RNA exosome complex. J Am Chem Soc 138:4069–4078
23. Gorle S, Pan Y, Sun Z, Shlyakhtenko LS, Harris RS, Lyubchenko YL, Vuković L (2017) Computational model and dynamics of monomeric full-length APOBEC3G. ACS Cent Sci 3:1180. https://doi.org/10.1021/acscentsci.7b00346
24. Gumbart JC, Roux B, Chipot C (2013) Standard binding free energies from computer simulations: what is the best strategy? J Chem Theory Comput 9:794–802
25. Phillips JC, Braun R, Wang W, Gumbart JC, Tajkhorshid E, Villa E, Chipot C, Skeel RD, Kale L, Schulten K (2005) Scalable molecular dynamics with NAMD. J Comput Chem 26:1781–1802

26. Case DA, Cerutti DS, Cheatham TE III et al (2017) AMBER 2017. University of California, San Francisco
27. Brooks BR, Brooks CL III, Mackerell AD Jr et al (2009) CHARMM: the biomolecular simulation program. J Comput Chem 30:1545–1614
28. Berendsen HJC, van der Spoel D, van Drunen R (1995) GROMACS: a message-passing parallel molecular dynamics implementation. Comp Phys Commun 91:43–56
29. Humphrey W, Dalke A, Schulten K (1996) VMD—visual molecular dynamics. J Mol Graph 14:33–38
30. The PyMOL Molecular Graphics System, Version 2.0 Schrödinger, LLC
31. Berman HM, Westbrook J, Feng Z, Gilliland G, Bhat TN, Weissig H, Shindyalov IN, Bourne PE (2000) The protein data bank. Nucleic Acids Res 28:235–242
32. Sali A, Blundell TL (1993) Comparative protein modelling by satisfaction of spatial restraints. J Mol Biol 234:779–815
33. Fiser A, Do RK, Sali A (2000) Modeling of loops in protein structures. Protein Sci 9:1753–1773
34. Best RB et al (2012) Optimization of the additive CHARMM all-atom protein force field targeting improved sampling of the backbone φ, ψ and side-chain χ1 and χ2 dihedral angles. J Chem Theory Comput 8:3257–3273
35. Darden DR, York TD, Pedersen L (1993) Particle mesh Ewald: an nlog(n) method for ewald sums in large systems. J Chem Phys 98:10089–10092
36. Ryckaert JP, Ciccotti G, Berendsen HJC (1977) Numerical integration of the Cartesian equations of motion of a system with constraints: molecular dynamics of n-alkanes. J Comput Phys 23:327–341
37. Liu P, Dehez F, Cai W, Chipot C (2012) A toolkit for the analysis of free-energy perturbation calculations. J Chem Theory Comput 8:2606–2616

Index

Yuri L. Lyubchenko (ed.), *Nanoscale Imaging: Methods and Protocols*, Methods in Molecular Biology, vol. 1814, https://doi.org/10.1007/978-1-4939-8591-3, © Springer Science+Business Media, LLC, part of Springer Nature 2018

C

D

G

H

I

J

K

L

M

N

O

P

Q

R

S

CPSIA information can be obtained
at www.ICGtesting.com
Printed in the USA
LVHW06*0829170718
583830LV00002BB/9/P